Trends in Logic

Volume 38

For further volumes:
http://www.springer.com/series/6645

TRENDS IN LOGIC
Studia Logica Library

VOLUME 38

SCOPE OF THE SERIES

The book series *Trends in Logic* covers essentially the same areas as the journal *Studia Logica*, that is, contemporary formal logic and its applications and relations to other disciplines. The series aims at publishing monographs and thematically coherent volumes dealing with important developments in logic and presenting significant contributions to logical research.

The series is open to contributions devoted to topics ranging from algebraic logic, model theory, proof theory, philosophical logic, non-classical logic, and logic in computer science to mathematical linguistics and formal epistemology. However, this list is not exhaustive; moreover, the range of applications, comparisons and sources of inspiration is open and evolves over time.

Christian Straßer

Adaptive Logics for Defeasible Reasoning

Applications in Argumentation, Normative Reasoning and Default Reasoning

Springer

Christian Straßer
Ghent University
Ghent
Belgium

ISSN 1572-6126 ISSN 2212-7313 (electronic)
ISBN 978-3-319-00791-5 ISBN 978-3-319-00792-2 (eBook)
DOI 10.1007/978-3-319-00792-2
Springer Cham Heidelberg New York Dordrecht London

Library of Congress Control Number: 2013940955

Printed on acid-free paper

Springer is part of Springer Science+Business Media (www.springer.com)

Dedicated to all the birds who cannot fly
Especially to Kiwis. . .

Sorry, Tweety, next time!

Preface

In this book I will study defeasible reasoning. There are many facts of reasoning that are captured under the term defeasible. Sometimes we argue on the basis of typicality, normality, sometimes we make inductive generalizations, etc. We "jump to conclusions" in different ways. It is not my intention to give an exhaustive characterization of all possible forms of defeasible reasoning. Hence I will paradigmatically examine various contexts in which defeasible reasoning is useful, such as default reasoning (Part II), reasoning in the context of argumentation (Part III), and normative reasoning (Part IV).

Still, my perspective is a unificatory one. It is gained by the choice of a specific formal logical framework. With the help of this framework I will develop logical models of forms of defeasible reasoning. The framework is that of *adaptive logics* which originates in the work of Diderik Batens. The standard format of adaptive logics provides a unified characterization of a class of logics that, as will be demonstrated and argued for in this manuscript, are decent tools in order to model defeasible reasoning.

The merits of the study offered in this book are two-fold.

First, it offers a deeper understanding of (forms of) defeasible reasoning. On the one hand, the logics that are introduced in this manuscript deepen our understanding of the formal properties (particular forms) of defeasible inferences, of retracting inferences, etc. On the other hand, formulating them in a unificatory framework offers possibilities to compare them and to identify formal properties they have in common.

Second, the book affirms and substantiates the status of adaptive logics as a generic formal framework for defeasible reasoning. It does so by offering case studies stemming from various contexts of defeasible reasoning. In addition, as will be shown, there are various metatheoretic advantages of adaptive logics compared to many other logics or logical frameworks that model defeasible reasoning.

The Structure of the Book

This book is structured as follows:

In Part I we begin with a general introduction into defeasible reasoning (Chap. 1). After that, adaptive logics (in short, ALs) are introduced (Chap. 2). It is demonstrated that they offer an intuitive and powerful framework to model defeasible reasoning. ALs are discussed in their *standard format*. It is argued that the standard format comes with an attractive meta-theory. In Chap. 3, it is shown how ALs can be combined. Chapter 4 contains joint work with Diderik Batens and Peter Verdée. We argue that ALs offer a transparent model for defeasible reasoning since elegant and intuitive criteria are available to decide whether (extensions of) premise sets are equivalent. Finally, in Chap. 5, it is demonstrated how the standard format can be generalized while keeping its metatheoretic merits intact. This is joint work with Frederik Van De Putte.

Part II contains two applications of ALs in the context of default reasoning. Let $A \rightsquigarrow B$ express that from A normally/usually/typically/etc. (depending on the application) follows B. Note that Modus Ponens is not unrestrictedly valid in such a context. This is due to cases of specificity. Where b stands for "being a bird" and f for "flying", we have $b \rightsquigarrow f$ ("Birds usually fly"). However, where p stands for "being a penguin", we also have $(p \wedge b) \rightsquigarrow \neg f$. Now suppose we have both premises, p and b. If Modus Ponens would be valid we would be able to derive both f and $\neg f$. Obviously this is not desired. In Chap. 6, I will propose a defeasible handling of Modus Ponens by means of ALs.

In [1] Lehmann, Magidor and Kraus tackle the question "What does a conditional knowledge base entail?" by means of a sophisticated semantic selection procedure, the so-called *Rational Closure* of a knowledge base. Chapter 7 offers an AL interpretation of Rational Closure. This way we gain a full logic for Rational Closure, one that is equipped with a (dynamic) proof theory. The semantic selection of [1] is very much in the spirit of Shoham's semantic selections (see e.g., [2]). Hence, the logic developed in Chap. 7 offers a paradigmatic demonstration that ALs are able to represent logics defined by semantic selections in the style of Shoham. This in turn substantiates the claim that ALs offer a very generic and unifying framework for defeasible reasoning.

In Part III, ALs are used for the modeling of argumentations. Dung presented in [3] a highly influential account of abstract argumentation. Arguments are represented as abstract entities and the relationships between arguments are modeled by an attack relation. The two elements define abstract argumentation systems. Dung offered a number of clear and intuitive semantics for selecting arguments from argumentation systems. Chapter 8 presents joint work with Dunja Šešelja in which we develop a unifying AL framework for abstract argumentation. Our family of logics models all the semantics proposed by Dung and moreover provides a dynamic proof-theory for each. In Chap. 9 I generalize the AL framework in accordance with Nielsen and Parsons' generalization of Dung's framework [4] in such a way that joint attacks are possible, i.e., attacks in which several arguments attack several arguments. This paradigmatically presents one of many possible enhancements to the systems introduced in Chap. 8.

Part IV features various applications of ALs in the context of deontic logics. Most of the systems presented in this part are heavily influenced by the work of Lou Goble. One of the main challenges for deontic logicians is to develop systems that are conflict-tolerant. That is to say, logics that do not exhibit explosive behavior when confronted with conflicting norms such as "You're obliged to bring about *A*" and "You're obliged to bring about not-*A*". Goble suggested an attractive way of tackling this problem, namely by restricting the so-called inheritance rule that allows to derive from the obligation to bring about *A* the obligation to bring about *B* in the case in which *A* necessitates *B*. Chapter 10 presents joint work with Joke Meheus and Mathieu Beirlaen in which we point out certain problems with Goble's systems and improve on them by strengthening them by means of ALs.

The remaining sections in Part IV feature applications in the context of conditional deontic logics. Chapter 11 generalizes and enhances the results of Chap. 10 for the conditional setting. In Chap. 12, I tackle a similar problem as in Chap. 6. The majority of conditional deontic logics does not allow for the factual detachment of conditional obligations. That is to say, given the commitment *A* under the condition *B* and the factual information *B*, in many circumstances it is desired that we derive the 'actual' and unconditional obligation to bring about *A*. However, similar as in the context of default reasoning, here we have to deal with cases of specificity as well. Moreover, we also have to take into consideration contrary-to-duty obligations. This motivates a defeasible handling of detachment. It is realized by means of ALs.

Ghent, March 2013 Christian Straßer

References

1. Lehmann, D.J., Magidor, M.: What does a conditional knowledge base entail? Artif. Intell. **55**(1), 1–60 (1992)
2. Shoham, Y.: A semantical approach to nonmonotonic logics. In: Ginsberg, M.L. (ed.) Readings in Non-Monotonic Reasoning, pp. 227–249. Morgan Kaufmann, Los Altos (1987)
3. Dung, P.M.: On the acceptability of arguments and its fundamental role in nonmonotonic reasoning, logic programming and n-person games. Artif. Intell. **77**, 321–358 (1995)
4. Nielsen, S.H., Parsons, S.: A generalization of Dung's abstract framework for argumentation: Arguing with sets of attacking arguments. In: Maudet, N., Parsons, S., Rahwan, I. (eds.) Argumentation in Multi-Agent Systems, Lecture Notes in Computer Science, vol. 4766, pp. 54–73. Springer, Heidelberg (2006)

Acknowledgments

I wish to thank the following persons:

- First and foremost my Ph.D.-supervisors Joke Meheus and Diderik Batens: I was able to learn a lot from them since I began working at the CLPS. Their work on adaptive logics has been deeply inspiring and forms the very foundation of this book. I am thankful for all the effort and time they have invested in guiding and assisting my research.
- Lou Goble for his fascinating and inspiring work on deontic logics on which a part of this book is based.
- Leon Van Der Torre, Lou Goble, Peter Verdée, and Erik Weber for their excellent feedback on my Ph.D.-thesis which was very useful for this manuscript.
- My co-authors and collaborators Mathieu Beirlaen, Dunja Šešelja, Frederik Van De Putte, and Peter Verdée: It has been and is a great pleasure to work with you. I couldn't wish for better collaborators.
- My great professors of philosophy Wilhelm Lütterfelds and Thomas Mohrs for evoking my fascination and sharpening my thinking in the subject, and for keeping me deeply inspired while struggling with philosophical problems. I couldn't have wished for better teachers.
- Two most remarkable professors of mathematics and computer sciences to whose teaching I was addicted since its crystal clearness shines like an archetype of formal thinking: Volker Weispfennig, for evoking my fascination for the beauty of algebra, mathematical logic, and model theory. Klaus Donner, for teaching me that the choice of the right formal mathematical tools can enable one to master even the most difficult problems with ease. The razor-sharp and perfectly ordered thinking of the both of them will keep on being an inspirational source and force to me.
- Thomas Sturm, from whom I learned a lot about writing a thesis properly. He would have been my Ph.D.-supervisor if I had stayed in Computer Science. If a day would have 48 hours I wouldn't have hesitated a second.
- My family for all the support they have given me when and since I moved to Belgium.

The Kiwi illustration on page v is designed by Mike Harding under the Creative Commons CC0 license. It can be freely obtained at http://thenounproject.com/noun/kiwi/#icon-No1010.

This book has been typeset in LaTeX in a GNU/Linux environment.

Content

Part I
Adaptive Logics as a Framework for Defeasible Logics

Chapter 1
Introduction

The purpose of this introduction is to familiarize the reader with defeasible reasoning. It will, on the one hand, answer the questions what defeasible reasoning is and how it is different from deductive reasoning. On the other hand, I will introduce some themes concerning the formalization of defeasible reasoning that will recur frequently in this book. I will close the section by indicating some gaps in the formal treatment of defeasible reasoning which will bridge to the central topic of this manuscript: the use of adaptive logics as a unifying formal framework for defeasible reasoning.

1.1 Defeasible Reasoning

Where we deductively infer from some premises $A_1, \ldots, A_n$ a conclusion C, the truth of the premises warrants the truth of the conclusion. The truth of the propositions (i) "n is a prime number." and (ii) "$n > 2$" guarantees the truth of the statement "n is odd." The situation is different in the case of defeasible reasoning. When we infer from (i) "Tweety is a bird." and (ii) "Birds fly." the conclusion "Tweety flies.", the truth of the premises does not warrant the truth of the conclusion. This is due to the fact that Tweety may be a kiwi or a penguin.

What distinguishes the supportive character of the premises in the deductive case from the defeasible case is that in the former we have an unconditional support. As a consequence, if a conclusion gained by deductive reasoning turns out to be false, then we can infer that one of the premises is false as well: the unconditional character of the support does not allow for exceptional circumstances. Not so in the defeasible case. All premises may be true and still the conclusion may be false. This may seem to undermine the status of these inferences as valid forms of reasoning. Hence, we arrive at the question what compensates for this lack of truth-conduciveness in order to make defeasible inferences rationally compelling? After all, following Toulmin [1], any argument relies on a "warrant" or inference license.

C. Straßer, *Adaptive Logics for Defeasible Reasoning*, Trends in Logic 38,
DOI: 10.1007/978-3-319-00792-2_1, © Springer International Publishing Switzerland 2014

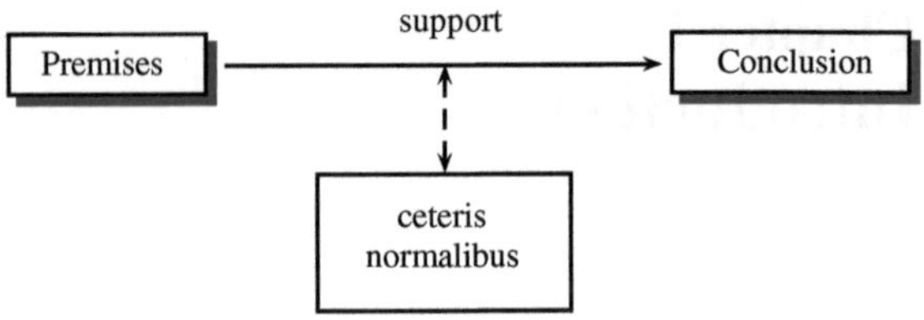

Fig. 1.1 A defeasible inference

Note first that there are many different types of defeasible reasoning. Different defeasible reasoning forms provide different rationales behind the support relation between the premises and the conclusion. We may for instance reason on the basis of normality, typicality, probability, etc.

Given such a defeasible inference type, although inferences do not always arrive at truthful conclusions from truthful premises, they are nevertheless "usually", "in most cases", "typically" or "normally" truth-conductive. Otherwise the given inference type would hardly be justified. Hence, the support provided by the premises has a certain, often tacit *ceteris normalibus* character (see Fig. 1.1). In the idealized "usual" resp. "normal" resp. "typical" etc. case, the inference is indeed truth-conductive. Hence, we may speak of an *idealized truth-conduciveness*, or a *truth-conduciveness ceteris normalibus*. The rationale given by a specific defeasible reasoning type explicates the exact character of the ceteris normalibus condition for the respective type. Let me give some examples.

The nature of the ceteris normalibus condition is obvious for defeasible reasoning on the basis of normality or typicality. The conclusion that Tweety flies is justified on the basis of our premises unless we also have the information that Tweety is a bird with rather exceptional or abnormal properties such as being a kiwi or a penguin, both of which cannot fly.

In inductive generalizations we may infer from the fact that a certain property common to a restricted number of samples of a certain class of objects is common to all objects in the class. What makes such an inference compelling is the tacit assumption that the sample class is normal in the sense that the homogeneity of the observed property does not just apply to the samples but rather that it is characteristic for the whole class.

In probabilistic reasoning we may apply what Pollock [2] dubbed "the statistical syllogism": "From 'X is an A and the probability of an A being a B is high', infer defeasibly, 'X is a B'". The tacit assumption here is that X is not abnormal or exceptional with respect to the given probabilities.

The fact that defeasible inferences are linked to a ceteris normalibus assumption that serves as its inference license is essentially connected to another feature of defeasible reasoning, namely its tentative character.[1] In strictly deductive logical systems the reasoning process has the following two distinctive properties.

[1] I agree with Blair that "an argument's warrant is not a premise, but instead an assumption" that warrants "inferences from *such* grounds to *such* conclusions": were the warrant a premise we would face a "vicious regress" ([3, p. 127] , see also Hitchcock [4]).

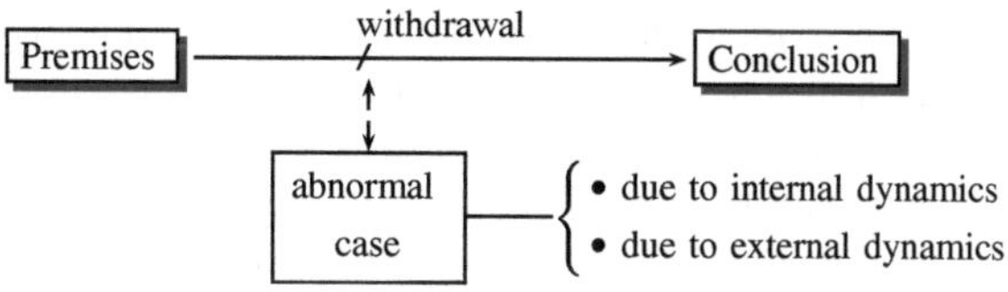

Fig. 1.2 Abnormalities and retraction

1. We infer in a monotonic manner, i.e. if we conclude A on the basis of a set of premises Γ, then we can deduce A also from the extended premise set $\Gamma \cup \Gamma'$.
2. We reason statically, i.e. if we infer at some point in the reasoning process A, we will not withdraw A due to the insights won in any extension of this reasoning process.

Defeasible reasoning differs from deductive reasoning at least in the latter, and usually in both aspects.

Note first that a defeasible inference is supported or justified as long as the information available to an agent does not give rise to certain abnormalities that interrupt the ceteris normalibus assumption and hence the support provided by the premises. In such cases defeasible inferences are withdrawn (see Fig. 1.2). For instance, Pollock points out that "[d]efeasibility arises from the fact that not all reasons are conclusive. Those that are not are prima facie *reasons*. Prima facie reasons create a presumption in favor of their conclusion, but it is defeasible" [5, p. 2]. Also Rescher argues that probative reasoning is "*presumptive* rather than deductively airtight" [6, p. 8] and the presumptions concerning "the usual, normal, customary course of things" (pp. 30–31) are "subject to defeat in being overthrown by sufficiently weighty countervailing considerations" whence "usually tentative and provisional" (p. 31).

The information available to an agent is on the one hand (a) explicit information presented to our agent by various external sources (such as perception, reliable agents, etc.) or on the other hand (b) the insight our agent has gained by analyzing and reasoning on the basis of this information.

It is important to notice that both factors, (a) and (b), have a certain dynamic character. Let us contrast the two types of dynamics with the characteristics of strictly deductive reasoning pointed out in 1. and 2. above.

1'. It is often the case that for defeasible reasoning processes the addition of new information leads to the retraction of previously drawn inferences. Pollock calls this the *synchronic defeasibility* of defeasible inferences [7]. If we obtain, for instance, the new information that Tweety is a penguin, we will withdraw the previous inference that Tweety flies, since, after all, penguins are exceptional to the default rule that birds fly. This corresponds to what we call the *external dynamics* of defeasible reasoning (see [8, 9]).

Hence, logics modeling defeasible reasoning are often nonmonotonic, i.e. what is derivable from a set of premises Γ may not be derivable from an enhancement of Γ with additional premises.[2]

2'. New information is not the only dynamic factor for defeasible reasoning. The growing insight into the given information may as well cause the withdrawal of previously drawn defeasible inferences even if no new information is available. Pollock dubs this the *diachronic defeasiblity* of defeasible inferences [7]. Often the given information is complex and interwoven. Only acute analyzing and reasoning may reveal new essential information with respect to previously drawn inferences. This corresponds to what we call the *internal dynamics* of defeasible reasoning (see [8, 9]).

1.2 Towards the Formalization of Defeasible Reasoning

The importance of defeasible reasoning for practical matters has already been noted and studied by Aristotle, for instance in the Topics. However, in the rise of non-defeasible mathematical logic, monotonicity provided a large bedrock for the mainstream of formal logic in the shadow of which defeasible enterprises were rather exotic and sparse for a long time. The term "defeasible" was introduced by Hart in 1948 in the philosophy of law [13]. But we have to wait until after the heyday of the Vienna Circle when the realization of the limitations of the purely deductive method grew widely, while at the same time the importance of the notion of defeasibility was emphasized. In epistemology defeasibility began to be a more widely discussed issue especially in the aftermath of Gettier's landmark article "Is true belief knowledge?" [14] (see for instance Lehrer and Paxson's [15], Annis' [16], Klein's [17], Swain's [18]). Two of the pioneers of defeasible reasoning are Chisholm (see e.g., [19]) and Pollock.

For instance Pollock's *Knowledge and Justification* [20] was one of the most successful landmarks in establishing defeasibility as a central notion in epistemology such that philosophical logicians could no longer turn a blind eye on it. Especially his distinction between two types of defeaters of defeasible arguments became paradigmatic. *Rebutting defeaters*, on the one hand, provide reasons supporting the belief in the negated conclusion of the given argument. On the other hand, there are *undercutting defeaters* that challenge the support that is provided by the premises of the given argument (see Fig. 1.3).

Soon researchers in artificial intelligence as well as philosophers began to develop first systems aiming at getting a formal grasp on defeasible reasoning. Nonmonotonic logics became its own research branch uniting scholars from philosophy, logic, artificial intelligence and computer science. Nowadays, about 30 years after the first

[2] There are exceptions such as the Weak Rescher-Manor inference relation [10] which is monotonic but which can be modeled by means of a dynamic proof theory that explicates internal dynamics (see point 2' below): this has been done in [11] and [12].

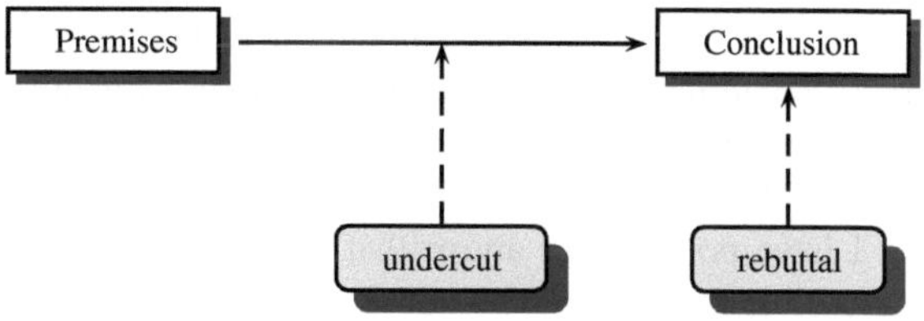

Fig. 1.3 Undercuts and Rebuttals

pioneering systems such as McCarthy's *circumscription* [21] or Reiter's *default logic* [22], we are facing a huge variety of systems. However, the evolutionary tree of nonmonotonic logics is multifarious and complicated. Most importantly it lacks a bedrock comparable to the one predicate logic provides for mathematical reasoning.

Nonmonotonicity is one of the central notions in the formal representation of defeasible reasoning. (Non)Monotonicity is a property of consequence relations. The latter are usually characterized as functions that map sets of formulas in a given language $\mathcal{L}$ into sets of formulas in $\mathcal{L}$. While some of the properties that are satisfied by the consequence relation of classical (propositional or predicate) logic, henceforth **CL**, are also desirable for consequence relations that model defeasible inferences, scholars agree that others have to be given up or have to be sufficiently altered. Let henceforth $Cn_{\mathbf{L}}(\Gamma)$ denote the set of **L**-consequences of the premise set Γ. For a nonmonotonic logic **L**, properties such as

$$\Gamma \subseteq Cn_{\mathbf{L}}(\Gamma) \qquad \text{(Reflexivity)}$$

$$Cn_{\mathbf{CL}}(\Gamma) \subseteq Cn_{\mathbf{L}}(\Gamma) \qquad \text{(Supraclassicality)}$$

are for many applications desirable. As has been already demonstrated by means of our Tweety example, the following central property of **CL** is abandoned in nonmonotonic systems:

$$Cn_{\mathbf{L}}(\Gamma) \subseteq Cn_{\mathbf{L}}\left(\Gamma \cup \Gamma'\right) \qquad \text{(Monotonicity)}$$

However, instead of totally abandoning the intuition behind monotonicity, it is one of the big challenges to develop a formal weakening of monotonicity that is adequate for certain forms of defeasible reasoning. Nowadays most scholars agree that the following cautious form of monotonicity is part of what forms the core properties of nonmonotonic consequence relations:

$$\text{If } \Gamma' \subseteq Cn_{\mathbf{L}}(\Gamma)\text{, then } Cn_{\mathbf{L}}(\Gamma) \subseteq Cn_{\mathbf{L}}\left(\Gamma \cup \Gamma'\right) \qquad \text{(CM)}$$

Cautious monotonicity is indeed an important and very intuitive notion. Suppose we have derived some A from a set of premises Γ. Adding A to our premise set Γ should in no way lead to less consequences than we have with Γ alone.

Such monotonicity-related properties of the consequence relation only reflect certain features of the external dynamics of the defeasible reasoning that is modeled.

Nonmonotonicity expresses that the arrival of new information, in some cases, gives rise to external dynamics. Properties such as cautious monotonicity resp. rational monotonicity (see Sect. 2.6.3.1) specify for a logic **L** enhancements of premise sets for which **L** does not give rise to external dynamics and hence behaves monotonic.

I have pointed out above that the internal dynamics plays an essential role in defeasible reasoning. The consequence relation itself does not give us any insight about the ways a logic models this aspect. After all, the consequence relation just informs us that a certain set of formulas Γ' is derivable from another set of formulas Γ if $\Gamma' \subseteq Cn_{\mathbf{L}}(\Gamma)$. The internal dynamics, however, concerns the path we take when we reason towards consequences. Due to the growing insight in the given premises we may retract certain defeasible inferences. The internal dynamics is not modeled by the consequence relation since the latter only offers a static view on the consequences of a given set of premises but doesn't inform us about the way in which we arrive at these consequences. The latter is offered by the proof theory of a logic. Static proofs[3] as they are usual for logics that model strictly deductive reasoning are obviously not able to display the internal dynamics of defeasible reasoning processes.

It is remarkable that most of the available systems that model defeasible reasoning lack a proof theory. That is to say, most of the proposed systems are not "full logics" in the sense of providing a proof theory that is complete with respect to a given semantics. Hence, it is also not surprising that scholars mostly emphasize the external dynamics and the nonmonotonic character of consequence relations while neglecting the explication of the internal dynamics of defeasible reasoning. Indeed, it is challenging to develop proof calculi that model defeasible reasoning. This is due to the internal and external dynamics of defeasible reasoning. The static proof format that is characteristic of classical logic is not apt to model the internal dynamics and has to be altered in favor of a dynamic format.

Another problem in the research done on the formalization of defeasible reasoning is the lack of a unificatory framework. Currently the various systems are formulated by means of a vast variety of formal frameworks. What is missing is a framework that is powerful enough to embed the given systems. This would be useful for various purposes. For instance, by bridging the different formalisms in which the proposals are expressed, we may get better means to compare or combine systems. Generic unificatory research programs and corresponding frameworks such as universal algebra or category theory in mathematics are still in a rather immature state for nonmonotonic logics.

⋆⋆⋆

This discussion marks the grounds and horizon from which this book emerges. In the next section I will introduce the reader to a logical framework, namely adaptive logics, that offers an, as I will argue, attractive contribution to the research on formalizations of defeasible reasoning forms. It has been developed by Diderik Batens in the early eighties. While the first application was to interpret (possibly) inconsistent

[3] A proof from some premises Γ is static if for any A that is derived on a line of it, A is a consequence. The reader will be introduced to the dynamic proof format of ALs in Part 2 of this book.

theories as consistently as possible [23], the family of adaptive logics has been growing rapidly and nowadays it covers many application contexts such as the modeling of induction, abduction, discussions, etc.[4]

It is a central claim and purpose of this manuscript to demonstrate

(a) that adaptive logics offer a unifying generic framework that is powerful enough to embed various well-known nonmonotonic systems;
(b) that adaptive logics are able to nonmonotonically strengthen, improve upon and enrich monotonic systems; and
(c) that adaptive logics offer a dynamic proof theory that explicates the dynamics of defeasible reasoning.

I will give examples from various fields, such as default reasoning, argumentative reasoning and normative reasoning.

References

1. Toulmin, S.E.: The Uses of Argument. Cambridge University Press, Cambridge (1958)
2. Pollock, J.: Defeasible reasoning. Cogn. Sci. **11**(4), 481–518 (1987)
3. Blair, A.: Informal logic and logic. In: Tindale, C.W. (ed.) Groundwork in the Theory of Argumentation, vol. 241, pp. 119–132
4. Hitchcock, D.: Toulmin's warrants. In: Eemeren F.H.V. (ed.) Anyone Who has a View: Theoretical Contributions to the Study of Argumentation, pp. 485–490. Springer, Dordrecht (2003)
5. Pollock, J.: How to reason defeasibly. Artif. Intell. **57**(1), 1–42 (1992)
6. Rescher, N.: Dialectics: a controversy-oriented approach to the theory of knowledge. SUNY Press, Albany (1977)
7. Pollock, J.: Defeasible reasoning. In: J.E. Adler, L.J. Rips (eds.) Reasoning: Studies of Human Inference and Its Foundations, pp. 451–470. Cambridge University Press, Cambridge (2008)
8. Batens, D.: The need for adaptive logics in epistemology. In: Gabbay, D., Rahman, S., Symons, J., Bendegem, J.P.V. (eds.) Logic, Epistemology Unity Science, pp. 459–485. Kluwer Academic Publishers, Dordrecht (2004)
9. Batens, D.: A universal logic approach to adaptive logics. Logica Universalis **1**, 221–242 (2007)
10. Rescher, N., Manor, R.: On inference from inconsistent premises. Theor. Decis. **1**, 179–217 (1970)
11. Batens, D., Vermeir, T.: Direct dynamic proofs for the Rescher-Manor consequence relations: the flat case. J. Appl. Non-Class Logics **12**, 63–84 (2002)
12. Meheus, J., Straßer, C., Verdée, P.: Which style of reasoning to choose in the face of conflicting information? J. Log. Comput. (2013, forthcoming)
13. Hart, H.L.A.: The ascription of responsibility and rights. In: Proceedings of the Aristotelian Society (1948)
14. Gettier, E.: Is justified true belief knowledge? Analysis **23**(6), 121–123 (1963)
15. Lehrer, K., Paxson Thomas, J.: Knowledge: undefeated justified true belief. J. Philos. **66**(8), 225–237 (1969)
16. Annis, D.: Knowledge and defeasibility. Philos. Stud. **24**(3) (1973)
17. Klein, P.D.: A proposed definition of propositional knowledge. J. Philos. **68**(16), 471–482 (1971)
18. Swain, M.: Epistemic defeasibility. Am. Philos. Q. **11**, 15 (1974)

[4] For a more thorough overview see Chap. 4.

19. Chisholm, R.M.: Theory of Knowledge. Prentice Hall, Englewood Cliffs, NJ (1966)
20. Pollock, J.: Knowledge and Justification. Princeton University Press, Princeton (1970)
21. McCarthy, J.: Circumscription—a form of non-monotonic reasoning. Artif. Intell. **13**, 27–29 (1980)
22. Reiter, R.: A logic for default reasoning. Artif. Intell. 1–2(13) (1980)
23. Batens, D.: Paraconsistent extensional propositional logics. Logique at Analyse **90–91**, 195–234 (1980)

Chapter 2
The Standard Format for Adaptive Logics

The purpose of this section is to introduce the reader to ALs with a special eye on the modeling of defeasible reasoning. The standard format of ALs has been introduced by Diderik Batens (see e.g. [1, 2] for a systematic study). As will be shown in the following, for the standard format a rich meta-theory is available which equips ALs with many desirable properties and at the same time provides a unifying framework to ALs.

2.1 The Standard Format

The basic idea behind ALs is to interpret a given set of premises "as normally as possible". Depending on the application this may have different meanings. Let me give some examples:

(i) In applications in which we are confronted with inconsistent information we may want to interpret the premises as consistently as possible.
(ii) In applications in which we are confronted with conflicting norms and obligations we may want to interpret the premises as non-conflicting as possible.

There are three elements that constitute ALs in the standard format:

1. the lower limit logic **LLL**,
2. the set of abnormalities Ω, and
3. the adaptive strategy: reliability or minimal abnormality.

$\mathbf{AL}^{\mathbf{r}}$ denotes the AL defined by $\langle \mathbf{LLL}, \Omega, \text{reliability} \rangle$ and $\mathbf{AL}^{\mathbf{m}}$ denotes the AL defined by $\langle \mathbf{LLL}, \Omega, \text{minimal abnormality} \rangle$. By **AL** I will refer to either of the two.

In the following sections I will introduce each element of the standard format, beginning with the lower limit logic.

C. Straßer, *Adaptive Logics for Defeasible Reasoning*, Trends in Logic 38,
DOI: 10.1007/978-3-319-00792-2_2, © Springer International Publishing Switzerland 2014

2.2 The Lower Limit Logic

ALs employ and strengthen a monotonic logic **LLL**, their so-called *lower limit logic*. This logic is a reflexive, transitive, monotonic and compact logic that has a characteristic semantics. Hence we have:

- *Reflexivity*: $\Gamma \subseteq Cn_{\mathbf{LLL}}(\Gamma)$.
- *Transitivity*: If $\Gamma' \subseteq Cn_{\mathbf{LLL}}(\Gamma)$ then $Cn_{\mathbf{LLL}}(\Gamma') \subseteq Cn_{\mathbf{LLL}}(\Gamma)$.
- *Monotonicity*: $Cn_{\mathbf{LLL}}(\Gamma) \subseteq Cn_{\mathbf{LLL}}(\Gamma \cup \Gamma')$.
- *Compactness*: If $A \in Cn_{\mathbf{LLL}}(\Gamma)$ then there is a finite $\Gamma' \subseteq \Gamma$ such that $A \in Cn_{\mathbf{LLL}}(\Gamma')$.

For instance in application (i) lower limit logics are of interest that are inconsistency-tolerant. That is to say, logics that do not validate the 'ex contradictione quodlibet' principle:

$$(A \wedge \neg A) \supset B \qquad \text{(ECQ)}$$

Were we to employ a logic as the lower limit logic that validates (ECQ) then **AL** would trivialize premise sets that contain $A \wedge \neg A$.

Candidates serving as lower limit logic are for instance **CLuN** (see [3]), **CLuNs** (see [4]) or da Costa's $\mathbf{C_i}$ systems (see [5, 6]). Note though that not all ALs that model reasoning on the basis of conflicting information are based on subclassical lower limit logics. Indeed, by translating the input Γ for instance to $\Gamma^{\Diamond} = \{\Diamond A \mid A \in \Gamma\}$ one can use classical modal logics as lower limits (see e.g., [7, 8]). We offer a more simple non-modal approach with a "dummy operator" that precedes premises in Sect. 2.4 and in [9].[1]

For application (ii) systems of interest are deontic logics that are conflict-tolerant, i.e. logics that do not cause deontic explosion given deontic conflicts. Where $\mathsf{O}A$ indicates the obligation to bring about A, the deontic explosion principle (D-EX) is given by

$$(\mathsf{O}A \wedge \mathsf{O}\neg A) \supset \mathsf{O}B \qquad \text{(D-EX)}$$

Examples of logics that do not validate (D-EX) are Lou Goble's **P** (see e.g. [10–12]) or his **DPM** systems (see e.g. [13–15]).

The lower limit logic constitutes the core of an AL in two senses. Semantically, an AL selects from the **LLL**-models of a given premise set models that are "sufficiently normal" according to a given standard of normality. The latter is characterized by the other two elements of ALs, the abnormalities and the adaptive strategy as will be demonstrated below.

Syntactically, all the rules of the proof theory of **LLL** are applicable. As a consequence, everything that is provable in **LLL** is also provable in the adaptive system. As will be explicated later, ALs enhance the static proof theory of **LLL** by a dynamic element, that in many cases allows for additional consequences.

[1] Note also that all lower limit logics used in applications in parts II-IV of this book are supraclassical.

Where **LLL** is defined over a language $\mathcal{L}$, we write $\mathcal{W}$ for the set of well-formed formulas in $\mathcal{L}$. The consequence relation of **LLL** is hence a mapping $\wp(\mathcal{W}) \rightarrow \wp(\mathcal{W})$.

For the adaptive meta-theory it is very useful to extend the language of **LLL** by classical connectives, written in a "checked way", e.g. $\check{\neg}$ and $\check{\vee}$. We denote the enriched language by $\mathcal{L}^+$ and the corresponding set of well-formed formulas by $\mathcal{W}^+$, where $\mathcal{W}^+$ is the $\langle \check{\neg}, \check{\vee}, \check{\wedge}, \check{\supset}, \check{\equiv} \rangle$-closure of $\mathcal{W}$. Note that this means that none of the "checked connectives" occurs within the scope of the connectives of $\mathcal{L}$. For instance, where $\rightarrow$ is a connective of $\mathcal{L}$, $\check{\neg}(A \rightarrow B)$ is a formula in $\mathcal{W}^+$, but $(\check{\neg} A) \rightarrow B$ is not.

Let **LLL**$^+$ be the logic that is the result of superimposing the classical symbols on **LLL**. Namely, **LLL**$^+$ takes over the axiomatization of **LLL** and restricts the rules and axioms of **LLL** to formulas in $\mathcal{W}$. Moreover, the classical axioms for the checked connectives are defined for all formulas in $\mathcal{W}^+$.[2] Semantically the internal structure of the **LLL**-models may be kept. Similarly as for the axiomatization, the semantic clauses of **LLL** are restricted to formulas of $\mathcal{L}$, while for the checked symbols we have $M \models \check{\neg} A$ iff $M \not\models A$, $M \models A \check{\vee} B$ iff $M \models A$ or $M \models B$, etc. Thus, it will not be necessary to formally distinguish between **LLL**-models and **LLL**$^+$-models.

In the adaptive meta-theory the derivability relation $\vdash_{\mathbf{LLL}^+}$ plays an essential role. However, it is customarily denoted by "$\vdash_{\mathbf{LLL}}$". Hence, the reader should not be surprised to find formulas in $\mathcal{W}^+ \setminus \mathcal{W}$ on the left- or right-hand-side of $\vdash_{\mathbf{LLL}}$. In order not to depart too much from the literature on ALs, I will adopt this convention while providing the reader unfamiliar with ALs with this warning.[3] Similarly there are two consequence relations corresponding to **LLL** and **LLL**$^+$. We define, where $\Gamma \subseteq \mathcal{W}$, $Cn^{\mathcal{L}}_{\mathbf{LLL}}(\Gamma) =_{\mathrm{df}} \{A \in \mathcal{W} \mid \Gamma \vdash_{\mathbf{LLL}} A\}$ and, where $\Gamma \subseteq \mathcal{W}^+$, $Cn^{\mathcal{L}^+}_{\mathbf{LLL}}(\Gamma) =_{\mathrm{df}} \{A \in \mathcal{W}^+ \mid \Gamma \vdash_{\mathbf{LLL}^+} A\}$. Where I skip the superscript either of the two readings may be applied.

2.3 The Abnormalities

In Sect. 1.1, I have characterized a defeasible inference as an inference that is supported by its premises 'ceteris normalibus' (cf. Fig. 1.1). The inference is warranted if and as long as there is no reason to suppose that certain abnormalities that violate the ceteris normalibus condition are the case (cf. Fig. 1.2). ALs formalize this principle.

[2] Often bridge principles need to be added. E.g., where $\vee$ is a classical disjunction in $\mathcal{L}$, the axiom $(A \check{\vee} B) \check{\equiv} (A \vee B)$ is added to ensure the equivalence between the two classical disjunctions.

[3] Note that the "checked" classical connectives are added even in the case that **LLL** already contains classical corresponding symbols. The reason is of a rather technical nature: it is to ensure that a formula is derivable already at a finite stage of an adaptive proof (cf. Section 2.7 and the discussion in Section 4.9.3 of [2]).

Abnormalities are characterized by a logical form F in the enriched language $\mathcal{L}^+$. Formulas of this form are supposed to be **LLL**-contingent, i.e. $\nvdash_{\mathbf{LLL}}$ F and $\nvdash_{\mathbf{LLL}} \check{\neg}$ F. By Ω we denote the set of all formulas of the form F.

For our application (i) abnormalities may have the form of inconsistencies, $A \wedge \neg A$. For application (ii) abnormalities may have the form of deontic conflicts, $\mathsf{O}A \wedge \mathsf{O}\neg A$.

To interpret the premises "as normally as possible" means to interpret the premises in such a way that as few abnormalities as possible are validated. We will see that semantically ALs select **LLL**-models of a given premise set that are "sufficiently normal" in terms of the abnormalities they validate. Proof-theoretically the idea is to apply certain rules conditionally, namely on the condition that certain abnormalities are false. These points are realized and disambiguated by the last element, the adaptive strategy.

2.4 The Adaptive Strategy

Adaptive strategies are the technically most involving aspect of ALs. Currently two strategies are part of the standard format: the minimal abnormality strategy and the reliability strategy. Together with the abnormalities they substantiate what it means to interpret premises as "normally as possible".

I will introduce a "toy" application in order to explicate the different intuitions behind the two strategies.

Let us model the defeasible reasoning of a detective. Suppose a murder happened. There are two witnesses. One states that the major suspect Mr. X entered the scene of crime right before the lethal shot was heard throughout the whole neighborhood. Another one states that he saw the major suspect leaving the scene of crime directly after the shot was heard. Moreover, our detective has the information that nobody else was at the scene of crime shortly before and shortly after the time of the killing.

We model the fact that there is evidence available for A by $\circ A$ (e.g., some witness may state A, A may be the result of forensic investigations, etc.). A $\circ$-less formula A expresses that A is a fact, or that there is definite proof for A, or that our detective accepts A as fact. Since we want to keep things simple we treat $\circ$ as a dummy operator and hence don't attach any logical properties to $\circ$. As a lower limit logic we employ classical propositional logic **CL** equipped with $\circ$. Let this logic be named $\mathbf{CL}_\circ$.[4] The semantics of $\mathbf{CL}_\circ$ is like the semantics for **CL**, just besides the usual assignment function v that assigns to each propositional letter a truth value, we also use an enhanced assignment function $v_\circ$ that (independently from v) associates each well-formed formula with a truth-value. Truth in a model M is defined as usual for the classical operators:

---

[4] In [9] we show that $\mathbf{CL}_\circ$ gives rise to very simple ALs that represent the Rescher-Manor consequence relations [16].

- $M \models A$ where A is a propositional letter iff $v(A) = 1$
- $M \models \neg A$ iff $M \not\models A$
- $M \models A \vee B$ iff $M \models A$ or $M \models B$
- and similar for the other classical connectives.

The $\circ$ operator is characterized by

- $M \models \circ A$ iff $v_\circ(A) = 1$.

The idea is that

(a) if our detective has evidence for A,— $\circ A$;
(b) and as long as there is no reason to assume that A is not the case,— $\neg \circ \neg A$,

then the detective is warranted to defeasibly infer that A is the case. Of course, $\mathbf{CL}_\circ$ is a monotonic system. We will in a moment strengthen it in a nonmonotonic adaptive way.

But let us return to our detective. Assume he has the following evidence:

- shortly before and shortly after the time of death nobody but the victim was at the scene of crime,— $\circ n$;
- that Mr. X entered the scene of crime alone right before the shot,— $\circ a$;
- that Mr. X left the scene of crime alone right after the shot,— $\circ b$.

Moreover, we presuppose that for some reason our detective accepts that if nobody else was at the scene of crime shortly before and shortly after the crime, but Mr. X entered the scene of crime alone right before the shot was heard, then he must be the murderer: $(a \wedge n) \supset c$. Similarly, $(b \wedge n) \supset c$.

What makes the situation more complicated is that our detective has definite proof that at least one of the witnesses has been bribed by one of Mr. X's enemies in order to fake a witness statement. Hence, since one of the witnesses lies, we have $\neg a \vee \neg b$. What should our detective conclude?[5]

2.4.1 The Reliability Strategy

If she takes a cautious stance, she will not conclude that Mr. X is the murderer since after all, both of the witnesses may be bribed. Let us elaborate a bit on this stance.

I have already mentioned that semantically ALs select from the lower limit logic models of the given premises the ones that are "sufficiently normal" with respect to a certain standard of normality. The latter is characterized by the abnormalities and the adaptive strategy.

The abnormalities for our application are cases where our detective has evidence for A but A is not the case. Hence $\Omega_\circ = \{\circ A \wedge \neg A\}$. Let henceforth $\mathbf{CL}_\circ{}^r$ be the AL defined by the triple:

---

[5] I do of course not claim that the modeling of the defeasible reasoning of our detective by $\mathbf{CL}_\circ$ is by any means optimal. It is however sufficiently intuitive and simple in order to serve as a toy application for introducing the basic concepts and mechanisms of ALs.

1. lower limit logic: $\mathbf{CL}_\circ$
2. abnormalities: $\Omega_\circ$
3. strategy: reliability

2.4.1.1 The Semantics

Let us first take a look at the semantics. What $\mathbf{CL}_\circ$-models of the premise set

$$\Gamma_1 = \{\circ n, (a \wedge n) \supset c, (b \wedge n) \supset c, \circ a, \circ b, \neg a \vee \neg b\}$$

should be selected according to the cautious rationale of our detective?

An important notion is the so-called *abnormal part* of a model. It consists of all the abnormalities validated by a given model M, in symbols

$$\mathrm{Ab}(M) = \{A \in \Omega \mid M \models A\}$$

For our applications the abnormal part of an $\mathbf{CL}_\circ$-model M is thus, $\mathrm{Ab}(M) = \{A \in \Omega_\circ \mid M \models A\}$. I will in the remainder of this section abbreviate abnormalities $\circ A \wedge \neg A$ by $!A$. Note that in $\mathbf{CL}_\circ$ we have the following:

$$\circ A, \circ B, \neg A \vee \neg B \vdash_{\mathbf{CL}_\circ} !A \check{\vee} !B$$

Hence, in every $\mathbf{CL}_\circ$-model of Γ_1 at least one of the abnormalities $!a$ and $!b$ is valid. Let us focus for our discussion on the following models of Γ_1:[6]

The abnormal part imposes a strict partial order $\sqsubset^{\Gamma}_{\mathrm{Ab}}$ on the lower limit logic models of a given premise set Γ where $M \sqsubset^{\Gamma}_{\mathrm{Ab}} M'$ iff $\mathrm{Ab}(M) \subset \mathrm{Ab}(M')$. Similarly, we define the partial order $\sqsubseteq^{\Gamma}_{\mathrm{Ab}}$ on the lower limit logic models of Γ by: $M \sqsubseteq M'$ iff $\mathrm{Ab}(M) \subseteq \mathrm{Ab}(M')$. For our six models this is illustrated in Fig. 2.1a.

Interpreting the premises "as normally as possible" first of all means that in cases in which we have no reason to suppose that an abnormality $!A$ occurs, we should

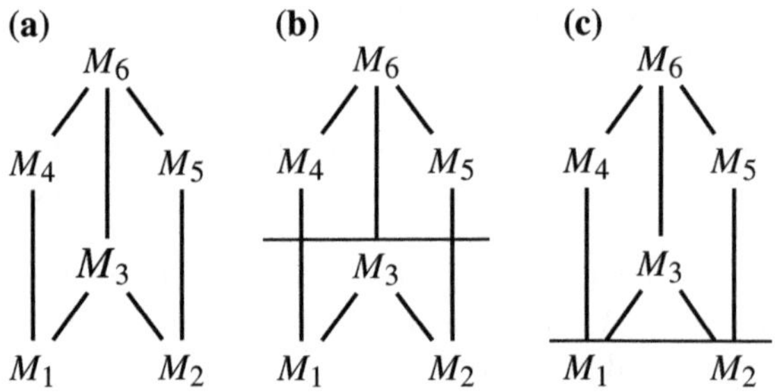

Fig. 2.1 **a** An excerpt of the partial order $\sqsubset^{\Gamma}_{\mathrm{Ab}}$ on the $\mathbf{CL}_\circ$-models of Γ_1; **b** under the line are reliable models; **c** under the line are minimal abnormal models

[6] I do not exhaustively characterize these models by means of what formulas they validate. However, it is obvious that models such as M_1 to M_6 exist.

presume that $!A$ is not the case. Take for instance our premise on. Since the premises give no reason for supposing $\neg n$ (we will make this formally precise in a moment) the semantic selection corresponding to the reasoning of our detective neglects models M_4, M_5 and M_6 since these models validate the abnormality $on \wedge \neg n$.

This can be made more precise by introducing another central notion for ALs: *minimal Dab-consequences*. Where $\Delta \subseteq \Omega$ is a finite and non-empty set of abnormalities, adaptive logicians use $\mathsf{Dab}(\Delta)$ as a notation for the classical disjunction of members in Δ: $\check{\bigvee}\Delta$. Where $\Delta = \emptyset$ the string '$\check{\vee}\,\mathsf{Dab}(\Delta)$' denotes the empty string. The minimal Dab-consequences derivable from a given premise set Γ are all $\mathsf{Dab}(\Delta)$ for which (i) $\Gamma \vdash_{\mathbf{LLL}} \mathsf{Dab}(\Delta)$ and (ii) there is no $\Delta' \subset \Delta$ such that $\Gamma \vdash_{\mathbf{LLL}} \mathsf{Dab}(\Delta')$. For a minimal Dab-consequence $\mathsf{Dab}(\Delta)$ we know that in each **LLL**-model of Γ at least one of the abnormalities in Δ is validated. Due to the minimality of Δ there is no $\Delta' \subset \Delta$ with the same property. Where $\mathsf{Dab}(\Delta_1)$, $\mathsf{Dab}(\Delta_2), \ldots$ are the minimal Dab-consequences, the set of *unreliable abnormalities* is $U(\Gamma) = \Delta_1 \cup \Delta_2 \ldots$

Indeed, there is no reason to assume that an abnormality is true in case it is not unreliable. After all, in this case it is not a disjunct of any minimal Dab-consequence. Of course, it may still be a disjunct of a *non*-minimal Dab-consequence. However, just as there is no reason to believe that it rains just because we can derive "It rains or it is windy" from "It is windy", there is no reason to believe that an abnormality is true just because by means of addition we can add it as a disjunct to a Dab-formula.

In our example the only minimal Dab-consequence is $!a \check{\vee} !b$. Hence, $U(\Gamma_1) = \{!a, !b\}$. Note that $!n \notin U(\Gamma_1)$. The idea is to select only lower limit models that validate only abnormalities in $U(\Gamma)$, i.e. models M that satisfy $\mathrm{Ab}(M) \subseteq U(\Gamma)$. We call these models the *reliable models of Γ*. Models M_1, M_2 and M_3 satisfy this requirement with respect to our premise set Γ_1 (see Fig. 2.1b).

As discussed above, the cautious rationale underlying the reliability strategy also takes into account the possibility that both of our witnesses have been bribed. Hence both abnormalities, $!a$ and $!b$ may be valid. Models M_1, M_2 and M_3 validate at least one of the two abnormalities. M_3 validates both of them. Note that in the model M_3, c is not validated. After all, the interpretation offered by M_3 treats both a and b as unreliable and thus in this interpretation neither $(a \wedge n) \supset c$ nor $(b \wedge n) \supset c$ can be used for deriving c. Hence, our cautious detective does not (tentatively) conclude that Mr. X is the murderer.

Generically the semantic consequence relation for the reliability strategy is defined as follows.

Definition 2.4.1. Where $\mathcal{M}_{\mathbf{AL}^{\mathrm{r}}}(\Gamma)$ is the set of all reliable **LLL**-models of Γ,

$$\Gamma \Vdash_{\mathbf{AL}^{\mathrm{r}}} A \text{ iff for all } M \in \mathcal{M}_{\mathbf{AL}^{\mathrm{r}}}(\Gamma)\,,\, M \models A.$$

Note that we have $\Gamma_1 \nVdash_{\mathbf{CL}_{\circ}{}^{\mathrm{r}}} c$ since the reliable model M_3 does not validate c.

Given the definition of reliable models we immediately get the following representational theorem (where $\Gamma^{\check{\neg}} =_{\mathrm{df}} \{\check{\neg} A \mid A \in \Gamma\}$):

Theorem 2.4.1. *Where* $\Gamma \subseteq \mathcal{W}^+$: $\Gamma \Vdash_{\mathbf{AL^r}} A$ *iff* $\Gamma \cup (\Omega \setminus U(\Gamma))^{\check{\neg}} \Vdash_{\mathbf{LLL}} A$.

By the compactness of **LLL** this implies:

Corollary 2.4.1. *Where* $\Gamma \subseteq \mathcal{W}^+$: $\Gamma \Vdash_{\mathbf{AL^r}} A$ *iff there is a* $\Delta \subseteq \Omega \setminus U(\Gamma)$ *such that* $\Gamma \Vdash_{\mathbf{LLL}} A \check{\vee} \mathsf{Dab}(\Delta)$.

2.4.1.2 The Proof Theory

Let me now show how the reliability strategy is realized by adaptive proofs. The adaptive proof format enhances the static proofs of the lower limit logic by an additional column in which conditions are attached to proof lines. Conditions are finite and possibly empty sets of abnormalities. A line in a proof consists of a line number, a formula, a justification, and a condition. The central feature of adaptive proofs is that they apply certain rules conditionally. Let me explicate this again by our example.

Note first that in $\mathbf{CL}_\circ$ the following rules are *not* valid:

$$\text{If } \circ A, \text{ then } A. \tag{2.1}$$

$$\text{If } \circ A \text{ and } A \supset B, \text{ then } B. \tag{2.2}$$

However, the following is valid[7]:

$$\circ A \vdash_{\mathbf{CL}_\circ} A \check{\vee} !A \tag{2.3}$$

$$\circ A, A \supset B \vdash_{\mathbf{CL}_\circ} B \check{\vee} !A \tag{2.4}$$

Hence, by (2.3), given $\circ A$ either A or the abnormality $!A$ is the case. Our AL enables conditional applications of rules (2.1) and (2.2). That is to say, from $\circ A$, A is derived "on the condition $\{!A\}$", or from $\circ A$ and $A \supset B$, B is derived "on the condition $\{!A\}$". Roughly the idea is to apply rules (2.1) and (2.2) on the condition that $!A$ can be considered not to be the case (see Fig. 2.2). This is still an ambiguous phrase and has different readings according to the two strategies.

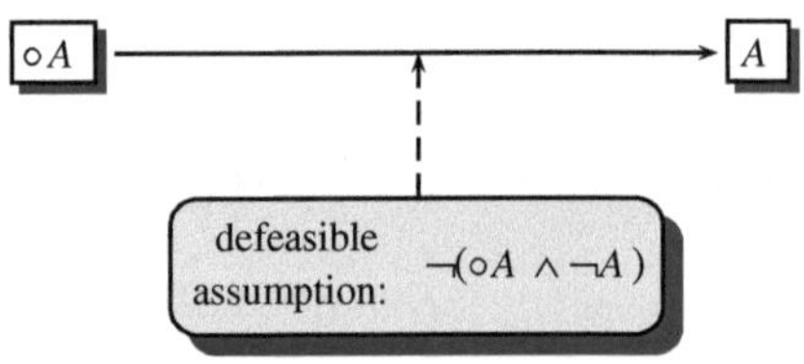

Fig. 2.2 Conditional inference

[7] In order to reduce notational clutter I will often omit set brackets on the left hand side of $\vdash$.

For the reliability strategy this is spelled out as follows: deriving A "on the condition Δ" means that A is derived on the condition that no member of Δ is unreliable. Let us have a look at a proof fragment:

1	$\circ n$	PREM	$\emptyset$
2	$(a \wedge n) \supset c$	PREM	$\emptyset$
3	$(b \wedge n) \supset c$	PREM	$\emptyset$
4	$\circ a$	PREM	$\emptyset$
5	$\circ b$	PREM	$\emptyset$
6	$\neg a \vee \neg b$	PREM	$\emptyset$
$^{10}7$	a	4; RC	$\{!a\}$
8	n	1; RC	$\{!n\}$
$^{10}9$	c	2, 7, 8; RU	$\{!a, !n\}$
10	$!a \check{\vee} !b$	4, 5, 6; RU	$\emptyset$

The first thing to notice is that, although for our applications we are interested in the adaptive consequence relation over the language $\mathcal{L}$ that characterizes our lower limit logic, the adaptive proofs are formulated in the enriched language $\mathcal{L}^+$. As the reader will see, this plays an important role in the modeling of defeasible reasoning in adaptive proofs. The proofs are governed by three generic rules: PREM, RU, and RC. Let us have a look at them separately.

At lines 1–6 premises are introduced. This is enabled by a generic premise introduction rule:

$$\text{If } A \in \Gamma: \quad \frac{\begin{matrix} \vdots & \vdots \end{matrix}}{\begin{matrix} A & \emptyset \end{matrix}} \tag{PREM}$$

Beside the premise introduction rule there are two other generic rules characterizing adaptive proofs: the unconditional rule RU and the conditional rule RC. Via RU the adaptive proofs come with all of the deductive power of the lower limit logic:

$$\text{If } A_1, \ldots, A_n \vdash_{\mathbf{LLL}} B: \quad \frac{\begin{matrix} A_1 & \Delta_1 \\ \vdots & \vdots \\ A_n & \Delta_n \end{matrix}}{\begin{matrix} B & \Delta_1 \cup \cdots \cup \Delta_n \end{matrix}} \tag{RU}$$

Note that the conditions of the used lines are carried forward.

The core and finesse of adaptive proofs comes with the conditional rule. It has been illustrated by means of the rules (2.1) and (2.2) above. In general the rule reads as follows[8]:

[8] Note that, as already mentioned earlier, I stick with the customary usage of $\vdash_{\mathbf{LLL}}$ in RU and RC as denoting the derivability relation $\vdash_{\mathbf{LLL}^+}$ characterizing the strengthened lower limit logic that operates on $\mathcal{L}^+$.

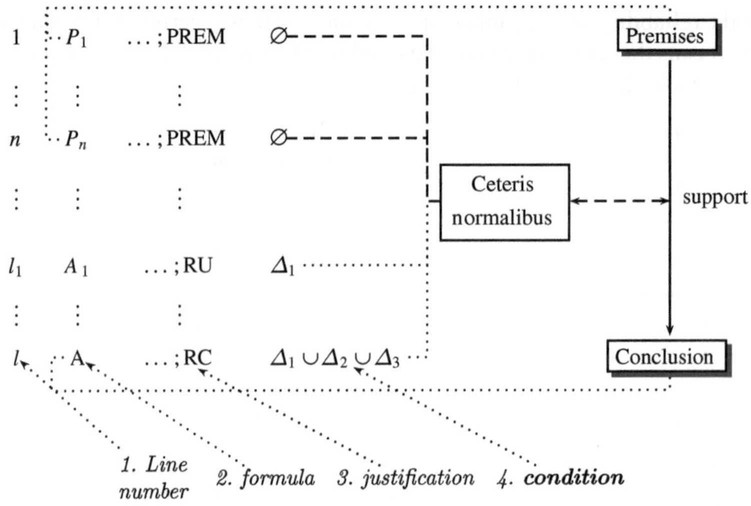

Fig. 2.3 Schematic illustration of an adaptive proof

$$\text{If } A_1, \ldots, A_n \vdash_{\mathbf{LLL}} B \check{\vee} \mathsf{Dab}(\Theta): \quad \begin{array}{ll} A_1 & \Delta_1 \\ \vdots & \vdots \\ A_n & \Delta_n \\ \hline B & \Delta_1 \cup \cdots \cup \Delta_n \cup \Theta \end{array} \qquad \text{(RC)}$$

At lines 7 and 8 we have conditional applications of rule (2.1). Take for instance line 7: the idea here is to derive defeasibly a from $\circ a$ on the condition $\{\circ a \wedge \neg a\}$. That is to say, from the fact that our detective has a good reason to assume a she derives a on the condition that not-a is not the case. The ceteris normalibus condition of this type of defeasible inference is that whenever there is a good reason to assume some a then, normally, $\neg a$ should not hold. In Fig. 2.3 our generic scheme for defeasible inferencing from Fig. 1.1 is related to the proof format of ALs.

At line 10 in our proof from Γ_1 the only minimal Dab-consequence is derived on the empty condition. At this point something important happens: the conditions of lines 7 and 9 are violated. After all, $!a$ turned out to be unreliable at line 10. In adaptive proofs, lines the conditions of which have been violated, are marked. The marking indicates that the second elements of these lines are not considered to be derived. Indeed, as long as the marking persists, the ceteris normalibus condition that guarantees the support from the premises is violated.

Before I give a formal definition of the marking, it is important to note that markings are dynamic. They may come and go. In order to see this, suppose for the moment that our detective has definite proof that the second witness has been bribed and thus has been lying. Where $\Gamma_2 = \Gamma_1 \cup \{\neg b\}$, we add the following lines to the proof from Γ_2:

11	$\neg b$	PREM	$\emptyset$
12	$!b$	5, 11; RU	$\emptyset$

What is remarkable here is that adding $\neg b$ to our premises leads to an alteration of the unreliable abnormalities. Now $!b$ is the only minimal Dab-consequence and $U(\Gamma_2) = \{!b\}$. Hence, the conditions of lines 7 and 9 can now be considered to be reliable. Consequently, these lines are unmarked at line 12.

At different stages of the proof the 'minimal Dab-formulas'[9] that are derivable are different. By analyzing a premise set in a proof, our insight in the premises grows and hence what is considered as an unreliable formula at a certain stage of the proof may change. Hence, in order to define the marking in such a way that it mirrors the dynamics of the defeasible reasoning that is modeled, we need to define the set of unreliable formulas such that it is relative to the stage of the current proof.

We say $\mathsf{Dab}(\Delta)$ is a *minimal Dab-formula at stage s* of a proof iff

(i) $\mathsf{Dab}(\Delta)$ has been derived on the empty condition at stage s, and
(ii) for all $\Delta' \subset \Delta$, $\mathsf{Dab}(\Delta')$ has not been derived on the empty condition at stage s.

Moreover, where $\mathsf{Dab}(\Delta_1)$, $\mathsf{Dab}(\Delta_2), \ldots$ are the minimal Dab-formulas at stage s, the set of unreliable formulas at stage s is $U_s(\Gamma) = \Delta_1 \cup \Delta_2 \cup \ldots$ The marking for the reliability strategy is defined as follows:

Definition 2.4.2 (Marking for the Reliability Strategy). Line i is marked at stage s iff, where Δ is its condition, $\Delta \cap U_s(\Gamma) \neq \emptyset$.

Note that, on the one hand, marked lines may be unmarked at a later stage of a proof. On the other hand, unmarked lines may be marked at a later stage. Suppose our detective has definite proof that also the first witness has been bribed. In this case the conditions of line 7 and 9 are violated again.

13	$\neg a$	PREM	$\emptyset$
14	$!a$	4, 13; RU	$\emptyset$

At this stage of the proof, $U_{14}(\Gamma_3) = \{!a, !b\}$, where $\Gamma_3 = \Gamma_2 \cup \{\neg a\}$. Hence, according to Definition 2.4.2, lines 7 and 9 are marked again at line 14.

Given a marking definition (the one for reliability introduced above or the one for minimal abnormality that is going to be introduced in the next section), the following definitions characterize the notion of derivation in adaptive dynamic proofs. The first definition concerns a dynamic notion of derivation:

Definition 2.4.3. A formula A has been *derived at stage s* of an adaptive proof, iff, at that stage, A is the second element of some unmarked line i.

In order to define a syntactic consequence relation we need a static, non-relative notion of derivability. This is provided by the following definition.

Definition 2.4.4 (Final derivability). A is *finally derived* from Γ on a finite line i of a proof at stage s iff

(i) A has been derived at stage s at line i;

[9] A precise meaning will be given to this notion in a moment.

(ii) every extension of the proof in which line i is marked may be further extended in such a way that line i is unmarked.

This definition can be interpreted in terms of an argumentation game where the proponent has a winning strategy in case her argument is able to withstand criticism (see [17]). Condition (i) says that the proponent is supposed to produce an argument for A by means of deriving it with an assumption that is not violated at some line l (otherwise the corresponding line would be marked). Now the opponent may respond and offer criticism. That is, he may derive Dab-formulas such that the proponent's argument is retracted (i.e., marked). However, our proponent is given the chance to reply: she repels the criticism in case she can further extend the proof such that her assumption is safe again and hence line l is unmarked. In case she is able to repel any possible criticism, she has a winning strategy and A is said to be finally derived.

This account fits in nicely with dialectical accounts of defeasible reasoning. For instance, Blair argued in his [18] that the view that "a valid inference is one whose justifying warrant can withstand criticism" (p. 116) and that "[t]he concepts of defeasibility and presumption are dialectical concepts" (p. 115) is common among many prominent theorists that deal with defeasible arguments such as Toulmin (see [19]), Wellman (see [20]), Rescher (see [21, chapter 3]), Pollock (see [22]), and Walton (see [23]).

Definition 2.4.5. $\Gamma \vdash_{\mathbf{AL^r}} A$ iff A is finally $\mathbf{AL^r}$-derivable from Γ.

Take for instance line 8 of our proof from Γ_1. There is no possible extension of the proof from Γ_1 that leads to the marking of this line. Hence, n is finally derivable from Γ_1. However, there is no way to finally derive a or b from Γ_1.

Note that for the reliability strategy the extensions referred to in point (ii) of Definition 2.4.4 can be restricted to the finite ones (see e.g. [2]).

The following theorem shows that A is derivable from Γ iff it is derivable on a condition Δ consisting of reliable formulas.

Theorem 2.4.2. *Where $\Gamma \subseteq \mathcal{W}$: $\Gamma \vdash_{\mathbf{AL^r}} A$ iff there is a $\Delta \subseteq \Omega$ for which $\Gamma \vdash_{\mathbf{LLL}} A \check{\vee} \mathsf{Dab}(\Delta)$ and $\Delta \cap U(\Gamma) = \emptyset$.*

I will not provide any meta-proofs for the theorems and lemmas in this chapter for the following two reasons. On the one hand, the meta-theory for the standard format that is presented in this chapter has been proven by Diderik Batens (e.g., in his seminal [1]). On the other hand, most of these results will follow as corollaries of the results presented in Chap. 5: there we introduce a generalization of the standard format and provide all the proofs for the meta-theory.

By making use of some basic properties of **LLL** we can alternatively characterize $\mathbf{AL}^{\mathbf{r}}$ as follows (where $\Gamma^{\check{\neg}} =_{\text{df}} \{\check{\neg} A \mid A \in \Gamma\}$)[10]:

Corollary 2.4.2. *Where* $\Gamma \subseteq \mathcal{W}$: $\Gamma \vdash_{\mathbf{AL}^{\mathbf{r}}} A$ *iff* $\Gamma \cup (\Omega \setminus U(\Gamma))^{\check{\neg}} \vdash_{\mathbf{LLL}} A$.

Finally, we have the following completeness and soundness result:

Theorem 2.4.3. *Where* $\Gamma \subseteq \mathcal{W}$: $\Gamma \vdash_{\mathbf{AL}^{\mathbf{r}}} A$ *iff* $\Gamma \Vdash_{\mathbf{AL}^{\mathbf{r}}} A$.

Although the derivability relation $\vdash_{\mathbf{AL}}$ is defined over $\mathcal{L}^+$, for applications we are mainly interested in the consequence set restricted to premises and consequences over the language $\mathcal{L}$ that characterizes our lower limit logic. However, for meta-theoretical insights also the enhanced consequence relation is of interest. Hence, we define, where $\Gamma \subseteq \mathcal{W}$, $Cn^{\mathcal{L}}_{\mathbf{AL}}(\Gamma) =_{\text{df}} \{A \in \mathcal{W} \mid \Gamma \vdash_{\mathbf{AL}} A\}$ and, where $\Gamma \subseteq \mathcal{W}^+$, $Cn^{\mathcal{L}^+}_{\mathbf{AL}}(\Gamma) =_{\text{df}} \{A \in \mathcal{W}^+ \mid \Gamma \vdash_{\mathbf{AL}} A\}$. I will also often omit the superscript, namely in cases in which both readings apply.

2.4.2 *The Minimal Abnormality Strategy*

We proceed analogous to the discussion of the reliability strategy: we first have a look at the semantics and then at the proof theory for the minimal abnormality strategy.

2.4.2.1 The Semantics

The minimal abnormality strategy is 'bolder' in comparison to the reliability strategy. Semantically the name is nearly self-explanatory. The *minimally abnormal models* are selected, i.e. the minimal elements of the partial order $\sqsubset^{\Gamma}_{\text{Ab}}$. In yet other words, all the **LLL**-models of a given premise set Γ that validate a minimal set of abnormalities. An **LLL**-model of Γ is a minimally abnormal model of Γ iff for all **LLL**-models M' of Γ, $\text{Ab}(M') \not\subset \text{Ab}(M)$. Note that $\text{Ab}(M_1)$, $\text{Ab}(M_2) \subset \text{Ab}(M_3)$ (see Fig. 2.1c). Hence, for the minimal abnormality strategy the reliable model M_3 is not selected.

For the minimal abnormality strategy "interpreting the premises as normally as possible" is read in a more rigorous way compared to the reliability strategy. The idea is to select $\mathbf{CL}_{\circ}$-models that validate as few abnormalities as possible. Given our (only) minimal Dab-consequence of Γ_1, $!a \check{\vee} !b$, models are selected that validate only one of the two unreliable abnormalities.

The semantic consequence relation for minimal abnormality is defined as follows.

[10] There is a $\Delta \subseteq \Omega \setminus U(\Gamma)$ for which $\Gamma \vdash_{\mathbf{LLL}} A \check{\vee} \text{Dab}(\Delta)$ iff [by the deduction theorem] there is a $\Delta \subseteq \Omega \setminus U(\Gamma)$ for which $\Gamma \cup \Delta^{\check{\neg}} \vdash_{\mathbf{LLL}} A$ iff [by the compactness and monotonicity of **LLL**] $\Gamma \cup (\Omega \setminus U(\Gamma))^{\check{\neg}} \vdash_{\mathbf{LLL}} A$.

Definition 2.4.6. Where $\mathcal{M}_{\mathbf{AL^m}}(\Gamma)$ is the set of all minimally abnormal **LLL**-models of Γ,

$$\Gamma \Vdash_{\mathbf{AL^m}} A \text{ iff for all } M \in \mathcal{M}_{\mathbf{AL^m}}(\Gamma),\, M \models A.$$

It is important to notice that the existence of minimally abnormal models is guaranteed.

Theorem 2.4.4. *$\sqsubset^{\Gamma}_{\mathrm{Ab}}$ is smooth (alias stoppered).*[11]

Immediate consequences of this are:

Corollary 2.4.3.

(i) If Γ has **LLL***-models then there are minimally abnormal models of Γ.* (Reassurance)

(ii) For every **LLL***-model M of Γ either M is minimally abnormal or there is an* **LLL***-model M' of Γ that is minimally abnormal and for which $Ab(M') \subset Ab(M)$.* (Strong Reassurance)

Moreover, it can be shown that every minimally abnormal model of Γ is also reliable. That is to say,

Theorem 2.4.5. $\mathcal{M}_{\mathbf{AL^m}}(\Gamma) \subseteq \mathcal{M}_{\mathbf{AL^r}}(\Gamma)$.

Hence, points (i) and (ii) in Corollary 2.4.3 also apply to reliable models.

Note that in our example all the minimally abnormal models of Γ_1 either validate $!a$ or $!b$ as the only abnormality. Hence, in all minimally abnormal models c is validated. This demonstrates that the minimal abnormality strategy is 'bolder' than the reliability strategy since $\Gamma_1 \Vdash_{\mathbf{CL_\circ^m}} c$ while $\Gamma_1 \nVdash_{\mathbf{CL_\circ^r}} c$.

Before I introduce the proof theory for minimal abnormality let me draw the reader's attention to a remarkable fact. Where $\mathsf{Dab}(\Delta_1)$, $\mathsf{Dab}(\Delta_2)$, ... are the minimal Dab-consequences of Γ, let $\Sigma(\Gamma) = \{\Delta_1, \Delta_2, \ldots\}$. A *choice set* of $\Sigma(\Gamma)$ is a set that contains a member from each Δ_i. Let $\Phi(\Gamma)$ be set of the minimal choice sets of $\Sigma(\Gamma)$, i.e. all choice sets $\varphi \subseteq \Omega$ of $\Sigma(\Gamma)$ such that there is no choice set $\varphi' \subseteq \Omega$ of $\Sigma(\Gamma)$ for which $\varphi' \subset \varphi$.[12]

The next theorem shows that each minimally abnormal model validates a minimal choice set as its abnormal part and vice versa, for each minimal choice set φ there is a minimally abnormal model that validates φ as its abnormal part.

Theorem 2.4.6. *Where $\Gamma \subseteq \mathcal{W}^+$ and $\mathcal{M}_{\mathbf{LLL}}(\Gamma)$ is non-empty.*

[11] A binary relation $\prec \subseteq X \times X$ is smooth (resp. stoppered) iff for every $a \in X$, either a is minimal or there is a $\prec$-minimal $b \in X$ for which $b \prec a$. The smoothness property will also play an important role when the standard format is generalized in Chap. 5 where we will—inter alia—prove this statement.

[12] Properties of choice sets that are useful in the context of ALs are inquired in the technical Appendix A.

(i) $\mathcal{M}_{\mathbf{AL^m}}(\Gamma) = \bigcup_{\varphi \in \Phi(\Gamma)}\{M \in \mathcal{M}_{\mathbf{LLL}}(\Gamma) \mid Ab(M) = \varphi\}$.
(ii) $\varphi \in \Phi(\Gamma)$ *iff there is an* $M \in \mathcal{M}_{\mathbf{AL^m}}(\Gamma)$ *for which* $Ab(M) = \varphi$.

This immediately implies a representational theorem:

Theorem 2.4.7. *Where* $\Gamma \subseteq \mathcal{W}^+$: $\Gamma \Vdash_{\mathbf{AL^m}} A$ *iff for each* $\varphi \in \Phi(\Gamma)$, $\Gamma \cup (\Omega \setminus \varphi)^{\check{\neg}} \Vdash_{\mathbf{LLL}} A$.

By the compactness of **LLL** this implies:

Corollary 2.4.4. *Where* $\Gamma \subseteq \mathcal{W}^+$: $\Gamma \Vdash_{\mathbf{AL^m}} A$ *iff for each* $\varphi \in \Phi(\Gamma)$ *there is a* $\Delta \subseteq \Omega \setminus \varphi$ *for which* $\Gamma \Vdash_{\mathbf{LLL}} A \check{\vee} \mathsf{Dab}(\Delta)$.

With the help of the minimally abnormal models we are able to give an alternative definition for the semantic selection for the reliability strategy.

Lemma 2.4.1. *Where* $\mathcal{M}$ *is a set of* **LLL***-models, define*

$$\Psi(\mathcal{M}) = \bigcup\{Ab(M) \mid M\, is\, minimally\, abnormal\, in\, \mathcal{M}\}.$$

Where $\Gamma \subseteq \mathcal{W}^+$: *M is a reliable* **LLL***-model of* Γ *iff* $Ab(M) \subseteq \Psi(\mathcal{M}_{\mathbf{LLL}}(\Gamma))$.

This characterization is attractive from a model-theoretic perspective since it is formulated independent of the consequence relation of the **LLL** which was used in the original definition in order to characterize the set $U(\Gamma)$. It is formulated only in terms of properties of the **LLL**-models of Γ, just like the definition of the semantic selection for the minimal abnormality strategy.

2.4.2.2 The Proof Theory

The proof theory for minimal abnormality differs from the one for reliability only with respect to the marking definition. We again employ the generic rules PREM, RU and RC.

As we have seen above, there is a direct link between the minimal choice sets (of $\Sigma(\Gamma)$) and the minimally abnormal interpretations of Γ provided by the minimally abnormal models. Also in the proof theory we will make use of this link. At any stage of the proof we are interested in the question which assumptions can be considered justified and which not. The information that we use in order to judge this is given by the minimal Dab-formulas that have been derived so far. While the reliability strategy considered each disjunct of a minimal Dab-formula as "unreliable", the minimal abnormality strategy is less skeptical. Let us first illustrate this by means of a simple example, and then make things more precise by making use of the notion of choice sets.

Suppose we have the following excerpt from a proof at some stage s (where we denote abnormalities by preceding them with "!"):

l	C	$\ldots$ $\{!A\}$
l'	C	$\ldots$ $\{!B\}$

Table 2.1 Possible interpretations of $\{!A \check{\vee} !B, !A \check{\vee} !C\}$

	$!A$	$!B$	$!C$
I_1	1	0	0
I_2	1	1	0
I_3	1	0	1
I_4	1	1	1
I_5	0	1	1

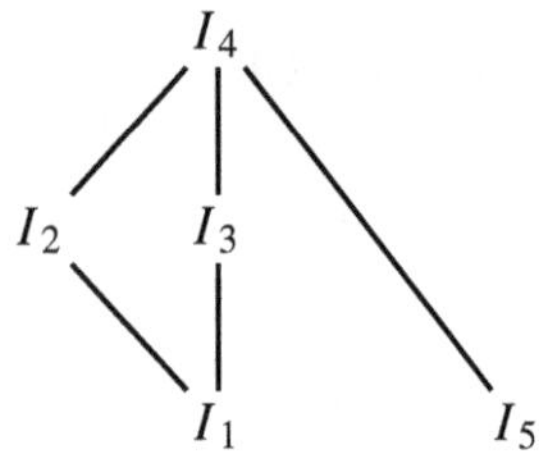

Fig. 2.4 Ordering of the interpretations in Table 2.1 in terms of abnormal parts

l'' $!A \check{\vee} !B$ $\quad\ldots\ \emptyset$
l''' $!A \check{\vee} !C$ $\quad\ldots\ \emptyset$

Suppose further that $!A \check{\vee} !B$ and $!A \check{\vee} !C$ are the only minimal Dab-formulas derived at stage s. The possible interpretations of these formulas are listed in Table 2.1. The corresponding ordering in terms of abnormal parts is illustrated in Fig. 2.4.

We have two minimally abnormal interpretations of these formulas: one I_1 according to which $!A$ is true, another one I_5 according to which $!B$ is true. Let us have a look at the formula C. Since both conditions on which it is derived contain unreliable abnormalities these lines are marked according to the reliability strategy. The situation is different for the minimal abnormality strategy. The reason is that the assumption expressed by the condition $\{!A\}$ is true in I_5 and the assumption expressed by the condition $\{!B\}$ is true in I_1. In other words, in each minimal abnormal interpretation of our minimal Dab-formulas derived so far C is justified.

Now, how does that relate to choice sets? Where $\mathsf{Dab}(\Delta_1), \mathsf{Dab}(\Delta_2), \ldots$ are the minimal Dab-formulas at stage s of a proof from Γ, the choice sets of $\Sigma_s(\Gamma) = \{\Delta_1, \Delta_2, \ldots\}$ give us exactly the possible interpretations of the minimal Dab-formulas derived so far. Hence, the minimal of these choice sets exactly correspond to the minimally abnormal interpretations of these minimal Dab-formulas.

In view of this, the marking of the minimal abnormality strategy will exactly mirror the idea of the semantics: we only take into account the minimally abnormal interpretations of the given premises—now contextualized to a given stage of the proof—and only claims that are justified in each of these interpretations are taken to be consequences at a given stage of the proof. This is realized by the following marking definition: where $\Phi_s(\Gamma)$ is the set of all minimal choice sets of $\Sigma_s(\Gamma)$ we define

Definition 2.4.7 (Marking for the Minimal Abnormality Strategy). Line i is marked at stage s iff, where A is derived on the condition Δ at line i,

(i) there is no $\varphi \in \Phi_s(\Gamma)$ such that $\varphi \cap \Delta = \emptyset$, or
(ii) for some $\varphi \in \Phi_s(\Gamma)$, there is no line at which A is derived on a condition Θ for which $\varphi \cap \Theta = \emptyset$.

Another way to interpret the marking definition is in terms of an argumentation game. Suppose the proponent derives a formula A on a line with condition Δ at stage s. Each minimal choice set $\varphi \in \Phi_s(\Gamma)$ represents a minimally abnormal interpretation of the Dab-formulas derived at stage s: each $B \in \varphi$ is true in this interpretation while each $B \in \Omega \setminus \varphi$ is false. Each minimal choice set φ thus represents a potential counter-argument against the defeasible assumption used by our proponent in order to derive A (namely that all members of Δ are false). φ is a counter-argument in case the defeasible assumption, i.e. the condition of line l, contains elements of φ. In this case the assumption of line l is not valid in the interpretation offered by φ.

In case there is no minimally abnormal interpretation φ in which the assumption holds (see point (i)), the proponent cannot defend herself and her inference to A is retracted in terms of being marked. But suppose there is a φ such that $\Delta \cap \varphi = \emptyset$. In this case there is at least one minimally abnormal interpretation in which the assumption of our proponent holds. But what about minimally abnormal interpretations in which the assumption does not hold, i.e. some $\varphi \in \Phi_s(\Gamma)$ for which $\varphi \cap \Delta \neq \emptyset$? In this case the proponent has to offer for each such φ another argument whose assumption is valid in φ (see point (ii)). If she is able to do so, i.e. if she is able to defend herself against all counter-arguments, then her argument is justified and hence line l is not marked at stage s.

In sum: suppose our proponent derived A on the assumption Δ at line l.

- *Is the argument at line l defensible?*
  Our proponent should be able to at least pinpoint one minimal abnormal interpretation of the Dab-formulas derived so far in which the assumption Δ holds.
- *Is the claim A justifiable?*
  For each counter-argument of our opponent, i.e. each minimally abnormal interpretation I of the Dab-formulas derived so far, she has to have an argument for A with an assumption that is valid in I.

If both questions are answered to the positive, our proponent wins the argumentation game at this stage. Otherwise, the opponent wins and line l is marked.[13]

[13] The terminological distinction between defensible and justified arguments is borrowed from abstract argumentation. Given a set of abstract entities (arguments) and an attack relation between them, there are various rationales according to which we can select arguments. (These rationales are called *extension types* in Part III.) If an argument is in all selections (that satisfy the criteria imposed by the rationale) it is called *justified*, if it is in some selection it is called *defensible*, if it is in no selection it is called *overruled*. See also the detailed discussion in [24]. The situation is analogous in our case: an argument for the claim A offered at a line l with an assumption expressed by the condition Δ is called justified if the assumption is valid in all minimally abnormal interpretations of the Dab-formulas (at the present stage), it is defensible if the assumption is valid in some minimally

Let us close this discussion by having another look at a proof from Γ_1, this time applying the marking definition for minimal abnormality.

1 $\circ n$	PREM	$\emptyset$
2 $(a \wedge n) \supset c$	PREM	$\emptyset$
3 $(b \wedge n) \supset c$	PREM	$\emptyset$
4 $\circ a$	PREM	$\emptyset$
5 $\circ b$	PREM	$\emptyset$
6 $\neg a \vee \neg b$	PREM	$\emptyset$
${}^{10}7$ a	4; RC	$\{!a\}$
${}^{10}8$ b	5; RC	$\{!b\}$
9 n	1; RC	$\{!n\}$
10 $!a \check{\vee} !b$	4, 5, 6; RU	$\emptyset$
11 c	2, 4, 9; RC	$\{!a, !n\}$
12 c	3, 5, 9; RC	$\{!b, !n\}$

Note that lines 11 and 12 are not marked as they would be according to the reliability strategy. For instance the condition of line 11 does (i) not intersect with all minimal choice sets in $\Phi_{12}(\Gamma_1) = \{\{!a\}, \{!b\}\}$ and (ii) it is not the case that there is a minimal choice set $\varphi \in \Phi_{12}(\Gamma_1)$ such that all conditions on which c has been derived intersect with φ. The reason for (ii) is that c is also derived on the condition $\{!b, !n\}$ at line 12. Indeed, c is valid in all minimally abnormal models of Γ_1.

A different situation occurs with respect to line 7. Its condition, and in fact all conditions on which a can be derived, intersect with the minimal choice set $\{!a\}$. Indeed, in the minimally abnormal model M_1 with abnormal part $\{!a\}$, a is not validated. An analogous argument applies to line 8.

Note that for our example the minimal choice sets $\Phi(\Gamma_1)$ are $\{!a\}$ and $\{!b\}$. Hence c is finally derivable.

The following theorem makes the link between the minimal choice sets and the adaptive consequences.

Theorem 2.4.8. *Where $\Gamma \subseteq \mathcal{W}$: $\Gamma \vdash_{\mathbf{AL^m}} A$ iff for every $\varphi \in \Phi(\Gamma)$ there is a $\Delta \subseteq \Omega$ for which $\Delta \cap \varphi = \emptyset$ and $\Gamma \vdash_{\mathbf{LLL}} A \check{\vee} \mathsf{Dab}(\Delta)$.*

Finally, we have the following completeness and soundness result:

Theorem 2.4.9. *Where $\Gamma \subseteq \mathcal{W}$: $\Gamma \vdash_{\mathbf{AL^m}} A$ iff $\Gamma \Vdash_{\mathbf{AL^m}} A$.*

(Footnote 13 continued)
abnormal interpretation of the Dab-formulas. The line is marked in case its argument is not justified. In Sect. 2.8 we present an alternative approach where the marking takes place in case an argument is not defensible and relate the two approaches to what is often called the skeptical and the credulous approach to defeasible reasoning.

2.4.3 A Special Case: The Simple Strategy

Sometimes we deal with cases in which both standard strategies, reliability and minimal abnormality, coincide. These are cases in which all minimal Dab-consequences of the lower limit logic **LLL** are abnormalities. That is to say, every minimal Dab-consequence $\mathsf{Dab}(\Delta)$ is such that Δ is a singleton. Let us call a premise set Γ for which all Dab-consequences are abnormalities, a *simple premise set.*

Where Γ is a simple premise set, it is straightforward to check that in this case $\Phi(\Gamma) = \{U(\Gamma)\}$ and, moreover, that in this case both strategies lead to the same consequence set.

Simple premise sets allow for a simplification of the adaptive strategy: the so-called *simple strategy.*

2.4.3.1 The Semantics

Let us first take a look at the semantics. Given a simple premise set Γ it is easy to see that all the minimally abnormal **LLL**-models M of Γ are such that $A \in \mathrm{Ab}(M)$ iff A is verified by every **LLL**-model of Γ. This is equivalent to: $\mathrm{Ab}(M) = \{A \in \Omega \mid \Gamma \vdash_{\mathbf{LLL}} A\}$ resp. $\mathrm{Ab}(M) = \{A \in \Omega \mid \Gamma \Vdash_{\mathbf{LLL}} A\}$ resp. $\mathrm{Ab}(M) = \bigcap_{M' \in \mathcal{M}_{\mathbf{LLL}}(\Gamma)} \mathrm{Ab}(M')$. The same holds for all the reliable **LLL**-models of Γ. This motivates the following definition:

Definition 2.4.8. An **LLL**-model M of Γ is *simple* iff $\mathrm{Ab}(M) = \{A \in \Omega \mid \Gamma \Vdash_{\mathbf{LLL}} A\}$.

Theorem 2.4.10. *Where Γ is a simple premise set, the following points are equivalent:*

(i) *A is verified by all simple models of Γ*
(ii) *A is verified by all reliable models of Γ*
(iii) *A is verified by all minimally abnormal models of Γ*

Definition 2.4.9. $\Gamma \Vdash_{\mathbf{AL}^s} A$ iff A is verified by all simple models of Γ.

2.4.3.2 The Proof Theory

Derivations are again governed by the generic rules PREM, RU, and RC. What changes and is simplified is the marking definition.

Definition 2.4.10 (Marking for the Simple Strategy). Line i is marked at stage s iff, where Δ is its condition, stage s contains a line on which an $A \in \Delta$ has been derived on the empty condition.

Final derivability is defined as for reliability and minimal abnormality. Hence, $\Gamma \vdash_{\mathbf{AL}^s} A$ iff A is finally derivable from Γ (with respect to the marking for the simple strategy).

Theorem 2.4.11. *Where* $\Gamma \subseteq \mathcal{W}$ *is a simple premise set,* $\Gamma \vdash_{\mathbf{AL}^r} A$ *iff* $\Gamma \vdash_{\mathbf{AL}^m} A$ *iff* $\Gamma \vdash_{\mathbf{AL}^s} A$.

Theorem 2.4.12. *Where* $\Gamma \subseteq \mathcal{W}$ *is a simple premise set,* $\Gamma \Vdash_{\mathbf{AL}^s} A$ *iff* $\Gamma \vdash_{\mathbf{AL}^s} A$.

2.5 Modeling Defeasibility in Adaptive Proofs

In this section we enhance our understanding of how ALs model defeasible reasoning. We start off with taking another look at dynamics in Sect. 2.5.1. Then, in Sect. 2.5.2, we compare the derivative power of the two strategies in view of so-called floating conclusions. Finally, in Sect. 2.5.3 we relate ALs to so-called plausible reasoning and a related problem concerning contraposition.

2.5.1 Internal and External Dynamics

As has been demonstrated above, formulas are derived conditionally in adaptive proofs. An unmarked line may be marked at a later stage of the proof and a marked line may be unmarked.[14] This is analogous to the tentative way of arriving at conclusions in defeasible reasoning, where we infer some A from some premises presuming that the circumstances satisfy some ceteris normalibus condition in order for the inference to be warranted. In ALs this is made explicit, on the one hand, by specifying what counts as an abnormality and, on the other hand, by specifying the exact nature of the normality condition by the adaptive strategy. In the adaptive proofs formulas are derived on conditions that are sets of abnormalities and the adaptive strategy specifies when the condition is met or violated. The marking definition that is characterized by the adaptive strategy determines when a formula counts as derived and when not.

We have distinguished between two types of dynamics. On the one hand, there is the internal dynamics according to which we may have to retract inferences in view of new insights gained by means of analyzing the given premises. On the other hand, there is the external dynamics according to which we may have to retract inferences in view of new information given by means of new premises.

The internal dynamics is modeled by the marking dynamics of AL proofs. We start off with a specific set of premises and analyze and reason on the basis of them with the help of the three generic rules PREM, RU, and RC. As we have seen, informed

[14] Note that when I speak of lines "being/getting marked" this should in no way be misunderstood as being an activity that is up to a decision by a user of the logic. The marking is characterized by the marking definition in a perfectly deterministic way.

by the minimal Dab-formulas derived at a specific stage, some inferences may be retracted by means of marking the corresponding line, while some inferences which were previously marked may be reinstated since the marking is removed. Since the retraction mechanism is fully determined by the analysis of the given premise set this is clearly an instance of the internal dynamics of defeasible reasoning.

As pointed out already, the external dynamics is mirrored by the nonmonotonicity of the consequence relation: sometimes new information may lead to the situation in which some formula that was previously a consequence is not anymore a consequence as soon as the new information is considered. I already discussed that the primary focus in the research on defeasible reasoning is on the external rather than the internal dynamics. ALs are nonmonotonic, so they obviously reflect the external dynamics as well. However, the question arises whether ALs add anything interesting when explicating the external dynamics which distinguishes them from other formal models. Here it is useful to distinguish between two ways in which a formal model **L** exhibits an external dynamics:

1. **L** is nonmonotonic: some previous output may not anymore be output given additional input. Hence, **L** can be said to *be externally dynamic.*
2. **L** *models the rationale* underlying the external dynamics by means of a procedural explication of the reasoning process that causes some previous consequences to cease to be consequences given new input.

My claim is that it is point 2 where ALs offer an essential contribution. Suppose our detective starts reasoning with the premise set $\Gamma_1 = \{\circ n, (a \wedge n) \supset c, (b \wedge n) \supset c, \circ a, \circ b\}$. The following proof $\mathcal{P}_1$ explicates her reasoning on the basis of the reliability strategy and Γ_1:

1	$\circ n$	PREM	$\emptyset$
2	$(a \wedge n) \supset c$	PREM	$\emptyset$
3	$(b \wedge n) \supset c$	PREM	$\emptyset$
4	$\circ a$	PREM	$\emptyset$
5	$\circ b$	PREM	$\emptyset$
6	c	1,2,4; RC	$\{!a, !n\}$
7	c	1,3,5; RC	$\{!b, !n\}$

Suppose at some point she gets new information which contains the definite proof that one of the witnesses was bribed and thus lied, she just doesn't know which one: $\neg a$ or $\neg b$. Instead of starting her reasoning process again from scratch from the enriched premise set $\Gamma_2 = \Gamma_1 \cup \{\neg a \vee \neg b\}$, she can continue her reasoning process $\mathcal{P}_1$ as follows in a proof $\mathcal{P}_2$ from Γ_2:

⋮	⋮	⋮	⋮
^{9}6	c	1,2,4; RC	$\{!a, !n\}$
^{9}7	c	1,3,5; RC	$\{!b, !n\}$
8	$\neg a \vee \neg b$	PREM	$\emptyset$
9	$!a \vee !b$	4,5,8; RU	$\emptyset$

The new information causes the marking of lines 6 and 7: while c was a consequence from Γ_1 it ceases to be a consequence given the new information $\neg a \vee \neg b$. Reusing and extending the proof $\mathcal{P}_1$ resulting in $\mathcal{P}_2$ explicates the reasoning process that leads to the retraction of the previous inferences resulting in c: hence it provides an understanding as to why our detective previously inferred c (given only Γ_1) and then she gave up on it (given Γ_2).

Moreover, ALs are also able to explicate cases of *reinstatements*: i.e., cases in which c is a consequence of Γ_1, ceases to be a consequence of Γ_2, and then is a consequence of Γ_3 again (where $\Gamma_1 \subset \Gamma_2 \subset \Gamma_3$). Let us demonstrate this by extending our example further.

Suppose that some informant provides our detective with the information that indeed the second witness has been bribed: $\neg b$. Hence, our premise set is now $\Gamma_3 = \Gamma_2 \cup \{\neg b\}$. Again, our detective can base her reasoning on the previous reasoning process and thus reuse $\mathcal{P}_2$ and extend it in the following way leading to a proof $\mathcal{P}_3$ from Γ_3:

	⋮	⋮	⋮	⋮
	6	c	1,2,4; RC	$\{!a, !n\}$
[11]	7	c	1,3,5; RC	$\{!b, !n\}$
	8	$\neg a \vee \neg b$	PREM	$\emptyset$
	9	$!a \vee !b$	4,5,8; RU	$\emptyset$
	10	$\neg b$	PREM	$\emptyset$
	11	$!b$	5,10; RU	$\emptyset$

Note that c at line 6 is reinstated in view of the new evidence. The reason is that $!a \vee !b$ is not anymore a minimal Dab-formula in view of $!b$ at line 11. Again, looking at the sequence $\mathcal{P}_1$, $\mathcal{P}_2$, $\mathcal{P}_3$ we see a detailed explication of the dynamics of her reasoning process: in $\mathcal{P}_1$ we see the rationale behind accepting the inference at line 6 as finally derived since the condition was reliable (meaning it only contained reliable abnormalities), in $\mathcal{P}_2$ the inference was retracted since the condition contained an unreliable abnormality, finally in $\mathcal{P}_3$ the inference is safe again since the condition is reliable again.

Note that where $\Gamma \subset \Gamma'$: an **AL**-proof from Γ is also an **AL**-proof from Γ'. This is the technical reason why our detective may reuse a previous proof (fragment) from Γ when reasoning on the basis of an enriched premise set Γ', as happened in the transition from $\mathcal{P}_1$ to $\mathcal{P}_2$ and from $\mathcal{P}_2$ to $\mathcal{P}_3$ in our example.

Note finally that, according to the given presentation, the way ALs explicate the external dynamics of defeasible reasoning is analogous to the way they explicate the internal dynamics: namely by a retraction mechanism that is implemented by means of (un-)marking lines. The difference is that in the case of the external dynamics we make a transition from a proof $\mathcal{P}$ from Γ to a proof $\mathcal{P}'$ from Γ' by reusing $\mathcal{P}$, while the internal dynamics occurs in one and the same proof. The analogous treatment is in no way surprising: after all, both dynamics are based on the fact that new insights may cause previous defeasible inferences to be retracted and the only difference concerns the source of the new insights. In the case of internal dynamics it is based on a better understanding of the given premises, while in the case of external dynamics it is based on new input. In practice both dynamics occur often as part of the same reasoning activities: think for instance of learning processes. Hence, the fact that there is a clear link between the nonmonotonicity of the consequence relation of ALs and the internal dynamics is an argument in favor of the unifying power of ALs as a formal model for defeasible reasoning.

2.5.2 Comparing the Strategies

We have seen that the standard format offers two strategies: the reliability and the minimal abnormality strategy. The latter offers for many examples a 'bolder type' of reasoning. That is to say, it offers a consequence relation that, in many examples, gives rise to more consequences compared to the one for reliability. This was illustrated by our example: while the reliability strategy corresponds to a rationale that refrains from drawing the conclusion that Mr. X is the murderer, according to the minimal abnormality strategy our detective concludes that Mr. X is the murderer.

We have distinguished the two strategies by means of their different handling of minimal Dab-consequences. For the reliability strategy it was sufficient that (a part of) the condition of a conditional application of a rule was unreliable, i.e. part of a minimal Dab-consequence, in order to invalidate the application. In contrast, for the bolder minimal abnormality strategy there are cases in which some A is derived on a condition Δ that involves unreliable abnormalities but is nevertheless not marked. Recall that by the minimal abnormality strategy our detective derives that Mr. X is the murderer. We have seen that in each minimally abnormal model she can rely on one of the two witnesses which is due to the fact that $a \vee b$ is valid in all minimally abnormal models. In contrast, the fact that $!a \check{\vee} !b$ is a minimal Dab-consequence of Γ_1 makes all the conditions on which c is derived unreliable and hence it is not derivable that Mr. X is the murderer according the the reliability strategy.

Scholars in defeasible reasoning sometimes distinguish between two basic types of conflicts:

1. a conflict between a defeasible inference and a "hard fact" (i.e., a premise) or any formula that can be inferred from the premises by means of non-defeasible rules;
2. a conflict between two defeasible inferences.

The first type of conflict is to be resolved by retracting the defeasible inference. Recall that in our proof from Γ_1 we derived n at line 8 by a defeasible inference on the basis of rule (2.1) on the condition $\{!n\}$:

8	n	1; RC	$\{!n\}$

Now suppose we introduce $\neg n$ as a hard fact by a new premise and let $\Gamma_4 = \Gamma_1 \cup \{\neg n\}$:

11	$\neg n$	PREM $\emptyset$
12	$!n$	1, 11; RU $\emptyset$

In this case line 8 gets marked. It is easy to see that this generalizes for all ALs in standard format. Say A has been derived conditionally at line i and some B has been derived on the empty condition. Suppose moreover that $B \vdash_{\mathbf{LLL}} \check{\neg} A$. Then line i is marked. This follows directly with the following derivable rule:

$$\frac{\begin{array}{ll} A & \Delta \\ \check{\neg} A & \Delta' \end{array}}{\mathsf{Dab}(\Delta \cup \Delta') \ \emptyset} \qquad \text{(RD)}$$

It is easy to see that, where Δ is the condition of a line l, and $\mathsf{Dab}(\Delta)$ is derived on the condition $\emptyset$ then l is marked according to both adaptive strategies.

RD is a consequence of the following lemma:

Lemma 2.5.1 (Conditions Lemma). *An* ***AL****-proof from Γ contains a line at which A is derived on the condition Δ iff $\Gamma \vdash_{\mathbf{LLL}} A \check{\vee} \mathsf{Dab}(\Delta)$.*[15]

The lemma gives immediately rise to the following rule:

$$\frac{A \qquad \Delta}{A \check{\vee} \mathsf{Dab}(\Delta) \ \emptyset} \qquad \text{(RA)}$$

We now discuss the second conflict type: conflicts between defeasible inferences. Again, a look at the derived rule RD helps us to understand how ALs handle such a conflict. It expresses that whenever we have a conflict between two claims, one derived on the condition Δ on line l and another one derived on the condition Δ' on line l', then we can derive (unconditionally) that one of the abnormalities in $\Delta \cup \Delta'$ is true. If there are no other minimal disjunctions of abnormalities in the proof and if there are no alternative arguments for our two claims, this means that according to both strategies both lines l and l' are retracted. However, the handling of such conflicts is not fully analogous with respect to the two strategies. This will be demonstrated in the following example.

Suppose a reliable although not infallible witness reports that

- Mr. X wore a long black coat in the bar in which he was seen half an hour before the murder. — $\circ l$

Another reliable although not infallible source however witnesses that

- Mr. X wore a short dark blue jacket and black trousers at the same time. — $\circ j$

Obviously $\neg(l \wedge j)$, since both cannot be the case. Moreover, we have

- If Mr. X was dressed in a long black coat, then he wore dark clothes. — $l \supset m$

[15] This is proven under the same name in [2, Chap. 4].

- If Mr. X was dressed in a short dark blue jacket and black trousers, then he wore dark clothes. — $j \supset m$

Let us have a look at a proof segment with the minimal abnormality strategy from $\Gamma_{\text{fc}} = \{\circ l, \circ j, \neg(l \wedge j), l \supset m, j \supset m\}$:

	1 $\circ l$	PREM	$\emptyset$
	2 $\circ j$	PREM	$\emptyset$
	3 $\neg(l \wedge j)$	PREM	$\emptyset$
	4 $l \supset m$	PREM	$\emptyset$
	5 $j \supset m$	PREM	$\emptyset$
12	6 l	1; RC	$\{!l\}$
	7 m	4, 6; RU	$\{!l\}$
12	8 j	2; RC	$\{!j\}$
	9 m	5, 8; RU	$\{!j\}$
12	10 $l \wedge j$	6, 8; RU	$\{!l, !j\}$
	11 $\check{\neg}(l \wedge j)$	3; RU	$\emptyset$
	12 $!l \check{\vee} !j$	10, 11; RD	$\emptyset$
	13 $l \vee j$	6; RU	$\{!l\}$
	14 $l \vee j$	8; RU	$\{!j\}$

Note that lines 7, 9, 13 and 14 are marked according to the reliability strategy, however they are unmarked according to the minimal abnormality strategy. Indeed, $l \vee j$ as well as m are finally derivable according to the minimal abnormality strategy. Note that for each choice set $\varphi \in \Phi_{14}(\Gamma_{\text{fc}}) = \{\{!l\}, \{!j\}\}$, $l \vee j$ is derived on a condition that has an empty intersection with φ. It is easy to see that there is no extension of the proof in which lines 13 and 14 are marked.

Conclusions such as m are often referred to as *floating conclusions*. Although no sequence of defeasible inferences leading to the conclusion m is valid in every selected model, in each of them at least one of these sequences is such that all the conditions of the rules constituting the sequence are valid. Note that there are two types of minimally abnormal models, one with abnormal part $\{!l\}$ and one with abnormal part $\{!j\}$. In the latter models none of the conditions of the sequence of inferences leading to the derivation of m explicated at lines 1, 4, 6 and 7 are violated. Similarly, in the former type of models none of the conditions of the sequence of inferences leading to the derivation of m explicated at lines 2, 5, 8 and 9 are violated.

In sum, according to the minimal abnormality strategy we get floating conclusions, while reliability blocks them.

2.5.3 Adaptive Logics and Plausible Reasoning

In this section we will demonstrate in which sense ALs model plausible reasoning and discuss a related problem that has to do with contraposition.

2.5.3.1 ALs Model Plausible Reasoning

As has become clear, ALs formally model defeasible reasoning by means of inferences based on assumptions. In the literature we can see two approaches to assumption-based reasoning:

(a) In the first approach the concrete assumption made in a defeasible inference is left unspecified or implicit. What is used is a defeasible inference rule. One way to realize this is for instance with a connective $A \rightarrow B$ to which a defeasible Modus Ponens rule is applicable so that B is defeasibly derived given A.
(b) In the other approach the assumptions that are associated with a defeasible inference are made explicit. Often this is expressed in the object language, e.g., $A \wedge \neg \mathrm{ab}_1 \supset B$. Given A and $\neg \mathrm{ab}_1$ we can apply Modus Ponens to derive B.

Moreover, various scholars (see [25–27]) make a difference between two types of reasoning:

(1) *Defeasible Reasoning* "as unsound (but still rational) reasoning on a solid basis" [27, p. 262]; and
(2) *Plausible Reasoning* "as sound (i.e., deductive) reasoning on an uncertain basis" [27, p. 262].

Hereby, (a) is often associated with (1), while (b) is associated with (2). The reason for the latter is that once we have explicit abnormality assumptions we can use the material implication as a conditional and Modus Ponens as an inference rule, whereas the (uncertain) abnormality assumptions are added as additional premises to the premise set. In the former case defeasible rules are applied to the premise set which is taken for granted (i.e., certain).

By now it is obvious that ALs belong to category (b): after all, normality assumptions are made explicit in the fourth column of adaptive proofs. The assumptions are generated by applications of the RC rule and stated in the fourth column of the proof. We have seen that the minimal Dab-consequences together with a rationale provided by the adaptive strategy determine which assumptions are considered safe and which not.

Let us now take a closer look at where ALs fall according to the second distinction.

Recall that the consequence relation of ALs is reflexive and yet (most frequently) nonmonotonic. This seems to indicate that we have a case of (1) where the reflexivity mirrors the "solid basis" and the nonmonotonicity mirrors the "unsound (but still rational)" reasoning.

But we should be more careful with our analysis. After all, the conditional inferences by means of the RC rule can be thought of as having the form of a classical deduction, i.e., of disjunctive syllogism: from $A \check{\vee} \mathsf{Dab}(\Delta)$ and the assumption $\check{\neg} \mathsf{Dab}(\Delta)$ derive by means of disjunctive syllogism A. Under this perspective ALs implement plausible reasoning in the following way. We have two premise sets, Γ and $\Omega^{\check{\neg}}$.[16] Γ provides a solid basis, while $\Omega^{\check{\neg}}$ is an uncertain basis consisting of normality

[16] Recall that $\Omega^{\check{\neg}} = \{\check{\neg} A \mid A \in \Omega\}$.

assumptions.[17] Whereas PREM only allows for the introduction of premises from the solid base, RC is a way of introducing premises from the uncertain base in such a way that (i) a record is held of the used uncertain premises in the fourth column of the proof, and (ii) the introduced normality assumptions are immediately applied in an instance of disjunctive syllogism (as described above). Viewed in this way, we only have a 'deductive' logic in which we formally distinguish between two types of premises. The adaptive marking then handles which parts of the uncertain basis may be considered safe in specific inferences and retracts inferences that are based on unsafe assumptions. Let us demonstrate this with a familiar example. On the left side we have a usual AL proof, on the right side a reconstruction that is more explicitly in the style of plausible reasoning and in which RC is replaced by an argument that makes use of disjunctive syllogism (DS) (where $!A =_{\rm df} \circ A \wedge \neg A$):

1	$\circ n$	PREM	$\emptyset$	$\circ n$		PREM1	$\emptyset$
2	$\circ a$	PREM	$\emptyset$	$\circ a$		PREM1	$\emptyset$
3	$\circ b$	PREM	$\emptyset$	$\circ b$		PREM1	$\emptyset$
4	$(a \wedge n) \supset c$	PREM	$\emptyset$	$(a \wedge n) \supset c$		PREM1	$\emptyset$
5	$(b \wedge n) \supset c$	PREM	$\emptyset$	$(b \wedge n) \supset c$		PREM1	$\emptyset$
6'				$n \check{\vee} !n$		1;RU	$\emptyset$
6"					$\check{\neg}!n$	PREM2	$\{\check{\neg}!n\}$
6	n	1; RC	$\{!n\}$	n		6',6";DS	$\{\check{\neg}!n\}$
7'				$c \check{\vee} !a \check{\vee} !n$		1,2,4;RU	$\emptyset$
✓ 7"					$\check{\neg}!a$	PREM2	$\{!a\}$
7	c	2,4,6; RC	$\{!a, !n\}$	c		6",7',7";DS	$\{\check{\neg}!a, \check{\neg}!n\}$
8'				$a \check{\vee} !a$		2; RU	$\emptyset$
✓ 8	a	2; RC	$\{!a\}$	a		7",8';DS	$\{\check{\neg}!a\}$
9'				$c \check{\vee} !b \check{\vee} !n$		1,3,5; RU	$\emptyset$
✓ 9"					$\check{\neg}!b$	PREM2	$\{\check{\neg}!b\}$
9	c	1,3,5; RC	$\{!b, !n\}$	c		6",9',9";DS	$\{\check{\neg}!b, \check{\neg}!n\}$
10	$\neg a \vee \neg b$	PREM	$\emptyset$	$\neg a \vee \neg b$		PREM1	$\emptyset$
11	$!a \check{\vee} !b$	2,3,10;RU	$\emptyset$	$!a \check{\vee} !b$		2,3,10;RU	$\emptyset$

On the right side we use two premise introduction rules: PREM1 for the premises in the solid base Γ and PREM2 for the premises in the uncertain premise set $\Omega^{\check{\neg}}$. We use an additional "boxed" column to introduce these premises for the sake of transparency. In the last column we keep a record of the used "uncertain" premises. RU is a generic rule for all the non-defeasible (i.e., deductive) inferences that stem from the lower limit logic. DS is disjunctive syllogism (we could have also just written RU since DS is valid in the lower limit logic enriched by the "checked connectives"). The question which parts of the uncertain premise set can be considered safe for a given inference is analogous to the determination of the marking of lines. For instance, according to the minimal abnormality strategy a line l with formula A and a record $\Delta \subset \Omega^{\check{\neg}}$ is marked at stage s iff, (i) there is no $\varphi \in \Phi_s(\Gamma)$ for which

[17] A similar distinction can be found for instance in the **ASPIC**$^+$-framework [27, 28] where we find an 'ordinary' knowledge base $\mathcal{K}_p$ that is uncertain and an 'axiomatic' solid knowledge base $\mathcal{K}_n$.

$\check{\varphi^{\neg}} \cap \Delta = \emptyset$, or (ii) for some $\varphi \in \Phi_s(\Gamma)$ there is no line l' with formula A and a record Θ such that $\check{\varphi^{\neg}} \cap \Theta = \emptyset$.

Altogether we now have a proof with only deductive inference steps and premise introduction where the marking retracts inferences based on unsafe premises in the uncertain premise set $\check{\Omega^{\neg}}$. Given this perspective ALs explicate plausible reasoning.

2.5.3.2 A Problem with Contraposition?

Formal models that explicate plausible reasoning have come under some criticism due to the fact that for the deductive rules which are used also their contraposition is available (most recently in Prakken [27] and Caminada in [29]). For instance, Prakken gives the following example (illustrated in Fig. 2.5a):

1. Birds normally fly: $b \wedge \neg \mathrm{ab}_1 \supset f$
2. Penguins normally don't fly: $p \wedge \neg \mathrm{ab}_2 \supset \neg f$
3. All penguins are birds: $p \supset b$
4. Penguins are abnormal birds with respect to flying: $p \supset \mathrm{ab}_1$
5. Tweety is observed as a penguin: o
6. Animals that are observed as penguins are normally penguins: $o \wedge \neg \mathrm{ab}_3 \supset p$

Now Prakken observes that we can construct an argument against applying 5 and 6 to infer p by means of applying contraposition to 4 and 6:

4'. $\neg \mathrm{ab}_1 \supset \neg p$
6'. $o \wedge \neg p \supset \mathrm{ab}_3$

Were contraposition not available this move would be blocked. Also Caminada states that given contraposition is available for the defeasible inference rules the principle "to keep the effects of possible conflicts as local as possible" [29, p. 113] (see also Hage [30, p. 109]) is violated. Note that besides the obvious conflict between f and $\neg f$, 4' also introduces a conflict between p and $\neg p$. While Caminada argues that contraposition should only be blocked in what he calls constitutive reasoning while it is "perfectly reasonable" in epistemic reasoning,[18] The example seems to indicate that Prakken would go further. He argues in [27] that contraposition is "a property which is too strong for default statements".[19]

Given the above analysis of ALs as a formal model for plausible reasoning we should expect a similar scenario. And indeed, contraposition is available for conditional inferences in ALs in the following sense. Suppose we can derive B from A defeasibly on the condition Δ by RC. That means: $A \vdash_{\mathbf{LLL}} B \check{\vee} \mathsf{Dab}(\Delta)$. But

[18] Caminada calls upon the distinction between epistemic and constitutive reasons in Hage [30, p. 60]: "Epistemic reasons are reasons for believing in facts that obtain independent of the reasons that plead for or against believing them. Constitutive reasons, on the contrary, influence the very existence of their conclusions."

[19] In [27] Prakken also argues against ad hoc solutions such as to strictly prioritize perceptual evidence since "the strength of perceptive inferences is highly context-dependent." (see his footnote 10) or to model perceptual inferences in a non-defeasible way.

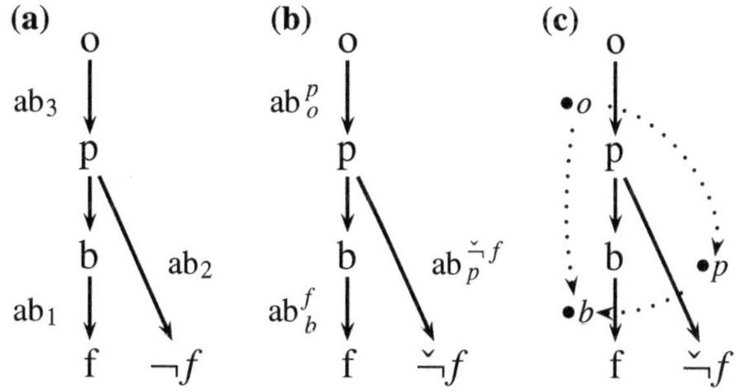

Fig. 2.5 Illustrations of the examples

then we also have $\check{\neg} B \vdash_{\mathbf{LLL}} \check{\neg} A \check{\vee} \mathsf{Dab}(\Delta)$ and we can thus derive $\check{\neg} A$ from $\check{\neg} B$ on the same condition Δ.

An obstacle in reconstructing examples as the one above by Prakken in an AL is that there is not one unique way to express it in ALs. Both, defaults and a defeasible Modus Ponens mechanism that models default inferencing, may be represented in various ways: for instance, in Part II we use a conditional that satisfies the so-called KLM properties. Alternatively we could use material implication $A \supset B$ preceded by a dummy operator $\circ(A \supset B)$ and adaptively activate them as much as possible by making use of the abnormality $\circ(A \supset B) \wedge \neg(A \supset B)$.[20] Let us thus stay on a more schematic level: suppose we have the following proof fragment from some premise set Γ (illustrated in Fig. 2.5b)

⋮	⋮	⋮	⋮
l_0	$p \supset b$	PREM	$\emptyset$
l_1	o	PREM	$\emptyset$
l_2	p	$\ldots, l_1$; RC	$\{\mathrm{ab}_o^p\}$
l_3	b	l_0, l_2; RU	$\{\mathrm{ab}_o^p\}$
l_4	$\neg f$	$\ldots, l_2$; RC	$\{\mathrm{ab}_o^p, \mathrm{ab}_p^{\neg f}\}$
l_5	f	$\ldots, l_3$; RC	$\{\mathrm{ab}_o^p, \mathrm{ab}_b^f\}$

It is easy to see that in view of the lines l_4 and l_5 we can derive

l_6	$\mathrm{ab}_o^p \check{\vee} \mathrm{ab}_p^{\neg f} \check{\vee} \mathrm{ab}_b^f$	l_4, l_5; RD	$\emptyset$

In view of line l_6 all our conditional derivations on lines $l_2, \ldots, l_5$ are marked. Moreover, in view of the proof fragment: $\Gamma \vdash_{\mathbf{LLL}} p \check{\supset} \left(\check{\neg} f \check{\vee} \mathrm{ab}_p^{\check{\neg} f}\right)$ and $\Gamma \vdash_{\mathbf{LLL}} b \check{\supset} \left(f \check{\vee} \mathrm{ab}_b^f\right)$. Since $p \supset b \in \Gamma$, $\Gamma \vdash_{\mathbf{LLL}} p \check{\supset} \left(f \check{\vee} \mathrm{ab}_b^f\right)$. By simple manipulations, $\Gamma \vdash_{\mathbf{LLL}} \check{\neg} p \check{\vee} \mathrm{ab}_p^{\check{\neg} f} \check{\vee} \mathrm{ab}_b^f$. Hence, we can produce the line

l_7	$\check{\neg} p$	$\ldots$	$\left\{\mathrm{ab}_p^{\check{\neg} f}, \mathrm{ab}_b^f\right\}$

In sum, the conflict between f and $\neg f$ is not isolated. Note that in view of line l_6 all the conditional inferences, including line l_2 with p, are marked. This shows that

---

[20] Both, Joke Meheus and Erik Weber independently suggested this in a conversation.

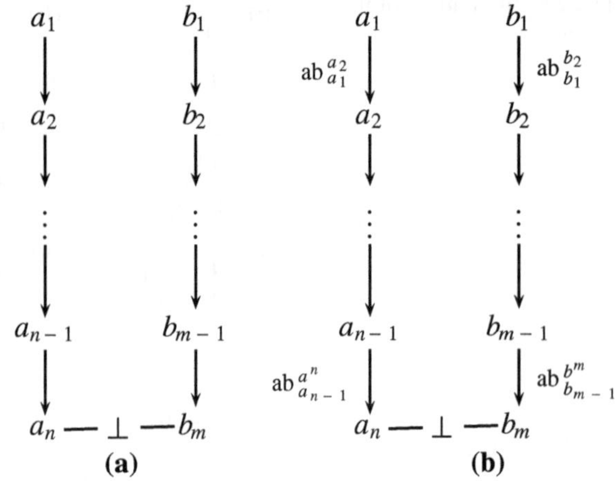

Fig. 2.6 Prakken's problem: abstract representation

the conflict between f and $\neg f$ spreads by effecting the defeasible inferences at lines l_2 and l_3 as well, since the corresponding abnormalities are involved in the minimal Dab-formula. Moreover, other conflicts are derivable such as the one between p and $\check{\neg} p$ (however, the corresponding lines are marked).

Let us conclude this discussion with various remarks.

1. In a more abstract phrasing the problem Prakken points out for plausible reasoning (that makes use of rules for which contraposition is available) is as follows. Suppose we have two sequences of rules $a_1 \to a_2 \to \cdots \to a_n$ and $b_1 \to b_2 \to \cdots \to b_m$ such that (i) $\vdash \neg(a_n \wedge b_m)$ and (ii) we have both a_1 and b_1 (see Fig. 2.6a). Due to the availability of contraposition we can construct an argument against any of the a_i (where $1 < i \leq n$): $b_1 \to b_2 \to \cdots \to b_m \to \neg a_n \to \neg a_{n-1} \to \ldots \to \neg a_i$. Hence, instead of the obvious conflict in a_n resp. b_m we suddenly end up with a conflict in each a_i. Since the same holds for all b_i (where $1 < i \leq m$) and since we are interested in a consistent consequence set we cannot—in view of symmetry—conclude any a_i nor any b_i (given a_1 and b_1).

We have shown by means of an example that whenever we have analogous sequences of conditional inferences in ALs (see Fig. 2.6b) we can (i) construct conditional arguments for each $\check{\neg} a_i$ and (ii) derive a Dab-formula which contains all the abnormalities in the conditions in the sequences:

$$\mathsf{Dab}\left(\left\{\mathrm{ab}_{a_1}^{a_2}, \ldots \mathrm{ab}_{a_{n-1}}^{a_n}, \mathrm{ab}_{b_1}^{b_2}, \ldots, \mathrm{ab}_{b_{m-1}}^{b_m}\right\}\right) \tag{2.5}$$

In case this Dab-formula is minimal and there are no alternative ways to obtain an inference for a_i, the conditional inference for a_i is marked and hence a_i is not a consequence. In this sense, Prakken's scenario is reproduced in ALs.

2. However, in many concrete ALs the problem is nevertheless avoided. Take for instance the ALs for default inferencing in Chap. 6. A default rule is represented by $a \rightsquigarrow b$ where $\rightsquigarrow$ is axiomatized by means of the KLM-properties. Moreover, the logic models a defeasible Modus Ponens as follows: from A and $A \rightsquigarrow B$ infer B unless we have $\bullet A$. The latter, $\bullet A$, expresses that the given circumstances are unusual for the proposition A which may be witnessed by the truth of some C that is less normal than A (this can be expressed by $(A \vee C) \rightsquigarrow \neg C$ in view of the KLM-properties).[21] A proof for Prakken's example may look as follows:

1	$o \rightsquigarrow p$	PREM $\emptyset$
2	$p \supset b$	PREM $\emptyset$
3	$p \rightsquigarrow \neg f$	PREM $\emptyset$
4	$b \rightsquigarrow f$	PREM $\emptyset$
5	o	PREM $\emptyset$
^{12}6	p	1,5; RC $\{\bullet o\}$
^{12}7	b	2,6; RU $\{\bullet o\}$
^{12}8	$\neg f$	3,6; RC $\{\bullet o, \bullet p\}$
^{12}9	f	4,7; RC $\{\bullet o, \bullet b\}$
10	$\bullet o \check{\vee} \bullet p \check{\vee} \bullet b$	8,9; RD $\emptyset$

So far it seems as if the problem is reproduced since in view of the minimal Dab-formula at line 12 our conditional inferences at lines 6–9 are marked.

The way the problem is avoided in this system is that $\bullet$ is 'inherited' along $\rightsquigarrow$-sequences: if $A \rightsquigarrow B$ and $\bullet A$ then $\bullet B$. Indeed, according to the KLM-properties, if $A \rightsquigarrow B$ then B is at least as normal as A. Hence, if A is excepted (i.e., we have an abnormal situation relative to A) then B is excepted as well (see Fig. 2.5c for an illustration: the dotted line indicates the 'inheritance'). Thus, the Dab-formula at line 12 is not minimal, but can be shortened to $\bullet p \check{\vee} \bullet b$ (and if we accept that $p \rightsquigarrow b$ we can further shorten it to $\bullet b$). In any case this will lead to the unmarking of lines 6 and 7 (resp. also to the unmarking of line 8).

We can conclude from this that although—in principle—the fact that contraposition is available for the conditional inferences in adaptive logics can cause the problem pointed out by Prakken, in concrete ALs it may nevertheless be avoided due to specific properties of the lower limit logic that may lead to the shortening of the Dab-formula (2.5).

3. Although we do get the "right" consequences in the example above (such as p), some may still argue that some of the inferences for which the logic allows (irrespective whether the corresponding lines are marked) are based on contrapositing default inferencing. E.g., the logic allows for the following inference:

13	$\neg p$	2–4; RU $\{\bullet p, \bullet b\}$

Of course, given our discussion above, this line gets marked. However, the mere fact that the logic allows for the inference may for some already be counter-intuitive. In view of this it is a research challenge to see whether the standard format for ALs can be adjusted in a way that allows for defeasible inferences that cannot be contraposed.

[21] For a detailed technical definition of the system see Chap. 6. An intuitive demonstration is enough for the purpose of the current section.

4. More research needs to be done on the question in which application contexts contraposition is a desired property of defeasible inferences. Caminada did an important step in clarifying this issue in [29]: it seems very plausible that in many contexts of epistemic reasoning contraposition is applicable while in many contexts of constitutive reasoning it is not. Nevertheless, examples such as the one by Prakken discussed above may indicate that the demarcation is not that smooth (see also [31]).

2.6 Some Properties of ALs in Standard Format

One of the merits of the standard format for ALs is that it comes along with many nice properties. For any concrete logic formulated in this format, these properties do not have to be proven since they have been shown generically to hold for any AL in standard format. Let me introduce some of these properties in this section. Later, in Chap. 4, I will point out some more specific properties related to premise sets. Most of the properties that are presented in this section will be shown to hold for a more general setting in Chap. 5. There and in the respective Appendix the reader can find meta-proofs.

2.6.1 Properties of the Adaptive Consequence Relations

The first result shows that the semantic and the syntactic consequence relations define identical relations. Indeed, by Theorem 2.4.3 and Theorem 2.4.9 we have the following soundness and completeness result for both adaptive strategies:

Theorem 2.6.1. (Soundness and Completeness of AL). *Where* $\Gamma \subseteq \mathcal{W}$*:* $\Gamma \vdash_{\mathbf{AL}} A$ *iff* $\Gamma \Vdash_{\mathbf{AL}} A$.

Soundness even holds for premise sets with "checked connectives"[22]:

Theorem 2.6.2. *Where* $\Gamma \subseteq \mathcal{W}^{+}$*:* $\Gamma \vdash_{\mathbf{AL}} A$ *implies* $\Gamma \Vdash_{\mathbf{AL}} A$.

The completeness doesn't hold for premise sets with "checked connectives", as is shown in Sect. 2.7.

By Definition 2.4.4, final derivability concerns finite stages of an adaptive proof. However, it is important to notice that it is essential for the minimal abnormality strategy that the extensions of the proof referred to in Definition 2.4.4ii may be infinite. Indeed, as demonstrated in [1, p. 229], there are premise sets for which it is true that for every way to finally derive some A at some line i there is an extension of the proof that leads to the marking of line i such that only an infinite further extension leads to the unmarking of line i.

22 We prove the corresponding theorem for the generalized standard format in Chap. 5 (see Corollary 5.4.3).

Although final derivability does not ensure the stability of a line i at which some A is finally derived with respect to its marking, for infinite proofs it is guaranteed that there is a stage from which on line i is unmarked and remains so.

Theorem 2.6.3. *Where $\Gamma \subseteq \mathcal{W}$: $\Gamma \vdash_{\mathbf{AL}} A$ iff A is derivable at an unmarked line of an* **AL***-proof from Γ that is stable with respect to that line.*[23]

The next theorem states certain properties concerning the strength of the adaptive consequence relation. It shows that **AL** is a reflexive and supraclassical (with respect to the enriched language) strengthening of **LLL**. Moreover, the 'bolder' minimal abnormality strategy leads indeed always to at least as many consequences (w.r.t. $\subset$) as the reliability strategy.

Theorem 2.6.4. *Where $\Gamma \subseteq \mathcal{W}^+$:*

(i) $\Gamma \subseteq Cn_{\mathbf{AL}}(\Gamma)$ (Reflexivity)
(ii) $Cn^{\mathcal{L}+}_{\mathbf{CL}}(\Gamma) \subseteq Cn^{\mathcal{L}+}_{\mathbf{AL}}(\Gamma)$ (Supraclassicality)
(iii) $Cn_{\mathbf{LLL}}(\Gamma) \subseteq Cn_{\mathbf{AL^r}}(\Gamma) \subseteq Cn_{\mathbf{AL^m}}(\Gamma)$.

The next theorem states some closure properties of the adaptive consequence set. The central result is that the adaptive consequences are a fixed point. If the AL is again applied to its own consequence set of some premise set Γ, nothing new will be derived. This is a desirable property. Suppose the idealized case that our detective at the end of the day reached all the final conclusions Γ' based on some premises Γ. It would be rather strange if next morning the same reasoning applied to $\Gamma \cup \Gamma'$ would lead her to new conclusions since she did not gather any new evidence. If the fixed point property would not hold she might never reach a final set of conclusions for her case.

Theorem 2.6.5. *Where $\Gamma \subseteq \mathcal{W}$:*

(i) $Cn_{\mathbf{LLL}}(Cn_{\mathbf{AL}}(\Gamma)) = Cn_{\mathbf{AL}}(\Gamma)$ (Redundancy of **LLL** with respect to **AL**)
(ii) $Cn_{\mathbf{AL}}(Cn_{\mathbf{LLL}}(\Gamma)) = Cn_{\mathbf{AL}}(\Gamma)$.
(iii) $\mathcal{M}_{\mathbf{AL}}(\Gamma) = \mathcal{M}_{\mathbf{AL}}(Cn_{\mathbf{AL}}(\Gamma))$ *and hence* $Cn_{\mathbf{AL}}(\Gamma) = Cn_{\mathbf{AL}}(Cn_{\mathbf{AL}}(\Gamma))$. (Fixed Point/Idempotence)

Beside **LLL** that defines the monotonic core and the lower limit of the adaptive strengthening, there is also an upper limit logic **ULL**. The upper limit logic explicates the standard of normality of an AL. An AL can be seen as interpreting a premise set in terms of its upper limit logic "as much as possible". For premise sets that do not give rise to abnormalities, i.e. premise sets Γ for which no Dab-formulas are in the **LLL**-consequence set of Γ, the **AL**-consequences are identical to the **ULL**-consequences. Such premise sets are called *normal*. This can be defined for instance in the following way: Γ is *normal* iff $U(\Gamma) = \emptyset$. Evidently, given a normal premise

[23] A proof from Γ is stable with respect to a line l iff the status of the marking (marked vs. unmarked) of line l remains the same for every possible extension of the proof. This is shown in Appendix B for the more generic setting in which n ALs are sequentially combined (see Corollary B.2.3): ALs in standard format are a border case in which $n = 1$.

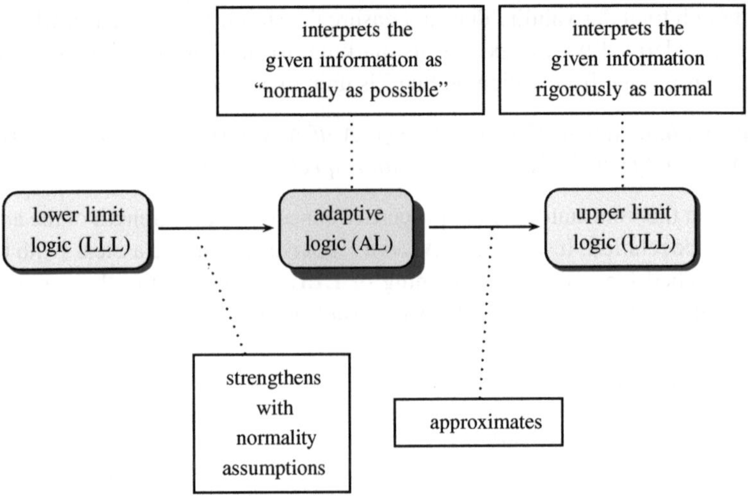

Fig. 2.7 The relationship between **LLL**, **AL**, and **ULL**

set, we expect an AL to realize its standard of normality. The upper limit logic is defined as follows:

Definition 2.6.1. Where $\Omega^{\check{\neg}} =_{\mathrm{df}} \left\{ \check{\neg} A \mid A \in \Omega \right\}$, **ULL** is characterized by the following consequence relation:

$$Cn^{\mathcal{L}^+}_{\mathbf{ULL}}(\Gamma) =_{\mathrm{df}} Cn^{\mathcal{L}^+}_{\mathbf{LLL}}\left(\Gamma \cup \Omega^{\check{\neg}}\right)$$

$$Cn^{\mathcal{L}}_{\mathbf{ULL}} =_{\mathrm{df}} \mathcal{W} \cap Cn^{\mathcal{L}^+}_{\mathbf{LLL}}\left(\Gamma \cup \Omega^{\check{\neg}}\right)$$

Moreover, $\mathcal{M}_{\mathbf{ULL}}(\Gamma) =_{\mathrm{df}} \mathcal{M}_{\mathbf{LLL}}\left(\Gamma \cup \Omega^{\check{\neg}}\right)$.

The following results show that **ULL** is indeed an upper limit to **AL**.

Theorem 2.6.6. *Where $\Gamma \subseteq \mathcal{W}^+$:*

(i) $Cn_{\mathbf{AL}}(\Gamma) \subseteq Cn_{\mathbf{ULL}}(\Gamma)$.
(ii) $\mathcal{M}_{\mathbf{ULL}}(\Gamma) \subseteq \mathcal{M}_{\mathbf{AL}}(\Gamma)$.
(iii) If Γ is normal, then $Cn_{\mathbf{AL}}(\Gamma) = Cn_{\mathbf{ULL}}(\Gamma)$ and $\mathcal{M}_{\mathbf{AL}}(\Gamma) = \mathcal{M}_{\mathbf{ULL}}(\Gamma)$.

Altogether we have (see also the illustration in Fig. 2.7),

Corollary 2.6.1. *Where $\Gamma \subseteq \mathcal{W}^+$:*

(i) $Cn_{\mathbf{LLL}}(\Gamma) \subseteq Cn_{\mathbf{AL^r}}(\Gamma) \subseteq Cn_{\mathbf{AL^m}}(\Gamma) \subseteq Cn_{\mathbf{ULL}}(\Gamma)$
(ii) $\mathcal{M}_{\mathbf{ULL}}(\Gamma) \subseteq \mathcal{M}_{\mathbf{AL^m}}(\Gamma) \subseteq \mathcal{M}_{\mathbf{AL^r}}(\Gamma) \subseteq \mathcal{M}_{\mathbf{LLL}}(\Gamma)$

The properties featured in the next theorem are "cautious" weakenings of properties that often characterize monotonic logics.

Theorem 2.6.7. *Where $\Gamma, \Gamma' \subseteq \mathcal{W}$:*

(i) *If $\Gamma' \subseteq Cn_{\mathbf{AL}}(\Gamma)$ then $Cn_{\mathbf{AL}}(\Gamma \cup \Gamma') \subseteq Cn_{\mathbf{AL}}(\Gamma)$.* (Cautious Cut / Cumulative Transitivity)

(ii) *If $\Gamma' \subseteq Cn_{\mathbf{AL}}(\Gamma)$ then $Cn_{\mathbf{AL}}(\Gamma) \subseteq Cn_{\mathbf{AL}}(\Gamma \cup \Gamma')$.* (Cautious Monotonicity / Cumulative Monotonicity)

(iii) *If $\Gamma' \subseteq Cn_{\mathbf{AL}}(\Gamma)$, then $Cn_{\mathbf{AL}}(\Gamma \cup \Gamma') = Cn_{\mathbf{AL}}(\Gamma)$.* (Cumulative Indifference / Cumulativity)

Cumulative indifference is a strengthening of the Fixed-Point property since it entails the latter (given the reflexivity of **AL**).[24] Moreover, it is a very desirable property itself. Suppose the case on which our detective is working is of a very complicated nature. Let the given evidence be Γ. Suppose further that in the evening she arrives at some—but due to the complicated nature of the case not all—final conclusions Γ'. That is to say, for every $A \in \Gamma'$ she is able to guarantee that no further analysis of Γ will lead to a withdrawal of A. For the adaptive proof this means A is finally derivable. Cumulative indifference guarantees that in the next morning she can reason on, on the basis of $\Gamma \cup \Gamma'$, i.e. she can rely on the insights she won the day before. This has the advantage that, technically speaking, once she established that some A is finally derivable, she doesn't have to keep track of the maybe very complicated conditions that enabled her to arrive at A, but rather she may introduce A as a premise in an adaptive proof from $\Gamma \cup \Gamma'$.

The nonmonotonicity of ALs can easily be demonstrated by the case of our detective that was introduced in Sect. 2.4. Let in the following $\mathbf{x} \in \{\mathbf{r}, \mathbf{m}\}$. Note that

$$\Gamma_1 \cup \{\neg b\} \vdash_{\mathbf{CL}_\circ^{\mathbf{x}}} c.$$

However, enhancing our premise set by $\{\neg a\}$, we have

$$\Gamma_1 \cup \{\neg b\} \cup \{\neg a\} \nvdash_{\mathbf{CL}_\circ^{\mathbf{x}}} c.$$

Also Cut/Transitivity is a property that does not hold for ALs in general. For instance we have $!a \vdash_{\mathbf{CL}_\circ^{\mathbf{x}}} c \vee !a$ and $c \vee !a \vdash_{\mathbf{CL}_\circ^{\mathbf{x}}} c$. However, $!a \nvdash_{\mathbf{CL}_\circ^{\mathbf{x}}} c$.

More generally the following can be proven.

Theorem 2.6.8. [25] *Where $\Gamma \subseteq \mathcal{W}$: If $Cn_{\mathbf{LLL}}(\Gamma) \subset Cn_{\mathbf{AL}}(\Gamma)$, then*

(i) **AL** *is nonmonotonic, and*

(ii) **AL** *is non-transitive.*

There are other properties that are often discussed in the context of defeasible reasoning and nonmonotonic logics: Rational Monotonicity and Rational Distributivity. The next section will demonstrate that these properties are not generically validated in ALs. At the same time it will demonstrate that this is rather an advantage of ALs since these properties are not without critical counter-examples.

---

[24] The proof is trivial and left to the reader.

[25] This is shown in [2].

2.6.2 Some Remarks on Computational Complexity

In this section I offer only some brief remarks on the computational complexity of ALs rather than providing a detailed survey of the given results. It would take significant space to spell out the technical preliminaries of complexity studies in the realm of the arithmetic hierarchy and hence lead us too far off the main course of the present study. I will instead provide pointers to the relevant literature for the interested reader.

While for most well-known formal accounts for nonmonotonic and defeasible reasoning there are thorough studies investigating complexity-related issues, such studies are sparse for ALs. Only rather recently some key results have been published. There is the critical study by Horsten and Welch [32] which caused two replies by adaptive logicians: [33] and [34]. Horsten and Welsh demonstrate that for some premise sets the consequence relation of the inconsistency-adaptive logic $\mathbf{CLuN^r}$ is Σ^0_3-hard in the arithmetic hierarchy. They argue that in view of this result the usefulness of ALs as a tool that explicates defeasible reasoning is put into question. In the technical study [33] Verdée proves that the minimal abnormality variant of the same inconsistency-adaptive logic $\mathbf{CLuN^m}$ falls into an even higher complexity class within the analytic hierarchy (he proves Π^1_1-completeness). Nevertheless, in a reply to the philosophical worries of Horsten and Welsh in [34] Batens et. al argue that such a high complexity class is to be expected from any serious formal attempt to capture the complexity of actual defeasible reasoning. It is not surprising then that many formal systems for defeasible reasoning fall in similar complexity classes (see e.g., [35, 36] for circumscription, [37, 38] for (generalized) closed world assumption).

Recently, Odintsov and Speranski contributed one paper [39] studying the complexity of some inconsistency-adaptive logics where they reaffirm and generalize some of the previous results. Finally, there is a forthcoming study [40] by them where these complexity results for the **CLuN**-based ALs are shown to hold generally for ALs in the standard format. For instance, the complexity upper bound Σ^0_3-for the reliability strategy and Π^1_1-for the minimal abnormality strategy are generalized for ALs in the standard format. The authors also investigate several interesting special cases (such as the case where $\Phi(\Gamma)$ is finite which is relevant for instance for our study of sequential combinations of ALs in Sect. 3.2.2).

2.6.3 Excursus on the Rational Properties

2.6.3.1 Rational Monotonicity

Besides cautious monotonicity there is another, in comparison stronger, weakening of monotonicity: rational monotonicity.

$$\text{If } A \in Cn_{\mathbf{L}}(\Gamma) \text{ and } A \notin Cn_{\mathbf{L}}(\Gamma \cup \{B\}), \text{ then } \check{\neg} B \in Cn_{\mathbf{L}}(\Gamma) \quad \text{(RM)}$$

The idea behind Rational Monotonicity is that, if adding B to the premise set Γ leads to nonmonotonicity, then $\check{\neg} B$ should be a consequence of Γ.

Rational Monotonicity is not a generic property of the consequence relation of ALs in standard format. Counter-examples are easily found. Rational Monotonicity, although an intuitive property in many cases, has also been criticized. In order to demonstrate the criticism and the fact that ALs do not in general validate Rational Monotonicity we "translate" an example by Stalnaker (see [41]) into the language of $\mathbf{CL}_\circ$. Suppose some reliable though not infallible source S_1 tells us that

- Bizet is a French composer,— $\circ f_B$;
- Satie is a French composer,— $\circ f_S$
- Verdi is an Italian composer,— $\circ i_V$.

Another reliable though not infallible source S_2 tells us that

- Verdi and Bizet are compatriots,— $\circ c_{V,B}$.

According to yet another reliable though not infallible source S_3,

- Verdi and Satie are compatriots,— $\circ c_{V,S}$.

Obviously the following is valid: $c_{V,B} \supset (\neg i_V \vee \neg f_B)$ and $c_{V,S} \supset (\neg i_V \vee \neg f_S)$. Let our premise set Γ_{RM} comprise sources S_1 and S_2 and thus be

$$\Gamma_{\mathsf{RM}} = \{\circ f_B, \circ f_S, \circ i_V, \circ c_{V,B}, c_{V,B} \supset (\neg i_V \vee \neg f_B), c_{V,S} \supset (\neg i_V \vee \neg f_S)\}.$$

The following proof fragment demonstrates how f_S can be derived from Γ_{RM}.

1	$\circ f_B$	PREM	$\emptyset$
2	$\circ f_S$	PREM	$\emptyset$
3	$\circ i_V$	PREM	$\emptyset$
4	$\circ c_{V,B}$	PREM	$\emptyset$
5	$c_{V,B} \supset (\neg i_V \vee \neg f_B)$	PREM	$\emptyset$
6	$c_{V,S} \supset (\neg i_V \vee \neg f_S)$	PREM	$\emptyset$
7	$c_{V,B} \supset (!i_V \vee !f_B)$	1, 3, 5; RU	$\emptyset$
8	$!c_{V,B} \check{\vee} !i_V \check{\vee} !f_B$	4, 7; RU	$\emptyset$
9	f_S	2; RC	$\{!f_S\}$

The minimal choice sets for Γ_{RM} are $\Phi(\Gamma_{\mathsf{RM}}) = \{\{!c_{V,B}\}, \{!i_V\}, \{!f_B\}\}$ and the set of unreliable abnormalities is $U(\Gamma_{\mathsf{RM}}) = \{!c_{V,B}, !i_V, !f_B\}$. Hence,

$$\Gamma_{\mathsf{RM}} \vdash_{\mathbf{CL}_\circ{}^{\mathbf{m}}} f_S, \text{ and} \quad (2.6)$$

$$\Gamma_{\mathsf{RM}} \vdash_{\mathbf{CL}_\circ{}^{\mathbf{r}}} f_S. \quad (2.7)$$

Note further that there is a minimal abnormal $\mathbf{CL}_\circ$-model M of Γ_{RM} such that $\mathrm{Ab}(M) = \{!i_V\}$ and $M \models \circ c_{V,S}$. Hence

$$\Gamma_{\mathsf{RM}} \nvdash_{\mathbf{CL}_{\circ}{}^{\mathbf{m}}} \check{\neg} \circ c_{V,S} \tag{2.8}$$
$$\Gamma_{\mathsf{RM}} \nvdash_{\mathbf{CL}_{\circ}{}^{\mathbf{r}}} \check{\neg} \circ c_{V,S} \tag{2.9}$$

Suppose now we take into account our source S_3 and add $\circ c_{V,S}$ to our premise set Γ_{RM}. In this case we add the following lines to a proof from $\Gamma_{\mathsf{RM}} \cup \{\circ c_{V,S}\}$:

10	$\circ c_{V,S}$	PREM	$\emptyset$
11	$c_{V,S} \supset (!i_V \vee !f_S)$	2, 3, 6; RU	$\emptyset$
12	$!c_{V,S} \check{\vee} !i_V \check{\vee} !f_S$	10, 11; RU	$\emptyset$

Note that at line 12 we have the following minimal choice sets of $\Gamma_{\mathsf{RM}} \cup \{\circ c_{V,S}\}$,

$$\Phi_{12}(\Gamma_{\mathsf{RM}} \cup \{\circ c_{V,S}\}) = \{\{!c_{V,B}, !c_{V,S}\}, \{!c_{V,B}, !f_S\}, \{!i_V\}, \{!c_{V,S}, !f_B\}\}.$$

Moreover, the set of unreliable abnormalities is

$$U_{12}(\Gamma_{\mathsf{RM}} \cup \{\circ c_{V,S}\}) = \{!c_{V,B}, !c_{V,S}, !f_S, !i_V, !f_B\}.$$

Hence, at this stage of the proof line 9 is marked according to both strategies. It is easy to see that there is no extension of the proof that leads to the unmarking of line 9. Since there is no other way to derive f_S we have

$$\Gamma_{\mathsf{RM}} \cup \{\circ c_{V,S}\} \nvdash_{\mathbf{CL}_{\circ}{}^{\mathbf{m}}} f_S, \text{ and} \tag{2.10}$$
$$\Gamma_{\mathsf{RM}} \cup \{\circ c_{V,S}\} \nvdash_{\mathbf{CL}_{\circ}{}^{\mathbf{r}}} f_S. \tag{2.11}$$

Altogether this shows that Rational Monotonicity is not valid in ALs. For reliability this is demonstrated by (2.7), (2.9) and (2.11), for minimal abnormality strategy it is demonstrated by (2.6), (2.8) and (2.10). As has been argued by Stalnaker, this is also the intuitive behavior.

2.6.3.2 Rational Distributivity

Similarly, ALs do not in general validate Rational Distributivity:

$$\text{If } A \notin Cn_{\mathbf{L}}(\Gamma \cup \{B\}) \text{ and } A \notin Cn_{\mathbf{L}}(\Gamma \cup \{C\}), \text{ then } A \notin Cn_{\mathbf{L}}(\Gamma \cup \{B \check{\vee} C\}) \tag{RD}$$

Consider the following example. A usually reliable, though not infallible source S_1 tells us that

- Peter had 6 points at the exam,— $\circ p_6$;
- Sue had 5 points at the exam,— $\circ s_5$;
- Anne had 4 points at the exam,— $\circ a_4$.

Another also reliable but not infallible source S_2 informs us that

- Peter was the worst in the exam,— $\circ p_w$.

Now suppose that yet another reliable but not infallible source S_3 states that

- Anne was the best in the exam,— $\circ a_b$.

Obviously we have $p_w \supset (\neg p_6 \vee (\neg s_5 \wedge \neg a_4))$ and $a_b \supset (\neg a_4 \vee (\neg p_6 \wedge \neg s_5))$. Let Γ_{RD} comprise only source S_1. Hence,

$$\Gamma_{\mathsf{RD}} = \{\circ p_6, \circ s_5, \circ a_4, p_w \supset (\neg p_6 \vee (\neg s_5 \wedge \neg a_4)), a_b \supset (\neg a_4 \vee (\neg p_6 \wedge \neg s_5))\}$$

The following proof fragment demonstrates that s_5 is not derivable from $\Gamma_{\mathsf{RD}} \cup \{\circ p_w\}$:

1	$\circ p_6$	PREM	$\emptyset$
2	$\circ s_5$	PREM	$\emptyset$
3	$\circ a_4$	PREM	$\emptyset$
4	$\circ p_w$	PREM	$\emptyset$
5	$p_w \supset (\neg p_6 \vee (\neg s_5 \wedge \neg a_4))$	PREM	$\emptyset$
6	$!p_w \vee !p_6 \vee (!s_5 \wedge !a_4)$	1, 2, 3, 4, 5; RU	$\emptyset$
7	$!p_w \check{\vee} !p_6 \check{\vee} !s_5$	6; RU	$\emptyset$
8	$!p_w \check{\vee} !p_6 \check{\vee} !a_4$	6; RU	$\emptyset$
^{7}9	s_5	2; RC	$\{!s_5\}$

Note that the minimal choice sets at this stage of the proof are $\Phi_9(\Gamma_{\mathsf{RD}} \cup \{\circ p_w\}) = \{\{!p_w\}, \{!p_6\}, \{!s_5, !a_4\}\}$. It is easy to see that $\Phi(\Gamma_{\mathsf{RD}} \cup \{\circ p_w\}) = \Phi_9(\Gamma_{\mathsf{RD}} \cup \{\circ p_w\})$. Since the only way to derive s_5 is on the condition $\{!s_5\}$, s_5 is not derivable. Thus,

$$\Gamma_{\mathsf{RD}} \cup \{\circ p_w\} \nvdash_{\mathbf{CL}_\circ^{\mathbf{m}}} s_5 \tag{2.12}$$

Similarly as above, $U(\Gamma_{\mathsf{RD}} \cup \{\circ p_w\}) = \{!p_w, !p_6, !a_4, !s_5\}$ and hence

$$\Gamma_{\mathsf{RD}} \cup \{\circ p_w\} \nvdash_{\mathbf{CL}_\circ^{\mathbf{r}}} s_5 \tag{2.13}$$

Analogously it can be shown that

$$\Gamma_{\mathsf{RD}} \cup \{\circ a_b\} \nvdash_{\mathbf{CL}_\circ^{\mathbf{m}}} s_5 \tag{2.14}$$

$$\Gamma_{\mathsf{RD}} \cup \{\circ a_b\} \nvdash_{\mathbf{CL}_\circ^{\mathbf{r}}} s_5 \tag{2.15}$$

The following proof fragment demonstrates that s_5 is derivable from $\Gamma_{\mathsf{RD}} \cup \{\circ p_w \check{\vee} \circ a_b\}$ for both adaptive strategies and hence that Rational Distributivity does not in general hold for ALs.

1	$\circ p_6$	PREM	$\emptyset$
2	$\circ s_5$	PREM	$\emptyset$
3	$\circ a_4$	PREM	$\emptyset$
4	$\circ a_b \check{\vee} \circ p_w$	PREM	$\emptyset$
5	$a_b \supset (\neg a_4 \vee (\neg p_6 \wedge \neg s_5))$	PREM	$\emptyset$
6	$p_w \supset (\neg p_6 \vee (\neg s_5 \wedge \neg a_4))$	PREM	$\emptyset$

7	$!a_b \vee !a_4 \vee (!p_6 \wedge !s_5) \vee !p_w \vee$ $!p_6 \vee (!s_5 \wedge !a_4)$	1, 2, 3, 4, 5, 6; RU	$\emptyset$
8	$!a_b \check{\vee} !a_4 \check{\vee} !p_w \check{\vee} !p_6$	7; RU	$\emptyset$
9	s_5	2; RC	$\{!s_5\}$

Note that $\Phi_9(\Gamma_{\mathsf{RD}} \cup \{\circ p_w \check{\vee} \circ a_b\}) = \{\{!a_b\}, \{!a_4\}, \{!p_w\}, \{!p_6\}\}$. Again, it is easy to show that $\Phi(\Gamma_{\mathsf{RD}} \cup \{\circ p_w \check{\vee} \circ a_b\}) = \Phi_9(\Gamma_{\mathsf{RD}} \cup \{\circ p_w \check{\vee} \circ a_b\})$. Moreover $U(\Gamma_{\mathsf{RD}} \cup \{\circ p_w \check{\vee} \circ a_b\}) = \{!a_b, !p_w, !a_4, !p_6\}$. Hence, line 9 is finally derived. Thus,

$$\Gamma_{\mathsf{RD}} \cup \{\circ p_w \check{\vee} \circ a_b\} \vdash_{\mathbf{CL}_\circ^{\mathbf{m}}} s_5 \tag{2.16}$$

$$\Gamma_{\mathsf{RD}} \cup \{\circ p_w \check{\vee} \circ a_b\} \vdash_{\mathbf{CL}_\circ^{\mathbf{r}}} s_5 \tag{2.17}$$

Note that (2.12), (2.14) and (2.16) show that Rational Distributivity does not hold for $\mathbf{CL}_\circ^{\mathbf{m}}$ and hence that it does not in general hold for ALs with minimal abnormality strategy. Moreover, (2.13), (2.15) and (2.17) show that it also does not hold for $\mathbf{CL}_\circ^{\mathbf{m}}$. Hence Rational Distributivity does not in general hold for ALs that employ the reliability strategy. The example shows that in some cases this is as desired. Although Rational Distributivity holds for a great variety of examples, there are some where it fails. In order for ALs to be a generic framework for defeasible reasoning it is desirable that ALs provide means to handle the latter cases in an intuitive way.

The fact that properties such as Rational Monotonicity and Rational Distributivity do no in general hold for ALs does not mean that ALs may not be used in order to characterize reasoning forms that explicate such properties. It only means that the characterization has to be realized under a translation (see Sect. 4.4).

2.7 The Necessity of Superimposing Classical Connectives

The reader may have the impression that, given a supraclassical lower limit logic **LLL**, the superimposing of the classical "checked" connectives is redundant.[26] Since all the ALs introduced in the following parts of this book are based on supraclassical lower limit logics it is important to avoid this confusion. For instance, some may think that Dab-formulas $\mathsf{Dab}(\Delta) = A_1 \check{\vee} \ldots \check{\vee} A_n$ can be simply expressed by $A_1 \vee \ldots \vee A_n$ where $\vee$ is the classical disjunction that is expressible in **LLL** (due to it being supraclassical). This impression may be further strengthened by the fact that in many papers that feature supraclassical lower limit logics checked symbols do not occur.

I was convinced of the redundancy of the checked connectives in cases in which **LLL** is supraclassical until I encountered the following example. It can be presented in a schematic and abstract form. We only need to presuppose that **LLL** is supraclassical and that abnormalities are denoted by $!A$ and formulated in $\mathcal{L}$. Let $\{!A_1, !A_2, \ldots\}$

[26] See Sect. 2.2.

be the set of all abnormalities in Ω. Let our premise set be $\Gamma = \Gamma_1 \cup \Gamma_2$, where

$$\Gamma_1 = \{!A_i \vee !A_j \mid 1 \leq i < j\}$$

$$\Gamma_2 = \left\{ \bigwedge_{1 \leq i < j \leq n} (!A_i \vee !A_j) \supset (A \vee !A_{n-1}) \mid 1 < n \right\}$$

Note that $\Phi(\Gamma) = \{\varphi_i \mid i \in \mathbb{N}\}$ where $\varphi_i = \Omega \setminus \{!A_i\}$. Moreover $\Gamma \vdash_{\mathbf{LLL}} A \vee !A_i$ for every $i \in \mathbb{N}$. Let M be a minimal abnormal model of Γ. By Theorem 2.4.6, there is a φ_i such that $\text{Ab}(M) = \varphi_i$. Hence $M \models \neg !A_i$. Since $M \models A \vee !A_i$, $M \models A$. Hence $\Gamma \Vdash_{\mathbf{AL^m}} A$.

In the following I will show that if formulas such as $!A_1 \vee !A_2$ are treated as Dab-formulas then the consequence set is not complete with respect to the semantics.

The reader is warned: In the following discussion I will incorrectly(!) treat formulas of the type $!A_1 \vee \ldots \vee !A_n$ as Dab-formulas.

The problem is the following: (†) A cannot be produced as the second element of a finite line i such that at some finite stage s line i is unmarked. In other words, at every finite stage s all conditional derivations of A are marked. Definition 2.4.4 requires (a) that A is the second element of a line i and (b) that there is a finite stage s at which line i is unmarked. Hence, A is not finally derived in any $\mathbf{AL^m}$-proof from Γ and thus $\Gamma \nVdash_{\mathbf{AL^m}} A$. Thus, $\mathbf{AL^m}$ is not complete for premise sets that contain checked connectives.

To illustrate (†) let me go a bit through a sample proof from Γ[27]:

	1 $!A_1 \vee !A_2$	PREM	$\emptyset$
	2 $(!A_1 \vee !A_2) \supset (A \vee !A_1)$	PREM	$\emptyset$
	3 $A \vee !A_1$	1, 2; RU	$\emptyset$
1	4 A	3; RC	$\{!A_1\}$
	5 $!A_1 \vee !A_3$	PREM	$\emptyset$
	6 $!A_2 \vee !A_3$	PREM	$\emptyset$
	7 $\bigwedge_{1 \leq i < j \leq 3} (!A_i \vee !A_j) \supset (A \vee !A_2)$	PREM	$\emptyset$
	8 $A \vee !A_2$	1, 5, 6, 7; RU	$\emptyset$
6	9 A	8; RC	$\{!A_2\}$

Note that at line 4 the minimal choice sets are $\{\{!A_1\}, \{!A_2\}\}$. Since there is a minimal choice set that intersects with all conditions on which A has been derived so far, namely $\{!A_1\}$ line 4 is marked.

[27] This example is formulated for the minimal abnormality strategy. A similar example was presented by Frederik Van De Putte in [42].

An analogous argument applies to line 9. The minimal choice sets are now $\{\{!A_1, !A_2\}, \{!A_1, !A_3\}, \{!A_2, !A_3\}\}$. Again, the choice set $\{!A_1, !A_2\}$ intersects with all the conditions of lines at which A has been derived, namely $\{!A_1\}$ and $\{!A_2\}$.

The problem is: it is only possible to derive A on the condition $\{!A_n\}$ at some stage s if all $!A_i \vee !A_j \in \Gamma$ where $1 \leq i, j \leq n+1$ have been introduced in the proof. But then some $\varphi_n \supseteq \{!A_1, \ldots, !A_n\}$ is a minimal choice set in $\Phi_s(\Gamma)$ and hence all the conditions of lines at which A has been introduced are marked since they intersect with φ_n.

This shows that at every finite stage of the proof every line that features A as second element is marked. By Definition 2.4.4, A is not finally derived and hence $\Gamma \nvdash_{\mathbf{AL^m}} A$.

When I confronted Diderik Batens with this "problem" and the proof fragment above, he reminded me of the role of the superimposed "checked" classical connectives.[28] Recall that (a) premises are supposed to be formulated in $\mathcal{L}$, and (b) Dab-formulas $\mathsf{Dab}(\Delta)$ are defined by $\check{\bigvee}\Delta$. As a consequence, lines 4 and 9 are unmarked in the proof above since for any stage s in the proof fragment above, $\Phi_s(\Gamma) = \{\emptyset\}$. Indeed, A is finally derived at line 4. In order to see this suppose line 4 is marked in an extension of the proof above. We extend the proof further in such a way that all formulas in Γ_1 are derived by PREM and that A is derived on any condition $!A_i$ where $i \in \mathbb{N}$. It is easy to see that (i) there is such an extension, (ii) that line 4 is unmarked at this stage, and (iii) that the marking remains stable from this stage on.

Of course, given a supraclassical **LLL**, whenever $\bigvee \Delta$ is produced at line l on the empty condition in an **AL**-proof from some premise set Γ' then also $\mathsf{Dab}(\Delta)$ is derivable on the empty condition, say on the next line l'. Hence, adaptive logicians often *conventionally* formulate object-level proofs in such a way that the marking is "shortcut": the marking is as if at line l the Dab-formula $\mathsf{Dab}(\Delta)$ has been derived and the derivation of the actual Dab-formula $\mathsf{Dab}(\Delta)$ is omitted in the presented proof. By treating formulas of the type $\bigvee \Delta$ as Dab-formulas, no $\check{\vee}$ connectives occur in the proofs which simplifies the presentation. In most cases, unlike the example above, this procedure is harmless in the sense that it produces the correct consequences. Obviously such proofs can be translated in a straightforward way into formally correct object-level proofs (by just adding a line l' featuring $\mathsf{Dab}(\Delta)$ on the empty condition, whenever at a line l, $\bigvee \Delta$ has been derived on the empty condition).

I will follow this convention throughout most of the following parts of this book.

Finally, it should be mentioned that **AL** is always sound and complete for any premise set $\Gamma \subseteq \mathcal{W}^+$ if we first close Γ under $\mathbf{LLL}^+$:

Theorem 2.7.1. *Where* $\Gamma = Cn^{\mathcal{L}^+}_{\mathbf{LLL}}(\Gamma)$*:* $\Gamma \Vdash_{\mathbf{AL}} A$ *iff* $\Gamma \vdash_{\mathbf{AL}} A$.

Proof. "$\Leftarrow$": this follows by Theorem 2.6.2.

[28] Indeed, he had already written a draft for a section for his forthcoming book (see [2, Part 4]) that discusses this problem with a similar example.

"$\Rightarrow$": Let $\Gamma \Vdash_{\mathbf{AL^r}} A$. Hence, for all $M \in \mathcal{M}_{\mathbf{AL^r}}(\Gamma)$, $M \models A$. Hence, for all $M \in \mathcal{M}_{\mathbf{LLL}}(\Gamma)$ for which $\mathsf{Ab}(M) \subseteq U(\Gamma)$, $M \models A$. Thus, $\Gamma \cup (\Omega \setminus U(\Gamma))^{\check{\neg}} \Vdash_{\mathbf{LLL}} A$. By the compactness of **LLL** there is a finite $\Delta \subseteq \Omega \setminus U(\Gamma)$ such that $\Gamma \cup \Delta^{\check{\neg}} \Vdash_{\mathbf{LLL}} A$. Thus, $\Gamma \Vdash_{\mathbf{LLL}} A \check{\vee} \mathsf{Dab}(\Delta)$. By the completeness of **LLL**, $A \check{\vee} \mathsf{Dab}(\Delta) \in Cn^{\mathcal{L}^+}_{\mathbf{LLL}}(\Gamma)$ and thus also $\check{\neg} A \check{\supset} \mathsf{Dab}(\Delta) \in Cn^{\mathcal{L}^+}_{\mathbf{LLL}}(\Gamma)$. We now prove A in an $\mathbf{AL^r}$-proof from $\Gamma = Cn^{\mathcal{L}^+}_{\mathbf{LLL}}(\Gamma)$ as follows: We introduce $\check{\neg} A \check{\supset} \mathsf{Dab}(\Delta)$ on line 1 by PREM. Then we derive A on the condition Δ by RC on line 2. Since no Dab-formulas are derived at this stage, line 2 is unmarked. Suppose line 2 is marked in an extension of the proof at some stage s. For each minimal Dab-formula Θ at stage s for which $\Delta \cap \Theta \neq \emptyset$ there is a $\Theta' \subset \Theta$ such that $\mathsf{Dab}(\Theta')$ is a minimal Dab-consequence of $Cn^{\mathcal{L}^+}_{\mathbf{LLL}}(\Gamma)$ and $\Theta' \cap \Delta = \emptyset$. This holds since $\Delta \subseteq \Omega \setminus U(\Gamma) = \Omega \setminus U(Cn^{\mathcal{L}^+}_{\mathbf{LLL}}(\Gamma))$. We extend the proof by introducing $\mathsf{Dab}(\Theta')$ for all these Θ's. Let the resulting stage be s'. Obviously, by the construction, $U_{s'}(\Gamma) \cap \Delta = \emptyset$.

The proof for minimal abnormality is similar and left to the reader. □

By Theorem 2.6.1 and Theorem 2.7.1 we immediately get:

Corollary 2.7.1. *Where* $\Gamma \subseteq \mathcal{W}$ *or* $\Gamma = Cn^{\mathcal{L}^+}_{\mathbf{LLL}}(\Gamma)$*:* $\Gamma \Vdash_{\mathbf{AL}} A$ *iff* $\Gamma \vdash_{\mathbf{AL}} A$.

2.8 Normal Selections: A 'Credulous' Strategy that is not in the Standard Format

The difference between the two standard strategies manifests itself in the fact that one, reliability, models a more 'cautious' and the other one, minimal abnormality, a 'bolder' style of defeasible reasoning. That is to say, the consequence relation for minimal abnormality is in many cases stronger than the one for reliability. However, there is also a more rigorous way of distinguishing between credulous and skeptical reasoning in the context of logics that model defeasible reasoning which can be found (under different names) in various well-known systems such as default logic, inheritance networks, abstract argumentation, Input/Output logic, the maximal consistent subset approach, etc.

Join Approach A is a *skeptical* consequence of Γ iff A is valid in/implied by/etc. *all* models/extensions/maximal consistent subsets/etc. of Γ

Meet Approach A is a *credulous* consequence of Γ iff A is valid in/implied by/etc. *some* interpretation/extension/maximal consistent subset/etc. of Γ.

Obviously, what is modeled by ALs in standard format (such as it is currently defined) is the former, *skeptical* notion. However, there is also an adaptive strategy that is in the spirit of the second, *credulous* notion.

According to the *normal selections strategy* A is a semantic consequence of Γ iff it is valid in a specific set of selected models of Γ. The latter sets are equivalence classes of **LLL**-models that have the same abnormal part. Where $M \sim M'$

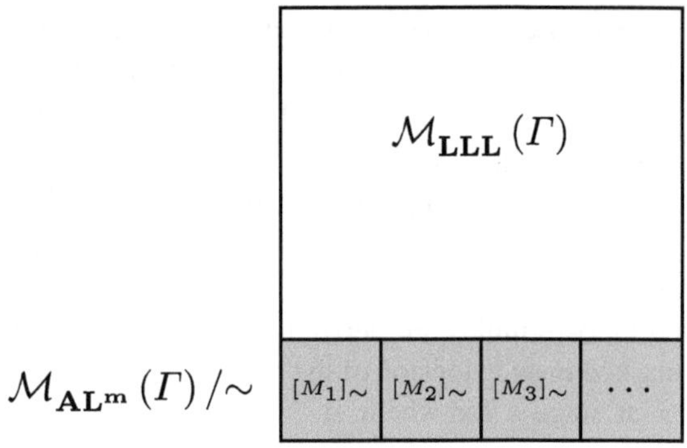

Fig. 2.8 The quotient structure $\mathcal{M}_{\mathbf{AL^m}}(\Gamma)/\sim$

Table 2.2 Equivalence class $[M]_\sim \in \mathcal{M}_{\mathbf{AL^m}}(\Gamma)/\sim$ represents a set of models

	M_1	M_2	M_3	M_4	M_5	M_6
$M \models$	$\neg a, b,$	$a, \neg b,$	$\neg a, \neg b,$	$\neg a, b,$	$a, \neg b,$	$\neg a, \neg b,$
	c, n	c, n	$\neg c, n$	$c, \neg n$	$c, \neg n$	$\neg c, \neg n$
$\mathrm{Ab}(M) =$	$\{!a\}$	$\{!b\}$	$\{!a, !b\}$	$\{!a, !n\}$	$\{!b, !n\}$	$\{!a, !b, !n\}$

iff $\mathrm{Ab}(M) = \mathrm{Ab}(M')$, $\mathcal{M}_{\mathbf{AL^m}}(\Gamma)/\sim$ is the quotient structure defined by the equivalence relation $\sim$ on the set of all $\mathbf{AL^m}$-models of Γ,[29] (see Fig. 2.8) and $[M]_\sim = \{M' \in \mathcal{M}_{\mathbf{AL^m}}(\Gamma) \mid M \sim M'\}$ we can define

Definition 2.8.1. $\Gamma \Vdash_{\mathbf{AL^n}} A$ iff there is a $[M]_\sim \in \mathcal{M}_{\mathbf{AL^m}}(\Gamma)/\sim$ such that for all $M' \in [M]_\sim$, $M' \models A$.

Alternatively this can be expressed by: $\Gamma \Vdash_{\mathbf{AL^n}} A$ iff there is a $M \in \mathcal{M}_{\mathbf{AL^m}}(\Gamma)$ such that for all $M' \in \mathcal{M}_{\mathbf{LLL}}(\Gamma)$ for which $\mathrm{Ab}(M') = \mathrm{Ab}(M)$, $M' \models A$ (Fig. 2.8).

Each equivalence class $[M]_\sim \in \mathcal{M}_{\mathbf{AL^m}}(\Gamma)/\sim$ represents a set of models that interpret the premise set Γ "as normally as possible". In each equivalence class this is realized in a different way. In our example M_1 and M_2 (see Table 2.2) belong to different equivalence classes. For instance a model M_7 for which $\mathrm{Ab}(M_7) = \{!a\}$ belongs to the same equivalence class as M_1. We have for instance $\Gamma_1 \Vdash_{\mathbf{CL_\circ^n}} a$ since for all $M \in [M_2]_\sim$, $M \models a$.

Each equivalence class offers a specific minimally abnormal interpretation of the given Dab-consequences. If we find one interpretation such that A is validated by all models that share this interpretation, A is considered a consequence. This distinguishes the normal selections strategy from both the reliability and the minimal abnormality strategy where A had to be valid in *all* models that offer sufficiently normal interpretations (so, the reliable resp. the minimally abnormal models). Thus, this makes the normal selections strategy more similar to the 'meet'-approach that is characteristic for credulous consequence relations, while the strategies of the stan-

[29] The fact that $\sim$ is a equivalence relation on $\mathcal{M}_{\mathbf{AL^m}}(\Gamma)$ can be easily shown and is left to the reader.

dard format are more similar to the 'join'-approach behind skeptical consequence relations. For instance a is valid in all models that have the (minimally) abnormal part $\{!b\}$ and hence it is a consequence according to the normal selections strategy. Note also that $\Gamma \not\Vdash_{\mathbf{CL}_\circ^{\mathrm{m}}} A$, i.e. a is not a consequence according to the minimal abnormality strategy. For instance the minimally abnormal model M_1 does not verify a.

With Theorem 2.4.6 we immediately get:

Theorem 2.8.1. *$\Gamma \Vdash_{\mathbf{AL}^{\mathrm{n}}} A$ iff there is a $\varphi \in \Phi(\Gamma)$ such that for all $M \in \{M' \in \mathcal{M}_{\mathbf{LLL}}(\Gamma) \mid Ab(M') = \varphi\}$, $M \models A$.*

Similarly, in the proof theory the idea is, that if A is derivable on an assumption that is not violated in some minimal abnormal interpretation of the Dab-consequences then A can be considered a consequence. This is realized by means of the following marking definition:

Definition 2.8.2 (Marking for normal selections, variant 1). A line l with condition Δ is marked at stage s, iff for all $\varphi \in \Phi_s(\Gamma)$, $\Delta \cap \varphi \neq \emptyset$.

In other words, a line with the condition Δ is *unmarked* in case there is a $\varphi \in \Phi(\Gamma)$ such that $\Delta \cap \varphi = \emptyset$. In the terminology of Sect. 2.4.2.2 a line l is unmarked in case the argument at line l is defensible.[30]

Otherwise the proof theory is the same as in the standard format: we again have the three generic rules PREM, RU, and RC.

The good news is that this marking condition can be simplified in a way that no reference need to be made to minimal choice sets:

Definition 2.8.3 (Marking for Normal Selections, variant 2). Line l is marked at stage s iff, where Δ is the condition of line l, $\mathsf{Dab}(\Delta')$ has been derived on the empty condition at stage s for some $\Delta' \subseteq \Delta$.[31]

In Appendix A we show (merely on the basis of set-theoretic insights into choice sets) that

Corollary 2.8.1. *Where $\Delta \subseteq \Omega$ is finite and $\Gamma \subseteq \mathcal{W}^+$:*

(i) there is a $\varphi \in \Phi_s(\Gamma)$ such that $\Delta \cap \varphi = \emptyset$ iff there is no minimal Dab*-formula* $\mathsf{Dab}(\Theta)$ *at stage s such that $\Theta \subseteq \Delta$;*
(ii) there is a $\varphi \in \Phi(\Gamma)$ such that $\Delta \cap \varphi = \emptyset$ iff there is no minimal Dab*-consequence* $\mathsf{Dab}(\Theta)$ *such that $\Theta \subseteq \Delta$.*

Note that (i) immediately implies the equivalence of the marking definitions.

---

[30] The distinction between the skeptical and the credulous approach has been discussed in relation to the distinction between justified and defensible arguments in [24, Sect. 4.3].

[31] Yet another way of phrasing the marking definition in such a way that it leads to the same adaptive consequences is by: Line l with condition Δ is marked at stage s iff $\mathsf{Dab}(\Delta)$ is derived on the empty condition at stage s. Obviously, if we can derive $\mathsf{Dab}(\Delta')$ for some $\Delta' \subseteq \Delta$ at stage s on the empty condition we can also derive $\mathsf{Dab}(\Delta)$ on the empty condition and so eventually mark line l according to the marking definition, variant 2.

Considering the second marking definition it is evident that once a line is marked, it will never be unmarked in a proof. Recall that this is unlike the marking in the standard format where a line may be marked at some point of the proof but get unmarked again at a later stage.

Let us have a simple demonstration by means of our detective case:

	1	$\circ n$	PREM	$\emptyset$
	2	$(a \wedge n) \supset c$	PREM	$\emptyset$
	3	$(b \wedge n) \supset c$	PREM	$\emptyset$
	4	$\circ a$	PREM	$\emptyset$
	5	$\circ b$	PREM	$\emptyset$
	6	$\neg a \vee \neg b$	PREM	$\emptyset$
	7	a	4; RC	$\{!a\}$
	8	b	5; RC	$\{!b\}$
	9	n	1; RC	$\{!n\}$
	10	$!a \check{\vee} !b$	4, 5, 6; RU	$\emptyset$
	11	c	2, 4, 9; RC	$\{!a, !n\}$
10	12	$a \wedge b$	7, 8; RU	$\{!a, !b\}$
	13	$\neg b$	6, 7; RU	$\{!a\}$
10	14	$b \wedge \neg b$	8, 13; RU	$\{!a, !b\}$

The first difference to the strategies of the standard format concerns lines 7 and 8: both are marked according to reliability and minimal abnormality but not according to normal selections. Similar as in the standard format lines 12 and 14 get marked: after all, the disjunction of the members of the condition of these lines has been derived at line 10 (cf. marking variant 2). What is most remarkable is that by means of normal selections we can derive both b (line 8) and $\neg b$ (line 13): for each respective condition Δ there is a minimal choice set (note that $\Phi_{14}(\Gamma) = \{\{!a\}, \{!b\}\}$) that has an empty intersection with Δ. However, line 14 with the formula $b \wedge \neg b$ gets marked. Obviously, there is no minimally abnormal interpretation which validates both abnormalities in the conditions: $!a$ and $!b$.

Final derivability is defined as usual (see Definition 2.4.4). Hence, we define $\Gamma \vdash_{\mathbf{AL^n}} A$ iff A is finally derivable in a $\mathbf{AL^n}$ proof from Γ.

Given the equivalence of our two marking definitions, it is not surprising that we get two corresponding representational theorems for the syntactic consequence relation.

Theorem 2.8.2. *Where $\Gamma \subseteq \mathcal{W}$ or $\Gamma = Cn^{\mathcal{L}^+}_{\mathbf{LLL}}(\Gamma)$: $\Gamma \vdash_{\mathbf{AL^n}} A$ iff there is a $\Delta \subseteq \Omega$ such that $\Gamma \vdash_{\mathbf{LLL}} A \check{\vee} \mathsf{Dab}(\Delta)$ and $\Gamma \nvdash_{\mathbf{LLL}} \mathsf{Dab}(\Delta)$.*

Theorem 2.8.3. *Where $\Gamma \subseteq \mathcal{W}$ or $\Gamma = Cn^{\mathcal{L}^+}_{\mathbf{LLL}}(\Gamma)$: $\Gamma \vdash_{\mathbf{AL^n}} A$ iff there is a $\Delta \subseteq \Omega$ such that (a) $\Gamma \vdash_{\mathbf{LLL}} A \check{\vee} \mathsf{Dab}(\Delta)$, and (b) for some $\varphi \in \Phi(\Gamma)$, $\varphi \cap \Delta = \emptyset$.*

Finally, this gives us soundness and completeness[32]:

[32] The proof is straight-forward in view of Theorem 2.8.1 and Theorem 2.8.3. In Chap. 5 we prove a generalized version of Theorem 2.8.4.

Theorem 2.8.4. *Where* $\Gamma \subseteq \mathcal{W}$ *or* $\Gamma = Cn^{\mathcal{L}^+}_{\mathbf{LLL}}(\Gamma)$*:* $\Gamma \vdash_{\mathbf{AL}^{\mathrm{n}}} A$ *iff* $\Gamma \Vdash_{\mathbf{AL}^{\mathrm{n}}} A$*.*

As a concluding remark it should be mentioned that the normal selections strategy can be represented by means of the simple strategy under a translation. This will be demonstrated in a future paper together with Joke Meheus.

References

1. Batens, D.: A universal logic approach to adaptive logics. Logica Universalis **1**, 221–242 (2007)
2. Batens, D.: Adaptive Logics and Dynamic Proofs. A Study in the Dynamics of Reasoning (201x)
3. Batens, D.: Inconsistency-adaptive logics. In: Orłowska, E. (ed.) Logic at Work. Essays Dedicated to the Memory of Helena Rasiowa, pp. 445–472. Physica Verlag (Springer), Heidelberg (1999)
4. Batens, D., De Clercq, K.: A rich paraconsistent extension of full positive logic. Log. Anal. **185–188**, 227–257 (2004). Appeared 2005
5. da Costa, N.C.: On the theory of inconsistent formal systems. Notre Dame J. Formal Logic **15**, 497–510 (1974)
6. Batens, D.: Adaptive $\mathbf{C}_n$ logics. In: Carnielli, W.A., Coniglio, M.E., Loffredo D'Ottaviano, I.M. (eds.) The Many Sides of Logic, Studies in Logic, vol. 21, pp. 27–45. College Publications, London (2009)
7. Batens, D., Meheus, J., Provijn, D., Verhoeven, L.: Some adaptive logics for diagnosis. Logic Log. Philos. **11/12**, 39–65 (2003)
8. Meheus, J.: An adaptive logic based on Jaśkowski's approach to paraconsistency. J. Philos. Logic **35**, 539–567 (2006)
9. Meheus, J., Straßer, C., Verdée, P.: Which style of reasoning to choose in the face of conflicting information? J. Logic Comput. (forthcoming)
10. Goble, L.: Preference semantics for deontic logic. Part II: Multiplex models. Log. Anal. **185–188**, 335–363 (2004)
11. Meheus, J., Beirlaen, M., Putte, F.V.D.: Avoiding deontic explosion by contextually restricting aggregation. In: Governatori, G., Sartor, G. (eds.) DEON (10th International Conference on Deontic Logic in Computer Science). Lecture Notes in Artificial Intelligence, vol. 6181, pp. 148–165. Springer, Berlin (2010)
12. Meheus, J., Beirlaen, M., Putte, F.V.D., Straßer, C.: Non-adjunctive deontic logics that validate aggregation as much as possible (forthcoming)
13. Goble, L.: A proposal for dealing with deontic dilemmas. In: A. Lomuscio, D. Nute (eds.) DEON. Lecture Notes in Computer Science, vol. 3065, pp. 74–113. Springer, Berlin (2004)
14. Straßer, C., Meheus, J., Beirlaen, M.: Tolerating deontic conflicts by adaptively restricting inheritance. Log. Anal. **219**, 477–506 (2012)
15. Straßer, C.: An adaptive logic framework for conditional obligations and deontic dilemmas. Logic Log. Philos. **19**(1–2), 95–128 (2010)
16. Rescher, N., Manor, R.: On inference from inconsistent premises. Theor. Decis. **1**, 179–217 (1970)
17. Batens, D.: Towards a dialogic interpretation of dynamic proofs. In: C. Dégremont et al. (ed.) Dialogues, Logics and Other Strange Things. Essays in Honour of Shahid Rahman, pp. 27–51. College Publications, London (2009)
18. Blair, A.: The "logic" of informal logic. In: Tindale, C.W. (ed.) Groundwork in the Theory of Argumentation: Selected Papers of J. Anthony Blair, pp. 101–118. Springer, Berlin (2012)
19. Toulmin, S.E.: The Uses of Argument. Cambridge University Press, Cambridge (1958)
20. Wellman, C.: Challenge and Response: Justification in Ethics. Southern Illinois University Press, Carbondale (1971)

21. Rescher, N.: Dialectics: A Controversy-Oriented Approach to the Theory of Knowledge. SUNY Press, New York (1977)
22. Pollock, J.: How to reason defeasibly. Artif. Intell. 57(1), 1–42 (1992)
23. Walton, D.: Argumentation Schemes for Presumptive Reasoning (Studies in Argumentation Theory). Routledge, London (1995)
24. Prakken, H., Vreeswijk, G.A.W.: Logics for Defeasible Argumentation, vol. 4, pp. 219–318. Kluwer, Dordrecht (2002)
25. Rescher, N.: Plausible Reasoning. An Introduction to the Theory and Practice of Plausibilistic Inference. Van Gorcum, Assen (1976)
26. Vreeswijk, G.A.W.: Studies in defeasible argumentation. Ph.D. thesis, Free University Amsterdam, Department of Computer Science (1993)
27. Prakken, H.: Some reflections on two current trends in formal argumentation. Logic Programs, Norms and Action, pp. 249–272 (2012)
28. Prakken, H.: An abstract framework for argumentation with structured arguments. Argument Comput. **3**1, 1–19 (2010)
29. Caminada, M.: On the issue of contraposition of defeasible rules. Front. Artif. Intell. Appl. **172**, 109–115 (2008)
30. Hage, J.: Reasoning with Rules: An Essay on Legal Reasoning and its Underlying Logic, vol. 27. Springer, Berlin (1996)
31. Yalcin, S.: A counterexample to modus tollens. J. Philos. Logic **41**(6), 1001–1024 (2012)
32. Horsten, L., Welch, P.: The undecidability of propositional adaptive logic. Synthese **158**, 41–60 (2007)
33. Verdée, P.: Adaptive logics using the minimal abnormality strategy are π_1^1-complex. Synthese **167**, 93–104 (2009)
34. Batens, D., De Clercq, K., Verdée, P., Meheus, J.: Yes fellows, most human reasoning is complex. Synthese **166**, 113–131 (2009)
35. Schlipf, J.S.: Decidability and definability with circumscription. Ann. Pure Appl. Logic **35**, 173–191 (1987)
36. Papalaskari, M.A., Weinstein, S.: Minimal consequence in sentential logic. J. Logic Program. **9**(1), 19–31 (1990)
37. Apt, K.R., Blair, H.A.: Arithmetic classification of perfect models of stratified programs. Fundam. Inform. **13**(1), 1–17 (1990)
38. Chomicki, J., Subrahmanian, V.: Generalized closed world assumption is π_2^0. Technical report complete. Technical Report TR-89-036, Department of Computer Science, University of North Carolina (1989)
39. Odintsov, S.P., Speranski, S.O.: On algorithmic properties of propositional inconsistency-adaptive logics. Logic Log. Philos. **21**(3), 209–228 (2012)
40. Odintsov, S.P., Speranski, S.O.: Computability issues for adaptive logics in expanded standard format. Presented as a keynote at Trends in Logic XI, Bochum, 2012 (forthcoming)
41. Stalnaker, R.F.: What is a nonmonotonic consequence relation? Fundam. Inform. **21**(1/2), 7–21 (1994)
42. Putte, F.V.D.: Generic formats for prioritized adaptive logics with applications in deontic logic, abduction and belief revision. Ph.D. thesis, Ghent University (March 28th, 2012)

Chapter 3
Sequential Combinations of ALs

There are various ways to combine ALs.[1] This section covers sequential combinations under a generic perspective, i.e., the kind of combinations that are used in this book. Sequential combinations are by far the most frequent combination type in the literature. In [3] the reader finds a study that compares sequential combinations of ALs to other formats for ALs such as lexicographic ALs [4].

I will consider the case that some given ALs in the standard format $\mathbf{AL}_{\mathbf{1}}^{\mathbf{x_1}}, \ldots, \mathbf{AL}_{\mathbf{n}}^{\mathbf{x_n}}$ are sequentially combined. In the remainder of this chapter, we have for all $i \in \{1, \ldots, n\}$, $\mathbf{AL}_{\mathbf{i}}^{\mathbf{x_i}} = \langle \mathbf{LLL}, \Omega_i, \text{strategy} \rangle$ and the strategy is reliability if $\mathbf{x_i} = \mathbf{r}$ and minimal abnormality if $\mathbf{x_i} = \mathbf{m}$. Adaptive logics in the standard format will sometimes be called *flat ALs* as opposed to the ALs that are the result of sequential combinations of ALs.

The idea is to define a consequence set of the logic that we dub **CAL** by first applying $\mathbf{AL}_{\mathbf{1}}^{\mathbf{x_1}}$ to the premise set Γ, then as a second step to apply $\mathbf{AL}_{\mathbf{2}}^{\mathbf{x_2}}$ to the $\mathbf{AL}_{\mathbf{1}}^{\mathbf{x_1}}$-consequence set of $\Gamma, \ldots,$ and finally as a nth step to apply $\mathbf{AL}_{\mathbf{n}}^{\mathbf{x_n}}$ to the $\mathbf{AL}_{\mathbf{n-1}}^{\mathbf{x_{n-1}}}$-consequences of the $\mathbf{AL}_{\mathbf{n-2}}^{\mathbf{x_{n-2}}}$-consequences of ... of the $\mathbf{AL}_{\mathbf{1}}^{\mathbf{x_1}}$-consequences of Γ. See Fig. 3.1 for an illustration.

Of course this raises the question whether such combinations are at all useful. Suppose for instance that we are interested in making inductive generalizations on the basis of possibly inconsistent information. This is a situation scientists sometimes face: different experiments and other empirical data may be inconsistent. Still we are interested in obtaining useful generalizations. In this situation we may first want to interpret the given data as consistent as possible and hence use an inconsistency-adaptive logic. Given this interpretation we proceed by means of an AL that helps us to make inductive generalizations.

In this chapter I will investigate some meta-theory of such combinations of ALs and hence stay on a technical and abstract level. This will serve as a theoretic

[1] Many of them are covered by Diderik Batens in [1]. Moreover, Frederik Van De Putte discusses some in [2].

C. Straßer, *Adaptive Logics for Defeasible Reasoning*, Trends in Logic 38,
DOI: 10.1007/978-3-319-00792-2_3, © Springer International Publishing Switzerland 2014

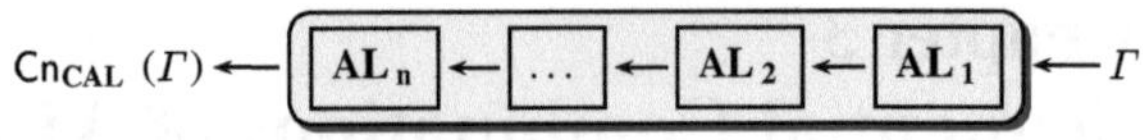

Fig. 3.1 Sequential combination of ALs

foundation for the concrete combinations of ALs that are presented later in this book. For examples and applications the reader is hence referred to:

- Part III, where combinations of ALs are used in order to model Dung's abstract argumentation framework.
- Part IV, where combinations of ALs are used in the context of deontic logics (see Sects. 11.6 and 12.7).

First, in Sect. 3.1 I will define a consequence relation for **CAL**. In Sect. 3.2 I will present some intuitive semantics. **CAL** will be shown to be sound and complete with respect to this semantics for certain cases, some of which are relevant for the applications in the later parts of this book. In Sect. 3.3 I will present a proof theory for the special case that the sets of abnormalities Ω_i are mutually disjoint.

3.1 The Consequence Relation of CAL

A first way to formally grasp the idea above is to define a consequence relation as follows:

$$Cn^{\mathcal{L}}_{\mathbf{CAL}}(\Gamma) =_{\mathrm{df}} Cn^{\mathcal{L}}_{\mathbf{AL}^{\mathbf{x_n}}_{\mathbf{n}}}\left(Cn^{\mathcal{L}}_{\mathbf{AL}^{\mathbf{x_{n-1}}}_{\mathbf{n-1}}}\left(\dots Cn^{\mathcal{L}}_{\mathbf{AL}^{\mathbf{x_1}}_{\mathbf{1}}}(\Gamma)\dots\right)\right) \tag{3.1}$$

$$Cn^{\mathcal{L}^+}_{\mathbf{CAL}}(\Gamma) =_{\mathrm{df}} Cn^{\mathcal{L}^+}_{\mathbf{AL}^{\mathbf{x_n}}_{\mathbf{n}}}\left(Cn^{\mathcal{L}^+}_{\mathbf{AL}^{\mathbf{x_{n-1}}}_{\mathbf{n-1}}}\left(\dots Cn^{\mathcal{L}^+}_{\mathbf{AL}^{\mathbf{x_1}}_{\mathbf{1}}}(\Gamma)\dots\right)\right) \tag{3.2}$$

where (3.1) describes the $\mathcal{L}$-consequences and (3.2) the $\mathcal{L}^+$-consequences. Of course, in order for this to be adequate we expect that

$$Cn^{\mathcal{L}}_{\mathbf{CAL}}(\Gamma) = Cn^{\mathcal{L}^+}_{\mathbf{CAL}}(\Gamma) \cap \mathcal{W} \tag{3.3}$$

As has been pointed out by Diderik Batens in [1] (see Part 6),[2] this is not in general warranted. Let me give an abstract example.

Example 3.1.1. Suppose we have two ALs in standard format: $\mathbf{AL}^{\mathbf{x_1}}_{\mathbf{1}}$ and $\mathbf{AL}^{\mathbf{x_2}}_{\mathbf{2}}$ where $\mathbf{x_1}, \mathbf{x_2} \in \{\mathbf{r}, \mathbf{m}\}$. Suppose **LLL** does not have means to express the classical negation, $\vee$ and $\supset$ have the classical meaning, $\circ$ is a dummy connective (as it was the case for $\mathbf{CL}_{\circ}$), and Modus Ponens (MP) is available in **LLL**. Let $\mathcal{W}$ be the $\langle \vee, \supset, \circ \rangle$-closure of the set of propositional atoms. Let moreover, $\Omega_1 = \{\circ A \mid A \in \mathcal{W}\}$ and $\Omega_2 = \{\check{\neg} A \mid A \in \Omega_1\}$.

[2] Diderik Batens remarks that Peter Verdée was the first to notice this problem.

We take a look at the premise set $\Gamma = \{p \vee \circ q, \circ q \supset r\}$ where $p, r \notin \Omega_1 \cup \Omega_2$. What is to be expected is that p is (finally) derivable by $\mathbf{AL_1^{x1}}$, since it can be derived on the condition $\{\circ q\}$. Evidently $\circ q$ is not unreliable with respect to $\mathbf{AL_1^{x1}}$ and hence $\check{\neg} \circ q \in Cn^{\mathcal{L}^+}_{\mathbf{AL_1^{x1}}}(\Gamma)$. Due to this we do not expect to derive r in $\mathbf{AL_2^{x2}}$ from the $\mathbf{AL_1^{x1}}$-consequences of Γ, since the only way to derive r is to apply MP to $\circ q \supset r$ and $\circ q$. However, as argued, the latter should not be available.

Note that indeed $\check{\neg} \circ q \in Cn^{\mathcal{L}^+}_{\mathbf{AL_1^{x1}}}(\Gamma)$ and $p \in Cn_{\mathbf{AL_1^{x1}}}(\Gamma)$. However,

$$\circ q \in Cn_{\mathbf{AL_2^{x2}}}\left(Cn^{\mathcal{L}}_{\mathbf{AL_1^{x1}}}(\Gamma)\right) \text{ since } \check{\neg} \circ q \notin Cn^{\mathcal{L}}_{\mathbf{AL_1^{x1}}}(\Gamma),$$

In order to illustrate this let us take a look at a proof. We start with applying $\mathbf{AL_1^{x1}}$ to Γ:

1	$p \vee \circ q$	PREM	$\emptyset$
2	$\circ q \supset r$	PREM	$\emptyset$
3	p	1; RC	$\{\circ q\}$
4	$\check{\neg} q$	-; RC	$\{\circ q\}$

There is of course no way to mark either lines 3 or 4 and hence their respective formulas are finally derivable. This shows that $p, \check{\neg} q \in Cn^{\mathcal{L}^+}_{\mathbf{AL_1^{x1}}}(\Gamma)$. Note however that $\check{\neg} q$ contains a "checked connective" and hence $\check{\neg} q \notin Cn^{\mathcal{L}}_{\mathbf{AL_1^{x1}}}(\Gamma)$. Thus, we have the following $\mathbf{AL_2^{x2}}$-proof from $Cn^{\mathcal{L}}_{\mathbf{AL_1^{x1}}}(\Gamma)$:

1	$\circ q \supset r$	PREM	$\emptyset$
2	$\circ q$	-; RC	$\{\check{\neg} \circ q\}$
3	r	1,2; RU	$\{\check{\neg} \circ q\}$

On the other hand, due to reflexivity of $Cn_{\mathbf{AL_2^{x2}}}$ we have:

$$\check{\neg} \circ q \in Cn^{\mathcal{L}^+}_{\mathbf{AL_2^{x2}}}\left(Cn^{\mathcal{L}^+}_{\mathbf{AL_1^{x1}}}(\Gamma)\right)$$

Hence we end up in an asymmetric situation:

$$r \in Cn_{\mathbf{AL_2^{x2}}}\left(Cn^{\mathcal{L}}_{\mathbf{AL_1^{x1}}}(\Gamma)\right) \text{ whereas } r \notin Cn_{\mathbf{AL_2^{x2}}}\left(Cn^{\mathcal{L}^+}_{\mathbf{AL_1^{x1}}}(\Gamma)\right)$$

One solution to the problem above is to use the cumbersome definition

$$Cn^{\mathcal{L}}_{\mathbf{CAL}}(\Gamma) =_{\mathrm{df}} \mathcal{W} \cap \left(Cn^{\mathcal{L}^+}_{\mathbf{AL_n^{xn}}}\left(Cn^{\mathcal{L}^+}_{\mathbf{AL_{n-1}^{x_{n-1}}}}\left(\ldots Cn^{\mathcal{L}^+}_{\mathbf{AL_1^{x1}}}(\Gamma)\ldots\right)\right)\right) \tag{3.4}$$

instead of (3.1). This way (3.3) is trivially satisfied.

The good news is that for supraclassical lower limit logics both definitions (3.1) and (3.4) are equivalent. In other words: given a supraclassical lower limit logic we can define the consequence relations as in (3.1) and (3.2) while (3.3) is warranted.

The key is that for each $\mathbf{AL}_{\mathbf{i}}^{\mathrm{x_i}}$ the $\mathcal{L}^+$-based consequence relation is identical to the closure of the $\mathcal{L}$-based consequence relation under $\mathbf{LLL}^+$:

$$\text{for all } \Gamma \subseteq \mathcal{W} : Cn_{\mathbf{LLL}}^{\mathcal{L}^+}\left(Cn_{\mathbf{AL}_{\mathbf{i}}^{\mathrm{x_i}}}^{\mathcal{L}}(\Gamma)\right) = Cn_{\mathbf{AL}_{\mathbf{i}}^{\mathrm{x_i}}}^{\mathcal{L}^+}(\Gamma) \tag{3.5}$$

Note that it is sufficient if (3.5) is satisfied for premise sets without "checked connectives".

We have seen in our example that (3.5) is not always satisfied. However, if (3.5) is guaranteed then we immediately get the equivalence of (3.1) and (3.4) (see Theorem 3.1.1 below), i.e., in that case we have:

$$Cn_{\mathbf{AL}_{\mathbf{n}}^{\mathrm{x_n}}}^{\mathcal{L}^+}\left(Cn_{\mathbf{AL}_{\mathbf{n-1}}^{\mathrm{x_{n-1}}}}^{\mathcal{L}^+}\left(\dots Cn_{\mathbf{AL}_{\mathbf{1}}^{\mathrm{x_1}}}^{\mathcal{L}^+}(\Gamma)\dots\right)\right) = Cn_{\mathbf{AL}_{\mathbf{n}}^{\mathrm{x_n}}}^{\mathcal{L}^+}\left(Cn_{\mathbf{AL}_{\mathbf{n-1}}^{\mathrm{x_{n-1}}}}^{\mathcal{L}}\left(\dots Cn_{\mathbf{AL}_{\mathbf{1}}^{\mathrm{x_1}}}^{\mathcal{L}}(\Gamma)\dots\right)\right) \tag{3.6}$$

and hence,

$$A \in \mathcal{W} \cap Cn_{\mathbf{AL}_{\mathbf{n}}^{\mathrm{x_n}}}^{\mathcal{L}^+}\left(Cn_{\mathbf{AL}_{\mathbf{n-1}}^{\mathrm{x_{n-1}}}}^{\mathcal{L}^+}\left(\dots Cn_{\mathbf{AL}_{\mathbf{1}}^{\mathrm{x_1}}}^{\mathcal{L}^+}(\Gamma)\dots\right)\right) \text{ iff } A \in Cn_{\mathbf{AL}_{\mathbf{n}}^{\mathrm{x_n}}}^{\mathcal{L}}\left(Cn_{\mathbf{AL}_{\mathbf{n-1}}^{\mathrm{x_{n-1}}}}^{\mathcal{L}}\left(\dots Cn_{\mathbf{AL}_{\mathbf{1}}^{\mathrm{x_1}}}^{\mathcal{L}}(\Gamma)\dots\right)\right) \tag{3.7}$$

Lemma 3.1.2 below shows that the supraclassicality of **LLL** indeed guarantees (3.5) and as a consequence (3.1) and (3.4) are equivalent whenever we are dealing with a supraclassical **LLL** (see Corollary 3.1.1). As a reminder: **LLL** is supraclassical in case $Cn_{\mathbf{LLL}}^{\mathcal{L}}(\Gamma)$ is at least as strong as classical logic. Note that this trivially holds for $\mathbf{LLL}^+$ (the lower limit logic **LLL** enriched by the "check connectives"). But, as our example shows, not all lower limit logics **LLL** are supraclassical. All the lower limit logics used in applications presented in this book are supraclassical while non-supraclassical lower limits are sometimes used for inconsistency-adaptive logics.

Theorem 3.1.1. *Where* $\Gamma \subseteq \mathcal{W}$*: If for every* $i < n$, (3.5) *holds, then* (3.6) *and* (3.7).

Proof. Suppose that for every $i \in \{1, \dots, n-1\}$, (3.5) holds. I show by induction that for all $i \in \{2, \dots, n\}$,

$$Cn^{\mathcal{L}^+}_{\mathbf{AL}_i^{x_i}}\left(Cn^{\mathcal{L}^+}_{\mathbf{AL}_{i-1}^{x_{i-1}}}\left(\ldots Cn^{\mathcal{L}^+}_{\mathbf{AL}_1^{x_1}}(\Gamma)\ldots\right)\right) = Cn^{\mathcal{L}^+}_{\mathbf{AL}_i^{x_i}}\left(Cn^{\mathcal{L}}_{\mathbf{AL}_{i-1}^{x_{i-1}}}\left(\ldots Cn^{\mathcal{L}}_{\mathbf{AL}_1^{x_1}}(\Gamma)\ldots\right)\right)$$

"$i = 2$": By Theorem 2.6.5ii and since $Cn^{\mathcal{L}}_{\mathbf{AL}_1^{x_1}}(\Gamma) \subseteq \mathcal{W}$,

$$Cn^{\mathcal{L}^+}_{\mathbf{AL}_2^{x_2}}\left(Cn^{\mathcal{L}}_{\mathbf{AL}_1^{x_1}}(\Gamma)\right) = Cn^{\mathcal{L}^+}_{\mathbf{AL}_2^{x_2}}\left(Cn^{\mathcal{L}^+}_{\mathbf{LLL}}\left(Cn^{\mathcal{L}}_{\mathbf{AL}_1^{x_1}}(\Gamma)\right)\right)$$

By (3.5) the latter is identical to $Cn^{\mathcal{L}^+}_{\mathbf{AL}_2^{x_2}}\left(Cn^{\mathcal{L}^+}_{\mathbf{AL}_1^{x_1}}(\Gamma)\right)$.

"$i-1 \Rightarrow i$": By Theorem 2.6.5ii, $Cn^{\mathcal{L}^+}_{\mathbf{AL}_i^{x_i}}\left(Cn^{\mathcal{L}}_{\mathbf{AL}_{i-1}^{x_{i-1}}}\left(\ldots Cn^{\mathcal{L}}_{\mathbf{AL}_1^{x_1}}(\Gamma)\ldots\right)\right) =$

$$Cn^{\mathcal{L}^+}_{\mathbf{AL}_i^{x_i}}\left(Cn^{\mathcal{L}^+}_{\mathbf{LLL}}\left(Cn^{\mathcal{L}}_{\mathbf{AL}_{i-1}^{x_{i-1}}}\left(\ldots Cn^{\mathcal{L}}_{\mathbf{AL}_1^{x_1}}(\Gamma)\ldots\right)\right)\right). \quad (\dagger)$$

($\dagger$) is by (3.5) identical to

$$Cn^{\mathcal{L}^+}_{\mathbf{AL}_i^{x_i}}\left(Cn^{\mathcal{L}^+}_{\mathbf{AL}_{i-1}^{x_{i-1}}}\left(Cn^{\mathcal{L}}_{\mathbf{AL}_{i-2}^{x_{i-2}}}\left(\ldots Cn^{\mathcal{L}}_{\mathbf{AL}_1^{x_1}}(\Gamma)\ldots\right)\right)\right) \quad (\ddagger)$$

By the induction hypothesis ($\ddagger$) is identical to

$$Cn^{\mathcal{L}^+}_{\mathbf{AL}_i^{x_i}}\left(Cn^{\mathcal{L}^+}_{\mathbf{AL}_{i-1}^{x_{i-1}}}\left(\ldots Cn^{\mathcal{L}^+}_{\mathbf{AL}_1^{x_1}}(\Gamma)\ldots\right)\right)$$ □

The following lemma will be crucial to establish in Lemma 3.1.2 that ALs with supraclassical lower limits satisfy (3.5).[3]

Lemma 3.1.1. *Where* $\Gamma = Cn^{\mathcal{L}^+}_{\mathbf{LLL}}(\Gamma)$ *or* $\Gamma \subseteq \mathcal{W}$:

$$Cn^{\mathcal{L}^+}_{\mathbf{AL}}(\Gamma) = Cn^{\mathcal{L}^+}_{\mathbf{LLL}}\left(Cn^{\mathcal{L}^+}_{\mathbf{AL}}(\Gamma)\right).$$

Proof. "$\subseteq$" follows by the reflexivity of **LLL**.

"$\supseteq$": Let $A \in Cn^{\mathcal{L}^+}_{\mathbf{LLL}}\left(Cn^{\mathcal{L}^+}_{\mathbf{AL}}(\Gamma)\right)$. By the compactness of **LLL**, there are $A_1, \ldots, A_n \in Cn^{\mathcal{L}^+}_{\mathbf{AL}}(\Gamma)$ such that $\{A_1, \ldots, A_n\} \vdash_{\mathbf{LLL}^+} A$. Hence, by Corollary 2.7.1, $\Gamma \Vdash_{\mathbf{AL}} A_1, \ldots, A_n$. Thus, for all $M \in \mathcal{M}_{\mathbf{AL}}(\Gamma)$, $M \models A_1, \ldots, A_n$.

[3] We prove a slightly more generic version than we need for Lemma 3.1.2 where we only make use of the case $\Gamma \subseteq \mathcal{W}$. However, the gained generality will be useful in the next Section.

Hence, since each such M is an **LLL**-model, for all $M \in \mathcal{M}_{\mathbf{AL}}(\Gamma)$, $M \models A$. Thus, $\Gamma \Vdash_{\mathbf{AL}} A$. Again by Corollary 2.7.1, $\Gamma \vdash_{\mathbf{AL}} A$. □

The next lemma shows that ALs with supraclassical lower limit logics do indeed satisfy (3.5).

Lemma 3.1.2. *Let* **LLL** *be supraclassical. Where* $\Gamma \subseteq \mathcal{W}$, *we have*

$$Cn^{\mathcal{L}^+}_{\mathbf{LLL}}\big(Cn^{\mathcal{L}}_{\mathbf{AL}}(\Gamma)\big) = Cn^{\mathcal{L}^+}_{\mathbf{LLL}}\big(Cn^{\mathcal{L}^+}_{\mathbf{AL}}(\Gamma)\big) = Cn^{\mathcal{L}^+}_{\mathbf{AL}}(\Gamma)$$

Proof. Since $Cn^{\mathcal{L}}_{\mathbf{AL}}(\Gamma) \subseteq Cn^{\mathcal{L}^+}_{\mathbf{AL}}(\Gamma)$, $Cn^{\mathcal{L}^+}_{\mathbf{LLL}}\big(Cn^{\mathcal{L}}_{\mathbf{AL}}(\Gamma)\big) \subseteq Cn^{\mathcal{L}^+}_{\mathbf{LLL}}\big(Cn^{\mathcal{L}^+}_{\mathbf{AL}}(\Gamma)\big)$ due to the monotonicity of $\mathbf{LLL}^+$. Moreover, by Lemma 3.1.1,

$$Cn^{\mathcal{L}^+}_{\mathbf{LLL}}\big(Cn^{\mathcal{L}^+}_{\mathbf{AL}}(\Gamma)\big) = Cn^{\mathcal{L}^+}_{\mathbf{AL}}(\Gamma) \tag{†}$$

Suppose **AL** employs the reliability strategy. Suppose further that $A \in Cn^{\mathcal{L}^+}_{\mathbf{LLL}}\big(Cn^{\mathcal{L}^+}_{\mathbf{AL}}(\Gamma)\big)$. Due to (†) there is by Theorem 2.4.2 a $\Delta \subseteq \Omega$ for which $A \check{\vee} \mathsf{Dab}(\Delta) \in Cn^{\mathcal{L}^+}_{\mathbf{LLL}}(\Gamma)$ and $\Delta \cap U(\Gamma) \neq \emptyset$. Let for a formula $B \in \mathcal{W}^+$, $B' \in \mathcal{W}$ be the result of replacing all occurrences of checked symbols by their classical equivalents in $\mathcal{L}$ (this is possible since **LLL** is supraclassical). Since $\vdash_{\mathbf{LLL}^+} A \check{\equiv} A'$, $A' \check{\vee} \mathsf{Dab}(\Delta) \in Cn^{\mathcal{L}^+}_{\mathbf{LLL}}(\Gamma)$ and hence $A' \in Cn^{\mathcal{L}}_{\mathbf{AL}}(\Gamma)$. Hence, $A \in Cn^{\mathcal{L}^+}_{\mathbf{LLL}}\big(Cn^{\mathcal{L}}_{\mathbf{AL}}(\Gamma)\big)$.

The case for minimal abnormality is similar and left to the reader. □

Corollary 3.1.1. *Where* $\Gamma \subseteq \mathcal{W}$ *and* **LLL** *is supraclassical we have* (3.6) *and hence* (3.7).

Proof. Follows by Lemma 3.1.2 and Theorem 3.1.1. □

Since the concrete sequential ALs that are introduced in Part III and Part IV of this book employ supraclassical lower limit logics, I will be able to stick with the more simple definition offered in (3.1) for the consequence relation of sequentially combined ALs. For the remainder of this Chapter however, $Cn^{\mathcal{L}}_{\mathbf{CAL}}$ is defined as in (3.4) by $\mathcal{W} \cap Cn^{\mathcal{L}^+}_{\mathbf{CAL}}$.

In order to reduce the notational clutter in the remainder of this chapter, we will from now on make use of the following abbreviation: where $i \leq n$,

$$Cn^{\mathcal{L}^+}_{\mathbf{CAL_i}}(\Gamma) =_{\mathrm{df}} Cn^{\mathcal{L}^+}_{\mathbf{AL}^{x_i}_{i}}\left(Cn^{\mathcal{L}^+}_{\mathbf{AL}^{x_{i-1}}_{i-1}}\left(\ldots Cn^{\mathcal{L}^+}_{\mathbf{AL}^{x_1}_{1}}(\Gamma)\ldots\right)\right)$$

We close this section with an observation that will be useful for proving the adequacy of the semantics and the adequacy of the proof theory in the following sections. Lemma 3.1.1 generalizes to **CAL**: i.e., the consequence relation of **CAL** is closed under **LLL**.

Lemma 3.1.3. *Where* $\Gamma \subseteq \mathcal{W}$: $Cn^{\mathcal{L}+}_{\mathbf{AL}_1^{x_1}}(\Gamma) = Cn^{\mathcal{L}+}_{\mathbf{LLL}}\left(Cn^{\mathcal{L}+}_{\mathbf{AL}_1^{x_1}}(\Gamma)\right)$ *and for each* $i \in \{2, \ldots, n\}$: $Cn^{\mathcal{L}+}_{\mathbf{CAL_i}}(\Gamma) = Cn^{\mathcal{L}+}_{\mathbf{LLL}}\left(Cn^{\mathcal{L}+}_{\mathbf{CAL_i}}(\Gamma)\right)$.

Proof. We show this by induction. "$i = 1$" has been shown in Lemma 3.1.1.
"$i \Rightarrow i{+}1$": By the induction hypothesis we have:

$$Cn^{\mathcal{L}+}_{\mathbf{AL}_{i+1}^{x_{i+1}}}\left(Cn^{\mathcal{L}+}_{\mathbf{CAL_i}}(\Gamma)\right) = Cn^{\mathcal{L}+}_{\mathbf{AL}_{i+1}^{x_{i+1}}}\left(Cn^{\mathcal{L}+}_{\mathbf{LLL}}\left(Cn^{\mathcal{L}+}_{\mathbf{CAL_i}}(\Gamma)\right)\right)$$

The rest follows with Lemma 3.1.1 and the idempotence of $\mathbf{LLL}^+$. □

This immediately shows that **CAL** is closed under $\mathbf{LLL}^+$.

Corollary 3.1.2. *Where* $\Gamma \subseteq \mathcal{W}$: $Cn^{\mathcal{L}+}_{\mathbf{CAL}}(\Gamma) = Cn^{\mathcal{L}+}_{\mathbf{LLL}}\left(Cn^{\mathcal{L}+}_{\mathbf{CAL}}(\Gamma)\right)$.

3.2 Semantics for Sequential ALs

The most intuitive approach to define a semantics for **CAL** is to first select all the adaptive (reliable resp. minimally abnormal) **LLL**-models of Γ with respect to $\mathbf{AL}_1^{x_1}$, then out of these the adaptive models with respect to $\mathbf{AL}_2^{x_2}$, and so on. This is formally defined in Definition 3.2.1 and illustrated in Fig. 3.2.

Definition 3.2.1. For an **LLL**-model M of Γ let $\mathrm{Ab}_i(M) = \{A \in \Omega_i \mid M \models A\}$. Given a premise set Γ, we define:

- $\mathcal{M}_1(\Gamma) = \mathcal{M}_1^{x_1}(\Gamma)$ where $\mathcal{M}_1^r(\Gamma)$ is the set of all reliable **LLL**-models of Γ (w.r.t. Ω_1) and $\mathcal{M}_1^m(\Gamma)$ is the set of all minimally abnormal **LLL**-models of Γ (w.r.t. Ω_1);

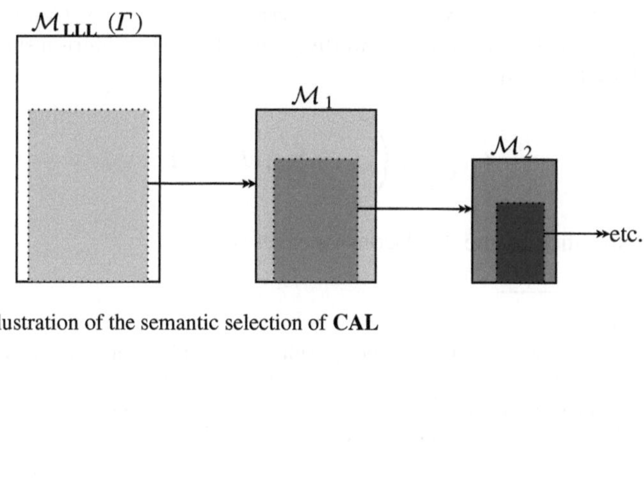

Fig. 3.2 Illustration of the semantic selection of **CAL**

- Where $1 < i \leq n$, $\mathcal{M}_i(\Gamma) = \mathcal{M}_i^{x_i}(\Gamma)$ where $\mathcal{M}_i^m(\Gamma)$ is the set of all minimally abnormal models in $\mathcal{M}_{i-1}(\Gamma)$ (w.r.t. Ω_i) and $\mathcal{M}_i^r(\Gamma)$ is the set of all models in $\mathcal{M}_{i-1}(\Gamma)$ for which $\mathrm{Ab}_i(M) \subseteq \Psi^i(\mathcal{M}_{i-1}(\Gamma))$ where[4]

$$\Psi^i\big(\mathcal{M}_{i-1}(\Gamma)\big) = \bigcup\big\{\mathrm{Ab}_i(M) \mid M \in \mathcal{M}_i^m(\Gamma)\big\}$$

Let $\mathcal{M}_{\mathbf{CAL}}(\Gamma) = \mathcal{M}_n(\Gamma)$. Let $\Vdash_{\mathbf{CAL}}$ be defined by: $\Gamma \Vdash_{\mathbf{CAL}} A$ iff for all $M \in \mathcal{M}_{\mathbf{CAL}}(\Gamma)$, $M \models A$.

Note that the consequence relation defined in (3.4) is neither sound nor complete with respect to the semantics above. Let me demonstrate this for completeness by an example Diderik Batens offers in [1].

Example 3.2.1. Let $\mathbf{AL_1^m}$ and $\mathbf{AL_2^x}$ (where $\mathbf{x} \in \{\mathbf{m}, \mathbf{r}\}$) be two ALs in standard format that employ the same lower limit logic **LLL** and the minimal abnormality strategy. Let $\Omega_1 = \{A_1^1, A_2^1, \ldots\}$ be the set of abnormalities of $\mathbf{AL_1^m}$ and $\Omega_2 = \{A_1^2, A_2^2, \ldots\}$ be the set of abnormalities of $\mathbf{AL_2^x}$. Let

$$\Gamma = \big\{A_i^1 \vee A_j^1 \mid i, j \in \mathbb{N}, i \neq j\big\} \cup \big\{B \vee A_i^1 \vee A_i^2 \mid i \in \mathbb{N}\big\}$$

It is easy to see that $B \notin Cn_{\mathbf{AL_2^x}}\big(Cn^{\mathcal{L}^+}_{\mathbf{AL_1^m}}(\Gamma)\big)$.

Let us take a look at the adaptive models. Note that for each minimally abnormal **LLL**-model (with respect to Ω_1) of Γ the abnormal part consists of all but one of the abnormalities in Ω_1. More formal, for each $M \in \mathcal{M}_{\mathbf{AL_1^m}}(\Gamma)$ there is an $i \in \mathbb{N}$ such that $\mathrm{Ab}_1(M) = \Omega_1 \setminus \{A_i^1\}$. Hence (†) for each of these models M there is an $i \in \mathbb{N}$ for which $M \models B \vee A_i^2$. The minimally abnormal resp. reliable models with respect to Ω_2 in $\mathcal{M}_{\mathbf{AL_1^m}}(\Gamma)$ are the ones that do no validate any abnormalities in Ω_2. Hence, due to (†), in all selected models B is valid.[5]

The example is stated for the minimal abnormality strategy. Indeed, things look better for the reliability strategy, as will be shown in Sect. 3.2.1 and for other special cases, as will be shown in Sect. 3.2.2.

In Theorem 3.2.1 I will now present a generic criterion that warrants soundness and completeness with respect to the intuitive semantics. The criterion states that for each $i < n$ the $\mathbf{AL_i^{x_i}}$-models of

$$Cn^{\mathcal{L}^+}_{\mathbf{AL_{i-1}^{x_{i-1}}}}\Big(\ldots Cn^{\mathcal{L}^+}_{\mathbf{AL_1^{x_1}}}(\Gamma)\ldots\Big)$$

are uniquely specified by the $\mathbf{AL_i^{x_i}}$-consequences, in signs

---

[4] See Lemma 2.4.1.

[5] Frederik Van De Putte was the first to come up with a very similar counter-example concerning the lack of soundness. Let $\Gamma' = \{A_i^1 \vee A_j^1 \mid i, j \in \mathbb{N}, i \neq j\} \cup \{B \vee A_i^1 \vee A_i^2 \mid i \in \mathbb{N}\}$ and take $\mathbf{AL_1^m}$ and $\mathbf{AL_2^m}$ from above. It can be shown that B is a syntactic consequence but not a semantic one with respect to the semantics defined above.

$$\mathcal{M}_{\mathbf{AL}_{\mathbf{i}}^{\mathbf{x_i}}}\left(Cn^{\mathcal{L}+}_{\mathbf{CAL_{i-1}}}(\Gamma)\right) = \mathcal{M}_{\mathbf{LLL}}\left(Cn^{\mathcal{L}+}_{\mathbf{CAL_{i}}}(\Gamma)\right) \tag{3.8}$$

In this case we get soundness and completeness: $\Gamma \Vdash_{\mathbf{CAL}} A$ iff $A \in Cn_{\mathbf{CAL}}(\Gamma)$.

Theorem 3.2.1. *If* (3.8) *then,* $\Gamma \Vdash_{\mathbf{CAL}} A$ *iff* $A \in Cn_{\mathbf{CAL}}(\Gamma)$.

We will prove this theorem below.

In our example we have seen that there are cases in which criterion (3.8) is violated. As mentioned, each $M \in \mathcal{M}_{\mathbf{AL}_{\mathbf{1}}^{\mathbf{m}}}(\Gamma)$ is such that $Ab_1(M) = \{A_i^1 \mid i \in \mathbb{N} \setminus \{j\}\}$ for some $j \in \mathbb{N}$. However, there are $M \in \mathcal{M}_{\mathbf{LLL}}\left(Cn^{\mathcal{L}+}_{\mathbf{AL}_{\mathbf{1}}^{\mathbf{m}}}(\Gamma)\right)$ such that $Ab_1(M) = \{A_i^1 \mid i \in \mathbb{N}\}$. The reason is that although some A_i^1 is false in every minimally abnormal model, we cannot express the infinite disjunction $\check{\bigvee}_{i \in \mathbb{N}} \check{\neg} A_i^1$ in our object language.

We will see in the next section that (3.8) is fulfilled for sequences of ALs that only use the reliability strategy.

Let us now prove Theorem 3.2.1. First we need a technical lemma:

Lemma 3.2.1. *If* (3.8) *then*

$$\mathcal{M}_{\mathbf{CAL}}(\Gamma) = \mathcal{M}_{\mathbf{AL}_{\mathbf{n}}^{\mathbf{x_n}}}\left(Cn^{\mathcal{L}+}_{\mathbf{CAL_{n-1}}}(\Gamma)\right) = \mathcal{M}_{\mathbf{LLL}}\left(Cn^{\mathcal{L}+}_{\mathbf{CAL}}(\Gamma)\right)$$

Proof. Suppose (3.8) holds. Let $\mathcal{N}_1^{x_1}(\Gamma) =_{\rm df} \mathcal{M}_{\mathbf{AL}_{\mathbf{1}}^{\mathbf{x_1}}}(\Gamma)$ and, where $1 < i \leq n$,

$$\mathcal{N}_i^{x_i}(\Gamma) =_{\rm df} \mathcal{M}_{\mathbf{AL}_{\mathbf{i}}^{\mathbf{x_i}}}\left(Cn^{\mathcal{L}+}_{\mathbf{CAL_{i-1}}}(\Gamma)\right)$$

I will show by induction that $\mathcal{M}_i^{x_i}(\Gamma) = \mathcal{N}_i^{x_i}(\Gamma)$ for all $i \in \{1, \ldots, n\}$.
"$i = 1$": This is so by definition.
"$i \Rightarrow i + 1$": Suppose $x_{i+1} = r$.

$$\mathcal{M}_{i+1}^r(\Gamma) = \left\{M \in \mathcal{M}_i^{x_i}(\Gamma) \mid Ab_{i+1}(M) \subseteq \Psi^{i+1}\left(\mathcal{M}_i^{x_i}(\Gamma)\right)\right\}$$

By the induction hypothesis,

$$\mathcal{M}_{i+1}^r(\Gamma) = \left\{M \in \mathcal{N}_i^{x_i}(\Gamma) \mid Ab_{i+1}(M) \subseteq \Psi^{i+1}\left(\mathcal{N}_i^{x_i}(\Gamma)\right)\right\} \tag{‡}$$

By Lemma 2.4.1, $\mathcal{N}_{i+1}^r(\Gamma) = \mathcal{M}_{\mathbf{AL}_{\mathbf{i+1}}^{\mathbf{r}}}\left(Cn^{\mathcal{L}+}_{\mathbf{CAL_i}}(\Gamma)\right) =$

$$\left\{M \in \mathcal{M}_{\mathbf{LLL}}\left(Cn^{\mathcal{L}+}_{\mathbf{CAL_i}}(\Gamma)\right) \middle| Ab_{i+1}(M) \subseteq \Psi^{i+1}\left(\mathcal{M}_{\mathbf{LLL}}\left(Cn^{\mathcal{L}+}_{\mathbf{CAL_i}}(\Gamma)\right)\right)\right\}$$

By (3.8) we have thus,

$$\mathcal{N}^r_{i+1}(\Gamma) = \{M \in \mathcal{N}^{x_i}_i(\Gamma) \mid \mathrm{Ab}_{i+1}(M) \subseteq \Psi^{i+1}(\mathcal{N}^{x_i}_i(\Gamma))\}$$

Hence, by (‡), $\mathcal{M}^r_{i+1}(\Gamma) = \mathcal{N}^r_{i+1}(\Gamma)$.

The case for $x_{i+1} = m$ is similar and left to the reader.

Thus, by our induction, $\mathcal{N}^{x_n}_n(\Gamma) = \mathcal{M}^{x_n}_n(\Gamma) = \mathcal{M}_{\mathbf{CAL}}(\Gamma)$. □

Proof (Proof of Theorem 3.2.1). Suppose (3.8). $A \in Cn^{\mathcal{L}^+}_{\mathbf{CAL}}(\Gamma)$ iff [by Corollary 3.1.2] $A \in Cn^{\mathcal{L}^+}_{\mathbf{LLL}}\left(Cn^{\mathcal{L}^+}_{\mathbf{CAL}}(\Gamma)\right)$ iff [by the soundness and completeness of **LLL**] $Cn^{\mathcal{L}^+}_{\mathbf{CAL}}(\Gamma) \Vdash_{\mathbf{LLL}^+} A$ iff [by Lemma 3.2.1 and (3.8)] $\Gamma \Vdash_{\mathbf{CAL}} A$. □

3.2.1 Soundness and Completeness for Reliability

In this section I will show that soundness and completeness is warranted for sequential combinations of ALs that employ the reliability strategy. As discussed in the previous section, it is sufficient to show that such sequences fulfill criterion (3.8).

In order to show this we first prove some useful lemmas.

Lemma 3.2.2. *Where* $\Gamma = Cn^{\mathcal{L}^+}_{\mathbf{LLL}}(\Gamma)$ *or* $\Gamma \subseteq \mathcal{W}$, $Cn^{\mathcal{L}^+}_{\mathbf{AL^r}}(\Gamma) = Cn^{\mathcal{L}^+}_{\mathbf{LLL}}(\Gamma \cup (\Omega \setminus U(\Gamma))^{\check{\neg}})$.

Proof. $\Gamma \vdash_{\mathbf{AL^r}} A$, iff [by Corollary 2.7.1], $\Gamma \Vdash_{\mathbf{AL^r}} A$, iff [by Theorem 2.4.1], $\Gamma \cup (\Omega \setminus U(\Gamma))^{\check{\neg}} \Vdash_{\mathbf{LLL}} A$, iff [by the soundness and completeness of **LLL**], $\Gamma \cup (\Omega \setminus U(\Gamma))^{\check{\neg}} \vdash_{\mathbf{LLL}} A$. □

Lemma 3.2.3. *Where* $\Gamma = Cn^{\mathcal{L}^+}_{\mathbf{LLL}}(\Gamma)$ *or* $\Gamma \subseteq \mathcal{W}$, $\mathcal{M}_{\mathbf{LLL}}\left(Cn^{\mathcal{L}^+}_{\mathbf{AL^r}}(\Gamma)\right) = \mathcal{M}_{\mathbf{AL^r}}(\Gamma)$.

Proof. Let $M \in \mathcal{M}_{\mathbf{LLL}}\left(Cn^{\mathcal{L}^+}_{\mathbf{AL^r}}(\Gamma)\right)$. Assume $M \models A$ for some $A \in U(\Gamma)$. By Lemma 3.2.2, $\check{\neg} A \in Cn^{\mathcal{L}^+}_{\mathbf{AL^r}}(\Gamma)$,—a contradiction. Hence, M is reliable. Let $M \in \mathcal{M}_{\mathbf{AL^r}}(\Gamma)$. By Corollary 2.7.1, $M \in \mathcal{M}_{\mathbf{LLL}}\left(Cn^{\mathcal{L}^+}_{\mathbf{AL^r}}(\Gamma)\right)$. □

The next result provides a representational theorem: it shows that the syntactic consequence relation of **CAL** can be expressed by means of **LLL** in case each AL in the sequence uses the reliability strategy.

Theorem 3.2.2. *Where* $\Gamma \subseteq \mathcal{W}$ *and*

$$\Gamma^1 =_{df} \Gamma \cup (\Omega \setminus U_1(\Gamma))^{\check{\neg}}$$

$$\textit{for all } i \in \{1, \ldots, n-1\} : \Gamma^{i+1} =_{df} \Gamma^i \cup (\Omega \setminus U_{i+1}(\Gamma^i))^{\check{\neg}}$$

$$\textit{we have: } Cn^{\mathcal{L}^+}_{\mathbf{AL^r_n}}\left(Cn^{\mathcal{L}^+}_{\mathbf{AL^r_{n-1}}}\left(\ldots Cn^{\mathcal{L}^+}_{\mathbf{AL^r_1}}(\Gamma)\ldots\right)\right) = Cn^{\mathcal{L}^+}_{\mathbf{LLL}}(\Gamma^n).$$

Proof. We show this by induction."$i = 1$": This is Corollary 2.4.2.
"$i \Rightarrow i{+}1$": By the induction hypothesis,

$$Cn^{\mathcal{L}+}_{\mathbf{AL}^{\mathbf{r}}_{\mathbf{i}}}\left(Cn^{\mathcal{L}+}_{\mathbf{AL}^{\mathbf{r}}_{\mathbf{i-1}}}\left(\dots Cn^{\mathcal{L}+}_{\mathbf{AL}^{\mathbf{r}}_{\mathbf{1}}}(\Gamma)\dots\right)\right) = Cn^{\mathcal{L}+}_{\mathbf{LLL}}\left(\Gamma^i\right).$$

Hence,

$$Cn^{\mathcal{L}+}_{\mathbf{AL}^{\mathbf{r}}_{\mathbf{i+1}}}\left(Cn^{\mathcal{L}+}_{\mathbf{AL}^{\mathbf{r}}_{\mathbf{i}}}\left(\dots Cn^{\mathcal{L}+}_{\mathbf{AL}^{\mathbf{r}}_{\mathbf{1}}}(\Gamma)\dots\right)\right) = Cn^{\mathcal{L}+}_{\mathbf{AL}^{\mathbf{r}}_{\mathbf{i+1}}}\left(Cn^{\mathcal{L}+}_{\mathbf{LLL}}(\Gamma^i)\right). \tag{3.9}$$

By Lemma 3.2.2 and the idempotence of $\mathbf{LLL}^+$,

$$\begin{aligned}
&Cn^{\mathcal{L}+}_{\mathbf{AL}^{\mathbf{r}}_{\mathbf{i+1}}}\left(Cn^{\mathcal{L}+}_{\mathbf{LLL}}(\Gamma^i)\right) = \\
&\quad Cn^{\mathcal{L}+}_{\mathbf{LLL}}\left(Cn^{\mathcal{L}+}_{\mathbf{LLL}}(\Gamma^i) \cup \left(\Omega_{i+1} \setminus U_{i+1}\left(Cn^{\mathcal{L}+}_{\mathbf{LLL}}(\Gamma^i)\right)\right)^{\check{\neg}}\right) = \\
&\qquad Cn^{\mathcal{L}+}_{\mathbf{LLL}}\left(\Gamma^i \cup \left(\Omega_{i+1} \setminus U_{i+1}(\Gamma^i)\right)^{\check{\neg}}\right) = Cn^{\mathcal{L}+}_{\mathbf{LLL}}\left(\Gamma^{i+1}\right).
\end{aligned}$$

Hence, together with (3.9) our induction step is finished. □

The next lemma shows that we indeed have (3.8) for sequences with the reliability strategy.

Lemma 3.2.4. *Where $\Gamma \subseteq \mathcal{W}$: $\mathcal{M}_{\mathbf{AL}^{\mathbf{r}}_{\mathbf{1}}}\Gamma = \mathcal{M}_{\mathbf{LLL}}\left(Cn^{\mathcal{L}+}_{\mathbf{AL}^{\mathbf{r}}_{\mathbf{1}}}(\Gamma)\right)$ and for each $i \in \{2,\dots,n\}$*

$$\begin{aligned}
&\mathcal{M}_{\mathbf{AL}^{\mathbf{r}}_{\mathbf{i}}}\left(Cn^{\mathcal{L}+}_{\mathbf{AL}^{\mathbf{r}}_{\mathbf{i-1}}}\left(\dots Cn^{\mathcal{L}+}_{\mathbf{AL}^{\mathbf{r}}_{\mathbf{1}}}(\Gamma)\dots\right)\right) = \\
&\qquad \mathcal{M}_{\mathbf{LLL}}\left(Cn^{\mathcal{L}+}_{\mathbf{AL}^{\mathbf{r}}_{\mathbf{i}}}\left(Cn^{\mathcal{L}+}_{\mathbf{AL}^{\mathbf{r}}_{\mathbf{i-1}}}\left(\dots Cn^{\mathcal{L}+}_{\mathbf{AL}^{\mathbf{r}}_{\mathbf{1}}}(\Gamma)\dots\right)\right)\right)
\end{aligned}$$

Proof. This follows immediately by Lemma 3.2.4, Theorem 3.2.1, and the idempotence of $\mathbf{LLL}^+$. □

The following corollary follows immediately with Lemma 3.2.4 and Theorem 3.2.1. It shows that we get soundness and completeness for sequences that only use the reliability strategy.

Corollary 3.2.1. *Where $x_i = r$ for all $i \leq n$ and $\Gamma \subseteq \mathcal{W}$: $A \in Cn^{\mathcal{L}+}_{\mathbf{CAL}}(\Gamma)$ iff $\Gamma \Vdash_{\mathbf{CAL}} A$.*

We also get soundness and completeness in case the outermost AL uses the minimal abnormality strategy, while all other ALs use reliability.

Theorem 3.2.3. *Where $\Gamma \subseteq \mathcal{W}$, $x_i = r$ for all $i < n$ and $x_n = m$: $A \in Cn^{\mathcal{L}+}_{\mathbf{CAL}}(\Gamma)$ iff $\Gamma \Vdash_{\mathbf{CAL}} A$.*

Proof. $A \in Cn^{\mathcal{L}^+}_{\mathbf{AL^m_n}}\left(Cn^{\mathcal{L}^+}_{\mathbf{AL^r_{n-1}}}\left(\ldots Cn^{\mathcal{L}^+}_{\mathbf{AL^r_1}}(\Gamma)\ldots\right)\right)$, iff [by the idempotence of $\mathbf{LLL}^+$, Theorem 3.2.2 and Corollary 2.7.1],

$$\text{for all } M \in \mathcal{M}_{\mathbf{AL^m_n}}\left(Cn^{\mathcal{L}^+}_{\mathbf{AL^r_{n-1}}}\left(\ldots Cn^{\mathcal{L}^+}_{\mathbf{AL^r_1}}(\Gamma)\ldots\right)\right), M \models A, \text{ iff,}$$
$$\text{for all } M \in \left\{M' \in \mathcal{M}_{\mathbf{LLL}}\left(Cn^{\mathcal{L}^+}_{\mathbf{AL^r_{n-1}}}\left(\ldots Cn^{\mathcal{L}^+}_{\mathbf{AL^r_1}}(\Gamma)\ldots\right)\right)\right| M' \text{ is minimally abnormal w.r.t. } \Omega_n\right\}, M \models A,$$

iff [by Lemma 3.2.4 and Lemma 3.2.1], for all $M \in \{M' \in \mathcal{M}_{n-1} \mid M'$ is minimally abnormal w.r.t. $\Omega_n\}$, $M \models A$, iff, for all $M \in \mathcal{M}_n$, $M \models A$, iff, $\Gamma \Vdash_{\mathbf{CAL}} A$. □

3.2.2 Soundness and Completeness for Some Other Special Cases

There are other applications in which completeness is guaranteed. For instance in case the sets of abnormalities are finite. Let me highlight some in the following.

Where $\Omega^* = \Omega_1 \cup \ldots \cup \Omega_n$ and $\Gamma \subseteq \mathcal{W}^+$ is some premise set, $U^*(\Gamma) \subseteq \Omega^*$ are the unreliable formulas with respect to Ω^* and **LLL**, and $\Phi^*(\Gamma) \subseteq \wp(\Omega^*)$ is the set of minimal choice sets with respect to Ω^* and **LLL**. We can specify specific special cases in view of the cardinality of $\Phi^*(\Gamma)$, the cardinality of the members of $\Phi^*(\Gamma)$, the cardinality of Ω^*, and the cardinality of $U^*(\Gamma)$. In Fig. 3.3 some relationships are illustrated. We show in this section that whenever any of these criteria is fulfilled, we get soundness and completeness.

In Appendix A we show by means of purely set-theoretic considerations about choice sets that:

Fact 3.2.1. *Where* $\Gamma \subseteq \mathcal{W}^+$:

(i) if every $\varphi \in \Phi^*(\Gamma)$ *is finite then* $\Phi^*(\Gamma)$ *is finite (Corollary A.3)*
(ii) every $\varphi \in \Phi^*(\Gamma)$ *is finite iff* $U^*(\Gamma)$ *is finite (Corollary A.4)*

As Fig. 3.3 illustrates, it is enough to show that we have soundness and completeness for the case that $\Phi^*(\Gamma)$ is finite, since it is implied by all other cases.

The following two lemmas will be useful in what follows. Both show that the set of $\mathbf{AL^m}$-models is uniquely determined by a specific set of formulas and **LLL** in case $\Phi(\Gamma)$ is finite.

Lemma 3.2.5. *Where* $\Gamma \subseteq \mathcal{W}^+$, $\Phi(\Gamma) = \{\varphi_1, \ldots, \varphi_n\}$, *and* $\Upsilon = \{\mathsf{Dab}(\Delta) \mid \Delta$ *is a choice set of* $\Phi(\Gamma)\}$: $\mathcal{M}_{\mathbf{AL^m}}(\Gamma) = \mathcal{M}_{\mathbf{LLL}}(\Gamma \cup \Upsilon)$

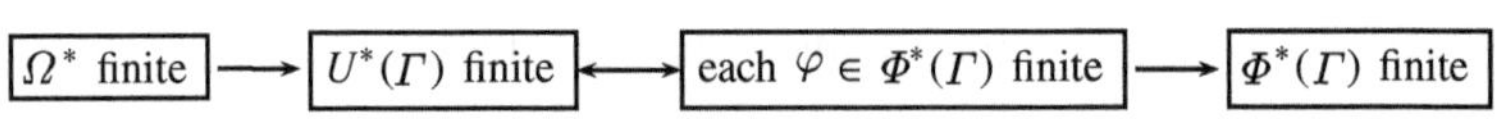

Fig. 3.3 Criteria that guarantee the soundness and completeness of **CAL**

Proof. Let $M \in \mathcal{M}_{\mathbf{AL^m}}(\Gamma)$. Hence, by Theorem 2.4.6, $\text{Ab}(M) = \varphi_i$ for some $i \leq n$. Hence, $M \models \Upsilon$. Thus, $M \in \mathcal{M}_{\mathbf{LLL}}(\Gamma \cup \Upsilon)$.

Let $M \in \mathcal{M}_{\mathbf{LLL}}(\Gamma \cup \Upsilon)$. Assume $M \notin \mathcal{M}_{\mathbf{AL^m}}(\Gamma)$. Hence, for all $\varphi_i \in \Phi(\Gamma)$, $\text{Ab}(M) \neq \varphi_i$. Hence, for each φ_i there is a $A_i \in \varphi_i$ such that $M \not\models A_i$. Note however that $\{A_i \mid i \leq n\} \in \Upsilon$,—a contradiction. □

Lemma 3.2.6. *Where* $\Gamma = Cn^{\mathcal{L}^+}_{\mathbf{LLL}}(\Gamma)$ *or* $\Gamma \subseteq \mathcal{W}$ *and* $\Phi(\Gamma)$ *is finite:* $\mathcal{M}_{\mathbf{LLL}}\left(Cn^{\mathcal{L}^+}_{\mathbf{AL}}(\Gamma)\right) = \mathcal{M}_{\mathbf{AL}}(\Gamma)$.

Proof. The case for reliability was already shown in Lemma 3.2.3. Suppose thus that **AL** uses minimal abnormality.

By Lemma 3.2.5, $\Gamma \Vdash_{\mathbf{AL}} \Upsilon$. Hence, by Lemma 3.2.2, (†) $\Upsilon \subseteq Cn^{\mathcal{L}^+}_{\mathbf{AL}}(\Gamma)$.

Let $M \in \mathcal{M}_{\mathbf{LLL}}\left(Cn^{\mathcal{L}^+}_{\mathbf{AL^m}}(\Gamma)\right)$. By (†), $M \in \mathcal{M}_{\mathbf{LLL}}(\Gamma \cup \Upsilon)$. By Lemma 3.2.5, $M \in \mathcal{M}_{\mathbf{AL^m}}(\Gamma)$. Let $M \in \mathcal{M}_{\mathbf{AL^m}}(\Gamma)$. Since by Lemma 3.2.2, $\Gamma \Vdash_{\mathbf{AL^m}} Cn^{\mathcal{L}^+}_{\mathbf{AL^m}}(\Gamma)$, also $M \in \mathcal{M}_{\mathbf{LLL}}\left(Cn^{\mathcal{L}^+}_{\mathbf{AL^m}}(\Gamma)\right)$. □

In the following we need a few more notational conventions.

Let $\boldsymbol{U^1}(\Gamma) \subseteq \Omega_1$ be the set of unreliable formulas with respect to $\mathbf{AL_1^{x_1}}$ and the premise set Γ, let for all $i \in \{2, \ldots, n\}$, $\boldsymbol{U^i}(\Gamma) \subseteq \Omega_i$ be the set of unreliable formulas with respect to $\mathbf{AL_i^{x_i}}$ and the premise set $Cn^{\mathcal{L}^+}_{\mathbf{AL}^{x_{i-1}}_{i-1}}\left(\ldots Cn^{\mathcal{L}^+}_{\mathbf{AL_1^{x_1}}}(\Gamma)\ldots\right)$.

Similarly let $\boldsymbol{\Phi^1}(\Gamma) \subseteq \wp(\Omega_1)$ be the set of minimal choice sets with respect to $\mathbf{AL_1^{x_1}}$ and the premise set Γ. For all $i \in \{2, \ldots, n\}$ let $\boldsymbol{\Phi^i}(\Gamma) \subseteq \wp(\Omega_i)$ be the set of minimal choice sets with respect to $\mathbf{AL_i^{x_i}}$ and the premise set $Cn^{\mathcal{L}^+}_{\mathbf{AL}^{x_{i-1}}_{i-1}}\left(\ldots Cn^{\mathcal{L}^+}_{\mathbf{AL_1^{x_1}}}(\Gamma)\ldots\right)$.

The following technical lemma is crucial in what follows.[6]

Lemma 3.2.7. *Where* $i \in \{1, \ldots, n\}$, $\Phi^*(\Gamma)$ *is finite, and* $\Gamma \subseteq \mathcal{W}$*:*

(i) $\Phi\left(Cn^{\mathcal{L}^+}_{\mathbf{AL_i^{x_i}}}\left(Cn^{\mathcal{L}^+}_{\mathbf{AL_{i-1}^{x_{i-1}}}}(\ldots Cn^{\mathcal{L}^+}_{\mathbf{AL_1^{x_1}}}(\Gamma))\right)\right) \subseteq \Phi^*(\Gamma)$

(ii) For all $\varphi \in \boldsymbol{\Phi^i}(\Gamma)$ *there is a* $\varphi' \in \Phi^*(\Gamma)$ *such that* $\varphi \subseteq \varphi'$.

(iii) $|\boldsymbol{\Phi^i}(\Gamma)| \leq |\Phi^*(\Gamma)|$

Corollary 3.2.2. *Where* $i \in \{1, \ldots, n\}$ *and* $\Gamma \subseteq \mathcal{W}$. *If* $\Phi^*(\Gamma)$ *is finite, then* $\boldsymbol{\Phi^i}(\Gamma)$ *is finite.*

The next theorem is the central result of this section: it shows that whenever $\Phi^*(\Gamma)$ is finite we get soundness and completeness.

First we show that (3.8) holds whenever $\Phi^*(\Gamma)$ is finite.

---

[6] The rather technical and lengthy proof can be found in Appendix B.1.

Lemma 3.2.8. *Where $\Phi^*(\Gamma)$ is finite:* (3.8) holds.

Proof. For $i = 1$ this follows directly with Corollary 3.2.2 and Lemma 3.2.6. "$i \Rightarrow i+1$": Note that

$$\mathcal{M}_{\mathbf{AL}_{\mathbf{i+1}}^{\mathbf{x_{i+1}}}}\left(Cn^{\mathcal{L}^+}_{\mathbf{CAL_i}}(\Gamma)\right) = \mathcal{M}_{\mathbf{LLL}}\left(Cn^{\mathcal{L}^+}_{\mathbf{CAL_{i+1}}}(\Gamma)\right)$$

by Corollary 3.2.2, Lemma 3.2.6 and since by Lemma 3.1.3,

$$Cn^{\mathcal{L}^+}_{\mathbf{CAL_i}}(\Gamma) = Cn^{\mathcal{L}^+}_{\mathbf{LLL}}\left(Cn^{\mathcal{L}^+}_{\mathbf{CAL_i}}(\Gamma)\right)$$

□

Theorem 3.2.4. *Where $\Phi^*(\Gamma)$ is finite and $\Gamma \subseteq \mathcal{W}$: $A \in Cn_{\mathbf{CAL}}(\Gamma)$ iff $\Gamma \Vdash_{\mathbf{CAL}} A$.*

Proof. This follows immediately with Lemma 3.2.8 and Theorem 3.2.1. □

As illustrated in Fig. 3.3 this result immediately applies to some other cases as well:

Corollary 3.2.3. *Where $\Gamma \subseteq \mathcal{W}$ and either of the following holds:*

1. *Ω^* is finite*
2. *$U^*(\Gamma)$ is finite*
3. *each $\varphi \in \Phi^*(\Gamma)$ is finite*

then: $A \in Cn_{\mathbf{CAL}}(\Gamma)$ iff $\Gamma \Vdash_{\mathbf{CAL}} A$.

In the last two sections I have presented cases in which completeness and soundness is warranted with respect to the intuitive semantics in terms of sequential selections that was explicated in the beginning of this section. As demonstrated in the example, not in all cases the consequence relation is complete with respect to this semantics. In such cases the semantic selection has to be defined in a more cumbersome way. For instance in the following way: Let $\Gamma \Vdash'_{\mathbf{CAL}} A$ iff $M \models A$ for all $M \in \mathcal{M}_{\mathbf{AL}_{\mathbf{n}}^{\mathbf{x_n}}}\left(Cn^{\mathcal{L}^+}_{\mathbf{AL}_{\mathbf{n-1}}^{\mathbf{x_{n-1}}}}\left(Cn^{\mathcal{L}^+}_{\mathbf{AL}_{\mathbf{n-2}}^{\mathbf{x_{n-2}}}}\left(\ldots Cn^{\mathcal{L}^+}_{\mathbf{AL}_{\mathbf{1}}^{\mathbf{x_1}}}(\Gamma)\ldots\right)\right)\right)$.

Theorem 3.2.5. *Where $\Gamma \subseteq \mathcal{W}$: $\Gamma \Vdash'_{\mathbf{CAL}} A$ iff $A \in Cn_{\mathbf{CAL}}(\Gamma)$.*

Proof. For $n = 1$ this holds by the soundness and completeness of ALs in the standard format. Let $n > 1$. By Lemma 3.1.3, $Cn^{\mathcal{L}^+}_{\mathbf{CAL_{n-1}}}(\Gamma) = Cn^{\mathcal{L}^+}_{\mathbf{LLL}}\left(Cn^{\mathcal{L}^+}_{\mathbf{CAL_{n-1}}}(\Gamma)\right)$. Hence, by Corollary 2.7.1, $A \in Cn^{\mathcal{L}^+}_{\mathbf{CAL}}(\Gamma)$ iff $Cn^{\mathcal{L}^+}_{\mathbf{CAL_{n-1}}}(\Gamma) \Vdash_{\mathbf{AL}_{\mathbf{n}}^{\mathbf{x_n}}} A$ iff $\Gamma \Vdash'_{\mathbf{CAL}} A$. □

3.3 A Proof Theory for CAL

So far we have only discussed the consequence relation and the semantics of **CAL**. While a consequence relation may inform us of the outcome of a reasoning process, it doesn't explicate the reasoning process. One way to do so is by means of a proof theory. We have already seen how the proof theory of ALs in the standard format explicates the internal dynamics of defeasible reasoning. Our task in this section is to introduce a similar proof theory for the sequential combinations of ALs **CAL**.

In this section I presuppose that the Ω_i's are disjoint, i.e., for all i, j for which $1 \leq i < j \leq n$, $\Omega_i \cap \Omega_j = \emptyset$.

The reader can find more generic—but also more involving—proof theories where this requirement is dropped in [5].

3.3.1 The Proof Format

The proof format of sequential superpositions which we present here is nearly identical to the one of flat ALs. Again, a line is a quadruple consisting of a line number, a formula, a justification and a condition. The only difference concerns the last element. While in the proof theory for flat ALs only formulas in the logical form of abnormalities with respect to a specific AL are part of the condition, in the sequential case abnormalities belonging to different Ω_i's can be part of the condition.

Suppose we have the following line in a proof[7]:

$$l \; A \qquad\qquad k_1, \ldots, k_n; \; R \; \Delta_1 \cup \Delta_2$$

where $\Delta_1 \subseteq \Omega_1$ and $\Delta_2 \subseteq \Omega_2$. The idea is that A is derived on the assumption that no abnormality in $\Delta_1 \cup \Delta_2$ is true. Hence, we make use of the defeasible reasoning forms represented by both $\mathbf{AL}_\mathbf{1}^{\mathbf{x_1}}$ and $\mathbf{AL}_\mathbf{2}^{\mathbf{x_2}}$. Moreover, in case A is finally derived at line l (see Definition 3.3.4 below), then A is a consequence of the superposition of $\mathbf{AL}_\mathbf{2}^{\mathbf{x_2}}$ on $\mathbf{AL}_\mathbf{1}^{\mathbf{x_1}}$, since no defeasible assumptions were made that correspond to ALs higher in the sequence of **SAL**.

In order to realize this idea we will again make use of three generic rules and marking definitions.

The generic rule for premise introduction, PREM, and the unconditional rule, RU, are the same as for flat ALs (see page 19).

In what follows we introduce the generic conditional rule RC. For each $\mathbf{AL}_\mathbf{i}^{\mathbf{x_i}}$ in the sequence it allows to make inferences that make use of the defeasible reasoning

[7] We use R as a metavariable for a generic inference rule.

corresponding to $\mathbf{AL}_{\mathbf{i}}^{\mathbf{x_i}}$. Where $i \leq n$, $\Theta \subseteq \Omega_i$, and for each Δ_j $(1 \leq j \leq m)$, $\Delta_j \subseteq \Omega_1 \cup \ldots \cup \Omega_i$, we have:

$$\text{If } A_1, \ldots, A_m \vdash_{\mathbf{LLL}} B \check{\vee} \mathsf{Dab}(\Theta): \quad \begin{array}{ll} A_1 & \Delta_1 \\ \vdots & \vdots \\ A_m & \Delta_m \\ \hline B & \Delta_1 \cup \ldots \cup \Delta_m \cup \Theta \end{array} \tag{RC}$$

Remark 3.3.1. Note that RCdoes not allow to introduce defeasible assumption corresponding to different $\mathbf{AL}_{\mathbf{i}}^{\mathbf{x_i}}$'s in the sequence in one inference step. After introducing the marking definitions we will give an example that illustrates why allowing for many defeasible assumptions in one inference step leads to trouble.

Hence the following is *not* a correct application of RC: Where $A_1 \in \Omega_1$ and $A_2 \in \Omega_2$,

l	$B \check{\vee} (A_1 \check{\vee} A_2)$	$\ldots$	$\emptyset$
l'	B	l; RC (!)	$\{A_1, A_2\}$

Neither is the *second* application of RC in the following proof correct:

l	$B \check{\vee} (A_1 \check{\vee} A_2)$	$\ldots$	$\emptyset$
l'	$B \check{\vee} A_1$	l; RC	$\{A_2\}$
l''	B	l'; RC (!)	$\{A_1, A_2\}$

B can be derived on the condition $\{A_1, A_2\}$, however the applications of RC have to respect the order of the sets of abnormalities in the sequence. The following proof is correct:

l	$B \check{\vee} (A_1 \check{\vee} A_2)$	$\ldots$	$\emptyset$
l'	$B \check{\vee} A_2$	l; RC	$\{A_1\}$
l''	B	l'; RC	$\{A_1, A_2\}$

More general, suppose we are able to derive $B \check{\vee} \mathsf{Dab}(\Theta_1 \cup \ldots \cup \Theta_m)$ in **LLL** from $A_1, \ldots, A_k$, where each $\Theta_i \subset \Omega_i$. In that case the proof theory allows us to defeasibly derive B from $A_1, \ldots, A_k$ step-wise:

l_1	$B \check{\vee} \mathsf{Dab}(\Theta_1 \cup \ldots \cup \Theta_m)$	$\vdots$	$\emptyset$
l_2	$B \check{\vee} \mathsf{Dab}(\Theta_2 \cup \ldots \cup \Theta_m)$	l_1; RC	Θ_1
$\vdots$	$\vdots$	$\vdots$	$\vdots$
l_m	$B \check{\vee} \mathsf{Dab}(\Theta_m)$	l_{m-1}; RC	$\Theta_1 \cup \ldots \cup \Theta_{m-1}$
l_{m+1}	B	l_m; RC	$\Theta_1 \cup \ldots \cup \Theta_m$

3.3.2 Preparing for the Marking Definitions

Of course, in order to explicate defeasible reasoning it is not enough to be able to apply certain rules conditionally. What is still missing is a mechanism that makes it possible to *retract* defeasible inferences. As in the case of flat ALs, lines in an **CAL**-proof are marked at a certain stage of the proof in order to signify that the corresponding inference is retracted at that stage.

The marking definitions reflect the hierarchical structure of the superposition. For each level $i \leq n$ we will state i-marking definitions. That a line is not i-marked for any $i \leq n$ indicates that we have no reason to suspect the assumption of this line. If a line in an **CAL**-proof is i-marked for an $i \leq n$, then this means the line is retracted at the given stage of the proof.

Since either $\mathbf{x_i} = \mathbf{r}$ or $\mathbf{x_i} = \mathbf{m}$, and since we also include superpositions of ALs with mixed strategies, we need to state i-marking definitions for both strategies. In order to do so it is useful to define sequential counter-parts to various notions that play a central role for the marking definitions in Chap. 2.

We first give a sequential account of minimal Dab-formulas resulting in the notion of a minimal Dab_i-formula for each $i \leq n$, i.e. a minimal disjunction of members of Ω_i. Just as the marking at stage s for flat ALs was determined by a set of minimal Dab-formulas relative to s, the i-marking in the sequential case will be determined by a set of Dab_i-formulas relative to s.

Definition 3.3.1. Let s be the stage of an **CAL**-proof from Γ and $i \leq n$.

- A proof line l with condition Δ is a $[\leq 0]$-*line* iff $\Delta = \emptyset$.
- A proof line l with condition Δ is a $[\leq i]$-*line* iff $\Delta \subseteq \Omega_1 \cup \ldots \cup \Omega_i$.
- A proof line l is an i-*line* iff it is a $[\leq i]$-line and not a $[\leq i-1]$-line.
- $\mathsf{Dab}(\Delta)$ is a Dab_i-*formula* iff $\Delta \subseteq \Omega_i$.
- Where $\Delta \subseteq \Omega_i$, $\mathsf{Dab}(\Delta)$ is a *minimal* Dab_i-*formula at stage* s in case

  (i) $\mathsf{Dab}(\Delta)$ is derived at some $[\leq i-1]$-line l at stage s,
  (ii) line l is not marked at stage s (see below for the marking definition), and
  (iii) for no $\Delta' \subset \Delta$, $\mathsf{Dab}(\Delta')$ is derived at an unmarked $[\leq i-1]$-line at stage s.

- Where $\{\mathsf{Dab}(\Delta_j) \mid j \in J\}$ is the set of the minimal Dab_i-formulas at stage s, let $\Sigma_s^i(\Gamma) =_{\text{df}} \{\Delta_j \mid j \in J\}$.
- $U_s^i(\Gamma) =_{\text{df}} \bigcup \Sigma_s^i(\Gamma)$
- A Dab_i-formula $\mathsf{Dab}(\Delta)$ is a *minimal* Dab_i-*consequence of* Γ iff $\Gamma \vdash_{\mathbf{LLL}} \mathsf{Dab}(\Delta)$ and for all $\Theta \subset \Delta$, $\Gamma \nvdash_{\mathbf{LLL}} \mathsf{Dab}(\Theta)$.

There is an important difference when comparing minimal Dab_i-formulas at some stage s to minimal Dab-formulas in proofs of flat ALs: while the latter are always derived on the condition $\emptyset$, the condition Δ on which a minimal Dab_i-formula is derived may be non-empty. However, the abnormalities in Δ are always lower than i.

3.3.3 The i-Marking for the Reliability Strategy

Now we are able to define the i-marking at a stage s. Let us begin with the marking definition for the reliability strategy.

Definition 3.3.2 (*i*-**marking for reliability**). An i-line l with condition Δ is i-marked at stage s iff $\Delta \cap U_s^i(\Gamma) \neq \emptyset$.

Note that the i-marking only concerns i-lines, i.e., lines for which the highest abnormalities in the condition are in Ω_i. Of course, it may be that an i-line l is derived from some j-line l' (where $j < i$) that is j-marked. In this case also the inference at line l is supposed to be retracted since it relies on a retracted inference. This is achieved by means of the following marking definition:

Definition 3.3.3 (inh-**marking of lines**). An i-line l with condition Δ and justification $l_1, \ldots, l_m$; R is inh-marked in case some l_j (where $1 \leq j \leq m$) is (i) k-marked for some $k < i$, or (ii) inh-marked.

We say a line is *marked* in case it is i-marked for some $i \leq n$ (see also the i-marking Definition 3.3.5 for minimal abnormality below) or it is inh-marked.

Before we turn to the i-marking definition for minimal abnormality, let us illustrate the generic inference rules and the above marking definition by means of a simple example.

Recall the logic $\mathbf{CL}_\circ$ from Chap. 2. Let us give a prioritized twist to it by enhancing its expressiveness: we prefix formulas with sequences of $\circ$'s in order to indicate the trustworthiness of the information resp. source. Where A is a formula without $\circ$'s, we indicate that A is preceded by i many $\circ$'s by means of $\circ^i A$. The more $\circ$'s that precede A the less trustworthy is the information resp. the source that states it. Of course, in case of conflicting statements we prefer the more trustworthy information. We realize this idea by means of a sequential combination of ALs. We combine the logics $\mathbf{CL}_{\circ_i}^{\mathbf{x_i}}$ where $\mathbf{CL}_{\circ_i}^{\mathbf{x_i}}$ is characterized by the lower limit $\mathbf{CL}_\circ$, the set of abnormalities $\Omega_i = \{\circ^i A \wedge \neg A \mid A$ is $\circ$-free, and the strategy $\mathbf{x_i} \in \{\mathbf{r}, \mathbf{m}\}$.

For the sake of the example let us combine $\mathbf{CL}_{\circ_1}^{\mathbf{r}}$ and $\mathbf{CL}_{\circ_2}^{\mathbf{r}}$. Now consider the premise set $\Gamma_{\mathsf{p1}} = \{\circ p, \circ\circ q, \circ\circ r, \neg p \vee \neg r\}$. According to this premise set, we have evidence for p, q and r, but p is more trustworthy than the other two propositions. However, we also know that either p or r is false. Hence we can expect that the prioritized logic will only allow us to finally derive p, and hence by disjunctive syllogism $\neg r$. Also, since q is not involved in the conflict, we expect it to be finally derivable. This can be done as follows.

We start by introducing the premises on the condition $\emptyset$:

1	$\circ p$	PREM $\emptyset$
2	$\circ\circ q$	PREM $\emptyset$
3	$\circ\circ r$	PREM $\emptyset$
4	$\neg p \vee \neg r$	PREM $\emptyset$

By the rule RC, we may subsequently derive p, q and r from the first three premises. In order to avoid notational clutter let us from now on abbreviate abnormalities $\circ^i A \wedge \neg A$ by $!^i A$. Note that $\Gamma_{\mathsf{p1}} \vdash_{\mathbf{CL}_\circ} p \check{\vee} !^1 p$, $\Gamma_{\mathsf{p1}} \vdash_{\mathbf{CL}_\circ} q \check{\vee} !^2 q$ and $\Gamma_{\mathsf{p1}} \vdash_{\mathbf{CL}_\circ} r \check{\vee} !^2 r$. Hence we can derive e.g. p on the assumption that $!^1 p$ is false. In the adaptive proof this means that we derive p on the condition $\{!^1 p\}$ and similar for q and r:

5	p	1; RC $\{!^1 p\}$
6	q	2; RC $\{!^2 q\}$
7	r	3; RC $\{!^2 r\}$

To understand the rule RU, consider the following continuation of the proof, in which the conditions of lines 5 and 6 are merged:

8	$p \wedge q$	5,6; RU $\{!^1 p, !^2 q\}$

Let us now turn to the marking. We use $\checkmark_i$ to denote that a line is i-marked. To avoid clutter, we only represent the marks at one stage: where k is the last line in the example proof the displayed marks represent the marking at stage k.

In order to render line 7 marked, we first have to derive the Dab_2-formula $!^2 r$. This is done as follows:

	⋮	⋮	⋮	⋮
	5	p	1; RC	$\{!^1 p\}$
	6	q	2; RC	$\{!^2 q\}$
$\checkmark_2$	7	r	3; RC	$\{!^2 r\}$
	8	$p \wedge q$	5,6; RU	$\{!^1 p, !^2 q\}$
	9	$!^1 p \check{\vee} !^2 r$	1,3,4; RU	$\emptyset$
	10	$!^2 r$	9; RC	$\{!^1 p\}$

Let us discuss the marking at stage 10 step by step. First of all, note that at stage 10, no Dab_1-formula has been derived on the condition $\emptyset$.[8] This means that $\Sigma^1_{10}(\Gamma_{\mathsf{p1}}) = \emptyset$, whence also $U^1_{10}(\Gamma_{\mathsf{p1}}) = \emptyset$. As a result, no line is 1-marked at stage 10.

Now consider line 10 and its formula $!^2 r$. This is a Dab_2-formula. Moreover, line 10 is not 1-marked. As a result, $!^2 r$ is a minimal Dab_2-formula at stage 10. This implies that $\Sigma^2_{10}(\Gamma_{\mathsf{p1}}) = \{\{!^2 r\}\}$, whence $U^2_{10}(\Gamma_{\mathsf{p1}}) = \{!^2 r\}$. As a result, line 7 is 2-marked at stage 10, as indicated by the symbol $\checkmark_2$.

We define final derivability for our proof theory exactly in the same way as it was defined for flat ALs in Definition 2.4.4.[9]

---

[8] The formula on line 9 is not a Dab_1-formula, since it contains the abnormality $!^2 r$ which is not a member of Ω_1.

[9] In case some $\mathbf{x_i} = \mathbf{m}$ this definition also makes reference to the i-marking for minimal abnormality which we define in Sect. 3.3.4.

Definition 3.3.4. *A is finally derived at a line l of a finite stage s in an* **CAL**-*proof, iff (i) line l is unmarked at stage s, and (ii) every extension of the proof in which l is marked can be further extended in such a way that l is unmarked.*

A is finally derivable from Γ in **CAL** *iff there is a proof from Γ in which A is finally derived. We write $\Gamma \vdash_{\mathbf{CAL}} A$ in case A is finally derivable from Γ.*

As a matter of fact, p, q and $p \wedge q$ are finally derived in the proof from Γ_{p1} above. Note that no Dab_1-consequence is derivable from this premise set, and the only minimal Dab_2-consequence that can be derived from Γ_{p1} is $!^2 r$. This means that in every extension of the proof, the marking of lines 1–10 remains unchanged.

3.3.4 The i-Marking for the Minimal Abnormality Strategy

The i-marking for minimal abnormality is slightly more complicated. Where $\Phi^i_s(\Gamma)$ is the set of minimal choice sets of $\Sigma^i_s(\Gamma)$ we define:

Definition 3.3.5 (i-marking for minimal abnormality). An i-line l with formula A and condition Δ is i-marked at stage s iff one of the following conditions hold:

(i) there is no $\varphi \in \Phi^i_s(\Gamma)$ such that $\Delta \cap \varphi \neq \emptyset$
(ii) for a $\varphi \in \Phi^i_s(\Gamma)$: there is no unmarked $[\leq i]$-line l' at stage s with formula A and condition Θ such that $\Theta \cap \varphi = \emptyset$.

Recall that final derivability as defined in Definition 3.3.4 also applies to superpositions that feature ALs with minimal abnormality. This completes the technical characterization of our first proof theory for **CAL**. In Appendix B we prove its adequacy:

Theorem 3.3.1. *Where $\Gamma \subseteq \mathcal{W}$: $\Gamma \vdash_{\mathbf{CAL}} A$ iff $A \in Cn_{\mathbf{CAL}}(\Gamma)$*

Let us in the remainder of this section illustrate the proof theory and discuss some noteworthy point concerning the marking for minimal abnormality.

Let us interpret Definition 3.3.5 in terms of an argumentation game (see also the discussion of Definition 2.4.7 on page 26). Suppose our proponent derives formula A on the condition Δ at stage s. The i-marking concerns the question whether the defeasible assumption that corresponds to level i in the superposition is feasible. The minimal choice sets of $\Sigma^i_s(\Gamma)$ offer minimally abnormal interpretations (in terms of abnormalities in Ω_i) of the premises with respect to the Dab_i-formulas at the given stage s. That is, they offer possible counter-arguments against the defeasible assumption Δ of line l. In case her assumption intersects with one of these minimally abnormal interpretations φ she has to offer an alternative argument for A whose assumption doesn't intersect with φ (condition (ii)). Moreover, there should be at least one φ such that her assumption doesn't intersect with φ (condition (i)). In Sect. 2.4.2.2 we said that her argument has to be justifiable (condition (ii)) and defensible (condition (i)). However, there is a slight complication involved.

The assumptions used in order to derive A may involve abnormalities of lower levels than i. Concerning the lower levels we adopt a bottom-up approach. In case one of the defeasible assumptions at a lower level is not feasible we rely on the marking corresponding to the lower level to retract the line. This is realized by (a) allowing only for one type of defeasible assumption in each inference step by means of RC, and (b) by using the inh-marking to "inherit" markings from a lower level line to a higher level line that calls upon the former in its justification. In this sense the i-marking procedure safely ignores the defeasible assumptions belonging to lower levels.

Let us demonstrate this by a simple example. As before, we use an $\mathbf{CL}_\circ$-based prioritized logic with only two levels of abnormalities. This time however, we consider the minimal abnormality-variant, i.e. characterized by the sequence $\langle \mathbf{CL}_{\circ 1}^{\mathbf{m}}, \mathbf{CL}_{\circ 2}^{\mathbf{m}} \rangle$.

Let $\Gamma_{\mathsf{p2}} = \{\circ p, \circ q, \circ\circ r, \circ\circ s, \neg p \vee \neg q, \neg p \vee \neg r, \neg q \vee \neg s\}$. Note that the following disjunctions of abnormalities are $\mathbf{CL}_\circ$-derivable from Γ_{p2}:

(i) $!^1 p \check{\vee} !^1 q$
(ii) $!^1 p \check{\vee} !^2 r$
(iii) $!^1 q \check{\vee} !^2 s$

However, (ii) and (iii) are neither Dab_1-formulas nor Dab_2-formulas. The following proof shows how we can derive Dab_2-formulas from Γ_{p2}:

	1 $\circ p$	PREM	$\emptyset$
	2 $\circ q$	PREM	$\emptyset$
	3 $\circ\circ r$	PREM	$\emptyset$
	4 $\circ\circ s$	PREM	$\emptyset$
	5 $\neg p \vee \neg q$	PREM	$\emptyset$
	6 $\neg p \vee \neg r$	PREM	$\emptyset$
	7 $\neg q \vee \neg s$	PREM	$\emptyset$
	8 $!^1 p \check{\vee} !^1 q$	1,2,5; RU	$\emptyset$
	9 $!^1 p \check{\vee} !^2 r$	1,3,6; RU	$\emptyset$
	10 $!^1 q \check{\vee} !^2 s$	2,4,7; RU	$\emptyset$
$\checkmark_1$	11 $!^2 r$	9; RC	$\{!^1 p\}$
$\checkmark_1$	12 $!^2 s$	10; RC	$\{!^1 q\}$
	13 $!^2 r \check{\vee} !^2 s$	11; RU	$\{!^1 p\}$
	14 $!^2 r \check{\vee} !^2 s$	12; RU	$\{!^1 q\}$

Note that $\Sigma^1_{14}(\Gamma_{\mathsf{p2}}) = \{\{!^1 p, !^1 q\}\}$, whence $\Phi^1_{14}(\Gamma_{\mathsf{p2}}) = \{\{!^1 p\}, \{!^1 q\}\}$. Hence, at the current stage of our proof there are two minimally abnormal interpretations with respect to the abnormalities in Ω_1: one according to which $!^1 p$ is the only true abnormality, and another one according to which $!^1 q$ is the only true abnormality. This means that we cannot finally derive $!^2 r$ on the condition $\{!^1 p\}$, since we cannot derive $!^2 r$ on an assumption that is valid in the minimally abnormal interpretation offered by means of the minimal choice set $\{!^1 q\}$ (see condition (ii) in Definition 3.3.5).

For the same reason, we cannot finally derive $!^2 s$. Both lines 11 and 12 are 1-marked. However, the disjunction of both level 2-abnormalities is finally derived at stage 14. This follows immediately from the fact that $\mathsf{Dab}(\Delta_1)$ where $\Delta_1 = \{!^1 p, !^1 q\}$ is the only minimal Dab_1-consequence of Γ_{p2}. Also, it can easily be verified that $\mathsf{Dab}(\Delta_2)$ where $\Delta_2 = \{!^2 r, !^2 s\}$ is the *only* minimal Dab_2-consequence of $Cn_{\mathbf{CL}_{\circ 1}^{\mathsf{m}}}(\Gamma_{\mathsf{p2}})$.

In view of the preceding, it is easy to see that the sets $\Sigma_s^1(\Gamma_{\mathsf{p2}})$ and $\Sigma_s^2(\Gamma_{\mathsf{p2}})$ remain stable from stage 14 on. Put differently, in every further stage s of the proof,

$$(\dagger 1)\ \Phi_s^1(\Gamma_{\mathsf{p2}}) = \Phi_{14}^1(\Gamma_{\mathsf{p2}}) = \{\{!^1 p\}, \{!^1 q\}\}$$
$$(\dagger 2)\ \Phi_s^2(\Gamma_{\mathsf{p2}}) = \Phi_{14}^2(\Gamma_{\mathsf{p2}}) = \{\{!^2 r\}, \{!^2 s\}\}$$

Let us now return to the iterative character of the conditional rule and the inheritance marking.

Were we to allow for a more generic RC-rule—let's call it RC′—that allows for the introduction of abnormalities that belong to different levels at the same time, we would be able to produce the following extension of our proof above in which the (arbitrarily chosen) formula t is derived:

Note that this is *not* a correct **CAL**-proof fragment.

9	$!^1 p \check{\vee} !^2 r$	1,3,6; RU	$\emptyset$
10	$!^1 q \check{\vee} !^2 s$	2,4,7; RU	$\emptyset$
⋮	⋮	⋮	⋮
15	$t \check{\vee} !^1 p \check{\vee} !^2 r$	9; RU	$\emptyset$
16	t	15; RC′	$\{!^1 p, !^2 r\}$
17	$t \check{\vee} !^1 q \check{\vee} !^2 s$	10; RU	$\emptyset$
18	t	17; RC′	$\{!^1 q, !^2 s\}$

Note that according to our marking Definition 3.3.5, neither line 16 nor line 18 is 1-marked or 2-marked. However, the arbitrarily chosen t is not a consequence. A proper **CAL**-proof in which t is derived looks as follows:

	9	$!^1 p \check{\vee} !^2 r$	1,3,6; RU	$\emptyset$
	10	$!^1 q \check{\vee} !^2 s$	2,4,7; RU	$\emptyset$
	⋮	⋮	⋮	⋮
	15	$t \check{\vee} !^1 p \check{\vee} !^2 r$	9; RU	$\emptyset$
$\checkmark_1$	16	$t \check{\vee} !^2 r$	15; RC	$\{!^1 p\}$
inh	17	t	16; RC	$\{!^1 p, !^2 r\}$
	18	$t \check{\vee} !^1 q \check{\vee} !^2 s$	10; RU	$\emptyset$
$\checkmark_1$	19	$t \check{\vee} !^2 s$	18; RC	$\{!^1 q\}$
inh	20	t	19; RC	$\{!^1 q, !^2 s\}$

Note that neither $t \check{\vee} !^2 r$ nor $t \check{\vee} !^2 s$ is a consequence of the first AL in the sequence. Accordingly, lines 16 and 19 both get 1-marked. Again, here it is important to notice that lines 17 and 20 are neither 1-, nor 2-marked. However, they get inh-marked due to the fact that they call upon a 1-marked line in their justification. This shows that the marking for minimal abnormality needs to be complemented by the inh-marking.[10]

Note that the proof theory proposed in [6] is not adequate with respect to the consequence relation of sequences of ALs with the minimal abnormality strategy and even trivializes some premise sets. The above example is one of those cases. As we have seen, this problem is solved by the current proof theory in terms of (a) allowing only for defeasible assumptions of one level to be introduced at a time by means of RC and (b) by supplementing the marking with the inh-marking.

3.3.5 An Alternative to the inh-Marking for Some Special Cases

In the special case where for an initial sequence of ALs in our combination—say $\langle \mathbf{AL}_1^{x_1}, \ldots, \mathbf{AL}_m^{x_m} \rangle$ where $m \leq n$—all ALs make use of the reliability strategy (i.e., $x_i = r$ for all $i \leq m$) we can do without the inh-marking for $[\leq m]$-lines. In order to do so we generalize the marking Definition 3.3.2 in the following way:

Definition 3.3.6 (**$[\leq i]$-marking for reliability**). A line l with condition Δ is $[\leq i]$-marked at stage s iff $\Delta \cap \bigcup_{j \leq i} U_s^j(\Gamma) \neq \emptyset$.

It is easy to see that the following holds:

Lemma 3.3.1. *Where $m \leq n$ and $x_i = r$ for all $i \leq m$:*

(i) a $[\leq m]$-line is $[\leq m]$-marked at stage s, iff it is either inh*-marked or i-marked for some $i \leq m$ at stage s,*
(ii) where $m < n$: a $(m+1)$-line is $[\leq m]$-marked at stage s iff it is inh*-marked at stage s.*

Proof. Let l be some line at stage s of the proof with condition Δ. Ad (i): Where l is a 0-line or a 1-line this is obvious. "$i \Rightarrow i+1$" where $i+1 \leq m$: Suppose l is an $(i+1)$-line. Suppose it is inh-marked. Hence, there is an $[\leq i]$-line l' that is called upon in the justification of l and that is marked. Hence, by the induction hypothesis l' is also $[\leq m]$-marked. Hence, where Δ' is the condition of line l', $\Delta' \cap \bigcup_{j \leq m} U_s^j(\Gamma) \neq \emptyset$. Since $\Delta' \subseteq \Delta$, also $\Delta \cap \bigcup_{j \leq m} U_s^j(\Gamma) \neq \emptyset$. Hence, l is also $[\leq m]$-marked. If l is $(i+1)$-marked then $\Delta \cap U_s^{i+1}(\Gamma) \neq \emptyset$ and hence, l is also $[\leq m]$-marked.

The other direction and the proof for (ii) is analogous and left to the reader. □

Some useful applications of this lemma are:

---

[10] In [5] we present another sequential proof theory where this is not necessary. However, the price to pay is that the marking definition for minimal abnormality is more complicated.

For the special case where each logic in the sequence uses the reliability strategy, we can use the $[\leq n]$-marking defined in Definition 3.3.6 instead of using i-markings for reliability for each $i \leq n$ and inh-marking.

For the special case in which only $x_n = m$ while $x_i = r$ for all $i < n$ we can use the $[\leq n]$-marking and n-marking according to minimal abnormality.

3.4 Normal Selections

In this section we investigate the case in which $\mathbf{AL_n}$ uses the normal selections strategy. Let us call this logic $\mathbf{CAL}^{\mathrm{ns}}$.[11] It has the following consequence relation:

$$Cn^{\mathcal{L}^+}_{\mathbf{CAL}^{\mathrm{ns}}}(\Gamma) =_{\mathrm{df}} Cn^{\mathcal{L}^+}_{\mathbf{AL}^{\mathrm{ns}}_{\mathbf{n}}}\left(Cn^{\mathcal{L}^+}_{\mathbf{AL}^{\mathbf{x_{n-1}}}_{\mathbf{n-1}}}\left(\ldots Cn^{\mathcal{L}^+}_{\mathbf{AL}^{\mathbf{x_1}}_{\mathbf{1}}}(\Gamma)\ldots\right)\right)$$

The meta-theory will be useful for instance in Chap. 8.

Let us first define the semantics. This is done analogously to the way we defined it in Sect. 3.2. Only the last selection is different. We define:

Definition 3.4.1. Where $\Gamma \subseteq \mathcal{W}^+$: $\Gamma \Vdash_{\mathbf{CAL}^{\mathrm{ns}}} A$ iff there is a $M \in \mathcal{M}^m_n$ such that for all $M' \in \mathcal{M}^m_n(\Gamma)$ for which $\mathrm{Ab}_n(M) = \mathrm{Ab}_n(M')$, $M' \models A$.

We have analogous soundness and completeness results as before.

Theorem 3.4.1. *If* (3.8) then, $\Gamma \Vdash_{\mathbf{CAL}^{\mathrm{ns}}} A$ iff $A \in Cn_{\mathbf{CAL}^{\mathrm{ns}}}(\Gamma)$.

Proof. Suppose (3.8). $A \in Cn^{\mathcal{L}^+}_{\mathbf{CAL}^{\mathrm{ns}}}(\Gamma)$, iff, $A \in Cn^{\mathcal{L}^+}_{\mathbf{AL}^{\mathrm{ns}}_{\mathbf{n}}}\left(Cn^{\mathcal{L}^+}_{\mathbf{CAL_{n-1}}}(\Gamma)\right)$, iff [since by Corollary 3.1.2. $Cn^{\mathcal{L}^+}_{\mathbf{CAL_{n-1}}}(\Gamma) = Cn^{\mathcal{L}^+}_{\mathbf{LLL}}\left(Cn^{\mathcal{L}^+}_{\mathbf{CAL_{n-1}}}(\Gamma)\right)$ and by Theorem 2.8.4], $Cn^{\mathcal{L}^+}_{\mathbf{CAL_{n-1}}}(\Gamma) \Vdash_{\mathbf{AL}^{\mathrm{ns}}_{\mathbf{n}}} A$, iff, there is a

$$M \in \mathcal{M}_{\mathbf{AL}^{\mathbf{m}}_{\mathbf{n}}}\left((Cn^{\mathcal{L}^+}_{\mathbf{CAL_{n-1}}}(\Gamma)\right)$$

such that for all $M' \in \mathcal{M}_{\mathbf{AL}^{\mathbf{m}}_{\mathbf{n}}}\left(Cn^{\mathcal{L}^+}_{\mathbf{CAL_{n-1}}}(\Gamma)\right)$ for which $\mathrm{Ab}_n(M') = \mathrm{Ab}_n(M)$, $M' \models A$, iff [by Lemma 3.2.1 and (3.8)], there is a $M \in \mathcal{M}^m_n(\Gamma)$ such that for all $M' \in \mathcal{M}^m_n(\Gamma)$ for which $\mathrm{Ab}_n(M') = \mathrm{Ab}_n(M)$, $M' \models A$, iff, $\Gamma \Vdash_{\mathbf{CAL}^{\mathrm{ns}}} A$. □

We also get soundness and completeness in case the outermost AL uses the normal selections strategy, while all other ALs use reliability.

[11] Usually the superscript ^{n} is used in order to indicate the normal selections strategy. However, since we use the subscript $_n$ in order to indicate the n-th logic in our sequence and in order to avoid needless ambiguities, we use ^{ns} for the normal selections strategy in this section.

Corollary 3.4.1. *Where $\Gamma \subseteq \mathcal{W}$, $x_i = r$ for all $i < n$: $\Gamma \vdash_{\mathbf{CAL}^{\mathrm{ns}}} A$ iff $\Gamma \Vdash_{\mathbf{CAL}^{\mathrm{ns}}} A$.*

This follows by Lemma 3.2.4 and Theorem 3.4.1.

Moreover, we have soundness and completeness in the finitary cases of Sect. 3.2.2:

Corollary 3.4.2. *Where $\Gamma \subseteq \mathcal{W}$ and either of the following holds:*

1. *Ω^* is finite*
2. *$U^*(\Gamma)$ is finite*
3. *each $\varphi \in \Phi^*(\Gamma)$ is finite*

then: $A \in Cn_{\mathbf{CAL}^{\mathrm{ns}}}(\Gamma)$ iff $\Gamma \Vdash_{\mathbf{CAL}^{\mathrm{ns}}} A$.

The follows by Lemma 3.2.8 and Theorem 3.4.1.

Let us close this section with a look at the proof theory for $\mathbf{CAL}^{\mathrm{ns}}$. The i-marking for $i < n$ is defined as before. We only alter the n-marking accordingly:

Definition 3.4.2 (**n-marking for normal selections**). An n-line l with condition Δ is n-marked iff there is an unmarked $[\leq n-1]$-line on which $\mathsf{Dab}(\Delta \cap \Omega_n)$ is derived.[12]

Finally derivability is defined as before. In Appendix B we prove that this proof theory is adequate:

Theorem 3.4.2. *Where $\Gamma \subseteq \mathcal{W}$: $\Gamma \vdash_{\mathbf{CAL}^{\mathrm{ns}}} A$ iff $A \in Cn_{\mathbf{CAL}^{\mathrm{ns}}}(\Gamma)$.*

3.5 Conclusion

In this chapter we have discussed sequential combinations of ALs. First we investigated some meta-theory for the syntactic consequence relation. Then we proposed an intuitive semantics and proved its adequacy for several special cases. We got soundness and completeness in case all the logics use the reliability strategy, or for cases in which $\Phi(\Gamma)$ is finite, or each member of $\Phi(\Gamma)$ is finite, or $U(\Gamma)$ is finite. Finally we presented a dynamic proof theory that is very much in the tradition of the dynamic proof theory for flat ALs that we discussed before.

Acknowledgments The research in this chapter is inspired by the work on combinations of ALs by Diderik Batens and by Frederik Van De Putte. I also thank both of them for valuable comments.

[12] Note that in Sect. 2.8 we present several variants of a marking definition for normal selections. We do not spell out the corresponding variants in this section, but the necessary adjustments are obvious.

References

1. Batens, D.: Adaptive Logics and Dynamic Proofs. A Study in the Dynamics of Reasoning (201x)
2. Putte, F.V.D.: Hierarchic adaptive logics. Logic J. IGPL **20**(1), 45–72 (2012)
3. Putte, F.V.D., Straßer, C.: Three formats of prioritized adaptive logics: a comparative study. Logic J. IGPL **2**(21), 127–159 (2013)
4. Putte, F.V.D., Straßer, C.: Extending the standard format of adaptive logics to the prioritized case. Logique et Analyse **55**(220), 601–641 (2012)
5. Straßer, C., Putte, F.V.D.: Proof theories for superpositions of adaptive logics (201x). Logique et Analyse (Forthcoming)
6. Batens, D.: A general characterization of adaptive logics. Logique et Analyse. **173–175**, 45–68 (2001)

Chapter 4
On the Transparency of Defeasible Logics: Equivalent Premise Sets, Equivalence of Their Extensions, and Maximality of the Lower Limit

For Tarski logics, there are simple criteria that enable one to conclude that two premise sets are equivalent. We shall show that the very same criteria hold for ALs, which is a major advantage in comparison to other approaches to defeasible reasoning forms.

A related property of Tarski logics is that the extensions of equivalent premise sets with the same set of formulas are equivalent premise sets. This does not hold for ALs. However a very similar criterion does.

We also shall show that every monotonic logic weaker than an AL is weaker than the lower limit logic of the AL or identical to it. This highlights the role of the lower limit for settling the adaptive equivalence of extensions of equivalent premise sets.

4.1 Formats for Logics for Defeasible Reasoning

This chapter has a specific and a more general aim. The specific aim is related to determining whether two premise sets are equivalent with respect to logics that explicate defeasible reasoning forms—henceforth *DRFs*. We shall show that ALs are superior to other formats in this respect. The more general aim is to highlight the advantages of the AL program with respect to other approaches to DRFs.

Let us compare the situation with Tarski logics, logics the consequence relation of which is Reflexive, Transitive and Monotonic. A variety of formulations has been developed: axiomatic, Fitch-style, Gentzen-style, etc. Each of these have their stronger points. The variety, however, is only apparent. First, there are relatively standard procedures that, for most logics, enable one to turn one formulation into another. Next, the different formulations are at best different ways to characterize

---

A previous version of this chapter has been published under the name "*On the Transparency of Defeasible Logics: Equivalent Premise Sets, Equivalence of Their Extensions, and Maximality of the Lower Limit*" in Logique et Analyse [1]. The paper is co-authored by Diderik Batens and Peter Verdée. Any mistakes in the new material that has been added are alone my responsibility.

C. Straßer, *Adaptive Logics for Defeasible Reasoning*, Trends in Logic 38,
DOI: 10.1007/978-3-319-00792-2_4,

the same basic entity, viz. the consequence relation, which assigns to every premise set a consequence set. There are some differences in semantic styles as well. Again, these may be reduced to each other, except that some logics require a more complex semantics than others.

The situation is drastically different for logics that explicate DRFs. Here a variety of syntactic formulations have been tried out, each of them often for some specific cases only. Many of these explications have no semantics, others require unusual techniques.[1] All this raises two central questions.

A first question is whether DRFs require a variety of formulations. It is indeed possible that the domain comprises reasoning forms that are so different from each other, that it is uninteresting or even impossible to forge them into the same format. Suppose, however, that it is possible to characterize all DRFs by the same type of logic or logical approach. Then, presumably, there will be several such approaches. If this is so, a second question should be raised: Which are the advantages and weaknesses of the different approaches?

It is the aim of the AL program to characterize all DRFs in terms of an AL in standard format (see Chap. 2). This was realized for a variety of DRFs, mostly by tackling such reasoning forms from scratch. Many DRFs have been decently described independently of the AL program. Quite a few of these were characterized by an AL in standard format—[5–8] for handling inconsistent knowledge bases as in [9–11]; [12] for the signed consequence relations from [13]; [14, 15] for default reasoning and circumscription,[2] documented in [16–18]; [19] for rational closure from [20]; [21] for abstract argumentation from [22, 23]; [24] for the belief merging protocols from [25]. Similarly for consequence relations not described as such in the literature—[26, 27] for question evocation from [28]; [29–33] for abduction as described in [34]; [35, 36] for diagnosis from [37]; [38] for the notion of empirical progress from [39]; [40] for belief revision. For several Tarski logics, an AL was developed to circumvent adding new premises (by tinkering)—[41] for the pragmatic structures from [42, 43]; [44] and [45] on causality as in [46]; [47, 48] for the deontic logics from [49, 50], and [51] for fuzzy logics. Those characterizations and extensions often require a translation to a different language. Where **L** is the original 'logic' and **AL** is an AL, the characterization may have the form: $\Gamma \vdash_{\mathbf{L}} A$ iff $f(\Gamma) \vdash_{\mathbf{AL}} f(A)$ where f is a function mapping formulas from the native language (for example the standard predicative language) to a different language (for example a modal language).

The successes on the adaptive side do not entail that the first question should be answered in the negative. All that follows is that adaptive logicians were successful where one attempted to find such a characterization. The attempts were not exhaustive with respect to the present literature and new forms of defeasible logics may be discovered in the future. So the situation seems to justify that adaptive logicians

---

[1] With respect to the semantics, useful unifying work was done by Shoham and associates, for example [2–4].

[2] These are older results, not in standard format, that soon will be improved upon.

continue their efforts, but it is possible that they will only be able to unify part of all DRFs.

Let us now turn to the second question. One of the arguments adduced in favour of characterizations in terms of ALs was precisely that this enterprise has a strong unifying effect, especially as the standard format provides ALs with a proof theory, a semantics, and all the interesting parts of the metatheory. But obviously, unification is not the only consideration that should be taken into account.

In the present chapter we shall consider a type of argument that is related to transparency. To be more precise, the argument concerns criteria for the equivalence of premise sets. This requires some explanation.

Theories may have different *formulations*: the *same* theory may be presented in different ways. To make the matter more precise, let a theory T be a couple $\langle \Gamma, \mathbf{L} \rangle$, in which Γ is a set of statements (the non-logical axioms of T) and $\mathbf{L}$ is a logic. The claims made by the theory are $Cn_{\mathbf{L}}(\Gamma) = \{A \mid \Gamma \vdash_{\mathbf{L}} A\}$. That $T = \langle \Gamma, \mathbf{L} \rangle$ and $T' = \langle \Gamma', \mathbf{L} \rangle$ are different formulations of the same theory obviously means that $Cn_{\mathbf{L}}(\Gamma) = Cn_{\mathbf{L}}(\Gamma')$. Similarly, people talking to each other about some subject may come to the conclusion that they fully agree on the topic. If they are serious about the matter, they mean to say that all one person believes on the subject is derivable from the statements made (or agreed to) by the other. We may safely take it that the agreeing parties share the underlying logic $\mathbf{L}$, at least in the context of their present communication. So their agreement may be formally expressed by a statement of the form $Cn_{\mathbf{L}}(\Gamma) = Cn_{\mathbf{L}}(\Gamma')$. Where this statement holds true, we shall say that Γ and Γ' are **L**-*equivalent premise sets*.

Sameness of theories and mutual agreement are important matters. If two theories are the same, everything proven from one of them may be carried over immediately to the other. If two people actually agree about some subject, they are able to predict everything the other believes about the subject and they may rely on this, for example in arguments about other topics. Yet, it is obvious that offering a *direct* proof of $Cn_{\mathbf{L}}(\Gamma) = Cn_{\mathbf{L}}(\Gamma')$ is out of the question. Put in a more precise way, it is impossible for humans to enumerate all members of $Cn_{\mathbf{L}}(\Gamma)$ and to demonstrate for each of them that it is also a member of $Cn_{\mathbf{L}}(\Gamma')$.[3] Humans rely on shortcuts in order to establish $Cn_{\mathbf{L}}(\Gamma) = Cn_{\mathbf{L}}(\Gamma')$.

In Sect. 4.2, we shall consider three common criteria for deciding that $Cn_{\mathbf{L}}(\Gamma) = Cn_{\mathbf{L}}(\Gamma')$. These criteria will be shown to be correct for Tarski logics. We shall show, however, that these criteria cannot be applied to certain popular formulations of DRFs and that no alternatives for the criteria seem available. This will lead to the question whether there are corresponding criteria for ALs. The answer is rather astonishing: the very same criteria may be applied in the case of ALs. This seems a strong argument in favour of the adaptive program.

In Sect. 4.5, we shall also consider a related question. Suppose that Γ and Γ' are **L**-equivalent. Does it follow that $\Gamma \cup \Delta$ and $\Gamma' \cup \Delta$ are **L**-equivalent premise sets? If two people study the same theory, but possibly a different formulation of it, and

[3] In the text, we neglect some border cases, which are irrelevant to the present discussion, for example the case in which $Cn_{\mathbf{L}}(\Gamma)$ is either empty or trivial.

both extend their formulation with the same set of statements, we might expect that the extensions are also **L**-equivalent. The answer to the question will be shown to be positive for Tarski logics, but negative for most defeasible logics presented in the literature. It will turn out that the answer is also negative for ALs. However, in the case of ALs, the answer is positive for a criterion that is extremely close to the considered one. Let $\mathbf{L}'$ be *weaker than* **L** iff $Cn_{\mathbf{L}'}(\Gamma) \subset Cn_{\mathbf{L}}(\Gamma)$ for some Γ and $Cn_{\mathbf{L}'}(\Gamma) \subseteq Cn_{\mathbf{L}}(\Gamma)$ for all Γ. We write $\mathbf{L}' \lhd \mathbf{L}$ to denote that **L** is weaker than $\mathbf{L}'$, and $\mathbf{L}' \unlhd \mathbf{L}$ to denote that $Cn_{\mathbf{L}'}(\Gamma) \subseteq Cn_{\mathbf{L}}(\Gamma)$ for all Γ.

The **L**-equivalence of the extensions is warranted if the two premise sets are $\mathbf{L}'$-equivalent, where $\mathbf{L}'$ is any Tarski logic weaker than **L**. We shall also present a criterion that is specific for ALs and comes very handy for many premise sets.

The lower limit logic of ALs is always a Tarski logic. As it is a constitutive element of the AL, it is natural to inquire whether it plays a specific role with respect to the criteria for equivalence of premise sets and for the equivalence of their extension. In Sect. 4.6, we shall show that the lower limit logic plays indeed a privileged role: if **L** is a monotonic logic and **L** is weaker than (or identical to) the AL or the adaptive consequence set is closed under **L**, then **L** is weaker than the AL's lower limit logic or identical to it.

The conclusion will be that ALs are not only attractive because of their unifying power, but also because they have certain properties which warrant a transparent handling of premise sets.

This chapter does not and cannot aim at establishing a final conclusion. As we said before, new DRFs may be discovered in the future. Still, the chapter offers a strong argument for ALs (and against some other approaches to DRFs) and the argument relies on the best present insights.

4.2 Equivalent Premise Sets

Let us start with some conventions. The set of closed formulas of the considered language will be called $\mathcal{W}$. A logic **L** is a function $\mathbf{L}\colon \wp(\mathcal{W}) \to \wp(\mathcal{W})$, in other words a logic **L** assigns to every premise set Γ a consequence set, which is denoted by $Cn_{\mathbf{L}}(\Gamma)$. We presuppose that $\mathcal{W}$ does not contain any "checked connectives" and denote—as before (see Sects. 2.2 and 2.7)—the set of wffs that are the result of superimposing the "check connectives" on $\mathcal{W}$ by $\mathcal{W}^+$.

A logic is a *Tarski logic* iff it fulfills the following three properties:

- Reflexivity: $\Gamma \subseteq Cn_{\mathbf{L}}(\Gamma)$.
- Transitivity: If $\Gamma' \subseteq Cn_{\mathbf{L}}(\Gamma)$ then $Cn_{\mathbf{L}}(\Gamma') \subseteq Cn_{\mathbf{L}}(\Gamma)$.
- Monotonicity: $Cn_{\mathbf{L}}(\Gamma) \subseteq Cn_{\mathbf{L}}(\Gamma \cup \Gamma')$.

Definition 4.2.1. Γ and Γ' are **L**-*equivalent premise sets* iff $Cn_{\mathbf{L}}(\Gamma) = Cn_{\mathbf{L}}(\Gamma')$. We write $\Gamma \sim_{\mathbf{L}} \Gamma'$.

We propose the following criteria for the **L**-equivalence of premise sets:

C1 If $\Gamma' \subseteq Cn_{\mathbf{L}}(\Gamma)$ and $\Gamma \subseteq Cn_{\mathbf{L}}(\Gamma')$, then $\Gamma \sim_{\mathbf{L}} \Gamma'$.

C2 Where $\mathbf{L}'$ is a Tarski logic for which $\mathbf{L}' \trianglelefteq \mathbf{L}$: $\Gamma \sim_{\mathbf{L}'} \Gamma'$ implies $\Gamma \sim_{\mathbf{L}} \Gamma'$.

C2(Props) Where $\mathbf{L}' \trianglelefteq \mathbf{L}$ and $\mathbf{L}'$ fulfills properties Props: $\Gamma \sim_{\mathbf{L}'} \Gamma'$ implies $\Gamma \sim_{\mathbf{L}} \Gamma'$.

C3 Where $\mathbf{L}$ is closed under a Tarski logic $\mathbf{L}'$ (viz. $Cn_{\mathbf{L}'}(Cn_{\mathbf{L}}(\Theta)) = Cn_{\mathbf{L}}(\Theta)$ for all $\Theta \subseteq \mathcal{W}$): $\Gamma \sim_{\mathbf{L}'} \Gamma'$ implies $\Gamma \sim_{\mathbf{L}} \Gamma'$.

C3(Props) Where $\mathbf{L}$ is closed under a logic $\mathbf{L}'$ that fulfills the properties Props: $\Gamma \sim_{\mathbf{L}'} \Gamma'$ implies $\Gamma \sim_{\mathbf{L}} \Gamma'$.

Note that obviously where Props implies Props' then Ci(Props) implies Ci(Props') (where $i \in \{2, 3\}$).

Criterion C1 states that, in order for Γ and Γ' to be $\mathbf{L}$-equivalent, it is sufficient that all members of Γ are $\mathbf{L}$-derivable from Γ' and vice versa. In terms of theories: if T and T' have the same underlying logic $\mathbf{L}$ and all axioms of T are $\mathbf{L}$-derivable from T' and vice versa, then the two axiom sets are $\mathbf{L}$-equivalent—T and T', if different, are different formulations of the same theory. A still different rendering proceeds in terms of mutual agreement. Suppose that two persons state their views about some subject in an exhaustive way—all one of them holds true about that subject is derivable from the statements made by this party. If each party then agrees with everything the other said on the subject, one may conclude that they have the same view on the subject. C1 is an immediate consequence of the Transitivity of $\mathbf{L}$.

Criterion C2 states that if two premise sets are equivalent with respect to a Tarski logic weaker than $\mathbf{L}$, then they are equivalent with respect to $\mathbf{L}$. It is easily seen that C2 holds for all Tarski logics $\mathbf{L}$. Suppose indeed that the antecedent of C2 is true. As $Cn_{\mathbf{L}'}(\Gamma) \subseteq Cn_{\mathbf{L}}(\Gamma)$, $Cn_{\mathbf{L}'}(\Gamma) \cup \Gamma \subseteq Cn_{\mathbf{L}}(\Gamma)$ by the reflexivity of $\mathbf{L}$ and hence $Cn_{\mathbf{L}}(Cn_{\mathbf{L}'}(\Gamma) \cup \Gamma) \subseteq Cn_{\mathbf{L}}(\Gamma)$ by the transitivity of $\mathbf{L}$. So, by the monotonicity of $\mathbf{L}$, $Cn_{\mathbf{L}}(Cn_{\mathbf{L}'}(\Gamma) \cup \Gamma) = Cn_{\mathbf{L}}(\Gamma)$. Finally, as $Cn_{\mathbf{L}'}(\Gamma) \cup \Gamma = Cn_{\mathbf{L}'}(\Gamma)$ by the reflexivity of $\mathbf{L}'$, $Cn_{\mathbf{L}}(Cn_{\mathbf{L}'}(\Gamma)) = Cn_{\mathbf{L}}(\Gamma)$. By the same reasoning $Cn_{\mathbf{L}}(Cn_{\mathbf{L}'}(\Gamma')) = Cn_{\mathbf{L}}(\Gamma')$. As $Cn_{\mathbf{L}'}(\Gamma) = Cn_{\mathbf{L}'}(\Gamma')$, $Cn_{\mathbf{L}}(\Gamma) = Cn_{\mathbf{L}}(\Gamma')$.

Criterion C3 is related to the fact that we expect operations under which $\mathbf{L}$-consequence sets are closed to define a logic that is weaker than $\mathbf{L}$ or identical to it, which triggers C2. If, for all Δ, $A \wedge B \in Cn_{\mathbf{L}}(\Delta)$ just in case $A \in Cn_{\mathbf{L}}(\Delta)$ and $B \in Cn_{\mathbf{L}}(\Delta)$, then we expect $\Gamma \cup \{p \wedge q\}$ and $\Gamma \cup \{p, q\}$ to be $\mathbf{L}$-equivalent premise sets.

Incidentally, that $\mathbf{L}'$ has certain basic properties and is not just any logic is essential for both C2 and C3. If $\mathbf{L}'$ were an arbitrary logic, these criteria would not hold. To see this, let $\mathcal{W}$ be the set of closed formulas of the standard language, let **CL** be classical logic and let **L**$'$ be defined by $Cn_{\mathbf{L}'}(\Gamma) = \{A \in \Gamma \mid \text{for all } B \in \mathcal{W}, B \notin Cn_{\mathbf{CL}}(\{A\})$ or $B \in \Gamma\}$. In words, the **L**$'$-consequence set of Γ are those members of Γ of which all **CL**-consequences are members of Γ. Obviously, it holds for all Δ that $Cn_{\mathbf{L}'}(\Delta) \subseteq Cn_{\mathbf{CL}}(\Delta)$ and also that $Cn_{\mathbf{L}'}(Cn_{\mathbf{CL}}(\Delta)) = Cn_{\mathbf{CL}}(\Delta)$. However, there are infinitely many Γ for which no $A \in \Gamma$ is such that $Cn_{\mathbf{CL}}(A) \subseteq \Gamma$. For all of them $Cn_{\mathbf{L}'}(\Gamma) = Cn_{\mathbf{L}'}(\emptyset)$ but $Cn_{\mathbf{CL}}(\Gamma) \neq Cn_{\mathbf{CL}}(\emptyset)$.

Note that, for instance, **L**$'$ does not satisfy reflexivity. Motivated by our example, we state a variant of each C2 and C3 that is formulated relative to certain properties of **L**$'$. We will see that for many logics **L**, **L**$'$ may have less properties than a Tarski logic and the variants of C2 and C3 still hold (see Theorem 4.3.4 and Corollary 4.3.6 in Sect. 4.3).

Obviously, C1 may be combined with C2 or C3. Thus if **L**$'$ is a Tarski logic weaker than **L**, $\Gamma' \subseteq Cn_{\mathbf{L}'}(\Gamma)$ and $\Gamma \subseteq Cn_{\mathbf{L}'}(\Gamma')$, then Γ and Γ' are **L**-equivalent.

Let us now turn to defeasible logics. Consider first the Strong (also called Inevitable) and Weak consequence relations from [9]—see also [10]. Given a possibly inconsistent set of premises Γ, $\Delta \subseteq \Gamma$ is a *maximal consistent subset* of Γ iff, for all $A \in \Gamma \setminus \Delta$, $\Delta \cup \{A\}$ is inconsistent. $\Gamma \vdash_{Strong} A$ iff A is a **CL**-consequence of every maximal consistent subset of Γ and $\Gamma \vdash_{Weak} A$ iff A is a **CL**-consequence of some maximal consistent subset of Γ.

It is easily seen that C1 does not hold for the Weak consequence relation. Here is an example: $\{p, q, \neg p\} \subseteq Cn_{Weak}(\{p \wedge q, \neg p\})$ and $\{p \wedge q, \neg p\} \subseteq Cn_{Weak}(\{p, q, \neg p\})$, but $\neg p \wedge q \in Cn_{Weak}(\{p, q, \neg p\})$ whereas $\neg p \wedge q \notin Cn_{Weak}(\{p \wedge q, \neg p\})$.

It is also easily seen that C3 does not hold for the Strong consequence relation. Let **LC** be the Tarski logic that consists, apart from the Premise rule, of the rules Adjunction and Simplification. All Strong consequence sets are closed under **LC**, viz. $Cn_{Strong}(\Gamma) = Cn_{\mathbf{LC}}(Cn_{Strong}(\Gamma))$ for all Γ. However, $Cn_{\mathbf{LC}}(\{p, q, \neg p\}) = Cn_{\mathbf{LC}}(\{p \wedge q, \neg p\})$ but $Cn_{Strong}(\{p, q, \neg p\}) \neq Cn_{Strong}(\{p \wedge q, \neg p\})$, for example $q \in Cn_{Strong}(\{p, q, \neg p\})$ whereas $q \notin Cn_{Strong}(\{p \wedge q, \neg p\})$.

For an example of a logic for which C2 does not hold, we shall remain close to the Rescher-Manor consequence relations, adding a (weak) Schotch-Jennings flavour—see for example [52]. A *partition* of Γ is a (possibly finite) set of sets $\{\Gamma_1, \Gamma_2, \ldots\}$ such that $\Gamma = \Gamma_1 \cup \Gamma_2 \cup \ldots$ and $\Gamma_i \cap \Gamma_j = \emptyset$ where $i \neq j$. A partition $\{\Gamma_1, \Gamma_2, \ldots\}$ of Γ is *consistent* iff every Γ_i is consistent. Obviously, Γ has a consistent partition iff all $A \in \Gamma$ are consistent. The *regular* partitions of Γ are the consistent ones or, if there are no consistent ones, all partitions of Γ. Define: $A \in Cn_{\mathbf{R}}(\Gamma)$ iff there is a regular partition $\{\Gamma_1, \Gamma_2, \ldots\}$ of Γ and an i such that $A \in Cn_{\mathbf{CL}}(\Gamma_i)$. Define $Cn_{\mathbf{Q}}(\Gamma) = Cn_{\mathbf{P}}(Cn_{\mathbf{R}}(\Gamma))$, in which **P** is full positive **CL**. If $\{\Gamma\}$ is a regular partition of Γ, $Cn_{\mathbf{Q}}(\Gamma) = Cn_{\mathbf{CL}}(\Gamma)$; if some $A \in \Gamma$ is inconsistent, $Cn_{\mathbf{Q}}(\Gamma)$ is trivial; if Γ is inconsistent but all $A \in \Gamma$ are consistent, $Cn_{\mathbf{Q}}(\Gamma)$ is inconsistent but non-trivial, border cases aside. Note that **P** is a Tarski logic and that it is weaker than **Q**, viz. $Cn_{\mathbf{P}}(\Gamma) \subseteq Cn_{\mathbf{Q}}(\Gamma)$ for all Γ.

C2 does not hold for the defeasible logic **Q**. Indeed, **P** is a Tarski logic weaker than **Q** and $Cn_{\mathbf{P}}(\{p, \neg p\}) = Cn_{\mathbf{P}}(\{p \wedge \neg p\})$, but $Cn_{\mathbf{Q}}(\{p \wedge \neg p\})$ is trivial whereas $Cn_{\mathbf{Q}}(\{p, \neg p\})$ is not.

These examples are rather 'generous' because the situation is actually worse for certain systems describing DRFs. For example for the many kinds of default logics the criteria C1–C3 should be reformulated in order to make a chance to be applicable. The set of defaults has to enter the picture and 'facts' and defaults are to some extend exchangeable. The situation is similar for many other logics that characterize DRFs, even for the very transparent pivotal-assumption consequences defined in [53].

4.3 Equivalent Premise Sets and ALs

It was proven that all ALs have the properties reflexivity (Theorem 2.6.4i), cumulativity (Theorem 2.6.7iii), and idempotency (Theorem 2.6.5iii). From this it is easily provable that C1–C3 hold for all of them. We will also show that some weakenings of C2 and C3 hold for ALs. Note that each of the three criteria greatly simplifies the identification of equivalent premise sets (or theories). An illustration of the following results is given in Fig. 4.1.

Theorem 4.3.1. *If* **L** *is cumulative, then C1 holds for* **L**.

Proof. Suppose that $\Gamma' \subseteq Cn_{\mathbf{L}}(\Gamma)$ and $\Gamma \subseteq Cn_{\mathbf{L}}(\Gamma')$. By cumulativity, $Cn_{\mathbf{L}}(\Gamma) = Cn_{\mathbf{L}}(\Gamma \cup \Gamma')$ and $Cn_{\mathbf{L}}(\Gamma') = Cn_{\mathbf{L}}(\Gamma \cup \Gamma')$. So $Cn_{\mathbf{L}}(\Gamma) = Cn_{\mathbf{L}}(\Gamma')$. □

Hence, since ALs are cumulative we immediately get:

Corollary 4.3.1. *C1 holds for all ALs.*

Before we continue with the other criteria it is important to point out that we are considering $Cn^{\mathcal{L}}_{\mathbf{AL}}$ as opposed to $itCn^{\mathcal{L}^+}_{\mathbf{AL}}$ in Corollary 4.3.1 and the results that follow below. Indeed, Theorem 4.3.1 is not applicable to $Cn^{\mathcal{L}^+}_{\mathbf{AL}}$ since the latter is not cumulative. We can show that C1 doesn't hold for $Cn^{\mathcal{L}^+}_{\mathbf{AL}}$ by means of the Γ from our example in Sect. 2.7. We have: $Cn^{\mathcal{L}^+}_{\mathbf{LLL}}(\Gamma) \subseteq Cn^{\mathcal{L}^+}_{\mathbf{AL^m}}(\Gamma)$ and $\Gamma \subseteq Cn^{\mathcal{L}^+}_{\mathbf{AL^m}}\left(Cn^{\mathcal{L}^+}_{\mathbf{LLL}}(\Gamma)\right)$. However, $A \in Cn^{\mathcal{L}^+}_{\mathbf{AL^m}}\left(Cn^{\mathcal{L}^+}_{\mathbf{LLL}}(\Gamma)\right) \setminus Cn^{\mathcal{L}^+}_{\mathbf{AL^m}}(\Gamma)$ as can easily be verified.[4]

The following technical fact will be useful for some of our proofs below:

Fact 4.3.1. *Where* **L** *is reflexive and* **L**$'$ *is monotonic:*

---

[4] One way to show this is as follows. We already argued in Sect. 2.7. that $\Gamma \Vdash_{\mathbf{AL^m}} A$ and that $A \notin Cn^{\mathcal{L}^+}_{\mathbf{AL^m}}(\Gamma)$. Obviously, $\mathcal{M}_{\mathbf{LLL}}(\Gamma) = \mathcal{M}_{\mathbf{LLL}}(Cn^{\mathcal{L}^+}_{\mathbf{LLL}}(\Gamma))$ and hence both premise sets have the same minimally abnormal models. This implies, $Cn^{\mathcal{L}^+}_{\mathbf{LLL}}(\Gamma) \Vdash_{\mathbf{AL^m}} A$. Thus, by Theorem 2.7.1, $A \in Cn^{\mathcal{L}^+}_{\mathbf{AL^m}}\left(Cn^{\mathcal{L}^+}_{\mathbf{LLL}}(\Gamma)\right)$.

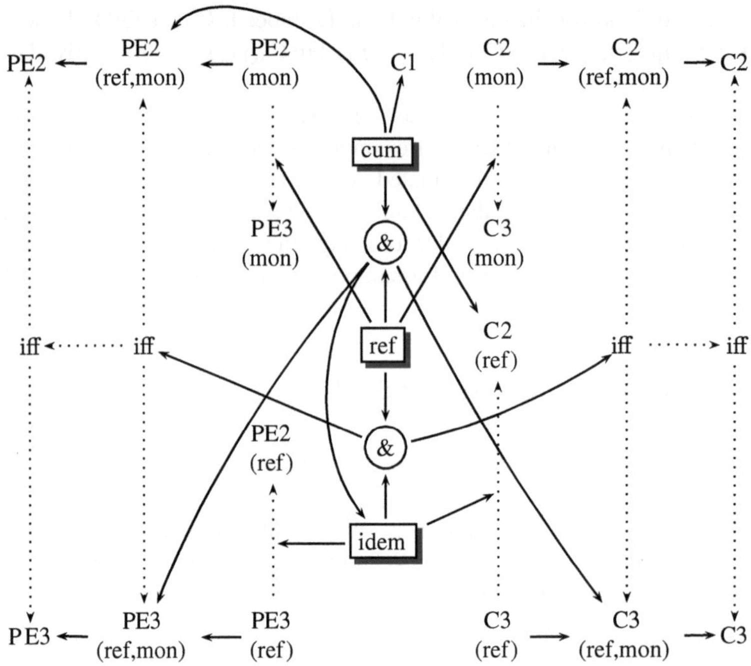

Fig. 4.1 Overview of the criteria with the abbreviations: ref = reflexivity, mon = monotonicity, cum = cumulativity, idem = idempotency. Omitted are for instance the *arrows* from PEi(Props) to Ci(Props) and from PEi to Ci (where $i \in \{2, 3\}$)

$$Cn_{\mathbf{L}'}(Cn_{\mathbf{L}}(\Gamma)) \subseteq Cn_{\mathbf{L}}(\Gamma) \text{ implies } Cn_{\mathbf{L}'}(\Gamma) \subseteq Cn_{\mathbf{L}}(\Gamma)\,.$$

Proof. Suppose that $\mathbf{L}$ is reflexive, $\mathbf{L}'$ is monotonic and $Cn_{\mathbf{L}'}(Cn_{\mathbf{L}}(\Gamma)) \subseteq Cn_{\mathbf{L}}(\Gamma)$. We have $\Gamma \subseteq Cn_{\mathbf{L}}(\Gamma)$ by the reflexivity of $\mathbf{L}$. Hence, $Cn_{\mathbf{L}'}(\Gamma) \subseteq Cn_{\mathbf{L}'}(Cn_{\mathbf{L}}(\Gamma))$ by the monotonicity of $\mathbf{L}'$. From this and the supposition follows that $Cn_{\mathbf{L}'}(\Gamma) \subseteq Cn_{\mathbf{L}}(\Gamma)$. □

Theorem 4.3.2. *Where* $\mathbf{L}$ *is reflexive*

(i) if Props *contains at least monotonicity, C2(*Props*) implies C3(*Props*) for* $\mathbf{L}$.
(ii) C2 implies C3 for $\mathbf{L}$.

Proof. Ad (i): Suppose $\mathbf{L}$ is reflexive, monotonicity is in Props and C2(Props) holds for $\mathbf{L}$. Now let $\mathbf{L}'$ be a monotonic logic for which $\Gamma \sim_{\mathbf{L}'} \Gamma'$ and for which $Cn_{\mathbf{L}'}(Cn_{\mathbf{L}}(\Theta)) = Cn_{\mathbf{L}}(\Theta)$ for all $\Theta \subseteq \mathcal{W}$. By Fact 4.3.1, $\mathbf{L}' \trianglelefteq \mathbf{L}$. By C2(Props), $\Gamma \sim_{\mathbf{L}} \Gamma'$. (ii) follows immediately. □

Fact 4.3.2. *Where* $\mathbf{L}$ *is idempotent and* $\mathbf{L}'$ *is reflexive*: $\mathbf{L}' \trianglelefteq \mathbf{L}$ *implies* $Cn_{\mathbf{L}'}(Cn_{\mathbf{L}}(\Gamma)) = Cn_{\mathbf{L}}(\Gamma)$ for all $\Gamma \subseteq \mathcal{W}$.

Proof. Let **L** be idempotent, **L**′ be reflexive, $\mathbf{L}' \trianglelefteq \mathbf{L}$, and $\Gamma \subseteq \mathcal{W}$ arbitrary. Thus, $Cn_{\mathbf{L}'}(Cn_{\mathbf{L}}(\Gamma)) \subseteq Cn_{\mathbf{L}}(Cn_{\mathbf{L}}(\Gamma))$. By the idempotency of **L**, $Cn_{\mathbf{L}'}(Cn_{\mathbf{L}}(\Gamma)) \subseteq Cn_{\mathbf{L}}(\Gamma)$. By the reflexivity of **L**′ $Cn_{\mathbf{L}'}(Cn_{\mathbf{L}}(\Gamma)) \supseteq Cn_{\mathbf{L}}(\Gamma)$. □

Theorem 4.3.3. *Where* **L** *is idempotent:*

(i) if Props *contains at least reflexivity then C3(*Props*) implies C2(*Props*) for* **L**.
(ii) C3 implies C2 for **L**.

Proof. Ad (i): Suppose **L** is idempotent, Props contains reflexivity, **L**′ is reflexive, $\mathbf{L}' \trianglelefteq \mathbf{L}$, and $\Gamma \sim_{\mathbf{L}'} \Gamma'$. By Fact 4.3.2, $Cn_{\mathbf{L}'}(Cn_{\mathbf{L}}(\Theta)) = Cn_{\mathbf{L}}(\Theta)$ for all $\Theta \subseteq \mathcal{W}$. Hence, by C3(Props), $\Gamma \sim_{\mathbf{L}} \Gamma'$. (ii) follows immediately. □

By Facts 4.3.1 and 4.3.2 we get:

Corollary 4.3.2. *Where* **L** *is reflexive and idempotent, and* **L**′ *is monotonic and reflexive:* $Cn_{\mathbf{L}'}(Cn_{\mathbf{L}}(\Gamma)) = Cn_{\mathbf{L}}(\Gamma)$ *for all* $\Gamma \subseteq \mathcal{W}$ *iff* $\mathbf{L}' \trianglelefteq \mathbf{L}$.

The following corollary establishes that C2 (resp. C2(Props)) and C3 (resp. C3(Props)) are coextensive whenever **L** is reflexive and idempotent (and Props contains at least monotonicity and reflexivity).

Corollary 4.3.3. *Where* **L** *is reflexive and idempotent:*

(i) if Props *contains at least monotonicity and reflexivity then C2(*Props*) and C3(*Props*) are coextensive for* **L**
(ii) C2 and C3 are coextensive for **L**.

By the reflexivity and idempotency of ALs we get:

Corollary 4.3.4. *(i) Where* Props *contains at least monotonicity and reflexivity: C2(*Props*) and C3(*Props*) are coextensive for* **AL**. *(ii) C2 and C3 are coextensive for* **AL**.

Theorem 4.3.4. *Where* **L** *is cumulative:*

(i) if Props *contains at least reflexivity, C2(*Props*) holds for* **L**.
(ii) C2 holds for **L**.

Proof. Ad (i): Suppose that $\mathbf{L}' \trianglelefteq \mathbf{L}$, $\Gamma \sim_{\mathbf{L}'} \Gamma'$, **L**′ is reflexive, and **L** is cumulative. By the reflexivity of **L**′, $\Gamma \subseteq Cn_{\mathbf{L}'}(\Gamma)$ and $\Gamma' \subseteq Cn_{\mathbf{L}'}(\Gamma')$. Hence, since $\Gamma \sim_{\mathbf{L}'} \Gamma'$ also $\Gamma \subseteq Cn_{\mathbf{L}'}(\Gamma')$ and $\Gamma' \subseteq Cn_{\mathbf{L}'}(\Gamma)$. By $\mathbf{L}' \trianglelefteq \mathbf{L}$, $\Gamma \subseteq Cn_{\mathbf{L}}(\Gamma')$ and $\Gamma' \subseteq Cn_{\mathbf{L}}(\Gamma)$. By the cumulativity of **L** and Theorem 4.3.1, $\Gamma \sim_{\mathbf{L}} \Gamma'$. (ii) follows immediately. □

Since ALs are cumulative we get:

Corollary 4.3.5. *(i) Where* Props *contains at least reflexivity:* **AL** *satisfies C2 (*Props*). (ii)* **AL** *satisfies C2.*

In other words, if L is a reflexive logic weaker than **AL** and Γ is **L**-equivalent with Γ', then Γ is also **AL**-equivalent with Γ'.

Fact 4.3.3. *Reflexivity and cumulativity imply idempotency.*

This can easily be seen. Suppose **L** is reflexive and cumulative. Then, since $Cn_{\mathbf{L}}(\Gamma) \subseteq Cn_{\mathbf{L}}(\Gamma)$, by cumulativity $Cn_{\mathbf{L}}(\Gamma) = Cn_{\mathbf{L}}(\Gamma \cup Cn_{\mathbf{L}}(\Gamma))$. By reflexivity, $Cn_{\mathbf{L}}(\Gamma) = Cn_{\mathbf{L}}(Cn_{\mathbf{L}}(\Gamma))$.

By Fact 4.3.3, Corollary 4.3.3 and Theorem 4.3.4 we immediately get:

Corollary 4.3.6. *Where* **L** *is cumulative and reflexive:*

(i) if Props *contains at least monotonicity and reflexivity, C3(*Props*) holds for* **L**.
(ii) C3 holds for **L**.

Of course, alternatively this can easily be shown directly:

Proof. Ad (i) Suppose **L** is cumulative and reflexive, **L**$'$ is monotonic and reflexive, that for all Θ, $Cn_{\mathbf{L}'}(Cn_{\mathbf{L}}(\Theta)) = Cn_{\mathbf{L}}(\Theta)$, and $\Gamma \sim_{\mathbf{L}'} \Gamma'$. Hence, $Cn_{\mathbf{L}'}(Cn_{\mathbf{L}}(\Gamma)) = Cn_{\mathbf{L}}(\Gamma)$ and $Cn_{\mathbf{L}'}\left(Cn_{\mathbf{L}}\left(\Gamma'\right)\right) = Cn_{\mathbf{L}}\left(\Gamma'\right)$. By Fact 4.3.1, $Cn_{\mathbf{L}'}(\Gamma) \subseteq Cn_{\mathbf{L}}(\Gamma)$ and $Cn_{\mathbf{L}'}\left(\Gamma'\right) \subseteq Cn_{\mathbf{L}}\left(\Gamma'\right)$. By the reflexivity of **L**$'$ and since $\Gamma \sim_{\mathbf{L}'} \Gamma'$, $\Gamma \subseteq Cn_{\mathbf{L}}\left(\Gamma'\right)$ and $\Gamma' \subseteq Cn_{\mathbf{L}}(\Gamma)$. By the cumulativity of **L** and Theorem 4.3.1, $\Gamma \sim_{\mathbf{L}} \Gamma'$. (ii) follows immediately. □

Since ALs are cumulative and reflexive we immediately get:

Corollary 4.3.7. *(i) Where* Props *contains at least reflexivity and monotonicity:* **AL** *satisfies C3(*Props*). (ii)* **AL** *satisfies C3.*

In other words, if **L** is a reflexive and monotonic logic under which **AL** is closed then: Γ is **L**-equivalent with Γ' implies Γ is also **AL**-equivalent with Γ'.

In order to give an example let us have a look at two typical examples of ALs, namely the inconsistency-ALs $\mathbf{CLuN}^r$ and $\mathbf{CLuN}^m$. They are defined as follows. The lower limit logic is **CLuN** (*C*lassical *L*ogic allowing for gl*u*ts with respect to *N*egation), viz. full positive **CL** with $(A \supset \neg A) \supset \neg A$ added as the only axiom for the standard negation, and extended with classical negation $\check{\neg}$.[5] While $A \vee \neg A$ is a **CLuN**-theorem, $A \wedge \neg A$ is **CLuN**-contingent. The set of abnormalities Ω comprises all formulas of the form $\exists(A \wedge \neg A)$ (the existential closure of $A \wedge \neg A$). The strategies are respectively reliability and minimal abnormality—see below. The resulting ALs will be called $\mathbf{CLuN}^r$ and $\mathbf{CLuN}^m$.

Note that, for every adaptive logic **AL**, **LLL** is a Tarski logic weaker than **AL**. So if two premise sets are **LLL**-equivalent, they are also **AL**-equivalent in view of C2. For some premise sets, however, one needs to rely directly on C1. An example is that $Cn_{\mathbf{CLuN}^m}(\{p\}) = Cn_{\mathbf{CLuN}^m}(\{p \vee (q \wedge \neg q)\})$. While $Cn_{\mathbf{CLuN}}(\{p\}) \neq Cn_{\mathbf{CLuN}}(\{p \vee (q \wedge \neg q)\})$, it is easy enough to show that $\{p\} \vdash_{\mathbf{CLuN}^m} p \vee (q \wedge \neg q)$ and that $\{p \vee (q \wedge \neg q)\} \vdash_{\mathbf{CLuN}^m} p$.

[5] Suitable axioms are $(A \supset \check{\neg} A) \supset \check{\neg} A$ and $A \supset (\check{\neg} A \supset B)$. The other classical symbols are stipulated to be identical to the corresponding standard symbols.

4.4 Characterizations Under a Translation

Some readers may be puzzled by our claim that criteria C1–C3 hold for the characterization of a DRF in terms of an AL while they do not hold for other characterizations—see Sect. 4.1. The reason is that the former characterizations are realized under a translation. Let us present two examples. Consider first the logic **Q** from Sect. 4.2, restricting the discussion to the propositional level.

Let $\rhd$ abbreviate $\Box\Diamond$. Where **T** is the well-known modal logic of Feys—see, for example, [54][6]—let the modal logic $\mathbf{T}^m$ be defined by (i) the lower limit **T**, (ii) the set of abnormalities $\Omega = \{(\rhd A \wedge \rhd B) \wedge \neg\rhd(A \wedge B) \mid A, B \in \mathcal{W}\}$, and (iii) minimal abnormality. It is provable that, where $B_1 \wedge \ldots \wedge B_n$ is the conjunctive normal form of A and $\Gamma^{\rhd} = \{\rhd C \mid C \in \Gamma\}$, $\Gamma \vdash_{\mathbf{Q}} A$ iff $\Gamma^{\rhd} \vdash_{\mathbf{T}^m} \rhd B_1 \wedge \ldots \wedge \rhd B_n$.

Note that every minimal Dab-formula that is **T**-derivable from $\Gamma^{\rhd}$ comprises only one disjunct.[7] This means that, for statements of the form $\Gamma^{\rhd} \vdash_{\mathbf{T}^m} \rhd B_1 \wedge \ldots \wedge \rhd B_n$, minimal abnormality and reliability boil down to the Simple strategy. Thus, the marking definition may be simplified to: a line is marked iff some member of its condition has been derived on the empty condition. Similarly, a **T**-model M of $\Gamma^{\rhd}$ is a $\mathbf{T}^m$-model of $\Gamma^{\rhd}$ iff $Ab(M) = \bigcap\{Ab(M') \mid M' \text{ is a } \mathbf{T}\text{-model of } \Gamma^{\rhd}\} = \{A \in \Omega \mid \Gamma^{\rhd} \vdash_{\mathbf{T}} A\}$.

Let us now turn to the fact that C2 does not hold for **Q**—for example **P**-equivalence does not warrant **Q**-equivalence—whereas C2 holds for $\mathbf{T}^m$ in view of Theorem 4.3.4. This opposition obviously derives from the fact that $\mathbf{T}^m$ distinguishes between $\rhd(p \wedge \neg p)$, which has no **T**-models, and $\rhd p \wedge \rhd\neg p$, which does, whereas **Q** blurs this distinction. For example $\{p, \neg p\}$ is **P**-equivalent to $\{p \wedge \neg p\}$, whereas $\{\rhd p, \rhd\neg p\}$ is not **T**-equivalent to $\{\rhd(p \wedge \neg p)\}$.

The situation is similar for the Strong and Weak consequence relations, which were employed to illustrate the non-applicability of C1 and C3. Here we only consider the Strong consequence relation. Let the premises be formulated with classical negation, $\check{\neg}$. Let $\Gamma^{\neg\check{\neg}} = \{\neg\check{\neg}A \mid A \in \Gamma\}$ and let $\mathcal{W}^{\not\neg}$ be the set of closed formulas that do not contain $\neg$ (but may contain $\check{\neg}$). It was proven in [5][8] that $Cn_{Strong}(\Gamma) = Cn_{\mathbf{CLuN}^m}\left(\Gamma^{\neg\check{\neg}}\right) \cap \mathcal{W}^{\not\neg}$. Note that, although Reflexivity (Theorem 2.6.4i), the fixed point property (Theorem 2.6.5iii) and cumulative indifference (Theorem 2.6.7iii) hold for the adaptive consequence relation $Cn_{\mathbf{CLuN}^m}(\Gamma)$, but that they do not hold for the consequence relation that maps Γ to $Cn_{\mathbf{CLuN}^m}\left(\Gamma^{\neg\check{\neg}}\right) \cap \mathcal{W}^{\not\neg}$. Thus if $p \wedge q, \check{\neg}p \in \Gamma$, neither of them will be in $Cn_{Strong}(\Gamma)$; $\neg\check{\neg}(p \wedge q)$ and $\neg\check{\neg}\check{\neg}p$ will be members of both $\Gamma^{\neg\check{\neg}}$ and $Cn_{\mathbf{CLuN}^m}\left(\Gamma^{\neg\check{\neg}}\right)$ but obviously not of $Cn_{\mathbf{CLuN}^m}\left(\Gamma^{\neg\check{\neg}}\right) \cap \mathcal{W}^{\not\neg}$. Note also that the required translation does not complicate the applicability of C1–C3.

---

[6] Except that, in order to define $\Gamma \vDash_{\mathbf{T}} A$, a **T**-model is defined as $M = \langle W, w_0, R, v\rangle$ with $w_0 \in W$ and M is said to verify A iff $v_M(A, w_0) = 1$.

[7] The property does not hold for all premise sets but is typical for premise sets $\Gamma^{\rhd}$ with Γ a set of modal-free formulas.

[8] The paraconsistent negation is there written as $\sim$ (here as $\neg$) and the classical negation as $\neg$ (here as $\check{\neg}$).

4.5 Extensions of Equivalent Premise Sets

Let us turn to the announced related problem: the equivalence of extensions of equivalent premise sets. We state similar criteria as in Sect. 4.2:

PE1 If $\Gamma \sim_{\mathbf{L}} \Gamma'$ implies $\Gamma \cup \Theta \sim_{\mathbf{L}} \Gamma' \cup \Theta$.

PE2 Where $\mathbf{L}'$ is a Tarski logic and $\mathbf{L}' \trianglelefteq \mathbf{L}$: $\Gamma \sim_{\mathbf{L}'} \Gamma'$ implies $\Gamma \cup \Theta \sim_{\mathbf{L}} \Gamma' \cup \Theta$.

PE2(Props) Where $\mathbf{L}'$ fulfills properties Props and $\mathbf{L}' \trianglelefteq \mathbf{L}$: $\Gamma \sim_{\mathbf{L}'} \Gamma'$ implies $\Gamma \cup \Theta \sim_{\mathbf{L}} \Gamma' \cup \Theta$.

PE3 Where $\mathbf{L}$ is closed under a Tarski logic $\mathbf{L}'$: $\Gamma \sim_{\mathbf{L}'} \Gamma'$ implies $\Gamma \cup \Theta \sim_{\mathbf{L}} \Gamma' \cup \Theta$.

PE3(Props) Where $\mathbf{L}$ is closed under a logic $\mathbf{L}'$ that fulfills the properties Props: $\Gamma \sim_{\mathbf{L}'} \Gamma'$ implies $\Gamma \cup \Theta \sim_{\mathbf{L}} \Gamma' \cup \Theta$.

PE1 is a strong criterion that is e.g. fulfilled by Tarski logics, but not by ALs.

Fact 4.5.1. *If* **L** *is a Tarski logic, then PE1 holds.*

It follows immediately from the following example that PE1 does not hold for ALs:

$$Cn_{\mathbf{CLuN}^m}(\{p\}) = Cn_{\mathbf{CLuN}^m}(\{p \vee (q \wedge \neg q)\}) \text{ but}$$
$$Cn_{\mathbf{CLuN}^m}(\{p, q \wedge \neg q\}) \neq Cn_{\mathbf{CLuN}^m}(\{p \vee (q \wedge \neg q), q \wedge \neg q\}).$$

Note that the example may be adjusted to any AL in which classical disjunction is present or definable. The example clearly indicates the most straightforward reason why the fact holds. The formula $q \wedge \neg q$ is an abnormality and hence is supposed to be false 'unless and until proven otherwise'. The original premise sets are equivalent because $p \vee (q \wedge \neg q)$ is the only premise of the second premise set and its minimally abnormal interpretation leads to p. If, however, $q \wedge \neg q$ is added to the premise sets, $\{p, q \wedge \neg q\}$ still gives us p because $\mathbf{CLuN}^m$ is reflexive, but p is not derivable from $\{p \vee (q \wedge \neg q), q \wedge \neg q\}$ because this extended premise set requires $q \wedge \neg q$ to be true and has the same $\mathbf{CLuN}^m$-consequences as $\{q \wedge \neg q\}$. To the negative fact corresponds a positive result which is very similar to it.

Theorem 4.5.1. *Where* **L** *is cumulative:*

(i) if Props *contains at least reflexivity and monotonicity:* **L** *satisfies PE2(*Props*).*
(ii) **L** *satisfies PE2.*

Proof. Ad (i): Suppose **L** is cumulative, $\mathbf{L}'$ is reflexive and monotonic, $\mathbf{L}' \trianglelefteq \mathbf{L}$, and $\Gamma_1 \sim_{\mathbf{L}'} \Gamma_2$. In view of the reflexivity of $\mathbf{L}'$, (i) $\Delta \subseteq Cn_{\mathbf{L}'}(\Gamma_1 \cup \Delta)$ and (ii) $Cn_{\mathbf{L}'}(\Gamma_1) = Cn_{\mathbf{L}'}(\Gamma_2)$ warrants that $\Gamma_2 \subseteq Cn_{\mathbf{L}'}(\Gamma_1)$. As $\mathbf{L}'$ is monotonic, it follows that $\Gamma_2 \cup \Delta \subseteq Cn_{\mathbf{L}'}(\Gamma_1 \cup \Delta)$. So $\Gamma_2 \cup \Delta \subseteq Cn_{\mathbf{L}}(\Gamma_1 \cup \Delta)$ in view of $Cn_{\mathbf{L}'}(\Gamma_1 \cup \Delta) \subseteq Cn_{\mathbf{L}}(\Gamma_1 \cup \Delta)$. By the same reasoning $\Gamma_1 \cup \Delta \subseteq Cn_{\mathbf{L}}(\Gamma_2 \cup \Delta)$. But then, in view of Theorem 4.3.1 and since **L** is cumulative, $Cn_{\mathbf{L}}(\Gamma_1 \cup \Delta) = Cn_{\mathbf{L}}(\Gamma_2 \cup \Delta)$. (ii) follows immediately. □

Since ALs are cumulative, we immediately get:

Corollary 4.5.1. *(i) Where* Props *contains at least reflexivity and monotonicity:* **AL** *satisfies PE2(*Props*). (ii)* **AL** *satisfies PE2.*

Theorem 4.5.2. *Where* **L** *is reflexive*

(i) if Props *contains at least monotonicity, PE2(*Props*) implies PE3(*Props*) for* **L**.
(ii) PE2 implies PE3 for **L**.

The proof is analogous to the proof for Theorem 4.3.2.

Theorem 4.5.3. *Where* **L** *is idempotent:*

(i) if Props *contains at least reflexivity then PE3(*Props*) implies PE2(*Props*) for* **L**.
(ii) PE3 implies PE2 for **L**.

The proof is analogous to the proof for Theorem 4.3.3.

By Theorem 4.5.2 and 4.5.3 we immediately get:

Corollary 4.5.2. *Where* **L** *is reflexive and idempotent:*

(i) if Props *contains at least monotonicity and reflexivity then PE2(*Props*) and PE3(*Props*) are coextensive for* **L**
(ii) PE2 and PE3 are coextensive for **L**.

By Fact 4.3.3, Theorem 4.5.1, and Corollary 4.5.2:

Corollary 4.5.3. *Where* **L** *is cumulative and reflexive:*

(i) if Props *contains at least reflexivity and monotonicity then* **L** *satisfies PE3 (*Props*).*
(ii) **L** *satisfies PE3.*

Of course, this can easily be proven directly.

Proof. Ad (i): Suppose that **L** is cumulative and reflexive, that $\mathbf{L}'$ is reflexive and monotonic, that for all Θ, $Cn_{\mathbf{L}'}(Cn_{\mathbf{L}}(\Theta)) = Cn_{\mathbf{L}}(\Theta)$, and that $\Gamma \sim_{\mathbf{L}'} \Gamma'$. Note first that due to the monotonicity and reflexivity of $\mathbf{L}'$, also $\Gamma \cup \Delta \subseteq Cn_{\mathbf{L}'}(\Gamma' \cup \Delta)$ and $\Gamma' \cup \Delta \subseteq Cn_{\mathbf{L}'}(\Gamma \cup \Delta)$. Since $Cn_{\mathbf{L}}$ is closed under $Cn_{\mathbf{L}'}$ and by Fact 4.3.1, $Cn_{\mathbf{L}'}(\Gamma \cup \Delta) \subseteq Cn_{\mathbf{L}}(\Gamma \cup \Delta)$ and $Cn_{\mathbf{L}'}(\Gamma' \cup \Delta) \subseteq Cn_{\mathbf{L}}(\Gamma' \cup \Delta)$. Altogether, $\Gamma \cup \Delta \subseteq Cn_{\mathbf{L}}(\Gamma' \cup \Delta)$ and $\Gamma' \cup \Delta \subseteq Cn_{\mathbf{L}}(\Gamma \cup \Delta)$ and hence by the cumulativity of **L** and Theorem 4.3.1, $\Gamma \cup \Delta \sim_{\mathbf{L}} \Gamma' \cup \Delta$. (ii) follows immediately. □

Note that this result improves on our previous result presented in Corollary 4.3.6 since obviously PE3(Props) implies C3(Props).

Corollary 4.5.4. *(i) Where* Props *contains at least reflexivity and monotonicity:* **AL** *satisfies PE3(*Props*). (ii)* **AL** *satisfies PE3.*

For ALs there is a weaker alternative for Fact 4.5.1. For this, we need another definition.

Definition 4.5.1. A set of formulas $\Theta \subseteq \mathcal{W}$ is an **AL**-*monotonic extension* of a set of formulas Γ iff $\Gamma \subset \Theta$ and $Cn_{\mathbf{AL}}(\Gamma) \subseteq Cn_{\mathbf{AL}}(\Theta)$.

Theorem 4.5.4. *If $\Gamma_1 \cup \Delta$ is an* **AL**-*monotonic extension of Γ_1 and $\Gamma_2 \cup \Delta$ is an* **AL**-*monotonic extension of Γ_2, then $Cn_{\mathbf{AL}}(\Gamma_1) = Cn_{\mathbf{AL}}(\Gamma_2)$ warrants that $Cn_{\mathbf{AL}}(\Gamma_1 \cup \Delta) = Cn_{\mathbf{AL}}(\Gamma_2 \cup \Delta)$*

Proof. Suppose $Cn_{\mathbf{AL}}(\Gamma_1) = Cn_{\mathbf{AL}}(\Gamma_2)$, $\Gamma_1 \cup \Delta$ is an **AL**-monotonic extension of Γ_1 and $\Gamma_2 \cup \Delta$ is an **AL**-monotonic extension of Γ_2. By Definition 4.5.1, the second supposition implies that

$$Cn_{\mathbf{AL}}(\Gamma_1) \subseteq Cn_{\mathbf{AL}}(\Gamma_1 \cup \Delta).$$

In view of the reflexivity of ALs

$$\Delta \subseteq Cn_{\mathbf{AL}}(\Gamma_1 \cup \Delta).$$

From the two previous results, one obtains immediately that

$$Cn_{\mathbf{AL}}(\Gamma_1) \cup \Delta \subseteq Cn_{\mathbf{AL}}(\Gamma_1 \cup \Delta),$$

and with Theorem 2.6.7iii

$$Cn_{\mathbf{AL}}(\Gamma_1 \cup \Delta) = Cn_{\mathbf{AL}}\left(Cn^{\mathcal{L}}_{\mathbf{AL}}(\Gamma_1) \cup \Delta \cup \Gamma_1\right).$$

In view of the reflexivity of **AL**, $\Gamma_1 \subseteq Cn_{\mathbf{AL}}(\Gamma_1)$, hence:

$$Cn_{\mathbf{AL}}(\Gamma_1 \cup \Delta) = Cn_{\mathbf{AL}}\left(Cn^{\mathcal{L}}_{\mathbf{AL}}(\Gamma_1) \cup \Delta\right).$$

With the same reasoning, the following is provable

$$Cn_{\mathbf{AL}}(\Gamma_2 \cup \Delta) = Cn_{\mathbf{AL}}\left(Cn^{\mathcal{L}}_{\mathbf{AL}}(\Gamma_2) \cup \Delta\right).$$

Since $Cn_{\mathbf{AL}}(\Gamma_1) = Cn_{\mathbf{AL}}(\Gamma_2)$,

$$Cn_{\mathbf{AL}}\left(Cn^{\mathcal{L}}_{\mathbf{AL}}(\Gamma_1) \cup \Delta\right) = Cn_{\mathbf{AL}}\left(Cn^{\mathcal{L}}_{\mathbf{AL}}(\Gamma_2) \cup \Delta\right).$$

Thus, $Cn_{\mathbf{AL}}(\Gamma_1 \cup \Delta) = Cn_{\mathbf{AL}}(\Gamma_2 \cup \Delta)$. □

There are criteria for deciding whether an extension is **AL**-monotonic. The criteria depend on the strategy, which is the third element of the adaptive logic **AL**. The criteria we introduce below may be not the sharpest possible ones, but it is obvious that they are correct. Let Γ be the original premise set and Γ' the extended premise set.

For the reliability strategy, the criterium reads: If $\Gamma \subset \Gamma'$ and $U(\Gamma') \subseteq U(\Gamma)$ then Γ' is an **AL**-monotonic extension of Γ. In words: if every abnormality that is unreliable with respect to Γ' is also unreliable with respect to Γ, then Γ' is an **AL**-monotonic extension of Γ. In terms of the proof theory, this means that every unmarked line in a proof from Γ remains unmarked if the premise set is extended to Γ'. This warrants that the final consequences of Γ are also final consequences of Γ'. Obviously, some lines that are marked in a proof from Γ may be unmarked in a proof from Γ'. The effect of this is that the latter premise set has more, but not less, consequences than the former.

For the minimal abnormality strategy, the criterion reads: If $\Gamma \subset \Gamma'$ and $\Phi(\Gamma') \subseteq \Phi(\Gamma)$, then Γ' is an **AL**-monotonic extension of Γ. This criterium is most easily understood from a semantic point of view. The antecedent warrants that every **AL**-model of Γ' is an **AL**-model of Γ and hence verifies every formula verified by all **AL**-models of Γ.

It is instructive to illustrate the difference between the two criteria in terms of $\mathbf{CLuN}^r$ and $\mathbf{CLuN}^m$. Let $\Gamma = \{(p \wedge \neg p) \vee (q \wedge \neg q), (p \wedge \neg p) \vee (r \wedge \neg r), s \vee (p \wedge \neg p), s \vee (q \wedge \neg q)\}$ and let $\Gamma' = \Gamma \cup \{q \wedge \neg q\}$. As $U(\Gamma) = U(\Gamma') = \{p \wedge \neg p, q \wedge \neg q, r \wedge \neg r\}$, Γ' is a $\mathbf{CLuN}^r$-monotonic extension of Γ. Note, however, that $\Phi(\Gamma) = \{\{p \wedge \neg p\}, \{q \wedge \neg q, r \wedge \neg r\}\}$ whereas $\Phi(\Gamma') = \{\{q \wedge \neg q, p \wedge \neg p\}, \{q \wedge \neg q, r \wedge \neg r\}\}$. So Γ' is not a $\mathbf{CLuN}^m$-monotonic extension of Γ and actually $\Gamma \vdash_{\mathbf{CLuN}^m} s$ whereas $\Gamma' \nvdash_{\mathbf{CLuN}^m} s$.

4.6 Maximality of the Lower Limit Logic

As **LLL** is a Tarski logic weaker than **AL**, Corollary 4.5.1 entails the following.

Corollary 4.6.1. *Where $Cn_{\mathbf{LLL}}(\Gamma_1) = Cn_{\mathbf{LLL}}(\Gamma_2)$: for all Δ,*

$$Cn_{\mathbf{AL}}(\Gamma_1 \cup \Delta) = Cn_{\mathbf{AL}}(\Gamma_2 \cup \Delta)\,.$$

We shall now prove that the lower limit logic **LLL** of an adaptive logic **AL** is not only a Tarski logic that is weaker than **AL** but that actually every monotonic logic **L** that is weaker than **AL** is weaker than **LLL** or identical to **LLL**. In the proof of the following theorem we rely on the compactness of **LLL**, but do not require **L** to be compact.

Lemma 4.6.1. *Where* $\Gamma' = \{\mathsf{Dab}(\Delta) \mid \Gamma \vdash_{\mathbf{LLL}} B \check{\vee} \mathsf{Dab}(\Delta)\}$. *If* $\Gamma \nvdash_{\mathbf{LLL}} B$, *then* $\Gamma \cup \Gamma' \nvdash_{\mathbf{LLL}} B$.

Proof. Suppose $\Gamma \nvdash_{\mathbf{LLL}} B$. Assume $\Gamma \cup \Gamma' \vdash_{\mathbf{LLL}} B$. Hence, by the compactness of **LLL**, there is a finite $\Gamma'' \subseteq \Gamma'$ such that $\Gamma \cup \Gamma'' \vdash_{\mathbf{LLL}} B$. Hence, $\Gamma \cup \{\check{\bigwedge} \Gamma''\} \vdash_{\mathbf{LLL}} B$. By the Deduction Theorem, $\Gamma \vdash_{\mathbf{LLL}} (\check{\bigwedge} \Gamma'') \check{\supset} B$. Hence,

$$\Gamma \vdash_{\mathbf{LLL}} \check{\neg} \left(\check{\bigwedge} \Gamma'' \right) \check{\vee} B \tag{4.1}$$

By the definition of Γ', for each $\mathsf{Dab}(\Delta) \in \Gamma'$, $\Gamma \vdash_{\mathbf{LLL}} \mathsf{Dab}(\Delta) \check{\vee} B$ and hence,

$$\Gamma \vdash_{\mathbf{LLL}} \left(\check{\bigwedge} \Gamma'' \right) \check{\vee} B \tag{4.2}$$

By (4.1) and (4.2), $\Gamma \vdash_{\mathbf{LLL}} B$,—a contradiction. □

Theorem 4.6.1. *Where* **L** *is a monotonic logic for which* $Cn^{\mathcal{L}^+}_{\mathbf{L}}(\Theta) \subseteq Cn^{\mathcal{L}^+}_{\mathbf{AL^m}}(\Theta)$ *for all* $\Theta \subseteq \mathcal{W}^+$: $Cn^{\mathcal{L}^+}_{\mathbf{L}}(\Gamma) \subseteq Cn^{\mathcal{L}^+}_{\mathbf{LLL}}(\Gamma)$ *for all* $\Gamma \subseteq \mathcal{W}^+$.

Proof. Suppose the antecedent holds and that there is a Γ and a B for which the following three hold.

$$\Gamma \nvdash_{\mathbf{LLL}} B \tag{4.3}$$
$$\Gamma \vdash_{\mathbf{L}} B \tag{4.4}$$
$$\Gamma \vdash_{\mathbf{AL^m}} B \tag{4.5}$$

Let $\Gamma' = \{\mathsf{Dab}(\Delta) \mid \Gamma \vdash_{\mathbf{LLL}} B \check{\vee} \mathsf{Dab}(\Delta)\}$. By Lemma 4.6.1, (4.3) entails (4.6); (4.7) follows from (4.4) by the monotonicity of **L**, and (4.8) follows from (4.7) by the supposition.

$$\Gamma \cup \Gamma' \nvdash_{\mathbf{LLL}} B \tag{4.6}$$
$$\Gamma \cup \Gamma' \vdash_{\mathbf{L}} B \tag{4.7}$$
$$\Gamma \cup \Gamma' \vdash_{\mathbf{AL^m}} B \tag{4.8}$$

Note first that due to (4.3) and (4.5), $\Gamma' \neq \emptyset$ (otherwise there is obviously no way to finally derive B in an $\mathbf{AL^m}$-proof from Γ). Thus, $\emptyset \notin \Phi(\Gamma \cup \Gamma')$. Moreover, by Theorem 2.4.4 and Theorem 2.4.6, $\Phi(\Gamma \cup \Gamma') \neq \emptyset$. By Theorem 2.6.2 and Corollary 2.4.4 there is a $\varphi \in \Phi(\Gamma \cup \Gamma')$ and a $\Delta_\varphi \subseteq \Omega \setminus \varphi$ such that $\Gamma \cup \Gamma' \vdash_{\mathbf{LLL}} B \check{\vee} \mathsf{Dab}(\Delta_\varphi)$.

In view of the compactness and monotonicity of **LLL** there are $\mathsf{Dab}(\Delta_1), \ldots, \mathsf{Dab}(\Delta_n) \in \Gamma'$ such that

$$\Gamma \cup \{\mathsf{Dab}(\Delta_1), \ldots, \mathsf{Dab}(\Delta_n)\} \vdash_{\mathbf{LLL}} B \check{\vee} \mathsf{Dab}(\Delta_\varphi)\,. \tag{4.9}$$

As $\Gamma \vdash_{\mathbf{LLL}} B \check{\vee} \mathsf{Dab}(\Delta_i)$ for every $i \in \{1, \ldots, n\}$,

$$\Gamma \vdash_{\mathbf{LLL}} B \check{\vee} \left(\check{\bigwedge}_{i=1}^{n} \mathsf{Dab}(\Delta_i) \right). \tag{4.10}$$

From (4.9) follows

$$\Gamma \cup \left\{ \check{\bigwedge}_{i=1}^{n} \mathsf{Dab}(\Delta_i) \right\} \vdash_{\mathbf{LLL}} B \check{\vee} \mathsf{Dab}(\Delta_\varphi), \tag{4.11}$$

whence, by the Deduction Theorem,

$$\Gamma \vdash_{\mathbf{LLL}} \left(\check{\bigwedge}_{i=1}^{n} \mathsf{Dab}(\Delta_i) \right) \check{\supset} (B \check{\vee} \mathsf{Dab}(\Delta_\varphi)). \tag{4.12}$$

From (4.10) and (4.12) follows

$$\Gamma \vdash_{\mathbf{LLL}} B \check{\vee} \mathsf{Dab}(\Delta_\varphi), \tag{4.13}$$

whence $\mathsf{Dab}(\Delta_\varphi) \in \Gamma'$. But then φ contains at least one member of Δ_φ,—a contradiction to the fact that $\Delta_\varphi \subseteq \Omega \setminus \varphi$. □

It follows from Theorem 2.6.iii that this result also holds when the third element of **AL** is reliability. Hence we obtain the following corollary.

Corollary 4.6.2. *Where* **L** *is a monotonic logic such that* $Cn^{\mathcal{L}^+}_{\mathbf{L}}(\Theta) \subseteq Cn^{\mathcal{L}^+}_{\mathbf{AL}}(\Theta)$ *for all* $\Theta \subseteq \mathcal{W}^+$*:* $Cn^{\mathcal{L}^+}_{\mathbf{L}}(\Gamma) \subseteq Cn^{\mathcal{L}^+}_{\mathbf{LLL}}(\Gamma)$ *for all* $\Gamma \subseteq \mathcal{W}^+$.

Fact 4.3.1 gives us a further corollary.

Corollary 4.6.3. *If* $Cn^{\mathcal{L}^+}_{\mathbf{AL}}(\Gamma)$ *is closed under a monotonic logic* **L**, *then* $Cn^{\mathcal{L}^+}_{\mathbf{L}}(\Gamma) \subseteq Cn^{\mathcal{L}^+}_{\mathbf{LLL}}(\Gamma)$ *for all* $\Gamma \subseteq \mathcal{W}^+$.

We close this section with a cautionary remark. Note that in Theorem 4.6.1 it is crucial that **L** is weaker or equal to **AL** relative to the language $\mathcal{L}^+$ that is enriched with the "checked connectives".[9] If we were to replace the antecedent of the theorem by

(†) "Where **L** is a monotonic logic for which $Cn^{\mathcal{L}}_{\mathbf{L}}(\Theta) \subseteq Cn^{\mathcal{L}}_{\mathbf{AL^m}}(\Theta)$ for all $\Theta \subseteq \mathcal{W}$"

the conclusion would not hold in general. In other words, (†) warrants

1. neither: $Cn^{\mathcal{L}^+}_{\mathbf{L}}(\Gamma) \subseteq Cn^{\mathcal{L}^+}_{\mathbf{LLL}}(\Gamma)$ for all $\Gamma \subseteq \mathcal{W}^+$
2. nor: $Cn^{\mathcal{L}}_{\mathbf{L}}(\Gamma) \subseteq Cn^{\mathcal{L}}_{\mathbf{LLL}}(\Gamma)$ for all $\Gamma \subseteq \mathcal{W}$

[9] It is for instance used in order to derive (4.8).

We can illustrate this by a simple example. Let $\mathcal{L}$ only consist of propositional atoms and a unary connective $\bullet$. Let **LLL** be the logic that only allows for premise introduction: $Cn^{\mathcal{L}}_{\mathbf{LLL}}(\Gamma) = \Gamma$. **L** is **LLL** enriched by the rule that allows to detach propositional atoms from $\bullet$: if $\bullet A$ then A, where A is a propositional atom. The consequence relations $Cn^{\mathcal{L}^+}_{\mathbf{LLL}}$ and $Cn^{\mathcal{L}^+}_{\mathbf{L}}$ are obtained in the usual way: we add the classical axioms and rules for the "checked connectives".

We now define an adaptive logic $\mathbf{AL}_\bullet$ by the triple $\langle \mathbf{LLL}, \Omega, x\rangle$ where $\Omega = \{\bullet A \check{\wedge} \check{\neg} A \mid A \text{ is a propositional atom}\}$ and x represents any of the two standard strategies.

Note that we have the following (where p is a propositional atom):

$$\text{for all } \Gamma \subseteq \mathcal{W} : Cn^{\mathcal{L}}_{\mathbf{L}}(\Gamma) \subseteq Cn^{\mathcal{L}}_{\mathbf{AL}_\bullet}(\Gamma) \tag{4.14}$$

$$Cn^{\mathcal{L}}_{\mathbf{LLL}}(\{\bullet p\}) \subset Cn^{\mathcal{L}}_{\mathbf{L}}(\{\bullet p\}) \tag{4.15}$$

$$Cn^{\mathcal{L}^+}_{\mathbf{LLL}}(\{\bullet p\}) \subset Cn^{\mathcal{L}^+}_{\mathbf{L}}(\{\bullet p\}) \tag{4.16}$$

Equation (4.14) holds since both **L** and $\mathbf{AL}_\bullet$ rigorously apply detachment to propositional atoms preceded by a $\bullet$. Indeed, **L** is the upper limit logic of $\mathbf{AL}_\bullet$ and any $\Gamma \subseteq \mathcal{W}$ is a normal premise set (see page 43). (4.15) and (4.16) are exemplified by $p \in Cn_{\mathbf{L}}(\{\bullet p\}) \setminus Cn_{\mathbf{LLL}}(\{\bullet p\})$.[10]

Now, does this give a counter-example to Theorem 4.6.1? Only if the antecedent of Theorem 4.6.1 is fulfilled by **L**. But it is not, since

$$Cn^{\mathcal{L}^+}_{\mathbf{AL}}\left(\{\bullet p, \check{\neg} p\}\right) \subset Cn^{\mathcal{L}^+}_{\mathbf{L}}\left(\{\bullet p, \check{\neg} p\}\right) = \mathcal{W}^+.$$

4.7 In Conclusion

We have proven that criteria C1–C3 and some weaker versions, which are standard for identifying equivalent premise sets with respect to Tarski logics, also apply to ALs. This is a major advantage of ALs in comparison to other formal approaches to defeasible reasoning forms because the criteria are transparent and easy to check.

With respect to extensions of equivalent premise sets, ALs do not behave like Tarski logics, but we have located a criterion that is simple and close to that for Tarski logics.

Let us summarize our central results. Where $\Gamma, \Gamma' \subseteq \mathcal{W}$ we have,

---

[10] Since it is easy to observe that $Cn^{\mathcal{L}}_{\mathbf{AL}_\bullet}$ is monotonic, the reader may conjecture Theorem 4.6.1 equipped with the antecedent (†) may hold at least for ALs for which $Cn^{\mathcal{L}}_{\mathbf{AL}}$ is non-monotonic. But this is not true either. Suppose we add two logical constants to our language $\mathcal{L}$: $\star$ and $\circ$. **LLL** and **L** are defined as before. We alter our **AL** by defining Ω by: $\{\bullet A \check{\wedge} \check{\neg} A \mid A \text{ is a propositional atom}\} \cup \{\star\} \cup \{\star \check{\vee} \check{\neg} \circ\}$. Note that $\emptyset \vdash_{\mathbf{AL}} \circ$ (each minimally abnormal model has the abnormal part $\emptyset$ and hence validates $\check{\neg}\star$ and $\circ$) but $\{\star\} \nvdash_{\mathbf{AL}} \circ$ (note that the minimally abnormal models all have the abnormal part $\{\star, \star \check{\vee} \check{\neg} \circ\}$ due to $\star$ being a premise: some also validate $\check{\neg}\circ$). This demonstrates that **AL** is non-monotonic (relative to $\mathcal{L}$). Furthermore, as before (4.14)–(4.16) hold.

(i) if $\Gamma \subseteq Cn_{\mathbf{AL}}(\Gamma')$ and $\Gamma' \subseteq Cn_{\mathbf{AL}}(\Gamma)$ then $Cn_{\mathbf{AL}}(\Gamma) = Cn_{\mathbf{AL}}(\Gamma')$
(ii) where **L** is a reflexive logic weaker than **AL** then $Cn_{\mathbf{L}}(\Gamma) = Cn_{\mathbf{L}}(\Gamma')$ implies $Cn_{\mathbf{AL}}(\Gamma) = Cn_{\mathbf{AL}}(\Gamma')$
(iii) where **L** is monotonic and reflexive and **AL** is closed under **L** (i.e., for all Θ, $Cn_{\mathbf{L}}(Cn_{\mathbf{AL}}(\Theta)) = Cn_{\mathbf{AL}}(\Theta)$), then $Cn_{\mathbf{L}}(\Gamma) = Cn_{\mathbf{L}}(\Gamma')$ implies $Cn_{\mathbf{AL}}(\Gamma \cup \Theta) = Cn_{\mathbf{AL}}(\Gamma' \cup \Theta)$
(iv) where **L** is a reflexive and monotonic logic weaker than **AL** then $Cn_{\mathbf{L}}(\Gamma) = Cn_{\mathbf{L}}(\Gamma')$ implies $Cn_{\mathbf{AL}}(\Gamma \cup \Theta) = Cn_{\mathbf{AL}}(\Gamma' \cup \Theta)$

We have also shown that the strongest monotonic logic weaker than an AL in standard format is its lower limit and that the lower limit logic is the strongest monotonic logic under which the adaptive consequence set is closed.

References

1. Batens, D., Straßer, C., Verdée, P.: On the transparency of defeasible logics: equivalent premise sets, equivalence of their extensions, and maximality of the lower limit. Logique et Analyse **52**(207), 281–304 (2009)
2. Shoham, Y.: A semantical approach to nonmonotonic logics. In: Ginsberg, M.L. (ed.) Readings in Non-Monotonic Reasoning, pp. 227–249. Morgan Kaufmann, Los Altos (1987)
3. Shoham, A.L.B.Y. Jr.: New results on semantical non-monotonic reasoning. In: Reinfrank, M., de Kleer, J., Ginsberg, M.L., Sandewall, E. (eds.) NMR. Lecture Notes in Computer Science, vol. 346, pp. 19–26. Springer (1988)
4. Lin, F., Shoham, Y.: Epistemic semantics for fixed-points non-monotonic logics. Morgan Kaufmann Publishers Inc., Pacific Grove, CA (1990)
5. Batens, D.: Towards the unification of inconsistency handling mechanisms. Logic Log. Philos. **8**, 5–31 (2000). Appeared 2002
6. Batens, D.: A strengthening of the Rescher-Manor consequence relations. Logique et Analyse **183–184**, 289–313 (2003). Appeared 2005
7. Batens, D., Vermeir, T.: Direct dynamic proofs for the Rescher-Manor consequence relations: the flat case. J. Appl. Non-Class. Logics **12**, 63–84 (2002)
8. Verhoeven, L.: Proof theories for some prioritized consequence relations. Logique et Analyse **183–184**, 325–344 (2003). Appeared 2005
9. Rescher, N., Manor, R.: On inference from inconsistent premises. Theor. Decis. **1**, 179–217 (1970)
10. Benferhat, S., Dubois, D., Prade, H.: Some syntactic approaches to the handling of inconsistent knowledge bases: a comparative study. Part 1: the flat case. Stud. Logica **58**, 17–45 (1997)
11. Benferhat, S., Dubois, D., Prade, H.: Some syntactic approaches to the handling of inconsistent knowledge bases: a comparative study. Part 2: The prioritized case. In: Orłowska, E. (ed) Logic at Work. Essays Dedicated to the Memory of Helena Rasiowa, pp. 473–511. Physica Verlag (Springer), Heidelberg (1999)
12. Batens, D., Meheus, J., Provijn, D.: An adaptive characterization of signed systems for paraconsistent reasoning (Unpublished).
13. Besnard, P., Schaub, T.: Signed systems for paraconsistent reasoning. J. Autom. Reason. **20**, 191–213 (1998)
14. De Clercq, K.: Two new strategies for inconsistency-adaptive logics. Logic Log. Philos. **8**, 65–80 (2000). Appeared 2002
15. Batens, D.: Inconsistency-adaptive logics and the foundation of non-monotonic logics. Logique et Analyse **145**, 57–94 (1994). Appeared 1996

16. Antoniou, G.: Nonmonotonic Reasoning. MIT Press, Cambridge (1996)
17. Brewka, G.: Nonmonotonic Reasoning: Logical Foundations of Commonsense. Cambridge University Press, Cambridge (1991)
18. Łukaszewicz, W.: Non-Monotonic Reasoning. Formalization of Commonsense Reasoning. Ellis Horwood, New York (1990)
19. Straßer, C.: An adaptive logic for rational closure. In: Carnielli, W., Coniglio, M.E., D'Ottaviano, I.M.L. (eds.) The Many Sides of Logic, pp. 47–67. College Publications (2009)
20. Lehmann, D.J., Magidor, M.: What does a conditional knowledge base entail? Artif. Intell. **55**(1), 1–60 (1992)
21. Straßer, C., Šešelja, D.: Towards the proof-theoretic unification of Dung's argumentation framework: an adaptive logic approach. J. Logic Comput. **21**, 133–156 (2010)
22. Dung, P.M., Son, T.C.: Nonmonotonic inheritance, argumentation and logic programming. In: Marek, V.W., Nerode, A. (eds.) Logic Programming and Nonmonotonic Reasoning, Proceedings. Lecture Notes in Computer Science. Third International Conference, LPNMR'95, Lexington, USA, June 26–28, 1995, vol. 928, pp. 316–329. Springer (1995)
23. Dung, P.M.: On the acceptability of arguments and its fundamental role in nonmonotonic reasoning, logic programming and n-person games. Artif. Intell. **77**, 321–358 (1995)
24. Primiero, G., Meheus, J.: Majority merging by adaptive counting. Synthese **165**, 203–223 (2008)
25. Konieczny, S., Pino Pérez, R.: Merging information under constraints: a logical framework. J. Logic Comput. **12**, 773–808 (2002)
26. Meheus, J.: Adaptive logics for question evocation. Logique et Analyse **173–175**, 135–164 (2001). Appeared 2003
27. Meheus, J.: Erotetic arguments from inconsistent premises. Logique et Analyse **165–166**, 49–80 (1999). Appeared 2002
28. Wiśniewski, A.: The Posing of Questions. Logical Foundations of Erotetic Inferences. Kluwer, Dordrecht (1995)
29. Meheus, J., Verhoeven, L., Van Dyck, M., Provijn, D.: Ampliative adaptive logics and the foundation of logic-based approaches to abduction. In: Magnani, L., Nersessian, N.J., Pizzi, C., et al (eds.) Logical and Computational Aspects of Model-Based Reasoning, pp. 39–71. Kluwer, Dordrecht (2002)
30. Meheus, J., Provijn, D.: Abduction through semantic tableaux versus abduction through goal-directed proofs. Theoria **22/3**, 295–304 (2007)
31. Meheus, J., Batens, D.: A formal logic for abductive reasoning. Logic J. IGPL **14**, 221–236 (2006)
32. Gauderis, T., Putte, F.V.D.: Abduction of generalizations. THEORIA Int. J. Theor. Hist. Found. Sci. **27**(3), (2012)
33. Gauderis, T.: Modelling abduction in science by means of a modal adaptive logic. Found. Sci. (2013)
34. Aliseda, A.: Abductive Reasoning. Logical Investigations into Discovery and Explanation. Springer, Dordrecht (2006)
35. Provijn, D., Weber, E.: Adaptive logics for non-explanatory and explanatory diagnostic reasoning. In: Magnani, L., Nersessian, N.J., Pizzi, C., et al (eds.) Logical and Computational Aspects of Model-Based Reasoning, pp. 117–142. Kluwer, Dordrecht (2002)
36. Batens, D., Meheus, J., Provijn, D., Verhoeven, L.: Some adaptive logics for diagnosis. Logic Log. Philos. **11/12**, 39–65 (2003)
37. Reiter, R.: A theory of diagnosis from first principles. Artif. Intell. **32**, 57–95 (1987)
38. Meheus, J.: Empirical progress and ampliative adaptive logics. In: Festa, R., Aliseda, A., Peijnenburg, J. (eds.) Confirmation, Empirical Progress, and Truth Approximation. Essays in Debate with Theo Kuipers. vol. 1, Poznan Studies in the Philosophy of the Sciences and the Humanities, vol. 83, pp. 193–217. Rodopi, Amsterdam/New York (2005)
39. Kuipers, T.A.F.: From Instrumentalism to Constructive Realism. On some Relations Between Confirmation, Empirical Progress, and Truth Approximation, Synthese Library, vol. 287. Kluwer, Dordrecht (2000)

40. Van De Putte, F., Verdée, P.: The dynamics of relevance: adaptive belief revision. Synthese **187**(1), 1–42 (2012)
41. Meheus, J.: An adaptive logic for pragmatic truth. In: Carnielli, W.A., Coniglio, M.E., Loffredo D'Ottaviano, I.M. (eds.) Paraconsistency. The Logical Way to the Inconsistent, pp. 167–185. Marcel Dekker, New York (2002)
42. Mikenberg, I., da Costa, N.C.A., Chuaqui, R.: Pragmatic truth and approximation to truth. J. Symb. Logic **51**, 201–221 (1986)
43. da Costa, N.C., Bueno, O., French, S.: The logic of pragmatic truth. J. Philos. Logic **27**, 603–620 (1998)
44. Van Dyck, M.: Causal discovery using adaptive logics. Towards a more realistic heuristics for human causal learning. Logique et Analyse **185–188**, 5–32 (2004). Appeared 2005
45. Leuridan, B.: Causal discovery and the problem of ignorance. An adaptive logic approach. J. Appl. Logic **7**, 188–205 (2009)
46. Pearl, J.: Causality. Models, Reasoning, and Inference. Cambridge University Press, Cambridge (2000)
47. Straßer, C.: An adaptive logic framework for conditional obligations and deontic dilemmas. Logic Log. Philos. **19**(1–2), 95–128 (2010)
48. Straßer, C., Meheus, J., Beirlaen, M.: Tolerating deontic conflicts by adaptively restricting inheritance. Logique Anal. **219**, 477–506 (2012)
49. Goble, L.: A proposal for dealing with deontic dilemmas. In: Lomuscio, A., Nute, D. (eds.) DEON Lecture Notes in Computer Science, vol. 3065, pp. 74–113. Springer (2004)
50. Goble, L.: A logic for deontic dilemmas. J. Appl. Logic **3**, 461–483 (2005)
51. Verdée, P., van der Waart van Gulik, S.: A generic framework for adaptive vague logics. Studia Logica **90**, 385–405 (2008)
52. Schotch, P.K., Jennings, R.E.: On detonating. In: Priest, G., Routley, R., Norman, J. (eds.) Paraconsistent Logic. Essays on the Inconsistent, pp. 306–327. Philosophia Verlag, München (1989)
53. Makinson, D.: Bridges from Classical to Nonmonotonic Logic, Texts in Computing, vol. 5. King's College Publications, London (2005)
54. Hughes, G., Cresswell, M.: An Introduction to Modal Logic. Methuen, New York (1972). First published 1968

46. Van De Putte, F., Verdée, P.: The dynamics of relevance: adaptive belief revision. Synthese 187(1), 1–42 (2012)

47. Meheus, J.: An adaptive logic for pragmatic truth. In: Carnielli, W.A., Coniglio, M.E., Loffredo D'Ottaviano, I.M. (eds.) [illegible]

48. [illegible] Pragmatic truth [illegible]

49. [illegible]

50. [illegible]

51. [illegible]

Chapter 5
Generalizing the Standard Format

5.1 Introduction

While the standard format for ALs provides a highly unifying framework for the modeling of defeasible reasoning, it has limitations. Let me mention two:

1. Recall that the basic idea behind ALs is to interpret a given set of premises "as normally as possible". Semantically this is realized by means of comparing and selecting models in view of their abnormal parts (i.e., the set of abnormalities they verify). Note that the underlying rationale behind these comparisons in the standard format is qualitative: a model M is preferable to another model M' in case the abnormal part of M is a subset of the abnormal part of M'. Besides these qualitative considerations also quantitative considerations are sometimes useful: e.g., the abnormal part of M may—according to this *quantitative rationale*—preferable to the abnormal part of M' in case it contains quantitatively less abnormalities. Indeed, there are ALs with so-called counting strategies (see e.g., [1–3]): however these systems are not in the standard format.
2. The standard format has also been criticized for offering a suboptimal framework to express *priorities among abnormalities*. In order to improve on this, a generalization of the standard format has been proposed in [4]: so-called lexicographic ALs. This format works with a stratified set of abnormalities $\Omega = \langle \Omega_i \rangle_I$ (where I is an index-set) where abnormalities in Ω_1 are more abnormal than abnormalities in Ω_2, and so on. Abnormal parts of models are compared by means of a lexicographic order. Lexicographic ALs have been applied to normative reasoning in [5], to belief revision [6], and to the handling of inconsistencies in [7]. Note however that lexicographic ALs also only model qualitative considerations (see point 1) and that the handling of priorities is limited to lexicographic comparisons of stratified sets of abnormalities.

In this chapter we generalize the standard format of ALs and thereby introduce an interesting larger class of ALs that can be characterized in a simple and intuitive way. We demonstrate that the new format overcomes the two shortcomings mentioned

C. Straßer, *Adaptive Logics for Defeasible Reasoning*, Trends in Logic 38,
DOI: 10.1007/978-3-319-00792-2_5, © Springer International Publishing Switzerland 2014

above. On the one hand, logics with both qualitative and quantitative rationales can be expressed in it. On the other hand, the format is expressive enough to allow for the handling of priorities in various ways.

We show that many ALs that have been considered in the literature fall within this larger class—for instance ALs in the standard format, ALs with counting strategies, lexicographic ALs—and that the characterization of this class offers many possibilities to formulate new logics.

One of the advantages of the format studied in this chapter is that a lot of meta-theory comes for free for any logic formulated in it. We show that ALs formulated in it are always sound and complete. Furthermore, many of the meta-theoretic properties that are usually associated with the standard format (such as cumulativity, fixed point property, (strong) reassurance, etc.) also hold for rich subclasses of logics formulated in the new format.

Recall that ALs in the standard format are characterized by a triple: (a) the lower limit logic **LLL** which is the monotonic, reflexive and transitive core logic, (b) a set of abnormalities Ω that is characterized by a (or some) logical form(s), and (c) an adaptive strategy.

Semantically speaking, the driving force behind ALs in the standard format is a selection semantics[1]:

1. We (partially) order the models of the lower limit logic according to their abnormal parts by means of set-inclusion $\subset$.
2. Then we select ("prefer") models whose abnormal part is lower than a given threshold that is determined by the adaptive strategy that is chosen. For instance, in case the strategy is minimal abnormality we simply choose the minimal models (i.e., the models whose abnormal part is minimal with respect to $\subset$): an **LLL**-model M of Γ is selected iff $\text{Ab}(M) \in \min_{\subset}\left(\text{Ab}^{\Gamma}_{\mathbf{LLL}}\right)$ where

$$\text{Ab}^{\Gamma}_{\mathbf{LLL}} =_{\text{df}} \left\{\text{Ab}(M') \mid M' \in \mathcal{M}_{\mathbf{LLL}}(\Gamma)\right\}$$

This idea can be generalized (see also Table 5.1).

1. First, instead of ordering **LLL**-models by means of their abnormal parts with respect to $\subset$ we may use another partial order $\prec$.[2] Throughout this chapter we will suppose that $\prec \subseteq \subset$. Obviously, if $\text{Ab}(M) \subset \text{Ab}(M')$ then M offers a more normal interpretation than M' which should be expressed by $\prec$.
2. Second, instead of determining the threshold for the selection in terms of minimally abnormal models or reliable models, we can specify other threshold functions Λ.

[1] The idea of selecting a certain set of models and then to define a semantic consequence relation on the basis of this selection is an integral part of many formal systems. Variants of it can be found in e.g., Shoham [8, 9], McCarthy [10], Schlechta [11], etc. Lindström [12] and Makinson [13] offer systematic overviews.

[2] In this chapter we will use $\prec$ to denote a *strict* partial order. Of course, one can easily define the corresponding non-strict $\preceq$ by $a \preceq b$ iff $a \prec b$ or $a = b$. Hence, this is a purely conventional choice.

Table 5.1 Generalizing the standard format for ALs

	Standard format	Generalization
Ordering of the abnormal parts of models w.r.t.	$\subset$	$\prec$
Threshold for the selection of models	$\min_{\subset}(\mathrm{Ab}^{\Gamma}_{\mathbf{LLL}})$ (*minimal abnormality*) $\{\varphi \mid \varphi \subseteq \bigcup \min_{\subset}(\mathrm{Ab}^{\Gamma}_{\mathbf{LLL}})\}$ (*reliability*)	$\Lambda(\mathrm{Ab}^{\Gamma}_{\mathbf{LLL}})$

The second point is especially important in cases in which the order $\prec$ that is imposed on the **LLL**-models is not smooth. In these cases there are models M for which there are no models M' with minimal abnormal part (relative to $\prec$) that are "better". We will also suggest different ways of specifying selection thresholds for such situations.

Besides the semantics we will introduce a dynamic proof theory for ALs in this generalized setting. Also the proof theory is a straight-forward generalization of the proof theory of ALs in the standard format. It is adequate with respect to the semantic consequence relation, equips us with a syntactic consequence relation, and mirrors the semantic selections in view of Λ by means of a defeasible retraction mechanism.

The chapter is structured as follows: In Sect. 5.2 we introduce several examples of ALs in the generalized setting. In Sect. 5.3 we present our generalized format: the selection semantics and the dynamic proof theory. Section 5.4 contains representational results for the semantic and syntactic consequence relations of our ALs including soundness and completeness. Section 5.5 contains a study of other meta-theoretic properties such as Cumulativity and Reassurance. In Sect. 5.6 we show that the proof theory can be simplified for a rich subclass of the generalized class of ALs. In Sect. 5.7 we show how the normal selections strategy can be expressed in the new format. Finally, in Sect. 5.8 we wrap things up and relate our meta-theoretic insights to the examples introduced in Sect. 5.2. Section 5.9 contains some final remarks including some additonal motivation of the new format and replies to possible objections. Many technical results are proven in the Appendix C.

5.2 Some Examples

In this section we will introduce some motivating examples for ALs that fall within the enriched class of logics whose meta-theory is studied in this paper.

There are three ways in which we make the presentation in this section more coherent and free of digressions. First, all the presented ALs are based on the logic $\mathbf{CL}_{\circ}$ (or simple variants of it). Recall that this logic is obtained by adding a "dummy" operator to classical propositional logic **CL**. Second, the applications we study in this paper are organized around discussive applications inspired e.g., by the research on Rescher-Manor consequence relations (see e.g., [14]). Third, in this section we will only focus on semantic considerations.

Most of the applications we mention would deserve a deeper discussion were we interested in developing a fully elaborated formal account for them. However, our aim is more modest: it is to make the reader aware of a variety of possibilities to define useful selection semantics. This in turn motivates the generic perspective on ALs which is introduced in the further run of this paper.

5.2.1 The Standard Format: Minimal Abnormality

Suppose we are to logically model a discussion. When some participant states A we represent this by $\circ A$. Hence, given a set of statements Γ we translate it to $\Gamma^\circ = \{\circ A \mid A \in \Gamma\}$. A statement $\circ A$ counts as accepted in a model M of Γ° in case A is valid in M. The idea is to select models in which as many statements as possible are accepted (while preserving consistency). Note that there are cases in which we cannot accept all given statements since some of these may be conflicting. We will give an example below.

We use the adaptive logic $\mathbf{AL}^\circ_{\min_\subset}$ that has the lower limit logic $\mathbf{CL}_\circ$ and the set of abnormalities

$$\Omega_\circ = \{\circ A \wedge \neg A \mid A \text{ has no occurrences of `} \circ \text{'}\}$$

Finally, we use the minimal abnormality strategy. We order the models of the lower limit logic with respect to their abnormal parts and according to $\subset$. Our semantic selection selects all $\mathbf{CL}_\circ$-models of Γ whose abnormal part is minimal with respect to $\subset$ and hence in

$$\min_\subset\left(\mathsf{Ab}^\Gamma_{\mathbf{CL}_\circ}\right)$$

where $\mathsf{Ab}^\Gamma_{\mathbf{CL}_\circ} =_{\text{df}} \left\{\text{Ab}(M) \mid M \in \mathcal{M}_{\mathbf{CL}_\circ}(\Gamma)\right\}$ and $\text{Ab}(M) =_{\text{df}} \{A \in \Omega_\circ \mid M \models A\}$.

The idea is very simple: a model M is "sufficiently normal" iff it is "minimally abnormal": i.e., there is no model whose abnormal part is a subset of the abnormal part of M.

We define the set of $\mathbf{AL}^\circ_{\min_\subset}$-models of Γ by

$$\mathcal{M}_{\mathbf{AL}^\circ_{\min_\subset}}(\Gamma) =_{\text{df}} \left\{M \in \mathcal{M}_{\mathbf{CL}_\circ}(\Gamma) \mid \text{Ab}(M) \in \min_\subset\left(\mathsf{Ab}^\Gamma_{\mathbf{CL}_\circ}\right)\right\}$$

The semantic consequence relation is then defined by

$$\Gamma \Vdash_{\mathbf{AL}^\circ_{\min_\subset}} A \text{ iff } M \models A \text{ for all } M \in \mathcal{M}_{\mathbf{AL}^\circ_{\min_\subset}}(\Gamma) \qquad (\star)$$

Suppose for instance that $\Gamma = \{p \wedge q, \neg p \wedge q, r\}$. Consider the following $\mathbf{CL}_\circ$-models of Γ°:

Model M	Ab(M)	$M \models$
M_1	$\circ(p \wedge q) \wedge \neg(p \wedge q), \circ r \wedge \neg r$	$\neg p, q, \neg r$
M_2	$\circ(\neg p \wedge q) \wedge \neg(\neg p \wedge q)$	p, q, r
M_3	$\circ(p \wedge q) \wedge \neg(p \wedge q)$	$\neg p, q, r$
M_4	$\circ(p \wedge q) \wedge \neg(p \wedge q), \circ(\neg p \wedge q) \wedge \neg(\neg p \wedge q)$	$p, \neg q, r$

The models are ordered with respect to their abnormal parts according to $\subset$ as follows:

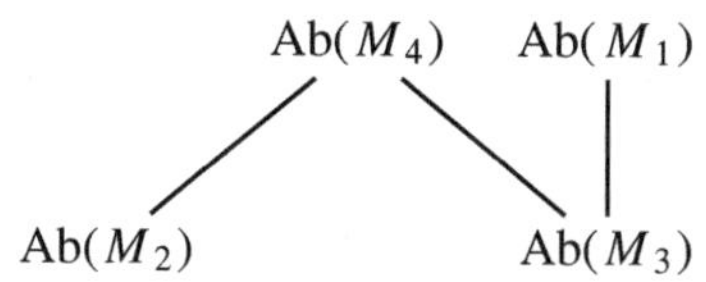

It is easy to see that all the models in $\mathcal{M}_{\mathbf{AL}^\circ_{\min\subset}}(\Gamma^\circ)$ have an abnormal part that is either identical to Ab(M_2) or identical to Ab(M_3). Thus $\Gamma^\circ \Vdash_{\mathbf{AL}^\circ_{\min\subset}} q, r$ but for instance $\Gamma^\circ \nVdash_{\mathbf{AL}^\circ_{\min\subset}} p, \neg p$.

Note that there is no conflict concerning r. Hence it is true in all minimal abnormal models. Second, the conflict in our statements concerns a disagreement in p. Since q is derivable from both of the conflicting statements and q is not conflicted otherwise, it is true in both minimally abnormal interpretations of our premises. Hence, it is a consequence.[3]

5.2.2 *Lexicographic ALs*

A recent example for ALs that are not in the standard format is the class of *lexicographic ALs* (see [4]). In these logics we work with a structured, i.e., prioritized set of abnormalities: $\Omega = \bigcup_I \Omega_i$ (where $I = \{1, \ldots, n\}$ or $I = \mathbb{N}$). The abnormalities in Ω_1 are considered to be "worst" and hence our priority is to avoid them. Abnormalities in Ω_2 are "second-worst", and so on. Lexicographic ALs have been applied to normative reasoning [5], to belief revision [6], and to reasoning with inconsistencies [7].[4]

For a simple example let us return to our logic $\mathbf{CL}_\circ$. However, we enhance our expressive powers slightly. We use sequences of $\circ$ in order to indicate the trustworthiness of the information. For instance, $\circ A$ indicates that the information A is provided by a most trust-worthy source. "$\circ\circ$" indicates a less trust-worthy source, etc. This way we can "prioritize" our set of abnormalities $\Omega = \bigcup_{\mathbb{N}} \Omega_i$. Where $\circ^i$ denotes a

[3] It can easily be shown that $\mathbf{AL}^\circ_{\min\subset}$ represents the universal Rescher-Manor consequence relation: A is derivable from all maximally consistent subsets of Γ iff $\Gamma^\circ \vdash_{\mathbf{AL}^\circ_{\min\subset}} A$. See [15].

[4] Lexicographic ALs have been compared to sequential combinations of ALs in the standard format and to hierarchical adaptive logics [16] in [17].

sequence of i-many $\circ$, we define

$$\Omega_i = \left\{\circ^i A \wedge \neg A \mid A \text{ is a formula without occurrences of `}\circ\text{'}\right\}$$

Now suppose we have the following premises: $\Gamma = \{\circ p, \circ\circ\neg p, \circ q, \circ\circ\circ r\}$. We consider four $\mathbf{CL}_\circ$-models of Γ:

Model M	Ab(M)	$M \models$
M_1	$\circ p \wedge \neg p$	$\neg p, q, r$
M_2	$\circ\circ\neg p \wedge \neg\neg p, \circ\circ\circ r \wedge \neg r$	$p, q, \neg r$
M_3	$\circ\circ\neg p \wedge \neg\neg p, \circ q \wedge \neg q$	$p, \neg q, r$
M_4	$\circ\circ\neg p \wedge \neg\neg p$	p, q, r

Now compare M_1 and M_4. Note that a more trust-worthy source states p than the source that states $\neg p$. Hence we should prefer p over $\neg p$. This makes M_4 preferable to M_1. By a similar argument M_2 is preferable to M_3.

One way of formally realizing this intuition is by means of a lexicographic order.

Definition 5.2.1 (Lexicographic order on $\wp(\Omega) \times \wp(\Omega)$). Where $\varphi, \psi \subseteq \Omega$ are sets of abnormalities, φ is preferable to ψ, in signs $\varphi \prec_{\mathsf{lex}} \psi$, iff, there is a $n \in \mathbb{N}$ for which

(a) $\varphi \cap \Omega_i = \psi \cap \Omega_i$ for all $i < n$ and
(b) $\varphi \cap \Omega_n \subset \psi \cap \Omega_n$.

This obviously imposes a partial order on the lower limit logic models (resp. on their abnormal parts) since now we can compare models with respect to their abnormal parts and $\prec_{\mathsf{lex}}$. Applying $\prec_{\mathsf{lex}}$ to our four models we get:

$$\text{Ab}(M_4) \prec_{\mathsf{lex}} \text{Ab}(M_2) \prec_{\mathsf{lex}} \text{Ab}(M_3), \text{Ab}(M_1)$$

As a threshold for our selection we use $\min_{\prec_{\mathsf{lex}}}\left(\mathsf{Ab}^{\Gamma}_{\mathbf{CL}_\circ}\right)$ and hence select all the $\prec_{\mathsf{lex}}$-minimally abnormal $\mathbf{CL}_\circ$-models of Γ. Thus, the set of selected models of the corresponding AL, we call it $\mathbf{AL}^{\circ}_{\prec_{\mathsf{lex}}}$, is defined by

$$\mathcal{M}_{\mathbf{AL}^{\circ}_{\prec_{lex}}}(\Gamma) =_{\text{df}} \left\{M \in \mathcal{M}_{\mathbf{CL}_\circ}(\Gamma) \mid \text{Ab}(M) \in \min_{\prec_{lex}}\left(\mathsf{Ab}^{\Gamma}_{\mathbf{CL}_\circ}\right)\right\}$$

The semantic consequence relation is defined analogous to ($\star$).

In our example all the minimal abnormal models M have the abnormal part $\text{Ab}(M) = \text{Ab}(M_4) = \{\circ\circ\neg p \wedge \neg\neg p\}$. Hence, we get $\Gamma \Vdash_{\mathbf{AL}^{\circ}_{\prec_{\mathsf{lex}}}} p, q, r$.

Since there is no conflict concerning q and r both are valid in our selected models and hence they are consequences. There is a conflict in p, however the more trustworthy source states p. Hence in our selected models p is the case, as desired.

5.2.3 Counting Abnormalities

Another class of ALs in which the abnormal parts of **LLL**-models are not compared by means of $\subset$ are ALs with so-called counting strategies (see e.g., [1–3]), or –more general–, ALs that use quantitative comparisons rather than qualitative ones.

Suppose we have a discussive application where we model possibly conflicting expert opinions. Instead of $\mathbf{CL}_{\circ}$ we use $\mathbf{CL}_{\circ}^{\star}$: instead of $\circ$ we now have a $\circ_i$ for each $i \in \mathbb{N}$. The idea is that each expert gets a number i, and everything she states is preceded by $\circ_i$.

In case some experts' opinions conflict, we do not prioritize between their expertise such as we did in the previous example where we distinguished the "trustworthiness of the source". However, we prefer expert opinions in proportion to how often they have been stated by different experts. E.g., if 6 experts state A and only 2 state $\neg A$, we prioritize A.

This can be realized as follows. We define

$$\Omega_* = \{\circ_i A \wedge \neg A \mid A \text{ is a } \circ\text{-free formula}\}$$

We compare the abnormal parts of $\mathbf{CL}_{\circ}^{\star}$-models by means of the order $\prec_c$:

Definition 5.2.2 (Counting order on $\wp(\Omega) \times \wp(\Omega)$). Where $\varphi, \psi \subseteq \Omega$ are sets of abnormalities, $\varphi \prec_c \psi$ iff, $|\varphi| < |\psi|$ or $\varphi \subset \psi$ where $|X|$ is the cardinality of X.[5]

So suppose we have the following scenario:

Expert	States
1	p, q
2	$\neg p, q$
3	p, q
4	$p, \neg q$

We can translate this into a premise set:

$$\Gamma_c = \{\circ_1 p, \circ_1 q, \circ_2 \neg p, \circ_2 q, \circ_3 p, \circ_3 q, \circ_4 p, \circ_4 \neg q\}$$

Now consider the following $\mathbf{CL}_{\circ}^{\star}$-models of Γ_c:

[5] Of course, for finite sets $\varphi \subset \psi$ implies $|\varphi| < |\psi|$. However, for infinite sets the comparison by means of the cardinality does not allow to prefer φ to ψ in case of $\varphi \subset \psi$, although the latter clearly indicates that φ is "better" ("less abnormal") than ψ.

Model	Ab(M)	$\lvert$Ab(M)$\rvert$	$M \models$
M_1	$\circ_2\neg p \wedge \neg\neg p, \circ_4\neg q \wedge \neg\neg q$	2	p, q
M_2	$\circ_2\neg p \wedge \neg\neg p, \circ_1 q \wedge \neg q, \circ_2 q \wedge \neg q, \circ_3 q \wedge \neg q$	4	$p, \neg q$
M_3	$\circ_1 p \wedge \neg p, \circ_3 p \wedge \neg p, \circ_4 p \wedge \neg p, \circ_4\neg q \wedge \neg\neg q$	4	$\neg p, q$
M_4	$\circ_1 p \wedge \neg p, \circ_3 p \wedge \neg p, \circ_4 p \wedge \neg p, \circ_1 q \wedge \neg q, \circ_2 q \wedge \neg q, \circ_3 q \wedge \neg q$	6	$\neg p, \neg q$

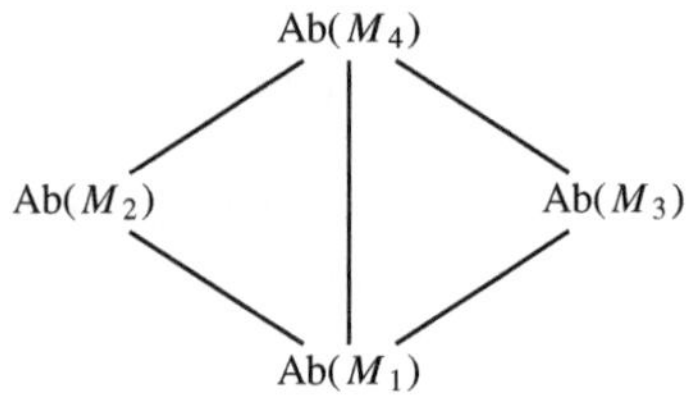

$\prec_c$ imposes the following partial order on our models:
Let $\mathbf{AL}^{\circ}_{\min_{\prec_c}}$ be the AL that is characterized by the lower limit logic $\mathbf{CL}^{\star}_{\circ}$, the set of abnormalities Ω_*, the order $\prec_c$ on the (abnormal parts) of our $\mathbf{CL}^{\star}_{\circ}$-models, and the threshold $\min_{\prec_c}$ for the semantic selection. Hence,

$$\mathcal{M}_{\mathbf{AL}^{\circ}_{\min_{\prec_c}}}(\Gamma) =_{\text{df}} \left\{M \in \mathcal{M}_{\mathbf{CL}^{\star}_{\circ}}(\Gamma) \mid \text{Ab}(M) \in \min_{\prec_c}\left(\mathsf{Ab}^{\Gamma}_{\mathbf{CL}^{\star}_{\circ}}\right)\right\}$$

The semantic consequence relation is defined analogous to ($\star$).

It is easy to see that in our example all selected models have the abnormal part of M_1. Hence, we get $\Gamma_c \Vdash_{\mathbf{AL}^{\circ}_{\min_{\prec_c}}} p, q$. This is as expected since p was stated by three experts, while $\neg p$ was stated by only one expert. An analogous argument applies to q.

5.2.4 A Most Skeptical Rationale: Reliability

The rationale behind the semantic selections presented thus far can be summarized as follows: given some partial order $\prec$, select all $\prec$-minimal models.

We will now depart from this rationale and focus on different threshold functions by considering ALs with the reliability strategy. Later –in Sect. 5.2.7– we take a look at two example with more refined quantitative rationales.

The reliability strategy is usually motivated by means of minimal Dab-consequences. Recall that a disjunction of abnormalities Dab(Δ) was called a Dab-*consequence of* Γ relative to a lower limit logic **LLL** iff $\Gamma \vdash_{\mathbf{LLL}}$ Dab(Δ). Dab(Δ) is a *minimal* Dab-consequence of Γ iff there is no $\Theta \subset \Delta$ such that Dab(Θ) is also a Dab-consequence of Γ.

The information provided by a minimal Dab-consequence Dab(Δ) is that (a) at least one of the abnormalities in Δ is true, and (b) there is no way of excluding any abnormality in Δ (due to the minimality of Δ).

The reliability strategy treats this situation in the most sceptical way possible: all abnormalities in a minimal Dab-consequence are considered unreliable. Hence, we also consider the worst case scenario in which each abnormality in Δ is true. For the semantic selection this means that we also select models in which all abnormalities in Δ are true. More general: where $\mathsf{Dab}(\Delta_1), \mathsf{Dab}(\Delta_2), \ldots$ is a list of all minimal Dab-consequences of Γ and $\Sigma(\Gamma) =_{\mathrm{df}} \{\Delta_1, \Delta_2, \ldots\}$, we select all **LLL**-models M for which $\mathrm{Ab}(M) \subseteq \bigcup \Sigma(\Gamma)$.

Note that an abnormality is a member of a minimal Dab-consequence iff it is validated by some $\subset$-minimally abnormal model. Hence, another way to express the selection for reliability is as follows: only select models M such that each abnormality validated by M is also validated by some $\subset$-minimally abnormal model. This means that our threshold function Λ is[6]:

$$\min{}^{\cup}_{\subset}(X) =_{\mathrm{df}} \left\{\varphi \in X \mid \varphi \subseteq \bigcup \min{}_{\subset}(X)\right\}$$

Let us exemplify this by means of the example from Sect. 5.2.1. Recall that the minimally abnormal models in our example were models with abnormal part $\mathrm{Ab}(M_2) = \{\circ(\neg p \wedge q) \wedge \neg(\neg p \wedge q)\}$ and models with abnormal part $\mathrm{Ab}(M_3) = \{\circ(p \wedge q) \wedge \neg(p \wedge q)\}$. Thus, $\min^{\cup}_{\subset}(\mathsf{Ab}^{\Gamma}_{\mathbf{CL}_\circ}) = \{\mathrm{Ab}(M) \in \mathsf{Ab}^{\Gamma}_{\mathbf{CL}_\circ} \mid \mathrm{Ab}(M) \subseteq \mathrm{Ab}(M_2) \cup \mathrm{Ab}(M_3)\} = \{\mathrm{Ab}(M) \in \mathsf{Ab}^{\Gamma}_{\mathbf{CL}_\circ} \mid \mathrm{Ab}(M) \subseteq \{\circ(\neg p \wedge q) \wedge \neg(\neg p \wedge q), \circ(p \wedge q) \wedge \neg(p \wedge q)\}\}$. Hence, also M_4 is selected since $\mathrm{Ab}(M_4) \subseteq \min^{\cup}_{\subset}(\mathsf{Ab}^{\Gamma}_{\mathbf{CL}_\circ})$. However, M_1 is not selected since $\circ r \wedge \neg r \in \mathrm{Ab}(M_1) \setminus \min^{\cup}_{\subset}(\mathsf{Ab}^{\Gamma}_{\mathbf{CL}_\circ})$.[7]

The fact that models like M_1 are not selected deserves some more discussion. The reason that M_1 is not selected is that it validates an abnormality that is not part of any minimal Dab-consequence. This is indeed reasonable in view of

(a) the fact that in ALs we assume an abnormality to be false "per default", i.e., unless there is a good reason to doubt this assumption
(b) as long as an abnormality is not involved in any minimal Dab-consequence there is indeed no reason to assume it to be true.

The fact that an abnormality is not involved in a minimal Dab-consequence indicates that it either is not part of any Dab-consequence (in case there are none), or that it was just added to a Dab-consequence by means of addition: but obviously this doesn't make it any more suspicious.

In other words: whatever our selection function is, it seems rational to require that it does not select models that validate abnormalities that are not part of any minimal Dab-consequence. As a consequence, for any threshold Λ used for the semantic selection, $\min^{\cup}_{\subset}$ should serve as an upper bound:

---

6 In the Appendix (Corollary C.2.1) we prove that $\min^{\cup}_{\subset}(\mathsf{Ab}^{\Gamma}_{\mathbf{LLL}}) = \mathsf{Ab}^{\Gamma'}_{\mathbf{LLL}}$ where $\Gamma' = \Gamma \cup (\Omega \setminus \bigcup \Sigma(\Gamma))^{\check{\neg}}$. This shows that the models in $\min^{\cup}_{\subset}(\mathsf{Ab}^{\Gamma}_{\mathbf{LLL}})$ are indeed exactly the models which do not validate any abnormalities which are not part of any minimal Dab-consequence.

7 In [15] we show that the AL $\mathbf{ACL}_\circ{}^{\mathbf{r}}$ in standard format that is characterized by the triple $\langle \mathbf{CL}_\circ, \Omega_\circ,$ reliability$\rangle$ represents the free Rescher-Manor consequence relation: A is a free Rescher-Manor consequence from Γ iff $\Gamma^\circ \Vdash_{\mathbf{ACL}_\circ{}^{\mathbf{r}}} A$.

$$\Lambda(\mathsf{Ab}^{\Gamma}_{\mathbf{LLL}}) \subseteq \min^{\cup}_{\subset}(\mathsf{Ab}^{\Gamma}_{\mathbf{LLL}})$$

In this sense, the reliability strategy of the standard format provides the most sceptical rationale for ALs.

5.2.5 Colexicographic ALs

Let us come back to our lower limit logic $\mathbf{CL}^{\star}_{\circ}$. We now focus on a single participant of the discussion and offer a different reading of $\circ_i A$. Namely, A has been uttered by our participant at time point i. We move from time point 1 on forward in time. We are interested in offering an account of the participant's changes of mind: in case she states $\neg A$ at time point 1 but changes her mind later on, say at time point 5 she states A, we expect to derive A. Hence, the consequences of our logic mirror the state of mind of our agent based on what she states at the latest time point she offers a statement.

Take as an example the following premise set

$$\Gamma_{co} = \{\circ_1 p, \circ_2 q, \circ_3 r, \circ_4 \neg q, \circ_5 \neg p\}$$

and consider the following models:

Model	Ab(M)	$M \models$
M_1	$\circ_5\neg p \wedge \neg\neg p, \circ_4\neg q \wedge \neg\neg q$	p, q, r
M_2	$\circ_5\neg p \wedge \neg\neg p, \circ_2 q \wedge \neg q$	$p, \neg q, r$
M_3	$\circ_1 p \wedge \neg p, \circ_2 q \wedge \neg q, \circ_3 r \wedge \neg r$	$\neg p, \neg q, \neg r$
M_4	$\circ_1 p \wedge \neg p, \circ_4\neg q \wedge \neg\neg q$	$\neg p, q, r$
M_5	$\circ_1 p \wedge \neg p, \circ_2 q \wedge \neg q$	$\neg p, \neg q, r$

The idea to prioritize later statements over former incompatible ones is realized by means of the following colexicographic order:

Definition 5.2.3 (Colexicographic order on $\wp(\Omega) \times \wp(\Omega)$). Where $\varphi, \psi \subseteq \Omega$, $\varphi \prec_{\mathsf{co}} \psi$ iff $\varphi \subset \psi$ or there is a $n \in \mathbb{N}$ for which

(a) $\varphi \cap \Omega_i = \psi \cap \Omega_i$ for all $i > n$, and
(b) $\varphi \cap \Omega_n \subset \psi \cap \Omega_n$

If we order our models by means of $\prec_{\mathsf{co}}$ with respect to their abnormal parts which are based on Ω_* we have:

$$\mathrm{Ab}(M_5) \prec_{\mathsf{co}} \mathrm{Ab}(M_3) \prec_{\mathsf{co}} \mathrm{Ab}(M_4) \prec_{\mathsf{co}} \mathrm{Ab}(M_2) \prec_{\mathsf{co}} \mathrm{Ab}(M_1)$$

Indeed, all models M with abnormal part in $\min_{\prec_{\mathsf{co}}}(\mathsf{Ab}^{\Gamma_{co}}_{\mathbf{CL}^{\star}_{\circ}})$ have the same abnormal part as our M_5.

Hence, where $\mathbf{AL}^{\circ}_{\text{min}_{\prec\mathsf{co}}}$ is characterized by the lower limit logic $\mathbf{CL}^{\star}_{\circ}$, the set of abnormalities Ω_{*}, and by the semantic selection based on $\text{min}_{\prec\mathsf{co}}$, we have $\Gamma \Vdash_{\mathbf{AL}^{\circ}_{\text{min}_{\prec\mathsf{co}}}} \neg p, \neg q, r$ (where $\Vdash_{\mathbf{AL}^{\circ}_{\text{min}_{\prec\mathsf{co}}}}$ is defined analogous to ($\star$)).[8]

5.2.6 *The Problem of Non-Smoothness*

Logics such as $\mathbf{AL}^{\circ}_{\text{min}_{\prec\mathsf{co}}}$ are troublesome as soon as the order $\langle \mathsf{Ab}^{\Gamma}_{\mathbf{CL}^{\star}_{\circ}}, \prec_{\mathsf{co}} \rangle$ is not smooth.[9,10] Take for instance the premise set (where p_i, q ($i \in \mathbb{N}$) are distinct propositional letters):

$$\Gamma_a = \{\neg p_i \vee \neg p_j \mid i < j\} \cup \{\circ p_i \mid i \in \mathbb{N}\} \cup \{\circ_1 q, \circ_1 r\} \cup \{\neg q \vee \neg r\}$$

There are infinitely descending sequences of "better and better and never best" models as illustrated in Fig. 5.1 where $\mho = \{\circ p_i \wedge \neg p_i \mid i \in \mathbb{N}\}$, $!^i A =_{\text{df}} \circ_i A \wedge \neg A$ and

$$\mathsf{Ab}(M^q_i) = \mho \setminus \{!^i p_i\} \cup \{!^1 q\}$$
$$\mathsf{Ab}(M^r_i) = \mho \setminus \{!^i p_i\} \cup \{!^1 r\}$$
$$\mathsf{Ab}(M^{q,r}_i) = \mho \setminus \{!^i p_i\} \cup \{!^1 q, !^1 r\}$$

Note that for instance,

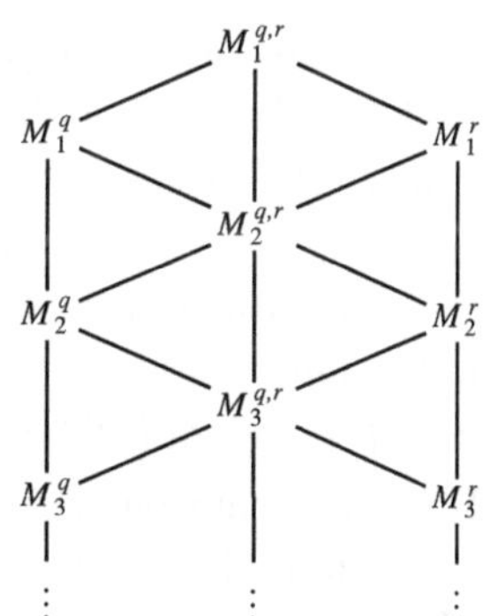

Fig. 5.1 An excerpt of the $\prec_{\mathsf{co}}$-order on the $\mathbf{CL}^{\star}_{\circ}$-models of Γ_a providing an illustration of non-smoothness

---

[8] Of course, we could define a logic on the basis of $\mathbf{CL}_{\circ}$ that realizes the same idea. Instead of e.g., $\circ_3 A$ we could use $\circ\circ\circ A$ in order to express that A is stated at time point 3.

[9] $\langle X, \prec\rangle$ is *smooth* iff for all $x \in X$ there is a $y \in \text{min}_{\prec}(X)$ such that $y \preceq x$.

[10] Non-smooth configurations similar to the following example have been discussed in the literature. See for instance Batens' discussion of Priest's $\mathbf{LP^m}$ in [18], or an example in the context of Circumscription discussed by Bossu and Siegel in [19].

$$\dots \prec_{\mathsf{CO}} M_n^q \prec_{\mathsf{CO}} \dots \prec_{\mathsf{CO}} M_2^q \prec_{\mathsf{CO}} M_1^q$$

Note further that $\min_{\prec_{\mathsf{CO}}}(\mathsf{Ab}^{\Gamma_a}_{\mathbf{CL}^\star_\circ}) = \emptyset$. This immediately implies that for any formula A, $\Gamma_a \Vdash_{\mathbf{AL}^\circ_{\min \prec_{\mathsf{CO}}}} A$. This is unfortunate since we do not want that our AL trivializes premise sets that are non-trivial in its lower limit logic. This problem can be avoided by slightly adjusting the selection procedure: instead of using as a threshold for our selection the set $\min_{\prec_{\mathsf{CO}}}(\mathsf{Ab}^{\Gamma}_{\mathbf{CL}^\star_\circ})$ we select[11]

$$\Psi_{\prec_{\mathsf{CO}}}(\mathsf{Ab}^{\Gamma}_{\mathbf{CL}^\star_\circ}) =_{\mathrm{df}} \Big\{\mathrm{Ab}(M) \in \min^{\cup}_{\subset}(\mathsf{Ab}^{\Gamma}_{\mathbf{CL}^\star_\circ}) \mid \\ \text{for all } \mathrm{Ab}(M') \in \min_{\prec_{\mathsf{CO}}}(\mathsf{Ab}^{\Gamma}_{\mathbf{CL}^\star_\circ}), \mathrm{Ab}(M') \not\prec_{\mathsf{CO}} \mathrm{Ab}(M)\Big\}$$

It follows immediately by the definition that $\min_{\prec_{\mathsf{CO}}}(\mathsf{Ab}^{\Gamma}_{\mathbf{CL}^\star_\circ}) \subseteq \Psi_{\prec_{\mathsf{CO}}}(\mathsf{Ab}^{\Gamma}_{\mathbf{CL}^\star_\circ})$. Hence all minimal models (if there are any) remain selected. The additional selected models are the ones in infinitely descending chains (for which there are no minimal models below).

However, there is still a problem. In our previous example each M_i^q and each M_i^r (where $i \in \mathbb{N}$) is selected according to this new selection. However, each $M_i^{q,r}$ is also selected, since there is no M' for which $\mathrm{Ab}(M') \in \min_{\prec_{\mathsf{CO}}}(\mathsf{Ab}^{\Gamma_a}_{\mathbf{CL}^\star_\circ})$ such that $\mathrm{Ab}(M') \prec_{\mathsf{CO}} \mathrm{Ab}(M_i^{q,r})$. This seems clearly counter-intuitive since for each level i, $\mathrm{Ab}(M_i^q), \mathrm{Ab}(M_i^r) \subset \mathrm{Ab}(M_i^{q,r})$ and

$$\min_{\prec_{\mathsf{CO}}}(\{\mathrm{Ab}(M_i^q), \mathrm{Ab}(M_i^r), \mathrm{Ab}(M_i^{q,r})\}) = \{\mathrm{Ab}(M_i^q), \mathrm{Ab}(M_i^r)\}$$

Note first that although according to $\prec_{\mathsf{CO}}$ the model $M_i^{q,r}$ is worse than all the models M_j^q and M_j^r (where $j \geq i$), $\prec_{\mathsf{CO}}$ doesn't offer a demarcation principle or rationale by means of which we select some M_j^q or M_j^r but not $M_i^{q,r}$. According to $\prec_{\mathsf{CO}}$, $M_i^{q,r}$ –just like M_j^q– is just another model in an infinitely descending chain for which there is no minimal model M' that is equal or better. There is just as much reason for (de-)selecting $M_i^{q,r}$ as for M_j^q. "Wait", one may say, "but $\mathrm{Ab}(M_j^q) \prec_{\mathsf{CO}} \mathrm{Ab}(M_i^{q,r})$. Isn't that a reason to select M_j^q and to deselect $M_i^{q,r}$?" However, analogously $\mathrm{Ab}(M_{j+1}^q) \prec_{\mathsf{CO}} \mathrm{Ab}(M_j^q)$ and $\mathrm{Ab}(M_{j+2}^q) \prec_{\mathsf{CO}} \mathrm{Ab}(M_{j+1}^q)$ and so forth. Hence, by applying this line of reasoning symmetrically, none of our M_j^q's would be selected. Altogether, we have an all-or-nothing choice: either we select all models in the infinitely descending chains or none. Every other choice would be asymmetric and hence ad hoc with respect to $\prec_{\mathsf{CO}}$.

[11] Batens mentioned this idea in the context of inconsistency-tolerant logics in [18]. Here it is applied in a more generic setting and we systematically investigate its meta-theory. For a motivation of the restriction of the selection to $\min^{\cup}_{\subset}$ see Sect. 5.2.4.

So, given that $\prec_{\mathsf{CO}}$ isn't doing any work in demarcating $M_i^{q,r}$ from the M_j^q's and the M_j^r's, is our intuition that we should rather de-select $M_i^{q,r}$ based on a confusion? Rather, we suggest, it is based on a second-order principle that we use on top of comparing models by means of $\prec_{\mathsf{CO}}$. The second order principle offers additional means of qualitatively demarcating some models from others. It helps us to express that some models that defer with respect to $\prec_{\mathsf{CO}}$ defer in a more significant sense than others. E.g., M_i^q and $M_i^{q,r}$ defer more significantly than M_{i+1}^q and M_i^q, although according to $\prec_{\mathsf{CO}}$ the situation is symmetric $\left(\mathrm{Ab}(M_{i+1}^q) \prec_{\mathsf{CO}} \mathrm{Ab}(M_i^q)\right.$ and $\left.\mathrm{Ab}(M_i^q) \prec_{\mathsf{CO}} \mathrm{Ab}(M_i^{q,r})\right)$. In this case the second order principle says: M_i^q and $M_i^{q,r}$ defer significantly because $\mathrm{Ab}(M_i^q) \subset \mathrm{Ab}(M_i^{q,r})$, while $\mathrm{Ab}(M_{i+1}^q) \not\subset \mathrm{Ab}(M_i^q)$ and hence these two models do not defer significantly.

Altogether, what we are interested in is another partial order $\prec'$ which emphasizes certain distinctions made within $\prec_{\mathsf{CO}}$ while neglecting others as less important. $\prec'$ should not introduce new distinctions that were not made by means of $\prec_{\mathsf{CO}}$ already. Hence, $\prec' \subset \prec_{\mathsf{CO}}$. Our discussion above motivates to use $\subset$ in order to implement our second order principle. The selection can then proceed by means of[12]:

$$\Psi_{[\prec_{\mathsf{CO}},\subset]}(\mathsf{Ab}^{\Gamma}_{\mathbf{CL}^{\star}_{\circ}}) =_{\mathrm{df}} \Psi_{\subset}(\Psi_{\prec_{\mathsf{CO}}}(\mathsf{Ab}^{\Gamma}_{\mathbf{CL}^{\star}_{\circ}}))$$

According to this selection only the models M_i^q and M_i^r are selected, while any model that validates both $\neg q$ and $\neg r$ (such as $M_i^{q,r}$) is de-selected. Hence, where $\mathbf{AL}^{\circ}_{\Psi[\prec_{\mathsf{CO}},\subset]}$ is the AL based on the semantic selection offered by $\Psi_{[\prec_{\mathsf{CO}},\subset]}$, we get $\Gamma_a \Vdash_{\mathbf{AL}^{\circ}_{\Psi[\prec_{\mathsf{CO}},\subset]}} q \vee r$.

5.2.7 More Refined Quantitative Examples

Let us come back to our application in the example of Sect. 5.2.3. There, models were selected in which statements that are offered the most frequent by our experts are validated. Now suppose we are not only interested in the models that validate the most frequently stated statements, but also some others which are "good enough". One way to do so would be by introducing a threshold value τ: instead of selecting models whose abnormal parts are in $\min_{\prec_c}(\mathsf{Ab}^{\Gamma}_{\mathbf{CL}^{\star}_{\circ}})$ we select models whose abnormal parts are in

$$\Lambda_c^1(\mathsf{Ab}^{\Gamma}_{\mathbf{CL}^{\star}_{\circ}}) =_{\mathrm{df}} \left\{\mathrm{Ab}(M) \in \min_{\subset}^{\cup}(\mathsf{Ab}^{\Gamma}_{\mathbf{CL}^{\star}_{\circ}}) \middle| |\mathrm{Ab}(M)| - \tau \leq |\varphi|\right\}$$

[12] This is not the same as simply defining another partial order $\prec'_{\mathsf{CO}}$ by "$\mathrm{Ab}(M) \prec'_{\mathsf{CO}} \mathrm{Ab}(M')$ iff $\mathrm{Ab}(M) \prec_{\mathsf{CO}} \mathrm{Ab}(M')$ or $\mathrm{Ab}(M) \subset \mathrm{Ab}(M')$" and then to use $\Psi_{\prec'_{\mathsf{CO}}}$. Note that since $\subset \subseteq \prec_{\mathsf{CO}}$, also $\prec'_{\mathsf{CO}} = \prec_{\mathsf{CO}}$.

where φ is an arbitrary element in $\min_{\prec_c}(\mathrm{Ab}^{\Gamma}_{\mathbf{CL}^{\star}_{\circ}})$.[13]

For instance, suppose $\tau = 3$. For our example based on Γ_c, both M_2 and M_3 would now be selected, while M_4 is still not selected. The value τ hence introduces some error tolerance: in this case we allow the majority to be mistaken about p (model M_3) or to be mistaken about q (model M_2) but not to be mistaken about both, p and q (model M_4 is not selected).

Another option is to use the following selection[14]:

$$\Lambda^2_c(\mathrm{Ab}^{\Gamma}_{\mathbf{CL}^{\star}_{\circ}}) =_{\mathrm{df}}$$
$$\left\{ \mathrm{Ab}(M) \in \min^{\cup}_{\subset}(\mathrm{Ab}^{\Gamma}_{\mathbf{CL}^{\star}_{\circ}}) \;\middle|\; |\mathrm{Ab}(M)| \leq \frac{\sum_{\mathrm{Ab}(M') \in \min_{\subset}(\mathrm{Ab}^{\Gamma}_{\mathbf{CL}^{\star}_{\circ}})} |\mathrm{Ab}(M')|}{|\min_{\subset}(\mathrm{Ab}^{\Gamma}_{\mathbf{CL}^{\star}_{\circ}})|} \right\}$$

Take for instance

$$\Gamma^2_c = \{\circ_i(p \wedge q), \circ_j(p \wedge \neg q), \circ_k p, \circ_{11}\neg p \mid i \in \{1,2,3\}, j \in \{4,5,6,7\}, k \in \{8,9,10\}\}$$

We have three types of models M with $\mathrm{Ab}(M) \in \min_{\subset}\left(\mathrm{Ab}^{\Gamma^2_c}_{\mathbf{CL}^{\star}_{\circ}}\right)$:

Model	Ab(M)	|Ab(M)|	$M \models$
M_1	$\circ_{11}\neg p \wedge \neg\neg p, \circ_i(p \wedge q) \wedge \neg(p \wedge q)$ $(i \in \{1,2,3\})$	4	$p, \neg q$
M_2	$\circ_{11}\neg p \wedge \neg\neg p, \circ_j(p \wedge \neg q) \wedge \neg(p \wedge \neg q)$ $(j \in \{4,5,6,7\})$	5	p, q
M_3	$\circ_i(p \wedge q) \wedge \neg(p \wedge q)$ $(i \in \{1,2,3\})$, $\circ_j(p \wedge \neg q) \wedge \neg(p \wedge \neg q)$, $(j \in \{4,5,6,7\})$, $\circ_k p \wedge \neg p$ $(k \in \{8,9,10\})$	10	$\neg p$

A model M is selected if its abnormal part $\mathrm{Ab}(M) \in \Lambda^2_c\left(\mathrm{Ab}^{\Gamma^2_c}_{\mathbf{CL}^{\star}_{\circ}}\right)$, i.e., if $|\mathrm{Ab}(M)| \leq (4+5+10)/3$. Hence, M_1 and M_2 are selected, while M_3 is not selected.

This selection is very contextual: an average value is calculated on the basis of how many statements are conflicted in the maximally consistent sets of statements. The models that conflict less than this average value are accepted. Suppose the distribution of conflicts in the example above would be different: instead of 4–5–10 ($|\mathrm{Ab}(M_1)|$, $|\mathrm{Ab}(M_2)$, $|\mathrm{Ab}(M_3)|$) we could have for instance 2–7–9. Then we would get the median value 6. In this case only M_1 would be selected.

---

[13] We prove that $\min_{\prec_c}(\mathrm{Ab}^{\Gamma}_{\mathbf{LLL}})$ (where **LLL** qualifies as a lower limit logic such as our $\mathbf{CL}^{\star}_{\circ}$) is non-empty for **LLL**-non-trivial Γ in Sect. 5.8.2 (see Fact 5.8.2 in combination with Lemma 5.3.2).

[14] Where $n \in \mathbb{N}^{\infty}$, we define $\infty/n =_{\mathrm{df}} \infty$. We show later that $\min_{\subset}(\mathrm{Ab}^{\Gamma}_{\mathbf{LLL}})$ is non-empty for **LLL**-non-trivial Γ (see Lemma 5.3.2 and Theorem 5.5.4): hence we do not have to worry about division by zero.

5.2.8 Summary

We have seen various examples for ALs that use orders different from set-inclusion in order to select models. Let us abstract away from the concrete examples and identify the formats of ALs we have discussed above. We characterize ALs in the following by triples: $\langle \mathbf{LLL}, \Omega, \Lambda \rangle$ where **LLL** is the lower limit logic, Ω is the set of abnormalities, and Λ determines the threshold for the selection of the models.[15]

We started with ALs in the standard format characterized by:

$$\langle \mathbf{LLL}, \Omega, \min_{\subset} \rangle \tag{5.1}$$

Selected are the **LLL**-models M of a premise set Γ whose abnormal part is in

$$\min_{\subset}\left(\mathsf{Ab}^{\Gamma}_{\mathbf{LLL}}\right) = \{\mathrm{Ab}(M) \mid M \in \mathcal{M}_{\mathbf{LLL}}(\Gamma), \quad \text{for all } M' \in \mathcal{M}_{\mathbf{LLL}}(\Gamma), \mathrm{Ab}(M') \not\subset \mathrm{Ab}(M)\}$$

Then we moved on to ALs with other partial orders $\prec$, but presupposing that

(†) for all premise sets Γ, $\langle \mathsf{Ab}^{\Gamma}_{\mathbf{LLL}}, \prec \rangle$ is smooth.

These logics can be characterized by:

$$\langle \mathbf{LLL}, \Omega, \min_{\prec} \rangle \tag{5.2}$$

The selected models of a premise set Γ are the ones whose abnormal part is in:

$$\min_{\prec}\left(\mathsf{Ab}^{\Gamma}_{\mathbf{LLL}}\right) = \{\mathrm{Ab}(M) \mid M \in \mathcal{M}_{\mathbf{LLL}}(\Gamma), \quad \text{for all } M' \in \mathcal{M}_{\mathbf{LLL}}(\Gamma), \mathrm{Ab}(M') \not\prec \mathrm{Ab}(M)\}$$

Dropping the smoothness requirement (†) for $\prec$ we first proposed the following format:

$$\langle \mathbf{LLL}, \Omega, \Psi_{\prec} \rangle \tag{5.3}$$

The selected models of a premise set Γ are the ones whose abnormal part is in:

$$\Psi_{\prec}(\mathsf{Ab}^{\Gamma}_{\mathbf{LLL}}) = \left\{\mathrm{Ab}(M) \in \min^{\cup}_{\subset}\left(\mathsf{Ab}^{\Gamma}_{\mathbf{LLL}}\right) \mid \text{there is no } \mathrm{Ab}(M') \in \min_{\prec}\left(\mathsf{Ab}^{\Gamma}_{\mathbf{LLL}}\right) \text{ such that } \mathrm{Ab}(M') \prec \mathrm{Ab}(M)\right\}$$

Since

Fact 5.2.1. *If* $\langle X, \prec \rangle$ *is smooth, then* $\Psi_{\prec}(X) = \min_{\prec}(X)$.

---

[15] We still focus on the semantic aspect of ALs in order not to open more doors than necessary at this point of the discussion. However, this should not distract from the fact that all these semantic features have a syntactic counter-part. We will investigate also the syntax of ALs beginning with the next section.

and given the presupposition (†) we used for the format (5.2), the semantic selection characterized by $\langle \mathbf{LLL}, \Omega, \min_{\prec} \rangle$ is identical to the one characterized by $\langle \mathbf{LLL}, \Omega, \Psi_{\prec} \rangle$.

As the example in Sect. 5.2.6 motivates, sometimes we are interested in basing our selection on more refined selections. On way to do so was by means of an order $\prec'$ where $\prec' \subseteq \prec$:

$$\langle \mathbf{LLL}, \Omega, \Psi_{[\prec, \prec']} \rangle \tag{5.4}$$

Here, the selection is refined by making use of the order $\prec'$ so that the selected models for a premise set Γ are the ones whose abnormal part is in

$$\Psi_{[\prec, \prec']}(\mathsf{Ab}^{\Gamma}_{\mathbf{LLL}}) = \Psi_{\prec'}(\Psi_{\prec}(\mathsf{Ab}^{\Gamma}_{\mathbf{LLL}}))$$

Of course, one may generalize this approach to a sequence of orders.

Definition 5.2.4. We call a $\prec'$ an *abstraction order of* $\prec$ iff $\prec' \subseteq \prec$. A sequence of partial orders $\langle \prec_1, \prec_2, \ldots, \prec_n \rangle$ is an *abstraction sequence* iff for each $i < n$, $\prec_{i+1}$ is an abstraction order of $\prec_i$.

Where $\langle \prec_1, \prec_2, \ldots, \prec_n \rangle$ is an abstraction sequence we define

$$\Psi_{[\prec_1, \ldots, \prec_n]}(\mathsf{Ab}^{\Gamma}_{\mathbf{LLL}}) =_{\text{df}} \Psi_{\prec_n}(\Psi_{\prec_{n-1}}(\ldots \Psi_{\prec_1}(\mathsf{Ab}^{\Gamma}_{\mathbf{LLL}})))$$

The examples presented in Sects. 5.2.4 and 5.2.7 explicated a different rationale according to which we are not anymore only interested in selecting minimal models (resp., in the case of $\Psi_{\prec}$, models for which there are no "better" minimal models). In this context we have argued that $\min^{\cup}_{\subset}$ is a reasonable upper bound for any threshold function Λ.

The most generic representation can be obtained when we abstract from the concrete way the threshold is obtained by an order $\prec$ and just focus on the threshold function Λ itself.

$$\langle \mathbf{LLL}, \Omega, \Lambda \rangle \tag{5.5}$$

Obviously, all the meta-theory that is available for the most generic format (5.5) straight-forwardly applies to the other formats as well.

In the next section we characterize the proof theory and semantics of this format which is followed by a study of its meta-theory.

5.3 The Generic Adaptive Logic $\mathbf{AL}_{\Lambda}$

We will now define adaptive logics $\mathbf{AL}_{\Lambda}$ characterized by our most generic presentation $\langle \mathbf{LLL}, \Omega, \Lambda \rangle$ where

1. the *Lower Limit Logic* **LLL** is a compact Tarski logic that has a characteristic semantics (see Sect. 2.2),[16]
2. the *set of abnormalities* Ω is characterized by a (or many) logical form(s) in a **LLL**-contingent way, i.e., there is no $\varphi \subseteq \Omega$ for which $\varphi \cup (\Omega \setminus \varphi)^{\check{\neg}}$ is **LLL**-trivial
3. $\Lambda : \wp(\wp(\Omega)) \rightarrow \wp(\wp(\Omega))$ is an **LLL**-*threshold function*, where

Definition 5.3.1. An **LLL**-*threshold function*[17] $\Lambda : \wp(\wp(\Omega)) \rightarrow \wp(\wp(\Omega))$ satisfies the following requirements for all X in the sub-domain $\Upsilon =_{\text{df}} \{\text{Ab}^{\Gamma}_{\mathbf{LLL}} \mid \Gamma \subseteq \mathcal{W}^+\}$

T1 Λ is inclusive in $\min^{\cup}_{\subset}$ (i.e., $\Lambda(X) \subseteq \min^{\cup}_{\subset}(X)$)
T2 $\Lambda(X)$ is a $\subset$-lower set[18] of X
T3 $\Lambda(X) = \emptyset$ implies $X = \emptyset$

Of course, $\Lambda(X)$ shall select sets in X and not introduce new sets of abnormalities. Moreover, as motivated in Sect. 5.2.4, $\min^{\cup}_{\subset}$ provides an upper bound for our selection Λ. Hence, **T1**. **T2** expresses that if some set φ of abnormalities is selected and hence deemed "sufficiently normal" in X according to Λ, then any $\psi \in X$ for which $\psi \subset \varphi$ shall also be selected: after all ψ contains even less abnormalities than φ. **T3** expresses that Λ makes a choice from X in case X is non-empty. The following properties are derivable:

Fact 5.3.1. *Where $X \in \Upsilon$: (i) $\Lambda(X) = \emptyset$ iff $X = \emptyset$, (ii) $\emptyset \in X$ iff $\{\emptyset\} = \Lambda(X)$.*

(i) follows by **T1** and **T3**. It expresses that no choice is made by means of Λ if and only if there is no choice available. (ii) follows by **T1** and **T3**. It expresses that the abnormality-free empty set is always selected if it is in X: there is indeed no reason to deem it not "sufficiently normal".

As is shown below, each of our examples from Sect. 5.2 is a threshold function and hence satisfies **T1**–**T3**, except for $\min_{\prec_{\text{co}}}$ since it doesn't satisfy **T3** as demonstrated in Sect. 5.2.6. We come back to our examples in more detail in Sect. 5.8.

In the following we will first present the semantics, then the dynamic proof theory of $\mathbf{AL}_\Lambda$. It will become evident that we hereby generalize the standard format. Hence, all the meta-theory for $\mathbf{AL}_\Lambda$ immediately applies to the standard format as well (see also the discussion in Sect. 5.8.2).

---

[16] In order to reduce notational clutter, we will in the following not distinguish between **LLL** and $\mathbf{LLL}^+$ and always write **LLL** in order to denote either. (Compare the discussion in Sect. 2.7). The context will always disambiguate this.

[17] We will in the remainder skip the reference to **LLL** whenever the context disambiguates.

[18] X is a $\prec$-*lower set of* Y iff for all $x \in X$ and all $y \in Y$, if $y \prec x$ then $y \in X$.

5.3.1 The Semantics

As in the standard format we use a selection semantics. The threshold function Λ determines which models we select from $\mathcal{M}_{\mathbf{LLL}}(\Gamma)$.

Definition 5.3.2. $\mathcal{M}_{\mathbf{AL}_\Lambda}(\Gamma) =_{\text{df}} \{M \in \mathcal{M}_{\mathbf{LLL}}(\Gamma) \mid \text{Ab}(M) \in \mathbf{\Lambda}(\mathsf{Ab}^{\Gamma}_{\mathbf{LLL}})\}$

Definition 5.3.3. $\Gamma \Vdash_{\mathbf{AL}_\Lambda} A$ iff for all $M \in \mathcal{M}_{\mathbf{AL}_\Lambda}(\Gamma)$, $M \models A$.

Remark 5.3.1. Note that with $\Lambda = \min_{\subset}$ we get exactly the semantics of the minimal abnormality strategy of the standard format, whereas with $\Lambda = \min^{\cup}_{\subset}$ we get the reliability strategy. We will come back to this in Sect. 5.8.2.

5.3.2 The Dynamic Proof Theory

The proof format of $\mathbf{AL}_\Lambda$ is identical to the proof format of the standard format. Each proof line has four elements: a line number, a formula, a justification, and a condition which is a finite set of abnormalities.

We again have the three generic rules: PREM, RU, and RC.

We only need to adjust the marking condition. Recall the marking definition of minimal abnormality (Definition 2.4.7, page 26). Our motivation was along the following lines.

Suppose we have derived A on the condition Δ_n at stage s on line l. That means our (defeasible) assumption for the argument that leads to line l is that each abnormality in Δ_n is false. Moreover, we may have also derived A already earlier in the proof on other conditions: $\Delta_1, \ldots, \Delta_{n-1}$. The fact whether line l is marked concerns two questions (compare our previous discussion in Sect. 2.4.2.2):

Q1 *Is the argument at line l defensible?*

Is the assumption of line l that neither of the abnormalities in Δ_n is true valid in some sufficiently normal interpretation of the disjunctions of abnormalities derived from the premises?

Q2 *Is the claim A of line l justified?*

Is it the case that for each sufficiently normal interpretation I of the Dab-formulas derived from the premises there is a Δ_i such that all the abnormalities in Δ_i are false in I?

Let us make this formally more precise. Recall that the minimal Dab-formulas derived from Γ at stage s were denoted by $\Sigma_s(\Gamma)$. Accordingly, interpretations of these Dab-formulas are represented by means of choice sets over $\Sigma_s(\Gamma)$. The set of all choice sets over $\Sigma_s(\Gamma)$ is denoted by $\Xi_s(\Gamma)$. More precisely, $\Xi_s(\Gamma)$ is the set of all set of abnormalities φ such that $\varphi \cap \Delta \neq \emptyset$ for all $\Delta \in \Sigma_s(\Gamma)$.

We have previously seen that the two strategies of the standard format provide different readings of what it means that a choice set is "sufficiently normal": e.g., for the minimal abnormality strategy we chose the $\subset$-minimal choice sets (denoted by $\Phi_s(\Gamma) = \min_{\subset}(\Xi_s(\Gamma))$). Now we use our threshold function Λ instead of $\min_{\subset}$. Let us write Λ_s^Γ for the choice sets that are selected at stage s by means of Λ.[19] Hence, formally the questions **Q1** and **Q2** have the following form:

Q1 Is there a $\varphi \in \Lambda_s^\Gamma$ such that $\Delta_n \cap \varphi = \emptyset$?
Q2 For each $\varphi \in \Lambda_s^\Gamma$: is there a Δ_i such that $\varphi \cap \Delta_i = \emptyset$?

This gives rise to the following marking definition which is analogous to the marking definition for minimal abnormality (Definition 2.4.7):

Definition 5.3.4 (Λ-**marking**). A line l with formula A and condition Δ is *marked at stage* s, iff $\Delta \neq \emptyset$ and

(i) there is no $\varphi \in \Lambda_s^\Gamma$ such that $\varphi \cap \Delta = \emptyset$, or
(ii) for a $\varphi \in \Lambda_s^\Gamma$ there is no line l' at stage s with formula A and condition Θ such that $\Theta \cap \varphi = \emptyset$.

Obviously, condition (i) is met in case the answer to **Q1** is negative, while condition (ii) is met in case the answer to **Q2** is negative.

There is a slight complication concerning the definition of Λ_s^Γ in the marking definition which is best demonstrated by means of an example.

Example 5.3.1. Let

$$\Lambda(X) = \left\{ \varphi \;\middle|\; |\varphi| \leq \frac{\sum_{\psi \in \min_{\subset}^{\cup}(X)} |\psi|}{\left|\min_{\subset}^{\cup}(X)\right|} \right\}$$

similar to Λ_c^2 from Sect. 5.2.7 and, where $!A =_{\mathrm{df}} \circ A \wedge A$, let

$$\Gamma = \{!p \vee !q, !p \vee !r, \neg !p \vee \neg !q, \neg !p \vee \neg !r, s \vee !q\}$$

We first look at the situation semantically. We have

$$\min_{\subset}\left(\mathrm{Ab}_{\mathbf{L}_\circ}^\Gamma\right) = \{\{!p\}, \{!q, !r\}\}$$

Due to the premises $\neg !p \vee \neg !q$ and $\neg !p \vee \neg !r$ we have:

$$\{!p, !q\}, \{!p, !r\}, \{!p, !q, !r\} \notin \mathrm{Ab}_{\mathbf{L}_\circ}^\Gamma$$

As a consequence:

[19] The reader may for the moment think of Λ_s^Γ as denoting the set $\Lambda(\Xi_s(\Gamma))$. However, we will have to make a slight adjustment below (see Definition 5.3.5).

$$\min^{\cup}_{\subset}\left(\mathrm{Ab}^{\Gamma}_{\mathbf{L}_\circ}\right) = \min_{\subset}\left(\mathrm{Ab}^{\Gamma}_{\mathbf{L}_\circ}\right)$$

Hence, since

$$\frac{\sum_{\psi\in\min^{\cup}_{\subset}(\mathrm{Ab}^{\Gamma}_{\mathbf{L}_\circ})}|\psi|}{\left|\min^{\cup}_{\subset}\left(\mathrm{Ab}^{\Gamma}_{\mathbf{L}_\circ}\right)\right|} = \frac{3}{2}$$

we have

$$\Lambda\left(\mathrm{Ab}^{\Gamma}_{\mathbf{L}_\circ}\right) = \{\{!p\}\} \tag{5.6}$$

Hence, since each selected model M has the abnormal part $\{!p\}$, also $M \models \neg !q$ and hence $M \models s$ due to the premise $s \vee !q$. Thus, s is a semantic consequence of the AL based on Λ.

Let us now look at a proof from Γ where, by way of trial, $\Lambda^{\Gamma}_{s} =_{\mathrm{df}} \Lambda(\Xi_s(\Gamma))$:

1	$!p \vee !q$	PREM	$\emptyset$
2	$!p \vee !r$	PREM	$\emptyset$
3	$\neg !p \vee \neg !q$	PREM	$\emptyset$
4	$\neg !p \vee \neg !r$	PREM	$\emptyset$
5	$\neg !p \vee \neg !q \vee \neg !r$	4; RU	$\emptyset$
6	$s \vee !q$	PREM	$\emptyset$
7	s	6; RC	$\{!q\}$

Members of $\Xi_7(\Gamma)$ are for instance $\{!p\}, \{!q, !r\}, \{!p, !q\}, \{!p, !r\}, \{!p, !q, !r\}$.[20] Moreover,

$$\min^{\cup}_{\subset}(\Xi_7(\Gamma)) = \left\{\{!p\}, \{!q, !r\}, \{!p, !q\}, \{!p, !r\}, \{!p, !q, !r\}\right\}$$

Note that

$$\frac{\sum_{\psi\in\min^{\cup}_{\subset}(\Xi_7(\Gamma))}|\psi|}{\left|\min^{\cup}_{\subset}(\Xi_7(\Gamma))\right|} = \frac{10}{5} = 2$$

Hence, we have:

$$\Lambda(\Xi_7(\Gamma)) = \left\{\{!p\}, \{!p, !q\}, \{!p, !r\}, \{!q, !r\}\right\} \tag{5.7}$$

According to this, line 7 is marked. Since we can only derive s on conditions containing $\{!q\}$ and there is no way to unmark line 7,[21] s is not derivable anymore on an unmarked line. But, recall that s is a semantic consequence. Something went wrong.

[20] For the sake of simplicity, we disregard in this discussion "checked connectives" (see Sect. 2.7).

[21] Note that we cannot introduce new minimal Dab-formulas in the proof.

Note that in view of (5.6), models with abnormal part $\{!q, !r\}$ are not selected. However, $\{!q, !r\}$ is below our Λ-threshold in the syntactic selection (5.7). The reason for the asymmetry is that for the calculation of the threshold on the syntactic side some choice sets are considered that do not correspond to abnormal parts of models since they are not satisfiable. Examples are $\{!p, !q\}$, $\{!p, !r\}$ and $\{!p, !q, !r\}$ which are due to the premises $\neg!p \vee \neg!q$ and $\neg!p \vee \neg!r$ not satisfiable and hence

$$\big\{\{!p, !q\}, \{!p, !r\}, \{!q, !r\}\big\} = \Lambda\big(\Xi_7(\Gamma)\big) \setminus \Lambda\big(\mathrm{Ab}^{\Gamma}_{\mathbf{L}_\circ}\big) \tag{5.8}$$

The good news is that this asymmetry can easily be avoided: instead of fixing the threshold on the basis of the choice sets of $\Sigma_s(\Gamma)$, i.e., on the basis of $\Xi_s(\Gamma)$, we calculate it on the basis of only the satisfiable choice sets of $\Sigma_s(\Gamma)$. This can be achieved by also paying attention to disjunctions of negated abnormalities that have been derived on the empty condition.

Let therefore $\mathsf{Dabn}(\Delta, \Theta) =_{\mathrm{df}} \check{\bigvee}\big(\Delta \cup \Theta^{\check{\neg}}\big)$ where $\Delta \cap \Theta = \emptyset$ and $\Delta \cup \Theta \subseteq \Omega$. We say that $\mathsf{Dabn}(\Delta, \Theta)$ is a *minimal* Dabn*-formula derived at stage* s iff $\mathsf{Dabn}(\Delta, \Theta)$ is derived at stage s on the condition $\emptyset$, and for all $\mathsf{Dabn}(\Delta', \Theta')$ derived at stage s on the condition $\emptyset$: if $\Delta' \subseteq \Delta$ and $\Theta' \subseteq \Theta$, then $\Delta' = \Delta$ and $\Theta' = \Theta$.

In analogy to $\Sigma_s(\Gamma)$ we define $\Sigma^{\mathrm{sat}}_s(\Gamma)$ to be the set of all $\Delta \cup \Theta^{\check{\neg}}$ such that $\mathsf{Dabn}(\Delta, \Theta)$ is a minimal Dabn-formula at stage s.

Instead of using $\Xi_s(\Gamma)$ that contains the choice sets of $\Sigma_s(\Gamma)$ we now use $\Xi^{\mathrm{sat}}_s(\Gamma)$ which contains all sets of abnormalities φ such that $\varphi \cup (\Omega \setminus \varphi)^{\check{\neg}}$ is a choice set of $\Sigma^{\mathrm{sat}}_s(\Gamma)$.

An alternative characterization of $\Xi^{\mathrm{sat}}_s(\Gamma)$ is given by the following Lemma. It mirrors nicely our informative reading that $\Xi^{\mathrm{sat}}_s(\Gamma)$ contains all choice sets in $\Xi_s(\Gamma)$ that are satisfiable.

Lemma 5.3.1. *Let* $\Gamma \subseteq \mathcal{W}^+$*:* $\Xi^{\mathrm{sat}}_s(\Gamma) = \Xi_s(\Gamma) \setminus \Xi^{\perp}_s(\Gamma)$*, where*

$$\Xi^{\perp}_s(\Gamma) =_{df} \big\{\varphi \in \Xi_s(\Gamma) \mid \textit{there are } \psi \subseteq \varphi \text{ and } \Delta \subseteq \Omega \setminus \varphi \textit{ such that } \mathsf{Dabn}(\Delta, \psi) \textit{ is a minimal } \mathsf{Dabn}\textit{-formula at stage } s\big\}$$

Proof. "$\supseteq$": Let $\varphi \in \Xi_s(\Gamma) \setminus \Xi^{\perp}_s(\Gamma)$. Assume there is a $\Delta \cup \Theta^{\check{\neg}} \in \Sigma^{\mathrm{sat}}_s(\Gamma)$ such that $\varphi \cap \Delta = \emptyset = (\Omega \setminus \varphi) \cap \Theta$. Hence, $\Theta \subseteq \varphi$ and $\Delta \subseteq \Omega \setminus \varphi$. But then, $\varphi \in \Xi^{\perp}_s(\Gamma)$,—a contradiction.

"$\subseteq$": Let (*) $\varphi \cup (\Omega \setminus \varphi)^{\check{\neg}}$ be a choice set of $\Sigma^{\mathrm{sat}}_s(\Gamma)$. Obviously, $\varphi \in \Xi_s(\Gamma)$. Assume $\varphi \in \Xi^{\perp}_s(\Gamma)$. Hence, there are $\psi \subseteq \varphi$ and $\psi' \subseteq \Omega \setminus \varphi$ such that $\mathsf{Dabn}(\psi', \psi)$ is a minimal Dabn-formula at stage s. Hence, $\psi' \cup \psi^{\check{\neg}} \in \Sigma^{\mathrm{sat}}_s(\Gamma)$. But this is a contradiction to (*). □

In our example we have $\Xi^{\mathrm{sat}}_7(\Gamma) \supseteq \big\{\{!p\}, \{!q, !r\}\big\}$,

$$\min^{\cup}_{\subset}\big(\Xi^{\mathrm{sat}}_7(\Gamma)\big) = \big\{\{!p\}, \{!q, !r\}\big\} \text{ and hence } \Lambda\big(\Xi^{\mathrm{sat}}_7(\Gamma)\big) = \Lambda\big(\mathrm{Ab}^{\Gamma}_{\mathbf{L}^*_\circ}\big)$$

Hence, if we let $\Lambda_s^\Gamma = \Lambda\big(\Xi_s^{\text{sat}}(\Gamma)\big)$ then line 7 is unmarked (and remains unmarked in any extension of the proof).

Note that for instance $\{!p, !r\} \notin \Xi_7^{\text{sat}}(\Gamma)$ since $\mathsf{Dabn}(\emptyset, \{!p, !r\}) = \neg!p \vee \neg!r$ is a minimal Dabn-formula derived at line 4. Thus, $\{!p, !r\} \in \Xi_7^{\perp}(\Gamma)$.

Let us sum up. What was demonstrated in the example is that some choice sets in $\Xi_s(\Gamma)$ may not be satisfiable. This causes two problems:

P1 *Redundancy*
First, there seem to be no good reasons why such choice sets should be selected by Λ or even be considered in the calculation of the threshold Λ.

P2 *Non-adequacy*
Moreover, in some configurations these non-satisfiable choice sets may alter the selection via Λ such that we get different outcomes in the syntactic and the semantic selection (see (5.8)).

The solution is to define

Definition 5.3.5. $\Lambda_s^\Gamma =_{\text{df}} \Lambda\big(\Xi_s^{\text{sat}}(\Gamma)\big)$

In Sect. 5.6 we will show that under specific conditions we can also use $\Lambda_s^\Gamma = \Lambda(\Xi_s(\Gamma))$ without running into the problem **P2**.

Note that in the limit $\Xi_s^{\text{sat}}(\Gamma)$ exactly corresponds to $\text{Ab}_{\mathbf{LLL}}^\Gamma$ as demonstrated in the following lemma. This result will be crucial to guarantee the soundness and completeness of $\mathbf{AL}_\Lambda$ (see Sect. 5.4). Where $\mathsf{Dabn}(\Delta, \Theta)$ is a Dabn-*consequence of* Γ iff $\Gamma \vdash_{\mathbf{LLL}} \mathsf{Dabn}(\Delta, \Theta)$, it is a *minimal* Dabn-*consequence of* Γ iff there are no Δ', Θ' such that $\Delta' \cup \Theta'^{\check{\neg}} \subset \Delta \cup \Theta^{\check{\neg}}$ and $\mathsf{Dabn}(\Delta', \Theta')$ is a Dabn-consequence of Γ. Let $\Sigma^{\text{sat}}(\Gamma)$ be the set of all $\Delta \cup \Theta^{\check{\neg}}$ such that $\mathsf{Dabn}(\Delta, \Theta)$ is a minimal Dabn-consequence of Γ. Finally, $\Xi^{\text{sat}}(\Gamma)$ is the set of all sets of abnormalities φ such that $\varphi \cup (\Omega \setminus \varphi)^{\check{\neg}}$ is a choice set of $\Sigma^{\text{sat}}(\Gamma)$.

Stage-independent versions of $\Xi_s(\Gamma)$ and $\Xi_s^{\perp}(\Gamma)$ are defined analogously on the basis of the Dab-consequences of Γ, resulting in $\Xi(\Gamma)$ and $\Xi^{\perp}(\Gamma)$.

Lemma 5.3.2. *Where* $\Gamma \subseteq \mathcal{W}^+$*:* $\Xi^{\text{sat}}(\Gamma) = \mathsf{Ab}_{\mathbf{LLL}}^\Gamma$

Proof. Let $\varphi \in \Xi^{\text{sat}}(\Gamma)$. Assume there is no $M \in \mathcal{M}_{\mathbf{LLL}}(\Gamma)$ such that $\text{Ab}(M) = \varphi$. Hence, $\mathcal{M}_{\mathbf{LLL}}\big(\Gamma \cup (\Omega \setminus \varphi)^{\check{\neg}} \cup \varphi\big) = \emptyset$. By the compactness of **LLL**, there are finite $\psi \subseteq \varphi$ and $\Delta \subseteq \Omega \setminus \varphi$ such that $\mathcal{M}_{\mathbf{LLL}}\big(\Gamma \cup \psi \cup \Delta^{\check{\neg}}\big) = \emptyset$. Hence, $\Gamma \Vdash_{\mathbf{LLL}} \mathsf{Dabn}(\Delta, \psi)$. By the completeness of **LLL**, $\Gamma \vdash_{\mathbf{LLL}} \mathsf{Dabn}(\Delta, \psi)$. But then there are $\Delta' \subseteq \Delta$ and $\psi' \subseteq \psi$ such that $\Delta' \cup \psi'^{\check{\neg}} \in \Sigma^{\text{sat}}(\Gamma)$,—a contradiction since $\varphi \cap \Delta' = \emptyset = \psi' \cap (\Omega \setminus \varphi)$ and $\varphi \in \Xi^{\text{sat}}(\Gamma)$.

Let $\varphi \in \mathsf{Ab}_{\mathbf{LLL}}^\Gamma$. Let $M \in \mathcal{M}_{\mathbf{LLL}}(\Gamma)$ such that $\varphi = \text{Ab}(M)$. Let $\Delta \cup \Theta^{\check{\neg}} \in \Sigma^{\text{sat}}(\Gamma)$. Hence, $\Gamma \Vdash_{\mathbf{LLL}} \mathsf{Dabn}(\Delta, \Theta)$. Thus, $M \models \mathsf{Dabn}(\Delta, \Theta)$. Hence, either $\varphi \cap \Delta \neq \emptyset$ or $(\Omega \setminus \varphi) \cap \Theta \neq \emptyset$. Thus, $\varphi \cup (\Omega \setminus \varphi)^{\check{\neg}}$ is a choice set of $\Sigma^{\text{sat}}(\Gamma)$ and hence $\varphi \in \Xi^{\text{sat}}(\Gamma)$. □

In view of Definitions 5.3.4 and 5.3.5 we can again define a stage-dependent and an absolute notion of derivability. The stage-dependent notion is as follows:

Definition 5.3.6. A formula A is *derived at stage* s of a proof iff there is a line l with formula A that is not marked at stage s.

Before we define the absolute notion of derivability, let us look at another example.

Example 5.3.2. We come back to our example from Sect. 5.2.5. We begin by introducing our first three premises:

1	$\circ_1 p$	PREM $\emptyset$
2	$\circ_2 q$	PREM $\emptyset$
3	$\circ_3 r$	PREM $\emptyset$

From this we can conditionally derive the following:

4	p	1; RC $\{\circ_1 p \wedge \neg p\}$
5	q	2; RC $\{\circ_2 q \wedge \neg q\}$
6	r	3; RC $\{\circ_3 r \wedge \neg r\}$

The proof gets more interesting when we introduce our fourth premise $\circ_4 \neg q$ since this leads to a conflict:

7	$\circ_4 \neg q$	PREM $\emptyset$
8	$(\circ_2 q \wedge \neg q) \vee (\circ_4 \neg q \wedge \neg\neg q)$	2, 7; RU $\emptyset$
9	$\neg q$	7; RC $\{\circ_4 \neg q \wedge \neg\neg q\}$

At this point we have

$$\{\circ_2 q \wedge \neg q\}, \{\circ_4 \neg q \wedge \neg\neg q\}, \{\circ_2 q \wedge \neg q, \circ_4 \neg q \wedge \neg\neg q\} \in \Xi_9^{\mathrm{sat}}(\Gamma_{co})$$

Note that

$$\{\circ_2 q \wedge \neg q\} \prec_{\mathsf{co}} \{\circ_4 \neg q \wedge \neg\neg q\} \prec_{\mathsf{co}} \{\circ_2 q \wedge \neg q, \circ_4 \neg q \wedge \neg\neg q\}$$

Indeed, $\Psi_{\prec_{\mathsf{co}}}\big(\Xi_9^{\mathrm{sat}}(\Gamma_{co})\big) = \big\{\{\circ_2 q \wedge \neg q\}\big\}$. Hence, at this point line 5 is marked while 9 is unmarked. This is as expected given the reading offered in Sect. 5.2.5. After all, $\neg q$ was stated at a later point than q and hence it is to be preferred.

Let us continue and introduce our last premise:

10	$\circ_5 \neg p$	PREM $\emptyset$
11	$(\circ_1 p \wedge \neg p) \vee (\circ_5 \neg p \wedge \neg\neg p)$	1, 10; RU $\emptyset$
12	$\neg p$	10; RU $\{\circ_5 \neg p \wedge \neg\neg p\}$

At this point we have four $\subset$-minimal choice sets in $\Xi_{12}^{\mathrm{sat}}(\Gamma_{co})$:

$\varphi_1 = \{\circ_2 q \wedge \neg q, \circ_1 p \wedge \neg p\}$
$\varphi_2 = \{\circ_2 q \wedge \neg q, \circ_5 \neg p \wedge \neg\neg p\}$
$\varphi_3 = \{\circ_4 \neg q \wedge \neg\neg q, \circ_1 p \wedge \neg p\}$
$\varphi_4 = \{\circ_4 \neg q \wedge \neg\neg q, \circ_5 \neg p \wedge \neg\neg p\}$

Note that $\varphi_1 \prec_{co} \varphi_3 \prec_{co} \varphi_2 \prec_{co} \varphi_4$. Indeed, $\Psi_{\prec_{co}}\left(\Xi^{\text{sat}}_{12}(\Gamma_{co})\right) = \{\varphi_1\}$. Hence, lines 4 and 5 are marked while lines 9 and 12 are unmarked.

In order to define the consequence relation of $\mathbf{AL}_{\mathbf{\Lambda}}$ we make use of a static notion of derivability. Final derivability is defined just as for the standard format:

Definition 5.3.7. *A* is *finally derived* at a line *l* at a finite stage *s* of a proof iff

(i) *A* is derived at line *l* at stage *s*,
(ii) for every further extension of the proof in which line *l* is marked there is yet another extension in which line *l* is unmarked.

Definition 5.3.8. $\Gamma \vdash_{\mathbf{AL}_{\mathbf{\Lambda}}} A$ iff *A* is finally derivable from Γ.

For a reading of final derivability in terms of an argumentation game between two players see page 22.

We are now in the position to define a syntactic consequence relation for $\mathbf{AL}_{\mathbf{\Lambda}}$.

Definition 5.3.9. $\mathrm{A} \in Cn_{\mathbf{AL}_{\mathbf{\Lambda}}}(\Gamma)$ iff $\Gamma \vdash_{\mathbf{AL}_{\mathbf{\Lambda}}} \mathrm{A}$

5.4 Representational Results for $\mathbf{AL}_{\mathbf{\Lambda}}$

In this Section we will offer various representational results for $\Vdash_{\mathbf{AL}_{\mathbf{\Lambda}}}$ and $\vdash_{\mathbf{AL}_{\mathbf{\Lambda}}}$. An overview is given in Fig. 5.2.

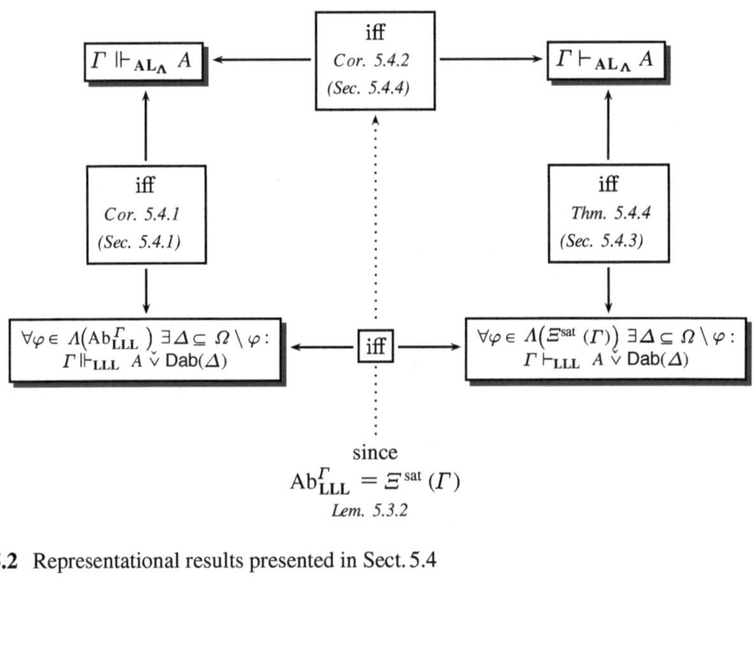

Fig. 5.2 Representational results presented in Sect. 5.4

5.4.1 A Representational Theorem for $\Vdash_{\mathbf{AL}_\Lambda}$

The semantic consequence relation $\Vdash_{\mathbf{AL}_\Lambda}$ can also be represented in terms of the consequence relation of the lower limit logic: $\Vdash_{\mathbf{LLL}}$.

Theorem 5.4.1. *Where $\Gamma \subseteq \mathcal{W}^+$: $\Gamma \Vdash_{\mathbf{AL}_\Lambda} A$ iff for all $\varphi \in \Lambda(\mathsf{Ab}^{\Gamma}_{\mathbf{LLL}})$, $\Gamma \cup (\Omega \setminus \varphi)^{\check{\neg}} \Vdash_{\mathbf{LLL}} A$.*

Proof. $\Gamma \Vdash_{\mathbf{AL}_\Lambda} A$ iff $M \models A$ for all M for which $\mathrm{Ab}(M) \in \Lambda(\mathsf{Ab}^{\Gamma}_{\mathbf{LLL}})$ iff $\Gamma \cup \mathrm{Ab}(M) \cup (\Omega \setminus \mathrm{Ab}(M))^{\check{\neg}} \Vdash_{\mathbf{LLL}} A$ for all $\mathrm{Ab}(M) \in \Lambda(\mathsf{Ab}^{\Gamma}_{\mathbf{LLL}})$ iff [by **T2**] $\Gamma \cup (\Omega \setminus \mathrm{Ab}(M))^{\check{\neg}} \Vdash_{\mathbf{LLL}} A$ for all $\mathrm{Ab}(M) \in \Lambda(\mathsf{Ab}^{\Gamma}_{\mathbf{LLL}})$. □

Hence, by the compactness of **LLL** and the deduction theorem, we get:

Corollary 5.4.1. *Where $\Gamma \subseteq \mathcal{W}^+$: $\Gamma \Vdash_{\mathbf{AL}_\Lambda} A$ iff for all $\varphi \in \Lambda(\mathsf{Ab}^{\Gamma}_{\mathbf{LLL}})$ there is a $\Delta \subseteq \Omega \setminus \varphi$ for which $\Gamma \Vdash_{\mathbf{LLL}} A \check{\vee} \mathsf{Dab}(\Delta)$.*

5.4.2 The Complete Stage of an Adaptive Proof

In the following it will be very useful to speak about the extension of a given (possibly empty) dynamic proof $\mathcal{P}$ from Γ in which A is derived on the condition Δ whenever $\Gamma \vdash_{\mathbf{LLL}} A \check{\vee} \mathsf{Dab}(\Delta)$. We dub a corresponding stage $g(\mathcal{P})$ a *complete stage*.

This stage exists and can be constructed along the following lines. Note that each well-formed formula has a Gödel-number. From this it follows immediately that $Cn_{\mathbf{LLL}}(\Gamma)$ is enumerable, e.g. $Cn_{\mathbf{LLL}}(\Gamma) = \{B_1, B_2, \ldots\}$. Moreover, due to the compactness of **LLL**, for each $B_i \in Cn_{\mathbf{LLL}}(\Gamma)$ there are some $A_1, \ldots, A_m$ such that $A_1, \ldots, A_m \vdash_{\mathbf{LLL}} B_i$. Hence, for each $B_i \in Cn_{\mathbf{LLL}}(\Gamma)$ we have the following proof $\mathcal{P}_i$:

l^i_1	A_1	PREM	$\emptyset$
$\vdots$	$\vdots$	$\vdots$	$\vdots$
l^i_m	A_m	PREM	$\emptyset$
l^i_{m+1}	B_i	$l^i_1, \ldots, l^i_m$; RU	$\emptyset$

In case B_i is of the form $A \check{\vee} \mathsf{Dab}(\Delta)$ we add a further line.

l^i_{m+2}	A	l^i_{m+1}; RC	Δ

Where $\mathcal{P}$ consists of lines $l^0_1, l^0_2, \ldots$, we now combine the proofs $\mathcal{P}, \mathcal{P}_1, \mathcal{P}_2, \ldots$ to a proof $\mathcal{P}'$ that extends $\mathcal{P}$ to the stage $g(\mathcal{P})$ by means of listing the respective lines as follows (and by renumbering the lines accordingly):

$$l^0_1, l^0_2, l^1_1, l^1_2, l^0_3, l^1_3, l^2_1, l^2_2, l^2_3, l^0_4, \ldots, l^2_4, l^3_1, \ldots, l^3_4, l^0_5, \ldots, l^3_5, l^4_1, \ldots, l^4_5, \ldots$$

Obviously, at the complete stage all formulas that can be derived on the empty condition from Γ are derived. Hence, all the **LLL**-consequences of Γ are derived on the empty condition. The following fact holds for the extension of a dynamic proof $\mathcal{P}$ to the stage $g(\mathcal{P})$:

Fact 5.4.1. *Where $\Gamma \subseteq \mathcal{W}^+$ and $A \in \mathcal{W}^+$, we have:*

(i) A is derived on the condition $\emptyset$ at stage $g(\mathcal{P})$ iff $\Gamma \vdash_{\mathbf{LLL}} A$
(ii) $\Sigma_{g(\mathcal{P})}(\Gamma) = \Sigma(\Gamma) = \Sigma_s(\Gamma)$ for any successor stage s of $g(\mathcal{P})$
(iii) $\Sigma^{\text{sat}}_{g(\mathcal{P})}(\Gamma) = \Sigma^{\text{sat}}(\Gamma) = \Sigma^{\text{sat}}_s(\Gamma)$ for any successor stage s of $g(\mathcal{P})$
(iv) $\Xi_{g(\mathcal{P})}(\Gamma) = \Xi(\Gamma) = \Xi_s(\Gamma)$ for any successor stage s of $g(\mathcal{P})$
(v) $\Xi^{\text{sat}}_{g(\mathcal{P})}(\Gamma) = \Xi^{\text{sat}}(\Gamma) = \Xi^{\text{sat}}_s(\Gamma)$ for any successor stage s of $g(\mathcal{P})$

Note that the marking at a stage is determined by $\Xi^{\text{sat}}_s(\Gamma)$ and Λ. Hence, the following fact follows immediately by (v).

Fact 5.4.2. *If a line l is marked at stage $g(\mathcal{P})$, then it is marked in every further extension. Hence, the markings remain stable from stage $g(\mathcal{P})$ on.*

5.4.3 A Representational Theorem For $\vdash_{\mathbf{AL}_\Lambda}$ and Related Results

The following theorem shows that $\mathbf{AL}_\Lambda$ is at least as strong as its lower limit logic:

Theorem 5.4.2. *Where $\Gamma \subseteq \mathcal{W}^+$: $\Gamma \vdash_{\mathbf{LLL}} A$ implies $\Gamma \vdash_{\mathbf{AL}_\Lambda} A$.*

Proof. By the compactness of **LLL** there is a finite set $\{A_1, \ldots, A_n\} \subseteq \Gamma$ such that $\{A_1, \ldots, A_n\} \vdash_{\mathbf{LLL}} A$. Hence we can construct a proof as follows: we introduce A_1 on line 1 be PREM, A_2 on line 2 by PREM, ..., A_n on line n by PREM. Finally we derive A on line $n + 1$ on the justification $1, \ldots, n$; RU and the condition $\emptyset$. This line stays unmarked in any extension of the proof. □

We call Γ **LLL**-*trivial* iff $\mathcal{M}_{\mathbf{LLL}}(\Gamma) = \emptyset$.

Lemma 5.4.1. *Where $\Gamma \subseteq \mathcal{W}^+$: Γ is* **LLL**-*trivial iff $\Xi^{\text{sat}}(\Gamma) = \Lambda(\Xi^{\text{sat}}(\Gamma)) = \emptyset$.*

Proof. $\Lambda(\Xi^{\text{sat}}(\Gamma)) = \emptyset$ iff [by Fact 5.3.1] $\Xi^{\text{sat}}(\Gamma) = \emptyset$ iff [by Lemma 5.3.2] $\text{Ab}^{\Gamma}_{\mathbf{LLL}} = \emptyset$ *iff Γ is* **LLL**-*trivial.* □

In case the premise set is **LLL**-trivial, **LLL** and $\mathbf{AL}_\Lambda$ define the same consequence relation:

Theorem 5.4.3. *If $\Gamma \subseteq \mathcal{W}^+$ is* **LLL**-*trivial: $\Gamma \vdash_{\mathbf{LLL}} A$ iff $\Gamma \vdash_{\mathbf{AL}_\Lambda} A$ iff $A \in \mathcal{W}^+$.*

Proof. Let Γ be **LLL**-trivial. Hence, by Lemma 5.4.1, $\Lambda(\Xi^{\text{sat}}(\Gamma)) = \emptyset$. "$\Rightarrow$": this is Theorem 5.4.2. "$\Leftarrow$": Let $\Gamma \vdash_{\mathbf{AL}_\Lambda} A$. Hence there is a finite proof $\mathcal{P}$ such that A is finally derived on a line l on a condition Δ at the corresponding stage s. We extend the proof to the complete stage $g(\mathcal{P})$. Note that $\Lambda(\Xi^{\text{sat}}(\Gamma)) = \Lambda\big(\Xi^{\text{sat}}_{g(\mathcal{P})}(\Gamma)\big) =$

$\Lambda(\Xi^{\text{sat}}_{s'}(\Gamma)) = \emptyset$ for all further stages s'. Hence, l is unmarked iff $\Delta = \emptyset$. Since, A was finally derived at stage s, $\Delta = \emptyset$. Hence, A is **LLL**-derivable. □

Lemma 5.4.2. *Where $\Gamma \subseteq \mathcal{W}^+$ and $\Gamma \vdash_{\mathbf{AL_\Lambda}} A$:*

(i) if Γ is **LLL***-non-trivial then A is derivable on a line l of a finite* $\mathbf{AL_\Lambda}$ *proof from Γ on a condition Δ such that $\Delta \cap \varphi = \emptyset$ for a $\varphi \in \Lambda(\Xi^{\text{sat}}(\Gamma))$.*
(ii) For every $\varphi \in \Lambda(\Xi^{\text{sat}}(\Gamma))$ there is a finite $\Delta \subseteq \Omega \setminus \varphi$ for which $\Gamma \vdash_{\mathbf{LLL}} A \check{\vee} \mathsf{Dab}(\Delta)$.

Proof. Suppose $\Gamma \vdash_{\mathbf{AL_\Lambda}} A$. Hence, there is a finite $\mathbf{AL_\Lambda}$-proof $\mathcal{P}$ from Γ such that (1) A is derived in this proof on an unmarked line l with a condition Δ, and (2) every extension of the proof in which l is marked can be further extended such that l is unmarked again. We now extend our proof $\mathcal{P}$ to the complete stage $\mathrm{g}(\mathcal{P})$. Note that $\Lambda(\Xi^{\text{sat}}(\Gamma)) = \Lambda(\Xi^{\text{sat}}_{\mathrm{g}(\mathcal{P})}(\Gamma)) = \Lambda(\Xi^{\text{sat}}_{s'}(\Gamma))$ for every later stage s'.

Ad (i): Since Γ is **LLL**-non-trivial and by Lemma 5.4.1, $\Lambda(\Xi^{\text{sat}}(\Gamma)) \neq \emptyset$. Assume there is no $\varphi \in \Lambda(\Xi^{\text{sat}}(\Gamma))$ such that $\Delta \cap \varphi = \emptyset$. By Definition 5.3.4, line l is marked at stage $\mathrm{g}(\mathcal{P})$ and hence at every later stage s'—a contradiction to (2). *Ad (ii):* Assume there is a $\varphi \in \Lambda(\Xi^{\text{sat}}(\Gamma))$ for which there is no $\Delta \subseteq \Omega$ such that $\Gamma \vdash_{\mathbf{LLL}} A \check{\vee} \mathsf{Dab}(\Delta)$ and $\Delta \cap \varphi = \emptyset$. By Definition 5.3.4.ii line l is marked at stage $\mathrm{g}(\mathcal{P})$ and hence at every later stage s'. This contradicts (2). □

Lemma 5.4.3. *Where $\Gamma \subseteq \mathcal{W}$ or $\Gamma = Cn^{\mathcal{L}^+}_{\mathbf{LLL}}(\Gamma)$: If $\Gamma \vdash_{\mathbf{LLL}} A \check{\vee} \mathsf{Dab}(\Delta)$ and $\Delta \cap \varphi = \emptyset$ for a $\varphi \in \Lambda(\Xi^{\text{sat}}(\Gamma))$, then there is a finite* $\mathbf{AL_\Lambda}$*-proof from Γ in which A is derived on the condition Δ at an unmarked line.*

Proof. Case $\Gamma \subseteq \mathcal{W}$: Suppose the antecedent holds. Due to the compactness of **LLL**, there is a $\Gamma' = \{A_1, \ldots, A_n\} \subseteq \Gamma$ for which $\Gamma' \vdash_{\mathbf{LLL}} A \check{\vee} \mathsf{Dab}(\Delta)$. We construct a proof as follows: At line 1 we introduce A_1 by the PREM rule, ..., at line n we introduce A_n by the PREM rule, and at line $n{+}1$ we derive A on the condition Δ by RC from lines $1, \ldots, n$. Let s be the corresponding stage of the proof. Since $\Gamma' \subseteq \Gamma \subseteq \mathcal{W}$, all Dab-formulas $B_1, \ldots, B_m$ that have been derived at stage s are members of Ω since no formulas with occurrences of $\check{\vee}$ have can be introduced by PREM. Thus, $\Sigma^{\text{sat}}_s(\Gamma) = \{\{B_1\}, \ldots, \{B_n\}\}$ and thus, $\min_{\subset}(\Xi^{\text{sat}}_s(\Gamma)) = \{\{B_1, \ldots, B_n\}\} = \min^{\cup}_{\subset}(\Xi^{\text{sat}}_s(\Gamma))$. Hence, by **T1** and **T3**, $\Lambda(\Xi^{\text{sat}}_s(\Gamma)) = \Lambda(\min^{\cup}_{\subset}(\Xi^{\text{sat}}_s(\Gamma))) = \{\{B_1, \ldots, B_n\}\}$.[22] Note that $\Gamma \vdash_{\mathbf{LLL}} B_i$ for all these abnormalities and hence $\{B_1, \ldots, B_m\} \subseteq \psi$ for all $\psi \in \Xi(\Gamma)$ and hence also for all $\psi \in \Xi^{\text{sat}}(\Gamma)$. Since $\varphi \in \Lambda(\Xi^{\text{sat}}(\Gamma))$ and since by **T1** Λ is inclusive, also $\varphi \in \Xi^{\text{sat}}(\Gamma)$. Hence, $\{B_1, \ldots, B_n\} \subseteq \varphi$. By the supposition $\Delta \cap \varphi = \emptyset$ and hence $\{B_1, \ldots, B_n\} \cap \Delta = \emptyset$. Hence, line $n{+}1$ is unmarked.

The case $\Gamma = Cn^{\mathcal{L}^+}_{\mathbf{LLL}}(\Gamma)$ is similar. Since $\Gamma \vdash_{\mathbf{LLL}} A \check{\vee} \mathsf{Dab}(\Delta)$ also $\Gamma \vdash_{\mathbf{LLL}} \check{\neg} A \check{\supset} \mathsf{Dab}(\Delta)$. Since Γ is a **LLL**-fixed point, also $\check{\neg} A \check{\supset} \mathsf{Dab}(\Delta) \in \Gamma$. Hence, we can introduce $\check{\neg} A \check{\supset} \mathsf{Dab}(\Delta)$ at line 1 of a proof by PREM and derive A on the

[22] Note that **T1**–**T3** are indeed applicable since $\Xi^{\text{sat}}_s(\Gamma)$, $\min^{\cup}_{\subset}(\Xi_s(\Gamma)) \in \Upsilon$ as we show in Lemma C.2.1 in the Appendix.

condition Δ by RC on line 2. Obviously this line is not marked at this stage since no Dab-formulas have been derived yet. □

Lemma 5.4.4. *Where $\Gamma \subseteq \mathcal{W}$ or $\Gamma = Cn^{\mathcal{L}+}_{\mathbf{LLL}}(\Gamma)$: If for every $\varphi \in \Lambda(\Xi^{\mathrm{sat}}(\Gamma))$ there is a finite $\Delta_\varphi \subseteq \Omega \setminus \varphi$ such that $\Gamma \vdash_{\mathbf{LLL}} A \check{\vee} \mathsf{Dab}(\Delta_\varphi)$, then $\Gamma \vdash_{\mathbf{AL}_\Lambda} A$.*

Proof. The case that Γ is **LLL**-trivial is covered by Theorem 5.4.3. Let hence Γ be **LLL**-non-trivial and thus by Lemma 5.4.1, $\Lambda(\Xi^{\mathrm{sat}}(\Gamma)) \neq \emptyset$.

Suppose the antecedent is true. By Lemma 5.4.3, for every Δ_φ there is a finite $\mathbf{AL}_\Lambda$-proof from Γ in which A is derived on the condition Δ_φ at an unmarked line l. Given any such proof, suppose the proof is extended to a stage s in which l is marked. Call this proof $\mathcal{P}$. We extend the proof further to the stage $\mathrm{g}(\mathcal{P})$. Note that for all $\varphi \in \Lambda(\Xi^{\mathrm{sat}}(\Gamma))$, A has been derived on the condition Δ_φ at this stage. By Definition 5.3.4, line l is unmarked at stage $\mathrm{g}(\mathcal{P})$. □

The following representational theorem characterizes the consequence relation of $\mathbf{AL}_\Lambda$ entirely by means of the consequence relation of **LLL** and the members of $\Lambda(\Xi^{\mathrm{sat}}(\Gamma))$. By Lemma 5.4.2 and 5.4.4 we immediately get:

Theorem 5.4.4. *Where $\Gamma \subseteq \mathcal{W}$ or $\Gamma = Cn^{\mathcal{L}+}_{\mathbf{LLL}}(\Gamma)$: $\Gamma \vdash_{\mathbf{AL}_\Lambda} A$ iff for every $\varphi \in \Lambda(\Xi^{\mathrm{sat}}(\Gamma))$ there is a $\Delta \subseteq \Omega \setminus \varphi$ for which $\Gamma \vdash_{\mathbf{LLL}} A \check{\vee} \mathsf{Dab}(\Delta)$.*

Due to the compactness of **LLL**, Theorem 5.4.4 can be alternatively expressed by:

Theorem 5.4.5. *Where $\Gamma \subseteq \mathcal{W}$ or $\Gamma = Cn^{\mathcal{L}+}_{\mathbf{LLL}}(\Gamma)$: $\Gamma \vdash_{\mathbf{AL}_\Lambda} A$ iff for every $\varphi \in \Lambda(\Xi^{\mathrm{sat}}(\Gamma))$, $\Gamma \cup (\Omega \setminus \varphi)^{\check{\neg}} \vdash_{\mathbf{LLL}} A$.*

Proof. In case Γ is **LLL**-trivial the theorem holds by Theorem 5.4.3. Let thus Γ be **LLL**-non-trivial. Hence, by Lemma 5.4.1, $\Lambda(\Xi^{\mathrm{sat}}(\Gamma)) \neq \emptyset$.

Suppose for every $\varphi \in \Lambda(\Xi^{\mathrm{sat}}(\Gamma))$, $\Gamma \cup (\Omega \setminus \varphi)^{\check{\neg}} \vdash_{\mathbf{LLL}} A$. Let $\varphi \in \Lambda(\Xi^{\mathrm{sat}}(\Gamma))$. Hence, $\Gamma \cup (\Omega \setminus \varphi)^{\check{\neg}} \vdash_{\mathbf{LLL}} A$. By the compactness of **LLL**, there is a finite $\Delta \subseteq \Omega \setminus \varphi$ such that $\Gamma \cup \Delta^{\check{\neg}} \vdash_{\mathbf{LLL}} A$. By the deduction theorem, $\Gamma \vdash_{\mathbf{LLL}} A \check{\vee} \mathsf{Dab}(\Delta)$. Hence, since φ was arbitrary in $\Lambda(\Xi^{\mathrm{sat}}(\Gamma))$ by Lemma 5.4.4, $\Gamma \vdash_{\mathbf{AL}_\Lambda} A$.

Let $\Gamma \vdash_{\mathbf{AL}_\Lambda} A$. By Lemma 5.4.2.ii, for every $\varphi \in \Lambda(\Xi^{\mathrm{sat}}(\Gamma))$ there is a $\Delta \subseteq \Omega \setminus \varphi$ for which $\Gamma \vdash_{\mathbf{LLL}} A \check{\vee} \mathsf{Dab}(\Delta)$. Assume for some $\varphi \in \Lambda(\Xi^{\mathrm{sat}}(\Gamma))$, $\Gamma \cup (\Omega \setminus \varphi)^{\check{\neg}} \nvdash_{\mathbf{LLL}} A$. By the monotonicity of **LLL**, there is no $\Delta \subseteq \Omega \setminus \varphi$ such that $\Gamma \cup \Delta^{\check{\neg}} \vdash_{\mathbf{LLL}} A$. Hence, there is no $\Delta \subseteq \Omega \setminus \varphi$ such that $\Gamma \vdash_{\mathbf{LLL}} A \check{\vee} \mathsf{Dab}(\Delta)$,—a contradiction. □

5.4.4 Soundness and Completeness

By our two representational results, Theorem 5.4.1 and 5.4.5 (or Corollary 5.4.1 and Theorem 5.4.4), and our Lemma 5.3.2 we immediately get soundness and completeness (see Fig. 5.2):

Corollary 5.4.2 (Soundness and Completeness of $\mathbf{AL}_\Lambda$). *Where $\Gamma \subseteq \mathcal{W}$ or $\Gamma = Cn^{\mathcal{L}^+}_{\mathbf{LLL}}(\Gamma)$: $\Gamma \vdash_{\mathbf{AL}_\Lambda} A$ iff $\Gamma \Vdash_{\mathbf{AL}_\Lambda} A$.*

Soundness even holds for premise sets with "checked connectives":

Corollary 5.4.3. *Where $\Gamma \subseteq \mathcal{W}^+$: $\Gamma \vdash_{\mathbf{AL}_\Lambda} A$ implies $\Gamma \Vdash_{\mathbf{AL}_\Lambda} A$.*

This follows immediately with Lemma 5.3.2, 5.4.2 and Corollary 5.4.1.

5.5 Other Meta-Theoretic Properties of $\mathbf{AL}_\Lambda$

In this section we will establish some crucial meta-theoretic properties of $\mathbf{AL}_\Lambda$. Due to the generality of our format, some properties only hold for a subclass of ALs. We will establish criteria for the threshold function Λ in order to distinguish some interesting classes.

5.5.1 Cumulativity and Equivalent Premise Sets

The following two criteria play an important role in what follows[23]:

CT where $X, Y \in \Upsilon$: $\Lambda(X) \subseteq Y \subseteq X$ implies $\Lambda(X) \subseteq \Lambda(Y)$
CM where $X, Y \in \Upsilon$: $\Lambda(X) \subseteq Y \subseteq X$ implies $\Lambda(X) \supseteq \Lambda(Y)$

The following theorem shows that **CT** and **CM** are sufficient to guarantee cautious transitivity resp. cautious monotonicity. Table 5.2 offers an overview for these and other criteria on Λ which are introduced in the following sections.

Theorem 5.5.1. *Where $\Gamma, \Gamma' \subseteq \mathcal{W}^+$ and $\Gamma \Vdash_{\mathbf{AL}_\Lambda} B$ for all $B \in \Gamma'$:*

(i) if Λ satisfies **CM** *then: $\Gamma \Vdash_{\mathbf{AL}_\Lambda} A$ implies $\Gamma \cup \Gamma' \Vdash_{\mathbf{AL}_\Lambda} A$.*
(ii) if Λ satisfies **CT** *then: $\Gamma \cup \Gamma' \Vdash_{\mathbf{AL}_\Lambda} A$ implies $\Gamma \Vdash_{\mathbf{AL}_\Lambda} A$.*

Proof. Since $\Gamma \Vdash_{\mathbf{AL}_\Lambda} B$ for all $B \in \Gamma'$, $\mathcal{M}_{\mathbf{AL}_\Lambda}(\Gamma) \subseteq \mathcal{M}_{\mathbf{LLL}}(\Gamma \cup \Gamma')$ and hence (1) $\Lambda(\mathsf{Ab}^{\Gamma}_{\mathbf{LLL}}) \subseteq \mathsf{Ab}^{\Gamma \cup \Gamma'}_{\mathbf{LLL}}$. Since by the monotonicity of **LLL**, $\mathcal{M}_{\mathbf{LLL}}(\Gamma \cup \Gamma') \subseteq \mathcal{M}_{\mathbf{LLL}}(\Gamma)$ also (2) $\mathsf{Ab}^{\Gamma \cup \Gamma'}_{\mathbf{LLL}} \subseteq \mathsf{Ab}^{\Gamma}_{\mathbf{LLL}}$.

Table 5.2 Overview of the criteria for Λ and their effects

CT	$\rightsquigarrow$	Cautious transitivity (Sect. 5.5.1)
CM	$\rightsquigarrow$	Cautious monotonicity (Sect. 5.5.1)
$\mathrm{RA}_\prec$	$\rightsquigarrow$	Strong reassurance w.r.t. $\prec$ (Sect. 5.5.3)
SIMP	$\rightsquigarrow$	Simplification of proof theory (Sect. 5.6)

[23] In [12] these criteria were first proposed for preferential semantics.

Ad (i): By **CM**, (1) and (2): $\Lambda(\mathsf{Ab}^{\Gamma}_{\mathbf{LLL}}) \supseteq \Lambda\left(\mathsf{Ab}^{\Gamma\cup\Gamma'}_{\mathbf{LLL}}\right)$. Thus, $\mathcal{M}_{\mathbf{AL}_{\Lambda}}(\Gamma) \supseteq \mathcal{M}_{\mathbf{AL}_{\Lambda}}(\Gamma \cup \Gamma')$. The rest follows immediately.

Ad (ii): By **CT**, (1) and (2): $\Lambda(\mathsf{Ab}^{\Gamma}_{\mathbf{LLL}}) \subseteq \Lambda\left(\mathsf{Ab}^{\Gamma\cup\Gamma'}_{\mathbf{LLL}}\right)$. Thus, $\mathcal{M}_{\mathbf{AL}_{\Lambda}}(\Gamma) \subseteq \mathcal{M}_{\mathbf{AL}_{\Lambda}}(\Gamma \cup \Gamma')$. The rest follows immediately. □

By Corollary 5.4.2 we immediately get:

Corollary 5.5.1. *Where $\Gamma, \Gamma' \subseteq \mathcal{W}$ and $\Gamma \vdash_{\mathbf{AL}_{\Lambda}} B$ for all $B \in \Gamma'$:*

(i) if Λ satisfies **CM** *then: $\Gamma \vdash_{\mathbf{AL}_{\Lambda}} A$ implies $\Gamma \cup \Gamma' \vdash_{\mathbf{AL}_{\Lambda}} A$.*
(ii) if Λ satisfies **CT** *then: $\Gamma \cup \Gamma' \vdash_{\mathbf{AL}_{\Lambda}} A$ implies $\Gamma \vdash_{\mathbf{AL}_{\Lambda}} A$.*

From this we immediately get:

Corollary 5.5.2. *Where $\Gamma \subseteq \mathcal{W}$ and Λ satisfies* **CT** *and* **CM***: $Cn_{\mathbf{AL}_{\Lambda}}(\Gamma) = Cn_{\mathbf{AL}_{\Lambda}}(Cn^{\mathcal{L}}_{\mathbf{AL}_{\Lambda}}(\Gamma))$.*

To see why this holds note that $Cn^{\mathcal{L}}_{\mathbf{AL}}(\Gamma) \subseteq Cn_{\mathbf{AL}_{\Lambda}}(\Gamma)$ and hence, by **CT**, **CM** and Corollary 5.5.1, $Cn_{\mathbf{AL}_{\Lambda}}(\Gamma) = Cn_{\mathbf{AL}_{\Lambda}}(Cn^{\mathcal{L}}_{\mathbf{AL}_{\Lambda}}(\Gamma) \cup \Gamma)$ By the reflexivity of $\mathbf{AL}_{\Lambda}$, $Cn_{\mathbf{AL}_{\Lambda}}(\Gamma) = Cn_{\mathbf{AL}_{\Lambda}}(Cn^{\mathcal{L}}_{\mathbf{AL}_{\Lambda}}(\Gamma))$.

Recall that cumulativity was crucial for various of the criteria for equivalent premise sets in Chap. 4. Hence, applying these results to our insights in this section we get:

Corollary 5.5.3. *Where $\Gamma, \Gamma' \subseteq \mathcal{W}$: If Λ satisfies* **CM** *and* **CT** *then*

(i) if $\Gamma \subseteq Cn_{\mathbf{AL}_{\Lambda}}(\Gamma')$ and $\Gamma' \subseteq Cn_{\mathbf{AL}_{\Lambda}}(\Gamma)$ then $Cn_{\mathbf{AL}_{\Lambda}}(\Gamma) = Cn_{\mathbf{AL}_{\Lambda}}(\Gamma')$
(ii) where **L** *is a reflexive logic weaker than* $\mathbf{AL}_{\Lambda}$ *then $Cn_{\mathbf{L}}(\Gamma) = Cn_{\mathbf{L}}(\Gamma')$ implies $Cn_{\mathbf{AL}_{\Lambda}}(\Gamma) = Cn_{\mathbf{AL}_{\Lambda}}(\Gamma')$*
(iii) where **L** *is monotonic and reflexive and* $\mathbf{AL}_{\Lambda}$ *is closed under* **L** *(i.e., for all Θ, $Cn_{\mathbf{L}}(Cn_{\mathbf{AL}_{\Lambda}}(\Theta)) = Cn_{\mathbf{AL}_{\Lambda}}(\Theta)$), then $Cn_{\mathbf{L}}(\Gamma) = Cn_{\mathbf{L}}(\Gamma')$ implies $Cn_{\mathbf{AL}_{\Lambda}}(\Gamma \cup \Theta) = Cn_{\mathbf{AL}_{\Lambda}}(\Gamma' \cup \Theta)$*
(iv) where **L** *is a reflexive and monotonic logic weaker than* $\mathbf{AL}_{\Lambda}$ *then $Cn_{\mathbf{L}}(\Gamma) = Cn_{\mathbf{L}}(\Gamma')$ implies $Cn_{\mathbf{AL}_{\Lambda}}(\Gamma \cup \Theta) = Cn_{\mathbf{AL}_{\Lambda}}(\Gamma' \cup \Theta)$*

Neither cautious monotonicity nor cautious transitivity hold for $\mathbf{AL}_{\Lambda}$ in general. We give two examples, starting with a counter-example for cautious transitivity.

Example 5.5.1. The logic $\mathbf{AL}_{\Lambda^2_{\mathbf{c}}}$ from Sect. 5.2.7 does not satisfy cautious transitivity. Recall that in our example there were three types of models M with $\mathsf{Ab}(M) \in \min_{\subset}\left(\mathsf{Ab}^{\Gamma^2_c}_{\mathbf{L}^*_\circ}\right)$, namely:

Model M	$\lvert\mathsf{Ab}(M)\rvert$	$M \models$
M_1	4	$p, \neg q$
M_2	5	p, q
M_3	10	$\neg p$

As demonstrated, selected are models with the abnormal part of M_1 or with the abnormal part of M_2. Thus, $\Gamma_c^2 \vdash_{\mathbf{AL}_{\mathbf{A}_c^2}} p$ and $\Gamma_c^2 \nvdash_{\mathbf{AL}_{\mathbf{A}_c^2}} \neg q$.

However, let us add p to our premise set: $\Gamma' = \Gamma_c^2 \cup \{p\}$. Now we have two types of models M with $\mathrm{Ab}(M) \in \min_{\subset}(\mathsf{Ab}^{\Gamma'}_{\mathbf{L}^*_\circ})$, namely models that have the abnormal part of M_1, and models that have the abnormal part of M_2. Hence, now we select models M for which: $|\mathrm{Ab}(M)| \leq \frac{4+5}{2}$. Models with the abnormal part of M_2 are thus not selected anymore, which leaves only models with the abnormal part of M_1. As a consequence, $\Gamma' \vdash_{\mathbf{AL}_{\mathbf{A}_c^2}} \neg q$. Hence, cautious transitivity does not hold.

Let us continue with a counter-example to cautious monotonicity.

Example 5.5.2. Let

$$\Lambda(X) = \begin{cases} \min_{\prec_c}(X) \text{ if } |\min_{\subset}(X)| \geq \omega \\ \min_{\subset}(X) \text{ else} \end{cases}$$

It is easy to see that Λ is a threshold function.[24] Suppose we have the following premise set (where p, q, r, s_i are propositional atoms ($i \in \mathbb{N}$)):

$$\Gamma = \{\neg p \vee (\neg q \wedge \neg r) \vee (\neg s_i \vee \neg s_j) \mid i < j\} \cup \{\circ p, \circ q, \circ r, \circ s_i \mid i \in \mathbb{N}\}$$

We have three types of $\subset$-minimally abnormal models (where $i \in \mathbb{N}$ and $!A =_{\mathrm{df}} \circ A \wedge \neg A$):

Model M	$\mathrm{Ab}(M)$	$\lvert\mathrm{Ab}(M)\rvert$	$M \models$
M_p	$\{!p\}$	1	$\neg p, q, r, s_j$ where $j \in \mathbb{N}$
$M_{q,r}$	$\{!q, !r\}$	2	$p, \neg q, \neg r, s_j$ where $j \in \mathbb{N}$
M_i	$\{!s_j \mid j \neq i\}$	ω	$p, q, r, s_i, \neg s_j$ where $j \neq i$

Since $\left|\min_{\subset}(\mathsf{Ab}^{\Gamma}_{\mathbf{L}_\circ})\right| \geq \omega$, only models with the abnormal part of M_p are selected since they are $\prec_c$-minimally abnormal. Hence, $\Gamma \vdash_{\mathbf{AL_A}} \neg p$ and thus also $\Gamma \vdash_{\mathbf{AL_A}} \neg p \vee (\neg q \wedge \neg r)$.

Now let $\Gamma' = \Gamma \cup \{\neg p \vee (\neg q \wedge \neg r)\}$. Then the $\subset$-minimally abnormal models have the abnormal part of either M_p or $M_{q,r}$. Hence, $\left|\min_{\subset}(\mathsf{Ab}^{\Gamma}_{\mathbf{L}_\circ})\right| < \omega$. But that means that also models with the abnormal part of $M_{q,r}$ are now selected: $\Lambda\left(\mathsf{Ab}^{\Gamma'}_{\mathbf{L}_\circ}\right) = \{\mathrm{Ab}(M_p), \mathrm{Ab}(M_{q,r})\}$. As a consequence, $\Gamma' \nvdash_{\mathbf{AL_A}} \neg p$. This shows that cautious monotonicity does not hold.

[24] In Theorem 5.5.4 and Fact 5.8.2 we show that $\langle X, \subset\rangle$ and $\langle X, \prec_c\rangle$ are smooth for all $X \in \Upsilon$ which ensures **T3**. In view of the definition of $\prec_c$, $\min_{\prec_c}(X) \subseteq \min_{\subset}(X)$ which ensures **T1**. **T2** is evident.

5.5.2 Some Classes of Cumulative Λ-Based ALs

In this section we will introducesome general criteria for the threshold function Λ which are sufficient for cumulativity.

The following notions will be useful in what follows:

Definition 5.5.1. Where $X \subseteq Y$: X is a *$\prec$-dense in Y* iff for all $y \in Y$ either $y \in X$ or there is a $x \in X$ for which $x \prec y$.

Definition 5.5.2. We say that Λ is *$\prec$-density invariant (on Υ)* iff for all $X, Y \in \Upsilon$ where X is a $\prec$-dense subset of Y, $\Lambda(X) = \Lambda(Y) \cap X$.

DI$_\prec$ Λ is $\prec$-density invariant (on Υ)
RA$_\prec$ for all $X \in \Upsilon$: $\Lambda(X)$ is $\prec$-dense in X

DI$_\prec$ expresses a certain invariance of our threshold: if φ is below our threshold in the context of X then it is also below the threshold in any Y in which X is $\prec$-dense, and vice versa. One may think of X as a filtered Y: elements are filtered out in such a way that it is guaranteed that for every element ψ that is filtered out a $\prec$-better one is preserved. Every element that is not filtered out and still below our threshold, should already be below our threshold in Y and vice versa, every element that is below our threshold in Y should still be below our threshold after a filtering process in which it is not filtered out.

RA$_\prec$ expresses that for every element that is above our threshold there should be a $\prec$-better one that is below the threshold. One motivation is as follows. If an element is not deemed to be good enough than this should be justified by means of pointing out a ($\prec$-) better element that is good enough.

The following fact shows that these two criteria are sufficient to guarantee cumulativity (Corollary 5.5.1) and the criteria for equivalent premise sets (Corollary 5.5.3).

Fact 5.5.1. *(**DI**$_\prec$ and **RA**$_\prec$) imply **CT** and **CM**.*

Proof. Let $\Lambda(X) \subseteq Y \subseteq X$ where $X, Y \in \Upsilon$. Since by **RA**$_\prec$ $\Lambda(X)$ is $\prec$-dense in X and $\Lambda(X) \subseteq Y$, also Y is $\prec$-dense in X. Hence, by **DI**$_\prec$, $\Lambda(Y) = \Lambda(X) \cap Y = \Lambda(X)$. □

We state also some weaker criteria in terms of $\min_\subset$.

C1 for all $X, Y \in \Upsilon$: $\min_\subset(X) = \min_\subset(Y)$ implies $\Lambda(X) \cap Y = \Lambda(Y) \cap X$

C1 is a similar invariance criterion as **DI**$_\prec$. Indeed, in the Appendix we show that **C1** is equivalent to **DI**$_\subset$ and that **DI**$_\prec$ implies **DI**$_\subset$ and hence also **C1**. The following criteria imply **C1** (with the exception of **C4**)[25]:

C2 for all $X \in \Upsilon$: $\Lambda(X) = f(\min_\subset(X))$ for some function f
C3 for all $X, Y \in \Upsilon$: $\min_\subset(X) = \min_\subset(Y)$ implies $\Lambda(X) = \Lambda(Y)$
C4 for all $X \in \Upsilon$: $\Lambda(X) \supseteq \min_\subset(X)$

[25] See Fact C.3.1.

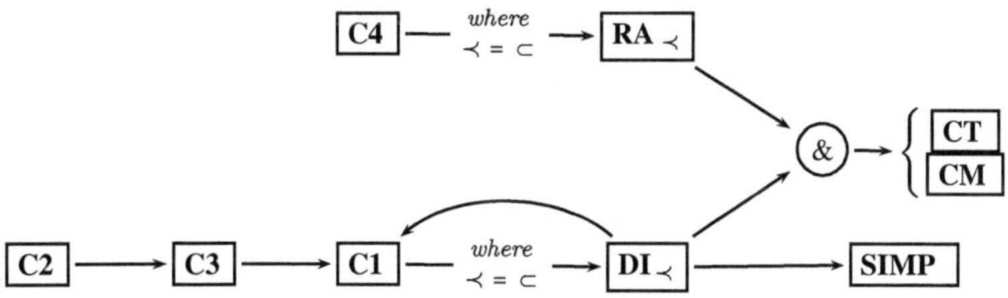

Fig. 5.3 Overview: criteria for Λ

These criteria express that Λ is determined by $\min_\subset$ to some degree. In the case of **C2** this is specified by a function f, while in the case of **C3** we have a weaker dependency: the sameness of the $\subset$-minimal elements of two sets guarantees that they have the same threshold. **C4** expresses that the threshold is always above or equal to $\min_\subset$.

Figure 5.3 gives an overview for the different criteria. They offer easy ways to check whether a given AL has certain properties. Take for instance the two threshold functions associated with the standard format: $\min_\subset$ and $\min_\subset^\cup$. Both obviously satisfy **C4** and **C2** and hence cumulativity follows immediately.

5.5.3 *Reassurance*

Reassurance means that whenever Γ is **LLL**-non-trivial, Γ is also $\mathbf{AL}_\Lambda$-non-trivial. In semantic terms: if $\mathcal{M}_{\mathbf{LLL}}(\Gamma) \neq \emptyset$ then $\mathcal{M}_{\mathbf{AL}_\Lambda}(\Gamma) \neq \emptyset$.

Theorem 5.5.2 **(Reassurance).** *If* Γ *is* **LLL**-*non-trivial then* Γ *is* $\mathbf{AL}_\Lambda$-*non-trivial.*

Proof. Suppose Γ is **LLL**-non-trivial. Hence, $\mathrm{Ab}^\Gamma_{\mathbf{LLL}} \neq \emptyset$. By **T3**, $\Lambda(\mathrm{Ab}^\Gamma_{\mathbf{LLL}}) \neq \emptyset$. Hence, Γ is $\mathbf{AL}_\Lambda$-non-trivial. □

The following lemma and its corollary are crucial to show that reassurance holds for ALs in the standard format (see also Sect. 5.8.2), or more general for ALs with threshold functions that satisfy **C4**.[26]

Lemma 5.5.1. *Let* X *be an enumerable set,* Σ *be a set of finite subsets of* X*, and* CS *denote the function that returns the choice sets of a set of sets. Where* $\varphi = \{A_i \mid i \in I\} \in \mathsf{CS}(\Sigma)$*, let* $\hat{\varphi} = \bigcap_{i \in I} \varphi_i$ *where* $\varphi_0 = \varphi$ *and (where* $i+1 \in I$*)*

$$\varphi_{i+1} = \begin{cases} \varphi_i & \text{if there is a } \Delta \in \Sigma \text{ such that } \varphi_i \cap \Delta = \{A_{i+1}\} \\ \varphi_i \setminus \{A_{i+1}\} & \text{else} \end{cases}$$

we have: $\hat{\varphi} \in \min_\subset(\mathsf{CS}(\Sigma))$.

[26] This and the following lemma are proven in Appendix A.

Hence, we immediately get:

Theorem 5.5.3. *Where $\Gamma \subseteq \mathcal{W}^+$: $\langle \Xi(\Gamma), \subset\rangle$ is smooth.*

Since also

Lemma 5.5.2. *Where $\Gamma \subseteq \mathcal{W}^+$ is* **LLL**-*non-trivial: $\Xi^{\text{sat}}(\Gamma)$ is $\subset$-dense in $\Xi(\Gamma)$.*

we get:

Theorem 5.5.4. *Where $\Gamma \subseteq \mathcal{W}^+$: $\langle \Xi^{\text{sat}}(\Gamma), \subset\rangle$ is smooth.*

This immediately shows that $\min_{\subset}$ and any function Λ that satisfies **C4** (e.g. $\min_{\subset}^{\cup}$) also satisfies reassurance and **T3**. Since $\min_{\subset}$ and $\min_{\subset}^{\cup}$ trivially satisfy **T1** and **T2**, this also suffices to show that $\min_{\subset}$ and $\min_{\subset}^{\cup}$ are threshold functions.

There is also a stronger form of reassurance. $\mathbf{AL}_{\Lambda}$ satisfies *strong reassurance* relative to a partial order $\prec$ iff, for each **LLL**-model M of Γ that is not selected there is a selected **LLL**-model M' of Γ that has a less abnormal part with respect to $\prec$: $\text{Ab}(M') \prec \text{Ab}(M)$. Obviously,

Fact 5.5.2. *If Λ satisfies $\mathbf{RA}_{\prec}$ then $\mathbf{AL}_{\Lambda}$ satisfies strong reassurance (relative to $\prec$).*

For instance, $\Psi_{\prec}$ and $\Psi_{[\prec_1,\ldots,\prec_n]}$ (where $\prec = \prec_1$) satisfy[27] $\mathbf{RA}_{\prec}$ and hence $\mathbf{AL}_{\Psi_{\prec}}$ and $\mathbf{AL}_{\Psi_{[\prec_1,\ldots,\prec_n]}}$ satisfy strong reassurance with respect to $\prec$.

Also, since **C4** implies $\mathbf{RA}_{\subset}$,[28] **C4** is by Fact 5.5.2 sufficient to guarantee strong reassurance (relative to $\subset$). For instance, $\min_{\subset}$ and $\min_{\subset}^{\cup}$ evidently satisfy **C4** and hence ALs in the standard format satisfy strong reassurance.

5.5.4 The Upper Limit Logic and the Maximality of the Lower Limit Logic

Recall that a premise set was called *normal* iff no Dab-formula is **LLL**-derivable from it. The upper limit logic **ULL** of an AL with abnormalities Ω was defined by $\Gamma \vdash_{\mathbf{ULL}} A$ iff $\Gamma \cup \check{\Omega}^{\neg} \vdash_{\mathbf{LLL}} A$. Semantically **ULL** is characterized by the set of models $\mathcal{M}_{\mathbf{ULL}}(\Gamma) =_{\text{df}} \{M \in \mathcal{M}_{\mathbf{LLL}}(\Gamma) \mid \text{Ab}(M) = \emptyset\}$.

It is easy to see that $\mathbf{AL}_{\Lambda}$ has exactly the same consequences as the upper limit logic in case Γ is normal.

Theorem 5.5.5. *Where $\Gamma \subseteq \mathcal{W}^+$ is normal: $\Gamma \vdash_{\mathbf{AL}_{\Lambda}} A$ iff $\Gamma \vdash_{\mathbf{ULL}} A$.*

Proof. In case Γ is normal, $\emptyset \in \text{Ab}^{\Gamma}_{\mathbf{LLL}}$ and hence by Fact 5.3.1, $\Lambda(\text{Ab}^{\Gamma}_{\mathbf{LLL}}) = \{\emptyset\}$. Hence, $\mathcal{M}_{\mathbf{AL}_{\Lambda}}(\Gamma) = \mathcal{M}_{\mathbf{ULL}}(\Gamma)$. □

Moreover, **ULL** is always an upper bound for $\mathbf{AL}_{\Lambda}$:

[27] This is shown in the Appendix: Fact C.3.4 and Fact C.3.11.

[28] See Fact C.3.1 in Appendix C.

Theorem 5.5.6. *Where $\Gamma \subseteq \mathcal{W}^+$: $Cn_{\mathbf{AL}_{\Lambda}}(\Gamma) \subseteq Cn_{\mathbf{ULL}}(\Gamma)$*

Proof. In case Γ is not normal, $Cn_{\mathbf{ULL}}(\Gamma)$ is trivial and hence the statement trivially holds. The case in which Γ is normal has been dealt with in Theorem 5.5.5. □

Hence, $\mathbf{AL}_{\Lambda}$ strengthens **LLL** and approximates **ULL** in the limit of normal premise sets. Altogether we have the following:

Corollary 5.5.4. *Where $\Gamma \subseteq \mathcal{W}^+$:*

$$\Gamma \subseteq Cn_{\mathbf{LLL}}(\Gamma) \subseteq Cn_{\mathbf{AL}_{\mathrm{min}^{\cup}_{\subset}}}(\Gamma) \subseteq Cn_{\mathbf{AL}_{\Lambda}}(\Gamma) \subseteq Cn_{\mathbf{ULL}}(\Gamma).$$

In Theorem 5.4.2 we have shown that $Cn_{\mathbf{LLL}}(\Gamma) \subseteq Cn_{\mathbf{AL}_{\mathrm{min}^{\cup}_{\subset}}}(\Gamma)$, in view of **T1** it directly follows that $Cn_{\mathbf{AL}_{\mathrm{min}^{\cup}_{\subset}}}(\Gamma) \subseteq Cn_{\mathbf{AL}_{\Lambda}}(\Gamma)$, and finally $Cn_{\mathbf{AL}_{\Lambda}}(\Gamma) \subseteq Cn_{\mathbf{ULL}}(\Gamma)$ follows by Theorem 5.5.6.

In Chap. 4 we have seen that the lower limit logic **LLL** is the maximal monotonic logic that is weaker or equal an adaptive logic based on it (see Theorem 4.6.1). The situation is analogous for $\mathbf{AL}_{\Lambda}$:

Theorem 5.5.7. *Where **L** is a monotonic logic for which $Cn^{\mathcal{L}^+}_{\mathbf{L}}(\Theta) \subseteq Cn^{\mathcal{L}^+}_{\mathbf{AL}_{\Lambda}}(\Theta)$ for all $\Theta \subseteq \mathcal{W}^+$: $Cn^{\mathcal{L}^+}_{\mathbf{L}}(\Gamma) \subseteq Cn^{\mathcal{L}^+}_{\mathbf{LLL}}(\Gamma)$ for all $\Gamma \subseteq \mathcal{W}^+$.*

Proof. The proof is nearly identical to the proof of Theorem 4.6.1. The only difference (besides the obvious replacing of occurrence of $\mathbf{AL^m}$ by $\mathbf{AL}_{\Lambda}$) concerns the paragraph after (4.8.) It is replaced by:

> Note first that due to (4.3) and (4.5), $\Gamma' \neq \emptyset$ (otherwise there is obviously no way to finally derive B in an $\mathbf{AL}_{\Lambda}$-proof from Γ). Thus, $\emptyset \notin \Phi(\Gamma \cup \Gamma')$. Note that by (4.6) $\Gamma \cup \Gamma'$ is **LLL**-non-trivial. Hence, $\mathsf{Ab}^{\Gamma \cup \Gamma'}_{\mathbf{LLL}} \neq \emptyset$ and thus by Lemma 5.3.2, $\Xi^{\mathrm{sat}}(\Gamma \cup \Gamma') \neq \emptyset$. By Lemma 5.4.2, there is a $\varphi \in \Lambda(\Xi^{\mathrm{sat}}(\Gamma \cup \Gamma'))$ and a $\Delta_{\varphi} \subseteq \Omega \setminus \varphi$ such that $\Gamma \cup \Gamma' \vdash_{\mathbf{LLL}} B \check{\vee} \mathsf{Dab}(\Delta_{\varphi})$.

The rest of the proof proceeds analogously. □

Similar as in Chap. 4 we obtain by Fact 4.3.1 a further corollary.

Corollary 5.5.5. *If $Cn^{\mathcal{L}^+}_{\mathbf{AL}_{\Lambda}}(\Gamma)$ is closed under a monotonic logic **L**, then $Cn^{\mathcal{L}^+}_{\mathbf{L}}(\Gamma) \subseteq Cn^{\mathcal{L}^+}_{\mathbf{LLL}}(\Gamma)$ for all $\Gamma \subseteq \mathcal{W}^+$.*

5.5.5 Further Properties

For a more thorough study of further meta-theoretic properties and their relationship to the threshold function Λ we refer the reader to Sten Lindström's [12]. Lindström's study concerns only selection semantics. Due to our soundness and completeness

Table 5.3 Overview of other meta-theoretic properties

Distributivity	$Cn_{\mathbf{AL}_\Lambda}(\Gamma) \cap Cn_{\mathbf{AL}_\Lambda}(\Gamma')$ $\subseteq$ $Cn_{\mathbf{AL}_\Lambda}(Cn_{\mathbf{LLL}}(\Gamma) \cap Cn_{\mathbf{LLL}}(\Gamma'))$	$\Lambda(X \cup Y)$ $=$ $\Lambda(X) \cup \Lambda(Y)$
Chernoff	$Cn_{\mathbf{AL}_\Lambda}(\Gamma \cup \Gamma')$ $\subseteq$ $Cn_{\mathbf{LLL}}(Cn_{\mathbf{AL}_\Lambda}(\Gamma) \cup \Gamma')$	$\Lambda(X) \cap Y$ $=$ $\Lambda(X \cap Y)$
Path independence	$Cn_{\mathbf{AL}_\Lambda}(Cn_{\mathbf{AL}_\Lambda}(\Gamma) \cap Cn_{\mathbf{AL}_\Lambda}(\Gamma'))$ $=$ $Cn_{\mathbf{AL}_\Lambda}(Cn_{\mathbf{LLL}}(\Gamma) \cap Cn_{\mathbf{LLL}}(\Gamma'))$	$\Lambda(\Lambda(X) \cup \Lambda(Y))$ $=$ $\Lambda(X \cup Y)$
Sen	If $Cn_{\mathbf{AL}_\Lambda}(\Gamma) \cup Cn_{\mathbf{AL}_\Lambda}(\Gamma')$ is **LLL**-non-trivial, then: $Cn_{\mathbf{AL}_\Lambda}(\Gamma) \cup Cn_{\mathbf{AL}_\Lambda}(\Gamma') \subseteq$ $Cn_{\mathbf{AL}_\Lambda}(\Gamma \cup \Gamma')$	$\Lambda(X) \cap \Lambda(Y) \neq \emptyset$ implies $\Lambda(X \cap Y) \subseteq$ $\Lambda(X) \cap \Lambda(Y)$
Arrow	If $Cn_{\mathbf{AL}_\Lambda}(\Gamma) \cup \Gamma'$ is **LLL**-non-trivial, then $Cn_{\mathbf{AL}_\Lambda}(\Gamma \cup \Gamma') =$ $Cn_{\mathbf{LLL}}(Cn_{\mathbf{AL}_\Lambda}(\Gamma) \cup \Gamma')$	$\Lambda(X) \cap Y \neq \emptyset$ implies $\Lambda(X \cap Y) = \Lambda(X) \cap Y$

result they apply to the semantic and syntactic consequence relations corresponding to $\mathbf{AL}_\Lambda$ as well. Table 5.3 provides a selective overview.

5.6 Simplifying the Proof-Theory for a Subclass of Λ-Based ALs

In the analysis of Example 5.3.1 we have seen that in general it is necessary to apply the threshold function Λ to only the satisfiable choice sets and hence to let $\Lambda_s^\Gamma = \Lambda(\Xi_s^{\text{sat}}(\Gamma))$ in the marking definition 5.3.4. We have seen in Sect. 2.4.2 that in the standard format we can use the more simple $\Lambda_s^\Gamma = \min_\subset(\Xi_s(\Gamma))$ for minimal abnormality (instead of $\Lambda_s^\Gamma = \min_\subset(\Xi_s^{\text{sat}}(\Gamma))$). This raises the question for a general criterion which ensures that we can use $\Lambda_s^\Gamma = \Lambda(\Xi_s(\Gamma))$ instead of $\Lambda_s^\Gamma = \Lambda(\Xi_s^{\text{sat}}(\Gamma))$. Such a criterion is:

SIMP $\Lambda(\Xi(\Gamma)) \setminus \Xi^\perp(\Gamma) = \Lambda(\Xi^{\text{sat}}(\Gamma))$

Let $\vdash_{\mathbf{AL}_\Lambda^*}$ be defined as $\vdash_{\mathbf{AL}_\Lambda}$ just that instead of $\Lambda_s^\Gamma = \Lambda(\Xi_s^{\text{sat}}(\Gamma))$ we use $\Lambda_s^\Gamma = \Lambda(\Xi_s(\Gamma))$ in the marking definition 5.3.4. We have[29]:

Theorem 5.6.1. *Where $\Gamma \subseteq \mathcal{W}$: If Λ satisfies* **SIMP**, *then $\Gamma \vdash_{\mathbf{AL}_\Lambda} A$ iff $\vdash_{\mathbf{AL}_\Lambda^*} A$ iff $\Gamma \Vdash_{\mathbf{AL}_\Lambda} A$.*

[29] This is proven in Appendix C.4.

This shows that in case Λ satisfies **SIMP** we can slightly simplify the proof theory while avoiding problem **P2** (see Sect. 5.3.2). Note though that problem **P1** is not avoided: still it may be the case that some choice sets are selected that are not satisfiable and in that sense redundant. Only, given **SIMP** these choice sets do not disturb the adequacy of the syntactic consequence relation with respect to the semantic consequence relation (as it was the case in Example 5.3.1).

In the Appendix we show that.[30]

Fact 5.6.1. $\mathbf{DI}_{\prec}$ *implies* **SIMP**.

For instance, the threshold functions $\min_{\subset}$, $\Psi_{\prec}$, $\Psi_{[\prec_1,\ldots,\prec_n]}$, and $\min_{\prec}$ (in case smoothness is guaranteed) satisfy $\mathbf{DI}_{\prec}$ (where $\prec\ =\ \prec_n$ in the case of $\Psi_{[\prec_1,\ldots,\prec_n]}$). Hence, these functions also satisfy **SIMP**.

5.7 Normal Selections: The Logic $\mathbf{AL}^{\mathbf{n}}_{\mathbf{\Lambda}}$

Recall the normal selections strategy from Sect. 2.8.

Syntactically the idea was that a conditional inference is not to be retracted in case it is defensible. We have called an argument on line l with a condition Δ defensible in case the assumption holds with respect to some sufficiently normal interpretation of the Dab-formulas derived from the premises at the corresponding stage. The sufficiently normal interpretations at stage s are given by $\Lambda^{\Gamma}_{s} = \Lambda(\Xi^{\mathrm{sat}}_{s}(\Gamma))$. This gives rise to the following marking definition:

Definition 5.7.1 (**Marking for Normal Selections**). A line l with condition Δ is marked at stage s iff for all $\varphi \in \Lambda^{\Gamma}_{s}$, $\varphi \cap \Delta \neq \emptyset$.

The syntactic consequence relation is defined as usual in terms of finally derivable formulas (where the definition of final derivability is exactly the same as for $\mathbf{AL}_{\mathbf{\Lambda}}$, see Definition 5.3.7).

Definition 5.7.2. $\Gamma \vdash_{\mathbf{AL}^{\mathbf{n}}_{\mathbf{\Lambda}}} A$ iff A is finally derivable in a $\mathbf{AL}^{\mathbf{n}}_{\mathbf{\Lambda}}$-proof from Γ.

We have an analogous representation theorem as in Sect. 2.8:

Theorem 5.7.1. *Where $\Gamma \subseteq \mathcal{W}$ or $\Gamma = Cn^{\mathcal{L}^+}_{\mathbf{LLL}}(\Gamma)$: $\Gamma \vdash_{\mathbf{AL}^{\mathbf{n}}_{\mathbf{\Lambda}}} A$ iff there is a $\varphi \in \Lambda(\Xi^{\mathrm{sat}}(\Gamma))$ and a $\Delta \subseteq \Omega \setminus \varphi$ such that $\Gamma \vdash_{\mathbf{LLL}} A \check{\vee} \mathsf{Dab}(\Delta)$.*

Proof. "$\Rightarrow$": Let $\Gamma \vdash_{\mathbf{AL}^{\mathbf{n}}_{\mathbf{\Lambda}}} A$. Hence, there is a finite stage of a proof $\mathcal{P}$ at which A is finally derived at some line l on a condition Δ. We claim that there is a $\varphi \in \Lambda(\Xi^{\mathrm{sat}}(\Gamma))$ such that $\Delta \cap \varphi = \emptyset$. To see that we extend the proof to the complete stage $\mathrm{g}(\mathcal{P})$. We have $\Xi^{\mathrm{sat}}(\Gamma) = \Xi^{\mathrm{sat}}_{\mathrm{g}(\mathcal{P})}(\Gamma) = \Xi^{\mathrm{sat}}_{s}(\Gamma)$ for any successor stage s. Were there no $\varphi \in \Lambda(\Xi^{\mathrm{sat}}(\Gamma))$ such that $\Delta \cap \varphi = \emptyset$ then line l would be marked

[30] See Theorem C.4.3.

and would remain marked from this stage on,—a contradiction to the fact that A was finally derived at line l.

"$\Leftarrow$": Suppose there is a $\varphi \in \Lambda(\Xi^{\text{sat}}(\Gamma))$ for which $\varphi \cap \Delta = \emptyset$ and $\Gamma \vdash_{\mathbf{LLL}} A \check{\vee} \mathsf{Dab}(\Delta)$. By Lemma 5.4.3, there is a finite proof such that A is derived on some line l on the condition Δ and line l is unmarked. Suppose line l is marked in an extension of the proof $\mathcal{P}$. We extend the proof further to the complete stage $\mathrm{g}(\mathcal{P})$. Since $\varphi \cap \Delta = \emptyset$ and $\varphi \in \Lambda\big(\Xi^{\text{sat}}(\Gamma)\big) = \Lambda\big(\Xi^{\text{sat}}_{\mathrm{g}(\mathcal{P})}(\Gamma)\big)$ line l is unmarked again. □

Semantically the idea was as follows: A is a semantic consequence of Γ if we can find a sufficiently normal **LLL**-model M of Γ such that in all **LLL**-models M' of Γ in which the same (or even more) normality assumptions hold as in M, A is true. In signs:

Definition 5.7.3. $\Gamma \Vdash_{\mathbf{AL}^{\mathbf{n}}_{\Lambda}} A$ iff there is a $M \in \mathcal{M}_{\mathbf{AL}_{\Lambda}}(\Gamma)$ such that for all $M' \in \mathcal{M}_{\mathbf{LLL}}(\Gamma)$ for which $\mathrm{Ab}(M') \subseteq \mathrm{Ab}(M)$, $M' \models A$.

In cases in which $\Lambda(\mathsf{Ab}^{\Gamma}_{\mathbf{LLL}}) \subseteq \min_{\subset}(\mathsf{Ab}^{\Gamma}_{\mathbf{LLL}})$, this comes down to identifying an $\sim$-equivalence class of selected models where $M \sim M'$ iff $\mathrm{Ab}(M) = \mathrm{Ab}(M')$. In that case it is immediately clear that our definition is equivalent to: $\Gamma \Vdash_{\mathbf{AL}^{\mathbf{n}}_{\Lambda}} A$ iff there is an $\mathcal{M} \in \mathcal{M}_{\mathbf{AL}_{\Lambda}}(\Gamma)/\sim$ such that for all $M \in \mathcal{M}$, $M \models A$.[31] This is the way we characterized the normal selections strategy in Sect. 2.8.

We have the following representation theorem that is analogous to the syntactic version:

Theorem 5.7.2. *Where $\Gamma \subseteq \mathcal{W}^{+}$: $\Gamma \Vdash_{\mathbf{AL}^{\mathbf{n}}_{\Lambda}} A$ iff there is a $\varphi \in \Lambda(\Xi^{\text{sat}}(\Gamma))$ such that $\Gamma \cup (\Omega \setminus \varphi)^{\check{\neg}} \Vdash_{\mathbf{LLL}} A$.*

Proof. $\Gamma \Vdash_{\mathbf{AL}^{\mathbf{n}}_{\Lambda}} A$, iff, there is a $M \in \mathcal{M}_{\mathbf{AL}_{\Lambda}}(\Gamma)$ such that for all $M' \in \mathcal{M}_{\mathbf{LLL}}(\Gamma)$ for which $\mathrm{Ab}(M) \supseteq \mathrm{Ab}(M')$, $M \models A$, iff, there is a $\varphi \in \Lambda(\mathsf{Ab}^{\Gamma}_{\mathbf{LLL}})$ such that for all $M \in \mathcal{M}_{\mathbf{LLL}}(\Gamma)$ for which $\mathrm{Ab}(M) \subseteq \varphi$, $M \models A$, iff, there is a $\varphi \in \Lambda(\mathsf{Ab}^{\Gamma}_{\mathbf{LLL}})$ such that $\Gamma \cup (\Omega \setminus \varphi)^{\check{\neg}} \Vdash_{\mathbf{LLL}} A$, iff [by Lemma 5.3.2], there is a $\varphi \in \Lambda(\Xi^{\text{sat}}(\Gamma))$ such that $\Gamma \cup (\Omega \setminus \varphi)^{\check{\neg}} \Vdash_{\mathbf{LLL}} A$. □

The corollary follows immediately by the compactness of **LLL**.

Corollary 5.7.1. *Where $\Gamma \subseteq \mathcal{W}^{+}$: $\Gamma \Vdash_{\mathbf{AL}^{\mathbf{n}}_{\Lambda}} A$ iff there is a $\varphi \in \Lambda(\Xi^{\text{sat}}(\Gamma))$ and a $\Delta \subseteq \Omega \setminus \varphi$ such that $\Gamma \vdash_{\mathbf{LLL}} A \check{\vee} \mathsf{Dab}(\Delta)$.*

Hence, we immediately get soundness and completeness:

Corollary 5.7.2. *Where $\Gamma \subseteq \mathcal{W}$ or $\Gamma = Cn^{\mathcal{L}^{+}}_{\mathbf{LLL}}(\Gamma)$: $\Gamma \vdash_{\mathbf{AL}^{\mathbf{n}}_{\Lambda}} A$ iff $\Gamma \Vdash_{\mathbf{AL}^{\mathbf{n}}_{\Lambda}} A$.*

---

[31] Or equivalently: $\Gamma \Vdash_{\mathbf{AL}^{\mathbf{n}}_{\Lambda}} A$ iff there is an $M \in \mathcal{M}_{\mathbf{AL}_{\Lambda}}(\Gamma)$ such that for all $M' \in \mathcal{M}_{\mathbf{LLL}}(\Gamma)$ for which $\mathrm{Ab}(M) = \mathrm{Ab}(M')$, $M' \models A$.

In the Appendix we show that $\min_{\subset}(\Xi^{\mathrm{sat}}(\Gamma)) = \min_{\subset}(\Xi(\Gamma))$ (where Γ is **LLL**-non-trivial, see Corollary C.1.1).[32] This immediately implies that $\mathbf{AL}^{\mathbf{n}}_{\min_{\subset}}$ has indeed the same consequence relation as $\mathbf{AL}^{\mathbf{n}}$ from Sect. 2.8.

5.8 Wrapping Things Up: Some Concrete Classes of Λ-Based ALs

In this chapter we wrap things up by relating our meta-theoretic insights to—among others—the examples presented in Sect. 5.2.

5.8.1 *The Threshold Functions* $\min_{\prec}$, $\Psi_{\prec}$, *and* $\Psi_{[\prec_1,\ldots,\prec_n]}$

The traditional perspective associated with preference semantics in the vein of Shoham is to select $\prec$-minimal models (of a given premise set) with respect to a partial order $\prec$.

One of the most serious problems behind this approach is the possibility of non-smoothness. A model may be such that there is no $\prec$-minimal model below. In the worst case there are no $\prec$-minimal models and we loose **T3** (see Sect. 5.2.6) in which case $\min_{\prec}$ is not a selection function according to our definition. Sometimes we may encounter asymmetric situations where for some models there are minimal models below them, while for other models there are none. Only selecting the minimal models in such cases can lead to rather counter-intuitive results. Suppose for instance that (a) in all minimal models C holds, and (b) in each infinitely descending chain of models there is a M such that in all models below M, C doesn't hold. This is illustrated in the following example.

Example 5.8.1. We use the logic $\mathbf{AL}^{\circ}_{\min_{\prec_{\mathsf{co}}}}$ from Sect. 5.2.6. Let the premise set be

$$\Gamma = \left\{!^1 q_1 \vee (!^i p_i \vee !^j p_j) \mid i, j \in \mathbb{N}, i < j\right\} \cup \\ \left\{!^i q_i \supset !^{i+1} q_{i+1} \mid i \in \mathbb{N}\right\} \cup \left\{!^1 q_1 \supset r, \check{\neg} !^1 q_1 \supset \neg r\right\}$$

where $!^i p =_{\mathrm{df}} \circ_i p \wedge \neg p$. Figure 5.4 represents an excerpt of the order on the (abnormal parts of the) $\mathbf{L}^*_{\circ}$-models of Γ by means of $\prec_{\mathsf{co}}$ where $\mathrm{Ab}(M_q) = \{!^i q_i \mid i \in \mathbb{N}\}$, $\mathrm{Ab}(M^p_i) = \{!^j p_j \mid j \neq i\}$ and $\mathrm{Ab}(M^p_q) = \mathrm{Ab}(M_q) \cup \{!^i p_i \mid i \in \mathbb{N}\}$.

It is not difficult to see that all models M in $\min_{\prec_{\mathsf{co}}}(\mathsf{Ab}^{\Gamma}_{\mathbf{L}^*_{\circ}})$ have the same abnormal part as M_q. Note that for all these models $M \models r$ (due to the premise $!^1 q_1 \supset r$).

[32] Note that $\min_{\subset}(\Xi(\Gamma))$ is just another way of writing $\Phi(\Gamma)$.

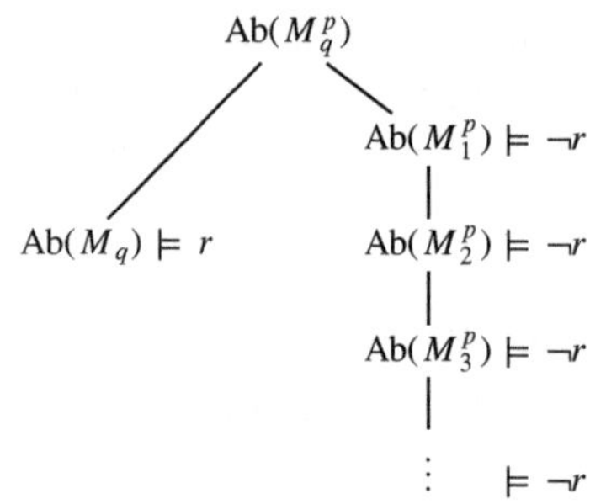

Fig. 5.4 Illustration with an infinite chain of models

Similarly, as indicated in the figure, $M_i^p \models \neg r$ for all $i \geq 1$. Altogether, since all selected models in $\min_{\prec^{co}}(\mathsf{Ab}^{\Gamma}_{\mathbf{L}^*_\circ})$ validate r we have $\Gamma \Vdash_{\mathbf{AL}^\circ_{\min_{\prec^{co}}}} r$.

We have presented a way to avoid such problems. The idea is to use $\Psi_\prec$ (or a more refined $\Psi_{[\prec_1,\ldots,\prec_n]}$) instead of $\min_\prec$.[33] Since $\min_\prec(\Xi(\Gamma)) = \Psi_\prec(\Xi(\Gamma))$ whenever $\langle \Xi(\Gamma), \prec\rangle$ is smooth, these logics are equivalent for all "non-problematic" Γ. Hence, whenever $\mathbf{RA}_\prec$ holds, the logics $\mathbf{AL}_{\min_\prec}$ and $\mathbf{AL}_{\Psi_\prec}$ lead to an identical consequence relation. However, for $\mathbf{AL}_{\Psi_\prec}$ (and for $\mathbf{AL}_{\Psi_{[\prec_1,\ldots,\prec_n]}}$) the problems pointed out above for non-smooth cases are avoided. Moreover, we get a rich meta-theory:

- Since $\Psi_\prec$ and $\Psi_{[\prec_1,\ldots,\prec_n]}$ satisfy **CT** and **CM**,[34] by Corollary 5.5.1 we get cumulativity, by Corollary 5.5.2 we get the fixed point property, and by Corollary 5.5.3 the same criteria for equivalent premise sets are satisfied as in the standard format (see Chap. 4).
- Since $\Psi_\prec$ and $\Psi_{[\prec_1,\ldots,\prec_n]}$ satisfy $\mathbf{RA}_\prec$, by Fact 5.5.2 we get strong reassurance.
- Finally, $\Psi_\prec$ and $\Psi_{[\prec_1,\ldots,\prec_n]}$ satisfy **SIMP** and hence we can use the simplified proof theory from Sect. 5.6.[35]

5.8.2 *The Standard Format*

ALs with the minimal abnormality strategy are representable in our format by means of the threshold function $\Lambda = \min_\subset$.[36] Since by Theorem 5.5.4 $\langle \Xi^{\text{sat}}(\Gamma), \min_\subset\rangle$ is smooth, we get the full meta-theory discussed in Sect. 5.8.1: cumulativity, fixed point, strong reassurance, etc.

---

[33] In the Appendix we show that $\Psi_\prec$ and $\Psi_{[\prec_1,\ldots,\prec_n]}$ are threshold functions (see Fact C.3.8).

[34] See Appendix: $\Psi_\prec$ satisfies both $\mathbf{DI}_\prec$ (Fact C.3.3) and $\mathbf{RA}_\prec$ (Fact C.3.4), and hence by Fact 5.5.1 also **CT** and **CM**. $\Psi_{[\prec_1,\ldots,\prec_n]}$ satisfies **CT** and **CM** by Fact C.3.12.

[35] By Fact C.3.3, $\Psi_\prec$ satisfies $\mathbf{DI}_\prec$ and hence by Fact 5.6.1 also **SIMP**. By Fact C.3.13, $\Psi_{[\prec_1,\ldots,\prec_n]}$ satisfies $\mathbf{DI}_{\prec_n}$ and hence by Fact 5.6.1 also **SIMP**.

[36] **T1** and **T2** are trivially satisfied, for **T3** see the discussion in Sect. 5.5.3.

In order to represent the reliability strategy we let[37]

$$\Lambda(X) = \min_{\subset}^{\cup}(X) = \left\{\varphi \in X \mid \varphi \subseteq \bigcup \min_{\subset}(X)\right\}$$

Note first that $\min_{\subset}^{\cup}$ is a threshold function (see footnote 36)

We get the following properties:

- Since $\min_{\subset}^{\cup}$ trivially satisfies **C4** we get[38] $\mathbf{RA}_{\subset}$ and hence by Fact 5.5.2 we have strong reassurance.
- Since we obviously have **C2**, we get $\mathbf{DI}_{\subset}$. Hence, since we also have $\mathbf{RA}_{\subset}$, by Fact 5.5.1 we get **CT** and **CM**. Thus, by Corollary 5.5.1 we get cumulativity, by Corollary 5.5.2 we get the fixed point property, and by Corollary 5.5.3 we get the criteria for equivalent premise sets.
- By $\mathbf{DI}_{\subset}$ and by Fact 5.6.3 we have **SIMP**. Thus, we can apply the simplified proof theory of Sect. 5.6.

5.8.3 (Co)-Lexicographic ALs

Lexicographic ALs as introduced in [4] are represented in our generic format by Λ_{lex} where $\prec_{\mathsf{lex}}$ is defined as in Definition 5.2.1 on the basis of a structured set of abnormalities $\Omega = \bigcup_I \Omega_i$. Note that

Fact 5.8.1. $\subset\ \subseteq \prec_{\mathsf{lex}}$

In the appendix of [4] we have shown that $\langle \Xi(\Gamma), \prec_{\mathsf{lex}}\rangle$ is smooth. Since $\Xi^{\mathrm{sat}}(\Gamma)$ is $\prec_{\mathsf{lex}}$-dense in $\Xi(\Gamma)$,[39] this means that also $\langle \Xi^{\mathrm{sat}}(\Gamma), \prec_{\mathsf{lex}}\rangle$ is smooth. Hence, we get all the properties discussed in Sect. 5.8.1 such as cumulativity, fixed point and strong reassurance, etc.

Our more generic perspective in this chapter gives rise to other lexicographic ALs which do not fall within the scope of the format presented in [4]. One may for instance define $\prec'_{\mathsf{lex}} \subseteq \wp(\Omega) \times \wp(\Omega)$ as follows on the basis of partial orders $\prec_i \subseteq \wp(\Omega_i) \times \wp(\Omega_i)$:

Definition 5.8.1. Where $\varphi, \psi \subseteq \Omega$ are sets of abnormalities, φ is preferable to ψ, in signs $\varphi \prec'_{\mathsf{lex}} \psi$, iff, there is an $n \in \mathbb{N}$ for which

(a) $\varphi \cap \Omega_i = \psi \cap \Omega_i$ for all $i < n$ and
(b) $\varphi \cap \Omega_n \prec_n \psi \cap \Omega_n$.

---

[37] The fact that this logic indeed characterizes the reliability strategy is proven in Theorem C.2.1 in the Appendix.

[38] For this and other relationships among the criteria see Fig. 5.3. For the proofs see Fact C.3.1 in the Appendix.

[39] See Lemma C.1.1 in the Appendix.

If we are able to guarantee the smoothness of $\langle \Xi(\Gamma), \prec'_{\mathsf{lex}} \rangle$ we can define Λ'_{lex} as before by $\min_{\prec'_{\mathsf{lex}}}$. Otherwise we define Λ'_{lex} by $\Psi_{\prec'_{\mathsf{lex}}}$. As pointed out in Sect. 5.8.1, we get the full meta-theory (soundness, completeness, cumulativity, fixed point, strong reassurance, etc.) for the resulting logic and can apply the simplified proof theory from Sect. 5.6.

In Sect. 5.2.5 we have seen another variant: instead of a lexicographic order we employed a colexicographic order. Unlike $\prec_{\mathsf{lex}}$, colexicographic orders sometimes give rise to non-smoothness. Hence, we used the selection function $\Psi_{\prec_{\mathsf{co}}}$ or the refined variant $\Psi_{[\prec_{\mathsf{co}}, \subset]}$. As discussed in Sect. 5.8.1, we get the full meta-theory for these logics (soundness, completeness, cumulativity, fixed point, and strong reassurance, etc.) and can apply the simplified proof theory from Sect. 5.6.

5.8.4 More on Quantitative Variants

5.8.4.1 Counting Strategies

In Sect. 5.2.3 we have introduced the order $\prec_c$. Note that

Fact 5.8.2. *Where $\Gamma \subseteq \mathcal{W}^+$: $\langle \Xi^{\mathrm{sat}}(\Gamma), \prec_c \rangle$ is smooth.*

Proof. *1. case:* $\Xi^{\mathrm{sat}}(\Gamma)$ only contains infinite sets. In this case $\min_{\prec_c}(\Xi^{\mathrm{sat}}(\Gamma)) = \min_{\subset}(\Xi^{\mathrm{sat}}(\Gamma))$ (recall that $\subset \subseteq \prec_c$). Hence we get smoothness by Theorem 5.5.4. *2. case:* $\Xi^{\mathrm{sat}}(\Gamma)$ contains finite sets. Let $n = \min_{<}(\{|\varphi| \mid \varphi \in \Xi^{\mathrm{sat}}(\Gamma)\})$. Note that $\min_{\prec_c}(\Xi^{\mathrm{sat}}(\Gamma)) = \{\varphi \in \Xi^{\mathrm{sat}}(\Gamma) \mid |\varphi| = n\} \neq \emptyset$. Let $\psi \in \Xi^{\mathrm{sat}}(\Gamma) \setminus \min_{\prec_c}(\Xi^{\mathrm{sat}}(\Gamma))$ and $\varphi \in \min_{\prec_c}(\Xi^{\mathrm{sat}}(\Gamma))$. Hence, $|\varphi| < |\psi|$ and thus $\varphi \prec_c \psi$. □

Hence, as pointed out in Sect. 5.8.1 we have the full meta-theory (soundness, completeness, cumulativity, fixed point, strong reassurance, etc.) for ALs based on the threshold function $\min_{\prec_c}$ and we can apply the simplified proof theory from Sect. 5.6.

Of course, one may for instance use more refined partial orders for quantitative comparisons, such as

Definition 5.8.2. $\varphi \prec'_c \psi$ iff $|\varphi \setminus \psi| < |\psi \setminus \varphi|$.

It is easy to see that $\prec_c \subseteq \prec'_c$. Note that $\prec'_c$ allows also to compare infinite sets that are not comparable by $\prec_c$. As an easy example apply the two partial orders to $\wp(\mathbb{N})$. Let $\varphi = E \cup \{1, \ldots, 10\}$ and $\psi = E \cup \{21, \ldots, 40\}$ where E is the set of even numbers in $\mathbb{N}$. Note that $\varphi \prec'_c \psi$ since $|\varphi \setminus \psi| = |\{1, 3, \ldots, 9\}| = 5 < 10 = |\{21, 23, \ldots, 39\}| = |\psi \setminus \varphi|$. However, φ and ψ are not comparable with respect to $\prec_c$.

5.8.4.2 More Involved Quantitative Approaches

We have already seen more involved examples that make use of a selection of models based on quantitative considerations in Sect. 5.2.7. Note that both Λ_c^1 and Λ_c^2 are threshold functions. Hence, by Corollary 5.4.2 we get soundness and completeness for both logics. Moreover, it can be easily shown that Λ_c^1 satisfies both **CT** and **CM** and hence we get cumulativity, while cumulative transitivity does not hold for the logic based on Λ_c^2 (see Example 5.5.1). We leave the proofs and examples to the interested reader and focus instead on yet another interesting quantitative approach that can be characterized in terms of a logic $\mathbf{AL}_{\mathbf{\Lambda}}$.

The driving idea behind prioritized ALs can be characterized by an iterative procedure:

- First we pick out models of the premises in which as less abnormalities in Ω_1 are validated as possible.
- Second, we refine the given selection from step 1 in such a way that models are selected that validate as less abnormalities in Ω_2 as possible.
- etc.

In some applications this procedure may be suboptimal. Take for instance default logics that are able to express the specificity order among defaults. Although in most cases it is more intuitive to prefer interpretations of a given set of defaults that violate more general defaults over interpretations that violate more specific defaults. For instance, given we know of Tweety that it is a penguin, and the following defaults, "Bird(X), then fly(X).", "Penguin(X), then not-fly(X)" we are usually inclined to choose an interpretation that violates the former default and we hence conclude that Tweety does not fly.

However, consider a case where we have the choice between violating one more specific default and, say, 20 slightly less specific defaults. In such cases, as has been pointed out by Goldszmidt, Morris and Pearl [20], it may be better to choose an interpretation that violates the more specific default. There are many weighing functions which can be employed for this. I will demonstrate the point with a very simple one.

Suppose that $\Omega = \Omega_1 \cup \ldots \cup \Omega_n$ where $\Omega_i = \{\circ_i A \wedge \neg A \mid A \text{ is } \circ\text{-free}\}$. Let

$$\mu(\Delta) = \begin{cases} \sum_{i=1}^{n} \frac{|\Delta \cap \Omega_i|}{i} & \text{if } \Delta \text{ is finite} \\ \mu(\Delta) = \infty & \text{else} \end{cases}$$

Define $\Delta \prec_\mu \Delta'$ iff $\mu(\Delta) < \mu(\Delta')$ or $\Delta \subset \Delta'$. We denote $\circ_i A \wedge \neg A$ by $!^i A$. In a similar way as in Fact 5.8.2 we can show that $\langle \Xi^{\text{sat}}(\Gamma), \prec_\mu \rangle$ is smooth for all $\Gamma \subseteq \mathcal{W}^+$. Hence, we can define the logic $\mathbf{AL}_{\min_{\prec_\mu}}$ and get the full meta-theory (soundness, completeness, cumulativity, fixed point, strong reassurance) and can apply the simplified proof theory from Sect. 5.6, as pointed out in Sect. 5.8.1.

Example 5.8.2. Suppose we have the situation that an expert of highest expertise states p. However, this is in conflict with the statements q, r, s, t and u which are stated by other experts of slightly less expertise. Hence, suppose our premise set is

$$\Gamma = \{\circ_1 p, \circ_3 q, \circ_2 r, \circ_3 s, \circ_3 t, \circ_3 u, \neg p \vee \neg q, \neg p \vee \neg r, \neg p \vee \neg s, \neg p \vee \neg t, \neg p \vee \neg u\}$$

Hence, we have

$$\Sigma(\Gamma) = \{\{!^1 p, !^3 q\}, \{!^1 p, !^2 r\}, \{!^1 p, !^3 s\}, \{!^1 p, !^3 t\}, \{!^1 p, !^3 u\}\}$$

which gives to the following $\subset$-minimal choice sets in $\Xi^{\text{sat}}(\Gamma)$: $\varphi = \{!^1 p\}$ and $\psi = \{!^3 q, !^2 r, !^3 s, !^3 t, !^3 u\}$. We have:

$$\mu(\varphi) = \frac{1}{1} = 1 < \mu(\psi) = 0 + \frac{1}{2} + \frac{4}{3} = 1\frac{5}{6}$$

and hence, $\varphi <_\mu \psi$. This means that with the AL characterized by the triple $\langle \mathbf{L}_\circ^*, \Omega, \min_{\prec_\mu} \rangle$, q, r, s and t are finally derivable while we don't get p. Hence, the expert opinions of the group of less expertise opposing our number one expert is prioritized by the logic.

In contrast, in a prioritized AL such as the one characterized by $\langle \mathbf{L}_\circ^*, \Omega, \min_{\prec_{\text{lex}}} \rangle$ the situation would be inverse: $\psi \prec_{\text{lex}} \varphi$ and hence we get the consequence p while q, r, s and u are not finally derivable. The reason is that the order $\prec_{\text{lex}}$ proceeds strictly stepwise: since $\psi \cap \Omega_1 \subset \varphi \cap \Omega_1$ the order $\prec_{\text{lex}}$ doesn't take into account any differences between the two choice sets concerning abnormalities of higher levels. The situation is different for $\prec_\mu$ since here the fact that on the higher levels 2 and 3 φ is less normal than ψ is the reason that $\varphi \prec_\mu \psi$ despite that fact that ψ fares better with respect to Ω_1.

Of course, in many cases both approaches will agree. Suppose for instance that our number one expert who states p is only opposed by one expert of lesser expertise who states $\neg p$. Hence, let $\Gamma' = \{\circ_1 p, \circ_2 \neg p\}$. Now we get $\Xi^{\text{sat}}(\Gamma') \supseteq \{\{!^1 p\}, \{!^2 \neg p\}\}$. We have $\mu(\{!^1 p\}) = 1 > \mu(\{!^2 \neg p\}) = 1/2$ and hence $\{!^2 \neg p\} \prec_\mu \{!^1 p\}$. We also have $\{!^2 \neg p\} \prec_{\text{lex}} \{!^1 p\}$. Hence, in both logics, p is a consequence.

5.9 Conclusion

In this chapter it was demonstrated that the standard format for ALs can be generalized in a natural way. Semantically speaking, in the standard format the models are compared with respect to the set of abnormalities they verify and set inclusion. The generalization allows for comparisons with respect to any partial order $\prec$. Also, we allow for a rich class of threshold functions Λ that select the models of the AL out of the models of the lower limit logic. We have shown that a huge class of ALs based on Λ has a strong meta-theory (soundness, completeness, cumulativity, fixed point, (strong) reassurance).

The generalization is natural since the main mechanisms of the standard format remain intact:

- As in the standard format we still have a selection semantics in which models are selected by virtue of their abnormal parts
- The dynamic proof format is the same as in the standard format. Formulas are derived conditionally where the conditions are sets of abnormalities. Derivations are executed by means of the familiar three generic rules PREM, RU, and RC. The marking is structurally analogous to the marking of the minimal abnormality strategy. The only significant difference is that instead of using the $\subset$-minimal choice sets with respect to the derived disjunctions of abnormalities, we use choice sets selected by the threshold function Λ.

We have given some special attention to the problem of non-smoothness. In the standard format where the abnormal parts of models are compared with respect to $\subset$ this problem does not appear: e.g., in the semantics the abnormal parts of the lower limit logic models of some premise set Γ are always guaranteed to be smooth with respect to $\subset$. However, as soon as we introduce other partial orders $\prec$ instead of $\subset$, this may not hold anymore. We have offered a way to deal with such situations: instead of using the threshold function $\min_{\prec}$ we introduced the threshold function $\Psi_{\prec}$. In this way the problems connected to non-smoothness are avoided while for premise sets in which smoothness is guaranteed we get equivalent consequence sets (compared to the logic based on $\min_{\prec}$).

Finally, one may argue that generalizations as the one offered in this chapter are useless formal overkills, it is shooting at flies with cannon-balls. For our defense let us point out the following.

First, we have hoped to convinced the reader that there are many applications that are intuitively represented with orders different from $\subset$. One class of applications concerns situations in which we have a structured set of abnormalities (e.g., they may be prioritized as in the lexicographic and the colexicographic format). The recently developed format of lexicographic ALs gave already rise to various useful formal systems. Similar developments can be expected for the many other possibilities that are offered within the new format. Another class of applications concerns quantitative approaches, that is logics in which we compare the abnormal parts of models by means of quantitative considerations (rather than qualitative ones such as in the standard format).

Once the reader agrees that there are application contexts in which a departure from the standard format is useful (or even needed), it is not difficult to further convince her of the usefulness of having a generic format with a rich associated meta-theory. On the one hand this reduces labor: once a logic is devised in this format there is no need anymore to check and prove many of the interesting meta-theoretic properties. We get them for free based on the research offered in this chapter. On the other hand, generic formats are fruitful for unification. Logics formulated in the same format can be easier compared, techniques used for one application context can be easier transferred to other application contexts, etc.

Finally, some may argue or at least conjecture that all ALs can be translated into the standard format. However, even if that were true, it is very likely that many of the translations will turn out to be technically very cumbersome and/or artificial on an intuitive level. Since one of the goals of ALs is to explicate reasoning processes, this seems counter-productive to this goal. In many cases switching to different orderings of the models (e.g., ones based on quantitative considerations) and to different threshold functions may offer a more natural and intuitive explication of reasoning processes than the best possible translations into the standard format can offer.

In this sense we hope that the research offered in this chapter proves useful and fruitful for one of the main goals behind the adaptive logic program: the intuitive and natural explication of defeasible reasoning processes.

Acknowledgments The results presented in this chapter are the product of joint research with Frederik Van De Putte.

References

1. Primiero, G., Meheus, J.: Majority merging by adaptive counting. Synthese **165**, 203–223 (2008)
2. Primiero, G., Meheus, J.: Quasi-merging and pure-arbitration on information for the family of adaptive logics ADM. In: Proceeding of the Workshop on Logic and Intelligent Interaction, ESSLLI, pp. 21–30. Hamburg (2008)
3. Primiero, G., Meheus, J.: Adaptive arbitration by variant counting on commuatative bases with weights. In: Proceedings of the Fusion 2008 Conference, Köln, 1–3 July, pp. 1374–1380. ISBN 978-3-00-024883-2 (2008)
4. Putte, F.V.D., Straßer, C.: Extending the standard format of adaptive logics to the prioritized case. Logique at Analyse **55**(220), 601–641 (2012)
5. Putte, F.V.D., Straßer, C.: A logic for prioritized normative reasoning. J. Logic Comput. **23**(3), 568–583 (2013)
6. Putte, F.V.D.: Generic formats for prioritized adaptive logics. with applications in deontic logic, abduction and belief revision. Ph.D. thesis, Ghent University (March 28th, 2012)
7. Verdée, P.: Combining a rich paraconsistent negation with classical negation by means of an infinitely valued logic (in preparation)
8. Shoham, Y.: A semantical approach to nonmonotonic logics. In: Ginsberg, M.L. (ed.) Readings in Non-Monotonic Reasoning, pp. 227–249. Morgan Kaufmann, Los Altos (1987)
9. Shoham, Y.: Reasoning about change: time and causation from the standpoint of artificial intelligence. Cambridge, MIT Press (1988)
10. McCarthy, J.: Circumscription—a form of non-monotonic reasoning. Artif. Intell. **13**, 27–29 (1980)
11. Schlechta, K.: Coherent Systems. Elsevier, Amsterdam (2004)
12. Lindström, S.: A semantic approach to nonmonotonic reasoning: inference operations and choice. Uppsala Prints and Reprints in, Philosophy, no 10, Department of Philosophy, University of Uppsala (1994)
13. Makinson, D.: General patterns in nonmonotonic reasoning. In: Handbook of Logic in Artificial Intelligence and Logic Programming, vol. III, pp. 35–110. Clarendon Press, Oxford (1994)
14. Rescher, N., Manor, R.: On inference from inconsistent premises. Theor. Decis. **1**, 179–217 (1970)

15. Meheus, J., Straßer, C., Verdée, P.: Which Style of Reasoning to Choose in the Face of Conflicting Information? J. Logic. Comput. (forthcoming)
16. Putte, F.V.D.: Hierarchic adaptive logics. Logic J. IGPL **20**(1), 45–72 (2012)
17. Putte, F.V.D., Straßer, C.: Three formats of prioritized adaptive logics: a comparative study. Logic J. IGPL **2**(21), 127–159 (2013)
18. Batens, D.: Minimally abnormal models in some adaptive logics. Synthese **125**, 5–18 (2000)
19. Bossu, G., Siegel, P.: Saturation, nonmonotonic reasoning and the closed-world assumption. Artif. Intell. **25**, 13–63 (1985)
20. Goldzsmidt, M., Morris, P., Pearl, J.: Amaximumentropy approach to nonmonotonic reasoning. IEEE Trans. Pattern Anal. Mach. Intell. **15**(3), 220–232 (1993)

Part II
Default Reasoning

Chapter 6
Adaptively Applying Modus Ponens in Conditional Logics of Normality

This chapter presents an adaptive logic enhancement of conditional logics of normality that allows for defeasible applications of Modus Ponens to conditionals. In addition to the possibilities these logics already offer in terms of reasoning about conditionals, this way they are enriched by the ability to perform default inferencing. The idea is to apply Modus Ponens defeasibly to a conditional $A \rightsquigarrow B$ and a fact A on the assumption that it is "safe" to do so concerning the factual and conditional knowledge at hand. It is, for instance, not safe if the given information describes exceptional circumstances: although birds usually fly, penguins are exceptional to this rule. The two adaptive standard strategies are shown to correspond to different intuitions, a skeptical and a credulous reasoning type, which manifest themselves in the handling of so-called floating conclusions.

6.1 Introduction

In this section I will first provide a brief introduction to reasoning on the basis of normality and then give an overview of this chapter.

6.1.1 Some Background

Since the early eighties, default reasoning, i.e., reasoning on the basis of what is normally or typically the case, has drawn much attention from philosophical logicians as well as scholars working in Artificial Intelligence. This is not surprising concerning the prominent role which reasoning on the basis of notions such as normality

---

A previous version of this chapter was published in the Journal of Applied Non-Classical Logic under the same title [1].

C. Straßer, *Adaptive Logics for Defeasible Reasoning*, Trends in Logic 38,
DOI: 10.1007/978-3-319-00792-2_6,

and typicality has. It clearly occupies a central place from everyday common sense reasoning to expert reasoning in many domains. Thus, logicians are urged to develop formal models which accurately explicate these reasoning forms.

In recent years the traditional formalisms of default reasoning such as presented in the landmark articles on default logic [2], on circumscription [3], and on autoepistemic logic [4] have been criticized and alternative conditional approaches have been developed.

In pioneering works on logics of conditionals the main interest was to model conditionals in everyday language which have the form "if ...then". Most of the research in this domain has been in the vein of the following influential conditional logics: Stalnaker [5] and Lewis [6] who offer an ontic interpretation of the conditional, Adams [7] who introduces probabilities in the discussion, and Gärdenfors' belief revision principles which are more concerned with acceptability than probability and truth [8].

There has been, especially since the late eighties, an increasing interest in making use of techniques and properties of conditional logics within the field of nonmonotonic reasoning, such as employed in default reasoning or reasoning with respect to prima facie obligations. The focus of this chapter is on conditional logics of normality that have been inspired by pioneering works such as [9–11]. There, a statement of the form $A \leadsto B$ is read as "From A normally/typically follows B" or "If A is the case then normally/typically also B is the case". We will call "$A \leadsto B$" a *conditional*, and a sequence of conditionals, written $A_1 \leadsto A_2 \leadsto \ldots \leadsto A_n$ as an abbreviation for $(A_1 \leadsto A_2) \wedge (A_2 \leadsto A_3) \wedge \ldots \wedge (A_{n-1} \leadsto A_n)$, an *argument*.

Conditional logics are attractive candidates for dealing with default reasoning for various reasons: First, the conditional $\leadsto$ does not have unwanted properties such as Strengthening the Antecedent, from $A \leadsto B$ infer $(A \wedge C) \leadsto B$, Transitivity, from $A \leadsto B$ and $B \leadsto C$ infer $A \leadsto C$, and Contraposition, from $A \leadsto B$ infer $\neg B \leadsto \neg A$. That the validity of any of these properties leads to undesired results in the context of reasoning on the basis of normality is well-known. Take, for instance, Strengthening the Antecedent: although birds usually fly, $b \leadsto f$, penguins do not, $(b \wedge p) \leadsto \neg f$. Thus $(b \wedge p) \leadsto f$ should not be derived. To find similar counterexamples for the other properties is left to the reader (see e.g., [9], p. 92.). Another advantage is the naturalness and simplicity of the representation of default knowledge by conditionals $A \leadsto B$ compared to the cumbersome representation by the classical approaches mentioned above. The latter use rules such as $A \wedge \pi(B) \supset B$ where $\pi(B)$ expresses for instance that we do not believe $\neg B$ in the case of autoepistemic logic, or that B can consistently be assumed in the case of default logic. Furthermore, certain disadvantages of the classical approach can be avoided in the framework of conditional logics. Boutilier for instance argues that certain paradoxes of material implication are inherited by the classical approaches due to the way default knowledge is represented in them (see [9], pp. 89–90).

Starting from the pioneering works such as [9–12] there has been vigorous research activity on conditional logics of normality. To mention a few: they have been applied to belief revision in [13, 14], strengthenings have been proposed for instance to give a more sophisticated account of Strengthening the Antecedent (see [15, 16]), a labeled natural deduction system has been introduced in [17], and various authors have investigated tableaux methods and sequent calculi for conditional logics (see e.g. [18, 19]). Furthermore, the influential work in [11] is greatly generalized in [20] by their plausible nonmonotonic consequence relations, and in [21] by their plausibility measures.

There is a remarkable agreement concerning fundamental properties for default reasoning in the various formal models. These properties have been dubbed *conservative core* by Pearl and Geffner [22] and are also commonly known as the *KLM-properties* (see [11]). Some of the most interesting and important problems in this field are, on the one hand, related to a proper treatment of irrelevant information (see [12]) and, on the other hand, to a proper treatment of specificity.

6.1.2 Contribution and Structure of this Chapter

This chapter tackles another important problem related to conditional logics of normality: while they are able to derive from conditional knowledge bases, i.e., sets of conditionals, other conditionals, their treatment of factual knowledge is mostly rather rudimentary. This concerns most importantly their treatment of Modus Ponens (MP), i.e., to derive B from A and $A \rightsquigarrow B$. We will also speak about detaching B from $A \rightsquigarrow B$ in case A is valid. Usually we do not only have a conditional knowledge base at hand but also factual information $\mathcal{F}$. In order to make use of the knowledge base, it is in our primary interest to derive, given $\mathcal{F}$, what normally should be the case. It goes without saying that for the practical usage of a conditional knowledge base this kind of application to factual information is essential and that the proper treatment of MP for conditionals is a central key to its modeling.

It is clear that full MP should not be applied unrestrictedly to conditional assertions: although birds usually fly, $b \rightsquigarrow f$, we should not deduce that a given bird flies if we also know that it is a penguin, since penguins usually do not fly, $p \rightsquigarrow \neg f$. However, if we do not know anything about it than the fact that it is a bird, MP should be applied to $b \rightsquigarrow f$ and b. Furthermore, it would be useful if this application is of a defeasible kind, since later we might learn that the bird in question is after all a penguin or a kiwi.

In this chapter a simple generic method is presented to enrich a given conditional logic of normality **L** by a defeasible MP. We consider **L** to consist at least of the core properties (see Sect. 6.2). We will refer to **L** as the *base logic*. As hinted above, there are several circumstances when we do not want to apply MP: cases of specificity such as the example with the penguin, or cases in which conditionals conflict, such as the well-known Nixon-Diamond. The central idea presented in this chapter is to apply MP conditionally, namely on the condition that it is safe to apply it. This idea will informally be motivated and outlined in Sect. 6.3. Formally, the conditional applications of MP are realized by ALs, namely $\mathbf{DLp^m}$ and $\mathbf{DLp^r}$ (see Definition 6.4.1). The idea of ALs is to interpret a premise set "as normally as possible" with respect to a certain standard of normality. They allow for some rules to be applied conditionally. In our case, as demonstrated in Sect. 6.4, MP is going to be applied as much as possible, i.e., as long as no cases of overriding via specificity or similar conflicts take place concerning the conditionals to which MP is going to be applied. That is to say, we are going to apply MP to $A \rightsquigarrow B$ and A on the condition that the other factual information at hand does not describe exceptional circumstances with respect to A. As a consequence, detachment from $b \rightsquigarrow f$ and b is for instance blocked if p is the case.

It will be demonstrated that choosing different adaptive strategies serves different intuitions: one corresponding to a more skeptical and the other one corresponding to a more credulous type of reasoning. This difference manifests itself in the handling of so-called floating conclusions.[1]

I will spend some time in demonstrating the modus operandi of the proposed logics and thereby their strengths by having a look at various benchmark examples. In Sect. 6.5 I highlight some advantages of the adaptive logic approach, compare it to other approaches, and discuss some other related issues. The semantics are investigated in Appendix D.

6.2 Conditional Logics, Their Core Properties and Related Work

Conditional logics are often presented in terms of extending classical propositional logic with a conditional operator $\rightsquigarrow$.[2] Our language is defined by the $(\wedge, \vee, \supset, \neg, \equiv)$-closure of the set of propositional variables and conditionals of the form $A \rightsquigarrow B$, where A and B are classical propositional formulas. Hence, to keep things simple we do not consider here nested occurrences of $\rightsquigarrow$ and focus on flat conditional logics. We refer to A as the *antecedent* and to B as the *conclusion* of the conditional. We write $\mathcal{W}$ for the set of all classical propositional formulas (i.e., formulas without occurrences of $\rightsquigarrow$). We abbreviate $(A \rightsquigarrow B) \wedge (B \rightsquigarrow A)$ by $A \sim B$ and $\neg(A \rightsquigarrow B)$ by $A \not\rightsquigarrow B$. Furthermore, we require that a conditional logic **L** satisfies the following core properties, where **CL** is classical propositional logic (see [11])[3]:

[1] A floating conclusion is a proposition that can be reached by two conflicting and equally strong arguments (see our discussion in Sect. 2.5).

[2] In some conditional logics of normality $\rightsquigarrow$ is not primitive. For instance in [9] it is defined by making use of Kripkean modal logic. There the core properties are shown to be equivalent to an extension of the modal logic **S4**. See [21] for a comparative study of various semantic systems for the core properties such as the preferential structures of [11], the ϵ-semantics of [23], the possibilistic structures of [24] and κ-rankings of [25, 26].

[3] We will use the name convention that is associated with conditional logics of normality (see [27, 28]) and not the one associated with nonmonotonic consequence relations which is used e.g. in [11].

$$\text{If } \vdash_{\mathbf{CL}} A \equiv B, \text{ then } \vdash (A \rightsquigarrow C) \equiv (B \rightsquigarrow C) \qquad \text{(RCEA)}$$

$$\text{If } \vdash_{\mathbf{CL}} B \supset C, \text{ then } \vdash (A \rightsquigarrow B) \supset (A \rightsquigarrow C) \qquad \text{(RCM)}$$

$$\vdash A \rightsquigarrow A \qquad \text{(ID)}$$

$$\vdash \big((A \rightsquigarrow B) \wedge ((A \wedge B) \rightsquigarrow C)\big) \supset (A \rightsquigarrow C) \qquad \text{(RT)}$$

$$\vdash \big((A \rightsquigarrow B) \wedge (A \rightsquigarrow C)\big) \supset ((A \wedge B) \rightsquigarrow C) \qquad \text{(ASC)}$$

$$\vdash \big((A \rightsquigarrow C) \wedge (B \rightsquigarrow C)\big) \supset ((A \vee B) \rightsquigarrow C) \qquad \text{(CA)}$$

The logic defined by these rules and axioms is **P**. Note that for instance the following properties are valid in **P**:[4]

$$\vdash \big((A \rightsquigarrow B) \wedge (A \rightsquigarrow C)\big) \supset (A \rightsquigarrow (B \wedge C)) \qquad \text{(CC)}$$

$$\vdash ((A \wedge B) \rightsquigarrow C) \supset (A \rightsquigarrow (B \supset C)) \qquad \text{(CW)}$$

$$\vdash \big((A \sim B) \wedge (B \rightsquigarrow C)\big) \supset (A \rightsquigarrow C) \qquad \text{(EQ)}$$

$$\vdash_{\mathbf{CL}} A \supset B, \text{ then } \vdash A \rightsquigarrow B \qquad \text{(CI)}$$

We consider these properties to be valid for all the conditional logics of normality in the remainder. Adding the following Rational Monotonicity principle to the core properties yields the logic **R** (see [15])[5]:

$$\vdash \big((A \rightsquigarrow C) \wedge (A \not\rightsquigarrow \neg B)\big) \supset ((A \wedge B) \rightsquigarrow C) \qquad \text{(RM)}$$

The core properties are not without criticism. On the one hand, it has been pointed out that certain principles of **P** resp. **R** are not always perfectly intuitive. For instance, Neufeld [29] has argued against (CA), Poole [30] against (CC), and Stalnaker [31] and Giordano et al. [32, 33] against (RM).[6]

On the other hand, the core properties have been criticized for being too weak. Many nonmonotonic strengthenings have been developed in order to overcome certain weaknesses.

Rational closure (see e.g. [15, 36, 37] and Chap. 7) for instance strengthens **R** by means of a Shoham-like preferential semantics [38, 39]. The idea is to assign natural numbers, i.e. ranks, to formulas. The rank indicates how exceptional a formula is. If for instance $(A \vee \neg A) \rightsquigarrow A$ then A has the lowest rank, 0. In our penguin example p is of a higher rank than b since after all $(p \vee b) \rightsquigarrow \neg b$. Each formula is ranked as low

---

[4] The proofs are fairly standard and can be found e.g. in [11].

[5] I adopt the names **P** and **R** for these logics from [19]. Although these are the same names as used for the systems in the pioneering KLM paper [11], the reader may be warned: the approach in terms of conditional logics differs from the KLM perspective which deals with rules of inference rather than with axioms. Also, strictly speaking, Rational Monotonicity as defined in [11] is a rule of inference whereas (RM) as defined above is an axiomatic counterpart to it.

[6] Some weakening or variants of Rational Monotonicity have been proposed: e.g. $\vdash ((A \rightsquigarrow B) \wedge ((A \wedge C) \not\rightsquigarrow \neg B)) \supset ((A \wedge C) \rightsquigarrow B)$ (IRR) in the context of Description Logic by Giordano et al. [34] or in the context of conditional deontic logics $\vdash ((A \rightsquigarrow B) \wedge (A \not\rightsquigarrow \neg(B \wedge C))) \supset ((A \wedge C) \rightsquigarrow B)$ (WRM) by Goble in [35].

as possible. A default $A \rightsquigarrow B$ is in the rational closure of a set of defaults D iff the rank of A is strictly less than the rank of $A \wedge \neg B$. In this way a significant problem of **P** and **R** is tackled, namely its suboptimal treatment of irrelevant information. For instance, the proposition "Tweety is a green bird.", g, will get the same rank as "Tweety is a bird.", b. Hence, the Rational Closure of $\{b \rightsquigarrow f, g \rightsquigarrow b\}$ contains the default $(b \wedge g) \rightsquigarrow f$, that green birds fly. The latter is neither entailed by **P** nor by **R**. Rational Closure has been shown to be equivalent to Pearl's system **Z** (see [16, 36]) which employs a probabilistic interpretation of defaults. These and similar approaches have been criticized for inheriting some of the weaknesses of the core properties (see e.g. [22]) resp. of rational monotonicity (see [33]) and for introducing new problems (see Example 6.4.7).

Giordano et al. introduce another preferential semantics based on **P** and a tableaux calculus for it. Their system $\mathbf{P}_{\min}$ selects models that minimize non-typical worlds with respect to a given set of formulas. Adding to our example the conditional $p \rightsquigarrow a$, that penguins live in the arctic, $\mathbf{P}_{\min}$ concludes nonmonotonically that there are no penguins that do not live in the arctic: $(p \wedge \neg a) \rightsquigarrow \bot$. This is not a consequence of **Z** resp. Rational Closure. However, $\mathbf{P}_{\min}$'s treatment of irrelevant information is suboptimal: unlike **Z** and Rational Closure $\mathbf{P}_{\min}$ does not lead to the consequence $(b \wedge g) \rightsquigarrow f$.

Lehmann's Lexicographic Closure (see [40, 41]) improves on some of the shortcomings of Rational Closure by strengthening it further.[7] On the one hand, it introduces a more rigorous approach to strengthening the antecedent and hence avoids the so-called Drowning Problem (we discuss this in more detail in Sect. 6.5). On the other hand, it makes sure that in cases of contradictory defaults quantitatively as many defaults as possible are satisfied. The policy is to strictly prefer more specific defaults over less specific ones. The quantitative aspect makes the Lexicographic Closure dependent on the way defaults are presented.

The maximum entropy approach of [42] is in the probabilistic tradition of the 1-entailment of system **Z**. It follows a similar intuition as Lexicographic Closure concerning conflicting defaults. One difference is, however, that in some cases the violation of a more specific default may lead to a higher overall entropy than the violation of some less specific defaults and may be thus preferred.

In his critical discussion of the core properties Delgrande [43] points out that there are two interpretations of conditionals $A \rightsquigarrow B$. Many approaches, such as the ones listed above, treat defaults as weak material implications that have a defeasible character, e.g. in specificity cases. He identifies several counter-intuitive instances where the core properties obtain contrapositives of defaults. This, so he argues, is a result of treating default conditionals in terms of material implications rather than in terms of inference rules. In the spirit of the latter perspective he develops a system based on a weakened core logic (in comparison to **P**). He demonstrates that his rule-based system has a lot of nice properties in terms of treating irrelevant informa-

[7] More precisely, Lexicographic Closure strengthens Rational Closure for all defaults with antecedents that have a finite rank: if A has finite rank and $A \rightsquigarrow B$ is in the rational closure of D, then $A \rightsquigarrow B$ is in the lexicographic closure of D.

tion and conflicting defaults. Another rule-based approach is e.g. presented by Dung and Son in [44].

The take on defaults in terms of weak material implications is very obvious in approaches that make abnormality assumptions explicit (see McCarthy's Circumscription [3], Geffner and Pearl's Conditional Entailment [22], as well as the one presented in this chapter). Here a default $A \rightsquigarrow B$ is presented by $A \wedge \alpha \supset B$ (or by both in the case of Conditional Entailment) where α expresses normality conditions that have to hold for this default. The interesting aspect of conditional entailment is that it extracts a priority order on the normality assumptions automatically from the knowledge base. The idea is to interpret a given knowledge base such that the normality assumptions of the defaults are validated "as much as possible". The priority order takes care that in case of conflicts more specific defaults are preferred where possible.

We conclude this section by noting that conditional logics have been successfully applied to various fields. For instance their relevance for belief revision has been investigated in [13, 14, 45]. The description logic $\mathcal{ALC}$ has been enhanced with a "typicality" operator in [34]. Similar to the logics that are going to be presented in the present chapter this system allows for inferences on the basis of factual information. However, in its current form the logic faces the problem of irrelevance pointed out above: given the information that typical birds fly the logic does not allow to infer that typical green birds fly. In order to deal with such problems the authors propose to integrate "a standard mechanism to reason about defaults" (p. 14) which is left for future research. Furthermore, recently conditional logics have been applied to access control and security in [46]. There the authors extend Garg and Abadi's access control logic **ICL** from [47] with intuitionistic conditional logic.

6.3 Modus Ponens in Conditional Logics of Normality

In this section I will informally motivate and outline the main idea behind the modeling of a defeasible MP in this chapter.

A naïve way to apply MP would be to use the unrestricted version

$$\vdash \big((A \rightsquigarrow B) \wedge A\big) \supset B \qquad \text{(MP}\rightsquigarrow\text{)}$$

However, this would lead to logical explosion whenever we are confronted with conflicting defaults, for instance in cases of *specificity*. Informally speaking, specificity occurs if a more specific argument overrides a more general one. One way to formalize this is as follows: if A is the case and $A \rightsquigarrow B \rightsquigarrow C$, as well as $A \rightsquigarrow \neg C$, then $B \rightsquigarrow C$ is overridden by $A \rightsquigarrow \neg C$, or in terms of arguments, $A \rightsquigarrow B \rightsquigarrow C$ is overridden by $A \rightsquigarrow \neg C$. The reader finds an illustration in Fig. 6.1a. The illustrations in the following figures have to be read in a similar way as inheritance networks (see [48]): nodes are in our case propositions, "$A \rightarrow B$" indicates $A \rightsquigarrow B$, "$A \cdot\!\cdot\!> B$"

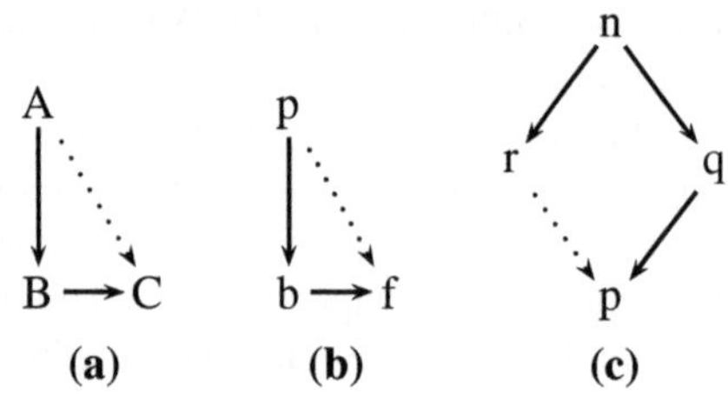

Fig. 6.1 Specificity and the Nixon Diamond

indicates $A \rightsquigarrow \neg B$, "$A \Rightarrow B$" indicates $A \vdash_{\mathbf{CL}} B$, and "$A \Longrightarrow B$" indicates $A \vdash_{\mathbf{CL}} \neg B$.

Example 6.3.1. A standard example illustrating a case of specificity is the following (see Fig. 6.1b):

- Birds normally fly.—$b \rightsquigarrow f$
- Penguins are (normally) birds.—$p \rightsquigarrow b$
- Penguins normally do not fly.—$p \rightsquigarrow \neg f$

The information represented by p is less specific or normal than the information represented by b. Thus, obviously the more specific $p \rightsquigarrow \neg f$ overrides $b \rightsquigarrow f$. This has an important consequence: Given $p \wedge b$ or p we do not want to apply MP to b and $b \rightsquigarrow f$. However, if we only have b as factual knowledge it would be justified on the basis of default reasoning to apply MP to b and $b \rightsquigarrow f$.

Since, as argued above, full MP is highly problematic in the context of default inferencing, we will in the remainder make use of a restricted MP. The idea is to restrict MP to "safe" antecedents. In order to express this, we introduce a unary operator $\bullet$ into our language which is applicable to propositional formulas. $\bullet A$ expresses that the given factual information is atypical or exceptional for A. Hence, in case $\bullet A$, MP should not be applied to conditionals with antecedent A. The following restricted MP realizes this idea.

$$\vdash \big((A \rightsquigarrow B) \wedge A \wedge \neg\bullet A\big) \supset B \tag{rMP}$$

Due to the restriction, MP is only applied in case we are able to derive that the factual information is not exceptional with respect to A, i.e., $\neg\bullet A$. The following is an immediate consequence of (rMP) and the core properties:

$$\vdash \big(A \wedge B \wedge (B \rightsquigarrow \neg A)\big) \supset \bullet B \tag{Spe1}$$

The antecedent of (Spe1) expresses that the default $B \rightsquigarrow \neg A$ is factually overridden since A is the case. If the factual information describes atypical circumstances for A and $A \rightsquigarrow B$, then we also have atypical circumstances for B, since after all A is at least as specific as B. This motivates the following axiom[8]:

[8] The name (Inh) indicates that the property of being exceptional is *inherited* along $\rightsquigarrow$-paths.

$$\vdash \big(\bullet A \wedge (A \rightsquigarrow B)\big) \supset \bullet B \tag{Inh}$$

Fact 6.3.1. (rMP), (Inh) *and the core properties entail*

$$\vdash \big(A \wedge (A \rightsquigarrow B \rightsquigarrow C) \wedge (A \rightsquigarrow \neg C)\big) \supset \bullet B \tag{Spe2}$$

The antecedent of (Spe2) describes a case of specificity: the default $B \rightsquigarrow C$ is overridden by the more specific default $A \rightsquigarrow \neg C$ and the fact A. Let us take a look at a proof fragment for our example:

1	$p \rightsquigarrow b$	PREM
2	$b \rightsquigarrow f$	PREM
3	$p \rightsquigarrow \neg f$	PREM
4	p	PREM
5	b	PREM
6	$\bullet b$	1,2,3,4; Spe2

Due to the fact that $\bullet b$ is derived at line 6, our restriction prevents MP of being applicable to b and $b \rightsquigarrow f$ in order to derive f. Indeed, due to p we are in atypical circumstances with respect to b. This is for instance witnessed by the fact that by the core properties $b \rightsquigarrow \neg p$ is derivable from our premise set, and p is a premise.

Note that something is still missing in order to model default inferencing properly. Due to the restricted MP we are able to block MP from being applied to excepted antecedents. However, we lack the ability to apply MP to $p \rightsquigarrow \neg f$ and p since we miss $\neg\bullet p$. This can be tackled by applying MP conditionally. More specifically, MP is applied to $A \rightsquigarrow B$ and A on the condition that the antecedent A can be assumed to be not excepted, i.e., on the condition that $\bullet A$ can be assumed not to be the case. This is technically realized by means of ALs.

I will introduce the ALs formally in Sect. 6.4 but let me sketch the main idea already now. In order to rewrite the proof above in the style of ALs, we need to add a fourth column containing sets of so-called abnormalities. In our case abnormalities are of the form $\bullet A$.

1	$b \rightsquigarrow f$	PREM	$\emptyset$
2	$p \rightsquigarrow b$	PREM	$\emptyset$
3	$p \rightsquigarrow \neg f$	PREM	$\emptyset$
4	b	PREM	$\emptyset$
5	p	PREM	$\emptyset$
^{7}6	f	1,4; RC	$\{\bullet b\}$
7	$\bullet b$	1,2,3,5; Spe2	$\emptyset$
8	$\neg f$	3,5; RC	$\{\bullet p\}$

At lines 6 and 8 MP is applied conditionally (indicated by RC for "rule conditional"). For instance at line 8 the condition is $\{\bullet p\}$. In other words, MP is applied to p and $p \rightsquigarrow \neg f$ on the condition that p can be assumed to be not excepted. Note that if $\neg\bullet p$ would be derivable, we would be able to apply (rMP) to p, $p \rightsquigarrow \neg f$ and $\neg\bullet p$ in order to detach $\neg f$. However, $\neg\bullet p$ is not derivable. Nevertheless, ALs offer the option to apply MP conditionally. Similarly, at line 6 MP is applied to $b \rightsquigarrow f$ and b on the condition that b is not excepted. However, at line 7, $\bullet b$ is derived. Note that at this point line 6 is marked by 7. The idea is that lines with "unsafe" conditions are marked and the formulas in the second column of marked lines are not considered as being derived. Of course, since f is derived on the condition that b is not excepted, this very condition cannot be considered safe anymore as soon as we derive that b is excepted at line 7. There are two adaptive strategies that specify what it exactly means that a condition of a line is "unsafe". For instance in case of the reliability strategy a line is marked at a given stage of the proof in case a member of its condition has been derived as part of a minimal disjunction of abnormalities (in our case a disjunction of formulas preceded by a $\bullet$) on the condition $\emptyset$.[9] Minimality means that no sub-formula of the disjunction has been derived. Since $\bullet p$ is not derivable as part of a disjunction of abnormalities, line 8 is not going to be marked. There is obviously no reason to treat its condition as unsafe.

In the following sections I will realize the idea that was informally presented in this section. In Sect. 6.4, the ALs for conditionally applying MP will be defined.

6.4 Applying Modus Ponens Conditionally

As discussed in Sect. 6.3, we use a unary operator '$\bullet$' in order to label propositional formulas for which MP should be blocked. These are propositions that are excepted by the information given in the premises. That is to say, the factual information at hand describes unusual circumstances concerning them.

We have seen that b, "Tweety is a bird.", is excepted if also p, "Tweety is a penguin.", is given. The second proposition describes an exceptional context for the first one due to the conditionals $b \rightsquigarrow f$, $p \rightsquigarrow b$ and $p \rightsquigarrow \neg f$ where f = "Tweety flies". Thus, f should not be detached from $b \rightsquigarrow f$ and b if p is the case: $b \rightsquigarrow f$ is overridden by the more specific $p \rightsquigarrow \neg f$.

The following fact shows that in various cases of specificity the least specific arguments are excepted.

Fact 6.4.1. *The core properties, (rMP) and (Inh) imply (Spe1), (Spe2) and the following:*

$$\textit{If } \vdash A \supset B, \textit{ then } \vdash \big(A \wedge (B \rightsquigarrow C) \wedge (A \rightsquigarrow \neg C)\big) \supset \bullet B \qquad \text{(sSpe)}$$

$$\vdash \big(A \wedge (A \rightsquigarrow B_1 \rightsquigarrow \ldots \rightsquigarrow B_n \rightsquigarrow C) \wedge (A \rightsquigarrow \neg C)\big) \supset \bullet B_n \qquad \text{(SpeG)}$$

$$\begin{aligned} &\vdash \big(A \wedge (A \rightsquigarrow B_1 \rightsquigarrow \ldots \rightsquigarrow B_n \rightsquigarrow D) \wedge \\ &(A \rightsquigarrow C_1 \rightsquigarrow \ldots \rightsquigarrow C_m \rightsquigarrow \neg D) \wedge (B_n \rightsquigarrow \ldots \rightsquigarrow C_m)\big) \supset \bullet C_m \end{aligned} \qquad \text{(PreE)}$$

$$\textit{If } \vdash \neg \bigwedge\nolimits_I D_i, \textit{ then } \vdash \Big(A \wedge \bigwedge\nolimits_I (A \rightsquigarrow \ldots \rightsquigarrow B_i \rightsquigarrow D_i)\Big) \supset \bigvee\nolimits_I \bullet B_i \qquad \text{(Conf)}$$

---

[9] A more precise notion of what it means that a condition is "unsafe" will be given in the next section by means of a marking definition.

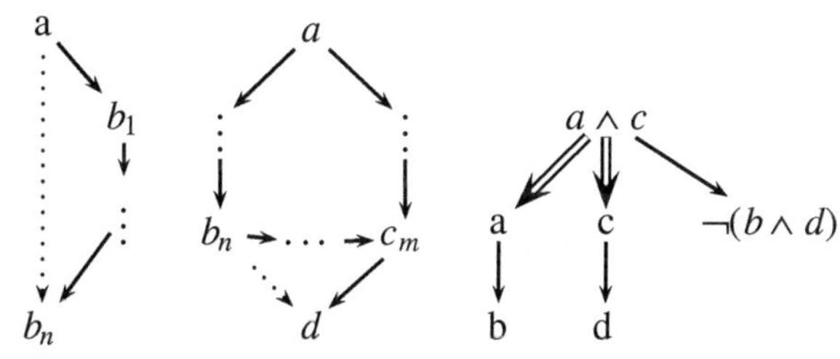

Fig. 6.2 Generalized specificity cases

(SpeG) is a generalization of (Spe2) (see Fig. 6.2a). The preemption rule (PreE) is a further generalization (see Fig. 6.2b).[10] (Conf) shows that if there are multiple conflicting arguments $A \rightsquigarrow \ldots \rightsquigarrow B_i \rightsquigarrow D_i$ then at least one of the B_i's is excepted.

Let in the remainder **Lp** be the base logic **L** enriched by (rMP) and (Inh). In this chapter we will focus on base logics $\mathbf{L} \in \{\mathbf{P}, \mathbf{R}\}$ (see Sect. 6.2).

Definition 6.4.1. We define $\mathbf{DLp^x}$ where $\mathbf{x} \in \{\mathbf{r}, \mathbf{m}\}$ as an AL in standard format by the following triple:

- the lower limit logic is **Lp**,
- the set of abnormalities is $\Omega = \{\bullet A \mid A \in \mathcal{W}\}$,
- the strategy is either **r**eliability (for $\mathbf{DLp^r}$) or **m**inimal abnormality (for $\mathbf{DLp^m}$).

To adaptively interpret premise sets "as normally as possible" means in our case to interpret the propositional formulas as not being excepted whenever possible, i.e., whenever this is consistent with the given premises. In turn, this allows us to apply MP as much as possible since the additional antecedents of (rMP), $\neg\bullet A$, are validated as much as possible. Note that due to (rMP) we have

$$\vdash_{\mathbf{Lp}} \big(A \wedge (A \rightsquigarrow B)\big) \supset \big(B \vee \bullet A\big)$$

Hence, by RC, B is derivable from A and $A \rightsquigarrow B$ on the condition $\{\bullet A\}$.

The (object-level) proofs presented in the following examples are for both ALs, $\mathbf{DLp^r}$, and $\mathbf{DLp^m}$, if not specified differently. I presume that $\mathbf{L} \in \{\mathbf{P}, \mathbf{R}\}$. Let us take a look at a simple case of specificity.

Example 6.4.1. We equip the conditional knowledge base in Example 6.3.1 (see Fig. 6.1b) with the factual knowledge $\{p\}$.

1	$p \rightsquigarrow b$	PREM	$\emptyset$
2	$b \rightsquigarrow f$	PREM	$\emptyset$
3	$p \rightsquigarrow \neg f$	PREM	$\emptyset$
4	p	PREM	$\emptyset$
5	b	1,4; RC	$\{\bullet p\}$

[10] Preemption plays an important role in the research on inheritance networks (see [48]).

[8]	6 f	2,5; RC	$\{\bullet p, \bullet b\}$
	7 $\neg f$	3,4; RC	$\{\bullet p\}$
	8 $\bullet b$	1,2,3,4; RU	$\emptyset$

At line 5, MP is applied to $p \rightsquigarrow b$ and p on the condition $\{\bullet p\}$. Similar conditional applications take place at lines 6 and 7. The desired $\neg f$ and b are (finally) derivable since the condition, $\bullet p$, is not part of any minimal Dab-consequence. Moreover, MP is blocked from $b \rightsquigarrow f$ and b since at line 8, $\bullet b$ is derived and hence line 6 is marked.

Example 6.4.2. Let us have a look at conflicting conditionals by means of the Nixon Diamond (see Fig. 6.1c) with the factual knowledge $\{n\}$ and the usual reading of q as 'being a Quaker', r 'being a Republican' and p as 'being a pacifist'.

	1 $n \rightsquigarrow q$	PREM	$\emptyset$
	2 $n \rightsquigarrow r$	PREM	$\emptyset$
	3 $q \rightsquigarrow p$	PREM	$\emptyset$
	4 $r \rightsquigarrow \neg p$	PREM	$\emptyset$
	5 n	PREM	$\emptyset$
	6 q	1,5; RC	$\{\bullet n\}$
	7 r	2,5; RC	$\{\bullet n\}$
[10]	8 p	3,6; RC	$\{\bullet n, \bullet q\}$
[10]	9 $\neg p$	4,7; RC	$\{\bullet n, \bullet r\}$
	10 $\bullet q \vee \bullet r$	1,2,3,4,5; RU	$\emptyset$

The logic proceeds as expected: r and q are derivable while the derivations of p and $\neg p$ get marked for both strategies. Note that the condition of line 6 and 7, namely $\{\bullet n\}$, is not part of any minimal Dab-consequence. In order to make the example more interesting let us introduce two more conditionals: $q \rightsquigarrow e$ and $r \rightsquigarrow e$ where e represents for instance 'being politically motivated'.

11 $q \rightsquigarrow e$	PREM	$\emptyset$
12 $r \rightsquigarrow e$	PREM	$\emptyset$
13 e	6,11; RC	$\{\bullet n, \bullet q\}$
14 e	7,12; RC	$\{\bullet n, \bullet r\}$

By the reliability strategy lines 13 and 14 are marked (due to the fact that $\bullet q \vee \bullet r$ at line 10 is a minimal Dab-consequence). They are not marked by the minimal abnormality strategy, since the minimal choice sets at line 14 are $\{\bullet q\}$ and $\{\bullet r\}$. It is easy to see that there is no way to extend the proof in a way such that lines 13 and 14 are marked according to the minimal abnormality strategy. This shows that the reliability strategy models a more skeptical reasoning in comparison to the bolder reasoning type modeled by the minimal abnormality strategy.

We have a similar scenario for the example depicted in Fig. 6.3d. By the minimal abnormality strategy p is derivable given the factual knowledge x. It is not derivable by the reliability strategy.

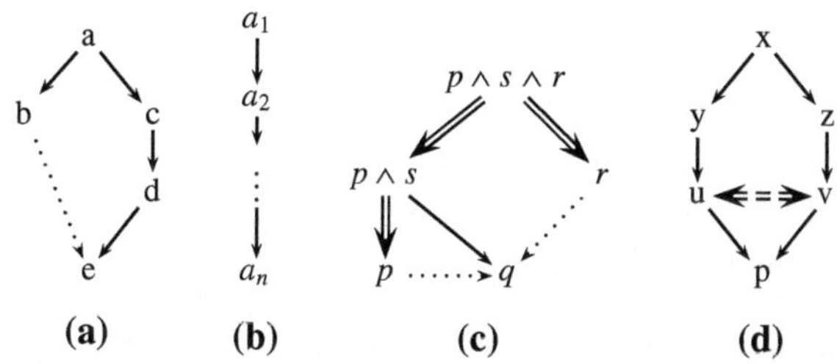

Fig. 6.3 Illustrations for our examples

Propositions such as p in Fig. 6.3d are commonly dubbed "floating conclusions". There is a vivid debate about whether such propositions should be accepted.[11] Instead of trying to have the final word on the discussion I want to point out that, as the example shows, the minimal abnormality strategy detaches floating conclusions, while the more skeptical reliability strategy rejects them. Different applications may ask for different strategies. The credulous character of the minimal abnormality strategy makes it interesting for applications in which "the value of drawing conclusions is high relative to the costs involved if some of those conclusions turn out not to be correct." ([48], p. 123). The reliability strategy on the other hand is, due to its more skeptical character, better "when the cost of error rises" (ibid.).

Example 6.4.3. Let our knowledge base be $\Gamma_{6.4.3} = \{a_i \rightsquigarrow a_{i+1} \mid 1 \leq i < n\}$ (see Fig. 6.3b) with factual knowledge $\{a_1\}$. Note that $\Gamma_{6.4.3} \nvdash_{\mathbf{P}} a_1 \rightsquigarrow a_j$ and $\Gamma_{6.4.3} \nvdash_{\mathbf{R}} a_1 \rightsquigarrow a_j$ where $2 < j \leq n$. However, our ALs are able to detach all the a_i's:

1	$a_1 \rightsquigarrow a_2$	PREM	$\emptyset$
⋮	⋮	⋮	$\emptyset$
$n-1$	$a_{n-1} \rightsquigarrow a_n$	PREM	$\emptyset$
n	a_1	PREM	$\emptyset$
$n+1$	a_2	n; RC	$\{\bullet a_1\}$
$n+2$	a_3	$n+1$; RC	$\{\bullet a_1, \bullet a_2\}$
⋮	⋮	⋮	⋮
$2n-1$	a_n	$2n-2$; RC	$\{\bullet a_1, \ldots, \bullet a_{n-1}\}$

Obviously none of the lines $n+1, \ldots, 2n-1$ can be marked by extending the proof. The fact that $\Gamma_{6.4.3} \cup \{a_1\} \vdash_{\mathbf{DLp^x}} a_i$, where $i \leq n$, $\mathbf{x} \in \{\mathbf{r}, \mathbf{m}\}$ and $\mathbf{L} \in \{\mathbf{P}, \mathbf{R}\}$, while $\Gamma_{6.4.3} \nvdash_{\mathbf{L}} a_1 \rightsquigarrow a_i$ demonstrates that our handling of MP overcomes certain weaknesses of the core logic in terms of the handling of transitive relations among conditionals.

Example 6.4.4. Let our factual knowledge be a. $b_1, \ldots, b_{n-1}$ and $\neg b_n$ are derivable from the knowledge base depicted in Fig. 6.2a by means of $\mathbf{DPp^x}$ (where $\mathbf{x} \in \{\mathbf{r}, \mathbf{m}\}$). We obtain e.g. the Dab-formula $\bullet b_{n-1}$ (by (SpeG)) and $\bullet b_n$ (by means of the former

[11] While Ginsberg [49], and Makinson and Schlechta [50] argue for the acceptance, Horty [51] argues against it.

and (Inh)). Note that no $\bullet b_i$ where $i < n-1$ is derivable as a part of a minimal Dab-consequence. Hence we can iteratively apply Modus Ponens conditionally to $a \rightsquigarrow b_1$ and $b_i \rightsquigarrow b_{i+1}$ where $i < n-1$ in such a way that the corresponding lines are unmarked. Note that $a \rightsquigarrow b_{n-1}$ is neither a **P**-consequence nor a **R**-consequence of the given premises, nor is it derivable by means of Rational Closure. However, it is entailed by $\mathbf{P}_{\min}$.[12]

The situation is slightly different in $\mathbf{DRp^x}$: besides $\bullet b_{n-1}$ and $\bullet b_n$ also $\bullet b_{n-2} \vee \bullet\neg b_{n-2}$ is **Rp**-derivable from the premises.[13] It is easy to see that due to this b_{n-1} is not $\mathbf{DRp^x}$-derivable since the only means of deriving b_{n-1} from the given premises is by detaching it from the conditional $b_{n-2} \rightsquigarrow b_{n-1}$ on the condition $\{\bullet b_{n-2}\}$. Yet, due to the minimal Dab-consequence $\bullet b_{n-2} \vee \bullet\neg b_{n-2}$ any such attempt gets marked in the proof. However, if we add the premise $b_{n-2} \not\rightsquigarrow b_n$ we get the consequences b_i for all $i < n$ and $\neg b_n$ just as for $\mathbf{DPp^x}$. It is easy to see that in this case $\bullet b_{n-2}$ is not anymore part of any minimal Dab-consequence.

Similarly, $b_1, \ldots, b_n, c_1, \ldots, c_m$ and $\neg d$ are $\mathbf{DPp^x}$-derivable from the knowledge base depicted in Fig. 6.2b. Analogous to the previous paragraph we need to add another premise, e.g. $c_{m-1} \not\rightsquigarrow a$, in order to get the same consequences for $\mathbf{DRp^x}$. The proofs are simple and left to the reader.

Example 6.4.5. Let us take a look at a variant of the Nixon Diamond (Fig. 6.3a) by means of the logic $\mathbf{DPp^x}$ (where $\mathbf{x} \in \{\mathbf{r}, \mathbf{m}\}$):

	1	$a \rightsquigarrow b$	PREM	$\emptyset$
	2	$a \rightsquigarrow c$	PREM	$\emptyset$
	3	$b \rightsquigarrow \neg e$	PREM	$\emptyset$
	4	$c \rightsquigarrow d$	PREM	$\emptyset$
	5	$d \rightsquigarrow e$	PREM	$\emptyset$
	6	a	PREM	$\emptyset$
	7	b	1,6; RC	$\{\bullet a\}$
12	8	$\neg e$	3,7; RC	$\{\bullet a, \bullet b\}$
	9	c	2,6; RC	$\{\bullet a\}$
	10	d	4,9; RC	$\{\bullet a, \bullet c\}$
12	11	e	5,10; RC	$\{\bullet a, \bullet c, \bullet d\}$
	12	$\bullet b \vee \bullet d$	1–6; RU	$\emptyset$

Note that neither is $a \rightsquigarrow d$ derivable by the core properties nor is it in the Rational Closure, nor is it entailed by Conditional Entailment.[14] Thus, in the given example our

---

[12] Note that in case we do not add $a \not\rightsquigarrow \bot$ to our premises, $\mathbf{P}_{\min}$ is rather rigorous and also entails $a \rightsquigarrow \bot$.

[13] The reason is as follows. Suppose first that $b_{n-1} \not\rightsquigarrow \neg b_{n-2}$. In this case by means of (RM) and since $b_{n-1} \rightsquigarrow b_n$ also $(b_{n-1} \wedge b_{n-2}) \rightsquigarrow b_n$. By (RT) and since $b_{n-2} \rightsquigarrow b_{n-1}$, $b_{n-2} \rightsquigarrow b_n$. But then since a, $a \rightsquigarrow \neg b_n$ and $a \rightsquigarrow \ldots \rightsquigarrow b_{n-2} \rightsquigarrow b_n$, by (SpeG), $\bullet b_{n-2}$. Now suppose $b_{n-1} \rightsquigarrow \neg b_{n-2}$. Since $\bullet b_{n-1}$ we get $\bullet\neg b_{n-2}$ by (Inh). Altogether, $\bullet b_{n-2} \vee \bullet\neg b_{n-2}$. Note that this argument does not hold in **Pp** since it makes essentially use of (RM).

[14] Note that $\mathbf{P}_{\min}$ entails $a \rightsquigarrow d$ and moreover $a \rightsquigarrow \bot$ (in case we do not manually add $a \not\rightsquigarrow \bot$ to the premises, see also footnote 12).

logic handles the transitive relations between defaults better than these systems, since (with both strategies) d is derivable following argument $a \rightsquigarrow c \rightsquigarrow d$. Furthermore, as desired, neither e nor $\neg e$ is derivable since there are conflicting arguments concerning e and $\neg e$.

The situation is different in $\mathbf{DRp^x}$ since by means of **Rp** also the minimal Dab-consequence $\bullet c \vee \bullet\neg c$ is derivable.[15] Hence, there d at line 10 is not finally derivable. However, by adding $c \not\rightsquigarrow \neg a$ to the premises also d is a $\mathbf{DRp^x}$-consequence of this premise set.

Example 6.4.6. We take a look at Fig. 6.2c with factual knowledge $\{a \wedge c\}$. This example illustrates a more complex case of specificity.

1	$a \wedge c \rightsquigarrow \neg(b \wedge d)$	PREM	$\emptyset$
2	$a \rightsquigarrow b$	PREM	$\emptyset$
3	$c \rightsquigarrow d$	PREM	$\emptyset$
4	$a \wedge c$	PREM	$\emptyset$
5	$\neg(b \wedge d)$	1,4; RC	$\{\bullet(a \wedge c)\}$
^{9}6	b	2,4; RC	$\{\bullet a\}$
^{9}7	d	3,4; RC	$\{\bullet c\}$
8	$\bullet a \vee \bullet c \vee \bullet(a \wedge c)$	1,2,3,4; RU	$\emptyset$
9	$\bullet a \vee \bullet c$	8; RU	$\emptyset$
10	$b \vee d$	6; RU	$\{\bullet a\}$
11	$b \vee d$	7; RU	$\{\bullet c\}$

Line 9 follows from line 8 in view of (Inh) and (CI). By the reliability strategy lines 10 and 11 are marked since both, $\bullet a$ and $\bullet c$, are unreliable formulas. Not so by the minimal abnormality strategy, since $b \vee d$ is derivable on both conditions, $\{\bullet a\}$ and $\{\bullet c\}$ (see Definition 2.4.7).

This example is interesting also in another respect. It features a more complex type of specificity. While none of the arguments $\mathsf{A}_1 = (a \wedge c) \rightsquigarrow a \rightsquigarrow b$ and $\mathsf{A}_2 = (a \wedge c) \rightsquigarrow c \rightsquigarrow d$ suffices in its own respect to cause a case of specificity with $(a \wedge c) \rightsquigarrow \neg(b \wedge d)$, both taken together do. Indeed, if we follow both lines of argument, A_1 and A_2, we arrive at b and d. However, the conjunction $b \wedge d$ contradicts $\neg(b \wedge d)$. Thus, $a \wedge c \rightsquigarrow \neg(b \wedge d)$ overrides the joint application of arguments A_1 and A_2 (see also the illustration in Fig. 6.2c).

Both, minimal abnormality and reliability strategy, validate $\neg(b \wedge d)$. Again, if we apply reliability we take a more skeptical route concerning A_1 and A_2, since both arguments are considered as being unreliable and thus neither argument is validated: we neither derive b, nor d, nor $b \vee d$. Minimal abnormality however validates one

---

[15] The reason is as follows. Suppose $\neg\bullet c$. Suppose (i) $b \rightsquigarrow \neg a$. Since also a, $a \rightsquigarrow b$ and $a \rightsquigarrow a$ (by (ID)) we get $\bullet b$ by (Spe2). Assume $b \not\rightsquigarrow \neg c$. By (RM), $(b \wedge c) \rightsquigarrow \neg a$. Since also a, $a \rightsquigarrow a$ and $a \rightsquigarrow (b \wedge c)$ (by (CC), $a \rightsquigarrow b$ and $a \rightsquigarrow c$) we have $\bullet(b \wedge c)$ by (Spe2). By (Inh) and (CI) also $\bullet c$,—a contradiction. Hence, $b \rightsquigarrow \neg c$. By (Inh) $\bullet\neg c$. Now suppose (ii) $b \not\rightsquigarrow \neg a$. Since $b \rightsquigarrow \neg e$ by (RM) $(a \wedge b) \rightsquigarrow \neg e$. Since $a \rightsquigarrow b$ by (RT) $a \rightsquigarrow \neg e$. By the latter, a, $\neg\bullet c$ and $a \rightsquigarrow c$ we have $c \not\rightsquigarrow e$ due to (Spe2). By a, $a \rightsquigarrow \neg e$ and $a \rightsquigarrow c \rightsquigarrow d \rightsquigarrow e$ we have $\bullet d$ due to (SpeG). Assume $d \not\rightsquigarrow \neg c$. Then by (RM) $(c \wedge d) \rightsquigarrow e$ and by (RT), $c \rightsquigarrow e$,—a contradiction. Hence $d \rightsquigarrow \neg c$ and by (Inh), $\bullet\neg c$. Altogether we get $\bullet c \vee \bullet\neg c$.

of the two arguments. Indeed, taken isolated from each other, neither A_1 nor A_2 is overridden by $a \wedge c \rightsquigarrow \neg(b \wedge d)$. Thus, the credulous reasoning provided by the minimal abnormality strategy validates $b \vee d$ and $\neg(b \wedge d)$.

Example 6.4.7. Given the factual knowledge $p \wedge s \wedge r$ and the defaults depicted in Fig. 6.3c we have the minimal Dab-consequence $\bullet(p \wedge s) \vee \bullet r$. That shows that neither q nor $\neg q$ is derivable. This is intuitive as pointed out by Geffner and Pearl in [22] since there are no reasons to prefer argument $(p \wedge s) \rightsquigarrow q$ over $r \rightsquigarrow \neg q$ or vice versa. Note however that the counter-intuitive $(p \wedge s \wedge r) \rightsquigarrow q$ is in the Rational and Lexicographic Closure, and it is entailed by the maximum entropy approach.[16]

6.5 Discussion

In this discussion section I will point out some advantages of the presented logics, also in comparison with other systems from the literature. Moreover I will comment on some other related and interesting points which were not mentioned so far.

6.5.1 Some Advantages of the Adaptive Approach

ALs offer a very generic framework enabling defeasible MP for conditional logics of normality since they can be applied to any conditional lower limit logic as long as it is reflexive, transitive, monotonic and compact. Depending on the application the reader is free to use any conditional logic of normality as **LLL** as long as it fulfills the mentioned requirements. Since ALs have shown great unifying power in representing nonmonotonic, defeasible logics, even conditional logics that do not fulfill the requirements may be represented by ALs.[17] By applying techniques of combining adaptive systems the framework developed in this chapter may be applicable also in such cases. Furthermore, similar techniques as presented here for defeasible MP in the context of default reasoning can be applied to conditional deontic logics (see Chap. 12 and [52]).

The meta-theory of ALs in standard format is well-researched (see [53, 54]). Many useful properties have been established generically. For instance, completeness and soundness of an AL are guaranteed by the completeness and soundness of its **LLL**, the consequence relation of an AL defines a fixed point and is cautious monotonic, etc.

Recall that Pollock distinguished in [55] between two types of dynamics that characterize defeasible reasoning: one based on *synchronic defeasibility* and another

[16] It is not entailed by $\mathbf{P}_{\text{min}}$ in case we add $(p \wedge s \wedge r) \not\rightsquigarrow \bot$.

[17] As will been shown, for instance, for Rational Closure in the next chapter.

one based on *diachronic defeasibility*. As I will discuss in the following, ALs are able to model both of them.

The *internal dynamics* of defeasible reasoning is caused by diachronic defeasibility. Often achieving a better understanding of the information at hand forces us to withdraw certain inferences even in cases in which no new information is available. This is modeled by the dynamic proof theory of ALs. For instance, if we (conditionally) apply MP to $b \rightsquigarrow f$ and b but at a later moment also derive p, $p \rightsquigarrow \neg f$ and $p \rightsquigarrow b$ from the same premises, we revise the former derivation. In the adaptive proof the line at which MP has been applied to $b \rightsquigarrow f$ and b is going to be marked and is hence considered not to be valid. Thus, while our insight in the given knowledge base—i.e., the premises—grows, we may consider revising some conclusions drawn before, especially if the knowledge base is of a complex nature. Hence, our treatment of common sense reasoning with factual information on the basis of conditional knowledge bases does not just reach intuitive results but the explication of the reasoning process itself is an integral part of the proof theory. This is an advantage compared to other systems which are able to model default inferencing such as Delgrande's [12], Lamarre's [56], or Geffner and Pearl's [22].

Lamarre in [56] presents a powerful approach based on semantic selection procedures on the models of a given conditional base logic, where the facts valid in all the selected models characterize the consequence set of his system.[18] What is missing, however, is a syntactical approach corresponding to it that mirrors our common sense reasoning by its proof theory. Delgrande's system [12] is syntactical in nature. The idea here is to iteratively enrich the given factual knowledge by further contingent information in order to form so-called maximal contingent extensions.[19] Special attention in building these extensions is given to cases of specificity: similar as in the presented approach, the world at hand is interpreted as non-exceptional as possible. Furthermore, in the construction of the extensions only relevant information is considered with respect to the knowledge base at hand. What is derivable by classical logic from these maximal contingent extensions corresponds to the factual consequences we draw via default reasoning. While Delgrande's assumptions concerning the normality of the actual world and his restriction to relevant information accord with a natural intuition concerning default reasoning, the way we arrive at the inferences by Delgrande's approach seems rather unnatural, i.e., the technical necessity to first built up all the maximal consistent factual extensions and then to infer from them by classical reasoning. This procedure does not model our actual default inferencing in an accurate way. Geffner and Pearl's Conditional Entailment has been already mentioned on Sect. 6.2. Although the authors provide a syntactic check-criterion for conditionally entailed propositions, they do not offer a deriva-

---

[18] As discussed in Sect. 6.2, ALs also employ semantic selections on the models of the **LLL**.

[19] Delgrande introduces in fact two equivalent proposals in this paper. The other one, which I do not discuss above, is based on forming maximal consistent extensions of the conditional knowledge base at hand (in contrast to the maximal consistent extensions of the factual knowledge which I discuss here). Note, however, that a similar criticism applies to both approaches.

tional procedure that mirrors our actual reasoning processes such as the dynamic proofs of ALs.[20]

As mentioned, another advantage of ALs is their ability to deal with the synchronic defeasibility that causes the *external dynamics* of reasoning processes (see also the discussion in Sect. 2.5.1). Often with the introduction of new information we are forced to withdraw certain inferences. Again, the markings of the dynamic proofs are able to model cases of specificity and conflicting arguments which might be caused by new information. In contrast, in Lamarre's approach the arrival of new information forces us to re-initiate the semantic selection procedure, and, similarly, for Delgrande's account we have to re-construct the maximal contingent extensions. In the adaptive approach, despite the fact that new information might force us to withdraw certain conclusions, the proof dynamics model in an accurate way the fact that we continue reasoning facing new information instead of beginning the reasoning process again from scratch.

6.5.2 The Drowning Problem

In Examples 6.4.3 and 6.4.4 it was demonstrated that the presented treatment of MP sometimes outgrows the abilities of the core system in terms of transitively closing $\rightsquigarrow$. However, there are limitations to it. To show this I extend Example 6.4.1 by a further conditional:

Example 6.5.1. We add to the conditionals of Example 6.4.1, $b \rightsquigarrow w$, where w stands for "having wings". The proof of Example 6.4.1 is extended in the following way:

9	$b \rightsquigarrow w$	PREM $\emptyset$
^{8}10	w	5,9; RC $\{\bullet p, \bullet b\}$

Note that the conditional derivation of w is not successful in the sense that it gets marked. This is due to the fact that b is excepted since we have p and $b \rightsquigarrow \neg p$. Indeed there is no way to derive w from the given premises. This is also due to the fact that in **P** and **R** neither $p \rightsquigarrow w$ nor $(p \wedge b) \rightsquigarrow w$ is derivable (neither are they in the Rational Closure). Note that if one of the latter would be derivable, w would be detachable from $p \rightsquigarrow w$ and p, or resp. $(p \wedge b) \rightsquigarrow w$ and $p \wedge b$. Thus, the limitation of the adaptive treatment of MP concerning excepted propositions mirrors a limitation of the base logic concerning conditional consequences.

This problem is commonly known as the Drowning Problem: suppose a default with antecedent A is excepted, then all other defaults with antecedent A are blocked from MP as well.

The first question to ask at this point is whether a "solution" to the drowning problem is at all desirable. Some scholars voice worries (see e.g. [14, 58–60]). For

[20] Computing Conditional Entailment is a pretty complex and challenging task. Hence, the authors only offer a computational approximation in terms of an assumption-based truth maintenance-like system (see [57]).

instance, Koons asserts that there are good reasons why we should not apply MP to defaults with excepted antecedents.

> Consider the following variant on the problem: birds fly, Tweety is a bird that doesn't fly, and birds have strong forelimb muscles. Here it seems we should refrain from concluding that Tweety has strong forelimb muscles, since there is reason to doubt that the strength of wing muscles is causally (and hence, probabilistically) independent of capacity for flight. Once we know that Tweety is an exceptional bird, we should refrain from applying other conditionals with *Tweety is a bird* as their antecedents, unless we know that these conditionals are independent of flight, that is, unless we know that the conditional with the stronger antecedent, *Tweety is a non-flying bird*, is also true. (see [60], Sect. 5.7)

Moreover, Lehmann in [40] points out that there are two perspectives on default reasoning. On the one hand, there is the *prototypical* reading where $b \rightsquigarrow f$ is understood as "Birds typically fly." On the other hand, according to the *presumptive* reading it is read as "Birds are presumed to fly unless there is evidence to the contrary." The former was proposed in [61] and Lehmann states that it is the intended reading for Rational Closure, whereas the presumptive reading is intended for the Lexicographic Closure. According to the prototypical reading the Drowning problem should not be solved. This is due to the fact that if there is an exception to some conditional with antecedent A then the situation is not typical with respect to A. However, defaults with antecedent A only account for typical situations (with respect to A). Hence, MP should not be applied to any conditional $A \rightsquigarrow B$ according to this view.

6.5.3 Taking Into Account Negative Knowledge

So far we focused on knowledge bases consisting on the one hand of conditionals and on the other hand of facts, i.e., facts expressed by propositions. It is interesting to enable the logic to also deal with knowledge bases including negative conditionals, i.e., formulas of the form $A \not\rightsquigarrow B$. Note that the framework proposed in this chapter is not able to deal with such knowledge bases in the case that our base system only consists of the core properties. Take for instance the simple penguin Example 6.4.1 and replace the premise $p \rightsquigarrow \neg f$ by $p \not\rightsquigarrow f$. Note that for the logics $\mathbf{DPp^x}$ (where $\mathbf{x} \in \{\mathbf{r}, \mathbf{m}\}$) the unwanted f is derivable for this premise set.

1	$p \rightsquigarrow b$	PREM	$\emptyset$
2	$p \not\rightsquigarrow f$	PREM	$\emptyset$
3	$b \rightsquigarrow f$	PREM	$\emptyset$
4	p	PREM	$\emptyset$
5	b	1,4; RU	$\{\bullet p\}$
6	f	3,5; RU	$\{\bullet p, \bullet b\}$

Note that there is no way to mark line 6 (in either of the strategies). However, as the following fact shows, the situation is different in case $\mathbf{R}$ is chosen as base system, i.e., for lower limit logic $\mathbf{Rp}$.

Fact 6.5.1. *The core properties, (RM), (rMP) and (Inh) imply*[21]

$$\vdash \big(A \wedge (A \rightsquigarrow B \rightsquigarrow C) \wedge (A \not\rightsquigarrow C)\big) \supset \bullet B \qquad \text{(Spe')}$$

$$\vdash \big(A \wedge (A \rightsquigarrow B_1 \rightsquigarrow \ldots \rightsquigarrow B_n \rightsquigarrow C) \wedge (A \not\rightsquigarrow C)\big) \supset \bullet B_n \qquad \text{(SpeG')}$$

$$\vdash \big(A \wedge (A \rightsquigarrow B_1 \rightsquigarrow \ldots \rightsquigarrow B_n) \wedge (B_n \not\rightsquigarrow D) \wedge$$
$$(A \rightsquigarrow C_1 \rightsquigarrow \ldots \rightsquigarrow C_m \rightsquigarrow D) \wedge (B_n \rightsquigarrow \ldots \rightsquigarrow C_m)\big) \supset \bullet C_m \qquad \text{(PreE')}$$

In $\mathbf{DRp^x}$ f is not derivable since line 6 is marked by the following extension of the proof:

7	$\bullet b$	1,2,3,4; Spe' ∅

6.6 Conclusion and Outlook

In this chapter an adaptive logic approach to Modus Ponens for conditional logics of normality was presented. By adaptively enhancing a given base logic we enrich it by the ability to model actual default inferencing. By means of benchmark examples it was demonstrated that the adaptive systems deal with specificity and conflicting arguments in an intuitive way. The two adaptive standard strategies have been shown to correspond to two different intuitions: a more skeptical and a more credulous one which gives rise to a different handling of floating conclusions.

Acknowledgments I thank Joke Meheus, Dunja Šešelja, and the anonymous reviewers of the Journal of Applied Non-Classical Logic for valuable comments concerning a previous version of this chapter.

References

1. Straßer, C.: Adaptively applying modus ponens in conditional logics of normality. J. Appl. Non-Class. Logic **22**(1–2), 125–148 (2012)
2. Reiter, R.: A logic for default reasoning. Artif. Intell. **13**(1–2), 81–132 (1980)
3. McCarthy, J.: Circumscription—a form of non-monotonic reasoning. Artif. Intell. **13**, 27–29 (1980)
4. Moore, R.C.: Possible-world semantics for autoepistemic logic. In: Proceedings of the Workshop on Non-monotonic Reasoning, pp. 344–354. AAAI (1984)
5. Stalnaker, R.F.: A theory of conditionals. In: Reischer, N. (ed.) Studies in Logical Theory. Basil Blackwell, Oxford (1968)
6. Lewis, D.: Counterfactuals. Blackwell, Oxford (2000)
7. Adams, E.W.: The Logic of Conditionals. D. Reidel Publishing Co., Dordrecht (1975)
8. Gärdenfors, P.: Conditionals and changes of belief. Acta Philos. Fennica **30** (2–4), 20 (1978)

[21] The proofs can be found in the Appendix.

9. Boutilier, C.: Conditional logics of normality: a modal approach. Artif. Intell. **68**(1), 87–154 (1994)
10. Lamarre, P.: S4 as the conditional logic of nonmonotonicity. In: Proceedings of the Second International Conference on the Principles of Knowledge Representation and Reasoning, pp. 357–367. Morgan Kaufmann, San Francisco (1991)
11. Kraus, S., Lehmann, D.J., Magidor, M.: Nonmonotonic reasoning, preferential models and cumulative logics. Artif. Intell. **44**, 167–207 (1990)
12. Delgrande, J.: An approach to default reasoning based on a first-order conditional logic: revised report. Artif. Intell. **36**(1), 63–90 (1988)
13. Boutilier, C.: Unifying default reasoning and belief revision in a modal framework. Artif. Intell. **68**(1), 33–85 (1994)
14. Wobcke, W.: Belief revision, conditional logic and nonmonotonic reasoning. Notre Dame J. Formal Logic **36**(1), 55–102 (1995)
15. Lehmann, D.J., Magidor, M.: What does a conditional knowledge base entail? Artif. Intell. **55**(1), 1–60 (1992)
16. Pearl, J.: System Z: a natural ordering of defaults with tractable applications to nonmonotonic reasoning. In: TARK '90: Proceedings of the 3rd Conference on Theoretical Aspects of Reasoning About Knowledge, pp. 121–135. Morgan Kaufmann Publishers Inc., San Francisco (1990)
17. Broda, K., Gabbay, D.M., Lamb, L.C., Russo, A.M.: Labelled natural deduction for conditional logics of normality. Logic J. IGPL **10**(2), 123–163 (2002)
18. Schröder, L., Pattinson, D., Hausmann, D.: Optimal tableaux for conditional logics with cautious monotonicity. In: Proceeding of the 2010 Conference on ECAI 2010: 19th European Conference on Artificial Intelligence, pp. 707–712. IOS Press, Amsterdam (2010)
19. Giordano, L., Gliozzi, V., Olivetti, N., Pozzato, G.: Analytic tableau calculi for KLM rational logic R. In: Proceedings of the 10th European Conference on Logics in Artificial Intelligence, pp. 190–202. ACM, New York (2006)
20. Arieli, O., Avron, A.: General patterns for nonmonotonic reasoning: from basic entailments to plausible relations. Logic J. IGPL **8**(2), 119–148 (2000)
21. Friedman, N., Halpern, J.Y.: Plausibility measures and default reasoning. J. ACM **48**, 1297–1304 (1996)
22. Geffner, H., Pearl, J.: Conditional entailment: bridging two approaches to default reasoning. Artif. Intell. **53**(2–3), 209–244 (1992)
23. Pearl, J.: Probabilistic semantics for nonmonotonic reasoning: a survey. In: Proceedings of the First International Conference on Principles of Knowledge Representation and Reasoning, pp. 505–516. Morgan Kaufmann Publishers Inc., San Francisco (1989)
24. Dubois, D., Prade, H.: Possibilistic logic, preferential models, non-monotonicity and related issues. In: Proceedings of the 12th International Joint Conference on Artificial Intelligence, pp. 419–424. Morgan Kaufmann, San Francisco (1991)
25. Goldszmidt, M., Pearl, J.: Rank-based systems: a simple approach to belief revision, belief update, and reasoning about evidence and actions. In: Proceedings of the Third International Conference on Knowledge Representation and Reasoning, pp. 661–672. Morgan Kaufmann, San Francisco (1992)
26. Spohn, W.: Ordinal conditional functions: a dynamic theory of epistemic states. In: W.L. Harper, B. Skyrms (eds.) Causation in Decision, Belief Change and Statistics, pp. 105–134. Springer, Heidelberg (1988)
27. Chellas, B.F.: Basic conditional logic. J. Philos. Logic **4**(2), 133–153 (1975)
28. Nute, D.: Topics in Conditional Logic. Philosophical Studies Series in Philosophy, vol. 20. D. Reidel Publishing. Co., Dordrecht (1980)
29. Neufeld, E.: Defaults and probabilities: extensions and coherence. In: Proceedings of the First International Conference on Principles of Knowledge Representation and Reasoning, pp. 312–323. Morgan Kaufmann Publishers Inc., San Francisco (1989)
30. Poole, D.: The effect of knowledge on belief: conditioning, specificity and the lottery paradox in default reasoning. Artif. Intell. **49**(1–3), 281–307 (1991)

31. Stalnaker, R.F.: What is a nonmonotonic consequence relation? Fundam. Inform. **21**(1/2), 7–21 (1994)
32. Giordano, L., Gliozzi, V., Olivetti, N., Pozzato, G.L.: Preferential vs rational description logics: which one for reasoning about typicality? In: Proceedings of the 2010 Conference on ECAI 2010: 19th European Conference on Artificial Intelligence, pp. 1069–1070. IOS Press (2010)
33. Giordano, L., Gliozzi, V., Olivetti, N., Pozzato, G.L.: A nonmonotonic extension of KLM preferential logic P. In: Proceedings of the 17th International Conference on Logic for Programming, Artificial Intelligence, and Reasoning, LPAR'10, pp. 317–332. Springer, Heidelberg (2010)
34. Giordano, L., Olivetti, N., Gliozzi, V., Pozzato, G.L.: ALC + T: a preferential extension of description logics. Fundam. Inform. **96**, 341–372 (2009)
35. Goble, L.: A proposal for dealing with deontic dilemmas. In: A. Lomuscio, D. Nute (eds.) DEON. Lecture Notes in Computer Science, vol. 3065, pp. 74–113. Springer, Heidelberg (2004)
36. Goldszmidt, M., Pearl, J.: On the relation between rational closure and system Z. In: Third International Workshop on Nonmonotonic Reasoning (South Lake Tahoe), pp. 130–140 (1990)
37. Freund, M.: Default extensions: dealing with computer information. Artif. Intell. **92**(1–2), 277–288 (1997)
38. Shoham, Y.: A semantical approach to nonmonotonic logics. In: Ginsberg, M.L. (ed.) Readings in Non-Monotonic Reasoning, pp. 227–249. Morgan Kaufmann, Los Altos (1987)
39. Shoham, Y.: Reasoning about Change: Time and Causation from the Standpoint of Artificial Intelligence. MIT Press, Cambridge (1988)
40. Lehmann, D.J.: Another perspective on default reasoning. Ann. Math. Artif. Intell. **15**(1), 61–82 (1995)
41. Benferhat, S., Cayrol, C., Dubois, D., Lang, J., Prade, H.: Inconsistency management and prioritized syntax-based entailment. In: IJCAI'93: Proceedings of the 13th International Joint Conference on Artificial Intelligence, pp. 640–645. Morgan Kaufmann Publishers Inc., San Francisco (1993)
42. Goldzsmidt, M., Morris, P., Pearl, J.: A maximum entropy approach to nonmonotonic reasoning. IEEE Trans. Pattern Anal. Mach. Intell. **15**(3), 220–232 (1993)
43. Delgrande, J.: On a rule-based interpretation of default conditionals. Ann. Math. Artif. Intell. **48**(3–4), 135–167 (2006)
44. Dung, P.M., Son, T.C.: An argument-based approach to reasoning with specificity. Artif. Intell. **133**(1–2), 35–85 (2001)
45. Booth, R.: The lexicographic closure as a revision process. J. Appl. Non-Class. Logics **11**(1–2), 35–58 (2001)
46. Genovese, V., Giordano, L., Gliozzi, V., Pozzato, G.L.: A conditional constructive logic for access control and its sequent calculus. In: Automated Reasoning with Analytic Tableaux and Related Methods. Lecture Notes in Computer Science, pp. 164–179. Springer, Heidelberg (2011)
47. Garg, D., Abadi, M.: A modal deconstruction of access control logics. In: Proceedings of the Theory and Practice of Software, 11th International Conference on Foundations of Software Science and Computational Structures, FOSSACS'08/ETAPS'08, pp. 216–230. Springer, Berlin (2008)
48. Horty, J.F.: Some direct theories of nonmonotonic inheritance. In: D. Gabbay, C.J. Hogger, J.A. Robinson (eds.) Handbook of Logic in Artificial Intelligence and Logic Programming, Vol. 3: Nonmonotonic Reasoning and Uncertain Reasoning, pp. 111–187. Oxford University Press, Oxford (1994)
49. Ginsberg, M.: Essentials of Artificial Intelligence. Morgan Kaufmann Publishers Inc., San Francisco (1994)
50. Makinson, D., Schlechta, K.: Floating conclusions and zombie paths: two deep difficulties in the "directly skeptical" approach to defeasible inheritance nets. Artif. Intell. **48**(2), 199–209 (1991)
51. Horty, J.F.: Skepticism and floating conclusions. Artif. Intell. **135**, 55–72 (2002)

52. Straßer, C.: A deontic logic framework allowing for factual detachment. J. Appl. Logic **9**(1), 61–80 (2010)
53. Batens, D.: A universal logic approach to adaptive logics. Log. Univers. **1**, 221–242 (2007)
54. Putte, F.V.D., Straßer, C.: Extending the standard format of adaptive logics to the prioritized case. Logique at Analyse **55**(220), 601–641 (2012)
55. Pollock, J.: Defeasible reasoning. Cogn. Sci. **11**(4), 481–518 (1987)
56. Lamarre, P.: A family of non-monotonic inference systems based on conditional logics. Tech. Rep., AAAI FS-93-01 (1993)
57. de Kleer, J.: An assumption-based TMS. In: Readings in Nonmonotonic Reasoning, pp. 280–298. Morgan Kaufmann, San Francisco (1987)
58. Jeffry, P.F., Renee, E.: On relevance in nonmonotonic reasoning: some empirical studies. In: Russ, G., Devika, S. (eds.) Relevance: AAAI 1994 Fall Symposium Series, pp. 64–67. AAAI Press, Palo Alto (1994)
59. Bonevac, D.: Deduction, Introductory Symbolic Logic, 2nd edn. Blackwell, Malden (2003)
60. Koons, R.: Defeasible reasoning. In: Zalta, E.N. (ed.) The Stanford Encyclopedia of Philosophy, Winter 2009 edn. Stanford University Press, Stanford (2009)
61. Reiter, R., Criscuolo, G.: On interacting defaults. In: IJCAI'81: Proceedings of the 7th International Joint Conference on Artificial Intelligence, pp. 270–276. Morgan Kaufmann, San Francisco (1981)

Chapter 7
An Adaptive Logic for Rational Closure

Monotonicity is an essential property of classical logic. For instance for a mathematician it would be rather inefficient (and demotivating) if a proof of a statement φ from a set of statements Γ would be invalidated by the addition of other statements to Γ. Of course, the situation in everyday life is quite different.The derivative power of human (and other intelligent) beings depends to a high degree on methods for drawing consequences which can be invalidated with the arrival of new information. Modeling ways of nonmonotonic reasoning in a formal framework is therefore very interesting for philosophers, as well as for the artificial intelligence community. Many systems were proposed in the literature to get a better grip on some of them, such as negation as failure [2], circumscription [3], default logic [4], or autoepistemic logic [5]. A more general view enabling comparative studies of nonmonotonic logics was introduced on the one hand by Gabbay [6], who focused on consequence relations, and on the other hand by Shoham (e.g. [7, 8]), who proposed a general model theory for nonmonotonic inference based on preference relations on models.

In [9] Kraus, Lehmann and Magidor give an account of nonmonotonic consequence relations by isolating a list of properties and putting them in the form of conditions on consequence relations. These properties have become well-known as the KLM-properties and form the basis of active research. They have been put in the context of a more generic framework in [10] and [11]. define, among other things, *preferential* consequence relations which prove to be a powerful tool for studying what could be considered as reasonable inference procedures. In [12] Lehmann and Magidor study a strong nonmonotonic, so-called *rational* consequence relation, which extends the preferential consequence relation of [9] by also validating the rule of rational monotonicity.

This is a substantially revised version of a paper that has been published under the name "*An adaptive logic for Rational Closure*" in "The Many Sides of Logic" in "Studies in Logic" Series, College Publications, London, 2009, [1].

C. Straßer, *Adaptive Logics for Defeasible Reasoning*, Trends in Logic 38,
DOI: 10.1007/978-3-319-00792-2_7, © Springer International Publishing Switzerland 2014

$$\frac{\alpha \mid\!\sim \gamma \quad \alpha \not\mid\!\sim \neg\beta}{\alpha \wedge \beta \mid\!\sim \gamma} \qquad \text{(Rational Monotonicity)}$$

The motivating question behind their study is as follows: given a conditional knowledge base **K** (i.e., a set of conditionals), what other conditionals can we infer from them. Hence, we are interested in extending the given conditional knowledge base by additional conditionals to $\mathbf{K}' = Cn(\mathbf{K})$ and this way to gain a notion of entailment. Moreover, we expect that our extended knowledge base $\mathbf{K}'$ satisfies all the properties of rational consequence relations such as Rational Monotonicity.

The most natural idea to achieve this goal seems to simply take the intersections of all extensions of **K** that satisfy all the properties of rational consequence relations. Obviously this will result in an extension of **K** since **K** is included in all of its extensions. The hope is that additionally this intersection also satisfies all the properties of rational consequence relations. However, it is ill-founded. This can easily be illustrated by an example. Suppose **K** only consists of $a \mid\!\sim b$. In the absence of $a \mid\!\sim \neg c$ one would—in view of Rational Monotonicity—expect that we get $(a \wedge c) \mid\!\sim b$. However, by intersection all extensions of **K** that satisfy the properties of rational consequence relations we don't get that. The reason is that on the one hand there are extensions of **K** where $a \mid\!\sim \neg c$ holds and $(a \wedge c) \mid\!\sim b$ does not hold (see Type 1 extensions in Table 7.1). As soon as we have $a \mid\!\sim \neg c$ Rational Monotonicity cannot be used to "enforce" $(a \wedge c) \mid\!\sim b$. On the other hand, there are extensions of **K** where $a \mid\!\sim \neg c$ does not hold and hence $(a \wedge c) \mid\!\sim b$ holds by Rational Monotonicity (see Type 2 extensions in Table 7.1). As a consequence of the existence of both types of extensions, $a \mid\!\sim b$ is in our intersection, while $a \mid\!\sim \neg c$ is not in our intersection and nevertheless $(a \wedge c) \mid\!\sim b$ is not in our intersection. Hence, our intersection violates Rational Monotonicity.

Note that given our knowledge base **K**, c is an irrelevant fact since it doesn't play a role in the conditional assertions of our knowledge base. Nevertheless, there are extensions in which c becomes relevant: for instance the type 1 extension in our Table 7.1.

Lehmann and Magidor propose a technique to gain a notion of rational entailment that avoids the problems of the ill-founded approach based on the intersection of extensions and that hence offers a better account of irrelevant factors such as c in our example. The idea is as follows. Given a rational consequence relation we can order propositional formulas by means of degrees of (ab)normality: $a \lessdot b$ iff $(a \vee b) \mid\!\sim \neg b$. In words: b is less normal than a in case $a \vee b$ usually implies that b is not the case. Hence, each rational consequence relation can be associated with such a normality order or a ranking in which each propositional formula is associated with a natural number indicating its degree of normality. The idea is now to pick out of all extensions of **K** that satisfy the properties of rational consequence relations the one in which every formula A is interpreted as normal as possible with respect to $\lessdot$. Indeed, they show that for a huge class of conditional knowledge bases there is such a unique minimally abnormal extension which they dub the *Rational Closure*.

Let us illustrate this with our example before we give a formally precise account in the next section. First, we want to interpret a is normal as possible: that means

Table 7.1 Various extensions of $\mathbf{K} = \{a \mid\!\sim b\}$

	Type 1	Type 2	Intersection
$a \mid\!\sim b$	✔	✔	✔
$a \mid\!\sim \neg c$	✔	✘	✘
$a \mid\!\sim c$	✘	✘	✘
$(a \wedge c) \mid\!\sim b$	✘	✔	✘

$\top \mid\!\sim \neg a$ (which is the same as $\top \lessdot a$) should not hold: nothing in our knowledge base suggests that a is usually not the case. Second, note that the type 1 interpretation satisfies $a \lessdot (a \wedge c)$ since by means of properties of rational consequence relations $a \mid\!\sim \neg c$ implies $a \vee (a \wedge c) \mid\!\sim \neg(a \wedge c)$. In the type 2 interpretation we have neither $a \lessdot (a \wedge c)$ nor $a \lessdot (a \wedge \neg c)$. In sum, the type 1 interpretation interprets $a \wedge c$ unnecessarily abnormal, while in type 2 interpretations the degree of normality of $a \wedge c$, $a \wedge \neg c$ and a is on par. This mirrors the fact that c is indeed irrelevant. In this sense extensions of type 2 offer a more normal interpretation than extensions of type 1 (other things being equal).

There are various ways to construct the rational closure of a knowledge base. For instance Lehmann and Magidor define a partial ordering on rational consequence relations that extend a given conditional knowledge base and the minimum of which represents the rational closure. Alternatively, there are set-theoretic constructions for a ranked model for the rational closure (see Sect. 7.1.4). Ranked models provide a semantics to rational consequence relations: we will discuss this in detail in the next section.

What is missing is a proof theory that enables one to derive all the conditionals from a conditional knowledge base $\mathbf{K}$ that are in the rational closure of $\mathbf{K}$. In this chapter I will propose such a proof theory by means of an AL for rational closure. As it has been shown in [13], rational closure is equivalent to *1-entailment* as defined in the context of Pearl's **Z**-system [14]. Therefore we additionally gain a proof theory for 1-entailment as well.

The idea is to realize the main rationale behind Rational Closure—to interpret each propositional formula as normally as possible—by means of dynamic proofs which allow for inferences that are equipped with normality assumptions. Lehmann and Magidor propose two perspectives on conditionals $a \mid\!\sim b$: on the one hand $\mid\!\sim$ may represent a consequence relation, on the other hand it may represent a conditional connective in the object language (just as $\leadsto$ in the previous chapter). Analogous to the previous chapter we will adopt the second perspective.[1] Moreover, it will be handy to express at the object level the degree of (ab)normality of a given formula. To sketch the main mechanism and to give the reader already a general idea of the proof theory, let me state an excerpt from a proof from $\mathbf{K} = \{a \mid\!\sim b\}$. In Sect. 7.2 we will explain the technical details.

[1] For representational reasons it will be more transparent to stick with the "$\mid\!\sim$"-notation of Lehmann and Magidor instead of using "$\leadsto$".

1 $a \mathrel{|\!\sim} b$	PREM	$\emptyset$
2 $\neg(\top \mathrel{|\!\sim} \neg(a \wedge c))$	RC	$\{l_0 \lessdot (a \wedge c)\}$
3 $\neg(a \mathrel{|\!\sim} \neg c)$	2; RU	$\{l_0 \lessdot (a \wedge c)\}$
4 $(a \wedge c) \mathrel{|\!\sim} b$	1,3; RU	$\{l_0 \lessdot (a \wedge c)\}$

The first line just introduces our premise. The assumption expressed in the condition of the second line is to be read as "the degree of abnormality of $a \wedge c$ is not worse than 0" (where 0 is the most normal degree and it gets less normal as we count up). That the degree of abnormality of $a \wedge c$ is 0 means that it is not usually the case that $\neg(a \wedge c)$. This is expressed by the formula at line 2. It is easy to see that $\neg(a \mathrel{|\!\sim} \neg c)$ follows from $\neg(\top \mathrel{|\!\sim} \neg(a \wedge c))$ by the KLM-properties. Hence, line 3. Finally we can derive $(a \wedge c) \mathrel{|\!\sim} b$ by Rational Monotonicity on line 4. In sum, in the AL proofs we interpret formulas as normal as possible by means of equipping lines with conditions in which these defeasible normality assumptions are expressed. The AL marking mechanism will realize the retraction of inferences whose associated assumptions are violated.

In this chapter I will proceed as follows. I will give a technically precise account of Rational Closure in Sect. 7.1. My original contribution, the AL for Rational Closure, is presented in Sect. 7.2. First I introduce the lower limit logic $\mathbf{R}^+$. The difference to the well-known $\mathbf{R}$ is that in $\mathbf{R}^+$ we can express the degree of (ab)normality of a formula in the object language. Then I introduce the AL $\mathbf{ARC^s}$ and shortly discuss the handling of negative conditional knowledge in Sect. 7.3. Finally I will conclude this chapter in Sect. 7.4.

7.1 Rational Closure

Let a propositional language $\mathcal{L}_p$ be characterized as follows. We denote the classical connectives by $\neg, \vee, \wedge, \supset$ and $\equiv$. Furthermore, we use the primitive symbols $\top$ and $\bot$. Lower case Greek letters α, β, γ and δ denote formulas in $\mathcal{V}_p$, the set of propositional formulas over the set of propositional letters $\mathcal{P} = \{p_1, \ldots, p_n\}$. In examples we will also often denote propositional variables by lower case Roman letters a, b, c, etc.

The authors in [9] introduce the symbol $\mathrel{|\!\sim}$ for conditional consequence relations. Statements of the form "$\alpha \mathrel{|\!\sim} \beta$", where $\alpha, \beta \in \mathcal{V}_p$, are called *conditional assertions* and can be interpreted as "from α sensibly conclude β" or "α usually implies β".

Definition 7.1.1 (see [9], **Def. 15).** A relation $\mathrel{|\!\sim} \subseteq \mathcal{V}_p \times \mathcal{V}_p$ is called a *preferential consequence relation* iff it satisfies the following properties[2]:

$$\alpha \mathrel{|\!\sim} \alpha \qquad \text{(Reflexivity)}$$

[2] This is the characterization presented in [12]. In [9] the authors give an alternative (equivalent) characterization replacing (And) with (Cut).

$$\frac{\vdash \alpha \equiv \beta \quad \alpha \mid\sim \gamma}{\beta \mid\sim \gamma} \qquad \text{(Left Logical Equivalence)}$$

$$\frac{\vdash \alpha \supset \beta \quad \gamma \mid\sim \alpha}{\gamma \mid\sim \beta} \qquad \text{(Right Weakening)}$$

$$\frac{\alpha \mid\sim \beta \quad \alpha \mid\sim \gamma}{\alpha \mid\sim \beta \wedge \gamma} \qquad \text{(And)}$$

$$\frac{\alpha \mid\sim \beta \quad \alpha \mid\sim \gamma}{\alpha \wedge \beta \mid\sim \gamma} \qquad \text{(Cautious Monotonicity)}$$

$$\frac{\alpha \mid\sim \gamma \quad \beta \mid\sim \gamma}{\alpha \vee \beta \mid\sim \gamma} \qquad \text{(Or)}$$

Another way to look at the properties is to interpret them as inferential rules where $\mid\sim$ is a logical connective and does therefore not represent a consequence relation. We call the system constituted by the above rules $\mathbf{P}_{\mathsf{KLM}}$.

Remarks 7.1.1. The following properties have been shown to be valid for preferential consequence relations in [9]:

$$\frac{\alpha \mid\sim \beta \quad \beta \mid\sim \alpha \quad \alpha \mid\sim \gamma}{\beta \mid\sim \gamma} \qquad \text{(Equivalence)}$$

$$\frac{\alpha \wedge \beta \mid\sim \gamma \quad \alpha \mid\sim \beta}{\alpha \mid\sim \gamma} \qquad \text{(Cut)}$$

$$\frac{\alpha \mid\sim \beta \supset \gamma \quad \alpha \mid\sim \beta}{\alpha \mid\sim \gamma} \qquad \text{(MPC)}$$

$$\frac{\alpha \wedge \beta \mid\sim \gamma}{\alpha \mid\sim \beta \supset \gamma} \qquad \text{(S)}$$

$$\frac{\alpha \wedge \neg\beta \mid\sim \gamma \quad \alpha \wedge \beta \mid\sim \gamma}{\alpha \mid\sim \gamma} \qquad \text{(D)}$$

In addition to Rational Monotonicity, the following properties are in general not valid for preferential consequence relations:

$$\frac{\alpha \wedge \gamma \not\mid\sim \beta \quad \alpha \wedge \neg\gamma \not\mid\sim \beta}{\alpha \not\mid\sim \beta} \qquad \text{(Negation Rationality)}$$

$$\frac{\alpha \not\mid\sim \gamma \quad \beta \not\mid\sim \gamma}{\alpha \vee \beta \not\mid\sim \gamma} \qquad \text{(Disjunctive Rationality)}$$

Remarks 7.1.2 (see [12], *Lem. 10, Lem. 12).* "Rational Monotonicity" implies "Disjunctive Rationality" and this implies "Negation Rationality".

Recall our ordering relation $\alpha \lessdot \beta =_{\text{df}} (\alpha \vee \beta \mid\sim \neg\beta)$.

Fact 7.1.1. *Where* $\mid\sim$ *is a preferential consequence relation*[3]*:*

[3] Proofs can be found in Appendix E.

(i) $\alpha \lessdot \beta$ *implies* $\alpha \mid\!\sim \alpha \wedge \neg\beta$, $\alpha \mid\!\sim \neg\beta$, *and* $\alpha \vee \beta \mid\!\sim \alpha$
(ii) $\alpha \mid\!\sim \beta$ *and* $\alpha \lessdot \gamma$ *imply* $\beta \lessdot \gamma$
(iii) $\alpha \lessdot \beta$ *and* $\beta \lessdot \gamma$ *imply* $\alpha \lessdot \gamma$

Definition 7.1.2 (see [12], Def. 13). A preferential consequence relation is a *rational consequence relation* if it satisfies rational monotonicity. We call the corresponding inferential system $\mathbf{R}_{\mathsf{KLM}}$.

A *(conditional) knowledge base* **K** is a set of conditional assertions. We call $\mid\!\sim$ a *preferential (resp. rational) extension of* **K** iff $\mid\!\sim$ is a preferential (resp. rational) consequence relation and it satisfies all the conditional assertions in **K**.

7.1.1 Preferential Entailment

Let us now tackle the question what conditional assertions a conditional knowledge base **K** entails. We first look at the simpler case concerning the rationale of preferential consequence relations. The question can be answered by means of intersecting all preferential extensions of **K**. Indeed, it can be shown that this intersection is itself a preferential consequence relation. This provides us with a notion of *preferential entailment*: $\alpha \mid\!\sim \beta$ is preferentially entailed by **K** iff it is in the intersection of all preferential extensions of **K**.

Alternatively and equivalently, the question can be answered by means of the inferential system $\mathbf{P}_{\mathsf{KLM}}$: $\alpha \mid\!\sim \beta$ is preferentially entailed by **K** iff it can be inferred from **K** by means of $\mathbf{P}_{\mathsf{KLM}}$.

Finally, we can answer the question in a semantic way. For this we have to first define a semantics for preferential consequence relations. A *preferential model* is a triple $\langle S, l, \prec\rangle$ where S is a set of states, l associates states with worlds (i.e., interpretations of propositional atoms) and $\prec$ is a partial order on S where $s \prec s'$ reads "state s is more normal than state s'". The validity of $\alpha \mid\!\sim \beta$ in a model M is then decided in view of the α-minimal states. A state s is α-minimal in case its associated world verifies α ($l(s) \models \alpha$) and there is no state s' below s ($s' \prec s$) such that $l(s') \models \alpha$. If in all the associated worlds of the α-minimal states β is also valid, the model validates $\alpha \mid\!\sim \beta$. We write $\alpha \mid\!\sim_M \beta$ if M validates $\alpha \mid\!\sim \beta$. Finally, we have the following *smoothness condition*: for each propositional formula α the set of states whose associated worlds validate α is $\prec$-smooth. Hence, for each state s for which $l(s) \models \alpha$ there is a state $s' \preceq s$ that is minimal with the property $l(s') \models \alpha$.[4]

Let us illustrate this by a preferential model M for $\mathbf{K} = \{p_1 \mid\!\sim p_2\}$. Suppose the set of propositional letters is $\{p_1, p_2, p_3\}$. We consider the worlds as listed in (Table 7.2):

Figure 7.1 illustrates the ordering $\prec$ of our model M:

[4] Recall that we used a similar condition for our threshold functions in the generalized standard format for ALs in Chap. 5. This ensures the strong reassurance property for ALs.

Table 7.2 Worlds for $\mathbf{K} = \{p_1 \mid\sim p_2\}$

	p_1	p_2	p_3
w_2	0	1	0
w_{123}	1	1	1
w_{12}	1	1	0
w_{13}	1	0	1

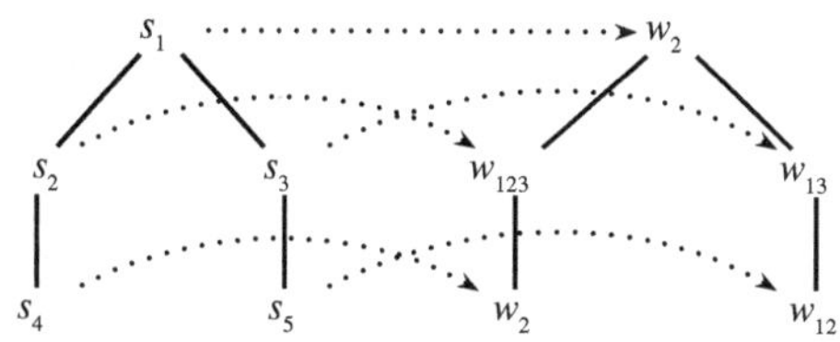

Fig. 7.1 *Left* the ordering on states in a preferential model. *Right* the corresponding ordering on the associated worlds

On the left hand side we have the ordering on the states. The dotted arrows illustrate our mapping l of states into worlds. On the right hand side we illustrate the ordering that is indirectly imposed on the worlds. This shouldn't be taken too literal but serves illustrative purposes. Note that w_2 appears twice: this is possible since different states may be mapped to the same world (like here s_1 and s_4 which are both mapped to w_2).

For instance, we have $p_1 \mid\sim_M p_2$ since in all minimal states whose associated worlds verify p_1 also p_2 is verified: these minimal states are s_2 and s_5. Note also that Rational Monotonicity is not valid in this model: we have $p_1 \mid\sim_M p_2$ and we don't have $p_1 \mid\sim_M \neg p_3$ but we also don't have $p_1 \wedge p_3 \mid\sim_M p_2$ as the reader can easily verify. (Note that s_3 is $p_1 \wedge p_3$-minimal.)

In [12] it was shown that $\mid\sim_M$ is a preferential consequence relation, and that each preferential consequence $\mid\sim$ relation can be represented by a preferential model M such that $\mid\sim = \mid\sim_M$.

Theorem 7.1.1 (see [12], Thm. 1). *A binary relation $\mid\sim \subseteq \mathcal{V}_p \times \mathcal{V}_p$ is a preferential consequence relation iff it is the consequence relation defined by some preferential model.*

Hence, our third way of expressing preferential entailment is as follows: **K** entails $\alpha \mid\sim \beta$ iff in all models M of **K**, $\alpha \mid\sim_M \beta$.

Given the equivalence of the three characterizations of preferential entailment it is arbitrary whether we define it in terms of the intersection of all preferential consequence relations that extend some knowledge base, or by means of the inferential system **P**, or in semantic terms. Here we go for the latter:

Definition 7.1.3 (see [12], Def. 6). A conditional assertion $\alpha \mid\sim \beta$ is *preferentially entailed* by a conditional knowledge base **K** iff it is satisfied by all preferential models of **K**. The set of all conditional assertions that are preferentially entailed by **K** is denoted by $\mathbf{K}^p$. The relation $\mathbf{K}^p$ is called the *preferential closure* of **K**.

As discussed, we have the following equivalences:

Theorem 7.1.2 (see [12], Thm. 2). *Let* **K** *be a conditional knowledge base. The following conditions are equivalent:*

1. $\alpha \mid\sim \beta \in \mathbf{K}^p$
2. *for all preferential models M of* **K**, $\alpha \mid\sim_M \beta$
3. $\alpha \mid\sim \beta$ *has a proof from* **K** *in* $\mathbf{P}_{\mathsf{KLM}}$
4. $\alpha \mid\sim \beta$ *for all preferential consequence relations* $\mid\sim$ *that extend* **K**

Note that the preferential closure satisfies all the KLM-properties:

Theorem 7.1.3 (see [12], Cor. 2). *For a conditional knowledge base* **K** *the preferential closure* $\mathbf{K}^p$ *defines a preferential consequence relation.*

7.1.2 Rational Entailment

We already mentioned in the beginning of this chapter that the idea to characterize rational entailment analogous to the notion of preferential entailment is ill-founded. The intersection of all rational extensions of a conditional knowledge base **K** is not warranted to be rational itself. Indeed, the intersection of all rational extensions of **K** yields exactly the same result as intersecting the preferential extensions of **K**: namely $\mid\sim_{\mathbf{K}^p}$ the preferential closure.

Theorem 7.1.4 (see [12], Thm. 6). *If an assertion* $\alpha \mid\sim \beta$ *is in all rational extensions of a conditional knowledge base* **K**, *then it is also in all preferential extensions of* **K**. *Furthermore,* $\alpha \mid\sim \beta$ *can be derived from* **K** *by the rules of* $\mathbf{R}_{\mathsf{KLM}}$ *iff it can be derived from* **K** *by the rules of* $\mathbf{P}_{\mathsf{KLM}}$.

In view of this, Magidor and Lehmann propose an alternative, more sophisticated characterization of rational entailment. For this purpose they introduce a partial order on rational consequence relations (not to be confused with the partial order on states that preferential models come with!):

Definition 7.1.4 (see [12], Def. 20). Let $\mid\sim_0$ and $\mid\sim_1$ be two rational consequence relations. We shall say that $\mid\sim_0$ is *preferable* to $\mid\sim_1$ ($\mid\sim_0 \sqsubset \mid\sim_1$) iff[5]:

1. there exists an assertion $\alpha \mid\sim_1 \beta$ such that $\alpha \not\mid\sim_0 \beta$ and for all γ for which $\gamma \lessdot \alpha$ for $\mid\sim_0$, and for all δ such that $\gamma \mid\sim_0 \delta$, we also have $\gamma \mid\sim_1 \delta$, and
2. for any assertion $\gamma \mid\sim_0 \delta$ such that $\gamma \not\mid\sim_1 \delta$, there are γ' and δ' for which $\gamma' \mid\sim_1 \delta'$, $\gamma' \not\mid\sim_0 \delta'$ and $\gamma' \lessdot \gamma$ for $\mid\sim_1$.

This can be interpreted as an argumentation game between two persons. Both persons come with a rational extension of a knowledge base. In order to attack the other person one may name a conditional assertion $\alpha \mid\sim \beta$ that is in the extension of the

[5] In [12] the authors use the "$\prec$" symbol for this relation. In order to disambiguate the usage of the various symbols for ordering relations in this chapter and therefore to serve readability, in this chapter "$\sqsubset$" is being used exclusively for the preferability relation of Definition 7.1.4.

other person but not in one's own. A defense consists of a counter-attack: it is to name a conditional assertion $\gamma \mid\!\sim \delta$ that is in the extension of the opponent but not in one's one and for which the antecedent γ is more normal than α according to the opponent's extension (hence the opponent has $\gamma \lessdot \alpha$ in her knowledge base). In order to win the proponent has to be able to attack the opponent in such a way that the opponent cannot offer a defense (criterion 1), and for every counter-attack of the opponent the proponent has to be able to respond (criterion 2).

The rational closure of a knowledge base **K** is the rational extension of **K** that is $\sqsubset$-preferable to all other rational extensions of **K**. Indeed, for instance for finite languages there is a unique rational extension that has this property.

Definition 7.1.5 (see [12], Def. 21). Let **K** be a knowledge base. If there is a rational extension $\mid\!\sim$ of **K** that is $\sqsubset$-preferable to all other rational extensions of **K**, then $\mid\!\sim$ is called the *rational closure* of **K**.

7.1.3 Ranked and Rational Models

Let us proceed by giving a semantics for rational consequence relations. In the next section we will use this semantic characterization in order to construct the rational closure of a knowledge base. The semantics is provided by a specific sub-class of preferential models, namely the ones for which the ordering on the states $\prec$ is modular.

Definition 7.1.6. A partial order $\prec$ over V is called *modular* iff it satisfies the following equivalent conditions:

- there is a totally ordered set Υ (the strict order on Υ is denoted by $\prec'$) and a function $\mathsf{r} : V \to \Upsilon$ (the *ranking function*) such that $x \prec y$ iff $\mathsf{r}(x) \prec' \mathsf{r}(y)$.
- for any $x, y, z \in V$ if $x \not\prec y$, $y \not\prec x$ and $z \prec x$, then $z \prec y$
- for any $x, y, z \in V$ if $x \prec y$, then either $z \prec y$ or $x \prec z$
- for any $x, y, z \in V$ if $x \not\prec y$, $y \not\prec z$, then $x \not\prec z$

In [12] (Def. 14) the authors define a *ranked model* as a preferential model $\langle S, l, \prec\rangle$ for which the strict partial order $\prec$ is modular. Therefore we can associate a ranking function r with each ranked model.

This is a good place to introduce some insights which help to simplify the technical requirements below. Using states and mapping states into worlds in the definition of a preferential model $M = \langle S, l, \prec\rangle$ allows for one and the same world w to be associated with different states, i.e., there are $s, t \in S$ such that $s \neq t$ and $l(s) = l(t)$.[6] However, if $\prec$ is modular, either both states have the same rank or one state is higher ranked than the other, e.g. $s \prec t$. Let $N = \langle S', l', \prec'\rangle$ where $S' = S \setminus \{t\}$, l' is l restricted to the domain S', and $\prec' = \prec \cap (S' \times S')$. Due to the modularity of $\prec$, it is obvious in the first case that $\mid\!\sim_N = \mid\!\sim_M$. In the second case t is such that for any

[6] See for instance the states s_1 and s_4 in Fig. 7.1.

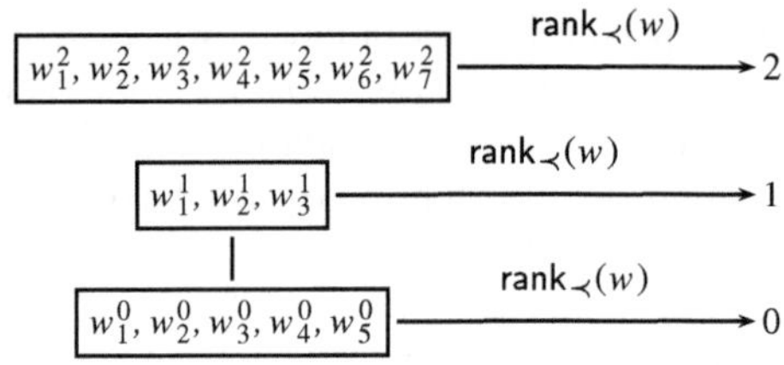

Fig. 7.2 Illustration of a rational model and its canonical ranking function

$\alpha \in \mathcal{V}_p$ for which $t \models \alpha$, t is not minimal in $\hat{\alpha}$ (since $s \prec t$ and $s \models \alpha$). What we have just seen is that, whenever for a ranked model two states are mapped into the same world, then we can do without one of them. But then, in case $\prec$ is modular, the talk about states is superfluous. Similar to [15] and to [16] we use therefore the following simplification:

Definition 7.1.7. A *rational model* M is a pair $\langle W, \prec\rangle$ where W is a set of worlds (i.e., interpretations of propositional atoms) and $\prec$ is a strict modular order on W. We define $\alpha \mid\!\sim_M \beta$ iff for all $w \in \min_\prec(\sigma_M(\alpha))$: $w \models \beta$, where $\sigma_M(\alpha) =_{\text{df}} \{w \in W : w \models \alpha\}$ and for all $W' \subseteq W$, $\min_\prec(W') =_{\text{df}} \{w \in W'$: there is no $w' \in W'$ such that $w' \prec w\}$.[7]

Note that due to the modularity of $\prec$ we have $w \prec w'$ for all $w \in \min_\prec(W')$ and for all $w' \in W' \setminus \min_\prec(W')$. Note further that the modularity of $\prec$ gives rise to equivalence classes of worlds which are mutually incomparable, namely the ones which have the same ranking. The equivalence classes themselves are totally ordered.

For a rational model $M = \langle W, \prec\rangle$ we define the *canonical ranking function* $\mathsf{rank}_\prec : W \to \mathbb{N}$ ($\mathbb{N}$ being the natural numbers incl. 0) as the unique epimorphism from $\langle W, \prec\rangle$ onto $\langle\{0, \ldots, n\}, <\rangle$ where $n \in \mathbb{N}$. Obviously for every equivalence class $[w]$ there is a unique natural number i assigned to it by $\mathsf{rank}_\prec$, i.e., for all $w' \in [w]$, $\mathsf{rank}_\prec(w') = \mathsf{rank}_\prec(w) = i$ and for each $w' \notin [w]$, $\mathsf{rank}_\prec(w') \neq i$. It is obvious that all minimal worlds in $\sigma_M(\alpha)$ belong to the same equivalence class. The canonical ranking of a rational model is illustrated in Fig. 7.2.

We define another useful related mapping $\mathsf{AbDeg}_\prec : \mathcal{V}_p \to \{0, \ldots, n, \omega\}$ on basis of $\mathsf{rank}_\prec$. We say $\mathsf{AbDeg}_\prec(\alpha)$ is the *abnormality degree of* α. Intuitively speaking, the abnormality degree of a formula α mirrors how surprised we would be if α were to be the case: higher degrees indicate a higher surprise factor.

For a $\alpha \in \mathcal{V}_p$, $\alpha \mapsto h$ iff $\sigma_M(\alpha) \neq \emptyset$ and $\mathsf{rank}_\prec(w) = h$ for any $w \in \min_\prec(\sigma_M(\alpha))$. If $\sigma_M(\alpha) = \emptyset$ then $\alpha \mapsto \omega$.

Suppose in Fig. 7.2, $\sigma_M(\alpha) = \{w_2^1, w_3^1, w_3^2, w_5^2, w_7^2\}$. Then $\min_\prec(\sigma_M(\alpha)) = \{w_2^1, w_3^1\}$ and $\mathsf{AbDeg}_\prec(\alpha) = 1$.

Just like preferential models adequately represented preferential consequence relations, rational models represent rational consequence relations:

[7] Since in this chapter we only deal with a finite language, the set of worlds W is finite as well. Hence, we don't need to add the smoothness condition that we used in the definition of preferential models.

Theorem 7.1.5 (see [12],Thm. 5). *A binary relation $\mid\!\sim$ is a rational consequence relation iff it is the consequence relation defined by some rational model.*

In the remainder of this section we present some properties that will be useful later on and that highlight the links between the various ways normality can be expressed in the present formal framework. Let in the following facts M be a rational model. We denote by $\alpha \lessdot_M \beta$ that M validates $\alpha \lessdot \beta$ (i.e., $\alpha \vee \beta \mid\!\sim_M \neg\beta$).

The normality order $\lessdot$ mirrors exactly the abnormality degrees: β is less normal than α in M in case the abnormality degree of α is lower than the one of β or both abnormality degrees are infinite (i.e., there are neither α-worlds nor β-words). Note that in the latter case $\alpha \vee \beta \mid\!\sim \gamma$ for any γ.

Fact 7.1.2. $\alpha \lessdot_M \beta$ *iff* ($\mathsf{AbDeg}_\prec(\alpha) < \mathsf{AbDeg}_\prec(\beta)$ *or* $\mathsf{AbDeg}_\prec(\alpha) = \mathsf{AbDeg}_\prec(\beta) = \omega$)

Proof. Let $\alpha \lessdot_M \beta$. Then $\alpha \vee \beta \mid\!\sim_M \neg\beta$. In case $\sigma_M(\alpha \vee \beta) \neq \emptyset$, for all $w \in \min_\prec(\sigma_M(\alpha \vee \beta))$, $w \models \neg\beta$. Hence $\min_\prec(\sigma_M(\alpha \vee \beta)) = \min_\prec(\sigma_M(\alpha \wedge \neg\beta))$. But then obviously for any $w \in \min_\prec(\sigma_M(\alpha))$ and any $w' \in \sigma_M(\beta)$, $\mathsf{rank}_\prec(w) < \mathsf{rank}_\prec(w')$ or $\sigma_M(\beta) = \emptyset$. Therefore $\mathsf{AbDeg}_\prec(\alpha) < \mathsf{AbDeg}_\prec(\beta)$. The case $\sigma_M(\alpha \vee \beta) = \emptyset$ is trivial, since then $\sigma_M(\alpha) = \sigma_M(\beta) = \emptyset$.

Let $\mathsf{AbDeg}_\prec(\alpha) < \mathsf{AbDeg}_\prec(\beta)$. Then $\sigma_M(\alpha) \neq \emptyset$. Furthermore, either $\sigma_M(\beta) = \emptyset$ or for all $w \in \sigma_M(\beta)$, $\mathsf{rank}_\prec(w) > \mathsf{rank}_\prec(w')$ where w' is any minimal world in $\sigma_M(\alpha)$. But then obviously $w' \models \neg\beta$. Hence $\min_\prec(\sigma_M(\alpha)) = \min_\prec(\sigma_M(\alpha \wedge \neg\beta)) = \min_\prec(\sigma_M(\alpha \vee \beta))$ and therefore $\alpha \vee \beta \mid\!\sim_M \neg\beta$. The case $\mathsf{AbDeg}_\prec(\alpha) = \mathsf{AbDeg}_\prec(\beta) = \omega$ is trivial, since then $\sigma_M(\alpha \vee \beta) = \emptyset$ and thus $\alpha \vee \beta \mid\!\sim \gamma$ for all γ. □

α normally implies β in M iff the abnormality degree of α is lower than the abnormality degree of $\alpha \wedge \neg\beta$ or the abnormality degree of α is infinite.

Fact 7.1.3. $\alpha \mid\!\sim_M \beta$ *iff* ($\mathsf{AbDeg}_\prec(\alpha) < \mathsf{AbDeg}_\prec(\alpha \wedge \neg\beta)$ *or* $\mathsf{AbDeg}_\prec(\alpha) = \omega$)

Proof. This is an immediate consequence of Fact 7.1.2 since $\alpha \mid\!\sim \beta$ is equivalent to $\alpha \lessdot \alpha \wedge \neg\beta$, as can easily be shown. □

If α is more normal than β then α normally implies $\neg\beta$.

Fact 7.1.4. *If* $\alpha \lessdot_M \beta$, *then* $\alpha \mid\!\sim_M \neg\beta$.

Proof. Since $\alpha \lessdot_M \beta$, $\alpha \vee \beta \mid\!\sim_M \neg\beta$. Suppose first $\sigma_M(\alpha \vee \beta) = \emptyset$. Then $\sigma_M(\alpha) = \emptyset$ and hence $\alpha \mid\!\sim_M \neg\beta$. Suppose now $\sigma_M(\alpha \vee \beta) \neq \emptyset$. Then, for all $w \in \min_\prec(\sigma_M(\alpha \vee \beta))$, $w \models \neg\beta$. Hence, since $w \models \alpha \vee \beta$, $w \models \alpha$. Hence, $\min_\prec(\sigma_M(\alpha)) = \min_\prec(\sigma_M(\alpha \vee \beta))$. Hence $\alpha \mid\!\sim_M \neg\beta$. □

$\lessdot_M$ is transitive.

Fact 7.1.5. *If* $\alpha \lessdot_M \beta$ *and* $\beta \lessdot_M \gamma$, *then* $\alpha \lessdot_M \gamma$.

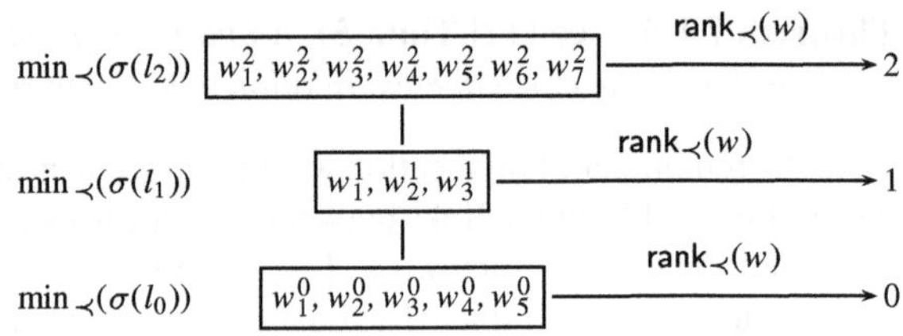

Fig. 7.3 Illustration of a rational$^+$ model and its canonical ranking function

Proof. Applying Fact 7.1.2 we have $\mathsf{AbDeg}_{\prec}(\alpha) < \mathsf{AbDeg}_{\prec}(\beta) \leq \mathsf{AbDeg}_{\prec}(\gamma)$ or $\mathsf{AbDeg}_{\prec}(\alpha) = \omega = \mathsf{AbDeg}_{\prec}(\beta) = \mathsf{AbDeg}_{\prec}(\gamma)$. The rest follows by Fact 7.1.2. □

We close our list of facts with two observations which immediately follow by the finiteness of our language:

Fact 7.1.6. *(i) For all $w \in W$ there is an α for which $w \in \min_{\prec}(\sigma(\alpha))$. (ii) For all $w \in W$, $\mathsf{rank}_{\prec}(w) \leq 2^{\mathsf{n}} - 1$.*

Proof. Ad (i): Take $\alpha = \bigwedge_{i \in I} p_i \wedge \bigwedge_{j \in \{1,\ldots,\mathsf{n}\} \setminus I} \neg p_j$ where w is the assignment $p_i \mapsto 1$ if $i \in I$ and $p_i \mapsto 0$ else.

Ad (ii): This follows from the fact that there are maximal 2^{n} worlds since there are only 2^{n} different assignments to the propositional letters $\{p_1, \ldots, p_{\mathsf{n}}\}$. □

7.1.4 Constructing a Model for Rational Closure

As shown in [12], the rational closure of a given knowledge base exists for a finite language $\mathcal{L}$. Furthermore, there are various model-theoretic approaches to construct a model for it (e.g. in [12], Sect. 5.7). The following proposal presented in [16] is interesting for our purposes. Let **K** be a conditional knowledge base. Let $\mathcal{M}$ be the set of all rational models of **K**. Define

$$U_0 =_{\text{df}} \bigcup_{\langle W, \prec \rangle \in \mathcal{M}} \{w \in W \mid \mathsf{rank}_{\prec}(w) = 0\}$$

$$U_i =_{\text{df}} \bigcup_{\langle W, \prec \rangle \in \mathcal{M}} \{w \in W \mid \mathsf{rank}_{\prec}(w) = i\} \setminus \bigcup_{j<i} U_j, \text{ where } i > 0$$

We are interested in $M = \langle U, \prec \rangle$ where $U = \bigcup_{i \in \mathbb{N}} U_i$ and $\prec$ is defined by: for all natural numbers i, j and for all w, w' for which $w \in U_i$ and $w' \in U_j$, $w \prec w'$ iff $i < j$. As proven in [16], M is a rational model defining the rational closure of **K** (Fig. 7.3).

The idea behind this construction is simple: each world gets a rank as low as possible. In terms of abnormality degrees of formulas this means that each formula

Table 7.3 Worlds for $\mathbf{K} = \{p_1 \mid\!\sim p_2, \neg p_1 \wedge \neg p_2 \mid\!\sim p_3\}$

	p_1	p_2	p_3		p_1	p_2	p_3
w_2	0	1	0	w_1	1	0	0
w_{123}	1	1	1	w_{23}	0	1	1
w_{12}	1	1	0	w_3	0	0	1
w_{13}	1	0	1	w_0	0	0	0

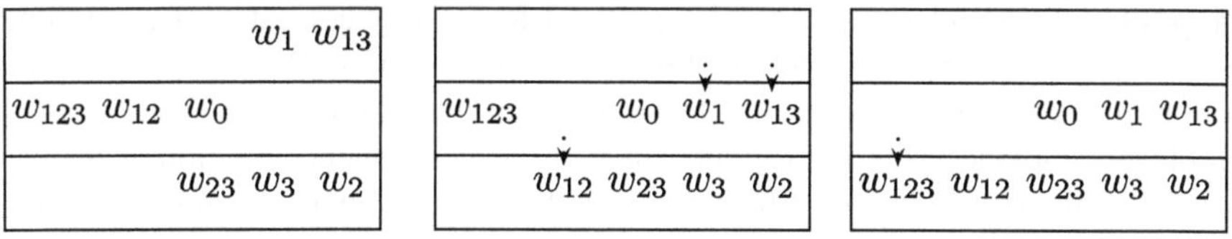

(a) model M_1 **(b)** model M_2 **(c)** model M_3

Fig. 7.4 The worlds are ordered by rank: bottom line is rank 0, second line is rank 1 and third line is rank 2

gets an abnormality degree as low as possible. Let us demonstrate this by a simple example where $\mathbf{K} = \{p_1 \mid\!\sim p_2, \neg p_1 \wedge \neg p_2 \mid\!\sim p_3\}$. Let the worlds be as in (Table 7.3):

In Fig. 7.4 we depict three models which we are going to compare now.

We see a sequence of models where worlds are ranked lower and lower from left to right (indicated by the dotted arrow). Indeed, model M_3 is the rational closure of **K** since no world can be ranked lower than it is ranked in M_3. This suggest we can visualize the construction as "dropping worlds" as low as possible. It is not difficult to see that "dropping worlds" in this manner leads to more and more preferable models: $\mid\!\sim_{M_3} \sqsubset \mid\!\sim_{M_2} \sqsubset \mid\!\sim_{M_1}$.

7.2 An AL for Rational Closure

We are now going to introduce our AL for rational closure. For this purpose we first define the logic **R**, then slightly enhance its expressive power in the logic $\mathbf{R}^+$. The latter will serve as our lower limit logic. What it adds to **R** is the ability to express in the object language the degree of (ab)normality of a formula. This will be handy for the AL $\mathbf{ARC^s}$ which is introduced in Sect. 7.2.3 since the abnormalities will be—informally speaking—of the form "α has an abnormality degree worse than n" (for each n). As a consequence $\mathbf{ARC^s}$ will interpret each formula as normally as possible relative to a given conditional knowledge base.

7.2.1 *The Logic* R

As our goal is to develop a logic, the consequences of which, for a given conditional knowledge base, correspond to the rational closure, the following logic **R** which generalizes the inferential system $\mathbf{R}_{\mathsf{KLM}}$ will be very useful. We are operating in a language $\mathcal{L}$ consisting of Boolean combinations of conditional assertions and propositional formulas (while the language of $\mathbf{R}_{\mathsf{KLM}}$ only consisted of conditional assertions). Let $\mathcal{V}_p$ be again the set of propositional formulas, $\mathcal{P} = \{p_1, \ldots, p_n\}$ the set of propositional letters, and $\mathcal{V}_{\mid\sim}$ the set of conditional assertions. $\mathcal{V}$ is the set of all Boolean combinations of the formulas in $\mathcal{V}_{\mid\sim}$. The consequence relation of **R** maps sets of formulas in $\mathcal{V}$ to sets of formulas in $\mathcal{V}$. Some notational conventions: as done so far we use α, β, γ and δ to refer to propositional formulas, φ and ψ are reserved for formulas in $\mathcal{V}$. Logic **R** consists of all axioms and rules of the propositional calculus together with the following axioms and rules:

$$\alpha \mid\sim \alpha \tag{ID}$$

$$\text{If } \vdash \alpha \equiv \beta, \text{ then } \vdash (\alpha \mid\sim \gamma) \supset (\beta \mid\sim \gamma) \tag{RCFA}$$

$$\text{If } \vdash \alpha \supset \beta, \text{ then } \vdash (\gamma \mid\sim \alpha) \supset (\gamma \mid\sim \beta) \tag{RCM}$$

$$\big((\alpha \mid\sim \beta) \wedge (\alpha \mid\sim \gamma)\big) \supset (\alpha \mid\sim \beta \wedge \gamma) \tag{CC}$$

$$\big((\alpha \mid\sim \gamma) \wedge (\beta \mid\sim \gamma)\big) \supset (\alpha \vee \beta \mid\sim \gamma) \tag{CA}$$

$$\big((\alpha \mid\sim \beta) \wedge (\alpha \mid\sim \gamma)\big) \supset (\alpha \wedge \beta \mid\sim \gamma) \tag{ASC}$$

$$\big((\alpha \mid\sim \beta) \wedge \neg(\alpha \mid\sim \neg\gamma)\big) \supset (\alpha \wedge \gamma \mid\sim \beta) \tag{RM}$$

The logic was more thoroughly studied in [10], for instance completeness and soundness was demonstrated for various semantics. In [15] a tableau calculus was developed for **R**. Like the authors in [15], we use a semantics based on rational models. For a rational model $M = \langle W, \prec \rangle$ we define $\models$ as follows: where $\alpha \mid\sim \beta \in \mathcal{V}_{\mid\sim}$: $M \models \alpha \mid\sim \beta$ iff $\alpha \mid\sim_M \beta$. Furthermore, for Boolean combinations $\models$ is defined in the usual way (e.g. $M \models \varphi \vee \psi$ iff $M \models \varphi$ or $M \models \psi$, etc.). We write $\Vdash_{\mathbf{R}}$ for the semantic consequence relation that is defined in the usual way by $\Gamma \Vdash_{\mathbf{R}} \varphi$ iff $M \models \varphi$ for all rational models of Γ.[8]

Remarks 7.2.1. Note that for instance the following properties hold in **R**:

$$\big((\alpha \mid\sim \beta) \wedge (\beta \mid\sim \alpha) \wedge (\alpha \mid\sim \gamma)\big) \supset (\beta \mid\sim \gamma) \tag{EQ}$$

$$\big((\alpha \wedge \beta \mid\sim \gamma) \wedge (\alpha \mid\sim \beta)\big) \supset (\alpha \mid\sim \gamma) \tag{RT}$$

$$\big((\alpha \wedge \beta) \mid\sim \gamma\big) \supset (\alpha \mid\sim (\beta \supset \gamma)) \tag{CW}$$

---

[8] In order to extend **R** such that its consequence relation maps formulas in the Boolean closure of $\mathcal{V}_{\mid\sim} \times \mathcal{V}_p$ to formulas in the Boolean closure of $\mathcal{V}_{\mid\sim} \times \mathcal{V}_p$ we can add an actual world @ to rational models and define $M \models p_i$ iff $@ \models p_i$. Conditional assertions and complex formulas get their truth values as before. However, here we will keeps things simple and only focus on Boolean combinations of conditional assertions.

7.2.2 The Lower Limit Logic $\mathbf{R}^+$

The main idea underlying the adaptive logic **ARC**$^{\mathbf{s}}$ for rational closure which is going to be introduced in this section is that propositional formulas are interpreted as normally as the given knowledge base allows for. Semantically this means that the worlds of rational models are ranked as low as possible (see Sect. 7.1.4). In order to make this description more precise and in order to define this logic we introduce an extended language $\mathcal{L}^+$ on basis of our language $\mathcal{L}$. Therefore we extend the given set of propositional letters $\mathcal{P} = \{p_1, \ldots, p_n\}$ with the set of propositional letters $\{l_i \mid i \in \mathcal{N}\}$ where $\mathcal{N} = \{0, \ldots, 2^n - 1\}$. We use the abbreviation m for $2^n - 1$. We define $\mathcal{P}^+ =_{\text{df}} \mathcal{P} \cup \{l_i \mid i \in \mathcal{N}\}$, $\mathcal{V}_p^+$ as the set of propositional formulas with propositional letters $\mathcal{P}^+$, and $\mathcal{V}_{\mid\sim}^+$ is the set of conditional assertions $\alpha \mid\sim \beta$ where $\alpha, \beta \in \mathcal{V}_p^+$. Finally, let $\mathcal{V}^+$ be the set of all Boolean combinations of formulas in $\mathcal{V}_{\mid\sim}^+$.

The idea is that the l_i's represent different "degrees of abnormality". $l_i \lessdot \alpha$ can be read as "α has a higher degree of abnormality than i", $\alpha \mid\sim l_i$ can be read as "α's degree of abnormality is at least i". The higher the index i, the higher is the degree of abnormality represented by l_i.

We call a rational consequence relation realizing this idea a *rational*$^+$ *consequence relation*. It is supposed to satisfy the following additional conditions, where $\alpha \in \mathcal{V}_p^+$:

$$\alpha \mid\sim l_0 \qquad (\mathrm{C}^+1)$$

$$\frac{l_{i-1} \lessdot \alpha}{\alpha \mid\sim l_j}, \text{ for all } j \leq i, \text{ where } i \in \mathcal{N} \setminus \{0\}, \qquad (\mathrm{C}^+2)$$

$$l_{i-1} \lessdot l_i, \text{ for all } i \in \mathcal{N} \setminus \{0\} \qquad (\mathrm{C}^+3)$$

$$\frac{l_{\mathsf{m}} \lessdot \alpha}{\alpha \mid\sim \bot} \qquad (\mathrm{C}^+4)$$

Note that due to the transitivity of $\lessdot$ (Fact 7.1.1.iv) and (C^+3) we immediately have $l_i \lessdot l_j$ for all $i, j \in \mathcal{N}$ for which $i < j$. (C^+1) ensures that each α has at least abnormality degree 0. By (C^+2), if α is more abnormal than $i - 1$, then it has at least abnormality degree j for $j \leq i$. Should α be more abnormal then m, then we consider it as maximally abnormal by (C^+4), i.e., $\alpha \mid\sim \bot$.

The following facts express that l_i works exactly as expected. Where $\mid\sim$ is a rational$^+$ consequence relation and $\alpha \in \mathcal{V}_p^+$, we have:

Fact 7.2.1. *Where* $i < j \leq \mathsf{m}$, $l_i \lessdot l_j$.

The l_i's are linearly ordered by $\lessdot$ according to their indexes. The fact holds by (C^+3) and Fact 7.1.1.iii.

Fact 7.2.2. *If $l_{i-1} \lessdot \alpha$ and $\alpha \lessdot l_j$, then (a) $\alpha \mid\!\sim l_{i'}$ for all $i' \leq i$ and $i' \leq \mathsf{m}$, and (b) $\alpha \mid\!\sim \neg l_{j'}$ for all $j' \geq j$ and $j' \leq \mathsf{m}$.*

If α has an abnormality degree worse than $i-1$ and better than j then (a) its abnormality degree is at least i' for all $i' \leq i$, and (b) its abnormality degree is not as bad as j' for all $j' \geq j$ that are lower or equal to m.

(a) holds due to (C^+2). Due to Fact 7.1.1.iii and (C^+3), $\alpha \lessdot l_{j'}$ for all $j' \geq j$ and $j' \leq \mathsf{m}$. Hence, by Fact 7.1.1.i, $\alpha \mid\!\sim \neg l_{j'}$.

Fact 7.2.3. *(i) $l_0 \mid\!\sim \top$ and $\top \mid\!\sim l_0$. (ii) $\top \mid\!\sim \alpha$ iff $l_0 \mid\!\sim \alpha$.*

$\top$ has abnormality degree 0 ("most normal").

$\top \mid\!\sim l_0$ holds by (C^+1) and the other direction is trivial. (ii) holds due to (Equivalence).

In order to avoid unnecessary confusions, we need to be very precise from now on concerning the language that is used in the context of models and consequence relations. Where $\hat{\mathcal{L}} \in \{\mathcal{L}, \mathcal{L}^+\}$, rational models are called rational $\hat{\mathcal{L}}$-models if the corresponding worlds are $\hat{\mathcal{L}}$-worlds, namely assignments of truth values to the propositional letters in $\hat{\mathcal{L}}$. Similarly we call a consequence relation $\mid\!\sim$ an $\hat{\mathcal{L}}$-consequence relation if it is defined over the propositional formulas in $\hat{\mathcal{L}}$. Finally, we call a knowledge base consisting of conditional assertions over propositional formulas in $\hat{\mathcal{L}}$ an $\hat{\mathcal{L}}$-conditional knowledge base.

Definition 7.2.1 (Rational$^+$ models). A *rational$^+$ $\mathcal{L}^+$-model* is a rational $\mathcal{L}^+$-model $M = \langle W, \prec \rangle$ that meets the following requirement (R):

(R) For each world $w \in W$, $\mathsf{rank}_\prec(w) \leq \mathsf{m}$ and where $\mathsf{rank}_\prec(w) = i$, (i) $w \models l_j$ for all $j \in \mathcal{N}$ such that $j \leq i$ and (ii) $w \models \neg l_j$ for all $j \in \mathcal{N}$ such that $j > i$.

A rational$^+$ $\mathcal{L}^+$-model is schematically illustrated in Fig. 7.3. As was already the case for rational models, each level of the modular order $\prec$ has an associated rank, i.e., $\mathsf{rank}_\prec(w^i_j) = i$. This partitions the worlds in equivalence classes of worlds with the same rank. Moreover, each equivalence class of rank i corresponds to the minimal worlds validating l_i. This way we can express in the object language the degree of abnormality of a formula β. As is shown in the following lemma, $\mathsf{AbDeg}_\prec(\beta) = i$ iff ($l_{i-1} \lessdot_M \beta$ and not $l_i \lessdot_M \beta$).

In the following we state some facts that highlight how rational$^+$ $\mathcal{L}^+$-models handle normality degrees.[9] Let $M = \langle W, \prec \rangle$ be a rational$^+$ $\mathcal{L}^+$-model.

Fact 7.2.4. $\alpha \mid\!\sim_M l_0$

Each α has abnormality degree of at best 0.

Fact 7.2.5. $\mathsf{AbDeg}_\prec(l_i) = i$ *iff* $\sigma_M(l_i) \neq \emptyset$, *else* $\mathsf{AbDeg}_\prec(l_i) = \omega$.

The abnormality degree of l_i is i except there are no l_i-worlds, in which case it is ω.

---

[9] The interested reader can find the proofs for these fact in Appendix E.

Fact 7.2.6. *Where $i, j \in \mathcal{N}$, if $i < j$ then $l_i \lessdot_M l_j$.*

The $\lessdot_M$-ordering of the l_i's reflects their indexes.

The next fact justifies why we speak of the "abnormality degree of α" in both cases, when we refer to $\mathsf{AbDeg}_{\prec}(\alpha)$ and when we refer to α's relation to the l_i's: there is a 1:1 between both notions.

Fact 7.2.7. *For all $\alpha \in \mathcal{V}_p^+$ and all $i \in \mathcal{N}$,*

(i) $i < \mathsf{AbDeg}_{\prec}(\alpha)$ *iff* $l_i \lessdot_M \alpha$.
(ii) $i \leq \mathsf{AbDeg}_{\prec}(\alpha)$ *iff* $\alpha \mid\!\sim_M l_i$.
(iii) where $i < \mathsf{m}$, $\mathsf{AbDeg}_{\prec}(\alpha) = i$ *iff,* $\alpha \mid\!\sim_M l_i$ *and* $\alpha \not\mid\!\sim l_{i+1}$
(iv) $\mathsf{AbDeg}_{\prec}(\alpha) = i$ *iff, not* $l_i \lessdot_M \alpha$ *and* ($l_{i-1} \lessdot_M \alpha$ *or* $i = 0$).

Ad (i): The abnormality degree as measured by $\mathsf{AbDeg}_{\prec}$ is worse than i iff the abnormality degree as measured by means of the l_j's is worse than i.

Ad (ii): The abnormality degree of α (as measured by $\mathsf{AbDeg}_{\prec}$) is at best i iff it is at best i as measured by the l_j's.

Ad (iii): If α's abnormality degree is i (in view of $\mathsf{AbDeg}_{\prec}$) then its abnormality degree (in view of the l_j's) is at best i and better than $i + 1$.

Ad (iv): If α's abnormality degree is i (in view of $\mathsf{AbDeg}_{\prec}$) then its abnormality degree is not worse than i and either it is 0 or it is worse than $i-1$.

Fact 7.2.8. *If $l_{i-1} \lessdot_M \alpha$, then $\alpha \mid\!\sim_M l_j$ for all $j \leq i$ and $j \leq \mathsf{m}$.*

Obviously, if α has an abnormality degree worse than $i-1$ then its abnormality degree is at best j where $j \leq i$ and $j \leq \mathsf{m}$.

Fact 7.2.9. *If $l_{\mathsf{m}} \lessdot_M \alpha$, then $\alpha \mid\!\sim_M \bot$.*

If α has an abnormality degree worse than m then there are no α-worlds and hence $\alpha \mid\!\sim_M \bot$.

It is in no way surprising that we have an analogous representation theorem for rational$^+$ consequence relations and rational$^+$ models as we had in the case of rational consequence relations and rational models. We first show that every rational$^+$-model defines a rational$^+$ consequence relation (Theorem 7.2.1), then that all rational models that define a rational$^+$ consequence relation are rational$^+$ (Theorem 7.2.2). This culminates in Corollary 7.2.1 which states the adequacy of rational$^+$ models for rational$^+$ consequence relations.

Theorem 7.2.1. *Let $M = \langle W, \prec \rangle$ be a rational$^+$ $\mathcal{L}^+$-model, then $\mid\!\sim_M$ is a rational$^+$ consequence relation.*

Proof. Follows by Fact 7.2.4, Fact 7.2.6, Fact 7.2.8, and Fact 7.2.9. □

Theorem 7.2.2. *For all rational $\mathcal{L}^+$-models $M = \langle W, \prec \rangle$ defining a rational$^+$ consequence relation $\mid\!\sim_M$, M satisfies requirement* (R).

Proof. Let $w \in W$. We prove the theorem by induction on $\mathsf{rank}_{\prec}(w)$. "$n = 0$": Let $\mathsf{rank}_{\prec}(w) = 0$. There is an $\alpha_0 \in \mathcal{V}_p^+$ such that $w \models \alpha_0$. Since $\mathsf{rank}_{\prec}(w) = 0$, $w \in \min_{\prec}(\sigma_M(\alpha_0))$. Since by ($C^+1$) $\alpha_0 \mathrel{|\!\sim}_M l_0$, we have $w \models l_0$. Since $\mathsf{rank}_{\prec}(w) = 0$, $w \in \min_{\prec}(\sigma_M(l_0)) = \min_{\prec}(\sigma_M(l_0 \vee l_i))$ for all $i \in \mathcal{N}$. Furthermore, since by Fact 7.2.1 $l_0 \lessdot l_i$, for all $i \in \mathcal{N} \setminus \{0\}$, $w \models \neg l_i$.

"$n \to n+1$": Let $\mathsf{rank}_{\prec}(w) = n+1$. By Fact 7.1.6 there is an $\alpha_{n+1} \in \mathcal{V}_p^+$ such that $w \in \min_{\prec}(\sigma_M(\alpha_{n+1}))$. By the induction hypothesis, $\mathsf{AbDeg}_{\prec}(l_i) = i$ for all $i < n+1$. Hence by Fact 7.1.2 we have $l_i \lessdot_M \alpha_{n+1}$ for all $i < n+1$. But then, due to (C^+2), $\alpha_{n+1} \mathrel{|\!\sim}_M l_i$ for all $i \leq n+1$. Therefore $\alpha_{n+1} \mathrel{|\!\sim}_M l_{n+1}$ and hence $w \models l_i$ for all $i \leq n+1$. It follows also immediately by the induction hypothesis that $w \in \min_{\prec}(\sigma_M(l_{n+1}))$. Since by Fact 7.2.1 $l_{n+1} \lessdot l_j$ for all $j \in \mathcal{N}$ for which $j > n+1$, $w \models \neg l_j$.

That $\mathsf{rank}_{\prec}(w) \leq \mathsf{m}$ follows immediately by (C^+4). □

Corollary 7.2.1. *A binary relation $\mathrel{|\!\sim}$ is a rational$^+$ $\mathcal{L}^+$-consequence relation iff it is the consequence relation defined by some rational$^+$ $\mathcal{L}^+$-model.*

Proof. Suppose $\mathrel{|\!\sim}$ is a rational$^+$ $\mathcal{L}^+$-consequence relation. Then, by Theorem 7.1.5, there is a rational $\mathcal{L}^+$-model M such that $\mathrel{|\!\sim} = \mathrel{|\!\sim}_M$. By Theorem 7.2.2 is M a rational$^+$ $\mathcal{L}^+$-model.

Suppose $\mathrel{|\!\sim}_M$ is such that M is a rational$^+$ $\mathcal{L}^+$-model. By Theorem 7.2.1, $\mathrel{|\!\sim}_M$ is a rational$^+$ $\mathcal{L}^+$-consequence relation. □

Corollary 7.2.2 is going to show that rational $\mathcal{L}$-models and rational$^+$ $\mathcal{L}^+$-models define the same rational $\mathcal{L}$-consequence relations. It is an immediate consequence that for each rational $\mathcal{L}$-consequence relation there is a rational$^+$ $\mathcal{L}^+$-model defining it and vice versa. Therefore, if a rational closure exists for a given $\mathcal{L}$-knowledge base, then there are rational$^+$ $\mathcal{L}^+$-models that define it. It will be the task of the adaptive logic $\mathbf{ARC^s}$ to pick out these models. In order to establish these results the following mappings λ and μ will be helpful.

Definition 7.2.2. We define a mapping λ from the rational $\mathcal{L}$-models into the rational$^+$ $\mathcal{L}^+$-models in the following way (see Fig. 7.5). For $M = \langle W, \prec \rangle$, $M \mapsto \langle \pi(W), \prec' \rangle$, where π maps $\mathcal{L}$-worlds into $\mathcal{L}^+$-worlds such that (i) for all $\alpha \in \mathcal{P}$, $w \models \alpha$ iff $\pi(w) \models \alpha$; and (ii) if $\mathsf{rank}_{\prec}(w) = i$ then $\pi(w) \models l_j$ for all $j \in \mathcal{N}$ for which $j \leq i$, and $\pi(w) \models \neg l_j$ for all $j \in \mathcal{N}$ for which $j > i$. Furthermore, define $\prec'$ in the following way, $w \prec w'$ iff $\pi(w) \prec' \pi(w')$.

The following lemma shows that the co-domain of λ is indeed the set of rational$^+$ $\mathcal{L}^+$-models (see (i) and (iv)) and that the mapping preserves conditional assertions (see (ii)) and abnormality degrees (see (iii)).

Lemma 7.2.1. *The following holds for the mapping λ:*[10]

(i) $\lambda(M)$ satisfies requirement (R).

[10] The proofs can be found in Appendix E.

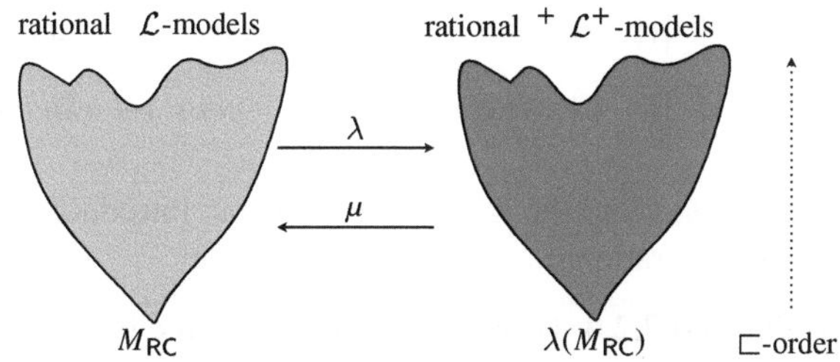

Fig. 7.5 The mappings λ (see Definition 7.2.2) and π (see Definition 7.2.3), M_{RC} is the model for the Rational Closure as constructed in Sect. 7.1.4

(ii) *For all* $\alpha, \beta \in \mathcal{V}_p$, $\alpha \mid\sim_M \beta$ *iff* $\alpha \mid\sim_{\lambda(M)} \beta$.
(iii) *For all* $w \in W$, $\mathsf{rank}_{\prec}(w) = \mathsf{rank}_{\prec'}(\pi(w))$.
(iv) $\lambda(M)$ *is a rational*$^+$ *model.*

Definition 7.2.3. We define a mapping μ from the rational$^+$ $\mathcal{L}^+$-models into the rational $\mathcal{L}$-models in the following way (see Fig. 7.5). For $M = \langle W, \prec \rangle$, $M \mapsto \langle \eta(W), \prec' \rangle$, where η maps $\mathcal{L}^+$-worlds into $\mathcal{L}$-worlds such that for all $\alpha \in \mathcal{P}$, $\eta(w) \models \alpha$ iff $w \models \alpha$. Furthermore, $\eta(w) \prec' \eta(w')$ iff for all $w'' \in \min_{\prec}(\eta^{-1}(\{\eta(w)\}))$ and for all $w''' \in \min_{\prec}(\eta^{-1}(\{\eta(w')\}))$, $w'' \prec w'''$.

The following lemma establishes that the co-domain of π is indeed the set of rational $\mathcal{L}$-models (see (i)), that the mapping preserves conditional assertions (see (ii)), and that the ranks of mapped worlds are at most their original rank. The reason why the ranks may not be equal is that for some $\mathcal{L}$-world w there may be different $\mathcal{L}^+$-worlds w' such for all $\mathcal{L}$-formulas α, $w \models \alpha$ iff $w' \models \alpha$. Hence, all the latter worlds are mapped by η to the same representative w which may have a lower rank in the resulting model $\mu(M)$ than some of the w' have in the original model M.

Lemma 7.2.2. *The following holds for the mapping* μ*:*[11]

(i) $\mu(M)$ *is a rational* $\mathcal{L}$*-model.*
(ii) *For all* $\alpha, \beta \in \mathcal{V}_p$, $\alpha \mid\sim_M \beta$ *iff* $\alpha \mid\sim_{\mu(M)} \beta$.
(iii) *For all* $w \in W$, $\mathsf{rank}_{\prec}(w) \geq \mathsf{rank}_{\prec'}(\eta(w))$.

Altogether the two lemmas show that the rational $\mathcal{L}$-models and the rational$^+$ $\mathcal{L}^+$-models represent the same rational consequence relations in $\mathcal{V}_p \times \mathcal{V}_p$.

Theorem 7.2.3. *(i) For each rational*$^+$ $\mathcal{L}^+$*-model* M *there is a rational* $\mathcal{L}$*-model* N *for which* $\mid\sim_M \cap (\mathcal{V}_p \times \mathcal{V}_p) = \mid\sim_N$. *And vice versa, (ii) for each rational* $\mathcal{L}$*-model* N *there is a rational*$^+$ $\mathcal{L}^+$*-model* M *for which* $\mid\sim_M \cap (\mathcal{V}_p \times \mathcal{V}_p) = \mid\sim_N$.

Proof. This follows by Lemma 7.2.1 and Lemma 7.2.2. □

Corollary 7.2.2. *Let* **K** *be an* $\mathcal{L}$*-knowledge base. We have,*

[11] The proof can be found in Appendix E.

$$\bigcap\{\mid\!\sim \mid\ \mid\!\sim \textit{ is a rational } \mathcal{L}\textit{-cons. rel. extending } \mathbf{K}\} =$$

$$\bigcap\{\mid\!\sim \mid\ \mid\!\sim \textit{ is a rational}^{+}\ \mathcal{L}^{+}\textit{-cons. rel. extending } \mathbf{K}\} \cap (\mathcal{V}_p \times \mathcal{V}_p).$$

Analogous to the rational case we can introduce a logic $\mathbf{R}^+$ on basis of the conditions for rational$^+$ consequence relations.

Definition 7.2.4 (Logic $\mathbf{R}^+$). We define the logic $\mathbf{R}^+$ by the axioms for $\mathbf{R}$ and the following:

$$\alpha \mid\!\sim l_0 \qquad (\mathrm{R}^+1)$$

$$(l_{i-1} \lessdot \alpha) \supset (\alpha \mid\!\sim l_j) \text{ for all } j \leq i, \text{ where } i \in \mathcal{N} \setminus \{0\} \qquad (\mathrm{R}^+2)$$

$$l_{i-1} \lessdot l_i \text{ for all } i \in \mathcal{N} \setminus \{0\} \qquad (\mathrm{R}^+3)$$

$$(l_{\mathsf{m}} \lessdot \alpha) \supset (\alpha \mid\!\sim \bot) \qquad (\mathrm{R}^+4)$$

A semantics for $\mathbf{R}^+$ is given by the rational$^+$ models. We define $\models$ and $\Vdash_{\mathbf{R}^+}$ for rational$^+$ models analogously to the way we defined $\models$ and $\Vdash_{\mathbf{R}}$ for rational models.

Theorem 7.2.4. $\mathbf{R}^+$ *is complete and sound with respect to rational*$^+$ *models.*

Proof. This can be easily shown on the basis of the soundness and completeness of $\mathbf{R}$. Let $\Gamma^+ = \Gamma \cup \{\alpha \mid\!\sim l_0 \mid \alpha \in \mathcal{V}_p^+\} \cup \{(l_{i-1} \lessdot \alpha) \supset (\alpha \mid\!\sim l_j) \mid \alpha \in \mathcal{V}_p^+, i \in \mathcal{N} \setminus \{0\}, j \leq i\} \cup \{l_{i-1} \lessdot l_i \mid i \in \mathcal{N} \setminus \{0\}\} \cup \{(l_{\mathsf{m}} \lessdot \alpha) \supset (\alpha \mid\!\sim \bot) \mid \alpha \in \mathcal{V}_p^+\}$. Note first that $\Gamma \vdash_{\mathbf{R}^+} \varphi$ iff $\Gamma^+ \vdash_{\mathbf{R}} \varphi$ since all instances of (R^+1)–(R^+4) are contained in Γ^+. By the soundness and completeness of $\mathbf{R}$, $\Gamma^+ \vdash_{\mathbf{R}} \varphi$ iff $\Gamma^+ \Vdash_{\mathbf{R}} \varphi$. Evidently, all rational models of Γ^+ define a rational$^+$ consequence relation. Hence, by Theorem 7.2.2, all rational models of Γ^+ are rational$^+$ models. Since, by definition, also every rational$^+$ model of Γ is a rational model of Γ^+, we have $\Gamma \Vdash_{\mathbf{R}^+} \varphi$ iff $\Gamma^+ \Vdash_{\mathbf{R}} \varphi$. Altogether, we get $\Gamma \vdash_{\mathbf{R}^+} \varphi$ iff $\Gamma \Vdash_{\mathbf{R}^+} \varphi$. □

7.2.3 The Adaptive Logic ARC$^{\mathbf{s}}$

We now define the adaptive logic $\mathbf{ARC^s}$ for rational closure. The lower limit logic is $\mathbf{R}^+$. Abnormalities are conditional assertions of the kind $l_i \lessdot \alpha$. The intention served by minimizing abnormalities of this form is to interpret a propositional formula α as normally as possible. Note here that for a rational$^+$ $\mathcal{L}^+$-model $M = \langle W, \prec\rangle$, $l_i \lessdot_M \alpha$ iff $i < \mathsf{AbDeg}_{\prec}(\alpha)$ (see Fact 7.2.7.i). Minimizing abnormalities can therefore in semantical terms be seen as ranking worlds as low as possible (see Sect. 7.1.4), hence as interpreting formulas as normally as possible.

Definition 7.2.5. $\mathbf{ARC^s}$ is the adaptive logic in standard format defined by the following triple:

- **LLL**: $\mathbf{R}^+$

- Abnormalities: $\Omega =_{\text{df}} \{l_i \lessdot \alpha \mid \alpha \in \mathcal{V}_p, i \in \mathcal{N}\}$
- Strategy: simple strategy

We write Ab(M) for the set $\{\varphi \in \Omega \mid M \models \varphi\}$ where M is an **LLL**-model of a given premise set.

We first prove the adequacy of **ARC^s** for characterizing the Rational Closure. Then we give some examples of dynamic proofs.

The following lemma establishes that, where M is the rational $\mathcal{L}$-model for which $\mid\!\sim_M$ is the rational closure of a knowledge base **K**, (i) $\lambda(M)$ is a minimally abnormal **R**$^+$-model of **K**, (ii) all minimally abnormal models of **K** have the same abnormal part, (iii) all formulas have the same normality degrees in the minimally abnormal models, (iv) all minimally abnormal models verify the same conditional assertions, namely the ones in the rational closure of **K**.

(iv) immediately implies that **ARC^s** is indeed adequate to represent the Rational Closure. (ii) allows us to use the simple strategy (see the discussion in Sect. 2.4.3).

Lemma 7.2.3. *Let $M = \langle W_M, \prec_M\rangle$ be the rational $\mathcal{L}$-model of the rational closure of an $\mathcal{L}$-knowledge base* **K** *constructed as in Sect. 7.1.4.*

(i) $\lambda(M) = \langle \pi(W_M), \prec_{\lambda(M)}\rangle$ is a minimally abnormal rational$^+$ $\mathcal{L}^+$-model of **K**;
(ii) for all minimally abnormal rational$^+$ $\mathcal{L}^+$-models N of **K**, *$Ab(N) = Ab(\lambda(M))$;*
(iii) for all minimally abnormal rational$^+$ $\mathcal{L}^+$-models $N = \langle W_N, \prec_N\rangle$ of **K** *and for all $\alpha \in \mathcal{V}_p$, $\mathsf{AbDeg}_{\prec_N}(\alpha) = \mathsf{AbDeg}_{\prec_M}(\alpha) = \mathsf{AbDeg}_{\prec_{\lambda(M)}}(\alpha)$;*
(iv) for all minimally abnormal rational$^+$ $\mathcal{L}^+$-models N of **K** *and for all conditional assertions $\alpha \mid\!\sim \beta$ (where $\alpha, \beta \in \mathcal{V}_p$), $\alpha \mid\!\sim_N \beta$ iff $\alpha \mid\!\sim_{\lambda(M)} \beta$ iff $\alpha \mid\!\sim_M \beta$.*

Proof. Ad (i) and (ii): Suppose there is a rational$^+$ $\mathcal{L}^+$-model $N = \langle W_N, \prec_N\rangle$ of **K** such that there is an $l_i \lessdot \alpha$ valid in $\lambda(M)$ but not in N (for $i \in \mathcal{N}, \alpha \in \mathcal{V}_p$). Hence, by Fact 7.2.7.i,

$$\mathsf{rank}_{\prec_N}(w) \leq i \tag{7.1}$$

where w is any world in $\min_{\prec_N}(\sigma_N(\alpha))$. By Lemma 7.2.2, $\mu(N) = \langle \eta(W_N), \prec_{\mu(N)}\rangle$ is a rational $\mathcal{L}$-model of **K**. Furthermore, by the construction of M in Sect. 7.1.4, $\eta(w) \in W_M$ and therefore $\pi(\eta(w)) \in \sigma_{\lambda(M)}(\alpha)$. Furthermore, since $l_i \lessdot_{\lambda(M)} \alpha$ and by Fact 7.2.7.i, $\mathsf{rank}_{\prec_{\lambda(M)}}(\pi(\eta(w))) > i$. By the construction of M we have

$$\mathsf{rank}_{\prec_M}(\eta(w)) \leq \mathsf{rank}_{\prec_{\mu(N)}}(\eta(w)) \tag{7.2}$$

Also, by Lemma 7.1.2.iii and Lemma 7.2.2.iii,

$$\mathsf{rank}_{\prec_{\lambda(M)}}(\pi(\eta(w)) = \mathsf{rank}_{\prec_M}(\eta(w)) \text{ and} \tag{7.3}$$

$$\mathsf{rank}_{\prec_{\mu(N)}}(\eta(w)) \leq \mathsf{rank}_{\prec_N}(w) \tag{7.4}$$

But (7.1), (7.2), (7.3) and (7.4) imply that $\mathsf{rank}_{\prec_{\lambda(M)}}(\pi(\eta(w)) \leq i$,—a contradiction.

Ad (iii): Assume for a $\alpha \in \mathcal{V}_p$, $i = \mathsf{AbDeg}_{\prec_{\lambda(M)}}(\alpha) < \mathsf{AbDeg}_{\prec_N}(\alpha)$. By Fact 7.2.7.i, $l_i \lessdot \alpha \notin$ Ab($\lambda(M)$) and $l_i \lessdot \alpha \in$ Ab(N). This contradicts (ii). Assume for a $\alpha \in \mathcal{V}_p$, $i = \mathsf{AbDeg}_{\prec_N}(\alpha) < \mathsf{AbDeg}_{\prec_{\lambda(M)}}(\alpha)$. But then, by an analogous argument as above, $l_i \lessdot \alpha \in \mathrm{Ab}(\lambda(M)) \setminus \mathrm{Ab}(N)$,—this contradicts (ii). Altogether, $\mathsf{AbDeg}_{\prec_{\lambda(M)}}(\alpha) = \mathsf{AbDeg}_{\prec_N}(\alpha)$. That $\mathsf{AbDeg}_{\prec_{\lambda(M)}}(\alpha) = \mathsf{AbDeg}_{\prec_M}(\alpha)$ follows by Lemma 7.2.1.iii.

Ad (iv): This is an immediate consequence of (iii) and Fact 7.1.3. □

Theorem 7.2.5. *Let* **K** *be an* $\mathcal{L}$*-conditional knowledge base. We have for all* $\alpha, \beta \in \mathcal{V}_p$*:* $\alpha \mathrel{|\!\sim} \beta$ *is in the rational closure of* **K** *iff* $\mathbf{K} \vdash_{\mathbf{ARC^s}} \alpha \mathrel{|\!\sim} \beta$.

Proof. Let M be the model for the rational closure of **K** constructed in Sect. 7.1.4.

"⇒": Suppose $\alpha \mathrel{|\!\sim} \beta$ is in the rational closure of **K**. Thus, $\alpha \mathrel{|\!\sim}_M \beta$. By Lemma 7.2.1.ii, $\alpha \mathrel{|\!\sim}_M \beta$ iff $\alpha \mathrel{|\!\sim}_{\lambda(M)} \beta$. Hence, by Lemma 7.2.3, all minimally abnormal models verify $\alpha \mathrel{|\!\sim} \beta$. Thus, $\mathbf{K} \Vdash_{\mathbf{ARC^s}} \alpha \mathrel{|\!\sim} \beta$ and by completeness $\mathbf{K} \vdash_{\mathbf{ARC^s}} \alpha \mathrel{|\!\sim} \beta$.

"⇐": Suppose $\mathbf{K} \vdash_{\mathbf{ARC^s}} \alpha \mathrel{|\!\sim} \beta$. Hence, by soundness $\mathbf{K} \Vdash_{\mathbf{ARC^s}} \alpha \mathrel{|\!\sim} \beta$. By Lemma 7.2.3, $\alpha \mathrel{|\!\sim}_M \beta$. □

Some examples are needed in order to demonstrate what we have achieved.

First some simple facts which will help us to shortcut the proofs:

Fact 7.2.10. *(i)* $\vdash_{\mathbf{R}^+} (\top \mathrel{|\!\not\sim} \neg\alpha) \vee (l_0 \lessdot \alpha)$
(ii) $\alpha \mathrel{|\!\sim} \beta \vdash_{\mathbf{R}^+} ((\alpha \wedge \gamma) \mathrel{|\!\sim} \beta) \vee (l_0 \lessdot (\alpha \wedge \gamma))$
(iii) $\{l_i \lessdot \alpha, \neg(l_{i+1} \lessdot (\alpha \wedge \beta)), \alpha \mathrel{|\!\sim} \gamma\} \vdash_{\mathbf{R}^+} \alpha \wedge \beta \mathrel{|\!\sim} \gamma$
(iv) $\{\alpha \mathrel{|\!\sim} \gamma, \beta \mathrel{|\!\sim} \neg\gamma, \beta \mathrel{|\!\sim} \alpha\} \vdash_{\mathbf{R}^+} \alpha \lessdot \beta$

Proof. Ad (i): Suppose $\top \mathrel{|\!\sim} \neg\alpha$. By (R^+1) $\top \mathrel{|\!\sim} l_0$ and since obviously $l_0 \mathrel{|\!\sim} \top$ we have by (EQ), $l_0 \mathrel{|\!\sim} \neg\alpha$. Note that also $l_0 \mathrel{|\!\sim} l_0 \vee \alpha$ and $l_0 \vee \alpha \mathrel{|\!\sim} l_0$. Hence, again by (EQ), $l_0 \vee \alpha \mathrel{|\!\sim} \neg\alpha$ which is the same as $l_0 \lessdot \alpha$.

Ad (ii): Suppose $\alpha \mathrel{|\!\sim} \beta$ and that $l_0 \lessdot \alpha \wedge \gamma$ is not the case. By (i), $\top \mathrel{|\!\not\sim} \neg(\alpha \wedge \gamma)$. Assume $\alpha \mathrel{|\!\sim} \neg(\alpha \wedge \gamma)$. Then, by (CW), $\top \mathrel{|\!\sim} (\alpha \supset \neg(\alpha \wedge \gamma))$ and hence $\top \mathrel{|\!\sim} \neg(\alpha \wedge \gamma)$,—a contradiction. Thus, $\alpha \mathrel{|\!\not\sim} \neg(\alpha \wedge \gamma)$. Since also $\alpha \mathrel{|\!\sim} \beta$, by (RM), $\alpha \wedge \gamma \mathrel{|\!\sim} \beta$.

Ad (iii): Suppose $\neg(l_{i+1} \lessdot (\alpha \wedge \beta))$, $l_i \lessdot \alpha$ and $\alpha \mathrel{|\!\sim} \gamma$. Since $l_i \lessdot \alpha$, by (R^+2), $\alpha \mathrel{|\!\sim} l_{i+1}$. Since $\neg(l_{i+1} \lessdot (\alpha \wedge \beta))$, by Fact 7.1.1.ii, $\neg(\alpha \lessdot (\alpha \wedge \beta))$. Thus, $\alpha \mathrel{|\!\not\sim} \neg(\alpha \wedge \beta)$. Since $\alpha \mathrel{|\!\sim} \gamma$, by (RM), $\alpha \wedge \beta \mathrel{|\!\sim} \gamma$.

Ad (iv): Suppose $\alpha \mathrel{|\!\sim} \gamma$, $\beta \mathrel{|\!\sim} \neg\gamma$, and $\beta \mathrel{|\!\sim} \alpha$. Note that by (CC) and (ID), $\beta \mathrel{|\!\sim} \beta \wedge \alpha$ and by (ID) and (RCM), $\beta \wedge \alpha \mathrel{|\!\sim} \beta$. Hence, by (EQ), $\beta \wedge \alpha \mathrel{|\!\sim} \neg\gamma$. Assume $\alpha \mathrel{|\!\not\sim} \neg\beta$. By (RM), $\alpha \wedge \beta \mathrel{|\!\sim} \gamma$,—a contradiction. Thus, $\alpha \mathrel{|\!\sim} \neg\beta$. Since $\alpha \mathrel{|\!\sim} \alpha$ and $\beta \mathrel{|\!\sim} \alpha$ by (CA), $\alpha \vee \beta \mathrel{|\!\sim} \alpha$. Since by (ID) and (RCM), $\alpha \mathrel{|\!\sim} \alpha \vee \beta$, we get by (EQ), $\alpha \vee \beta \mathrel{|\!\sim} \neg\beta$ and thus $\alpha \lessdot \beta$. □

Example 7.2.1. Let our knowledge base be $\Gamma = \{b \mathrel{|\!\sim} f, p \mathrel{|\!\sim} \neg f, p \mathrel{|\!\sim} b\}$: Birds usually fly, while penguins usually don't fly. Moreover, Penguins are usually birds. We first introduce our premises:

1	$b \mathrel{|\!\sim} f$	PREM $\emptyset$
2	$p \mathrel{|\!\sim} \neg f$	PREM $\emptyset$
3	$p \mathrel{|\!\sim} b$	PREM $\emptyset$

Now we interpret b as normally as possible. This means that we assume that $\top \mathrel{|\!\sim} \neg b$ is not the case since this would imply that b is not normal:

4 $\top \not\mathrel{|\!\sim} \neg b$	RC	$\{l_0 \lessdot b\}$

We have made use of Fact 7.2.10.i. We proceed symmetrically for $\neg b$:

5 $\top \not\mathrel{|\!\sim} b$	RC	$\{l_0 \lessdot \neg b\}$

Now we introduce an irrelevant factor w. By Fact 7.2.10.ii we know that in case we can safely assume that $b \wedge w$ has abnormality degree 0 we can derive $b \wedge w \mathrel{|\!\sim} f$ from $b \mathrel{|\!\sim} f$:

6 $b \wedge w \mathrel{|\!\sim} f$	1; RC	$\{l_0 \lessdot (b \wedge w)\}$

We now defeasibly interpret p as having abnormality degree 0:

^{10}7 $\top \not\mathrel{|\!\sim} \neg p$	RC	$\{l_0 \lessdot p\}$
8 $b \mathrel{|\!\sim} l_0$	RU	$\emptyset$
9 $b \lessdot p$	1,2,3; RU	$\emptyset$
10 $l_0 \lessdot p$	8,9; RU	$\emptyset$

Line 8 follows by ($\mathrm{R}^{+}1$). Line 9 follows by Fact 7.2.10.iv and Line 10 follows with Fact 7.1.1.ii. It shows that our assumption at line 7 was mistaken and hence we retract the inference at line 7 by means of marking it.

Finally we derive $p \wedge w \mathrel{|\!\sim} \neg f$ from $p \mathrel{|\!\sim} \neg f$ in an analogous way as we derived $b \wedge w \mathrel{|\!\sim} f$ from $b \mathrel{|\!\sim} f$:

11 $\neg(l_1 \lessdot (p \wedge w))$	RC	$\{l_1 \lessdot (p \wedge w)\}$
12 $(p \wedge w) \mathrel{|\!\sim} \neg f$	2,10,11; RU	$\{l_1 \lessdot (p \wedge w)\}$

Line 12 follows with Fact 7.2.10.iii. All formulas on unmarked lines are finally derivable and hence in the rational closure of **K**, as can easily be verified.

Example 7.2.2. We consider the so-called *Nixon diamond* example where $\Gamma = \{r \mathrel{|\!\sim}\} \neg p, q \mathrel{|\!\sim} p.$[12] The formulas can be interpreted by:

- Being a republican usually implies not being a pacifist. ($r \mathrel{|\!\sim} \neg p$)
- Being a Quaker usually implies being a pacifist. ($q \mathrel{|\!\sim} p$)

It would for example be desirable to derive $q \wedge w \mathrel{|\!\sim} p$ (where w can be read as "being a worker").

1 $r \mathrel{|\!\sim} \neg p$	PREM	$\emptyset$
2 $q \mathrel{|\!\sim} p$	PREM	$\emptyset$
3 $\top \not\mathrel{|\!\sim} \neg(w \wedge q)$	RU	$\{l_0 \lessdot w \wedge q\}$
4 $\top \not\mathrel{|\!\sim} \neg w \vee \neg q$	3; RU	$\{l_0 \lessdot w \wedge q\}$
5 $(\top \not\mathrel{|\!\sim} \neg w \vee \neg q) \supset (q \not\mathrel{|\!\sim} \neg w)$	RU	$\emptyset$
6 $q \not\mathrel{|\!\sim} \neg w$	4,5; RU	$\{l_0 \lessdot w \wedge q\}$
7 $((q \mathrel{|\!\sim} p) \wedge (q \not\mathrel{|\!\sim} \neg w)) \supset (q \wedge w \mathrel{|\!\sim} p)$	RU	$\emptyset$
8 $q \wedge w \mathrel{|\!\sim} p$	2,6,7; RU	$\{l_0 \lessdot w \wedge q\}$

[12] This has been presented e.g. in [12].

At line 3 we make use of Fact 7.2.10.i, at line 5 we arrive by the contraposition to (CW), line 7 is an instance of (RT). It is easy to see that there is no way to derive $l_0 \lessdot w \wedge q$ on the empty condition,—$q \wedge w \mid\sim p$ is therefore finally derivable.

7.3 Including Negative Knowledge

Knowledge bases were so far restricted to contain only the positive knowledge an agent may have. However, an agent might also have statements of the following kind in his knowledge base: if α then not normally β, i.e., $\alpha \not\mid\sim \beta$. Adding statements of this kind to a knowledge base requires special attention since such enriched knowledge bases might not be consistent anymore. Suppose $\mathbf{K} = \{a \mid\sim b\}$ and we add $a \not\mid\sim \neg c$ and $a \wedge c \not\mid\sim b$. Applying rational monotonicity to $a \mid\sim b$ and $a \not\mid\sim \neg c$ leads to a contradiction.

Let a *general knowledge* base be a conditional knowledge base which may contain negated conditional assertions. As shown in [16], for a consistent general conditional knowledge base $\mathbf{K}$, $\mid\sim_M$ is the rational closure of $\mathbf{K}$, where M is the model constructed in Sect. 7.1.4. By Theorem 7.2.3 we get following result:

Corollary 7.3.1. *Let* $\mathbf{K}$ *be a consistent general* $\mathcal{L}$*-conditional knowledge base. We have for all* $\alpha, \beta \in \mathcal{V}_p$*:* $\alpha \mid\sim \beta$ *is in the rational closure of* $\mathbf{K}$ *iff* $\mathbf{K} \vdash_{\mathbf{ARC^s}} \alpha \mid\sim \beta$.

7.4 Conclusion

The preferential closure of a conditional knowledge base $\mathbf{K}$ can be obtained by intersecting all supersets of $\mathbf{K}$ satisfying the conditions for preferential consequence relations. The proof theory of preferential closure is therefore simply given by a logic defined by these conditions, interpreted as rules. Adding Rational Monotonicity as a further condition, on the other hand, defines a family of consequence relations which is not closed under intersection. Although Lehmann and Magidor are able to characterize it semantically by a selection on ranked models, a proof theory for it was missing. This syntactical gap has been filled in this chapter by the adaptive logic $\mathbf{ARC^s}$ for finite propositional languages for which the rational closure is guaranteed to exist.

In [12] it has been shown that the rational closure exists for a larger class of knowledge bases and infinite languages, namely the so-called admissible knowledge bases (e.g. knowledge bases for which a well-founded preferential model exists). If we want to proceed in a similar manner as in this chapter we are in need of a set-theoretic construction for the rational closure of a knowledge base in this more general case. This is no trivial exercise: the one we used in Sect. 7.1.4 (cp. [16]) is restricted to finite languages. The authors in [12] presented another procedure.

However, this construction requires another severe restriction: it only works on the basis of so-called well-founded preferential consequence relations.[13] Unfortunately, as has been shown in [12], even finite knowledge bases can give rise to a non well-founded preferential closure, and the existence of a well-founded preferential model M does not ensure that the consequence relation defined by M is well-founded.

It is a future challenge to develop a proof theory allowing for admissible knowledge bases in general and furthermore to take into account the predicative case (cp. [17]).

Acknowledgments I am thankful to Diderik Batens and Joke Meheus for many constructive comments which helped to improve this paper.

References

1. Straßer, C.: An adaptive logic for rational closure. In: Carnielli, W. Coniglio, M.E., D'Ottaviano I.M.L. (eds.) The Many Sides of Logic, pp. 47–67. College Publications (2009)
2. Clark, K.L.: Negation as failure. In: Logic and Data Bases, pp. 293–322 (1977)
3. McCarthy, J.: Circumscription—a form of non-monotonic reasoning. Artif. Intell. **13**, 27–29 (1980)
4. Reiter, R.: A logic for default reasoning. Artif. Intell. **13**(1-2), 81–132 (1980)
5. Moore, R.C.: Possible-world semantics for autoepistemic logic. In: Proceedings of the Workshop on Non-monotonic Reasoning, pp. 344–354. AAAI (1984)
6. Gabbay, D.M.: Theoretical foundations for non-monotonic reasoning in expert systems. In: Logics and Models of Concurrent Systems, pp. 439–457. Springer-Verlag, New York, (1985)
7. Shoham, Y.: Reasoning About Change: Time and Causation from the Standpoint of Artificial Intelligence. M.I.T Press, Cambridge (1988)
8. Shoham, Y.: A semantical approach to nonmonotonic logics. In: Ginsberg, M.L. (ed.) Readings in Non-Monotonic Reasoning, pp. 227–249. Morgan Kaufmann, Los Altos (1987)
9. Kraus, S., Lehmann, D.J., Magidor, M.: Nonmonotonic reasoning, preferential models and cumulative logics. Artif. Intell. **44**, 167–207 (1990)
10. Friedman, N., Halpern, J.Y.: Plausibility measures and default reasoning. J. ACM **48**, 1297–1304 (1996)
11. Arieli, O., Avron, A.: General patterns for nonmonotonic reasoning: from basic entailments to plausible relations. Logic J. IGPL **8**(2), 119–148 (2000)
12. Lehmann, D.J., Magidor, M.: What does a conditional knowledge base entail? Artif. Intell. **55**(1), 1–60 (1992)
13. Goldszmidt, M., Pearl, J.: On the relation between rational closure and system Z. In: Third International Workshop on Nonmonotonic Reasoning South Lake Tahoe, pp. 130–140 (1990)
14. Pearl, J.: System Z: a natural ordering of defaults with tractable applications to nonmonotonic reasoning. In: TARK '90: Proceedings of the 3rd Conference on Theoretical Aspects of Reasoning About Knowledge, pp. 121–135. Morgan Kaufmann Publishers Inc., San Francisco, (1990)
15. Giordano, L., Gliozzi, V., Olivetti, N., Pozzato, G.: Analytic tableau calculi for KLM rational logic R. In: Proceedings of the 10th European Conference on Logics in Artificial Intelligence, pp. 190–202. ACM, New York (2006)

[13] The well-foundedness of a preferential consequence relation $\mid\!\sim$ should not be mistaken for well-foundedness of preferential models, e.g. a model defining $\mid\!\sim$.

16. Booth, R., Paris, J.: A note on the rational closure of knowledge bases with both positive and negative knowledge. J. Logic Lang. Inform. **7**(2), 165–190 (1998)
17. Lehmann, D., Magidor, M.: Preferential logics: the predicate calculus case. In: TARK '90: Proceedings of the 3rd Conference on Theoretical Aspects of Reasoning About Knowledge, pp. 57–72. Morgan Kaufmann Publishers Inc., San Francisco, (1990)

Part III
Argumentation Theory

Chapter 8
Towards the Proof-Theoretic Unification of Dung's Argumentation Framework: An Adaptive Logic Approach

This chapter presents a unifying adaptive logic framework for abstract argumentation. It consists of a core system for abstract argumentation and various ALs based on it. These logics represent in an accurate sense all standard extensions defined within Dung's abstract argumentation framework (see [2, 3]) with respect to skeptical and credulous acceptance. The models of our logics correspond exactly to specific extensions of given argumentation frameworks. Additionally, the dynamics of adaptive proofs mirror the argumentative reasoning of a rational agent. In particular, the presented logics allow for external dynamics, i.e., they are able to deal with the arrival of new arguments and are therefore apt to model open-ended argumentations by providing provisional conclusions.[1]

8.1 Introduction

Theories of argumentation have been the subject of intensive research within the fields of logic, philosophy, artificial intelligence and computer science. Bench-Capon and Dunne [4] speak of a "core study within Artificial Intelligence", the relation of argumentation to nonmonotonic logics has been pointed out and researched (see [3, 5, 6]), argumentation has been discovered as a powerful tool within logic programming (see [7, 8]), while in the field of practical reasoning many specific applications have been developed, e.g., the analysis of legal discourse (see [9, 10]). One of the most influential approaches to argumentation systems is Dung's account of argumentation frameworks (see [2, 3]). The importance of Dung's approach derives from the fact that it abstracts away from the nature of arguments and argumentation rules: the framework consists of a set of arguments, which are taken to be abstract entities represented by alphabetical letters, and the binary (so-called *attack*) relation defined

---

[1] A former version of the content of this chapter has been published under the name "*Towards the Proof-Theoretic Unification of Dung's Argumentation Framework: An Adaptive Logic Approach*" in the "Journal of Logic and Computation", 2010, [1]. The paper is co-authored by Dunja Šešelja.

C. Straßer, *Adaptive Logics for Defeasible Reasoning*, Trends in Logic 38,
DOI: 10.1007/978-3-319-00792-2_8, © Springer International Publishing Switzerland 2014

over this set. Such a framework is capable of formalizing various approaches to nonmonotonic inferencing in the fields of AI, logic programming and human reasoning, which also "suggests that it is meaningful and interesting to incorporate the idea of argumentation into nonmonotonic logic" ([2], p. 855). For a survey on abstract argumentation frameworks see [4, 11] or [5].

The key notion of Dung's account is the acceptability of arguments. Dependent on the criteria for acceptance, it is possible to formulate different semantics. These define a number of extensions representing sets of acceptable arguments, such as *admissible*, *grounded*, *complete*, *(semi)-stable* and *preferred* extensions. Dung's system has been extended and generalized in various respects. We name just a few: preferences [12], values [13] and audiences [14] in the sense of Perelman [15] have been introduced, joint attacks have been enabled [16], the system has been used for an improved account to default reasoning [6, 17], it has been applied to multi-agent systems [18, 19], it has been used for applications in the philosophy of science [20], new semantics/extensions have been presented [21, 22], game-theoretic approaches have been developed [23], etc.

This chapter offers an adaptive logic framework with a specific core axiomatic system, on the basis of which we define logics for obtaining all the standard extension types of Dung's account with respect to the skeptical and the credulous acceptability of arguments. We have seen that the main idea of ALs is to interpret a premise set "as normally as possible", given a certain standard of normality. Based on the lower limit logic (**LLL**), they select certain **LLL**-models of a given premise set which satisfy the standard of normality. Syntactically they enrich the derivative power of the **LLL** by allowing for certain rules to be applied conditionally. In case a condition turns out to be unsafe, formulas derived on this very condition are marked in the proof and thus not considered as being derived anymore. Markings in ALs come and go while we reason on. Indeed, adaptive proofs are dynamic in two ways: internally in the sense that during the reasoning process certain conditions might turn out to be unsafe/safe (again) while we get more insight into the given premises, externally in the sense that the introduction of new information in the form of new premises may alter our treatment of certain conditions and thus alter the markings in the proof. In argumentation the situation is similar: as rational debaters we introduce an argument a in such a way that we are willing to withdraw it under certain circumstances. For instance in the case that it is conflicting with other arguments, or in the case that we cannot defend it against certain counterarguments. However, at a later point, a new argument which defends the attacked a, might enter the scene, and cause the acceptance of the latter one again. Therefore, argumentation is in a similar way dynamic as adaptive logic proofs: internally in the sense that the progressing analyses of the relationship of given arguments might alter our choice for accepted arguments, and externally in the sense that the introduction of new arguments might make us reconsider the acceptance of some arguments.

As we have already mentioned, Dung defined various extension types which select certain subsets of (non-conflicting) arguments with respect to given criteria. This mirrors the semantic selection of ALs: while the **LLL** defines minimal criteria that have to be fulfilled in every model (such as the absence of conflicts between the validated arguments and the property that every validated argument is defended by the other validated arguments against all possible attacks), the adaptive enhancements refine the semantic selection by modeling the criteria given by the various extension types. Furthermore, the dynamic adaptive proofs model the reasoning process leading to these selections. In summary, ALs are very suitable for providing a unifying logical framework for abstract argumentation. Thus, the logics for abstract argumentation which will be presented in this chapter are ALs. They employ the core set of axioms as **LLL** and define different standards of normality. In this way we can obtain logics for *admissible, complete, preferred, (semi-)stable* as well as *grounded* extensions.

One of the main advantages of our approach compared to other proposals for proof theories for abstract argumentation (see [24–26]) lies in its unifying power. A single framework is able to capture all standard semantics/extensions with respect to both skeptical and credulous acceptance, which makes it a decent logical surrounding for their comparison, further elaboration, enhancements and generalizations.

In addition, our system represents a contribution to the research done in applications of ALs to different dialogical contexts (for a survey see [27]). Furthermore, this chapter confirms the claim that the adaptive logic program offers a general and unifying framework for nonmonotonic and defeasible logics (as has recently been argued for in [28]).

The chapter is structured in the following way. First we present Dung's abstract argumentation framework in Sect. 8.2. In Sect. 8.3 we introduce the reader step-wise into our adaptive logic framework for abstract argumentation. In Sect. 8.4 we define all the ALs for the various extension types with respect to skeptical acceptance and state the corresponding representational results. Section 8.5 features all the logics for credulous acceptance and the respective representational results. In Sect. 8.6 we localize the adaptive logic approach within the field of logical representations of abstract argumentation and point out some advantages. The Appendix contains all the meta-proofs for our results.

8.2 Dung's Argumentation Framework: Key Terms

We will use lower case letters $a_1, a_2, a_3, \ldots$ for arguments and lower case fraktur letters $\mathfrak{a}, \mathfrak{b}, \mathfrak{c}, \ldots$ as meta-variables for arguments. Let

$$\mathcal{A}_n =_{\mathrm{df}} \{a_1, a_2, \ldots, a_n\}.$$

In [3] Dung defined his abstract argumentation frameworks as follows[2]:

Definition 8.2.1. A *finite argumentation framework (AF)* is a pair $\langle \mathcal{A}, \rightarrow \rangle$ where $\mathcal{A} \subseteq \mathcal{A}_n$ is a finite set of arguments, and $\rightarrow \subseteq \mathcal{A} \times \mathcal{A}$ is a relation between arguments. The expression $\mathfrak{a} \rightarrow \mathfrak{b}$ is pronounced as "$\mathfrak{a}$ attacks $\mathfrak{b}$".

[2] We restrict the discussion in this chapter to the finite case, i.e., to argumentation frameworks with a finite number of propositional letters.

Given an AF $\langle \mathcal{A}, \rightarrow \rangle$ we are particularly interested in giving an account of reasonable choices of arguments in $\mathcal{A}$: a minimal criterion is, for instance, that no argument in a selection S should attack another argument in S. Of course, more interesting selection types can be defined:

Definition 8.2.2. Given an argumentation framework $\mathsf{A} = \langle \mathcal{A}, \rightarrow \rangle$ we define the following notions.

(i) An argument $\mathfrak{a}$ is *attacked* by a set of arguments $B \subseteq \mathcal{A}$ iff there is a $\mathfrak{b} \in B$ such that $\mathfrak{b} \rightarrow \mathfrak{a}$.
(ii) An argument $\mathfrak{a}$ is *acceptable* with respect to a set of arguments $C \subseteq \mathcal{A}$, iff every attacker of $\mathfrak{a}$ is attacked by C. It is said that C *defends* $\mathfrak{a}$.
(iii) A set of arguments $S \subseteq \mathcal{A}$ is *conflict-free* iff S doesn't attack any argument in S.
(iv) A conflict-free set of arguments $S \subseteq \mathcal{A}$ is *admissible* iff each argument in S is acceptable with respect to S.
(v) A set of arguments $S \subseteq \mathcal{A}$ is a *preferred extension* iff it is a maximal (w.r.t. $\subseteq$) admissible set.
(vi) A conflict-free set of arguments $S \subseteq \mathcal{A}$ is a *stable extension* iff it attacks every argument in $\mathcal{A} \setminus S$.
(vii) An admissible set of arguments $S \subseteq \mathcal{A}$ is a *complete extension* iff $F(S) = S$, where $F(S) =_{\text{df}} \{\mathfrak{c} \mid S \text{ defends } \{\mathfrak{c}\}\}$.
(viii) A set of arguments $S \subseteq \mathcal{A}$ is a *grounded extension* iff it is the minimal (w.r.t. $\subseteq$) complete extension.
(ix) A complete extension $S \subseteq \mathcal{A}$ is a *semi-stable extension* [3] iff $S \cup S^+$ is maximal (w.r.t. $\subseteq$), where S^+ is the set of all arguments in $\mathcal{A} \setminus S$ which are attacked by S.
(x) A set of arguments $S \subseteq \mathcal{A}$ is *credulously accepted* according to preferred [(semi)-stable, complete or grounded] semantics (w.r.t. A) iff it is contained in at least one preferred [(semi)-stable, complete or grounded] extension of A.
(xi) A set of arguments $S \subseteq \mathcal{A}$ is *skeptically accepted* according to preferred [(semi)-stable, complete or grounded] semantics (w.r.t. A) iff it is contained in every preferred [(semi)-stable, complete or grounded] extension of A.

Suppose we select a conflict-free set $E \subseteq \mathcal{A}$. There are two types of arguments in $\mathcal{A} \setminus E$ which are not selected. On the one hand, arguments in E^+ which are attacked by the selected arguments and on the other hand the ones that are not attacked by E, i.e., arguments in $\mathcal{A} \setminus (E \cup E^+)$. We call the former arguments *defeated*, since they are attacked by at least some of our selected arguments E. Admissibility requires that the set of defeated arguments for a given selection of arguments S consists at least of all the attackers of S. Opposite to attacks, we only speak of a defeat in view of a given selection of arguments: $\mathfrak{a}$ attacks $\mathfrak{b}$ iff $\mathfrak{a} \rightarrow \mathfrak{b}$, while $\mathfrak{b}$ is defeated iff there is a selected argument $\mathfrak{a}$ that attacks $\mathfrak{b}$. It would be misleading to confuse attack and

[3] Semi-stable semantics were defined by Caminada in [22] and are equivalent to Verhijs' *admissible stage extensions* in [29].

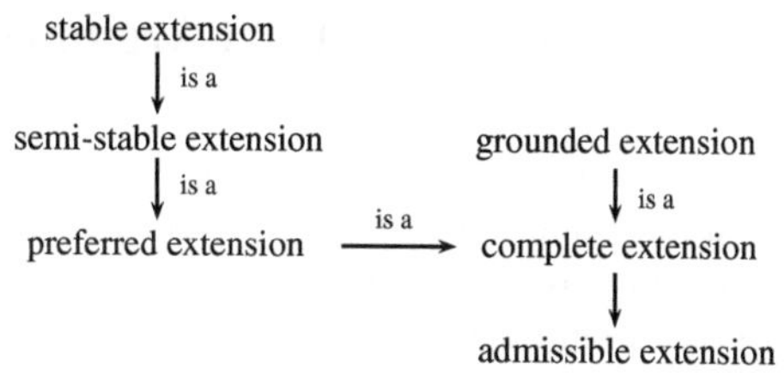

Fig. 8.1 The relationship between the extensions types

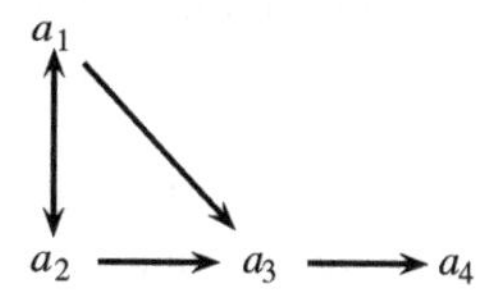

Fig. 8.2 An attack-diagram

defeat: an agent may (argumentatively) attack another agent, but we only consider the attack as a defeat if the argument used for the attack is considered as valid.[4]

Figure 8.1 illustrates some basic relationships between the various extension types. For a more thorough study of them we refer the reader to the rich literature mentioned in the introduction.

Argumentation frameworks $\langle \mathcal{A}, \rightarrow \rangle$ are often represented by directed graphs, the so-called *attack-diagrams* (see Fig. 8.2). The nodes are arguments in $\mathcal{A}$ and there is an edge from $\mathfrak{a}$ to $\mathfrak{b}$ iff $(\mathfrak{a}, \mathfrak{b}) \in \rightarrow$.

Example 8.2.1. We will demonstrate the concepts just introduced with the attack-diagram in Fig. 8.2. The following table lists the extensions belonging to the extension types introduced in Definition 8.2.2.

Admissible	Preferred	Semi-stable	Complete	Grounded
$\emptyset, \{a_1\}, \{a_2\}, \{a_1, a_4\}, \{a_2, a_4\}$	$\{a_1, a_4\}, \{a_2, a_4\}$	$\{a_1, a_4\}, \{a_2, a_4\}$	$\emptyset, \{a_1, a_4\}, \{a_2, a_4\}$	$\emptyset$

Stable extensions are, unlike the other extension types, not guaranteed to exist. Semi-stable extensions (see [22]) improve on that: they are guaranteed to exist, and in case stable extensions exists the semi-stable extensions are identical to them.[5] Moreover, there is one unique grounded extension.

---

[4] Our notion of defeat differs from the way defeat is defined in various preference or value based enhancements of Dung's abstract argumentation framework. Defeat is there usually defined as a binary relation between arguments which is a subset of the attack relation: a_1 defeats a_2 iff a_1 attacks a_2 and a_2 is not 'preferable' to a_1. The preferability of one argument over another is modeled in different ways: in terms of a preference relation between arguments in [30], by allowing for arguments to attack an attack in [31], or in terms of mapping arguments into partially ordered values in [19].

[5] The following fact offers an alternative definition of semi-stable extensions in terms of admissible

8.3 A Logic for Abstract Argumentation

In this chapter we are going to present propositional logics for abstract argumentation. We will present for each extension type $\mathcal{E}$ (such as admissible, complete, preferred, etc.) a corresponding logic $\mathbf{L}_{\mathcal{E}}$. The major idea is that $\mathbf{L}_{\mathcal{E}}$ derives for a given argumentation framework $\mathsf{A} = \langle \mathcal{A}, \rightarrow \rangle$ all skeptically (resp. credulously) accepted arguments and that the models represent the extensions of type $\mathcal{E}$.

This section will make this idea more precise. It will introduce basic notions and the modus operandi of our logical framework for abstract argumentation. First, in Sect. 8.3.1, we will propose a formal language and the core set of rules for our logics. Section 8.3.2 offers a way to represent AFs as premise sets. We give a precise account of the representational requirements for our logics in Sect. 8.3.3. In the remainder of this section we will introduce the reader into the main ideas of ALs by focusing on example cases, paradigmatically for preferred extensions. Section 8.4 will contain the definitions for all the other variants and the representational results.

8.3.1 Language and Rules

In order to represent a given AF A as a premise set we need a formal language which allows us to express the basic notions of abstract argumentation. The idea is, on the one hand, to represent arguments by propositional letters and, on the other hand, to enrich the language of classical propositional logic by a binary logical operator $\twoheadrightarrow$ where $\alpha \twoheadrightarrow \beta$ means that α attacks β. Since we represent arguments by propositional letters only, we restrict our language in such a way that only propositional letters are arguments of $\twoheadrightarrow$.[6] Formally the set of well-formed formulas $\mathcal{W}_n$ (where n is a natural number) is defined in the following way:

$$\begin{aligned} \mathcal{V}_n \; &:= p_1 \mid p_2 \mid p_3 \mid \ldots \mid p_n \\ \mathcal{W}_n^{\twoheadrightarrow} &:= \langle \mathcal{V}_n \rangle \twoheadrightarrow \langle \mathcal{V}_n \rangle \mid \bot \twoheadrightarrow \langle \mathcal{V}_n \rangle \\ \mathcal{W}_n \; &:= \bot \mid \langle \mathcal{V}_n \rangle \mid \langle \mathcal{W}_n^{\twoheadrightarrow} \rangle \mid \neg \langle \mathcal{W}_n \rangle \mid \langle \mathcal{W}_n \rangle \wedge \langle \mathcal{W}_n \rangle \mid \\ &\qquad \langle \mathcal{W}_n \rangle \vee \langle \mathcal{W}_n \rangle \mid \langle \mathcal{W}_n \rangle \supset \langle \mathcal{W}_n \rangle \end{aligned}$$

$\mathcal{V}_n$ are the propositional letters of our language. We will in the remainder abbreviate $\neg(\alpha \twoheadrightarrow \beta)$ by $\alpha \not\twoheadrightarrow \beta$.

Let us introduce the rules characterizing our core logic for abstract argumentation. First of all, it is obvious that if α is valid and it attacks β, $\alpha \twoheadrightarrow \beta$, then β should not be a consequence of our logic:

(Footnote 5 continued)
sets S for which $S \cup S^+$ is maximal: Let $\mathsf{A} = \langle \mathcal{A}, \rightarrow \rangle$ be an AF and $S \subseteq \mathcal{A}$. S is a semi-stable extension iff S is an admissible set of arguments for which there is no admissible set of arguments $T \subseteq \mathcal{A}$ such that $T \cup T^+ \supset S \cup S^+$. The statement is proven in Appendix F, Fact F.3.1.

[6] We also allow for $\bot$ on the left hand side of $\twoheadrightarrow$. We will comment on this in a moment.

$$\frac{\alpha \quad \alpha \twoheadrightarrow \beta}{\neg\beta} \tag{R$\twoheadrightarrow$}$$

This rule guarantees the conflict-freeness of our consequences since by (R$\twoheadrightarrow$) we immediately get

$$\vdash (\alpha \twoheadrightarrow \beta) \supset (\neg\alpha \vee \neg\beta)$$

So, whenever $\alpha \twoheadrightarrow \beta$, either α or β is considered to be invalid. Moreover, in our language it is easy to express that an argument has been defeated by one or more arguments. We define:

$$\mathsf{def}\,\beta =_{\mathrm{df}} \bigvee_{\alpha \in \mathcal{V}_n} \big(\alpha \wedge (\alpha \twoheadrightarrow \beta)\big) \tag{Def}$$

It is easy to verify that the following properties are immediate consequences of this definition and rule (R$\twoheadrightarrow$):

$$\vdash \mathsf{def}\,\alpha \supset \neg\alpha$$

$$\vdash \big(\alpha \wedge (\alpha \twoheadrightarrow \beta)\big) \supset \mathsf{def}\,\beta$$

The first property guarantees that, if an argument has been defeated, then it is supposed not to be validated by the logic. The second property assures that, if α and $\alpha \twoheadrightarrow \beta$ have been derived then β is defeated.

Remember that the idea behind admissible extensions is that a selected set of arguments S is required to defend itself. That is to say, in case an argument $\mathfrak{a}$ in S is attacked by another argument $\mathfrak{b}$, $\mathfrak{b} \rightarrow \mathfrak{a}$, then there is an argument $\mathfrak{c}$ in S which attacks $\mathfrak{b}$ or, in yet other words, $\mathfrak{b}$ is defeated. For our logics we can express this as follows:

$$\frac{\alpha \quad \beta \twoheadrightarrow \alpha}{\mathsf{def}\,\beta} \tag{Rad}$$

If we have α and $\beta \twoheadrightarrow \alpha$, then β is supposed to be defeated. Note that it would be insufficient to replace the conclusion $\mathsf{def}\,\beta$ of rule (Rad) by $\neg\beta$. In case an argument $\mathfrak{a}$ is selected and $\mathfrak{b}$ attacks $\mathfrak{a}$, $\mathfrak{b} \rightarrow \mathfrak{a}$, then it is not enough simply not to select $\mathfrak{b}$. By the requirement of admissible extensions the selected arguments have to defend themselves against all attackers. Thus, what is required in terms of our language is $\mathsf{def}\,\beta$. This ensures that there is an argument γ which attacks and thus defeats β. This is guaranteed by $\mathsf{def}\,\beta$ due to its definition $\bigvee_{\delta \in \mathcal{V}_n} (\delta \wedge (\delta \twoheadrightarrow \beta))$. The existence of a defeating argument γ of β would not be guaranteed were we to replace $\mathsf{def}\,\beta$ by $\neg\beta$ in the conclusion of (Rad).

An attentive reader might have noticed that our language also allows for $\bot \twoheadrightarrow \alpha$. This is helpful in order to express that a given propositional letter α corresponds to an argument in the given argumentation framework $\langle \mathcal{A}, \rightarrow \rangle$. The cardinality of $\mathcal{V}_n$ might be higher than the cardinality of $\mathcal{A}$ and thus there might be propositional letters which do not correspond to the given arguments $\mathcal{A}$. We express the fact that a propositional letter represents an argument by $\bot \twoheadrightarrow \alpha$. All propositional letters that do not represent an argument are guaranteed not to be validated by the following rule:

$$\frac{\bot \not\twoheadrightarrow \alpha}{\neg\alpha} \tag{R$\bot$}$$

Let us have a look at one last rule that will help us to represent complete extensions. The main idea behind this extension type is that any argument that is defended by a given selected set of arguments S is supposed to be in the selection S. That is to say, if S defends argument $\mathfrak{a}$, then $\mathfrak{a} \in S$. This can be expressed by the following rule:

$$\frac{\bot \twoheadrightarrow \beta \quad \bigwedge_{\alpha \in \mathcal{V}_n} \big((\alpha \twoheadrightarrow \beta) \supset \mathsf{def}\,\alpha\big)}{\beta} \tag{RCo}$$

That β is defended by the set of validated arguments is expressed by $\bigwedge_{\alpha \in \mathcal{V}_n}((\alpha \twoheadrightarrow \beta) \supset \mathsf{def}\,\alpha)$: for every attacker α of β, α is supposed to be defeated. We also add the condition that β is actually representing an argument, $\bot \twoheadrightarrow \beta$, since, if it did not, then β would have no attackers and thus the other antecedent, $\bigwedge_{\alpha \in \mathcal{V}_n}(\alpha \twoheadrightarrow \beta \supset \mathsf{def}\,\alpha)$, would be valid. But, as already pointed out, we want to keep propositional letters which do not represent arguments out of the consequence sets of our logics.

The presented rules enable us to define the following two logics which will serve as lower limit logics for our adaptive systems.

Definition 8.3.1. $\mathbf{L_A}$ is classical propositional logic enriched by the rules (R$\twoheadrightarrow$), (Rad), and (R$\bot$). $\mathbf{L_C}$ is $\mathbf{L_A}$ enriched by (RCo).

We define the semantics for logics $\mathbf{L} \in \{\mathbf{L_A}, \mathbf{L_C}\}$ via an assignment function $v : \mathcal{V}_n \cup \mathcal{W}_n^{\twoheadrightarrow} \to \{0, 1\}$ and an $\mathbf{L}$-valuation $v_M^{\mathbf{L}} : \mathcal{W}_n \to \{0, 1\}$ determined by the assignment. We use an extended assignment function $v : \mathcal{V}_n \cup \mathcal{W}_n^{\twoheadrightarrow} \to \{0, 1\}$ that assigns truth values to both, propositional letters and 'attacks', i.e., formulas in $\mathcal{W}_n^{\twoheadrightarrow}$. A model M is defined by an assignment function v. The following definitions are useful in order to define the $\mathbf{L_A}$-valuation based on v:

$$\begin{aligned}
v_{\mathrm{Rral}} &=_{\mathrm{df}} 1 - \max_{\alpha,\beta \in \mathcal{V}_n} \big(\min(v(\alpha), v(\alpha \twoheadrightarrow \beta), v(\beta))\big)\\
v_{\mathrm{Rbot}} &=_{\mathrm{df}} 1 - \max_{\alpha \in \mathcal{V}_n} \big(\min(v(\alpha), 1 - v(\bot \twoheadrightarrow \alpha))\big)\\
v_{\mathrm{Rad}} &=_{\mathrm{df}} 1 - \max_{\alpha,\beta \in \mathcal{V}_n} \Big(\min\big(v(\alpha), v(\beta \twoheadrightarrow \alpha), 1 - \max_{\gamma \in \mathcal{V}_n}\big(\min(v(\gamma), v(\gamma \twoheadrightarrow \beta))\big)\big)\Big),\\
v_{\mathsf{i}}^{\mathbf{L_A}} &=_{\mathrm{df}} \min(v_{\mathrm{Rral}}, v_{\mathrm{Rbot}}, v_{\mathrm{Rad}})
\end{aligned}$$

Note that v_{Rral} corresponds to our syntactical rule (R$\twoheadrightarrow$) in the sense that $v_{\mathrm{Rral}} = 1$ iff the assignment satisfies the semantic counterpart to (R$\twoheadrightarrow$). That is to say, $v_{\mathrm{Rral}} = 1$ iff v satisfies

$$\text{If } v(\alpha) = v(\alpha \twoheadrightarrow \beta) = 1, \text{ then } v(\beta) = 0. \tag{S$\twoheadrightarrow$}$$

The situation is analogous for v_{Rbot} and v_{Rad} with respect to the following properties:

$$\text{If } v(\alpha) = 1, \text{ then } v(\bot \twoheadrightarrow \alpha) = 1. \tag{S$\bot$}$$

$$\begin{array}{l}\text{If } v(\alpha) = v(\beta \twoheadrightarrow \alpha) = 1, \text{ then}\\ \quad \text{there is a } \gamma \in \mathcal{V}_n \text{ for which } v(\gamma) = v(\gamma \twoheadrightarrow \beta) = 1.\end{array} \tag{Sad}$$

We call an assignment $\mathbf{L_A}$-*intended* iff $v_{\mathsf{i}}^{\mathbf{L_A}} = 1$. In Appendix F it is shown that an assignment v is $\mathbf{L_A}$-intended iff v satisfies (S$\twoheadrightarrow$), (Sad) and (S$\bot$).

We define the valuation function $v_M^{\mathbf{L}} : \mathcal{W}_n \to \{0, 1\}$ paradigmatically for $\mathbf{L} = \mathbf{L_A}$. The one for $\mathbf{L_C}$ is defined in a similar way and can be found in Appendix F. Where $\alpha, \beta \in \mathcal{V}_n$ and $\varphi, \varphi_1, \varphi_2 \in \mathcal{W}_n$ we define:

$$v_M^{\mathbf{L}}(\bot) =_{\text{df}} 0 \tag{s$\bot$}$$

$$v_M^{\mathbf{L}}(\alpha \twoheadrightarrow \beta) =_{\text{df}} v(\alpha \twoheadrightarrow \beta) \tag{s$\twoheadrightarrow$}$$

$$v_M^{\mathbf{L}}(\bot \twoheadrightarrow \alpha) =_{\text{df}} v(\bot \twoheadrightarrow \alpha) \tag{s$\bot\twoheadrightarrow$}$$

$$v_M^{\mathbf{L}}(\alpha) =_{\text{df}} \min(v_{\mathsf{i}}^{\mathbf{L_A}}, v(\alpha)) \tag{sPA}$$

$$v_M^{\mathbf{L}}(\varphi_1 \wedge \varphi_2) =_{\text{df}} \min(v_M^{\mathbf{L}}(\varphi_1), v_M^{\mathbf{L}}(\varphi_2)) \tag{s$\wedge$}$$

$$v_M^{\mathbf{L}}(\varphi_1 \vee \varphi_2) =_{\text{df}} \max(v_M^{\mathbf{L}}(\varphi_1), v_M^{\mathbf{L}}(\varphi_2)) \tag{s$\vee$}$$

$$v_M^{\mathbf{L}}(\varphi_1 \supset \varphi_2) =_{\text{df}} \max(1 - v_M^{\mathbf{L}}(\varphi_1), v_M^{\mathbf{L}}(\varphi_2)) \tag{s$\supset$}$$

$$v_M^{\mathbf{L}}(\neg\varphi) =_{\text{df}} 1 - v_M^{\mathbf{L}}(\varphi) \tag{s$\neg$}$$

Obviously, by (s$\twoheadrightarrow$) and (s$\bot\twoheadrightarrow$) the valuation inherits the truth values for 'attacks' in $\mathcal{W}_n^{\twoheadrightarrow}$ from the assignment function. Note that although (sPA) is of a rather complex form, it is fully determined by the assignment v. In the case $v_{\mathsf{i}}^{\mathbf{L_A}} = 1$, i.e., in the case that the assignment is $\mathbf{L_A}$-intended, the valuation takes over all truth values from the assignment for all formulas in $\mathcal{V}_n$. However, if $v_{\mathsf{i}}^{\mathbf{L_A}} = 0$, the valuation assigns to all propositional letters the truth value 0. Note that for a given AF A the empty selection is always an admissible extension. Thus, the valuation on the basis of a non-intended assignment corresponds to the empty extension. In Appendix F it is shown that $\mathbf{L_A}$-valuations satisfy (S$\twoheadrightarrow$), (Sad) and (S$\bot$).

Model validity and the semantic consequence relation are defined in the usual way. Where $\mathbf{L} \in \{\mathbf{L_A}, \mathbf{L_C}\}$, we define $M \models_{\mathbf{L}} \varphi$ iff $v_M^{\mathbf{L}}(\varphi) = 1$. We say a model M is an $\mathbf{L}$-model of $\Gamma \subseteq \mathcal{W}_n$ iff $M \models_{\mathbf{L}} \varphi$ for all $\varphi \in \Gamma$. We write $\mathcal{M}_{\mathbf{L}}(\Gamma)$ for the set of all $\mathbf{L}$-models of Γ. The semantic consequence relations $\Vdash_{\mathbf{L}}$ are defined in the usual way: $\Gamma \Vdash_{\mathbf{L}} \varphi$ iff for all $\mathbf{L}$-models M of Γ, $M \models_{\mathbf{L}} \varphi$.

Completeness and soundness for both logics, $\mathbf{L_A}$ and $\mathbf{L_C}$ are proven in Appendix F.

8.3.2 Representing AFs as Premise Sets

Let us now see how to represent AFs in terms of premise sets. There is an easy and intuitive way to do so:

- First, we need to map the arguments of a given AF $\mathsf{A} = \langle \mathcal{A}, \rightarrow \rangle$, where $\mathcal{A} \subseteq \mathcal{A}_n$, into the set of propositional letters $\mathcal{V}_n$. Of course, we need at least as many propositional letters as we have arguments. A canonical way to do so is by $\lambda_n : \mathcal{A}_n \rightarrow \mathcal{V}_n, a_i \mapsto p_i$ for all $1 \leq i \leq n$. We can say that p_i represents, or corresponds to an argument a_i iff $a_i \in \mathcal{A}$.
- Second, we need to represent the attack relation. This can be simply done by adding to the premise set $p_i \twoheadrightarrow p_j$ iff $(a_i, a_j) \in \rightarrow$.
- It is furthermore important to indicate which propositional letters belong to the AF in question. We do this by adding $\bot \twoheadrightarrow p_i$ to the premise set iff $a_i \in \mathcal{A}$.

For premise sets constructed in this way we write Γ^n_{A}.

Example 8.3.1. For instance, the AF A from Example 8.2.1 is represented by the premise set $\Gamma^n_{\mathsf{A}} = \{p_1 \twoheadrightarrow p_2, p_2 \twoheadrightarrow p_1, p_1 \twoheadrightarrow p_3, p_2 \twoheadrightarrow p_3, p_3 \twoheadrightarrow p_4\} \cup \{\bot \twoheadrightarrow p_1, \bot \twoheadrightarrow p_2, \bot \twoheadrightarrow p_3, \bot \twoheadrightarrow p_4\}$ where $n \geq 4$.

For many applications it is interesting to choose a language that has more propositional letters than an argumentation framework $\langle \mathcal{A}, \rightarrow \rangle$ that is initially modeled. Some examples:

1. In order to model the argumentative reasoning of intelligent agents a system has to deal with more and more information in form of new arguments coming in. The initial setup is thus iteratively enriched as the argumentation proceeds. An argumentation can thus be seen as a sequence of argumentation frameworks $\mathsf{A}^1, \mathsf{A}^2, \ldots, \mathsf{A}^m$ where $\mathsf{A}^{i+1} = \langle \mathcal{A}^{i+1}, \rightarrow^{i+1} \rangle$ is an enhancement of $\mathsf{A}^i = \langle \mathcal{A}^i, \rightarrow^i \rangle$, i.e., $(\mathcal{A}^{i+1} \cup \rightarrow^{i+1}) \supset (\mathcal{A}^i \cup \rightarrow^i)$.
2. Abstract argumentation is a promising framework for applications such as machine learning (see [32]), belief revision (see [33] for a survey), or decision theory (see [34]) since knowledge/belief bases may be represented by or with the help of argumentation frameworks.

For such applications it is obviously important to have enough propositional letters available in order to represent the successive stages $\mathsf{A}^1, \mathsf{A}^2, \ldots, \mathsf{A}^m$.

Furthermore, a logic has to be able to deal with new information arriving, resulting in the transition from A^i to A^{i+1}. That is to say, it has to allow for *external dynamics* (see our discussion in Sect. 2.5.1. To simply apply a given algorithm producing the accepted argument in question again from scratch to A^{i+1} is cumbersome, especially since it doesn't model in any way the rationale of the rational agent going through this very transition. We will thus offer a dynamic proof procedure which by dynamic markings is able to model the reasoning of the agent in question throughout the sequence of updates she is exposed to. We will discuss this feature more in Sect. 8.3.6.

8.3.3 Representational Requirements

Given an extension type $\mathcal{E}$ (such as admissible, complete, preferred, etc.) and an AF A, what are our requirements for a logic for abstract argumentation $\mathbf{L}_{\mathcal{E}}$? What should its consequences look like, what should its models represent?

Let us presuppose that for the AFs $\langle \mathcal{A}, \rightarrow \rangle$ under consideration, $\mathcal{A} \subseteq \mathcal{A}_n$, and that $\mathbf{L}_{\mathcal{E}}$ is formulated in the language $\mathcal{W}_n$. This simply makes sure that we have enough propositional letters to logically represent the AFs in question.

We have two straightforward and intuitive representational requirements for a complete and sound logic $\mathbf{L}_{\mathcal{E}}$: a syntactical one and a semantical one. Let $\mathsf{A} = \langle \mathcal{A}, \rightarrow \rangle$ be an AF:

1. *Syntactic adequacy for skeptical (resp. credulous) acceptance*: We require that $\Gamma^n_{\mathsf{A}} \vDash_{\mathbf{L}_{\mathcal{E}}} p_i$ iff $a_i \in \mathcal{A}$ and a_i is skeptically (resp. credulously) accepted according to $\mathcal{E}$. Informally this simply means that a propositional letter is derivable iff it represents a skeptically (resp. credulously) acceptable argument (according to $\mathcal{E}$).
2. *Semantic adequacy*: Let $\mathcal{M}_{\mathbf{L}_{\mathcal{E}}}(\Gamma^n_{\mathsf{A}})$ be the set of all $\mathbf{L}_{\mathcal{E}}$-models of Γ^n_{A}. We require that,

   (a) For each $\mathcal{E}$-extension $E \subseteq \mathcal{A}$ of A there is a model $M \in \mathcal{M}_{\mathbf{L}_{\mathcal{E}}}(\Gamma^n_{\mathsf{A}})$ for which $M \vDash_{\mathbf{L}_{\mathcal{E}}} p_i$ iff $a_i \in E$,
   (b) and vice versa, for each model $M \in \mathcal{M}_{\mathbf{L}_{\mathcal{E}}}(\Gamma^n_{\mathsf{A}})$ there is an $\mathcal{E}$-extension E of A for which $M \vDash_{\mathbf{L}_{\mathcal{E}}} p_i$ iff $a_i \in E$.

   Thus, the models of the logic correspond exactly to the $\mathcal{E}$-extensions in the sense that they validate exactly the propositional letters representing the arguments in the extensions.

For instance, an adequate logic for preferred extensions for our Example 8.2.1 would have two types of models: one validating p_1 and p_4, representing the preferred extension $\{a_1, a_4\}$, and another one validating p_2 and p_4, representing the preferred extension $\{a_2, a_4\}$. For skeptical acceptance, the only propositional letter derivable is supposed to be p_4, since the only skeptically acceptable argument is a_4. Due to completeness, $p_1 \vee p_2$ is obviously also a consequence. This is also intuitive since in every preferred extension either a_1 or a_2 is valid.

Definition 8.3.2. Let $\mathcal{E}$ be an extension type. If a logic $\mathbf{L}$ fulfills requirement (1) for $\mathcal{E}$ and for all AFs $\mathsf{A} = \langle \mathcal{A}, \rightarrow \rangle$ where $\mathcal{A} \subseteq \mathcal{A}_n$, then we say that $\mathbf{L}$ *syntactically represents extension type $\mathcal{E}$ with respect to skeptical (resp. credulous) acceptance for argumentation frameworks with at most n arguments*. If a logic $\mathbf{L}$ fulfills the requirements in (2) for $\mathcal{E}$ and for all AFs $\mathsf{A} = \langle \mathcal{A}, \rightarrow \rangle$ where $\mathcal{A} \subseteq \mathcal{A}_n$, then we say that $\mathbf{L}$ *semantically represents extension type $\mathcal{E}$ for argumentation frameworks with at most n arguments*.

8.3.4 Interpreting a Premise Set as Normally as Possible

The logics we are going to present belong to the class of ALs. The essential feature of ALs is that they interpret a premise set "as normally as possible" given certain criteria for normality. Semantically speaking, ALs select from all **LLL**-models the ones that "are normal enough" and hence satisfy a certain standard of normality. We will give a technically precise explication of this in a moment. Translated in the context of our application and given an AF $\mathsf{A} = \langle \mathcal{A}, \rightarrow \rangle$ this means, for instance,

- that, in the case of preferred extensions, in each selected **LLL**-model of Γ^n_{A} as many arguments as possible[7] are validated while the criteria for admissible extensions are satisfied,
- that, in the case of grounded extensions, in each selected **LLL**-model of Γ^n_{A} as few arguments as possible are validated while the criteria for complete extensions are satisfied,
- that, in the case of semi-stable extensions, in each selected **LLL**-model of Γ^n_{A} as many arguments as possible are validated and at the same time as many arguments as possible are defeated while the criteria for complete extensions are satisfied.

The adaptive strategy defines what it means for a model to satisfy the standard of normality. For all the ALs in this chapter it will mean that the models should be as "normal as possible", or in other words, "minimally abnormal". The abnormalities define what is considered as abnormal for these models. For instance, in case of preferred extension

$$\Omega_P = \{\neg p_i \mid i \leq n\}$$

is a good choice for abnormalities. Assume for the moment that, given an AF A, the **LLL**-models of Γ^n_{A} correspond to the admissible extensions of A.[8] The minimal abnormal models are the **LLL**-models in which as few negated propositional letters $\neg p_i$ are validated as possible. Inversely that means that as many propositional letters are validated as possible. Then these models correspond exactly to the preferred extensions, since these are the maximal admissible extensions.

More generally, where Ω is the set of abnormalities and $\mathbf{L}$ is a logic, we define the abnormal part of an **L**-model M as follows, $\mathrm{Ab}^{\mathbf{L}}_{\Omega}(M) = \{\varphi \in \Omega \mid M \models_{\mathbf{L}} \varphi\}$. We say that an **L**-model M of Γ is an *Ω-minimally abnormal* **L**-*model of* Γ iff for all **L**-models M' of Γ, $\mathrm{Ab}^{\mathbf{L}}_{\Omega}(M') \not\subset \mathrm{Ab}^{\mathbf{L}}_{\Omega}(M)$. The ALs presented in this chapter select all the Ω-minimally abnormal **LLL**-models of a given premise set where the exact nature of Ω depends on the extension type under consideration.

In terms of proofs, this idea is realized by allowing for certain lines to be added to the proof conditionally while the **LLL**-rules are unconditionally applicable. For

[7] More precisely we would have to express this by "as many propositional letters as possible that represent arguments of the given AF".

[8] Actually, as we will see in Sect. 8.3.5, the $\mathbf{L}_{\mathsf{A}}$-models of Γ^n_{A} are a superset of the models corresponding to the admissible extensions of A. Due to this we will perform a pre-selection on the $\mathbf{L}_{\mathsf{A}}$-models before selecting the Ω_P-minimally abnormal models (see Sect. 8.3.6).

instance in case of preferred extensions we are interested in adding an argument α on the condition that the assumption that $\neg\alpha$ is not the case is safe.

But let us exemplify the notions just introduced by having a look at a proof for the simple AF given by $\mathsf{A}^1 = \langle\{a_1, a_2\}, \rightarrow^1\rangle$ where $\rightarrow^1 = \{(a_1, a_2)\}$. As discussed above, the premise set corresponding to A^1 for language $\mathcal{W}_3$ is given by $\Gamma^3_{\mathsf{A}^1} = \{p_1 \twoheadrightarrow p_2, \bot \twoheadrightarrow p_1, \bot \twoheadrightarrow p_2\}$. We use more than two propositional letters (namely three) since we are later going to enhance A^1. The only preferred extension for A^1 is $\{a_1\}$. Thus, what is expected from the logic for preferred extensions is to derive p_1. A good choice for an **LLL**is our core system $\mathbf{L_A}$.

1	$p_1 \twoheadrightarrow p_2$	PREM	$\emptyset$
2	p_1	RC	$\{\neg p_1\}$
3	$\neg p_2$	1, 2; R$\twoheadrightarrow$	$\{\neg p_1\}$

As discussed above, the idea for preferred extensions is to derive as many arguments as possible. Technically, this is made possible by allowing for the conditional introduction of arguments, i.e., propositional letters. For instance at line 2, p_1 is introduced on the condition $\{\neg p_1\}$. The elements of conditions are abnormalities, in our case members of Ω_P.

Now, once p_1 is considered to be valid, we know that p_2 cannot be valid since $p_1 \twoheadrightarrow p_2$. Indeed, at line 3 we derive $\neg p_2$ by rule (R$\twoheadrightarrow$). Note that the condition of line 2 is carried forward to line 3 since the derivational step performed at line 3 uses also line 2.

In our example proof we write the rules of the lower limit logic which are used for unconditional derivations instead of writing "RU" for the unconditional generic rule. This is for the sake of transparency. E.g., we derive $\neg p_2$ at line 3 by the **LLL**-rule (R$\twoheadrightarrow$) with antecedents in lines 1 and 2. The conditions of these lines, namely $\emptyset$ and $\{\neg p_1\}$, are carried forward to line 3.

The essential strength of ALs comes with the rule RC. It enables us to derive formulas conditionally. Since $\vdash p_1 \vee \neg p_1$ is a theorem of propositional logic, we derive p_1 by RC at line 2 on the condition $\{\neg p_1\}$.

Note that adaptive proofs are not yet fully characterized by the generic rules PREM, RU and RC. What is missing are means to invalidate lines which are derived on conditions that have to be considered as unsafe. The marking definition of our ALs will give a precise account of when a condition is considered as unsafe. We will come to that at the end of the next subsection.

8.3.5 The Problem of an Interpretative Surplus

It was indicated above that an intuitive semantic selection procedure is to select all Ω_P-minimally abnormal $\mathbf{L_A}$-models of $\Gamma^3_{\mathsf{A}^1}$. This section will show that this idea, although it is on the right path, gives rise to a problem. Namely, some of the models selected by the proposed procedure validate attacks, $p_i \twoheadrightarrow p_j$, and propositional letters, p_k, that do not correspond to attacks or arguments in the given AF A, i.e.,

$(a_i, a_j) \notin \rightarrow$ and $a_k \notin \mathcal{A}$. Thus, some models 'interpret too much into the given AF'. After explicating the problem in this subsection we will propose a solution by refining our semantic selection (Sect. 8.3.6).

Note that, in our example, neither is a_1 a member of all admissible extensions of A nor is p_1 derivable by $\mathbf{L_A}$. Argument a_1 constitutes the unique preferred extension $\{a_1\}$ and thus p_1 should be derivable by a logic for preferred extensions and furthermore, it should be the only propositional letter derivable. Also, since $(p_1 \twoheadrightarrow p_2) \supset (\neg p_1 \vee \neg p_2)$, p_1 and p_2 are never both valid in the same model. This is as desired since a_1 attacks a_2. Furthermore, it can be easily shown that there is an Ω_P-minimally abnormal $\mathbf{L_A}$-model of Γ^3_{A} that verifies p_1 and $\neg p_2$. There are however two problems:

(1) As can easily be verified, all the Ω_P-minimally abnormal $\mathbf{L_A}$-models of $\Gamma^3_{\mathsf{A}^1}$ also validate p_3. Since a_3 is not part of the AF in question, A^1, this is undesired and is not in accordance with our adequacy requirements.
(2) It is easy to see that there are other Ω_P-minimally abnormal $\mathbf{L_A}$-models which verify p_2 and $\neg p_1$. Some of these models verify $p_2 \twoheadrightarrow p_1$, which enables p_2 to defend itself against the attack from p_1. However, a_2 does not attack a_1 in our AF A^1.

Thus, the problem is that, in order to validate as many arguments as possible, the logic selecting Ω_P-minimally abnormal $\mathbf{L_A}$-models, (a), validates propositional letters that do not correspond to arguments of the given AF, and, (b), in some models validates attacks which are not part of the given premises. To interpret a premise set as normally as possible the logic should thus always also take care, (a), that all the arguments that do not correspond to arguments in the given AF are not validated, and, (b), that no additional attacks are derived or validated in the models.

One way to do so would be to directly enhance the premise set for AF $\langle \mathcal{A}, \rightarrow \rangle$ by $p_i \not\twoheadrightarrow p_j$ iff $(a_i, a_j) \notin \rightarrow$ and by $\bot \not\twoheadrightarrow p_i$ iff $a_i \notin \mathcal{A}$. In our case the enriched premise set is

$$\Gamma'_{\mathsf{A}^1} = \Gamma^3_{\mathsf{A}^1} \cup \bigcup_{i=1}^{3} \{p_i \not\twoheadrightarrow p_1\} \cup \bigcup_{i=2}^{3} \{p_i \not\twoheadrightarrow p_2\} \cup \bigcup_{i=1}^{3} \{p_i \not\twoheadrightarrow p_3\} \cup \{\bot \not\twoheadrightarrow p_3\}.$$

However, to enhance the premise set in this way has disadvantages. Some of them are rather obvious: our first proposal for the representation of AFs in terms of premise sets as exemplified by $\Gamma^3_{\mathsf{A}^1}$ is more intuitive, simple and elegant compared to the enhanced presentation exemplified by Γ'_{A^1}. Furthermore, such enhanced premise sets can have a very high cardinality (namely, $n! + n$) already for small AFs. Instead of manually adding all the additional formulas to the premise set $\Gamma^3_{\mathsf{A}^1}$ it would be better if the logic were able to derive them on its own.

Additionally, for some applications the enhancement of the premise set proposed above is counterproductive. Suppose we are to model the argumentative reasoning of intelligent agents: argumentation is a dynamic process, new information in form of new arguments and new attack relations might come in (see also [35]). In terms of argumentation frameworks that means that the initial state of an argumentation might be given by A^1, while, at a later stage a new argument might enter the scene. For instance an agent might, in order to defend a_2 argue that a new argument a_3 attacks a_1. Thus, A^1 is extended to $\mathsf{A}^2 = \langle\{a_1, a_2, a_3\}, \{(a_1, a_2), (a_3, a_1)\}\rangle$. Now if we represent A^1 by the premise set Γ'_{A^1}, then there is no way anymore to introduce $p_3 \twoheadrightarrow p_1$ at a later point of the proof, since this contradicts the premise $p_3 \not\twoheadrightarrow p_1 \in \Gamma'_{\mathsf{A}^1}$. Were we to add $p_3 \twoheadrightarrow p_1$, this would lead to explosion. Similar applications requiring from the logic the ability to deal with new information on-the-fly would be in the fields of belief revision or machine learning (see p. 218). Fortunately, ALs offer a way to avoid these difficulties. The following subsection explores how.

8.3.6 A Better Solution: Going Adaptive and Enabling External Dynamics

Instead of enriching the premise set in the way demonstrated above, it would thus be preferable to have a logic that, (a), has the virtue of dealing with such cases of external dynamics, that is to say a logic which is able to deal with the addition of new arguments and new attacks at any point during the proof without exhibiting explosive behaviour, and, (b), can deal with the intuitive and simple representation Γ^n_{A} of AFs A as premise sets as defined in Sect. 8.3.2 without exhibiting the problems explicated in Sect. 8.3.5.

This is where the strengths of ALs can again be of use. The idea is now to interpret the relation between two arguments a_i and a_j as non-attacking as long as the premise $p_i \twoheadrightarrow p_j$ has not been introduced, and to treat p_i as not representing an argument as long as the premise $\bot \twoheadrightarrow p_i$ has not been introduced. For our example that means that as long as our agent doesn't introduce p_3, the logic should, (a), treat the relation between p_3 and p_i for all $p_i \in \mathcal{V}_n$ as non-attacking and thus derive $p_3 \not\twoheadrightarrow p_i$, and, (b), derive $\bot \not\twoheadrightarrow p_3$ and hence by (R$\bot$) $\neg p_3$. As a result, as long as p_3 has not been introduced, the only argument in the consequence set should be p_1 since a_1 constitutes the unique preferred extension of A^1. However, as soon as p_3 and $p_3 \twoheadrightarrow p_1$ have been introduced, we are interested in deriving p_2 and p_3 as only arguments. The reason is that $\{a_2, a_3\}$ constitutes the unique preferred extension of A^2. As the reader might have already guessed, the way to achieve this behaviour via an AL is to define abnormalities by the logical form $\alpha \twoheadrightarrow \beta$. Let thus

$$\Omega_{\twoheadrightarrow} =_{\text{df}} \{\alpha \twoheadrightarrow \beta \mid \alpha \in \mathcal{V}_n \cup \{\bot\}, \beta \in \mathcal{V}_n\}$$

The idea is to ensure in this way that $p_i \not\twoheadrightarrow p_j$ and $\bot \not\twoheadrightarrow p_k$ are derivable whenever $p_i \twoheadrightarrow p_j$ and $\bot \twoheadrightarrow p_k$ are not part of the premise set.

We have seen in Sect. 8.3.5 that the Ω_P-minimally abnormal $\mathbf{L_A}$-models of $\Gamma^3_{\mathsf{A}^1}$ do not correspond to the preferred extensions since they validate arguments and attacks which are not part of A^1. In order to improve on that we employ a sequential combination of adaptive logics.[9] In semantic terms the idea is realized in two steps:

(1) First, we pre-select the set $\mathcal{M}_{\twoheadrightarrow}$ of $\mathbf{L_A}$-models of $\Gamma^3_{\mathsf{A}^1}$ which validate only the attacks that are actually a part of the given AF and which invalidate all propositional letters that do not represent arguments.
(2) Second, from our preselection $\mathcal{M}_{\twoheadrightarrow}$ we select the Ω_P-minimally abnormal $\mathbf{L_A}$-models.

Now we have all tools at hand, at least semantically, to introduce our logic for preferred extensions:

$$\mathbf{AL_P} = \langle \mathbf{L_A}, [\Omega_{\twoheadrightarrow}, \Omega_P], [\text{simple strategy}, \text{ minimal abnormality strategy}]\rangle$$

The first element, $\mathbf{L_A}$, is the lower limit logic. The second element lists the abnormalities for the first and second selection. The third element lists the strategies used for the semantic selections , or syntactically, for the markings in the proof. We will comment on the simple strategy more in a moment; what is now important is that semantically both strategies select minimally abnormal models.[10] $\mathbf{AL_P}$ is semantically characterized by the two steps of the selection procedure which have just been introduced. Only performing the first step characterizes another, flat AL that can be shown to represent admissible extensions:

$$\mathbf{AL_A} = \langle \mathbf{L_A}, \Omega_{\twoheadrightarrow}, \text{ simple strategy}\rangle$$

It is easy to prove that, for a given AF $\mathsf{A} = \langle \mathcal{A}, \rightarrow\rangle$, the models selected by the first selection, i.e., the $\mathbf{AL_A}$-models of Γ^n_{A},

(a) validate $p_i \twoheadrightarrow p_j$ iff $(a_i, a_j) \in \rightarrow$, and,
(b) for all $a_i \notin \mathcal{A}$ validate $\bot \not\twoheadrightarrow p_i$ and thus, by (s⊥), $\neg p_i$.

---

[9] See Chap. 3 for a detailed discussion of their meta-theory.

[10] The reader should not be confused by the fact that for both strategies, simple strategy resp. minimal abnormality, we apply the same semantic selection, namely the selection of minimally abnormal $\mathbf{L_A}$-models with respect to the abnormalities in $\Omega_{\twoheadrightarrow}$ resp. Ω_P. The reason for this is that the simple strategy is equivalent to the minimal abnormality strategy for a lower limit logic **LLL**, abnormalities Ω and a class of premise sets $\boldsymbol{\Gamma}$ if the following fact holds:

($F\star$): For all $\Gamma \in \boldsymbol{\Gamma}$ and all finite and non-empty $\Delta \subseteq \Omega$, $\Gamma \vdash_{\mathbf{LLL}} \mathsf{Dab}(\Delta)$, then there is a $\varphi \in \Delta$ such that $\Gamma \vdash_{\mathbf{LLL}} \varphi$.

This is the case for our $\mathbf{L_A}$, $\Omega_{\twoheadrightarrow}$ and premise sets defined by Γ^n_{A} (as shown in Appendix F). Hence, in this case the simple strategy, as we will see, allows for a simplified marking strategy (see Definition 8.3.3) compared to the one for minimal abnormality (which is defined in Sect. 8.3.8, Definition 8.3.4). Of course, due to ($F\star$) the semantic selection for the simple strategy can also be characterized as follows: selected are all $\mathbf{L_A}$-models of Γ^n_{A} that validate only those abnormalities in $\Omega_{\twoheadrightarrow}$ that are $\mathbf{L_A}$-derivable from Γ^n_{A} (or equivalently, that are validated by all other $\mathbf{L_A}$-models of Γ^n_{A}). Note, that in the case that fact ($F\star$) does not hold, such models are not guaranteed to exist. See also the discussion in Sect. 2.4.3.

This obviously solves the problem of Sect. 8.3.5. Indeed, the $\mathbf{AL_A}$-models of $\Gamma^3_{\mathsf{A}^1}$ correspond to the admissible extensions of A^1.[11] This is as expected, since the models in our second selection, the Ω_P-minimally abnormal models from these $\mathbf{AL_A}$-models, are expected to correspond to the maximal admissible sets. We will see in Sect. 8.4 that the same sequential selection procedure is applied to other extension types, only the abnormalities for the second selection have to be adjusted.

We have talked a lot about semantics. Let us now take a look at a continuation of our proof from p. 221 and see how the ideas presented above are applied syntactically.

1	$p_1 \twoheadrightarrow p_2$	PREM	$\emptyset$
2	p_1	RC	$\{\neg p_1\}$
3	$\neg p_2$	1,2; R$\twoheadrightarrow$	$\{\neg p_1\}$
$^{13}4$	p_2	RC	$\{\neg p_2\}$
$^{13}5$	$\mathsf{def}\ p_1$	1,4; Rad	$\{\neg p_2\}$
$^{13}6$	$\neg p_1$	1,4; R$\twoheadrightarrow$	$\{\neg p_2\}$
$^{13}7$	$\bigvee_{i=1}^{3}(p_i \wedge (p_i \twoheadrightarrow p_1))$	5; Def	$\{\neg p_2\}$
$^{13}8$	$\left[\begin{array}{l}(p_2 \wedge (p_2 \twoheadrightarrow p_1))\vee \\ \quad (p_3 \wedge (p_3 \twoheadrightarrow p_1))\end{array}\right.$	6,7; RU	$\{\neg p_2\}$
$^{15}9$	p_3	RC	$\{\neg p_3\}$
10	$\left[\begin{array}{l}\neg p_2 \vee (p_2 \wedge (p_2 \twoheadrightarrow p_1)) \\ \quad \vee (p_3 \wedge (p_3 \twoheadrightarrow p_1))\end{array}\right.$	8; RA	$\emptyset$
11	$p_2 \not\twoheadrightarrow p_1$	RC	$\{p_2 \twoheadrightarrow p_1\}$
12	$p_3 \not\twoheadrightarrow p_1$	RC	$\{p_3 \twoheadrightarrow p_1\}$
13	$\neg p_2$	10,11,12; RU	$\{p_2 \twoheadrightarrow p_1, p_3 \twoheadrightarrow p_1\}$
14	$\bot \not\twoheadrightarrow p_3$	RC	$\{\bot \twoheadrightarrow p_3\}$
15	$\neg p_3$	14; R$\bot$	$\{\bot \twoheadrightarrow p_3\}$

What is happening in the proof segment above? At line 4 we conditionally introduce p_2. This gives rise to p_1 being defeated under the same condition at line 5. Furthermore, at line 8 we derive that either p_2 or p_3 has to be the defeater of p_1. We introduce p_3 conditionally at line 9. What we expect from the proof is that lines 4–9 get invalidated, since it is not in our interest to derive p_2 and p_3, as neither is a part of the unique preferred extension $\{a_1\}$.

At lines 11, 12 and 14 we realize the ideas from above, namely that two propositional letters p_i and p_j should be considered to not attack each other as long as no premise $p_i \twoheadrightarrow p_j$ has been introduced, and that a propositional letter p_i should be considered as not valid unless one of the introduced premises states that it is part of the AF under consideration, i.e., $\bot \twoheadrightarrow p_i$. Thus, we derive $p_2 \not\twoheadrightarrow p_1$, $p_3 \not\twoheadrightarrow p_1$ and $\bot \not\twoheadrightarrow p_3$ conditionally at line 11, 12 and 14. Since $p_2 \twoheadrightarrow p_1$, $p_3 \twoheadrightarrow p_1$ and $\bot \twoheadrightarrow p_3$ are not part of the premise set $\Gamma^3_{\mathsf{A}^1}$, these lines are not going to be marked in our proof for A^1. Using these lines we are able to derive $\neg p_2$ at line 13 on the condition $\{p_2 \twoheadrightarrow p_1, p_3 \twoheadrightarrow p_1\}$ as well as $\neg p_3$ at line 15 on condition $\{\bot \twoheadrightarrow p_3\}$. Now something very important happens. Note that lines 4–8 have been derived on

[11] The representational results are stated in Sect. 8.4 (see Theorem 8.4.1 and Corollary 8.4.1) and proven in Appendix F.

the very condition that $\neg p_2$ is not valid. However, now we have derived $\neg p_2$ and thus all lines which were derived on this condition should be considered as invalid derivations and thus have to be marked. Similarly, by introducing $\bot \not\twoheadrightarrow p_3$ at line 14 we derive $\neg p_3$ at line 15 which causes the marking of line 9. The idea behind the marking is thus to invalidate lines on conditions that have to be considered as unsafe. What is considered as unsafe depends on the adaptive strategy used. Recall that we have defined two types of abnormalities: Ω_P and $\Omega_{\twoheadrightarrow}$. Each of these come with their own marking definition. The marking definition for abnormalities in $\Omega_{\twoheadrightarrow}$ is very simple, after all it is based on the adaptive strategy called the *simple strategy* (see Sect. 2.4.3).

Definition 8.3.3 (Marking for the simple strategy). A line with condition Δ is marked at stage s if a $\alpha \twoheadrightarrow \beta \in \Delta \cap \Omega_{\twoheadrightarrow}$ has been derived on the empty condition.

Suppose for a moment that new information comes in: one agent, in order to defend a_2, voices a_3 which attacks a_1. In this case we would introduce $p_3 \twoheadrightarrow p_1$ and $\bot \twoheadrightarrow p_3$ by PREM. Note that lines 12–15 would get marked. These lines would not anymore be considered to be derived since they rely on the condition that $p_3 \twoheadrightarrow p_1$ and resp. $\bot \twoheadrightarrow p_3$ are not derivable. This behavior is obviously intuitive.

The marking conditions for Ω_P are technically a bit more complicated. We will introduce them later in Sect. 8.3.8 in order not to complicate things more than necessary at this point. However, let us make another important remark.

So far we have discussed the prioritized aspect of ALs only in terms of the semantic selection. Of course, this has a syntactic equivalent to it. This is illustrated in the proof, for instance, at line 13: here we derive an abnormality $\neg p_2 \in \Omega_P$ at an unmarked line on a condition $\{p_2 \twoheadrightarrow p_1, p_3 \twoheadrightarrow p_1\} \subset \Omega_{\twoheadrightarrow}$. This causes the marking of all lines that have $\neg p_2$ as a part of the condition. Similarly, at line 15 we derive $\neg p_3$ on the condition $\{\bot \twoheadrightarrow p_3\}$ which causes the marking of line 9. Hence, lines are considered as (un)safe due to conditions in Ω_P on the basis of abnormalities in Ω_P (and their disjunctions, as we will see in Sect. 8.3.8) derived on unmarked lines on the empty condition or on conditions which are subsets of $\Omega_{\twoheadrightarrow}$. In contrast, the marking condition for $\Omega_{\twoheadrightarrow}$ requires that, in order to mark a line with condition Δ, a $\alpha \twoheadrightarrow \beta \in \Delta$ has to be derived on the empty(!) condition. It is not enough to derive $\alpha \twoheadrightarrow \beta$ on a condition $\Delta' \subseteq \Omega_P$. This evidently mirrors syntactically the prioritized aspect of the two semantic selections.

8.3.7 External Dynamics: Letting New Information In

Let us now proceed from A^1 to A^2. The new information in $\Gamma^3_{\mathsf{A}^2} \setminus \Gamma^3_{\mathsf{A}^1}$ is introduced at lines 16 and 17. We restate lines 2, 4, 9, 13–15. Let $\Theta = \{p_1 \twoheadrightarrow p_3, p_2 \twoheadrightarrow p_3, p_3 \twoheadrightarrow p_3\}$.

^{22}2	p_1	RC	$\{\neg p_1\}$
$\vdots$	$\vdots$	$\vdots$	$\vdots$
4	p_2	RC	$\{\neg p_2\}$
$\vdots$	$\vdots$	$\vdots$	$\vdots$
9	p_3	RC	$\{\neg p_3\}$
$\vdots$	$\vdots$	$\vdots$	$\vdots$
^{16}13	$\neg p_2$	10,11,12; RU	$\{p_2 \twoheadrightarrow p_1, p_3 \twoheadrightarrow p_1\}$
^{17}14	$\bot \not\twoheadrightarrow p_3$	RC	$\{\bot \twoheadrightarrow p_3\}$
^{17}15	$\neg p_3$	14; R$\bot$	$\{\bot \twoheadrightarrow p_3\}$
16	$p_3 \twoheadrightarrow p_1$	PREM	$\emptyset$
17	$\bot \twoheadrightarrow p_3$	PREM	$\emptyset$
18	$\mathsf{def}\, p_3$	2, 16; Rad	$\{\neg p_1\}$
19	$\bigvee_{i=1}^{3}(p_i \wedge (p_i \twoheadrightarrow p_3))$	18; Def	$\{\neg p_1\}$
20	$\neg p_1 \vee \bigvee_{i=1}^{3}(p_i \wedge (p_i \twoheadrightarrow p_3))$	19; RA	$\emptyset$
21	$\bigwedge_{i=1}^{3}(p_i \not\twoheadrightarrow p_3)$	RC	Θ
22	$\neg p_1$	20, 21; RU	Θ

Due to the new information in A^2, p_3 now corresponds to an argument, namely a_3. Furthermore a_3 attacks a_1. Thus we introduce at lines 16 and 17 the new premises $p_3 \twoheadrightarrow p_1$ and $\bot \twoheadrightarrow p_3$. This immediately leads to the marking of line 13 since this line was derived on the condition that p_3 does not attack p_1, and to the marking of lines 15 and 15 since these lines were derived on the condition $\{\bot \twoheadrightarrow p_3\}$, i.e., that a_3 does not belong to the AF in question. Furthermore, the new information enables us to derive $\neg p_1$ at line 22 on the condition $\Theta \subset \Omega_{\twoheadrightarrow}$, which leads to the marking of line 2. Moreover, line 4 is now unmarked, since the line which caused it to be marked before, namely 13, is now itself marked. Analogously for line 9: line 15 which caused it to be marked is now marked itself and hence, 9 is unmarked.

8.3.8 The Minimal Abnormality Strategy and Final Derivability

We postponed the exact marking definition for the minimal abnormality strategy so far. Recall that this strategy is used for the abnormalities in Ω_P. The marking definition can be better motivated if we take a look at another simple example. Let $\mathsf{A} = \langle\{a_1, a_2\}, \{(a_1, a_2), (a_2, a_1)\}\rangle$. The two preferred extensions are $\{a_1\}$ and $\{a_2\}$. Thus, there should be two types of Ω_P-minimally abnormal $\mathbf{AL_A}$-models: on the one hand a model verifying p_1 and $\neg p_2$ and on the other hand a model verifying p_2 and $\neg p_1$. That means that we expect from our logic to derive $p_1 \vee p_2$ since either p_1 or p_2 is valid in each minimally abnormal model. Let us take a look at a proof for $\Gamma_{\mathsf{A}}^4 = \{p_1 \twoheadrightarrow p_2, p_2 \twoheadrightarrow p_1, \bot \twoheadrightarrow p_1, \bot \twoheadrightarrow p_2\}$ and the language $\mathcal{W}_4$.

1	$p_1 \twoheadrightarrow p_2$	PREM	$\emptyset$
2	$p_2 \twoheadrightarrow p_1$	PREM	$\emptyset$

3 $\bot \twoheadrightarrow p_1$	PREM	$\emptyset$
4 $\bot \twoheadrightarrow p_2$	PREM	$\emptyset$
5 p_1	RC	$\{\neg p_1\}$
6 p_2	RC	$\{\neg p_2\}$
7 $p_1 \vee p_2$	5; RU	$\{\neg p_1\}$
8 $p_1 \vee p_2$	6; RU	$\{\neg p_2\}$

With the analyses given above, what we expect from our logic is that lines 5 and 6 are marked, while line 7 and 8 are not marked and hence $p_1 \vee p_2$ is considered as being derived. Note that with line 1 we can derive the following disjunction of abnormalities:

9 $\neg p_1 \vee \neg p_2$	1; R$\twoheadrightarrow$	$\emptyset$

It is important to notice that neither $\neg p_1$ nor $\neg p_2$ can be derived on a condition $\Delta \subseteq \Omega_{\twoheadrightarrow}$ (including the empty condition $\emptyset$). All we know is that either p_1 is false or p_2 or both. If our rationale is to interpret the premises "as normal as possible" then we will not opt for the latter but assume that at most one of the two propositional letters is false. However, if at most one of them is false, either p_1 or p_2 will be true. In that case at least on of the two assumptions $\{\neg p_1\}$ or $\{\neg p_2\}$ is not mistaken. Since $p_1 \vee p_2$ is derived on both assumptions, it should be considered a safe defeasible inference at this point of the proof. We will now make this insight technically precise.

In order to define our marking conditions we need to introduce some terminology. Let us do this in a more general setting for a logic

$$\mathbf{AL} = \langle \mathbf{L_X}, [\Omega_{\twoheadrightarrow}, \Omega], [\text{simple strategy, minimal abnormality strategy}] \rangle,$$

where $\mathbf{X} \in \{\mathbf{A}, \mathbf{C}\}$. Obviously our $\mathbf{AL_P}$ is such a logic for $\Omega = \Omega_P$. Where $\Delta \subseteq \Omega$ and Δ is finite and non-empty, we say that $\mathsf{Dab}(\Delta)$ is a *Ω-minimal Dab-formula at a stage s* of the proof iff it is the formula of an unmarked line with a condition $\Delta_{\twoheadrightarrow} \subseteq \Omega_{\twoheadrightarrow}$ and no $\mathsf{Dab}(\Delta')$, where $\Delta' \subset \Delta$, is the formula of an unmarked line with a condition $\Delta'_{\twoheadrightarrow} \subseteq \Omega_{\twoheadrightarrow}$. A *choice set* of $\Sigma = \{\Delta_1, \Delta_2, \dots\}$ is a set that contains an element out of each member of Σ. A *minimal choice set* of Σ is a choice set of Σ of which no proper subset is a choice set of Σ.[12] Where $\mathsf{Dab}(\Delta_1), \dots, \mathsf{Dab}(\Delta_n)$ are the Ω-minimal Dab-formulas at stage s for a premise set Γ, $\Phi_s(\Gamma)$ is the set of minimal choice sets of $\{\Delta_1, \dots, \Delta_n\}$.

With this terminology we can define marking conditions for the abnormalities Ω and the minimal abnormality strategy. Let Γ be a premise set.

Definition 8.3.4 (Marking for the minimal abnormality strategy (with respect to $\boldsymbol{\Omega}$)). Line i is marked at stage s if, where φ is derived on the condition Δ at line i,

(i) there is no $\Delta' \in \Phi_s(\Gamma)$ such that $\Delta' \cap \Delta = \emptyset$, or
(ii) for some $\Delta' \in \Phi_s(\Gamma)$, there is no line at which φ is derived on a condition Θ for which $\Delta' \cap \Theta = \emptyset$.

[12] Let for instance $\Sigma = \{\{1, 2\}, \{1, 3\}\}$. Choice sets are $\{1\}, \{1, 2\}, \{1, 3\}, \{2, 3\}$ and $\{1, 2, 3\}$. Minimal are $\{1\}$ and $\{2, 3\}$.

Let us return to our example. Note that at this stage of the proof $\neg p_1 \vee \neg p_2$ at line 9 is a Ω_P-minimal Dab-formula. Thus, the minimal choice sets at this stage of the proof are $\{\neg p_1\}$ and $\{\neg p_2\}$. By (ii) lines 5 and 6 are marked. This is as desired, since after all, neither is a_1 nor is a_2 a skeptically accepted argument. However, the situation is different for lines 7 and 8. Note that neither (i) nor (ii) apply, due to the fact that we are able to derive $p_1 \vee p_2$ on condition $\{\neg p_1\}$ and(!) on condition $\{\neg p_2\}$.

Since, as we have seen with our examples, markings come and go in adaptive proofs, we need a stable criterion for derivability in order to define a consequence relation.

Definition 8.3.5. φ is *finally derived* from Γ on line i of a proof at stage s iff

(i) φ is the second element of line i,
(ii) line i is not marked at stage s and
(iii) for every extension of the proof in which line i is marked there is a further extension in which line i is unmarked.

The definition states that a formula derived at an unmarked line is finally derived in the case that there is no way anymore to mark it by extending the proof. For instance in our proof above it is easy to see that $p_1 \vee p_2$ is finally derived since there is no way of extending the proof in such a way that lines 7 and 8 get marked. This is due to the fact that neither $\neg p_1$ nor $\neg p_2$ are derivable on a condition $\Delta \subseteq \Omega_{\twoheadrightarrow}$.

Let us close this section by introducing $p_1 \twoheadrightarrow p_3, p_2 \twoheadrightarrow p_3, p_3 \twoheadrightarrow p_4, \bot \twoheadrightarrow p_3, \bot \twoheadrightarrow p_4$ to our last example so that we arrive at the AF from Example 8.2.1. Hence our new premise set is $\{p_1 \twoheadrightarrow p_2, p_2 \twoheadrightarrow p_1, p_1 \twoheadrightarrow p_3, p_2 \twoheadrightarrow p_3, p_3 \twoheadrightarrow p_4, \bot \twoheadrightarrow p_1, \bot \twoheadrightarrow p_2, \bot \twoheadrightarrow p_3, \bot \twoheadrightarrow p_4\}$. In the following proof we use the abbreviations:

$$\Theta_1 = \{p_1 \twoheadrightarrow p_1, p_3 \twoheadrightarrow p_1, p_4 \twoheadrightarrow p_1\}$$
$$\Theta_2 = \{p_2 \twoheadrightarrow p_2, p_3 \twoheadrightarrow p_2, p_4 \twoheadrightarrow p_2\}$$

The proof is as follows:

10	$p_1 \twoheadrightarrow p_3$	PREM	$\emptyset$
11	$p_2 \twoheadrightarrow p_3$	PREM	$\emptyset$
12	$p_3 \twoheadrightarrow p_4$	PREM	$\emptyset$
13	$p_1 \not\twoheadrightarrow p_1 \wedge p_3 \not\twoheadrightarrow p_1 \wedge p_4 \not\twoheadrightarrow p_1$	RC	Θ_1
14	$\mathsf{def}\, p_1 \supset p_2$	13; RU	Θ_1
15	$p_2 \not\twoheadrightarrow p_2 \wedge p_3 \not\twoheadrightarrow p_2 \wedge p_4 \not\twoheadrightarrow p_2$	RC	Θ_2
16	$\mathsf{def}\, p_2 \supset p_1$	15; RU	Θ_2
17	$p_3 \supset \mathsf{def}\, p_1$	10; Rad	$\emptyset$
18	$p_3 \supset \mathsf{def}\, p_2$	11; Rad	$\emptyset$
19	$p_3 \supset (\mathsf{def}\, p_1 \wedge \mathsf{def}\, p_2)$	17,18; RU	$\emptyset$
20	$p_3 \supset (p_1 \wedge p_2)$	14,16,19; RU	$\Theta_1 \cup \Theta_2$
21	$\neg p_3$	9,20; RU	$\Theta_1 \cup \Theta_2$
22	p_4	RC	$\{\neg p_4\}$

It is easy to see that there is no way to mark line 22. Thus, as desired, $p_1 \vee p_2$ and p_4 are finally derivable. Recall that a_4 is the only accepted argument with respect to preferred extensions and that in each preferred extension there is either a_1 or a_2 (but obviously never both simultaneously). Thus, p_1 and p_2 are not derivable, but $p_1 \vee p_2$ and $\neg p_1 \vee \neg p_2$ are.

After having introduced a logic for preferred extensions it is time to introduce the other logics for abstract argumentation in the next section.

8.4 The AL Framework for Skeptical Acceptance

In this section we will introduce ALs for all the standard extension types for abstract argumentation. After our discussion in the previous section this can be done very smoothly. Let us first recapitulate the three characteristic elements of ALs that were introduced in the previous section:

The lower limit logic (**LLL**) While all the rules of **LLL** are valid in **AL**, the latter additionally allows for certain rules to be applied conditionally. This strengthens **LLL**as it allows to derive at least as much as **LLL** and in most cases even more. Thus, the consequence set of **AL** is a superset of the consequence set of **LLL**: $Cn_{\mathbf{AL}}(\Gamma) \supseteq Cn_{\mathbf{LLL}}(\Gamma)$ for all premise sets Γ. In semantic terms ALs select a subset of "sufficiently normal" **LLL**-models. In the case of our logics, models are "minimally abnormal" in the sense that they validate as few abnormalities as possible. This brings us to the next point:

The abnormalities The set of abnormalities is defined by a logical form F.[13] In the last section we, for instance, used abnormalities of the form $\neg\alpha$ where α is a propositional letter. Thus, one set of abnormalities was characterized by the set $\Omega_P = \{\neg p_i \mid i \leq n\}$ for the language $\mathcal{W}_n$.

The strategy Together with the abnormalities the strategy gives an exact account of what it means to interpret a premise set "as normally as possible". In the previous section we, for instance, employed the minimal abnormality strategy. In semantic terms this strategy selects all **LLL**-models M of a given premise set Γ for which there are no **LLL**-models that validate less abnormalities (w.r.t. $\subset$). In syntactic terms strategies are realized by a marking definition. In the last section we have demonstrated that adaptive proofs are dynamic: the markings invalidate lines, however if new information is introduced and/or while we reason along, markings may come and go. In this respect adaptive proofs resemble human reasoning.

We have introduced the following notation to define ALs (where $\mathbf{X} \in \{\mathbf{A}, \mathbf{C}\}$):

$$\mathbf{AL_1} = \langle \mathbf{L_X}, \Omega_{\twoheadrightarrow}, \text{simple strategy} \rangle$$
$$\mathbf{AL_2} = \langle \mathbf{L_X}, [\Omega_{\twoheadrightarrow}, \Omega], [\text{simple strategy, minimal abnormality}] \rangle$$

[13] F is considered to be **LLL**-contingent, i.e., neither $\vdash_{\mathbf{LLL}} \mathsf{F}$ nor $\vdash_{\mathbf{LLL}} \neg\, \mathsf{F}$.

We have already seen examples for both cases: the logic $\mathbf{AL_A}$ is a flat AL for admissible extensions. On the other hand, the logic $\mathbf{AL_P}$ for preferred extensions is a prioritized AL.

The idea of the prioritization can be easily put in semantic terms: first the set of minimally abnormal $\mathbf{L_X}$-models $\mathcal{M}_{\twoheadrightarrow}$ with respect to $\Omega_{\twoheadrightarrow}$ is selected, and then, from these selected models in $\mathcal{M}_{\twoheadrightarrow}$ the minimally abnormal models with respect to Ω are selected. This is mirrored in the proof dynamics: the marking definition for the minimal abnormality strategy (w.r.t. Ω) is such that not only Ω-minimal Dab-formulas (i.e., minimal disjunctions of abnormalities) derived on the empty condition are considered for the marking procedure, but also the ones derived at unmarked lines with conditions $\Delta \subseteq \Omega_{\twoheadrightarrow}$ (see Definition 8.3.4).

We presuppose the language $\mathcal{W}_n$ throughout this and the next section for an arbitrary natural number n. Let us give a general account of the consequence relations for the flat and prioritized logics that we are going to introduce in this chapter:

Definition 8.4.1. Let $\mathcal{M}_{\mathbf{AL_1}}(\Gamma) =_{\text{df}} \{M \in \mathcal{M}_{\mathbf{L_X}}(\Gamma) \mid \text{there is no } M' \in \mathcal{M}_{\mathbf{L_X}}(\Gamma) \text{ such that } \text{Ab}^{\mathbf{L_X}}_{\Omega_{\twoheadrightarrow}}(M') \subset \text{Ab}^{\mathbf{L_X}}_{\Omega_{\twoheadrightarrow}}(M)\}$.

We define $\Gamma \Vdash_{\mathbf{AL_1}} \varphi$ iff for all $M \in \mathcal{M}_{\mathbf{AL_1}}(\Gamma)$, $M \models_{\mathbf{L_X}} \varphi$. Furthermore, $\Gamma \vdash_{\mathbf{AL_1}} \varphi$ iff φ is finally derivable in terms of the marking conditions for the simple strategy defined in Definition 8.3.3.

We define $\Gamma \models_{\mathbf{AL_2}} \varphi$ iff for all Ω-minimally abnormal $\mathbf{AL_1}$-models M of Γ, $M \models_{\mathbf{L_X}} \varphi$. Furthermore, $\Gamma \vdash_{\mathbf{AL_2}} \varphi$ iff φ is finally derivable in terms of the marking conditions for the simple strategy in Definition 8.3.3 and for the minimal abnormality strategy in Definition 8.3.4.

Since our logics are in the standard format for ALs, a lot of meta-theory for them has already been investigated in other places (see e.g., [36]). Thus, for instance, completeness and soundness follow immediately with the completeness and soundness of our lower limit logics $\mathbf{L_A}$ and $\mathbf{L_C}$ (see Appendix F for proofs). The consequence set of the sequential AL $\mathbf{AL_2}$ in Definition 8.4.1 is given by $Cn_{\mathbf{AL_2}}(\Gamma) = Cn_{\mathbf{AL'_2}}(Cn_{\mathbf{AL_1}}(\Gamma))$ where[14]

$$\mathbf{AL'_2} = \langle \mathbf{LLL}, \Omega, \text{minimal abnormality strategy} \rangle$$

We are now able to define our adaptive logic framework for abstract argumentation. The idea is that we first define flat, non-prioritized ALs for admissible and complete extensions. While $\mathbf{AL_A}$ is the logic for admissible extensions, the logic for complete extensions is the strengthening of $\mathbf{AL_A}$ by the rule (RCo):

$$\mathbf{AL_C} = \langle \mathbf{L_C}, \Omega_{\twoheadrightarrow}, \text{simple strategy} \rangle$$

[14] We have not characterized the marking conditions for minimal abnormality for logics that employ the minimal abnormality strategy for the flat case such as $\mathbf{AL'_2}$. They are a straightforward specification of our Definition 8.3.4. See also the characterization of sequential ALs in Chap. 3 where the proof theory is presented in generic terms.

In the second step we define the ALs for all the other extension types (preferred, grounded and semi-stable) by simply adding an "adaptive layer" to the flat adaptive logics $\mathbf{AL_A}$, resp. $\mathbf{AL_C}$. We have seen this already in Sect. 8.3.6 for preferred extensions: we added a second level to the AL for admissible extensions resulting in the logic $\mathbf{AL_P}$.

For grounded extensions the idea is similar. There are three differences compared to preferred extensions:

- While preferred extensions are a specific selection of admissible extensions, the grounded extension is a specific complete extension. Thus, instead of using $\mathbf{L_A}$ as **LLL**, we now use $\mathbf{L_C}$.
- While preferred extensions were *maximal* admissible extensions, the grounded extension is the *minimal* complete extension. Thus, instead of verifying as many propositional letters as possible we now verify as few as possible. Hence, instead of defining the abnormalities as $\Omega_P = \{\neg p_i \mid i \leq n\}$ we now define them as

$$\Omega_G = \{p_i \mid i \leq n\}$$

- While there may be many preferred extensions, the grounded extension is always *unique*. This allows for a simplification, namely to use instead of the minimal abnormality strategy the simple strategy.[15]

Thus, we define the AL for grounded extensions as follows[16]:

$$\mathbf{AL_G} = \langle \mathbf{L_C}, [\Omega_{\twoheadrightarrow}, \Omega_G], \text{[simple strategy, simple strategy]}\rangle$$

The AL for semi-stable extensions shouldn't come as a surprise anymore: instead of maximizing the number of arguments validated, we now maximize not only the number of arguments validated but also the number of defeated arguments. Thus, our abnormalities are defined by

$$\Omega_S = \{\neg p_i \wedge \neg \mathsf{def}\, p_i \mid i \leq n\}$$

For as many propositional letters as possible the logic is supposed to derive $\neg(\neg p_i \wedge \neg \mathsf{def}\, p_i)$. This is equivalent to $p_i \vee \mathsf{def}\, p_i$: either p_i is valid or it is defeated. We define our AL for semi-stable extensions as follows[17]:

$$\mathbf{AL_S} = \langle \mathbf{L_C}, [\Omega_{\twoheadrightarrow}, \Omega_S], \text{[simple strategy, minimal abnormality strategy]}\rangle$$

---

[15] As is well-known in the adaptive logic research, in case all minimally abnormal models validate the same set of abnormalities, the minimal abnormality strategy and the simple strategy are equivalent (cf. Footnote 10). See Sect. 2.4.3.

[16] In view of our discussion it is straightforward to define the marking conditions for the simple strategy for Ω_G in $\mathbf{AL_G}$: A line with condition Δ is marked at stage s if a $p_i \in \Delta \cap \Omega_G$ has been derived at an unmarked line on a condition $\Delta' \subseteq \Omega_{\twoheadrightarrow}$.

[17] In accordance with Footnote 5, $\mathbf{AL_S}$ can easily be shown to be equivalent to $\langle \mathbf{L_A}, [\Omega_{\twoheadrightarrow}, \Omega_S]$, [simple strategy, minimal abnormality strategy]$\rangle$.

The following results show that the logics defined above satisfy our representational requirements from Sect. 8.3.3.

Theorem 8.4.1. $\left\{\begin{array}{c} (i)\ \mathbf{AL_A} \\ (ii)\ \mathbf{AL_C} \\ (iii)\ \mathbf{AL_P} \\ (iv)\ \mathbf{AL_G} \\ (v)\ \mathbf{AL_S} \end{array}\right\}$ *semantically represents* $\left\{\begin{array}{l} \textit{admissible} \\ \textit{complete} \\ \textit{preferred} \\ \textit{grounded} \\ \textit{semi-stable} \end{array}\right\}$ *extensions for argumentation frameworks with at most n arguments.*

Due to our generic soundness and completeness results in Chap. 3 we immediately get:

Corollary 8.4.1. $\left\{\begin{array}{c} (i)\ \mathbf{AL_A} \\ (ii)\ \mathbf{AL_C} \\ (iii)\ \mathbf{AL_P} \\ (iv)\ \mathbf{AL_G} \\ (v)\ \mathbf{AL_S} \end{array}\right\}$ *syntactically represents* $\left\{\begin{array}{l} \textit{admissible} \\ \textit{complete} \\ \textit{preferred} \\ \textit{grounded} \\ \textit{semi-stable} \end{array}\right\}$ *extensions with respect to skeptical acceptance for argumentation frameworks with at most n arguments.*

8.5 Adaptive Logics for Credulous Acceptance

So far we have presented logics modeling skeptical acceptance. The current section will deal with credulous acceptance. In the skeptical case we were interested in arguments located in the intersection of all extensions of a given type. Now we are focusing on their union. The so-called normal selections strategy will prove to be very useful for this purpose.[18] We will see that, given the systems for skeptical acceptance, everything which has to be done in order to model credulous acceptance is to use the normal selections strategy instead, resp. on top of the minimal abnormality strategy.

The reason for this can be easily understood when we take a look at the normal selections strategy from a semantic point: like the minimal abnormality strategy, the normal selections strategy selects minimally abnormal **LLL**-models. However, semantic consequences are not defined in terms of the intersection of the models but in terms of their union. This obviously mirrors the difference between skeptical and credulous acceptance, where the former is defined with respect to the intersection of all models of a certain extension type while the latter is defined in terms of the union of these extensions.

We use in this section the language $\mathcal{W}_n$ for an arbitrary natural number n. The semantic consequence relation for our prioritized logics for credulous acceptance is defined as follows:

---

[18] The normal selections strategy was first introduced in [37]. See also [38] and Sect. 2.8 for a more elaborated representation. In Sect. 3.4 normal selections are introduced in sequential combinations of ALs and in Sect. 5.7 normal selections are presented in a more general way.

Definition 8.5.1. Where $\mathbf{X} \in \{\mathbf{A}, \mathbf{C}\}$ let

$$\mathbf{AL^n} = \langle \mathbf{L_X}, [\Omega_{\twoheadrightarrow}, \Omega], [\text{simple strategy, normal selections}]\rangle.$$

Where $\mathcal{M}_{\mathbf{AL_X}}(\Gamma)$ is the set of all $\mathbf{L_X}$-models of Γ that are $\Omega_{\twoheadrightarrow}$-minimally abnormal and $\mathcal{M}_{\mathbf{AL^n}}(\Gamma)$ is the set of all Ω-minimally abnormal $\mathbf{L_X}$-models in $\mathcal{M}_{\mathbf{AL_X}}(\Gamma)$, we define the semantic consequence relation as follows: $\Gamma \Vdash_{\mathbf{AL^n}} \varphi$ iff there is a $M \in \mathcal{M}_{\mathbf{AL^n}}(\Gamma)$ for which $M \models_{\mathbf{L_X}} \varphi$.[19]

Thus, we are going to define, for instance, a logic for preferred extensions with respect to credulous acceptance by

$$\mathbf{ALC_P} = \langle \mathbf{L_A}, [\Omega_{\twoheadrightarrow}, \Omega_P], [\text{simple strategy, normal selections}]\rangle.$$

Note that each of the selected models from Definition 8.5.1 exactly corresponds to a preferred extension.

We are still lacking a syntactic characterization of the normal selections strategy. The marking conditions are technically straightforward. The following definition covers the generic case for $\mathbf{AL^n}$ from Definition 8.5.1:

Definition 8.5.2 Marking for normal selections (w.r.t. Ω). Line i is marked at stage s if, where Δ is the condition of line i, $\mathsf{Dab}(\Delta \cap \Omega)$ has been derived at an unmarked line on a condition Δ' for which $\Delta' \cap \Omega = \emptyset$.

Definition 8.5.3. $\Gamma \vdash_{\mathbf{AL^n}} \varphi$ iff φ is finally derivable with respect to the marking conditions for simple selections (w.r.t. $\Omega_{\twoheadrightarrow}$) and for normal selections (w.r.t. Ω).

Let us again take a look at the AF A from our Example 8.2.1 for the logic $\mathbf{ALC_P}$ and with the language $\mathcal{W}_4$. Recall that the premise set is $\Gamma_{\mathsf{A}}^4 = \{p_1 \twoheadrightarrow p_2, p_2 \twoheadrightarrow p_1, p_1 \twoheadrightarrow p_3, p_2 \twoheadrightarrow p_3, p_3 \twoheadrightarrow p_4, \bot \twoheadrightarrow p_1, \bot \twoheadrightarrow p_2, \bot \twoheadrightarrow p_3, \bot \twoheadrightarrow p_4\}$. In the proof we use the following abbreviation: $\Theta = \{p_1 \twoheadrightarrow p_1, p_3 \twoheadrightarrow p_1, p_4 \twoheadrightarrow p_1, p_2 \twoheadrightarrow p_2, p_3 \twoheadrightarrow p_2, p_4 \twoheadrightarrow p_2\}$.

1	$p_1 \twoheadrightarrow p_2$	PREM	$\emptyset$
2	$p_2 \twoheadrightarrow p_1$	PREM	$\emptyset$
3	$p_1 \twoheadrightarrow p_3$	PREM	$\emptyset$
4	$p_2 \twoheadrightarrow p_3$	PREM	$\emptyset$

[19] Usually the semantic consequence relation has to be defined in terms of equivalence classes of Ω-minimally abnormal $\mathbf{AL_X}$-models. For two $\mathbf{AL_X}$-models $M \sim N$ iff $\mathrm{Ab}_{\Omega}^{\mathbf{L_X}}(M) = \mathrm{Ab}_{\Omega}^{\mathbf{L_X}}(N)$. The semantic consequence relation is then defined by $\Gamma \Vdash_{\mathbf{AL^n}} \varphi$ iff there is an Ω-minimally abnormal $\mathbf{AL_X}$-model M of Γ such that for all Ω-minimally abnormal $\mathbf{AL_X}$-models N of Γ for which $N \sim M$, $N \models_{\mathbf{L_X}} \varphi$ (see Definition 2.8.1). However, the nature of our abnormalities and of our premise sets allows for the simplification in Definition 8.5.1 since it can easily be shown that for all AFs A and for all Ω-minimally abnormal $\mathbf{AL_X}$-models of Γ_{A}^n, M and N,

$$(M \sim N) \text{ iff (for all } \varphi \in \mathcal{W}_n, M \models_{\mathbf{L_X}} \varphi \text{ iff } N \models_{\mathbf{L_X}} \varphi)$$

The simplification is explicated in a more detailed way in Appendix F.4.

5	$p_3 \twoheadrightarrow p_4$	PREM	$\emptyset$
6	p_1	RC	$\{\neg p_1\}$
7	$\neg p_2 \wedge \mathsf{def}\, p_2$	1, 6; R$\twoheadrightarrow$,Def	$\{\neg p_1\}$
8	p_2	RC	$\{\neg p_2\}$
9	$\neg p_1 \wedge \mathsf{def}\, p_1$	2, 8; R$\twoheadrightarrow$,Def	$\{\neg p_2\}$
^{13}10	$p_1 \wedge \neg p_1$	6, 9; RU	$\{\neg p_1, \neg p_2\}$
^{13}11	$p_2 \wedge \neg p_2$	7, 8; RU	$\{\neg p_1, \neg p_2\}$
^{13}12	$p_1 \wedge p_2$	6, 8; RU	$\{\neg p_1, \neg p_2\}$
13	$\neg p_1 \vee \neg p_2$	1; R$\twoheadrightarrow$	$\emptyset$
14	$\bigwedge_{i \in \{1,3,4\}} p_i \not\twoheadrightarrow p_1$	RC	$\{p_1 \twoheadrightarrow p_1, p_3 \twoheadrightarrow p_1, p_4 \twoheadrightarrow p_1\}$
15	$\mathsf{def}\, p_1 \supset p_2$	14; Def	$\{p_1 \twoheadrightarrow p_1, p_3 \twoheadrightarrow p_1, p_4 \twoheadrightarrow p_1\}$
16	$\bigwedge_{i \in \{2,3,4\}} p_i \not\twoheadrightarrow p_2$	RC	$\{p_2 \twoheadrightarrow p_2, p_3 \twoheadrightarrow p_2, p_4 \twoheadrightarrow p_2\}$
17	$\mathsf{def}\, p_2 \supset p_1$	16; Def	$\{p_2 \twoheadrightarrow p_2, p_3 \twoheadrightarrow p_2, p_4 \twoheadrightarrow p_2\}$
18	$p_3 \supset \mathsf{def}\, p_1$	3; Rad	$\emptyset$
19	$p_3 \supset \mathsf{def}\, p_2$	4; Rad	$\emptyset$
20	$p_3 \supset (\mathsf{def}\, p_1 \wedge \mathsf{def}\, p_2)$	18, 19; RU	$\emptyset$
21	$p_3 \supset (p_1 \wedge p_2)$	15, 17, 20; RU	Θ
22	$\neg p_3$	13, 21; RU	Θ
23	p_4	RC	$\{\neg p_4\}$
24	$\bot \twoheadrightarrow p_4$	PREM	$\emptyset$

Although the proof is very similar to the one presented in Sect. 8.3.8, there are important differences. Let us take a closer look. At lines 1–5 we introduce some premises. At line 6 we conditionally derive p_1. Next, due to p_1 being conditionally derived, p_2 gets defeated at line 7 on the condition $\{\neg p_1\}$. This branch of the proof corresponds to the preferred extension $\{a_1, a_4\}$ in the sense that the proof proceeds under the condition that p_1 is valid. The only Ω_P-minimally abnormal $\mathbf{AL_A}$-model validating p_1 is the one also validating p_4 and $\neg p_2, \neg p_3$. Analogously lines 8–9 correspond to the preferred extension $\{p_2, p_4\}$. At line 13 the Ω_P-minimal Dab-formula $\neg p_1 \vee \neg p_2$ is derived. At line 22 another Ω_P-minimal Dab-formula is derived, namely $\neg p_3$.

Note that, unlike the marking procedure for the logic $\mathbf{AL_P}$ (see Definition 8.3.4), at line 13 we do not mark lines 6–9. The reason for this is that, concerning the marking conditions for normal selections, we would have to derive $\neg p_1$ (resp. $\neg p_2$) at an unmarked line on a condition $\Delta \subseteq \Omega_{\twoheadrightarrow}$ in order to mark lines derived on the condition $\{\neg p_1\}$ (resp. $\{\neg p_2\}$).

In our example we have two minimal choice sets at line 24, namely $\{\neg p_1, \neg p_3\}$ and $\{\neg p_2, \neg p_3\}$. Therefore we have two Ω_P-minimally abnormal $\mathbf{AL_A}$-models: M_a and M_b where $\mathrm{Ab}^{\mathbf{L_A}}_{\Omega_P}(M_a) = \{\neg p_2, \neg p_3\}$ and $\mathrm{Ab}^{\mathbf{L_A}}_{\Omega_P}(M_b) = \{\neg p_1, \neg p_3\}$. It is easy to see that $M_a \models_{\mathbf{L_A}} p_1, p_4, \neg p_2, \neg p_3$ and $M_b \models_{\mathbf{L_A}} p_2, p_4, \neg p_1, \neg p_3$. M_a (resp. M_b) therefore corresponds to the preferred extension $\{a_1, a_4\}$ (resp. corresponds to the other preferred extension $\{a_2, a_4\}$).

What is important to notice in this example is that we have

$$\Gamma^4_{\mathsf{A}} \vdash_{\mathbf{ALC_P}} p_1$$
$$\Gamma^4_{\mathsf{A}} \vdash_{\mathbf{ALC_P}} p_2$$
$$\Gamma^4_{\mathsf{A}} \vdash_{\mathbf{ALC_P}} \neg p_1$$
$$\Gamma^4_{\mathsf{A}} \vdash_{\mathbf{ALC_P}} \neg p_2$$

However, we also have for instance

$$\Gamma^4_{\mathsf{A}} \nvdash_{\mathbf{ALC_P}} p_1 \wedge p_2 \tag{8.1}$$
$$\Gamma^4_{\mathsf{A}} \nvdash_{\mathbf{ALC_P}} p_1 \wedge \neg p_1 \tag{8.2}$$

Note that lines 10–12 are marked. While p_1 and p_2 are credulously acceptable in their own respect, as both corresponding arguments—a_1 and a_2—are members of preferred extensions, their conjunction is not. Indeed, there is no preferred extension in which both appear simultaneously. This justifies (8.1). As for (8.2), note that this prevents an explosion. Also, as a_1 is a member of one preferred extension, it is credulously acceptable. Still, a_1 is defeated in another preferred extension and therefore the corresponding p_1 is false in the corresponding model M_b. Of course, we do not want this to lead to the credulous acceptance of '$p_1 \wedge \neg p_1$'. This reasonable behaviour of the logics for credulous acceptance causes that they, unlike the logics for skeptical acceptance, are not in the standard format of ALs.[20]

Let us take a look at the other extension types. Admissible, complete, and preferred extensions share the same credulously accepted arguments since an argument is credulously accepted with respect to preferred extensions iff it is in a maximal admissible (resp. complete) extension iff it is in an admissible (resp. complete) extension iff it is credulously accepted with respect to admissible (resp. complete) extensions. Thus, logic $\mathbf{ALC_P}$ also models credulous acceptance for admissible (resp. complete) extensions.

Also for grounded extensions we know that an argument is credulously accepted iff it is a member of the unique grounded extension iff it is skeptically accepted. Thus, logic $\mathbf{AL_G}$ also models credulous acceptance.

For semi-stable extensions we proceed in a similar way as for preferred extensions:

$$\mathbf{ALC_S} = \langle \mathbf{L_C}, [\Omega_{\twoheadrightarrow}, \Omega_S, \Omega_P], \text{[simple strategy, minimal abnormality strategy, normal selections]}\rangle$$

The semantic consequence relation for $\mathbf{ALC_S}$ is defined similar to Definition 8.5.1 (details can be found in Appendix F.4.4).

[20] In particular they lack e.g. the following properties which hold for ALs in the standard format:

(i) fixed-point property – $Cn_{\mathbf{LLL}}(Cn_{\mathbf{AL}}(\Gamma)) = Cn_{\mathbf{AL}}(\Gamma)$,
(ii) closure of the consequence set with respect to the **LLL** – $Cn_{\mathbf{LLL}}(Cn_{\mathbf{AL}}(\Gamma)) = Cn_{\mathbf{AL}}(\Gamma)$,

Theorem 8.5.1. $\left\{\begin{array}{c} (i)\ \mathbf{ALC_P} \\ (ii)\ \mathbf{ALC_P} \\ (iii)\ \mathbf{ALC_S} \end{array}\right\}$ *semantically represents* $\left\{\begin{array}{l} \textit{max. complete} \\ \textit{preferred} \\ \textit{semi-stable} \end{array}\right\}$ *extensions for argumentation frameworks with at most n arguments.*

Theorem 8.5.2. $\left\{\begin{array}{c} (i)\ \mathbf{ALC_P} \\ (ii)\ \mathbf{ALC_P} \\ (iii)\ \mathbf{ALC_P} \\ (iv)\ \mathbf{AL_G} \\ (v)\ \mathbf{ALC_S} \end{array}\right\}$ *syntactically represents* $\left\{\begin{array}{l} \textit{admissible} \\ \textit{complete} \\ \textit{preferred} \\ \textit{grounded} \\ \textit{semi-stable} \end{array}\right\}$ *extensions with respect to credulous acceptance for argumentation frameworks with at most n arguments.*

8.6 Discussion

In this discussion section we will localize our results within the context of logical representations of abstract argumentation and highlight some of its advantages.

There are basically two types of logical approaches to argumentation: a meta-level and an object-level approach (see [39]).

The meta-level approaches are often framed in terms of modal logics (see [39, 40]). Where **L** is such a modal logic, argumentation frameworks are models of **L** and arguments are possible worlds. This way extension types and other key properties of argumentation theory can be expressed in terms of the validity of certain formulas in the models.[21]

Object-level approaches "model argumentation from within" ([39], p. 134). An AF is represented in terms of premises and arguments as atoms. The logic is supposed to derive acceptable arguments with respect to a given extension type. Obviously, this is the way we motivated our logical framework. Caminada and Gabbay [39] present such a (modal) system for grounded extensions. Our system is clearly more unifying in the sense that it is able to represent all standard extensions of Dung's framework. It is to our knowledge the most unifying object-level logical modeling of argumentation in this sense. This is due to the fact that the adaptive logic framework offers easily adjustable and thus powerful mechanisms that make it possible to obtain a generic proof-theoretic framework for all the different extensions, i.e., with the same representation of AFs as premise sets and only slight variations in the abnormalities and strategies.

It is important to point out that our logical (object-level) approach to abstract argumentation has a variety of advantages that go beyond the capabilities of a simple algorithmic framework that produces skeptically resp. credulously accepted arguments. Some were already mentioned before. For instance, the defeasible character

[21] A different meta-level approach is [41]. Here, arguments are presented by propositions and extensions are presented by primitives. The authors in [39] offer, besides the modal systems, also a classical logic meta-level approach (using circumscription).

of our modeling allows for the addition of new elements to an AF A^1 on-the-fly, resulting in A^2 (see especially Sect. 8.3.7 for the technical details). Traditional algorithms have to be applied first to A^1 and then again, from scratch, to A^2. However, our proof theory adjusts to new situations by updating the markings while the argumentation goes on, providing provisional consequences for each step. In this way the dynamics and the rationale of an ongoing argumentation are modeled.

Furthermore, the consequence set of our logics applied to an AF contains more useful information beside the acceptable arguments. Take our Example 8.2.1. As expected, our logic for preferred extensions does derive the only skeptically acceptable p_4. Moreover, the following formulas are derivable: $\varphi_1 = p_1 \vee p_2$, $\varphi_2 = \mathsf{def}\, p_1 \vee \mathsf{def}\, p_2$, $\varphi_3 = \mathsf{def}\, p_3$.[22] The first formula, φ_1, expresses that either a_1 or a_2 is valid in every preferred extension, φ_2 expresses that either a_1 or a_2 is defeated in every preferred extension, and φ_3 expresses that a_3 is defeated in every preferred extension.

Moreover, in some cases the user may take some arguments, X, in A for granted and is thus only interested in, say, preferred extensions that cohere with X, i.e., all preferred extensions E of A such that $E \supseteq X$. In this case the premise set Γ^n_{A} may be enriched by $\{p_i \mid a_i \in X\}$. Our logic for preferred extensions for instance derives inter alia all skeptically acceptable arguments with respect to this subset of all preferred extensions. E.g., for $X = \{a_1\}$ we get the following consequences: $p_4, \mathsf{def}\, p_2, \neg p_2, \mathsf{def}\, p_3, \neg p_3$. This expresses that a_4 is skeptically acceptable in the discussed sense, that a_2 and a_3 are defeated and thus not part of any of the preferred extensions of interest.

Furthermore, we are able to introduce proof theoretic techniques developed for ALs which are interesting for abstract argumentation. For instance, the interpretation of adaptive proofs in terms of argumentation games[23] can be used to model a debate between two parties. Thus, we gain a view on all the standard extension types for abstract argumentation in terms of dialogical games for free on the basis of the presented proof dynamics.

Another advantage of our framework is that it is easily extendable. We give two examples. Often it is not a single argument but rather a bundle of arguments that together attack another argument (see also [16]). Such joint attacks can easily be introduced in our framework by, for instance, allowing for formulas such as $\left(\bigwedge_I p_i\right) \twoheadrightarrow p_j$ expressing that arguments $\{a_i \mid i \in I\}$ attack argument a_j. Our rules (R$\twoheadrightarrow$), (Rad), (R$\bot$), (RCo) and the definition (Def) can be adjusted in a straightforward way. This is presented in the next Chapter. On the other hand, it is interesting to allow for arguments attacking attacks rather than arguments (see e.g., [31]). One such reason is to express a preference for one argument, for instance a_1 attacks $a_2 \rightarrow a_3$ since a_3 is preferred compared to a_2. For our modeling this means that we allow for nested occurrences of the attack operator: $p_1 \twoheadrightarrow (p_2 \twoheadrightarrow p_3)$.[24] It is important to notice that

---

22 Note that neither of the following is derivable: $p_1, p_2, \mathsf{def}\, p_1, \mathsf{def}\, p_2$.

23 See [42] for different variants of argumentation games proposed for ALs, and for instance [24] for the interpretation of abstract argumentation in form of games.

24 We will present the technical details in a future paper.

enhancements to abstract argumentation and their combinations can be modeled by minor natural adjustments to our framework and that the modeling is very intuitive.

We have at the end of Sect. 8.3.2 already pointed out that the external dynamics enabled by the AL representation of abstract argumentation not only opens the possibility to model open, on-going argumentations, but may also be useful for applications in machine learning, belief revision and decision theory.

Moreover, the interpretation of abstract argumentation in terms of ALs offers the possibility to combine the strengths of both frameworks in the modeling of various reasoning forms. Scholars have pointed out the strength of both systems to model for instance defeasible reasoning (see e.g., [17, 28]) or abduction (see e.g., [43, 44]). It remains open for future research to explore these options.

8.7 Conclusion

In this chapter we have presented an adaptive logic characterization of abstract argumentation. Our framework is unifying in the sense that adaptive logic enhancements of one core logic are able to represent all standard extension types for skeptical and credulous acceptance. Skeptically as well as credulously accepted arguments with respect to a given extension type are represented syntactically via the consequence set and semantically in the sense that the models correspond to the extensions. The logics differ only insofar as different strategies and different sets of abnormalities are employed.

Moreover, the presented family of logics is apt for the modeling of open-ended argumentations. The logics are able to derive provisional conclusions at different stages of an ongoing discussion. Thus, they explicate the rationale underlying the acceptance of arguments. This is mirrored also by their dynamic proof theory.

It is interesting to notice that the proof dynamics of these logics can be interpreted in terms of argumentation games (see [42]). Finally, it should be mentioned that the logics can easily be extended by preferences [12], values [13], audiences [14] as well as joint attacks [16] (see Chap. 9).

Acknowledgments I am thankful to Diderik Batens, Joke Meheus and the anonymous referees of the Journal of Logic and Computation for valuable comments which helped to improve the paper.

References

1. Straßer, C., Šešelja, D.: Towards the proof-theoretic unification of Dung's argumentation framework: an adaptive logic approach. J. Logic Comput. **21**, 133–156 (2010)
2. Dung, P.M.: On the acceptability of arguments and its fundamental role in nonmonotonic reasoning and logic programming. In: International Joint Conference on Artificial Intelligence, Proceedings, pp. 852–859 (1993)

3. Dung, P.M.: On the acceptability of arguments and its fundamental role in nonmonotonic reasoning, logic programming and n-person games. Artif. Intell. **77**, 321–358 (1995)
4. Bench-Capon, T.J.M., Dunne, P.E.: Argumentation in artificial intelligence. Artif. Intell. **171**(10–15), 619–641 (2007)
5. Chesevar, C.I., Maguitman, A.G., Loui, R.P.: Logical models of argument. ACM Comput. Surv. **32**(4), 337–383 (2000)
6. Dung, P.M., Son, T.C.: An argumentation-theoretic approach to reasoning with specificity. In: Proceedings of the Fifth International Conference on Principles of Knowledge Representation and Reasoning (KR'96). Morgan Kaufmann Publischers, Inc., Cambridge, MA (1996)
7. Kakas, A.C., Toni, F.: Computing argumentation in logic programming. J. Logic Comput. **9**(4), 515–562 (1999)
8. Dung, P.M.: An argumentation-theoretic foundation for logic programming. J. Logic Program. **22**(2), 151–171 (1995)
9. Atkinson, K., Bench-Capon, T.J.M.: Legal case-based reasoning as practical reasoning. Artif. Intell. Law **13**, 93–131 (2005)
10. Bench-Capon, T.J.M., Atkinson, K., Chorley, A.: Persuasion and value in legal argument. J. Logic Comput. **15**, 1075–1097 (2005)
11. Vreeswijk, G.A.W.: Abstract argumentation systems. Artif. Intell. **90**, 225–279 (1997)
12. Amgoud, L., Cayrol, C.: On the acceptability of arguments in preference-based argumentation. In: Cooper, G.F., Moral, S. (eds.) UAI, pp. 1–7. Morgan Kaufmann (1998)
13. Bench-Capon, T.J.M.: Value based argumentation frameworks. CoRR cs.AI/0207059 (2002)
14. Bench-Capon, T.J.M., Doutre, S., Dunne, P.E.: Audiences in argumentation frameworks. Artif. Intell. **171**(1), 42–71 (2007)
15. Perelman, C., Olbrechts-Tyteca, L.: The New Rhetoric: A Treatise on Argumentation. University of Notre Dame Press, Indiana (1969)
16. Nielsen, S.H., Parsons, S.: A generalization of Dung's abstract framework for argumentation: arguing with sets of attacking arguments. In: Maudet, N., Parsons, S., Rahwan, I. (eds.) Argumentation in Multi-Agent Systems. Lecture Notes in Computer Science, vol. 4766, pp. 54–73. Springer (2006)
17. Bondarenko, A., Dung, P.M., Kowalski, R.A., Toni, F.: An abstract, argumentation-theoretic approach to default reasoning. Artif. Intell. **93**, 63–101 (1997)
18. Coste-Marquis, S., Devred, C., Konieczny, S., Lagasquie-Schiex, M.C., Marquis, P.: On the merging of dung's argumentation systems. Artif. Intell. **171**, 730–753 (2007)
19. Bench-Capon, T.J.M.: Persuasion in practical argument using value-based argumentation frameworks. J. Logic Comput. **13**, 429–448 (2003)
20. Šešelja, D., Straßer, C.: Abstract argumentation and explanation applied to scientific debates. Synthese **190**(12), 2195–2217 (2013)
21. Dung, P.M., Mancarella, P., Toni, F.: Computing ideal sceptical argumentation. Artif. Intell. **171**, 642–674 (2007)
22. Caminada, M.: Semi-stable semantics. In: Computational Models of Argument, pp. 121–132. IOS Press (2006)
23. Dunne, P.E., Bench-Capon, T.J.M.: Two party immediate response disputes: properties and efficiency. Artif. Intell. **149**(2), 221–250 (2003)
24. Vreeswijk, G.A.W., Prakken, H.: Credulous and Sceptical Argument Games for Preferred Semantics. Lecture Notes in Computer Science, vol. 1919, pp. 239–253 (2000)
25. Cayrol, C., Doutre, S., Mengin, J.: On decision problems related to the preferred semantics for argumentation frameworks. J. Logic Comput. **13**(3), 377–403 (2003)
26. Dung, P.M., Mancarella, P., Toni, F.: Argumentation-based proof procedures for credulous and sceptical non-monotonic reasoning. Lecture Notes in Artificial Intelligence, vol. 2408, pp. 289–310 (2002)
27. Batens, D.: Aspects of the dynamics of discussions and logics handling them. Informal Publication. (2001)
28. Batens, D., Straßer, C., Verdée, P.: On the transparency of defeasible logics: equivalent premise sets, equivalence of their extensions, and maximality of the lower limit. Logique at Analyse **52**(207), 281–304 (2009)

29. Verheij, B.: Two approaches to dialectical argumentation: admissible sets and argumentation stages. In: Proceedings of the Biannual International Conference on Formal and Applied Practical Reasoning (FAPR) Workshop, pp. 357–368 (1996)
30. Amgoud, L., Cayrol, C.: A reasoning model based on the production of acceptable arguments. Ann. Math. Artif. Intell. **34**, 197–215 (2002)
31. Modgil, S.: Reasoning about preference in argumentation frameworks. Artif. Intell. **173**(9–10), 901–934 (2009)
32. Možina, M., Žabkar, J., Bratko, I.: Argument based machine learning. Artif. Intell. **171**(10–15), 922–937 (2007)
33. Falappa, M.A., Kern-Isberner, G., Simari, G.R.: Belief revision and argumentation theory. In: Argumentation in Artificial Intelligence, Chap. 17, pp. 341–360. Springer US (2009)
34. Amgoud, L., Vesic, S.: On revising argumentation-based decision systems. In: Sossai, C., Chemello, G. (eds.) Symbolic and Quantitative Approaches to Reasoning with Uncertainty, vol. 5590, Chap. 8, pp. 71–82. Springer, Berlin, (2009)
35. Cayrol, C., de Saint-Cyr, F.D., Lagasquie-Schiex, M.C.: Revision of an argumentation system. In: Brewka, G., Lang, J. (eds.) KR, pp. 124–134. AAAI Press (2008)
36. Batens, D.: A universal logic approach to adaptive logics. Logica Universalis **1**, 221–242 (2007)
37. Batens, D.: Towards the unification of inconsistency handling mechanisms. Logic Log. Philos. **8**, 5–31 (2000). Appeared 2002
38. Batens, D., Meheus, J., Provijn, D.: An adaptive characterization of signed systems for paraconsistent reasoning (to appear)
39. Caminada, M., Gabbay, D.: A logical account of formal argumentation. Studia Logica **93**(2), 109–145 (2009)
40. Grossi, D.: Doing argumentation theory in modal logic. Tech. Rep, ILLC Technical Report (2009)
41. Boella, G., Hulstijn, J., van der Torre, L.W.N.: A logic of abstract argumentation. In: Parsons, S., Maudet, N., Moraitis, P., Rahwan, I. (eds.) Argumentation in Multi-Agent Systems. Lecture Notes in Computer Science, vol. 4049, pp. 29–41. Springer (2005)
42. Batens, D.: Towards a dialogic interpretation of dynamic proofs. In: Dégremont, C., et al. (eds.) Dialogues, Logics and Other Strange Things. Essays in Honour of Shahid Rahman, pp. 27–51. College Publications, London (2009)
43. Neophytos, D., Antonis, K.: Argumentation with abduction. In: Proceedings of the Fourth Panhellenic Symposium of Logic (2003)
44. Meheus, J., Batens, D.: A formal logic for abductive reasoning. Logic J. IGPL **14**, 221–236 (2006)

Verheij, B.: Two approaches to dialectical argumentation: admissible sets and argumentation stages. In: Proceedings of the Biannual International Conference on Formal and Applied Practical Reasoning (FAPR'96), pp. 357–368 (1996)

[illegible]

Chapter 9
Allowing for Joint Attacks

In the last chapter I have announced that the adaptive modeling of Dung's abstract argumentation framework can easily be enhanced. In order to demonstrate this I will focus in this chapter on a useful generalization of Dung's framework proposed by Nielsen and Parsons in [1].

9.1 Motivation

In [1] Nielsen and Parsons argue that Dung's abstract argumentation framework [2] has a major drawback, namely its inability to model *joint attacks* on arguments. An argumentative setting in which more than one argument, say $a_1, \ldots, a_n$, attack another argument b cannot be modeled in a straightforward way since the attack relation is only defined for pairs of arguments. Hence, Nielsen and Parsons argue that "if his [Dung's] framework is expected to be able to model all possible kinds of attack, there is an implicit assumption that the underlying language contains a logical 'and' connective" ([1], p. 55). Namely, in order to model the setting above, we need to presuppose that there is an argument c that represents the conjunction of arguments $a_1, \ldots, a_n$ and $c \rightarrow b$.

But are such joint attacks occurring in argumentations? Nielsen and Parsons offer various examples.

Example 9.1.1. [1] Three persons, A, B, and C have the following argumentation:

a_1 "Peter's Porsche looks purple, so he drives a purple car."
b_1 "Peter drives an Austin Montego, but not a Porsche."
b_2 "Since Peter's car looked purple to me when I saw it yesterday evening, he drives a purple car."
c_1 "According to my catalogue, Austin Montego has never produced purple cars."

[1] This is a slight modification of an examples offered by Nielsen and Parsons in [1], p. 57.

C. Straßer, *Adaptive Logics for Defeasible Reasoning*, Trends in Logic 38,
DOI: 10.1007/978-3-319-00792-2_9,

a_2 "Actually, in 1969 Austin Montego produced a limited special edition for a movie that was purple."

The setting suggests the following attacks: b_1 attacks a_1, c_1 together with b_1 attacks b_2, c_1 together with b_2 attacks b_1, and a_2 attacks c_1.

Nielsen and Parsons have argued that, in order to model for instance the attack of c_1 together with b_1 on b_2 in terms of Dung's argumentation framework, "C would have to repeat a previously stated argument [namely the argument b_1 which was uttered by B], which is not only inelegant, but also forces C to implicitly acknowledge [b_1]" (p. 58). This is problematic since C does not necessarily have to agree with b_1 in order to make her argumentative move, the aim of which may be to show that b_1 and b_2 cannot hold at the same time. This would not be reflected, were C's argumentative move be modeled by a pseudo-argument p stating the conjunction of c_1 and b_1.

Furthermore, the attack of argument a_2 on c_1 would have to be modeled by $a_2 \rightarrow p$ which does not adequately represent the fact that a_2 is directed against c_1.

In order to give an adequate representation of joint attacks Nielsen and Parsons generalize the attack relation so that it does not only hold between single arguments, but between sets of arguments and sets of arguments:

$$\{\mathfrak{a}_1, \ldots, \mathfrak{a}_n\} \rightarrow \{\mathfrak{b}_1, \ldots, \mathfrak{b}_m\}$$

Following Verheij [3] there are two types of interpretation to such joint attacks[2]:

Collective Attack From the validity of $\mathfrak{a}_1, \ldots, \mathfrak{a}_n$ follows that neither of the $\mathfrak{b}_i$'s is valid. That is to say, each of the $\mathfrak{b}_i$'s is considered to be jointly attacked by the set of arguments $\{\mathfrak{a}_1, \ldots, \mathfrak{a}_n\}$.

Indeterministic Attack From the validity of $\mathfrak{a}_1, \ldots, \mathfrak{a}_n$ follows that at least one of the $\mathfrak{b}_i$'s is invalid. That is to say, the fact that all the $\mathfrak{b}_i$'s hold is jointly attacked by the set of arguments $\{\mathfrak{a}_1, \ldots, \mathfrak{a}_n\}$.

Of course, if a set of arguments jointly attacks a singleton then both interpretations are equivalent.

9.2 Complex Argumentation Frameworks

Definition 9.2.1. A *complex argumentation framework (CAF)* is a triple

$$\langle \mathcal{A}, \rightarrow_{\mathsf{c}}, \rightarrow_{\mathsf{i}} \rangle$$

[2] Verheij speaks of collective and indeterministic *defeats*. Since *defeat* is a specific technical term in our logical modeling of Dung's argumentation framework that is going to be generalized in this chapter, I choose the term *attack* instead of *defeat*.

where $\mathcal{A}$ is a finite[3] set of arguments, $\rightarrow_{\mathsf{c}}$ is the collective attack relation where $\rightarrow_{\mathsf{c}} \subseteq (\wp(\mathcal{A}) \setminus \{\emptyset\}) \times (\wp(\mathcal{A}) \setminus \{\emptyset\})$, and $\rightarrow_{\mathsf{i}}$ is the indeterministic attack relation where $\rightarrow_{\mathsf{i}} \subseteq (\wp(\mathcal{A}) \setminus \{\emptyset\}) \times (\wp(\mathcal{A}) \setminus \{\emptyset\})$.

Nielsen and Parsons argue that both notions, collective and indeterministic attacks, can be modeled by restricting the attack relation to hold between sets of arguments and single arguments, $\{\mathfrak{a}_1, \ldots, \mathfrak{a}_n\} \rightarrow \mathfrak{b}$.

First, a collective attack of a set of arguments $\{\mathfrak{a}_1, \ldots, \mathfrak{a}_n\}$ on a set of arguments $\{\mathfrak{b}_1, \ldots, \mathfrak{b}_m\}$ can be reformulated as a series of attacks

$$\begin{gathered}\{\mathfrak{a}_1, \ldots, \mathfrak{a}_n\} \rightarrow \mathfrak{b}_1 \\ \vdots \\ \{\mathfrak{a}_1, \ldots, \mathfrak{a}_n\} \rightarrow \mathfrak{b}_m\end{gathered}$$

Evidently, if the arguments $\mathfrak{a}_1, \ldots, \mathfrak{a}_n$ are valid, each of the $\mathfrak{b}_i$'s is invalid.

An indeterministic attack of a set of arguments $\{\mathfrak{a}_1, \ldots, \mathfrak{a}_n\}$ on a set of arguments $\{\mathfrak{b}_1, \ldots, \mathfrak{b}_m\}$ is reformulated in the following way:

$$\begin{gathered}\{\mathfrak{a}_1, \ldots, \mathfrak{a}_n, \mathfrak{b}_2, \ldots, \mathfrak{b}_m\} \rightarrow \mathfrak{b}_1 \\ \{\mathfrak{a}_1, \ldots, \mathfrak{a}_n, \mathfrak{b}_1, \mathfrak{b}_3, \ldots, \mathfrak{b}_m\} \rightarrow \mathfrak{b}_2 \\ \vdots \\ \{\mathfrak{a}_1, \ldots, \mathfrak{a}_n, \mathfrak{b}_1, \ldots, \mathfrak{b}_{m-1}\} \rightarrow \mathfrak{b}_m\end{gathered}$$

Given an indeterministic attack of $\{\mathfrak{a}_1, \ldots, \mathfrak{a}_n\}$ on $\{\mathfrak{b}_1, \ldots, \mathfrak{b}_m\}$, given the validity of all the $\mathfrak{a}_i$'s and all but one of the $\mathfrak{b}_i$'s, say $\mathfrak{b}_j$, evidently $\mathfrak{b}_j$ cannot be valid.

Accepting Nielsen and Parson's reformulations of collective and indeterministic attacks in terms of attacks of sets of arguments on single arguments each CAF has a canonical representation as an argumentation framework that only features an attack relation between (non-empty) sets of arguments and single arguments.

Definition 9.2.2. A *simplified complex argumentation framework (sCAF)* is a pair $\langle \mathcal{A}, \rightarrow \rangle$ where $\mathcal{A}$ is a finite set of arguments, and $\rightarrow \subseteq (\wp(\mathcal{A}) \setminus \{\emptyset\}) \times \mathcal{A}$ is the attack relation.

Evidently, every CAF can be transformed into a unique sCAF by the procedure indicated above.

Example 9.2.1. Let $\mathsf{A} = \langle \mathcal{A}, \rightarrow_{\mathsf{c}}, \rightarrow_{\mathsf{i}} \rangle$ be a CAF, where $\mathcal{A} = \{a_1, a_2, a_3, a_4, a_5\}$,

$$\begin{aligned}\rightarrow_{\mathsf{c}} &= \{(\{a_1\}, \{a_2\}), (\{a_2\}, \{a_1, a_3\})\}, \text{ and} \\ \rightarrow_{\mathsf{i}} &= \{(\{a_1, a_5\}, \{a_3, a_4\}), (\{a_2, a_5\}, \{a_3, a_4\})\}\end{aligned}$$

[3] As in Chap. 8 I restrict the discussion to the finite case.

The canonical representation of A as a sCAF is given by $\mathsf{A}' = \langle \mathcal{A}, \rightarrow \rangle$ where

$$\rightarrow = \big\{ (\{a_1\}, a_2), (\{a_2\}, a_1), (\{a_2\}, a_3), (\{a_1, a_3, a_5\}, a_4), \\ (\{a_1, a_4, a_5\}, a_3), (\{a_2, a_3, a_5\}, a_4), (\{a_2, a_4, a_5\}, a_3) \big\}$$

Example 9.2.2. Our Example 9.1.1 is for instance represented by the CAF $\mathsf{A} = \langle \{a_1, a_2, b_1, b_2, c_1\}, \rightarrow_{\mathsf{c}}, \rightarrow_{\mathsf{i}} \rangle$ where $\rightarrow_{\mathsf{c}} = \big\{ (\{b_1\}, \{a_1\}), (\{a_2\}, \{c_1\}) \big\}$ and $\rightarrow_{\mathsf{i}} = \big\{ (\{c_1\}, \{b_1, b_2\}) \big\}$.

The corresponding sCAF is $\langle \{a_1, a_2, b_1, b_2, c_1\}, \rightarrow \rangle$ where

$$\rightarrow = \big\{ (\{b_1\}, a_1), (\{a_2\}, c_1), (\{b_1, c_1\}, b_2), (\{b_2, c_1\}, b_1) \big\}$$

9.3 Extension Types

We alter the definition of our standard extension types as follows:

Definition 9.3.1. Given a CAF $\mathsf{A} = \langle \mathcal{A}, \rightarrow_{\mathsf{c}}, \rightarrow_{\mathsf{i}} \rangle$ we define the following notions on the basis of its canonical representation as a sCAF $\langle \mathcal{A}, \rightarrow \rangle$.

(i) A set of arguments A is *attacked* by a set of arguments $B \subseteq \mathcal{A}$ iff there is a $B' \subseteq B$ and a $\mathfrak{a} \in A$ such that $B' \rightarrow \mathfrak{a}$.
(ii) A set of arguments A is *acceptable* with respect to a set of arguments $C \subseteq \mathcal{A}$, iff every set of attackers of A is attacked by C. It is said that C *defends* A.
(iii) A set of arguments $S \subseteq \mathcal{A}$ is *conflict-free* iff S does not attack itself.
(iv) A conflict-free set of arguments $S \subseteq \mathcal{A}$ is *admissible* iff S is acceptable with respect to S.
(v) A set of arguments $S \subseteq \mathcal{A}$ is a *preferred extension* iff it is a maximal (w.r.t. $\subseteq$) admissible set.
(vi) A conflict-free set of arguments $S \subseteq \mathcal{A}$ is a *stable extension* iff it attacks every set of arguments in $\mathcal{A} \setminus S$.
(vii) An admissible set of arguments $S \subseteq \mathcal{A}$ is a *complete extension* iff $\bigcup F(S) = S$, where $F(S) =_{\mathrm{df}} \{C \mid S \text{ defends } C\}$.
(viii) A set of arguments $S \subseteq \mathcal{A}$ is a *grounded extension* iff it is the minimal (w.r.t. $\subseteq$) complete extension.
(ix) A complete extension $S \subseteq \mathcal{A}$ is a *semi-stable extension*[4] iff $S \cup \bigcup S^+$ is maximal (w.r.t. $\subseteq$), where S^+ is the set of sets of arguments that are subsets of $\mathcal{A} \setminus S$ and that are attacked by S.
(x) A set of arguments $S \subseteq \mathcal{A}$ is *credulously accepted* according to preferred [(semi)-stable, complete or grounded] semantics (w.r.t. A) iff it is contained in at least one preferred [(semi)-stable, complete or grounded] extension of A.

[4] Semi-stable semantics were defined by Caminada in [4] for AFs and are equivalent to Verhijs' *admissible stage extensions* in [5].

(xi) A set of arguments $S \subseteq \mathcal{A}$ is *skeptically accepted* according to preferred [(semi)-stable, complete or grounded] semantics (w.r.t. A) iff it is contained in every preferred [(semi)-stable, complete or grounded] extension of A.

Example 9.3.1. Let us return to our CAF A from Example 9.2.1. We have the following preferred and (semi)-stable extensions:

$$\{a_1, a_3, a_5\}, \{a_1, a_4, a_5\}, \{a_2, a_4, a_5\}$$

Note that $\{a_1, a_3, a_4\}$ is not admissible since it does not defend itself from the attacker $\{a_1, a_4, a_5\}$. The grounded extension is $\{a_5\}$.

Example 9.3.2. For our Example 9.1.1 we have the preferred and at the same time (semi-)stable extension $\{a_2, b_1, b_2\}$. Moreover, this is also the grounded extension as can easily be seen.

9.4 Allowing for Joint Attacks in the Lower Limit Logics

In the remainder I will generalize the ALs for Dung's argumentation framework that have been presented in Chap. 8 in such a way that they are able to model CAFs (resp. sCAFs).

In the remainder we allow for conjunctions of atoms on the left hand side of $\twoheadrightarrow$ and we allow for disjunctions and conjunctions of atoms on the right hand side of $\twoheadrightarrow$. Thus, for instance,

$$(p_1 \wedge p_2 \wedge p_3) \twoheadrightarrow (p_4 \vee p_5)$$
$$(p_1 \wedge p_2 \wedge p_3) \twoheadrightarrow (p_4 \wedge p_5)$$

are now well-formed formulas.

Where $\bigwedge\{p_i\} = \bigvee\{p_i\} = p_i$, we define $\mathcal{V}_n^{\wedge}$ to be the set of all conjunctions of atoms in $\mathcal{V}_n = \{p_1, \ldots, p_n\}$, $\mathcal{V}_n^{\wedge} = \{\bigwedge_I p_i \mid I \subseteq \{1, \ldots, n\}, I \neq \emptyset\}$, and $\mathcal{V}_n^{\vee}$ to be the set of all disjunctions of atoms in $\mathcal{V}_n$, $\mathcal{V}_n^{\vee} = \{\bigvee_I p_i \mid I \subseteq \{1, \ldots, n\}, I \neq \emptyset\}$. Let

$$\mathcal{W}_n^{\twoheadrightarrow} := \langle \mathcal{V}_n^{\wedge} \rangle \twoheadrightarrow \langle \mathcal{V}_n^{\wedge} \rangle \mid \langle \mathcal{V}_n^{\wedge} \rangle \twoheadrightarrow \langle \mathcal{V}_n^{\vee} \rangle \mid \bot \twoheadrightarrow \langle \mathcal{V}_n \rangle$$
$$\mathcal{W}_n := \bot \mid \langle \mathcal{V}_n \rangle \mid \langle \mathcal{W}_n^{\twoheadrightarrow} \rangle \mid \neg \langle \mathcal{W}_n \rangle \mid \langle \mathcal{W}_n \rangle \wedge \langle \mathcal{W}_n \rangle \mid \langle \mathcal{W}_n \rangle \vee \langle \mathcal{W}_n \rangle \mid \langle \mathcal{W}_n \rangle \supset \langle \mathcal{W}_n \rangle$$

As noted above, collective attacks $\{\mathfrak{a}_i \mid i \in I\} \rightarrow_{\mathsf{c}} \{\mathfrak{a}_j \mid j \in J\}$ can be expressed by a series of attacks, namely by $\{\mathfrak{a}_i \mid i \in I\} \rightarrow \mathfrak{a}_j$ for all $j \in J$. We will express collective attacks by $\left(\bigwedge_I p_i\right) \twoheadrightarrow \left(\bigvee_J p_j\right)$ and characterize them in accordance with the discussion in Sect. 9.2 as follows:

$$\left(\left(\bigwedge_{i\in I} p_i\right) \twoheadrightarrow \left(\bigvee_{j\in J} p_j\right)\right) \equiv \left(\bigwedge_{j\in J}\left(\left(\bigwedge_{i\in I} p_i\right) \twoheadrightarrow p_j\right)\right) \tag{CA}$$

Indeterministic attacks $\{\mathfrak{a}_i \mid i \in I\} \rightarrow_{\mathrm{i}} \{\mathfrak{a}_j \mid j \in J\}$ will be expressed by $\left(\bigwedge_I p_i\right) \twoheadrightarrow \left(\bigwedge_J p_j\right)$. As noted above, also indeterministic attacks can be translated into a series of attacks of single arguments, namely into $\left\{\mathfrak{a}_i \mid i \in I \cup J \setminus \{j\}\right\} \twoheadrightarrow p_j$ for all $j \in J$. This motivates the following equivalence:

$$\left(\left(\bigwedge_{i\in I} p_i\right) \twoheadrightarrow \left(\bigwedge_{j\in J} p_j\right)\right) \equiv \left(\bigwedge_{j\in J}\left(\left(\bigwedge_{i\in I\cup J\setminus\{j\}} p_i\right) \twoheadrightarrow p_j\right)\right) \tag{IA}$$

Where $\bigwedge_I \beta_i \in \mathcal{V}_n^{\wedge}$, I alter the definition of def in the following way:

$$\mathsf{def} \bigwedge_{i\in I} \beta_i =_{\mathrm{df}} \bigvee_{i\in I} \bigvee_{\alpha\in\mathcal{V}_n^{\wedge}} \left(\alpha \wedge (\alpha \twoheadrightarrow \beta_i)\right)$$

The other rules are defined as in Chap. 8, just that we relax the constraint that only atoms attack atoms:

$$\frac{\alpha \quad \alpha \twoheadrightarrow \beta}{\neg\beta} \text{ where } \alpha \in \mathcal{V}_n^{\wedge}, \beta \in \mathcal{V}_n \tag{R↠C}$$

$$\frac{\alpha \quad \beta \twoheadrightarrow \alpha}{\mathsf{def}\,\beta} \text{ where } \alpha \in \mathcal{V}_n, \beta \in \mathcal{V}_n^{\wedge} \tag{RadC}$$

$$\frac{\bot \not\twoheadrightarrow \alpha}{\neg\alpha} \text{ where } \alpha \in \mathcal{V}_n \tag{R⊥C}$$

These rules are motivated and discussed in Sect. 8.3.1.

For complete extensions we add the following rule:

$$\frac{\bot \twoheadrightarrow \beta \quad \bigwedge_{\alpha\in\mathcal{V}_n^{\wedge}}\left((\alpha \twoheadrightarrow \beta) \supset \mathsf{def}\,\alpha\right)}{\beta} \text{ where} \beta \in \mathcal{V}_n \tag{RCoC}$$

Definition 9.4.1. $\mathbf{CL_A}$ is classical propositional logic enriched by rules (R↠C), (RadC), (R⊥C), (CA), and (IA). $\mathbf{CL_C}$ is $\mathbf{CL_A}$ enriched by (RCoC).

The semantics for these logics and all the proofs for our meta-theory can be found in Appendix 9.

Theorem 9.4.1. *The following holds, where* $\mathbf{X} \in \{\mathbf{A}, \mathbf{C}\}$*:*

$$\vdash_{\mathbf{CL_X}} (\alpha \wedge (\alpha \twoheadrightarrow \beta)) \supset \neg\beta, \text{ where } \alpha \in \mathcal{V}_n^{\wedge} \text{ and } \beta \in \mathcal{V}_n^{\vee} \cup \mathcal{V}_n^{\wedge} \tag{gR↠C}$$

$$\vdash_{\mathbf{CL_X}} (\alpha \wedge (\beta \twoheadrightarrow \alpha)) \supset \mathsf{def}\,\beta, \text{ where } \alpha \in \mathcal{V}_n^{\vee} \cup \mathcal{V}_n^{\wedge} \text{ and } \beta \in \mathcal{V}_n^{\wedge} \tag{gRadC}$$

$$\vdash_{\mathbf{CLc}} \big((\bigwedge_I \bot \twoheadrightarrow \beta_i) \wedge \bigwedge_{\alpha \in \mathcal{V}_n^{\wedge}} ((\alpha \twoheadrightarrow \beta) \supset \mathsf{def}\,\alpha)\big) \supset \beta,$$

$$\textit{where } \beta \in \{\bigwedge_I \beta_i, \bigvee_I \beta_i \mid \beta_i \in \mathcal{V}_n\} \qquad \text{(gRCoC)}$$

Proof. I paradigmatically prove (gR↠C). Suppose $\alpha \wedge (\alpha \twoheadrightarrow \beta)$. Suppose $\beta = \bigvee_I \beta_i \in \mathcal{V}_n^{\vee}$. By (CA), $\bigwedge_I \alpha \twoheadrightarrow \beta_i$. By (R↠C), $\neg\beta_i$ for each $i \in I$. Hence, by aggregation, $\bigwedge_I \neg\beta_i$. Thus, $\neg \bigvee_I \beta_i$.

Suppose now $\beta = \bigwedge_I \beta_i$. Hence, by (IA), $\bigwedge_{i \in I}\big((\alpha \wedge \bigwedge_{j \in I \setminus \{i\}} \beta_j) \twoheadrightarrow \beta_i\big)$. By (R↠C), $\big((\bigwedge_I \beta_i) \wedge \bigwedge_{i\in I}\big((\alpha \wedge \bigwedge_{j \in I\setminus\{i\}} \beta_j) \twoheadrightarrow \beta_i\big) \wedge \alpha\big) \supset (\bigwedge_I \neg\beta_i)$. Hence, $\big((\alpha \wedge \bigwedge_{j\in I\setminus\{i\}} \beta_j) \twoheadrightarrow \beta_i\big) \wedge \alpha\big) \supset \big((\bigwedge_I \neg\beta_i) \vee \neg(\bigwedge_I \beta_i)\big)$. Thus, $\big((\alpha \wedge \bigwedge_{j\in I\setminus\{i\}} \beta_j) \twoheadrightarrow \beta_i\big) \wedge \alpha\big) \supset \neg(\bigwedge_I \beta_i)$. By MP, $\neg \bigwedge_I \beta_i$ and hence, $\neg\beta$.

The proofs for (gRadC) and (gRCoC) are similar and left to the reader. □

Suppose for instance that $(p_1 \wedge p_2) \twoheadrightarrow (p_3 \wedge p_4)$. This represents the indeterministic attack of $\{a_1, a_2\}$ on $\{a_3, a_4\}$. Suppose now, that we have $p_1 \wedge p_2$. By (gR↠C) (and Modus Ponens), we get $\neg(p_3 \wedge p_4)$ and hence $\neg p_3 \vee \neg p_4$. This is as expected since due to the nature of indeterministic attacks at least one of the two arguments a_3 and a_4 is supposed to be invalid.

Similar for the collective attack that is represented by $(p_1 \wedge p_2) \twoheadrightarrow (p_3 \vee p_4)$. Given $p_1 \wedge p_2$ we are able to derive $\neg(p_3 \vee p_4)$ by (gR↠C) (and Modus Ponens). The latter is equivalent to $\neg p_3 \wedge \neg p_4$. Again, this is as desired, since due to the validity of a_1 and a_2 both arguments, a_3 and a_4, are expected to be invalid.

9.5 Representing Complex Argumentation Frameworks as Premise Sets

The representation of a given CAF or a given sCAF in terms of a premise set is again straightforward and very similar to the way AFs were presented in terms of premise sets in Sect. 8.3.2.

Given a CAF $\mathsf{A} = \langle \mathcal{A}, \rightarrow_{\mathsf{c}}, \rightarrow_{\mathsf{i}} \rangle$, where $\mathcal{A} \subseteq \mathcal{A}_n$, we define the premise set Γ_{A}^n as follows:

$$\Gamma_{\mathsf{A}}^n = \Big\{(\bigwedge_I p_i) \twoheadrightarrow (\bigvee_J p_j) \mid \{a_i \mid i \in I\} \rightarrow_{\mathsf{c}} \{a_j \mid j \in J\}\Big\} \cup$$
$$\Big\{(\bigwedge_I p_i) \twoheadrightarrow (\bigwedge_J p_j) \mid \{a_i \mid i \in I\} \rightarrow_{\mathsf{i}} \{a_j \mid j \in J\}\Big\} \cup \big\{\bot \twoheadrightarrow p_i \mid a_i \in \mathcal{A}\big\}$$

Although our primary interest is to give a logical modeling of CAFs, it is technically straightforward to represent sCAFs as well. Given an sCAF $\mathsf{A} = \langle \mathcal{A}, \rightarrow \rangle$, where $\mathcal{A} \subseteq \mathcal{A}_n$, we define Γ_{A}^n as follows:

$$\Gamma_{\mathsf{A}}^n = \Big\{(\bigwedge_I p_i) \twoheadrightarrow p_j \mid \{a_i \mid i \in I\} \rightarrow a_j\Big\} \cup \big\{\bot \twoheadrightarrow p_i \mid a_i \in \mathcal{A}\big\}$$

Given a CAF A and its canonical representation as a sCAF, A′, the corresponding representations as premise sets are equivalent with respect to our logics $\mathbf{CL_A}$ and $\mathbf{CL_C}$.

Theorem 9.5.1. *Where* A *is a CAL and* A′ *is its canonical representation as an sCAL and* $\mathbf{X} \in \{\mathbf{A}, \mathbf{C}\}$, $\Gamma^n_{\mathsf{A}} \vdash_{\mathbf{CL_X}} \Gamma^n_{\mathsf{A}'}$ *and* $\Gamma^n_{\mathsf{A}'} \vdash_{\mathbf{CL_X}} \Gamma^n_{\mathsf{A}}$.

Proof. This is so by the definition of A′ and axioms (CA) and (IA). □

Example 9.5.1. The presentation of our CAF A from Example 9.2.1 as a premise set is as follows:

$$\Gamma^5_{\mathsf{A}} = \{p_1 \twoheadrightarrow p_2, p_2 \twoheadrightarrow (p_1 \vee p_3), (p_1 \wedge p_5) \twoheadrightarrow (p_3 \wedge p_4), (p_2 \wedge p_5) \twoheadrightarrow (p_3 \wedge p_4)\} \\ \cup\{\bot \twoheadrightarrow p_i \mid i \in \{1, \ldots, 5\}\}$$

9.6 Going Adaptive

The adaptive systems for the various extension types are defined analogous to the way they were defined in Sect. 8.4.

Definition 9.6.1. Where $\mathbf{X} \in \{\mathbf{A}, \mathbf{C}\}$ and $\Omega^C_{\twoheadrightarrow} = \{\alpha \twoheadrightarrow \beta \mid \alpha \in \mathcal{V}^{\wedge}_n \cup \{\bot\}, \beta \in \mathcal{V}_n\}$ we define:

$$\begin{aligned}
\mathbf{ACL_X} &= \langle \mathbf{CL_X}, \Omega^C_{\twoheadrightarrow}, \text{simple strategy}\rangle, \\
\mathbf{ACL_P} &= \langle \mathbf{CL_A}, [\Omega^C_{\twoheadrightarrow}, \Omega_P], [\text{simple strategy, minimal abnormality}]\rangle, \\
\mathbf{ACL_G} &= \langle \mathbf{CL_C}, [\Omega^C_{\twoheadrightarrow}, \Omega_G], [\text{simple strategy, simple strategy}]\rangle, \\
\mathbf{ACL_S} &= \langle \mathbf{CL_C}, [\Omega^C_{\twoheadrightarrow}, \Omega_S], [\text{simple strategy, minimal abnormality}]\rangle.
\end{aligned}$$

The sets of abnormalities Ω_P, Ω_G and Ω_S are defined exactly as in Chap. 8.

We have the following representational results:

Theorem 9.6.1. $\left\{\begin{array}{c} (i)\ \mathbf{ACL_A} \\ (ii)\ \mathbf{ACL_C} \\ (iii)\ \mathbf{ACL_P} \\ (iv)\ \mathbf{ACL_G} \\ (v)\ \mathbf{ACL_S} \end{array}\right\}$ *semantically represents*[5] $\left\{\begin{array}{c} \textit{admissible} \\ \textit{complete} \\ \textit{preferred} \\ \textit{grounded} \\ \textit{semi-stable} \end{array}\right\}$ *extensions for CAFs and sCAFs with at most n arguments.*

[5] See Definition 8.3.2.

Corollary 9.6.1. $\left\{\begin{array}{c}(i)\ \mathbf{ACL_A}\\(ii)\ \mathbf{ACL_C}\\(iii)\ \mathbf{ACL_P}\\(iv)\ \mathbf{ACL_G}\\(v)\ \mathbf{ACL_S}\end{array}\right\}$ *syntactically represents*[6] $\left\{\begin{array}{l}\textit{admissible}\\\textit{complete}\\\textit{preferred}\\\textit{grounded}\\\textit{semi-stable}\end{array}\right\}$ *extensions with respect to skeptical acceptance for CAFs and sCAFs with at most n arguments.*

Example 9.6.1. Let me demonstrate for instance the logic $\mathbf{ACL_P}$ by means of our Example 9.1.1. Let us first take a look at the situation before the introduction of argument a_2. Let

$$\mathsf{A}^1 = \langle \{a_1, b_1, b_2, c_1\}, \{(\{b_1\}, \{a_1\})\}, \{(\{c_1\}, \{b_1, b_2\})\}\rangle.$$

Hence, our premise set is:

$$\Gamma^5_{\mathsf{A}^1} = \{p^1_b \twoheadrightarrow p^1_a, p^1_c \twoheadrightarrow (p^1_b \wedge p^2_b)\} \cup \{\bot \twoheadrightarrow p^1_a, \bot \twoheadrightarrow p^1_b, \bot \twoheadrightarrow p^2_b, \bot \twoheadrightarrow p^1_c\},$$

where $p^1_a = p_1$, $p^2_a = p_2$, $p^1_b = p_3$, $p^2_b = p_4$, $p^1_c = p_5$. Let moreover $\Theta_c = \{\varphi \twoheadrightarrow p^1_c \mid \varphi \in \mathcal{V}^{\wedge}_5\}$.

$^3 1$	p^1_a	RC	$\{\neg p^1_a\}$
2	$p^1_b \twoheadrightarrow p^1_a$	PREM	$\emptyset$
3	$\neg p^1_a \vee \neg p^1_b$	2; RU	$\emptyset$
$^3 4$	p^1_b	RC	$\{\neg p^1_b\}$
$^9 5$	p^2_b	RC	$\{\neg p^2_b\}$
6	$p^1_c \twoheadrightarrow (p^1_b \wedge p^2_b)$	PREM	$\emptyset$
7	$\bigwedge_{\varphi \in \mathcal{V}^{\wedge}_5} \varphi \twoheadrightarrow p^1_c$	RC	Θ_c
8	$\neg\,\mathsf{def}\, p^1_c$	7; RU	Θ_c
9	$\neg p^1_b \vee \neg p^2_b$	6,8; RU	Θ_c
10	$p^1_b \vee (p^2_b \wedge p^1_a)$	4; RU	$\{\neg p^1_b\}$
11	$p^1_b \vee (p^2_b \wedge p^1_a)$	1,5; RU	$\{\neg p^1_a, \neg p^2_b\}$
12	p^1_c	RC	$\{\neg p^1_c\}$

At line 1 we conditionally introduce argument a_1. Since b_1 attacks a_1 (line 2), at least one of them is not valid (line 3). At this point of the proof $\neg p^1_a \vee \neg p^1_b$ is a Ω_P-minimal Dab-formula. Hence, line 1 is marked. After introducing the indeterministic attack of c_1 on $\{b_1, b_2\}$ (line 6), we derive that c_1 is un-defeated on the condition that none of the given arguments attacks c_1 (line 8). Since c_1 attacks $\{b_1, b_2\}$ indeterministically, at least one of the two has to be invalid (line 9). Note that neither line 7, nor line 8 nor line 9 can be marked in any extension of the proof since our premise set $\Gamma^5_{\mathsf{A}^1}$ does not feature any attack on c_1. Since $\neg p^1_b \vee \neg p^2_b$ has been derived at an unmarked line on the condition $\Theta_c \subset \Omega_{\twoheadrightarrow}$, lines 4 and 5 are

---

[6] See Definition 8.3.2.

marked. Note that the minimal choice sets at line 11 are $\{\neg p_b^1\}$ and $\{\neg p_a^1, \neg p_b^2\}$. Since $p_b^1 \vee (p_b^2 \wedge p_a^1)$ is derived on both conditions, $\{\neg p_b^1\}$ and $\{\neg p_a^1, \neg p_b^2\}$, lines 10 and 11 are not marked according to the marking definition for minimal abnormality. It is easy to see that $\Gamma^5_{\mathsf{A}^1} \vdash_{\mathbf{ACL_P}} p_c^1, p_b^1 \vee (p_b^2 \wedge p_a^1)$, while $\Gamma^5_{\mathsf{A}^1} \nvdash_{\mathbf{ACL_P}} p_b^1, p_b^2, p_a^1$. This corresponds to the two preferred extensions $\{a_1, b_2, c_1\}$ and $\{b_1, c_1\}$.

Let us now look at the situation when argument a_2 enters the scene. Our premise set is enriched to $\Gamma^5_{\mathsf{A}^2} = \Gamma^5_{\mathsf{A}^1} \cup \{p_a^2 \twoheadrightarrow p_c^1, \bot \twoheadrightarrow p_a^2\}$ where A^2 is the (full) CAF from Example 9.1.1. Let $\Theta_a = \{\varphi \twoheadrightarrow p_a^2 \mid \varphi \in \mathcal{V}_5^{\wedge}\}$ and $\Theta_b = \{\varphi \twoheadrightarrow p_b^1 \mid \varphi \in \mathcal{V}_5^{\wedge}\{p_b^2 \wedge p_c^1\}\}$.

$^{19}1$	p_a^1	RC	$\{\neg p_a^1\}$
2	$p_b^1 \twoheadrightarrow p_a^1$	PREM	$\emptyset$
3	$\neg p_a^1 \vee \neg p_b^1$	2; RU	$\emptyset$
4	p_b^1	RC	$\{\neg p_b^1\}$
5	p_b^2	RC	$\{\neg p_b^2\}$
6	$p_c^1 \twoheadrightarrow (p_b^1 \wedge p_b^2)$	PREM	$\emptyset$
$^{13}7$	$\bigwedge_{\varphi \in \mathcal{V}_5^{\wedge}} \varphi \twoheadrightarrow p_c^1$	RC	Θ_c
$^{13}8$	$\neg\,\mathsf{def}\, p_c^1$	7; RU	Θ_c
$^{13}9$	$\neg p_b^1 \vee \neg p_b^2$	6,8; RU	Θ_c
10	$p_b^1 \vee (p_b^2 \wedge p_a^1)$	4; RU	$\{\neg p_b^1\}$
$^{19}11$	$p_b^1 \vee (p_b^2 \wedge p_a^1)$	1,5; RC	$\{\neg p_a^1, \neg p_b^2\}$
$^{15}12$	p_c^1	RC	$\{\neg p_c^1\}$
13	$p_a^2 \twoheadrightarrow p_c^1$	PREM	$\emptyset$
14	$\neg\,\mathsf{def}\, p_a^2$	RC	Θ_a
15	$\neg p_c^1$	13,14; RU	Θ_a
16	p_a^2	RC	$\{\neg p_a^2\}$
17	$\mathsf{def}\, p_c^1$	13,16; RU	$\{\neg p_a^2\}$
18	$\neg\,\mathsf{def}\, p_b^1$	6,15; RC	$\Theta_a \cup \Theta_b$
19	$\neg p_a^1$	2,18; RU	$\Theta_a \cup \Theta_b$

At line 13 we introduce the new attack of a_2 on c_1. This leads to the marking of lines 7, 8 and 9, since these lines were derived under the condition that c_1 is not attacked by any of the given arguments. Under the condition that none of the given arguments attacks a_2 we derive that a_2 is not defeated at line 14. But that means that c_1 cannot be the case since c_1 is not defended from a_2 (line 15). This leads to the marking of our conditional derivation of c_1 at line 12. At line 16 we conditionally derive a_2. If a_2 is valid, c_1 is defeated since a_2 attacks c_1 (line 17). Under the condition that only $\{b_2, c_1\}$ attacks b_1 we can derive from the fact that c_1 is not valid (line 15), that b_1 is un-defeated (line 18). In this case a_1 cannot be valid, since a_1 would be undefended from the attack of b_1 (line 19). It is easy to check that p_b^1 at line 4, p_b^2 at line 5, and p_a^2 at line 16 are finally derivable. Altogether, $\Gamma^5_{\mathsf{A}^2} \vdash_{\mathbf{ACL_P}} p_b^1, p_b^2, p_a^2, \mathsf{def}\, p_a^1, \mathsf{def}\, p_c^1$. This corresponds to the only preferred extension $\{a_2, b_1, b_2\}$.

9.7 Conclusion

In this chapter I have demonstrated that the AL framework for abstract argumentation of Chap. 8 can easily be enhanced to model a more generic setting in which joint attacks of sets of arguments are allowed.

The results were explicated with respect to the skeptical acceptance type. Note that the credulous acceptance type can be easily represented by employing the normal selections strategy analogous to the logics that were introduced in Sect. 8.5.

References

1. Nielsen, S.H., Parsons, S.: A generalization of Dung's abstract framework for argumentation: arguing with sets of attacking arguments. In: Maudet, N., Parsons, S., Rahwan, I. (eds.) Argumentation in Multi-Agent Systems. Lecture Notes in Computer Science, vol. 4766, pp. 54–73. Springer, Berlin (2006)
2. Dung, P.M.: On the acceptability of arguments and its fundamental role in nonmonotonic reasoning, logic programming and n-person games. Artif. Intell. **77**, 321–358 (1995)
3. Verheij, B.: Rules, reasons, arguments. formal studies of argumentation and defeat. Ph.D. Thesis, Universiteit Maastricht (1996)
4. Caminada, M.: Semi-stable semantics. In: Computational Models of Argument, pp. 121–132. IOS Press (2006)
5. Verheij, B.: Two approaches to dialectical argumentation: admissible sets and argumentation stages. In: Proceedings of the Biannual International Conference on Formal and Applied Practical Reasoning (FAPR) Workshop, pp. 357–368 (1996)

9.7 Conclusion

In this chapter I have demonstrated that the AL framework for abstract argumentation of Chapter [illegible] can easily be enhanced to model a more general setting in which also attacks on sets of arguments are allowed.

[illegible]

References

[illegible]

Part IV
Deontic Logics

Chapter 10
Avoiding Deontic Explosion by Contextually Restricting Modal Inheritance

In order to deal with the possibility of deontic conflicts Lou Goble developed a group of logics (**DPM**) that are characterized by a restriction of the inheritance principle. While they approximate the deductive power of standard deontic logic, they do so only if the user adds certain statements to the premises. By adaptively strengthening the **DPM** logics, this chapter presents logics that overcome this shortcoming. Furthermore, these ALs are capable of modeling the dynamic and defeasible aspect of our normative reasoning by their dynamic proof theory. This way they enable us to have a better insight into the relations between obligations and thus to localize deontic conflicts.

10.1 Introduction

Recent work in deontic logics has shown a growing interest in systems that are able to deal with deontic conflicts (e.g., [2–11]). A deontic conflict between obligations occurs when the obligations cannot be jointly realized. Note that deontic conflicts are not just an abstruse philosophical notion, but that they occur quite commonly in our every-day moral lives (see e.g. [12, 13]). This has for instance to do with the fact that different obligations and behavioral codices may stem from different moral systems and institutions. Sartre famously reports on one of his students who found himself in an unfortunate situation. On the one hand, he felt obliged to support the French army in their resistance against Nazi Germany. On the other hand, however, there was the obligation to stay at home in order to support his ill mother. Obviously, it was not possible for him to fulfill both obligations simultaneously.

In deontic logics a modal operator O is used where $\mathsf{O}A$ expresses the obligation to bring about A. In order to accommodate deontic conflicts systems that tolerate

---

A former version of the content of this chapter has been elaborated in the article "*Avoiding Deontic Explosion by Contextually Restricting Modal Inheritance*" [1]. It is co-authored by Joke Meheus and Mathieu Beirlaen.

C. Straßer, *Adaptive Logics for Defeasible Reasoning*, Trends in Logic 38,
DOI: 10.1007/978-3-319-00792-2_10,

them need to be developed, i.e., systems that do not lead to triviality when applied to conflicting obligations. Formally, conflict-tolerant deontic logics do not validate the following principle of deontic explosion:

$$\vdash (\mathsf{O}A \wedge \mathsf{O}\neg A) \supset \mathsf{O}B \qquad \text{(DEX)}$$

Note that standard deontic logic (**SDL**)[1] is not conflict-tolerant. One reason is that it validates the principle (D), $\vdash \mathsf{O}A \supset \neg\mathsf{O}\neg A$. Thus, $\neg(\mathsf{O}A \wedge \mathsf{O}\neg A)$ is a theorem of **SDL** and all conflicts of the form $\mathsf{O}A \wedge \mathsf{O}\neg A$ lead to explosion.

There are various proposals for conflict-tolerant deontic logics. First, one could restrict or reject the *ex contradictione quodlibet* principle ($(A \wedge \neg A) \vdash B$, for any B), i.e., go paraconsistent (see e.g. [2, 3, 9]). Another approach is to restrict the aggregation principle (if $\mathsf{O}A$ and $\mathsf{O}B$, then $\mathsf{O}(A \wedge B)$) or to abandon it (see [4, 8, 10, 14]).

Yet another approach is given by Goble's logics **DPM** (see [5, 6]). They prevent deontic explosion by restricting the inheritance principle

$$\text{If } \vdash A \supset B, \text{then} \vdash \mathsf{O}A \supset \mathsf{O}B \qquad \text{(RM)}$$

Note that any system that validates full aggregation, full inheritance as well as *ex contradictione quodlibet* leads to explosion when applied to conflicts of the form $\mathsf{O}A \wedge \mathsf{O}\neg A$. By aggregation $\mathsf{O}(A \wedge \neg A)$ is derivable from $\mathsf{O}A \wedge \mathsf{O}\neg A$ and, in view of *ex contradictione quodlibet*, $\mathsf{O}B$ follows from $\mathsf{O}(A \wedge \neg A)$ by (RM).

We will argue in this chapter that, although Goble's **DPM** logics are conflict-tolerant with respect to conflicting obligations, they are suboptimal in other respects. In order to overcome this, we will present adaptive strenghtenings of the **DPM** logics. The idea behind ALs (see [15, 16]) is to interpret a given premise set "as normally as possible". In our case obligations are interpreted as non-conflicting as possible. It will be demonstrated that the adaptive systems are significantly stronger than the **DPM** logics and approximate **SDL**. For instance, for premise sets that are conflict-free, the adaptive versions of the **DPM** systems lead to exactly the same consequence set as **SDL**.

Let us outline the structure of this chapter. In Sect. 10.2, we introduce Goble's **DPM** systems and explain their semantics in Sect. 10.3. We show that the **DPM** systems have some shortcomings in Sect. 10.4. Motivated by the limitations of the **DPM** systems, we suggest ALs as a way to tackle the given problems in Sect. 10.5. In Sects. 10.6 and 10.7 we present the adaptive strengthenings **ADPM.1** and **ADPM.2′**. We list some meta-theoretical properties of the ALs in Sect. 10.8. In Sect. 10.9 we discuss some shortcomings of our logics and relate them to other systems. Finally, in Sect. 10.10 we offer a conclusion and in Sect. 10.11 an outlook. Appendix H features the proofs of our results.

[1] **SDL** is obtained by adding the principle (D), $\vdash \mathsf{O}A \supset \neg\mathsf{O}\neg A$ to the normal modal logic **K**. There are various alternative axiomatizations of **SDL**, cfr. footnote 5.

10.2 Dealing with Deontic Conflicts by Restricting Inheritance

In the remainder we work with a propositional language enriched by a monadic obligation operator O. Where $\mathcal{S}$ is the set of sentential letters, our set of well-formed formulas $\mathcal{W}$ is given by the $\langle \neg, \wedge, \vee, \supset, \mathsf{O} \rangle$-closure of $\mathcal{S}$ with the usual rules for brackets. We define $A \equiv B$ by $(A \supset B) \wedge (B \supset A)$ and the permission operator $\mathsf{P}A$ by $\neg\mathsf{O}\neg A$.

The idea behind Goble's **DPM** systems is to restrict the inheritance principle via permission statements. The full inheritance principle (RM) is replaced by the following 'rule of permitted inheritance'

$$\text{If } \vdash A \supset B, \text{ then } \vdash \mathsf{P}A \supset (\mathsf{O}A \supset \mathsf{O}B) \tag{RPM}$$

What the rule (RPM) comes to is this: if A is obligatory and A entails B, then B is also obligatory *provided that* it is explicitly stated that A is permitted, or what comes to the same, that the obligation to bring about A is unconflicted.[2] Thus, $\mathsf{O}B$ follows neither from $\Gamma_1 = \{\mathsf{O}(A \wedge B), \mathsf{O}\neg(A \wedge B)\}$ nor from $\Gamma_2 = \{\mathsf{O}(A \wedge B)\}$, but it does follow from $\Gamma_3 = \{\mathsf{O}(A \wedge B), \mathsf{P}(A \wedge B)\}$.

Classical propositional logic enriched with the rules (RPM), and

$$\text{If } \vdash A \equiv B, \text{ then } \vdash \mathsf{O}A \equiv \mathsf{O}B \tag{RE}$$

and the axioms

$$\vdash \mathsf{O}\top \tag{N}$$

$$\vdash (\mathsf{O}A \wedge \mathsf{O}B) \supset \mathsf{O}(A \wedge B) \tag{AND}$$

defines the system **DPM.1**. More precisely, **DPM.1** is the least set of formulas containing all classical tautologies of formulas of $\mathcal{W}$, plus all instances of (N) and (AND), that is closed under Modus Ponens, (RE), and (RPM) with '$\vdash$' indicating membership in **DPM.1**. We define in a canonical way, $\vdash_{\mathbf{DPM.1}} A$ iff A is a member of **DPM.1**. Furthermore, where $\Gamma \subseteq \mathcal{W}$, $\Gamma \vdash_{\mathbf{DPM.1}} A$ iff for some $B_1, \ldots, B_n \in \Gamma$ we have $\vdash_{\mathbf{DPM.1}} (B_1 \wedge \cdots \wedge B_n) \supset A$.[3]

Besides **DPM.1** Goble presented another system, **DPM.2**, that also employs the restricted inheritance principle (RPM), but that moreover restricts aggregation. We have motivated the restriction of the inheritance principle and of the aggregation principle as a way to gain conflict-tolerant deontic logics. As will be stated in Theorem 10.2.1, **DPM.1** does not validate (DEX). Hence, since **DPM.1** is already a conflict-tolerant deontic logic, the question arises concerning the use of this further restriction. Let us give some reasons. First, it is not clear that aggregation should hold unrestrictedly. For instance, should aggregation be applied to conflict-

---

[2] In view of the definition of $\mathsf{P}A$, $\mathsf{O}A \wedge \mathsf{P}A$ expresses that the obligation OA is unconflicted.

[3] See also [17] where the authors define consequence relations for rank-1 modal logics in this way and prove strong completeness.

ing obligations? Example: do we want to derive $\mathsf{O}(A \wedge B)$ from $\{\mathsf{O}A,\ \mathsf{O}\neg A,\ \mathsf{O}B\}$? Analogously, should aggregation be applied in cases where it leads to (additional) deontic conflicts? For instance, should one allow that $\mathsf{O}(A \wedge B)$ is derivable from $\{\mathsf{O}A,\ \mathsf{O}B,\ \mathsf{O}\neg(A \wedge B)\}$, thus creating an additional conflict? A negative answer to these questions motivates the restriction of aggregation. Secondly, principle

$$\neg\mathsf{O}\bot \tag{P}$$

has quite some intuitive appeal. Obviously it is impossible to bring about $\bot$. The Kantian principle 'ought implies can' says that we are not obliged to bring about things that are impossible to realize. However, allowing for unrestricted aggregation in the presence of a conflict $\mathsf{O}A \wedge \mathsf{O}\neg A$ leads to $\mathsf{O}(A \wedge \neg A)$ and hence to $\mathsf{O}\bot$. Thus, adding (P) as an axiom to **DPM.1** leads to explosion when applied to deontic conflicts. This can be avoided by restricting aggregation.

Due to the fact that there are various conflict-tolerant deontic logics that *only* restrict (or abandon) aggregation, the reader may still wonder why in **DPM.2** both principles are restricted. One reason is, as Goble pointed out in his [6], that many systems that restrict (but do not abandon) aggregation are not conflict-tolerant *enough*. In his critical analysis he elaborated various refined explosion principles. Besides the very strict notion of deontic explosion that underlies (DEX), namely situations in which all obligations are derivable, there are weaker notions. Take for instance the following explosion principle[4]:

$$\text{If } \nvdash \neg B \text{ then } \mathsf{O}A, \mathsf{O}\neg A \vdash \mathsf{O}B \tag{DEX-1}$$

Another notion of deontic explosion is given if, for every B, $\mathsf{O}B \vee \mathsf{O}\neg B$ is derivable. Semantically speaking this corresponds to the case where all models are such that for every B there is either the obligation to bring about B or there is the obligation to bring about not-B. Although weaker than (DEX) it is equally counter-intuitive that $\mathsf{O}B \vee \mathsf{O}\neg B$ is derivable from $\{\mathsf{O}A \wedge \mathsf{O}\neg A\}$. Hence, we expect from conflict-tolerant deontic logics that they do not validate the following explosion principle:

$$\mathsf{O}A, \mathsf{O}\neg A \vdash \mathsf{O}B \vee \mathsf{O}\neg B \tag{DEX-2}$$

This may be weakened further. Facing a deontic conflict, $\mathsf{O}A \wedge \mathsf{O}\neg A$ as well as an unconflicted obligation $\mathsf{O}C \wedge \neg\mathsf{O}\neg C$, it would be undesired that, for every B, the formula $\mathsf{O}B \vee \mathsf{O}\neg B$ would be derivable. This is expressed as follows:

$$\mathsf{O}A, \mathsf{O}\neg A, \mathsf{O}C, \neg\mathsf{O}\neg C \vdash \mathsf{O}B \vee \mathsf{O}\neg B \tag{DEX-3}$$

[4] We slightly adjusted the criteria (DEX-1)–(DEX-3) (the latter two will be introduced in a moment) offered by Goble since his criteria were formulated in terms of theoremhood while we focus on the consequences of premise sets.

Validating (DEX-3) is counter-intuitive, since for some arbitrary B the conflict $\mathsf{O}A \wedge \mathsf{O}\neg A$ together with the other, otherwise unrelated and unproblematic obligation $\mathsf{O}C$ does not entail that we are either obliged to bring about B or to bring about $\neg B$: $\mathsf{O}B \vee \mathsf{O}\neg B$.

By restricting aggregation along with inheritance the various advantages can be combined. In this way we gain systems that follow the Kantian intuition 'ought implies can', that hence validate (P), and that are strongly conflict-tolerant such that they do not validate any of the explosion principles (DEX), (DEX-1)–(DEX-3).

In order to achieve such a conflict-tolerant logic, Goble uses the following permitted aggregation principle:

$$\vdash (\mathsf{O}A \wedge \mathsf{O}B \wedge \mathsf{P}(A \wedge B)) \supset \mathsf{O}(A \wedge B) \qquad \text{(PAND')}$$

The idea is to apply aggregation to $\mathsf{O}A$ and $\mathsf{O}B$, *provided that* $A \wedge B$ is explicitly permitted. Goble's logic **DPM.2** is defined by (RPM), (RE), (N), (P), and (PAND'). The consequence relation $\vdash_{\mathbf{DPM.2}}$ is defined analogous to $\vdash_{\mathbf{DPM.1}}$.

There is an alternative way of restricting aggregation that offers several advantages over (PAND'), namely:

$$\vdash (\mathsf{O}A \wedge \mathsf{O}B \wedge \mathsf{P}A \wedge \mathsf{P}B) \supset \mathsf{O}(A \wedge B) \qquad \text{(PAND')}$$

Here, the idea is to apply aggregation to $\mathsf{O}A$ and $\mathsf{O}B$ provided that *both* A and B are explicitly permitted.

The logic **DPM.2**$'$ is defined by (RPM), (RE), (N), (P), and (PAND'). The consequence relation $\vdash_{\mathbf{DPM.2'}}$ is defined analogous to $\vdash_{\mathbf{DPM.1}}$.

Henceforth we will use **DPM** as a generic term for **DPM.1**, **DPM.2** and **DPM.2**$'$.

As announced already, **DPM** is sufficiently conflict-tolerant not to validate any of the introduced explosion principles.

Theorem 10.2.1. *Where* $\mathbf{L} \in \{\mathbf{DPM.1}, \mathbf{DPM.2}, \mathbf{DPM.2'}\}$, $\mathbf{L}$ *does not validate any of the explosion principles* (DEX), (DEX-1–DEX-3).

Lou Goble argued in [5] in favor of the following criterion of adequacy for conflict-tolerant deontic logics:

($\star$) A conflict-tolerant deontic logic should be such that the result of adding (D), namely $\vdash \mathsf{O}A \supset \neg\mathsf{O}\neg A$, as an axiom leads to the same consequence relation as **SDL**.[5]

Theorem 10.2.2. *Where* $\alpha \in \{1, 2'\}$, $\mathbf{DPM}.\alpha$ *satisfies* ($\star$).

The logic **DPM.2**$'$ has several advantages as compared to Goble's **DPM.2**. First, the restricted aggregation principle of **DPM.2**$'$, i.e. (PAND'), coheres better with the idea underlying (RPM) than (PAND'). Note that the idea underlying (RPM) was to

[5] Goble axiomatizes **SDL** by adding (D), (N), (RE), (RM), and (AND) to full propositional logic. A consequence relation $\vdash_{\mathbf{SDL}}$ can be defined analogous to $\vdash_{\mathbf{DPM.1}}$. We slightly adjusted Goble's ($\star$) since he is mainly interested in theoremhood, while we focus on consequence relations.

restrict inheritance to those obligations that are explicitly permitted, or what comes to the same, are explicitly unconflicted. This idea is applied to the aggregation principle by (PAND')—aggregation can only be applied if both obligations are explicitly unconflicted. In contrast, the idea underlying (PAND') is to apply aggregation to $\mathsf{O}A$ and $\mathsf{O}B$ provided that the *outcome* of the aggregation, $A \wedge B$, is explicitly permitted. Thus in the case of (PAND'), but not in the case of (PAND'), $\mathsf{O}(A \wedge B)$ is derivable from $\{\mathsf{O}A, \mathsf{O}\neg A, \mathsf{O}B, \mathsf{P}(A \wedge B)\}$.[6] Second, **DPM.2**$'$ satisfies Goble's criterion ($\star$) while **DPM.2** does not. Third, by choosing **DPM.2**$'$ instead of **DPM.2** as a basis for the adaptive strengthenings that are introduced in Sect. 10.5 we will avoid some technical problems.

Before we take a look at some of the shortcomings of the **DPM** logics, let us introduce the semantics.

10.3 The Semantics of DPM

The semantics that we introduce in this section are very similar to Goble's neighborhood semantics for his **DPM** logics in [5]. The only difference is that we employ an actual world. This makes the semantics philosophically more intuitive for our application, since we are not only interested in modeling theoremhood but also in defining a semantic consequence relation.

One of the basic ideas for the neighborhood semantics is that propositions are interpreted in terms of sets of worlds. Moreover, each world has associated with it propositions, i.e., sets of worlds. The idea is that an obligation $\mathsf{O}A$ is true at a world w, in case A is one of its associated propositions.

Let $\wp(X)$ be the power-set of some set X. A neighborhood frame F is a tuple $\langle W, \mathcal{O} \rangle$ where W is a set of points and $\mathcal{O} : W \to \wp(\wp(W))$. We call elements of W worlds. Thus, $\mathcal{O}$ assigns to each world $w \in W$ a set of propositions, i.e., $\mathcal{O}(w) \subseteq \wp(W)$. We write from now on $\mathcal{O}_w$ instead of $\mathcal{O}(w)$. An F-model M on a frame F is a triple $\langle F, v, @ \rangle$ where $@ \in W$ is called the actual world and $v : \mathcal{S} \to \wp(W)$. A propositional atom is mapped by v into the set of worlds in which it is supposed to hold. Where $w \in W$ and $|A|_M =_{\text{df}} \{w \in W \mid M, w \models A\}$, we define:

(M-$\mathcal{P}$) $M, w \models A$ iff $w \in v(A)$, where $A \in \mathcal{S}$
(M-$\mathcal{O}$) $M, w \models \mathsf{O}A$ iff $|A|_M \in \mathcal{O}_w$
(M-$\neg$) $M, w \models \neg A$ iff $M, w \not\models A$
(M-$\vee$) $M, w \models A \vee B$ iff $M, w \models A$ or $M, w \models B$
(M-$\wedge$) $M, w \models A \wedge B$ iff $M, w \models A$ and $M, w \models B$
(M-$\supset$) $M, w \models A \supset B$ iff $M, w \models \neg A \vee B$
(M-$\top$) $M, w \models \top$
(M-$\bot$) $M, w \not\models \bot$

---

[6] A restricted inheritance principle following the intuition of (PAND') would be: If $\vdash A \supset B$, then $\vdash \mathsf{P}B \supset (\mathsf{O}A \supset \mathsf{O}B)$. Inheritance is applied to $\mathsf{O}A$ in order to derive $\mathsf{O}B$ if it does not result in a deontic conflict $\mathsf{O}B \wedge \mathsf{O}\neg B$.

Furthermore, $M \models A$ iff $M, @ \models A$. Where $\Gamma \subseteq \mathcal{W}$, we say that M is an F-model of Γ iff M is an F-model and $M \models A$ for all $A \in \Gamma$.

We define the following requirements on frames $F = \langle W, \mathcal{O} \rangle$. For all $w \in W$:

(a) $W \in \mathcal{O}_w$
(b) If $X \in \mathcal{O}_w$ and $Y \in \mathcal{O}_w$, then $X \cap Y \in \mathcal{O}_w$
(b') If $X \in \mathcal{O}_w$; $Y \in \mathcal{O}_w$; $W \setminus X \notin \mathcal{O}_w$; and $W \setminus Y \notin \mathcal{O}_w$, then $X \cap Y \in \mathcal{O}_w$
(b") If $X \in \mathcal{O}_w$; $Y \in \mathcal{O}_w$; $W \setminus (X \cap Y) \notin \mathcal{O}_w$, then $X \cap Y \in \mathcal{O}_w$
(c) If $X \subseteq Y$; $X \in \mathcal{O}_w$ and $W \setminus X \notin \mathcal{O}_w$, then $Y \in \mathcal{O}_w$
(d) $\emptyset \notin \mathcal{O}_w$

Condition (a) corresponds to (N), (b) corresponds to (AND), (b') corresponds to (PAND'), (b") corresponds to (PAND'), (c) corresponds to (RPM), and (d) corresponds to (P). We call the class of all frames that satisfy (a), (b) and (c) the **DPM.1**-frames, the ones that satisfy (a), (b'), (c) and (d) the **DPM.2**$'$-frames, and the ones that satisfy (a), (b"), (c) and (d) the **DPM.2**-frames.

Let $\Gamma \subseteq \mathcal{W}$. A semantic consequence relation can be defined as follows. Where F is a frame, $\Gamma \Vdash_F A$ iff for all F-models M of Γ, $M \models A$. Moreover, where $\alpha \in \{1, 2, 2'\}$, $\Gamma \Vdash_{\mathbf{DPM}.\alpha} A$ iff $\Gamma \Vdash_F A$ for all **DPM**.α-frames F.

Theorem 10.3.1. *Where* $\alpha \in \{1, 2, 2'\}$ *and* $\Gamma \subseteq \mathcal{W}$*:*

$$\Gamma \vdash_{\mathbf{DPM}.\alpha} A \textit{ iff } \Gamma \Vdash_{\mathbf{DPM}.\alpha} A.$$

10.4 Some Shortcomings of DPM

In order to apply the weakened inheritance principle (resp. also the weakened aggregation principle in the case of **DPM.2** and **DPM.2**$'$) the user has to "manually" add permission statements. For instance, in order to apply the restricted inheritance principle (RPM) to $\mathsf{O}A$ we also need $\mathsf{P}A$. In cases in which $\mathsf{P}A$ is not derivable from the premises by means of **DPM**, the user has to add manually $\mathsf{P}A$ to the premises. This is suboptimal for various reasons.[7]

1. For all interesting cases, determining which permission statements can safely be added to a set of premises (that is, in such a way that no explosion follows) requires *reasoning*. This kind of reasoning falls entirely outside the scope of the **DPM** systems and is therefore left to the user of the **DPM** systems. So, the **DPM** systems are inadequate to fully explicate the reasoning processes that are

[7] In [11], van der Torre and Tan presented a sequential system which, in a first phase, disables the application of (RM) and allows for the application of a restricted aggregation rule. In a second phase, it disables this aggregation rule and allows for the application of (RM). Although this system overcomes this problem, it can do so only by introducing two different O-operators and by requiring that (RM) is never applied before the restricted aggregation rule. As the authors themselves admit, this is rather strange from an intuitive point of view (see also [6], pp. 470–471).

needed to apply the **DPM** systems in a sensible way (that is, in a way that modal inheritance is applied "as much as possible").

2. The fact that permissions have to be added manually is especially problematic in cases where the relationship between the premises is interwoven. For instance in complicated setups it might not be obvious at all that $\mathsf{O}A \wedge \mathsf{O}\neg A$ is derivable. However, suppose that in this case the user naïvely added $\mathsf{P}A$ to the premises in order to apply (RPM) to $\mathsf{O}A$. Since $\mathsf{P}A$ is equivalent to $\neg\mathsf{O}\neg A$ the user caused in this way an explosion.
3. For most premise sets the **DPM** systems are rather weak. Recall that in order to achieve the deductive strength of **SDL** we had to add (D) to the axiomatization of **DPM.1** (resp. **DPM.2**′). Suppose we accept **SDL** as the normative standard for the modeling of non-conflicting obligations. It would be desirable then that conflict-tolerant logics apply all rules of **SDL** to non-conflicting obligations. For instance, given a premise set $\Gamma \subseteq \mathcal{W}$ that is conflict-free, we expect a deontic logic to lead to a consequence set that is the same as that of **SDL** without the need of strengthening the premise set manually by adding instances of (D), or by adding premises. Thus, if the premises are conflict-free, the logic should apply all the rules of **SDL** unrestrictedly.

The discussion above motivates the following strengthening of Goble's requirement ($\star$).[8]

($\star\star$) For **SDL**-consistent premise sets[9] $\Gamma \subseteq \mathcal{W}$ a conflict-tolerant deontic logic should lead to the same consequence set as **SDL**.

Note that neither of the introduced **DPM** logics satisfies ($\star\star$). We will in the next section adaptively strengthen **DPM.1** and **DPM.2**′ so that they satisfy criterion ($\star\star$).

As should have become evident from the discussion above, there is a certain trade-off for monotonic deontic logics such as **DPM**. In order to offer conflict-tolerance certain **SDL**-principles such as (D) and inheritance have to be restricted or abandoned. In return, this weakens the logics even in cases in which it would be unproblematic to apply the principles in question.

The adaptive strengthenings introduced in the next sections will overcome this trade-off. On the one hand, applying the logic does not involve any user interference. On the other hand, by interpreting a premise set as non-conflicting as possible, principles such as inheritance, (D) and aggregation in the case of the adaptive strengthening of **DPM.2**′ will be applied as much as possible.

---

[8] To stay in line with Goble's ($\star$), we formulate the strengthened requirement in terms of **SDL**. If one's preferred logic is different from **SDL**, the requirement may easily be adapted. The basic idea is that, where one's preferred deontic logic (for conflict-free premise sets) is **L**, one expects from a conflict-tolerant deontic logic on the basis of **L** that it leads to the same consequence set as **L** for all **L**-consistent premise sets. This is exactly what ALs allow for.

[9] Γ is **SDL**-consistent iff $\Gamma \nvdash_{\mathbf{SDL}} \bot$.

10.5 Adaptive Logics

As discussed before, the main feature of ALs is that they interpret a given premise set "as normally as possible". The standard of normality depends on the application. For instance, there are inconsistency-ALs that, while allowing for classical inconsistencies, interpret a given premise set as consistently as possible (see e.g. [16, 18]). In our case the idea is to interpret premise sets "as non-conflicting as possible". We will in the following give a precise meaning to this vague notion.

We will in this chapter use **DPM.1** or **DPM.2**′ as lower limit logics.[10] Let henceforth $\alpha \in \{1, 2'\}$.

We will focus in this chapter on the minimal abnormality strategy. It gives rise to a stronger consequence relation since it allows for the derivation of $\mathsf{O}(A \vee B)$ from two incompatible obligations $\mathsf{O}A$ and $\mathsf{O}B$ while reliability does not. We will give an example of this later on (see Example 10.7.1). Of course, the corresponding logics with reliability can easily be devised, which is left to the reader.

Since our aim is to interpret a given premise set Γ as non-conflicting as possible, we define our abnormalities to be deontic conflicts, $\mathsf{O}A \wedge \mathsf{O}\neg A$.

Now we have the three elements needed to define the ALs that are explicated in this chapter:

1. the lower limit logic is either **DPM.1** or **DPM.2**′,
2. the logical form of the abnormalities is $\mathsf{O}A \wedge \mathsf{O}\neg A$ and Ω is the set of all abnormalities,[11]
3. the strategy is minimal abnormality.

We dub these systems **ADPM.1** and **ADPM.2**′.

We define the semantics as usual: a certain well-defined set of **DPM**.α-models of a given premise set Γ is selected, namely the ones that are "minimally abnormal". The models are selected with respect to their abnormal part, i.e. the abnormalities they verify: where M is a **DPM**.α-model, $\text{Ab}(M) = \{A \in \Omega \mid M \models A\}$. For a given logic **L**, we write $\mathcal{M}_{\mathbf{L}}(\Gamma)$ for the set of **L**-models verifying all members of Γ.

Definition 10.5.1. *A* **DPM**.α*-model* $M \in \mathcal{M}_{\mathbf{DPM}.\alpha}(\Gamma)$ *is* minimally abnormal *iff there is no* **DPM**.α*-model* $M' \in \mathcal{M}_{\mathbf{DPM}.\alpha}(\Gamma)$ *such that* $Ab(M') \subset Ab(M)$. *We write* $\mathcal{M}_{\mathbf{ADPM}.\alpha}(\Gamma)$ *for all the minimal abnormal* **DPM**.α*-models of* Γ.

$\Gamma \Vdash_{\mathbf{ADPM}.\alpha} A$ *(A is an* **ADPM**.α-semantic consequence of Γ*) iff A is verified by all* $M \in \mathcal{M}_{\mathbf{ADPM}.\alpha}(\Gamma)$.

Hence all the selected models are such that they validate a minimal amount of conflicts (in the set-theoretical sense). This justifies our claim that the ALs interpret the premises as non-conflicting as possible.

Let us proceed with the syntactic counter-part to the semantic selection. It is realized by dynamic adaptive proofs. While all the rules of **DPM**.α are valid, a

[10] We will discuss the case of **DPM.2** being a lower limit logic shortly in Sect. 10.7.

[11] In Section 10.6 in the context of the lower limit logic **DPM.1** we will have to make a slight adjustment when defining Ω. However, the main idea stays the same.

key feature of adaptive proofs is that they allow for certain additional rules to be applied *conditionally*. In our case the idea is to apply the inheritance principle and principle (D) conditionally. Recall that **DPM**.α, in order to avoid deontic explosions, only validates a restricted version of (RM). This led to the problem (discussed in Sect. 10.4) that in many cases the user needs to add manually permission statements. Note that the following is valid in **DPM**.α:

$$\text{If } \vdash_{\mathbf{DPM}.\alpha} A \supset B, \text{ then } \quad \mathsf{O}A \vdash_{\mathbf{DPM}.\alpha} \mathsf{O}B \vee (\mathsf{O}A \wedge \mathsf{O}\neg A).$$

The underlying idea of the restriction is that inheritance is applicable to $\mathsf{O}A$ if there is no deontic conflict concerning $\mathsf{O}A$. In the AL we make use of this: the inheritance principle is applied to $\mathsf{O}A$ on the condition $\{\mathsf{O}A \wedge \mathsf{O}\neg A\}$. That is to say, on the condition that there is no reason to suppose that there is a deontic conflict concerning $\mathsf{O}A$.

This is still very vague, but we will make it more precise in a moment. Suppose $\mathsf{O}A$ is one of the premises. A fragment of an adaptive proof may look as follows:

1	$\mathsf{O}A$	PREM	$\emptyset$
2	$\mathsf{O}(A \vee B)$	1; RC	$\{\mathsf{O}A \wedge \mathsf{O}\neg A\}$

Line 2 contains a conditional application of the inheritance rule to $\mathsf{O}A$. This is indicated by the generic rule RC and by a fourth column in which the conditions of the respective lines are contained. The condition of line 1 is empty since it is the result of a premise introduction. The condition of line 2 is $\{\mathsf{O}A \wedge \mathsf{O}\neg A\}$. Now suppose we are able to derive the following disjunction of abnormalities from the given premises at a line l:

$$l\ (\mathsf{O}A \wedge \mathsf{O}\neg A) \vee (\mathsf{O}C \wedge \mathsf{O}\neg C) \quad \ldots;\ \text{RU}\ \emptyset$$

By the generic rule RU we indicate the (unconditional) applications of the **DPM**.α rules. Note that the disjunction of abnormalities that has been derived at line l also features the condition of line 2. This gives us reasons to suspect that $\mathsf{O}A$ may after all be part of a deontic conflict. In this case line 2 is marked according to the marking definition. Formulas that are the second element of marked lines are not considered to be derived at that stage. Note however that markings may come and go. Assume for instance that $\mathsf{O}C \wedge \mathsf{O}\neg C$ has been derived at a later stage of the proof. In this case there is no reason anymore to suspect that $\mathsf{O}A$ is part of a deontic conflict and the conditional application of inheritance at line 2 can again be considered as valid. The marking will thus be defined on basis of the minimal disjunctions of abnormalities that have been derived at a given stage of the proof.

Besides the lower limit logic there is also an upper limit logic for each AL. It is the strengthening **UDPM**.α of **DPM**.α that trivializes abnormalities. It is defined by, $\Gamma \vdash_{\mathbf{UDPM}.\alpha} A$ iff $\Gamma \cup \{\neg B \mid B \in \Omega\} \vdash_{\mathbf{DPM}.\alpha} A$.

The next result follows immediately by Theorem 2.6.6 (recall that there premise sets that are consistent with respect to the upper limit logic are called "normal").

Corollary 10.5.1. *For* **UDPM**.α*-consistent premise sets* Γ, **ADPM**.α *leads to the same consequence set as* **UDPM**.α.

10.6 The Adaptive Logic ADPM.1

In this section we introduce a concrete adaptive system on the basis of Lou Goble's **DPM.1**. **ADPM.1** is defined by the triple

$$\langle \textbf{DPM.1}, \Omega, \text{minimal abnormality}\rangle,$$

where $\Omega = \{!A \mid A \in \mathcal{W}\}$ is the set that contains for each propositional formula A and its set of subformulas Sub(A) (including A itself) the formula $!A =_{\text{df}} \bigvee_{B \in \text{Sub}(A)} \mathsf{O}B \wedge \mathsf{O}\neg B$. This is a slight complication to our discussion in Sect. 10.5: a formula A behaves abnormal if it gives rise to a deontic conflict, or (and this is the complication) if any of its subformulas gives rise to a deontic conflict. We will comment some more on this in Example 10.6.1. below.

This logic will allow us to apply the inheritance principle to $\mathsf{O}A$ on the condition $\{!A\}$. Moreover, also (D) may be applied conditionally. Note that the following is valid:

$$\mathsf{O}A \vdash_{\textbf{DPM}} \mathsf{P}A \vee (\mathsf{O}A \wedge \mathsf{O}\neg A)$$
$$\text{and hence } \mathsf{O}A \vdash_{\textbf{DPM}} \mathsf{P}A \vee !A$$

This allows to derive $\mathsf{P}A$ from $\mathsf{O}A$ on the condition $\{!A\}$ by the generic rule RC.

Example 10.6.1. Let us take a look at a first concrete example of a proof in **ADPM.1** (where A, B, and C are atoms):

1 $\mathsf{O}A$	PREM	$\emptyset$
2 $\mathsf{O}\neg A$	PREM	$\emptyset$
3 $\mathsf{O}(B \wedge C)$	PREM	$\emptyset$
4 $\mathsf{O}B$	3; RC	$\{!(B \wedge C)\}$
5 $\mathsf{P}B$	4; RC	$\{!(B \wedge C), !B\}$

Note that $\mathsf{O}A, \mathsf{O}\neg A, \mathsf{O}(B \wedge C) \nvdash_{\textbf{DPM.1}} \mathsf{O}B$ and $\mathsf{O}A, \mathsf{O}\neg A, \mathsf{O}(B \wedge C) \nvdash_{\textbf{DPM.1}} \mathsf{P}B$. In Sect. 10.4 we have pointed out that this is suboptimal. A deontic logic should apply the rules of **SDL** to non-conflicting parts of the premise set and there should be no need for the user to add permission statements. In order to derive $\mathsf{O}B$ from $\mathsf{O}(B \wedge C)$ the user of **DPM.1** would have to manually add $\mathsf{P}(B \wedge C)$ to the premises. Moreover, there is no way of deriving $\mathsf{P}B$ from the given premises in **DPM.1**.

In contrast, the adaptive logic **ADPM.1** applies inheritance conditionally to $\mathsf{O}(B \wedge C)$ in order to derive $\mathsf{O}B$ at line 4. Moreover, (D) is applied conditionally to $\mathsf{O}B$ in order to derive $\mathsf{P}B$ at line 5. Note that $\mathsf{O}(B \wedge C)$ and $\mathsf{O}B$ are unrelated to the deontic conflict $\mathsf{O}A \wedge \mathsf{O}\neg A$, and hence, as discussed in Sect. 10.4, inheritance and (D) should be applicable to them. It can easily be seen that lines 4 and 5 are finally derived.

We still need to explain why we need the complication in the definition of the set of abnormalities Ω. Indeed, in the original formulation [1], Ω was defined by $\{\mathsf{O}A \wedge \mathsf{O}\neg A \mid A \in \mathcal{W}\}$. However, Lou Goble [19] pointed out that the resulting system is suboptimal. Take our example. Were we to use $\mathsf{O}A \wedge \mathsf{O}\neg A$ as our form of abnormalities we would derive $\mathsf{O}B$ at line 4 on the condition $\{\mathsf{O}(B \wedge C) \wedge \mathsf{O}\neg(B \wedge C)\}$. However, we have e.g., $\{\mathsf{O}A, \mathsf{O}\neg A, \mathsf{O}(B \wedge C)\} \vdash_{\mathbf{DPM.1}} \bigvee \Delta$ where $\Delta = \big\{\mathsf{O}(B \wedge C) \wedge \mathsf{O}\neg(B \wedge C), \mathsf{O}(A \wedge C) \wedge \mathsf{O}\neg(A \wedge C), \mathsf{O}(\neg A \wedge C) \wedge \mathsf{O}\neg(\neg A \wedge C)\big\}$. Deriving $\bigvee \Delta$ on the empty condition leads to the marking of line 4. Moreover, for all $\Delta' \subset \Delta$, $\{\mathsf{O}A, \mathsf{O}\neg A, \mathsf{O}(B \wedge C)\} \nvdash_{\mathbf{DPM.1}} \bigvee \Delta'$ which means that $\bigvee \Delta$ is minimal and in view of this it is not difficult to see that indeed $\mathsf{O}B$ would not follow.

Because of this complication, Goble proposed the solution to redefine Ω as presented above. This avoids the problem from the previous paragraph for the simple reason that we can derive $!A$ and hence also $!(A \wedge C)$ and $!(\neg A \wedge C)$. As a consequence, there is no way to involve $!(B \wedge C)$ in a minimal disjunction of abnormalities.

In the remainder we will frequently make use of the following fact.

Fact 10.6.1. *Where $\alpha \in \{1, 2'\}$, the following holds in* **DPM.**α[12]*:*

$$\textit{If } \vdash_{\mathbf{CL}} \neg(A \wedge B), \textit{ then } \mathsf{O}A, \mathsf{O}B \vdash_{\mathbf{DPM}.\alpha} !A \vee !B.$$

The fact shows that a pair of conflicting obligations gives rise to a disjunction of abnormalities in the **DPM** logics, even in case they are not directly conflicting (such as $\mathsf{O}A$ and $\mathsf{O}\neg A$).

Example 10.6.2. In the example above it was demonstrated that **ADPM.1** delivers all the consequences of a premise ($\mathsf{O}(B \wedge C)$) that is consistent with the other two directly conflicting premises ($\mathsf{O}A, \mathsf{O}\neg A$). In the following example we have two conflicting premises: $\mathsf{O}\neg A$ and $\mathsf{O}(A \wedge B)$. Note that $\mathsf{O}B$ is –via unrestricted inheritance– a consequence of $\mathsf{O}(A \wedge B)$ and that it is not conflicting with the premise $\mathsf{O}\neg A$. This poses the question whether $\mathsf{O}B$ is a desired consequence of these premises. We now demonstrate that **ADPM.1** takes a skeptical stance and does not lead to the consequence $\mathsf{O}B$. Let us take a look at an adaptive proof (where A and B are atoms):

1	$\mathsf{O}\neg A$	PREM	$\emptyset$
2	$\mathsf{O}(A \wedge B)$	PREM	$\emptyset$
^{4}3	$\mathsf{O}B$	2; RC	$\{!(A \wedge B)\}$
4	$!(A \wedge B)$	1,2; RU	$\emptyset$

[12] This can easily be seen: from $\vdash_{\mathbf{CL}} \neg(A \wedge B)$, $\mathsf{O}A$ and restricted inheritance we get $\mathsf{O}\neg B \vee !A$. The latter together with $\mathsf{O}B$ results in $!A \vee !B$.

Although $\mathsf{O}B$ is conditionally derivable at line 3, it is marked since the condition of line 3 is part of the minimal Dab-consequence derived at line 4. Note that by Fact 10.6.1 we get $!A \vee !(A \wedge B)$ and hence also $!(A \wedge B)$ from lines 1 and 2.

What motivates the skeptical stance according to which $\mathsf{O}B$ is not a consequence of the given premises is that there may be certain causal connections between A and B. This is for instance the case where A stands for taking a train to London and B for buying the ticket for it. If there is also the obligation not to take the train, then despite the fact that there is no direct conflict between $\neg A$ and B in terms of being jointly realizable, there is nevertheless clearly a tension. As long as we don't have reasons to prefer the obligation to take the train there is no reason to derive the obligation to buy the ticket.

Example 10.6.3. Let us alter the example slightly by adding $\mathsf{O}A$ to the premises. This way we make the conflict between $\mathsf{O}A$ and $\mathsf{O}\neg A$ explicit in the premises.

	1 $\mathsf{O}A$	PREM	$\emptyset$
	2 $\mathsf{O}\neg A$	PREM	$\emptyset$
	3 $\mathsf{O}(A \wedge B)$	PREM	$\emptyset$
^{6}	4 $\mathsf{O}B$	3; RC	$\{!(A \wedge B)\}$
	5 $!A$	1,2; RU	$\emptyset$
	6 $!(A \wedge B)$	5; RU	$\emptyset$

Evidently, we can again derive $!(A \wedge B)$ on the empty condition and mark line 4. Hence, **ADPM.1** is also in this example coherent to its skeptical stance that was illustrated already in the example above.

Example 10.6.4. Examples 10.6.2 and 10.6.3 have demonstrated that in some cases we cannot derive non-conflicting conjuncts of conflicting premises by means of **ADPM.1**. This opens the question whether $\mathsf{O}B$ is derivable if it is part of each conflicting premise. Where A and B are atoms, we have the following proof:

	1 $\mathsf{O}(A \wedge B)$	PREM	$\emptyset$
	2 $\mathsf{O}(\neg A \wedge B)$	PREM	$\emptyset$
^{7}	3 $\mathsf{O}A$	1; RC	$\{!(A \wedge B)\}$
^{7}	4 $\mathsf{O}\neg A$	2; RC	$\{!(\neg A \wedge B)\}$
	5 $\mathsf{O}B$	1; RC	$\{!(A \wedge B)\}$
	6 $\mathsf{O}B$	2; RC	$\{!(\neg A \wedge B)\}$
	7 $!(A \wedge B) \vee !(\neg A \wedge B)$	1,2; RU	$\emptyset$

As the proof indicates, $\mathsf{O}B$ is derivable from $\mathsf{O}(A \wedge B)$ and $\mathsf{O}(\neg A \wedge B)$. The reason why lines 5 and 6 are unmarked is that for each minimal choice set $\{!(A \wedge B)\}$ and $\{!(\neg A \wedge B)\}$ at this stage, $\mathsf{O}B$ is derivable on a condition which has an empty intersection with it. As can easily be checked, $\mathsf{O}B$ is finally derived at line 6.

Example 10.6.5. Let us generalize the example above. We take a look at the premises $\mathsf{O}(A \wedge B)$ and $\mathsf{O}(\neg A \wedge C)$. Although $B \vee C$ is not a conjunct of any of the two premises, it is a consequence of both $A \wedge B$ and $\neg A \wedge C$. In coherence with the example above we expect thus $\mathsf{O}(B \vee C)$ to be an **ADPM.1**-consequence. On the other hand we expect $\mathsf{O}B$ and $\mathsf{O}C$ not to be a consequence. Where A, B, and C are atoms, we have the following proof:

	$1\ \mathsf{O}(A \wedge B)$	PREM	$\emptyset$
	$2\ \mathsf{O}(\neg A \wedge C)$	PREM	$\emptyset$
^{7}	$3\ \mathsf{O}B$	1; RC	$\{!(A \wedge B)\}$
	$4\ \mathsf{O}(B \vee C)$	1; RU	$\{!(A \wedge B)\}$
^{7}	$5\ \mathsf{O}C$	2; RC	$\{!(\neg A \wedge C)\}$
	$6\ \mathsf{O}(B \vee C)$	2; RU	$\{!(\neg A \wedge C)\}$
	$7\ !(A \wedge B) \vee !(\neg A \wedge C)$	1,2; RU	$\emptyset$
	$8\ !\top$	1,2; RU	$\emptyset$

Indeed, the conditional derivation of $\mathsf{O}B$ and $\mathsf{O}C$ is marked, while $\mathsf{O}(B \vee C)$ is derivable since it is derived on both conditions, $\{!(A \wedge B)\}$ and $\{!(\neg A \wedge C)\}$. Since the minimal choice sets in this example are $\{!\top, !(A \wedge B)\}$ and $\{!\top, !(\neg A \wedge C)\}$ this ensures that $\mathsf{O}(B \vee C)$ is finally derived.[13]

Example 10.6.6. The following example of an **ADPM.1**-proof features a more complex setup. Where A, B, C, and D are atoms, we have:

	$1\ \mathsf{O}A$	PREM	$\emptyset$
	$2\ \mathsf{O}B$	PREM	$\emptyset$
	$3\ \mathsf{O}(C \vee D)$	PREM	$\emptyset$
	$4\ \mathsf{O}\neg(A \wedge C)$	PREM	$\emptyset$
	$5\ \mathsf{O}\neg(B \wedge D)$	PREM	$\emptyset$
	$6\ \mathsf{O}(A \wedge \neg C)$	1,4; RU	$\emptyset$
	$7\ \mathsf{O}(B \wedge \neg D)$	2,5; RU	$\emptyset$
^{16}	$8\ \mathsf{P}A \wedge \mathsf{P}B \wedge \mathsf{P}(C \vee D)$	1,2,3; RC	$\{!A, !B, !(C \vee D)\}$
	$9\ \mathsf{O}(A \vee \neg C) \vee !A$	1; RU	$\emptyset$
	$10\ \mathsf{O}(B \vee \neg D) \vee !B$	2; RU	$\emptyset$
	$11\ \mathsf{O}\neg C \vee !A$	4,9;RU	$\emptyset$
^{16}	$12\ \mathsf{O}\neg C$	11; RC	$\{!A\}$
	$13\ \mathsf{O}\neg D \vee !B$	5,10; RU	$\emptyset$
^{16}	$14\ \mathsf{O}\neg D$	13; RC	$\{!B\}$
	$15\ \mathsf{O}\neg(C \vee D) \vee !A \vee !B$	11,13; RU	$\emptyset$
	$16\ !A \vee !B \vee !(C \vee D)$	3,15; RU	$\emptyset$
	$17\ \neg\mathsf{P}A \vee \neg\mathsf{P}B \vee \neg\mathsf{P}(C \vee D)$	15; RU	$\emptyset$

Without engaging in more advanced reasoning processes it is for a user of **DPM.1** in no way clear which permission statements may be added without causing deontic conflicts or explosion. We have pointed out this problem in Sect. 10.4 (point 2). Would (s)he, for instance, add $\mathsf{P}A$, $\mathsf{P}B$, and $\mathsf{P}(C \vee D)$ it would cause explosion, since via **DPM.1**, $\neg\mathsf{P}A \vee \neg\mathsf{P}B \vee \neg\mathsf{P}(C \vee D)$ is derivable at line 17. Note that in the adaptive proof line 8 is marked in view of line 16 and does therefore not cause any harm. Thus, the AL identifies given deontic conflicts and blocks undesired consequences

[13] Note that line 8 follows by (N) and applying aggregation to lines 1 and 2.

from them. For instance, the unintuitive derivations of $\mathsf{O}\neg C$ and $\mathsf{O}\neg D$ at lines 12 and 14 are marked.

10.7 The Adaptive Logic ADPM.2′

We have already pointed out various advantages of our **DPM.2**′ logic over Goble's **DPM.2**. Besides these points **DPM.2**′ is also more apt as a lower limit logic.

One idea to define an AL on the basis of **DPM.2** would be in terms of the triple $\langle$**DPM.2**, Ω, minimal abnormality$\rangle$. However, this has a severe shortcoming. Suppose our premises are $\Gamma = \{\mathsf{O}A, \mathsf{O}B\}$. Since these premises do not give rise to any deontic conflicts we expect from the logic to apply aggregation to them. However, $\Gamma \nvdash_{\mathbf{ADPM.2}} \mathsf{O}(A \wedge B)$. Since **DPM.2** also restricts aggregation beside inheritance it is desirable that the logic is able to apply aggregation conditionally in a similar way as **ADPM.1** applies inheritance and (D) conditionally. The way aggregation is restricted in **DPM.2**, namely by (PAND'), does not allow for utilizing the same set of abnormalities as for **ADPM.1**. A way around this problem is to define the abnormalities in a different way, for instance by $\Omega' = \{\mathsf{O}A \wedge \mathsf{O}B \wedge \mathsf{O}\neg(A \wedge B) \mid A, B \in \mathcal{W}\}$. This would allow to apply the aggregation principle conditionally. However, this logic is very weak. For instance given two incompatible obligations $\mathsf{O}A, \mathsf{O}B$, where $\mathsf{O}\neg(A \wedge B)$ expresses their incompatibility, we are not able to derive $\mathsf{O}(A \vee B)$. The upshot is that, with **DPM.2** as the lower limit logic, we were not able to define a set of abnormalities in such a way that the resulting AL is sufficiently strong and aggregation is conditionally applicable.

The situation is different if we employ **DPM.2**′ as lower limit logic. In that case, we can use the set of abnormalities $\Omega = \{\mathsf{O}A \wedge \mathsf{O}\neg A \mid A \in \mathcal{W}\}$. This contributes to a more unifying adaptive framework. Let **ADPM.2**′ be defined by the triple $\langle$**DPM.2**′, Ω, minimal abnormality$\rangle$. Note that in **DPM.2**′ the following is a consequence of (PAND'):

$$\vdash_{\mathbf{DPM.2'}} \big(\mathsf{O}A \wedge \mathsf{O}B\big) \supset \big(\mathsf{O}(A \wedge B) \vee \big((\mathsf{O}A \wedge \mathsf{O}\neg A) \vee (\mathsf{O}B \wedge \mathsf{O}\neg B)\big)\big)$$

This makes it possible to apply aggregation to $\mathsf{O}A \wedge \mathsf{O}B$ on the condition $\{\mathsf{O}A \wedge \mathsf{O}\neg A, \mathsf{O}B \wedge \mathsf{O}\neg B\}$ by the rule RC. Similarly, inheritance and (D) are applied to $\mathsf{O}A$ on the condition $\{!A\}$ where now $!A$ abbreviates $\mathsf{O}A \wedge \mathsf{O}\neg A$.

Note that we simplified the form of the abnormalities in **ADPM.2**′ compared to **ADPM.1**. The reason why there was a need for the complication in **ADPM.1** was the presence of the unrestricted aggregation principle (AND). For instance, it caused that the unconflicted formula $\mathsf{O}(B \wedge C)$ became involved in a minimal disjunction of normative conflicts in our Example 10.6.1. It is easy to see that the only minimal disjunctions of deontic conflicts derivable via **DPM.2**′ from $\{\mathsf{O}A, \mathsf{O}\neg A, \mathsf{O}(B \wedge C)\}$ are $\mathsf{O}A' \wedge \mathsf{O}\neg A'$ for all A' for which $\vdash_{\mathbf{DPM.2'}} A' \equiv A$. This immediately implies that we can finally derive $\mathsf{O}B$ from $\mathsf{O}(B \wedge C)$ on the condition $\{\mathsf{O}(B \wedge C) \wedge \mathsf{O}\neg(B \wedge C)\}$ in **ADPM.2**′.

In order to demonstrate the modus operandi of **ADPM.2**′ we take a look at a concrete example.

Example 10.7.1. Let us come back to Sartre's unfortunate student. On the one hand, he has the obligation to stay with his sick mother ($\mathsf{O}M$). However, on the other hand he has the obligation to fight at the front against Nazi Germany ($\mathsf{O}F$). Due to the lack of alethic modalities we have to find a way to model the fact that both obligations are not simultaneously realizable. This can be done by $\mathsf{O}\neg(M \wedge F)$. We add another premise which is independent of the first two: as a good citizen he is obliged to pay taxes and to vote, $\mathsf{O}(T \wedge V)$.[14]

	1 $\mathsf{O}M$	PREM	$\emptyset$
	2 $\mathsf{O}F$	PREM	$\emptyset$
	3 $\mathsf{O}\neg(M \wedge F)$	PREM	$\emptyset$
	4 $\mathsf{O}(T \wedge V)$	PREM	$\emptyset$
	5 $\mathsf{O}T$	4; RC	$\{!(T \wedge V)\}$
7	6 $\mathsf{O}(M \wedge F)$	1,2; RC	$\{!M, !F\}$
	7 $!(M \wedge F) \vee !M \vee !F$	1,2,3; RU	$\emptyset$
	8 $\mathsf{O}(M \vee F)$	1; RC	$\{!M\}$
	9 $\mathsf{O}(M \vee F)$	2; RC	$\{!F\}$

As discussed in Sect. 10.4, we expect that all rules of **SDL** can be applied to non-conflicting obligations such as $\mathsf{O}(T \wedge V)$. Indeed, after having introduced the premises at lines 1–4, inheritance is applied to $\mathsf{O}(T \wedge V)$ in order to derive $\mathsf{O}T$, the student's duty to pay taxes. Furthermore, the application of aggregation to $\mathsf{O}M$ and $\mathsf{O}F$ at line 6 gets marked. This accords with the fact that (PAND') is a rule that isolates deontic conflicts, i.e. it is not applicable to conflicting obligations.

ADPM.2′ realizes our design requirements. On the one hand, it blocks rules from being applied to conflicting obligations (such as aggregation at line 6). On the other hand it allows for applications of rules to non-conflicting premises (such as inheritance at line 5 to $\mathsf{O}(T \wedge V)$) without requiring the manual addition of permission statements.

Moreover, note that it is desired to derive $\mathsf{O}(M \vee F)$, the student's obligation to either stay with his mother or to fight the Nazis. Indeed, since $\mathsf{O}(M \vee F)$ is derivable on both conditions, $\{\mathsf{O}M \wedge \mathsf{O}\neg M\}$ and $\{\mathsf{O}F \wedge \mathsf{O}\neg F\}$, lines 8 and 9 are not marked. Similarly for instance $\mathsf{O}(T \wedge (M \vee F))$ and $\mathsf{O}(V \wedge (M \vee F))$ are derivable.

In contrast, using one of the **DPM** logics, the user would have to add manually $\mathsf{P}(T \wedge V)$ to the premises in order to derive $\mathsf{O}T$. Furthermore, it is unclear how to derive $\mathsf{O}(M \vee F)$ in **DPM**. Only by adding either $\mathsf{P}M$ or $\mathsf{P}F$ permitted inheritance is applicable to $\mathsf{O}M$ or resp. $\mathsf{O}F$. However, there is no reason to prefer $\mathsf{P}M$ over $\mathsf{P}F$ or vice versa. Moreover, would the user add both of them, it may lead to explosion. Consider the case where also $\mathsf{P}\neg(M \wedge F)$ is a premise. It can be argued that, since M and F cannot be mutually realized an agent is allowed either to bring about not-M or to bring about not-F. However, in view of the final derivability of the formula at

[14] The proofs of $\mathsf{O}(M \vee F)$ and $\mathsf{O}T$ for **ADPM.1** are left to the reader.

line 7, this leads to the derivability of $(\mathsf{O}M \wedge \mathsf{O}\neg M) \vee (\mathsf{O}F \wedge \mathsf{O}\neg F)$. In case the user would have added both, $\mathsf{P}M$ and $\mathsf{P}F$, explosion would result.

Example 10.7.2. **ADPM.2**$'$ leads to similar results in examples 10.6.1, 10.6.2, 10.6.4 and 10.6.5 as **ADPM.1**:

$$\mathsf{O}A, \mathsf{O}\neg A, \mathsf{O}(B \wedge C) \nvdash_{\mathbf{ADPM.2}'} \mathsf{O}B, \mathsf{P}B$$
$$\mathsf{O}\neg A, \mathsf{O}(A \wedge B) \nvdash_{\mathbf{ADPM.2}'} \mathsf{O}B$$
$$\mathsf{O}(A \wedge B), \mathsf{O}(\neg A \wedge B) \nvdash_{\mathbf{ADPM.2}'} \mathsf{O}B$$
$$\mathsf{O}(A \wedge B), \mathsf{O}(\neg A \wedge C) \nvdash_{\mathbf{ADPM.2}'} \mathsf{O}B, \mathsf{O}C$$
$$\mathsf{O}(A \wedge B), \mathsf{O}(\neg A \wedge C) \nvdash_{\mathbf{ADPM.2}'} \mathsf{O}(B \vee C)$$

Example 10.7.3. A difference between **ADPM.2**$'$ and **ADPM.1** is in the handling of the premise set consisting of $\mathsf{O}A$, $\mathsf{O}\neg A$ and $\mathsf{O}(A \wedge B)$ (where A and B are atoms):

1 $\mathsf{O}A$	PREM	$\emptyset$
2 $\mathsf{O}\neg A$	PREM	$\emptyset$
3 $\mathsf{O}(A \wedge B)$	PREM	$\emptyset$
4 $\mathsf{O}A \wedge \mathsf{O}\neg A$	1,2; RU	$\emptyset$
5 $\mathsf{O}B$	3; RC	$\{!(A \wedge B)\}$

In **ADPM.1** an obligation behaves abnormal as soon as one of its subformulas is involved in a deontic conflict. As a result and due to the conflict in A, it is not possible to finally derive $\mathsf{O}B$ from $\mathsf{O}(A \wedge B)$ in **ADPM.1** in Example 10.6.3. The situation is different in **ADPM.2**$'$ where an obligation is only considered abnormal if it is itself involved in a conflict. There is no way to involve $!(A \wedge B)$ in a minimal Dab-consequence in this example. Responsible is also the restricted aggregation principle. Given (AND), we can involve $!(A \wedge B)$ in a minimal Dab-consequence since $\{\mathsf{O}A, \mathsf{O}\neg A, \mathsf{O}(A \wedge B)\} \vdash_{\mathbf{DPM.1}} (\mathsf{O}(A \wedge B) \wedge \mathsf{O}\neg(A \wedge B)) \vee (\mathsf{O}(\neg A \wedge B) \wedge \mathsf{O}\neg(\neg A \wedge B))$. To see this suppose the first disjunct $!(A \wedge B)$ does not hold. Then, inheritance is applicable to $\mathsf{O}(A \wedge B)$ to derive for instance (1) $\mathsf{O}(A \vee \neg B)$ and (2) $\mathsf{O}(A \vee B)$. Via (AND), (2) and $\mathsf{O}\neg A$ we get $\mathsf{O}(\neg A \wedge B)$. By (1) we get the second disjunct $!(\neg A \wedge B)$. In absence of (AND) this derivation is blocked in **DPM.2**$'$ since we would also need for instance $\neg\mathsf{O}A$ in order to apply (PAND') to (2) and $\mathsf{O}\neg A$ to arrive at $\mathsf{O}(\neg A \wedge B)$. However, we have $\mathsf{O}A$ in our premises.

10.8 Some Meta-Theoretical Properties

This section features some meta-theoretic properties of the introduced logics. The meta-theory of ALs in standard format equips our systems immediately with soundness and completeness. As desired, the adaptive strengthenings are conflict-tolerant.

Theorem 10.8.1. *Where $\alpha \in \{1, 2'\}$, none of Goble's explosion principles (DEX), (DEX-1)–(DEX-3) is valid in* **ADPM**.α.

Theorem 10.8.2. *Where $\alpha \in \{1, 2'\}$, the upper limit logic of* **ADPM**.α *is* **SDL**.

Corollary 10.8.3. *Where $\alpha \in \{1, 2'\}$,* **ADPM**.α *satisfies* ($\star$).

We introduced another, in a sense more demanding criterion, ($\star\star$). For **SDL**-consistent premise sets the given logics should have the same derivative power as **SDL**. This criterion is not fulfilled by the **DPM** logics. However, as the following corollary shows, it is fulfilled by our adaptive strengthenings. The corollary is a direct consequence of Corollary 10.5.1 and Theorem 10.8.2.

Corollary 10.8.4. *Where $\alpha \in \{1, 2'\}$,* **ADPM**.α *satisfies* ($\star\star$).

10.9 Shortcomings and Related Work

The starting point and main motivation of this paper was to present a technically straightforward way to equip Goble's **DPM** systems with a defeasible mechanism that strengthens the logics in a non-monotonic fashion. In this section, we point out some parallels and differences between the **ADPM**.α-systems resulting from our enterprise on the one hand, and existing approaches for dealing with deontic conflicts on the other hand.

The behavior of **ADPM.1** explicated in examples 10.6.1–10.6.5 may suggest that the consequence set of **ADPM.1** can be characterized in the following way: $\mathsf{O}B$ is a consequence of a set of obligations $\{\mathsf{O}A \mid A \in \Gamma\}$ (where Γ is a set of propositions) iff A is (classically) derivable from each maximal consistent subset[15] of Γ. This characterization is similar to approaches making use of maximally consistent subsets with a 'skeptical' operator for obligation, e.g. [7, 20] or the "full meet constraint output" of Input/Output logics in [21].[16] However, there are more complex settings which show that **ADPM.1** is too weak.[17] It is a question for future research to find ways to strengthen this logic so that the rationale above is realized and a defeasible

[15] A subset Γ' of Γ is maximally consistent iff Γ' is consistent and for every consistent $\Gamma'' \subseteq \Gamma$, if $\Gamma' \subseteq \Gamma''$ then $\Gamma'' = \Gamma'$.

[16] In the deontic logic literature on maximally consistent subsets, skeptical operators are contrasted with more 'credulous' operators. The latter allow one to derive the obligation $\mathsf{O}A$ as soon as A is derivable from some maximal consistent subset of Γ (see e.g. [20, 22] or the "full join constraint output" of Input/Output logics in [21]) while the former allow to derive $\mathsf{O}A$ iff A is derivable in all maximal consistent subsets of Γ.

[17] One such counterexample is given by the premise set $\{\mathsf{O}A, \mathsf{O}B, \mathsf{O}\neg(A \wedge B), \mathsf{O}(\neg(A \wedge B) \wedge D), \mathsf{O}(C_1 \wedge C_2)\}$. Here it is not possible to derive $\mathsf{O}C_1$ by means of **ADPM.1** or **ADPM.2**$'$ since $!(C_1 \wedge C_2)$ is involved in a minimal Dab-consequence. Note that we can derive $\mathsf{O}C_1$ on the condition that $\mathsf{O}(C_1 \wedge C_2)$ is not conflicted. By applying aggregation we get $\mathsf{O}(C_1 \wedge (A \wedge B))$ (in **ADPM.2**$'$ we also need the condition that both $\mathsf{O}A$ and $\mathsf{O}B$ are not conflicted). Similarly, we can derive $\mathsf{O}\neg(C_1 \wedge (A \wedge B))$ from $\mathsf{O}(\neg(A \wedge B) \wedge D)$ by means of inheritance on the condition that the latter

proof theory is obtained for the well-known conflict-tolerant approach in terms of maximal consistent subsets.

We have mentioned that for examples 10.6.1, 10.6.2, 10.6.4, and 10.6.5 **ADPM.2′** has a similar behavior as **ADPM.1**. This may suggest that the consequences of **ADPM.2′** can be characterized as follows: $\mathsf{O}B$ is a consequence of a set of obligations $\{\mathsf{O}A \mid A \in \Gamma\}$ iff A is in Γ or it is (classically) derivable from each maximal consistent subset of Γ. However, again there are examples in which the consequence set of **ADPM.2′** is too weak (see Footnote 17).

Moreover, we have already mentioned the rather peculiar asymmetry of **ADPM.2′** in the handling of the premise sets $\Gamma_1 = \{\mathsf{O}A, \mathsf{O}\neg A, \mathsf{O}(A \wedge B)\}$ and $\Gamma_2 = \{\mathsf{O}\neg A, \mathsf{O}(A \wedge B)\}$ (see Example 10.7.3). While in the first case $\mathsf{O}B$ is derivable, in the second case it isn't. One way to make the behavior of **ADPM.2′** more coherent is to restrict inheritance in the lower limit logic more severely by also blocking it from obligations that display "sub-conflicts", i.e. conflicts concerning subformulas of two obligations. In our example inheritance should be blocked from $\mathsf{O}(A \wedge B)$ since this conflicts with $\mathsf{O}\neg A$. In the adaptive system this would be mirrored by a type of abnormality that is not only restricted to direct conflicts but also considers conflicts such as the one between $\mathsf{O}(A \wedge B)$ and $\mathsf{O}\neg A$. A deontic adaptive logic that takes into account such conflicts can be found in [23].

The treatment of Γ_1 points also to another direction. Namely to design an adaptive logic for which $\mathsf{O}B$ is derivable from a set of obligations $\{\mathsf{O}A \mid A \in \Gamma\}$ iff A is (classically) derivable from some maximal consistent subset of Γ and A is consistent with all maximal consistent subsets of Γ.[18] One way to realize this is to adjust the restricted inheritance principle in the following way. Instead of applying inheritance if the antecedent is consistent the idea is to apply inheritance if the consequent is consistent:

$$\text{If } \vdash A \supset B, \text{ then } \mathsf{O}A \supset (\mathsf{P}B \supset \mathsf{O}B).$$

We will investigate in these systems in the future.

Recently various other adaptive deontic systems have been presented. In [8] and the forthcoming [23] strictly non-aggregative deontic logics were adaptively strengthened. These systems are based on the lower limit logic $\mathbf{P_eSDL_a}$ from [14]. Two modalities are used: O_e and O_a where the former e.g. expresses obligations that stem from some authority while the latter expresses obligations upon which all authorities agree. These systems validate full inheritance but have no aggregation for the O_e-operator. In [27] these systems are formulated in the prioritized format for adaptive logics from [28]. This way they are able to deal with preferences among obligations.

is not conflicted. Altogether this shows that $!(C_1 \wedge C_2)$, $!(\neg(A \wedge B) \wedge D)$ and $!(C_1 \wedge (A \wedge B))$ are involved in a Dab-consequence. It is not difficult to see that they are indeed involved in a minimal Dab-consequence.

[18] In a non-deontic setting, this approach was taken up by Rescher & Manor in their definition of an 'argued' consequence relation [24]. Adaptive characterizations of Rescher & Manor's consequence relations are given in [25, 26].

The abnormalities used in all of the systems mentioned in this paragraph are of a more complex nature since they also take into account conflicts between subformulas of obligations. One advantage of the systems presented in the present paper is that by means of **ADPM.1** and **ADPM.2**$'$ we are able to derive $\mathsf{O}(A \vee B)$ from $\mathsf{O}A$, $\mathsf{O}B$ and $\mathsf{O}\neg(A \wedge B)$. This 'disjunctive solution' has been defended for instance in [7, 29]. Moreover, deontic adaptive logics have been devised that are based on paraconsistent modal logics [2, 30]. This approach has also been applied to a multi-agent setting [31, 32].

10.10 Conclusion

In this chapter we introduced the adaptive strengthenings **ADPM.1** and **ADPM.2**$'$ of Goble's conflict-tolerant logic **DPM.1** and of our variant **DPM.2**$'$ of Goble's **DPM.2**. We have demonstrated various advantages of the adaptive systems.

ADPM.α (where henceforth $\alpha \in \{1, 2'\}$) is significantly stronger than **DPM.**α. It is not just the case that adding (D) to the logic leads to equivalent systems to **SDL**. For any **SDL**-consistent premise set, **ADPM.**α proves to be equivalent to **SDL**. Moreover, **ADPM.**α applies restricted inheritance "as much as possible". In contrast, in Goble's system permission statements have to be added by the user in order to apply the inheritance principle. The needed permission statements are generated in the adaptive systems automatically. This brings us to another point.

The adaptive systems **ADPM.**α have the design virtue that the logics model all the reasoning for the user. In contrast, in **DPM.**α the user not just has to interfere in order to derive as much consequences as possible (by adding permission statements). Moreover, finding out which permission statements are harmless and may be added to a given premise set involves advanced reasoning by the user. This is especially the case for complicated setups.

The meta-theory of ALs in standard format is well-established. Many key-features do not have to be proven (anymore) for the **ADPM.**α logics since they have been shown to be valid for all ALs in standard format. For instance, the completeness and soundness of **ADPM.**α follows immediately from the completeness and soundness of **DPM.**α.

The dynamic adaptive proofs mirror actual reasoning processes. While the insight in a given premise set grows, some lines of the proof may get marked, others unmarked due to the fact that their conditions turn out to be (not) trustworthy. Furthermore, the adaptive proofs are able to deal with new information in the form of new premises and thus to handle the defeasibility that comes with it.

10.11 Outlook

While we presented in this chapter ways to apply inheritance, (D), and aggregation conditionally by ALs, the approach can be fruitfully applied in other contexts of deontic logics. Let me give some examples.

10.11.1 Obligation-Permission Conflicts: Towards More Conflict-Tolerant Deontic Logics

It is remarkable that scholars have been mainly focusing on only one type of deontic conflicts, namely conflicts between obligations (henceforth, OO-conflicts) such as for instance between $\mathsf{O}A$ and $\mathsf{O}\neg A$. However, another type of conflict has been neglected, namely conflicts between obligations and permissions (henceforth, OP-conflicts) such as for instance between $\mathsf{O}A$ and $\mathsf{P}\neg A$. In this respect the numerous proposed conflict-tolerant deontic logics (henceforth, CTDLs) are not "fully" conflict tolerant since they validate the principle

$$(\mathsf{O}A \wedge \mathsf{P}\neg A) \supset \mathsf{O}B \tag{OP-EX}$$

This creates an unnecessary and moreover unmotivated asymmetry in the modeling offered by these logics.

This is a severe defect of CTDLs. However, it is not a fatal one. In [33] Mathieu Beirlaen and I propose a transformation procedure that turns a given monotonic CDTL **L** into a monotonic CTDL $\mathbf{L}\star$ that is also conflict-tolerant concerning OP-conflicts. The mainspring of our transformation is to give up the interdefinability between obligations and permission that is characteristic for most deontic logics. Usually, either $\mathsf{P}A$ is defined by $\neg\mathsf{O}\neg A$ or there is an axiom which enforces the equivalence. In the transformed logic $\mathbf{L}\star$ only one direction of the equivalence holds, i.e. $\mathbf{L}\star$ does not validate

$$\mathsf{P}A \supset \neg\mathsf{O}\neg A \tag{DfP1}$$

The logics gained by our transformation procedure offer many advantages. First and foremost, acknowledging the so far neglected conflict type does not necessitate the development of entirely new deontic systems, but many of the already proposed systems can be transformed in a way that preserves their strengths and modeling features while making them at the same time more conflict-tolerant.

Giving up (DfP1) weakens the transformed logic compared to **L**. In the first instance this sounds like a shortcoming. However, this makes it possible to add a very intuitive principle that is usually not validated by CTDLs, namely axiom (D) that allows to derive from the obligation to bring about A, the permission to bring about A. Given the intuitive appeal of (D) we argue in [33] that this is indeed a very good compensation for losing (DfP1). We have paradigmatically demonstrated this transformation, inter alia, by means of Goble's **DPM**.α (where $\alpha \in \{1, 2'\}$). The transformed systems are denoted by **DPM**.$\alpha\star$.

One of the drawbacks of our transformations in [33] is that we had to give up on (DfP1) in order to gain conflict-tolerance with respect to OP-conflicts. The problem is that, as some may argue, in non-conflicting cases (DfP1) has quite some intuitive appeal. Hence, what is needed is a logic that applies (DfP1) whenever it is unproblematic. Given that the aim is to interpret a given set of norms as non-conflicting and as coherent as possible, it is desirable to derive $\neg\mathsf{O}\neg A$ from $\mathsf{P}A$ whenever there is no conflicting situation with respect to $\mathsf{P}A$. Similarly, whenever there is no conflicting situation with respect to a $\mathsf{O}B$ it is desirable to derive $\neg\mathsf{P}\neg B$ from $\mathsf{O}B$.

In [33] we demonstrate that ALs offer a very intuitive solution to this problem. We define ALs, inter alia on the basis of **DPM**.$\alpha\star$, that interpret a given premise set Γ as non-conflicting as possible. That is to say, Γ is interpreted in such a way that as few deontic conflicts are the case as is coherent with Γ. As a consequence the ALs validate (DfP1) as much as possible. Thus, we are able to define deontic logics that are OO- and OP-conflict tolerant, that validate the intuitive principle (D), and that interpret obligations and permissions in the usual way according to the equivalence $\mathsf{P}A \equiv \neg\mathsf{O}\neg A$ as much as is coherent with a given premise set.

The way this is technically realized is deeply inspired by the adaptive systems presented in this Chapter. Instead of the lower limit logic **DPM**.α there we use **DPM**.$\alpha\star$, instead of using OO-conflicts $\mathsf{O}A \wedge \mathsf{O}\neg A$ as abnormalities we use there OP-conflicts $\mathsf{O}A \wedge \mathsf{P}\neg A$ as abnormalities. The strategy is again minimal abnormality. Classical logic and (D) warrants

$$\begin{aligned} \mathsf{O}A &\vdash_{\mathbf{DPM}.\alpha\star} \neg\mathsf{P}\neg A \vee (\mathsf{O}A \wedge \mathsf{P}\neg A) \\ \mathsf{P}A &\vdash_{\mathbf{DPM}.\alpha\star} \neg\mathsf{O}\neg A \vee (\mathsf{O}\neg A \wedge \mathsf{P}A) \\ \mathsf{O}A &\vdash_{\mathbf{DPM}.\alpha\star} \neg\mathsf{O}\neg A \vee (\mathsf{O}A \wedge \mathsf{P}\neg A) \\ \text{If } \vdash_{\mathbf{DPM}.\alpha\star} A \supset B\text{, then }\quad \mathsf{O}A &\vdash_{\mathbf{DPM}.\alpha\star} \mathsf{O}B \vee (\mathsf{O}\neg A \wedge \mathsf{P}A) \end{aligned}$$

Hence, in the adaptive systems we are able to derive $\neg\mathsf{P}\neg A$ from $\mathsf{O}A$ conditionally, resp. to derive $\neg\mathsf{O}\neg A$ from $\mathsf{P}A$ conditionally, resp. to derive $\neg\mathsf{O}\neg A$ from $\mathsf{O}A$ conditionally, resp. to apply inheritance conditionally. Semantically, selected are the **DPM**.$\alpha\star$ models that validate the least OP-conflicts. Since by (D) every OO-conflict $\mathsf{O}A \wedge \mathsf{O}\neg A$ gives rise to two OP-conflicts, $\mathsf{O}A \wedge \mathsf{P}\neg A$ and $\mathsf{P}A \wedge \mathsf{O}\neg A$, the selected models are also models that validate the least OO-conflicts.

10.11.2 Going Conditional

While we have so far focused on monadic deontic operators, the techniques and ideas introduced in this chapter can also be applied in conditional deontic logics.

Lou Goble presented in [5] conditional versions of his **DPM** systems. Also there, he uses a restricted version of the inheritance principle. The adaptive handling of inheritance and aggregation introduced in this chapter can be applied in the conditional context straightforwardly (see Sect. 11.5).

One problem for conditional deontic logics is related to the detachment of conditional obligations. Given a conditional obligation to bring about A if B is the case, written $\mathsf{O}(A \mid B)$, and given that the condition B is fulfilled, one may want to derive the 'actual obligation' to bring about A. However, detachment is not universally valid. Note that being served a meal and given that the meal is asparagus, we do not want to detach the obligation not to eat with our fingers, although its condition, that a meal is served, is fulfilled. This is due to the fact that being served asparagus we are in exceptional circumstances. Again, applying adaptively detachment to $\mathsf{O}(A \mid B)$ and B "as much as possible" leads to interesting solutions to this problem (see Chap. 12). Furthermore, the semantics with an actual world introduced in this chapter can easily be generalized to the conditional case. This way factual premises can be handled semantically. Similarly an adaptive approach can be used in order to apply strengthening the antecedent (if $\mathsf{O}(A \mid B)$, then $\mathsf{O}(A \mid B) \wedge C)$ "as much as possible", i.e., to apply it whenever the factual premises do no describe an exceptional situation (see Sect. 11.5.2).

Altogether, the work presented in this chapter is inspiring and transferable to the tackling of other important problems within the context of deontic logics.

References

1. Straßer, C., Meheus, J., Beirlaen, M.: Tolerating deontic conflicts by adaptively restricting inheritance. Logique at Analyse **219**, 477–506 (2012)
2. Beirlaen, M., Straßer, C., Meheus, J.: An inconsistency-adaptive deontic logic for normative conflicts. J. Philos. Logic **2**(42), 285–315 (2013)
3. Da Costa, N., Carnielli, W.: On paraconsistent deontic logic. Philosophia **16**(3), 293–305 (1986)
4. Fraassen, B.C.V.: Values and the heart's command. J. Philos. **70**(1), 5–19 (1973)
5. Goble, L.: A proposal for dealing with deontic dilemmas. In: Lomuscio, A., Nute, D. (eds.) DEON. Lecture Notes in Computer Science, vol. 3065, pp. 74–113. Springer, Heidelberg (2004)
6. Goble, L.: A logic for deontic dilemmas. J. Appl. Logic **3**, 461–483 (2005)
7. Horty, J.F.: Reasoning with moral conflicts. Noûs **37**(4), 557–605 (2003)
8. Meheus, J., Beirlaen, M., Putte, F.V.D.: Avoiding deontic explosion by contextually restricting aggregation. In: Governatori, G., Sartor, G. (eds.) DEON (10th International Conference on Deontic Logic in Computer Science). Lecture Notes in Artificial Intelligence, vol. 6181, pp. 148–165. Springer, Heidelberg (2010)
9. Priest, G.: In Contradiction: A Study of the Transconsistent. Oxford University Press, Oxford (2006)
10. Schotch, P.K., Jennings, R.E.: Non-kripkean deontic logic. In: Hilpinen, R. (ed.) New Studies in Deontic Logic: Norms, Actions, and the Foundations of Ethics, pp. 149–162. Reidel Publishing Company, Dordrecht (1981)
11. van der Torre, L., Tan, Y.H.: Two-phase deontic logic. Logique at Analyse **171–172**, 411–456 (2000)
12. Gowans, C.W. (ed.): Moral Dilemmas. Oxford University Press, New York (1987)
13. Sinnott-Armstrong, W.: Moral Dilemmas. Basil Blackwell, Oxford (1988)

14. Goble, L.: Multiplex semantics for deontic logic. Nord. J. Philos. Logic **5**, 113–134 (2000)
15. Batens, D.: A universal logic approach to adaptive logics. Logica Universalis **1**, 221–242 (2007)
16. Batens, D.: Adaptive Logics and Dynamic Proofs. A Study in the Dynamics of Reasoning. (201x)
17. Schröder, L., Pattinson, D.: Rank-1 modal logics are coalgebraic. J. Logic Comput. **5**(20), 1113–1147 (2010)
18. Batens, D.: Towards the unification of inconsistency handling mechanisms. Logic Log. Philos. **8**, 5–31 (2000). Appeared 2002
19. Goble, L.: Deontic logic (adapted) for normative conflicts. Logic Journal of IGPL (2013). Online First
20. Hansen, J.: Conflicting imperatives and dyadic deontic logic. J. Appl. Logic **3**, 484–511 (2005)
21. Makinson, D., van der Torre, L.: Constraints for input/output logics. J. Philos. Logic **30**(2), 155–185 (2001)
22. Van Fraassen, B.C.: Values and the heart's command. J. Philos. **70**, 5–19 (1973)
23. Meheus, J., Beirlaen, M., Putte, F.V.D., Straßer, C.: Non-adjunctive deontic logics that validate aggregation as much as possible (forthcoming)
24. Rescher, N., Manor, R.: On inference from inconsistent premises. Theor. Decis. **1**, 179–217 (1970)
25. Batens, D.: A strengthening of the Rescher-Manor consequence relations. Logique at Analyse **183–184**, 289–313 (2003). Appeared 2005
26. Batens, D., Vermeir, T.: Direct dynamic proofs for the Rescher-Manor consequence relations: the flat case. J. Appl. Non-Class. Logics **12**, 63–84 (2002)
27. Putte, F.V.D., Straßer, C.: A logic for prioritized normative reasoning. J. Logic Comput. (2012)
28. Putte, F.V.D., Straßer, C.: Extending the standard format of adaptive logics to the prioritized case. Logique at Analyse **55**(220), 601–641 (2012)
29. Brink, D.: Moral conflict and its structure. Philos. Rev. **103**, 215–247 (1994)
30. Beirlaen, M., Straßer, C.: Two adaptive logics of norm-propositions. J. Appl. Logic **11**(2), 147–148 (2013)
31. Beirlaen, M., Straßer, C.: A paraconsistent multi-agent framework for dealing with normative conflicts. In: Leite, J., Torroni, P., Agotnes, T., Boella, G., van der Torre, L. (eds.) Computational Logic in Multi-Agent Systems. Lecture Notes in Computer Science, pp. 312–329. Springer, Heidelberg (2011)
32. Beirlaen, M., Straßer, C.: Nonmonotonic reasoning with normative conflicts in multi-agent deontic logic. J. Logic Comput. (2013)
33. Straßer, C., Beirlaen, M.: Towards more conflict-tolerant deontic logics by relaxing the interdefinability between obligations and permissions (in preparation)

Chapter 11
An Adaptive Logic Framework for Conditional Obligations and Deontic Dilemmas

Lou Goble proposed powerful conditional deontic logics (**CDPM**) in [1, 2] that are able to deal with deontic conflicts by restricting the inheritance principle. One of the central problems for dyadic deontic logics is to properly treat the restricted applicability of the principle "strengthening the antecedent". In most cases it is desirable to derive from an obligation A under condition B, that A is also obliged under condition B and C. However, there are important counterexamples. Goble proposed a weakened rational monotonicity principle to tackle this problem. This solution is suboptimal as it is for some examples counter-intuitive or even leads to explosion. The chapter identifies also other problems of Goble's systems. For instance, to make optimal use of the restricted inheritance principle, in many cases the user has to manually add certain statements to the premises.

An adaptive logic framework on basis of **CDPM** is proposed which is able to tackle these problems. It allows for certain rules to be applied as much as possible. In this way counter-intuitive consequences as well as explosion can be prohibited and no user interference is required. Furthermore, for non-conflicting premise sets the ALs are equivalent to Goble's dyadic version of standard deontic logic.

11.1 Introduction

Recent work in ALs has shown growing interest in (monadic) deontic systems which are able to deal with deontic conflicts (see e.g., [3–6]). A deontic conflict between obligations occurs when the obligations cannot be mutually realized. As has been argued in Chap. 10, deontic conflicts are not just an abstruse philosophical notion, but that they occur quite commonly in our every-day moral lives.

A former version of the content of this chapter has been published under the name "*An Adaptive Logic Framework for Conditional Obligations and Deontic Dilemmas*" in the Journal "Logic and Logical Philosophy", 2010, [1].

C. Straßer, *Adaptive Logics for Defeasible Reasoning*, Trends in Logic 38,
DOI: 10.1007/978-3-319-00792-2_11,

Unlike standard deontic logic **SDL**, Goble's logics **DPM** (see [1, 2, 7] and Chap. 10) prevent deontic explosions in such cases by means of restricting the inheritance principle ("if $\vdash A \supset B$ then $\vdash \mathsf{O}A \supset \mathsf{O}B$"), while having the same range of desired consequences for non-conflicting premise sets. Developing adaptive versions of **DPM** in [3] the authors were able to improve them in various aspects (see Chap. 10).

It is well known that attempts to model conditional obligations in terms of monadic ought-operators (e.g. $\mathsf{O}(A \supset B)$ or $A \supset \mathsf{O}B$) have several shortcomings. This has led to various approaches based on dyadic ought-operators $\mathsf{O}(A \mid B)$—"if B is the case you are obliged to do/bring about A". Goble in [1, 2] developed conditional versions of his conflict-tolerant **DPM** systems (**CDPM**) that are also based on a restricted inheritance principle.

One of the most difficult problems for dyadic deontic logics is to handle cases in which the principle 'strengthening the antecedent' ($\vdash \mathsf{O}(A \mid B) \supset \mathsf{O}(A \mid B \wedge C)$) has to be restricted. Paradigmatic instances are settings in which exceptions and/or violations of general obligations occur, as for example (cp. [8])[1]:

- You ought not to eat with your fingers: $\mathsf{O}(\neg f \mid \top)$
- You ought to put your napkin on your lap: $\mathsf{O}(n \mid \top)$
- If you are served asparagus, you are allowed to eat it with your fingers: $\mathsf{P}(f \mid a)$

By adding some intuitive permission statements, Goble's preferred conditional logic **CDPM.2c** is able to derive all the desired obligations (e.g. $\mathsf{O}(\neg f \wedge n \mid \top)$, $\mathsf{O}(n \mid a)$). Note, however, that given the intuitive permission $\mathsf{P}(\neg f \wedge a \mid \top)$, also the counter-intuitive $\mathsf{O}(\neg f \mid a)$ is derivable. This leads to triviality since the latter is equivalent to $\neg\mathsf{P}(f \mid a)$,—a severe shortcoming. Obviously nothing is wrong with the permission not to eat with your fingers and to eat asparagus, it is perfectly coherent with our three premises.

We will also identify other shortcomings of Goble's logics. For instance, in order to make optimal use of the restricted inheritance principle, in many cases the user needs to manually add certain statements to the premise set. Furthermore, some of the rules of Goble's **CDPM** logics do not behave well together and cause undesired consequences.

This chapter presents ALs based on Goble's **CDPM** logics which are able to tackle these problems. They allow for certain rules to be applied as much as possible. In this way counter-intuitive consequences as well as explosion can be prohibited and no user interference is required. For instance, it will be shown that for the adaptive approach there is no need to explicitly add premises in order to make use of the restricted inheritance rule. The proof dynamics of the ALs takes care of this as part of the reasoning process which is explicated by the proof. In addition, the dynamic aspect of our moral reasoning is nicely recaptured by the dynamic proof theory. This also enables us to have a better insight in the relations between obligations/permissions and thus to localize the deontic conflicts as well as violations and exceptions of obligations as the product of an actual reasoning process. Furthermore, for

[1] The permission operator is as usually defined by $\mathsf{P}(A \mid B) =_{\mathrm{df}} \neg\mathsf{O}(\neg A \mid B)$.

non-conflicting premise sets the ALs are equivalent to Goble's dyadic version of standard deontic logic.

In Appendix I the interested reader can find semantics for the introduced logics and proofs for the (meta)-theorems presented in this chapter.

11.2 The Problem With Strengthening the Antecedent

Most of our moral or behavioral norms are in a conditional form. For instance "Being in an airplane you ought to turn off your mobile phone." One proposal to model this in deontic logics is to use $p \supset \mathsf{O}m$. A disadvantage of this form is that it offers unrestricted "strengthening the antecedent" (SA): from $p \supset \mathsf{O}m$, $(p \wedge b) \supset \mathsf{O}m$ is derivable. In many cases this is as expected. But consider the following case:

1. In general we're supposed not to eat with our fingers.
2. Eating asparagus we're allowed to eat with our fingers.

Modeling (1) by $\top \supset \mathsf{O}\neg f$, $a \supset \mathsf{O}\neg f$ is derivable. This obviously is in conflict with (2) $a \supset \mathsf{P}f$ which is equivalent to $a \supset \neg\mathsf{O}\neg f$. Similarly problematic is to use $\mathsf{O}(C \supset B)$ to represent the obligation B under condition C. In the example we can derive by the inheritance principle the counter-intuitive $\mathsf{O}(a \supset \neg f)$ from $\mathsf{O}(\top \supset \neg f)$. Also restricting the inheritance principle as proposed by Goble doesn't help, as nothing speaks against adding the harmless premise $\mathsf{P}(\top \supset \neg f)$. However, this again enables the derivation of $\mathsf{O}(a \supset \neg f)$ by (RPM) from $\mathsf{O}(\top \supset \neg f)$ (and $\mathsf{P}(\top \supset \neg f)$).

Monadic approaches also were object of other kinds of criticism. The most prominent class of problems have to do with paradoxes such as the Chisholm Paradox (see [9]) or the Gentle Murderer Paradox (see [10]). It is commonly agreed that dyadic approaches are in general better suited to deal with these kind of problems.

Modeling conditional obligations with dyadic obligation operators also allows for a more subtle approach to (SA). We use $\mathsf{O}(A \mid B)$ in order to express that "under condition B it ought to be that A". As usual, the permission operator $\mathsf{P}(A \mid B)$ is defined by $\neg\mathsf{O}(\neg A \mid B)$. We use a propositional language. In contrast to the monadic approach presented in Chap. 10 we enrich the language now by a dyadic obligation operator. Our set of well-formed formulas $\mathcal{W}_2$ is defined as follows:

$$\mathcal{W}_2 ::= \bot \mid \top \mid \langle\mathcal{S}\rangle \mid \langle\mathcal{W}_2\rangle \wedge \langle\mathcal{W}_2\rangle \mid \langle\mathcal{W}_2\rangle \vee \langle\mathcal{W}_2\rangle \mid \\ \langle\mathcal{W}_2\rangle \supset \langle\mathcal{W}_2\rangle \mid \neg\langle\mathcal{W}_2\rangle \mid \mathsf{O}\langle\mathcal{W}_2\rangle\langle\mathcal{W}_2\rangle$$

In this way we can formalize our example by:

$$\mathsf{O}(\neg f \mid \top) \tag{A1}$$

$$\mathsf{P}(f \mid a) \tag{A2}$$

An unrestricted (SA)

$$\vdash \mathsf{O}(B \mid A) \supset \mathsf{O}(B \mid A \wedge C) \tag{SA}$$

would lead to the counter-intuitive $\mathsf{O}(\neg f|a)$.

As in the case of the restricted inheritance principle (RPM) (see Chap. 10), a possible way of restricting (SA) is to require certain permission statements. Inspired by Kraus, Lehmann and Magidor's work (see [11, 12]) on nonmonotonic consequence relations a candidate can be found in the principle of *Rational Monotonicity*

$$\vdash \big(\mathsf{O}(B \mid A) \wedge \mathsf{P}(C \mid A)\big) \supset \mathsf{O}(B \mid A \wedge C) \tag{RatMono}$$

Lou Goble proposed different axiomatizations of a standard dyadic deontic logic (which are equivalent to van Fraassen's **CD** in [13], David Lewis's **VN** in [14] and his own **SDDL** in [15]). Let **RSDDL** be the logic characterized by all classical tautologies, (RatMono) and the following rules

$$\text{If } \vdash A \equiv B \text{ then } \vdash \mathsf{O}(C \mid A) \equiv \mathsf{O}(C \mid B) \tag{RCE}$$

$$\text{If } \vdash B \supset C \text{ then } \vdash \mathsf{O}(B \mid A) \supset \mathsf{O}(C \mid A) \tag{RCM}$$

and axioms

$$\vdash \mathsf{O}(\top \mid \top) \tag{CN}$$

$$\vdash \neg\mathsf{O}(\bot \mid A) \tag{CP}$$

$$\vdash \big(\mathsf{O}(B \mid A) \wedge \mathsf{O}(C \mid A)\big) \supset \mathsf{O}(B \wedge C \mid A) \tag{CAND}$$

$$\vdash \mathsf{O}(B \mid A) \supset \mathsf{O}(A \mid A) \tag{QR}$$

$$\vdash \mathsf{O}(C \mid A \wedge B) \supset \mathsf{O}(B \supset C \mid A) \tag{S}$$

But employing (RatMono) instead of (SA) is problematic as well. In our example it is in no way counter-intuitive to add the statement $\mathsf{P}(a \mid \top)$, "It is in general allowed to eat asparagus", to the premise set. But now it is again possible to derive $\mathsf{O}(\neg f \mid a)$. A further restriction is needed.

11.3 Lou Goble's CDPM

Lou Goble proposes in [1, 2] the following *Weak Rational Monotonicity* principle:

$$\vdash \big(\mathsf{O}(B \mid A) \wedge \mathsf{P}(B \wedge C \mid A)\big) \supset \mathsf{O}(B \mid A \wedge C) \tag{WRM}$$

On basis of his **DPM** systems he presents the following conditional logics:

Definition 11.3.1. Enriching classical propositional logic by (RCE), (CN), (CAND) and the rules

$$\text{If } \vdash B \equiv C \text{ then } \vdash \mathsf{O}(B \mid A) \equiv \mathsf{O}(C \mid A) \tag{CRE}$$

$$\text{If } \vdash B \supset C \text{ then } \vdash \mathsf{P}(B \mid A) \supset \big(\mathsf{O}(B \mid A) \supset \mathsf{O}(C \mid A)\big) \tag{RCPM}$$

results in logic **CDPM.1**. Logic **CDPM.1c** is **CDPM.1** enriched by (WRM), (QR) and (S).

The idea behind the restricted inheritance principle (RCPM) is analogous to the monadic case (see Chap. 10): inheritance is only applied to non-conflicting obligations. That is to say, it is applied to $\mathsf{O}(A \mid B)$ only in case we also have $\mathsf{P}(A \mid B)$.

Definition 11.3.2. **CDPM.2′** is defined analogous to **CDPM.1**, just add (CP) and replace (CAND) by[2]

$$\vdash \big(\mathsf{O}(A \mid C) \wedge \mathsf{O}(B \mid C) \wedge \mathsf{P}(A \mid C) \wedge \mathsf{P}(B \mid C)\big) \supset \mathsf{O}(A \wedge B \mid C) \tag{CPAND′}$$

Logic **CDPM.2′c** is like **CDPM.1c** with exception of (CAND) which is replaced by (CPAND′). Furthermore (CP) is added.

Definition 11.3.3. For the remainder it is useful to introduce some writing conventions: We write $\oplus(\{P_1, \ldots, P_n\}, \mathbf{L})$ for the logic $\mathbf{L}'$ that is defined as $\mathbf{L}$ with the addition of principles $P_1, \ldots, P_n$ where each $P_i \in \mathcal{P}$ and[3]

$$\mathcal{P} =_{\text{df}} \{\text{WRM}, \text{WRM}_\star, \text{CAND}, \text{CPAND}', \text{QR}, \text{PS}, \text{S}, \text{CP}, \text{AWRM}_\star, \text{CD}\}.$$

Define $\mathcal{L} =_{\text{df}} \{\oplus(\Psi, \mathbf{CDPM}^-) \mid \Psi \in \wp(\mathcal{P})\}$ where $\mathbf{CDPM}^-$ is defined by (RCE), (CN), (CRE) and (RCPM). Let $\mathbf{L} = \oplus(\{P_1, \ldots, P_n\}, \mathbf{CDPM}^-) \in \mathcal{L}$. We write $\ominus(\{P_i \mid i \in I\}, \mathbf{L})$ (where $I \subseteq \{1, \ldots, n\}$) for the logic $\mathbf{L}' = \oplus(\{P_i \mid i \in \{1, \ldots, n\} \setminus I\}, \mathbf{CDPM}^-)$. We write $\oplus_P \mathbf{L}$ (resp. $\ominus_P \mathbf{L}$) instead of $\oplus(\{P\}, \mathbf{L})$ (resp. $\ominus(\{P\}, \mathbf{L})$). Further define for $\mathbf{L} = \oplus(\{P_i \mid i \in I\}, \mathbf{CDPM}^-) \in \mathcal{L}$, $\downarrow\mathbf{L}$ as the set of all sub-logics of $\mathbf{L}$ in $\{\ominus(\Psi, \mathbf{L}) \mid \Psi \in \wp(\{P_i \mid i \in I\})\}$.

None of the following "deontic explosion principles" are valid in **CDPM1.c** and **CDPM.2′c**[4]:

$$\text{If } \nvdash \neg B \text{ then } \mathsf{O}(A \mid C), \mathsf{O}(\neg A \mid C) \vdash \mathsf{O}(B \mid C) \tag{CDEX-1}$$

$$\mathsf{O}(A \mid C), \mathsf{O}(\neg A \mid C), \mathsf{P}(B \mid C) \vdash \mathsf{O}(B \mid C) \tag{CDEX-2}$$

---

[2] As in the monadic case in Chap. 10, we use a slight variation of Goble's **CDPM.2** which employs $\vdash \big(\mathsf{O}(A \mid C) \wedge \mathsf{O}(B \mid C) \wedge \mathsf{P}(A \wedge B \mid C)\big) \supset \mathsf{O}(A \wedge B \mid C)$ (CPAND) instead of our (CPAND′). Using **CDPM.2′c** instead of **CDPM.2c** as lower limit logic leads to technically more elegant ALs. Furthermore, in contrast to **CDPM.2c**, **CDPM.2′c** fulfills criterion $(C\star)$, that is going to be introduced in a moment.

[3] Some of the principles in $\mathcal{P}$ will be defined later on (namely $\text{WRM}_\star$, PS, CD and $\text{AWRM}_\star$).

[4] We slightly adjusted the criteria offered by Goble since his criteria were formulated in terms of theoremhood, while we focus on the consequences of premise sets. Models validating counter-instances of the criteria can be found in the proof of Theorem I.2.1 in Appendix I.

$$\mathsf{O}(D \mid C), \mathsf{P}(D \mid C), \mathsf{O}(A \mid C), \mathsf{O}(\neg A \mid C), \mathsf{P}(B \mid C) \vdash \mathsf{O}(B \mid C) \quad \text{(CDEX-3)}$$

Goble defined a dyadic variant of convention ($\star$) which is satisfied by both logics, **CDPM.1c** and **CDPM.2′c**:

(C$\star$) A dyadic deontic logic for dilemmas should be such that the result of adding (CD), namely $\mathsf{O}(B|A) \supset \neg\mathsf{O}(\neg B|A)$, as an axiom results in a logic that has the same consequence relation as **(R)SDDL**.

Theorem 11.3.1. *Where* $\alpha \in \{1, 2'\}$, **CDPM.**α**c** *satisfies* (C$\star$).

11.4 A Critical Analysis

In the following we will give a critical analysis of the **CDPM** logics which will indicate several problems. This in turn will motivate to strengthen these systems within the AL framework.

11.4.1 CDPM.1c *Validates (PRatMono)*

As has been demonstrated by Goble, one severe shortcoming of **CDPM.1c** is that

$$\vdash \big(\mathsf{O}(B \mid A) \wedge \mathsf{P}(C \mid A) \wedge \mathsf{P}(B \mid A)\big) \supset \mathsf{O}(B \mid A \wedge C) \quad \text{(PRatMono)}$$

is derivable from it. However, this is suboptimal. If just the harmless premise $\mathsf{P}(\neg f \mid \top)$ in addition to $\mathsf{P}(a \mid \top)$ is added to (A1) and (A2) then again $\mathsf{O}(\neg f \mid a)$ is derivable.

11.4.2 Explosive Behavior and Other Problems With (WRM)

We saw above that different restricted versions of (SA), such as (RatMono) and (PRatMono), are counter-intuitive as soon as we add further harmless premises to our asparagus example. Is the weak rational monotonicity principle proposed by Lou Goble robust to criticism of this kind? Goble demonstrates that it is not. It is not conflicting with our moral intuitions to add the premise $\mathsf{P}(\neg f \wedge a \mid \top)$, "in general it is allowed not to eat with your fingers and also to eat asparagus", to the premise set consisting of (A1) and (A2). But in this case, by applying (WRM) to $\mathsf{O}(\neg f \mid \top)$ and $\mathsf{P}(\neg f \wedge a \mid \top)$, we arrive at $\mathsf{O}(\neg f \mid a)$. This causes not just a deontic conflict, but a full-fledged explosion, as we also have $\mathsf{P}(f \mid a)$ which is equivalent to $\neg\mathsf{O}(\neg f \mid a)$.

Note also that the following counter-intuitive statement is a consequence of (WRM):

$$\vdash \big(\mathsf{O}(\neg f \mid \top) \wedge \mathsf{P}(f \mid a)\big) \supset \neg\mathsf{P}(\neg f \wedge a \mid \top)$$

Obviously the general obligation not to eat with your fingers and the fact that in case asparagus is served it is allowed to eat with your fingers do in no way entail that it is in general not allowed not to eat with your fingers and to eat asparagus.

Example 11.4.1. We add another problematic example:

(1) In a hospital you ought not to smoke—$\mathsf{O}(\neg s \mid h)$;
(2) If you're in a smoking room, you're allowed to smoke—$\mathsf{P}(s \mid r)$;
(3) If you're in a hospital, you are allowed to be in a smokers room and not to smoke—$\mathsf{P}(r \wedge \neg s \mid h)$;
(4) If you're in a smoking room (and) in a hospital, you're allowed to smoke—$\mathsf{P}(s \mid h \wedge r)$.

Note that by (WRM) we can derive $\mathsf{O}(\neg s \mid h \wedge r)$ from (1) and (3). In face of (2) and (4) this is counter-intuitive and contradictory.

11.4.3 Solving the Problems by Further Restricting (WRM)

There is a way of tackling our problems in a non-adaptive, monotonic way by further restricting (WRM):

$$\vdash \big(\mathsf{O}(B \mid A) \wedge \mathsf{P}(B \wedge C \mid A) \wedge \neg\mathsf{P}(\neg B \wedge A \mid C)\big) \supset \mathsf{O}(B \mid A \wedge C) \quad (\text{WRM}_\star)$$

The idea is to strengthen the antecedent A of obligation B by C only if $\neg B \wedge A$ is not allowed in the context described by C. Looking back at the asparagus example this obviously blocks the unwanted derivation. By (WRM$_\star$) we have

$$\vdash \big(\mathsf{O}(\neg f \mid \top) \wedge \mathsf{P}(\neg f \wedge a \mid \top) \wedge \neg\mathsf{P}(f \mid a)\big) \supset \mathsf{O}(\neg f \mid a)$$

Note that $\mathsf{P}(f \mid a)$ is a premise and thus $\mathsf{O}(\neg f \mid a)$ is not derivable by (WRM$_\star$). Also in the case of the second example we have no means to derive $\mathsf{O}(\neg s \mid h \wedge r)$. We would need $\neg\mathsf{P}(s \wedge h \mid r)$, which is obviously counter-intuitive.

As (PRatMono) was derivable in **CDPM.1c** while not being derivable in **CDPM.2′c**, this lead to an undesirable asymmetry. However, this asymmetry disappears in case of $\oplus_{\text{WRM}_\star}\ominus_{\text{WRM}}$**CDPM.1c** and $\oplus_{\text{WRM}_\star}\ominus_{\text{WRM}}$**CDPM.2′c**. In both systems neither (PRatMono) nor (WRM) is derivable. This is clearly as desired.

However, the price to pay for this is that neither $\oplus_{\text{WRM}_\star}\ominus_{\text{WRM}}$**CDPM.1c** nor $\oplus_{\text{WRM}_\star}\ominus_{\text{WRM}}$**CDPM.2′c** is equivalent to **(R)SDDL** if we add (CD) as an axiom. Hence neither of the two systems satisfies requirement ($C\star$).

Both systems, $\oplus_{\text{WRM}_\star}\ominus_{\text{WRM}}$**CDPM.1c** and $\oplus_{\text{WRM}_\star}\ominus_{\text{WRM}}$**CDPM.2′c**, do not validate (CDEX-1)–(CDEX-3) and are therefore sufficiently robust with respect to deontic conflicts.

Note that in **(R)SDDL** all instances of

$$\vdash \big(\mathsf{O}(B \mid A) \wedge \mathsf{P}(B \wedge C \mid A)\big) \supset \mathsf{O}(A \supset B \mid C) \qquad (\text{AWRM}_\star)$$

are valid. Hence, whenever the antecedent of (WRM) is fulfilled in **(R)SDDL**, then also the antecedent of ($\text{WRM}_\star$) is fulfilled. Where $\alpha \in \{1, 2'\}$, if all instances of (CD) are added to $\oplus(\{\text{WRM}_\star, \text{AWRM}_\star\}, \ominus_{\text{WRM}}\mathbf{CDPM}.\alpha\mathbf{c})$ then the resulting logic has the same consequence relation as **(R)SDDL**.

Theorem 11.4.1. *Where* $\alpha \in \{1, 2'\}$,
$\oplus(\{\text{WRM}_\star, \text{AWRM}_\star\}\ominus_{\text{WRM}}\mathbf{CPDM}.\alpha\mathbf{c})$ *satisfies* ($C\star$).

One of the major problems with restricting (WRM) is analogous to the problem we already pointed out in connection with the restricted inheritance principle (RCPM): in order to apply ($\text{WRM}_\star$) the user has to manually add permission statements. However, on the one hand, if the relationships between various obligations are of a complicated nature, the manual addition of permission statements might lead to undesired results such as explosion. On the other hand we would like to delegate as much reasoning as possible from the user to the logic itself. We will therefore in Sect. 11.5.2 propose an AL which applies (WRM) "as much as possible" without the need of user interference.

11.4.4 A Problem with Aggregation and (S)

Example 11.4.2. Consider the following strict version of the asparagus example:

$$\mathsf{O}(\neg f \mid \top) \qquad (\text{SA1})$$

$$\mathsf{O}(f \mid a) \qquad (\text{SA2})$$

$$\mathsf{P}(a \mid \top) \qquad (\text{SA3})$$

$$\mathsf{P}(\neg f \wedge \neg a \mid \top) \qquad (\text{SA4})$$

From (SA2) we get by (S),

$$\mathsf{O}(a \supset f \mid \top) \qquad (\text{SA5})$$

Note that it is a consequence of (RCPM) that $\vdash \mathsf{P}(C \mid B) \supset (\mathsf{P}(C' \mid B) \vee \mathsf{O}(C' \mid B))$ where $C \vdash C'$. Therefore, by (SA3), $\mathsf{P}(a \vee f \mid \top) \vee \mathsf{O}(a \vee f \mid \top)$. Since (SA4) is equivalent to $\neg\mathsf{O}(a \vee f \mid \top)$ we arrive at

$$\mathsf{P}(a \vee f \mid \top) \qquad (\text{SA6})$$

In case we apply (CAND) to (SA1) and (SA5) we get $\mathsf{O}(\neg f \wedge (a \supset f) \mid \top)$ and therefore, by (CRE), $\mathsf{O}(\neg f \wedge \neg a \mid \top)$. But this is equivalent to $\neg\mathsf{P}(a \vee f \mid \top)$,—a contradiction with (SA6). Employing the weaker aggregation principle (CPAND′) doesn't help either. In this case we add the harmless statements $\mathsf{P}(\neg f \mid \top)$ and $\neg\mathsf{O}(a \wedge \neg f \mid \top)$ to the premises. By (CPAND′), $\mathsf{O}(\neg f \wedge \neg a \mid \top)$ is again derivable (as the reader can easily verify herself),—in contradiction with (SA6).

Thus, the example shows that (CAND) and (CPAND′) do not behave well together with (S). One possible solution is to use a restricted version of (S), namely[5]

$$\vdash \big(\mathsf{P}(C \mid A) \wedge \mathsf{O}(C \mid A \wedge B)\big) \supset \mathsf{O}(B \supset C \mid A) \tag{PS}$$

Note that in $\oplus_{\mathrm{PS}}\ominus_{\mathrm{S}}$**CDPM.1c** and $\oplus_{\mathrm{PS}}\ominus_{\mathrm{S}}$**CDPM.2′c** the derivation of (SA5) is blocked since we would need $\mathsf{P}(f \mid \top)$, but we have $\mathsf{O}(\neg f \mid \top)$ which is equivalent to $\neg\mathsf{P}(f \mid \top)$. It is also worth mentioning that, where $\alpha \in \{1, 2'\}$, $\oplus(\{\mathrm{CD}, \mathrm{PS}\}, \ominus_{\mathrm{S}}$**CDPM**.$\alpha$**c**) has the same consequence relation as $\oplus_{\mathrm{CD}}$**CDPM**.α**c** and therefore it also has the same consequence relation as **(R)SDDL**.

Theorem 11.4.2. *Where* $\alpha \in \{1, 2'\}$, $\oplus_{\mathrm{PS}}\ominus_{\mathrm{S}}$**CDPM**.$\alpha$**c** *satisfies* ($C\star$).

Furthermore, in view of Theorem (11.4.1) it is not surprising that in case we replace (WRM) with ($\mathrm{WRM}_\star$) we need to also add ($\mathrm{AWRM}_\star$) in order to get (C⋆):

Theorem 11.4.3. *Where* $\alpha \in \{1, 2'\}$,
$\oplus(\mathrm{WRM}_\star, \mathrm{AWRM}_\star, \mathrm{PS}, \ominus(\{\mathrm{WRM}, \mathrm{S}\},$ **CPDM**.α**c**)) *satisfies (C⋆).*

11.5 Going Adaptive – The Conditional Case

After having located various problems of **CDPM.1c** and **CDPM.2′c** we are now going to introduce an adaptive logic framework that can deal with these problems. One of our goals is to develop conflict-tolerant logics that are able to derive from the non-conflicting 'parts' of a given premise set as much as possible without the need of manually adding premises. Due to this, criterion (C⋆) is not adequate anymore since it measures the derivative power of a logic in view of adding all instances of (CD) to it. We alter it the following way:

(C‡) For all premise sets for which **(R)SDDL** is non-explosive, a dyadic deontic logic for dilemmas should have the same consequence set as **(R)SDDL**.

Note that logics satisfying this criterion are in a sense stronger than logics only satisfying (C⋆) since, in order to achieve the same consequences as **(R)SDDL**, (CD) does not have to be added to the former ones.

[5] (PS) was proposed by Goble in [1] in connection with another problem with respect to (RatMono).

11.5.1 Applying Inheritance Conditionally

In order to apply the dyadic inheritance principle (RCPM) conditionally we proceed in the same way as in the monadic case (see Chap. 10).

Definition 11.5.1. $\mathbf{A_c}(\mathbf{LLL})$ is defined by the triple

$$\langle \mathbf{LLL}, \Omega_d^c, \text{minimal abnormality} \rangle$$

where $\mathbf{LLL} \in \downarrow\mathbf{CDPM.1c} \cup \downarrow\mathbf{CDPM.2'c}$, $\Omega_d^c =_{\mathrm{df}} \{!_c\mathsf{O}(A \mid B) \mid A, B \in \mathcal{W}_2\}$, and $!_c\mathsf{O}(\neg A \mid B)$ abbreviates

- $\mathsf{O}(A \mid B) \wedge \mathsf{O}(\neg A \mid B)$ in case **LLL** does not validate (CAND), and otherwise
- $\bigvee_{C \in \mathrm{Sub}(A)} \mathsf{O}(C \mid B) \wedge \mathsf{O}(\neg C \mid B)$ (where $\mathrm{Sub}(A)$ is the set of subformulas of A including A).

In Chapter 10 we have already explained why in the presence of an unrestricted aggregation principle we need a slightly more complicated logical form to characterize our abnormalities. We have:

$$\text{If } \vdash_{\mathbf{L}} A \supset B, \text{ then } \mathsf{O}(A \mid C) \vdash_{\mathbf{L}} \mathsf{O}(B \mid C) \vee !_c\mathsf{O}(A \mid C)$$
$$\mathsf{O}(D \mid F), \mathsf{O}(E \mid F) \vdash_{\mathbf{L.2}} \mathsf{O}(D \wedge E \mid F) \vee \left(!_c\mathsf{O}(D \mid F) \vee !_c\mathsf{O}(E \mid F)\right)$$

where $\mathbf{L}, \mathbf{L.2} \in \downarrow\mathbf{CDPM.1c} \cup \downarrow\mathbf{CDPM.2'c}$, $!_c$ has any of the two interpretations stated above, and $\mathbf{L.2}$ validates (CPAND$'$). Thus, it is possible

- to derive $\mathsf{O}(B \mid C)$ from $\mathsf{O}(A \mid C)$ on the condition $\{!_c\mathsf{O}(A \mid C)\}$, and
- to derive $\mathsf{O}(D \wedge E \mid F)$ from $\mathsf{O}(D \mid F)$ and $\mathsf{O}(E \mid F)$ on the condition $\{!_c\mathsf{O}(D \mid F), !_c\mathsf{O}(E \mid F)\}$.

Example 11.5.1. As we discussed the monadic variant already in Chap. 10, we are only going to take a look at a very simple example for $\mathbf{A_c}(\mathbf{CDPM.2'c})$. Let as usual f express 'eating with your fingers', n 'using a napkin' and let b stand for 'belching at the table'.

1	$\mathsf{O}(\neg f \wedge \neg b \mid \top)$	PREM	$\emptyset$
2	$\mathsf{O}(\neg b \mid \top)$	1; RC	$\{!_c\mathsf{O}(\neg f \wedge \neg b \mid \top)\}$
3	$\mathsf{O}(n \mid \top)$	PREM	$\emptyset$
4	$\mathsf{O}(\neg f \wedge \neg b \wedge n \mid \top)$	1, 3; RC	$\{!_c\mathsf{O}(\neg f \wedge \neg b \mid \top), !_c\mathsf{O}(n \mid \top)\}$

In line 2 we have a conditional application of the inheritance principle, in line 4 one of the aggregation principle.

Similar as in the monadic case (see Corollary 10.8.4) we have:

Theorem 11.5.1. *Where $\alpha \in \{1, 2'\}$, we have:*

(i) $\mathbf{A_c}(\mathbf{CDPM.}\alpha\mathbf{c})$ *satisfies (C‡).*
(ii) $\mathbf{A_c}(\oplus_{\mathrm{PS}}\ominus_{S}\mathbf{CDPM.}\alpha\mathbf{c})$ *satisfies (C‡).*

Theorem 11.5.2. *Where* **LLL** $\in \downarrow$**CDPM.1c** $\cup \downarrow$**CDPM.2′c**, $\mathbf{A_c}$(**LLL**) *does not validate* (CDEX-1), (CDEX-2), *and* (CDEX-3).

11.5.2 Applying Weak Rationality Conditionally

Furthermore, our logics can be equipped with the ability to conditionally apply (WRM). Compared with **CDPM.1c** and **CDPM.2′c** the advantage is that in case an application of (WRM) leads to unwanted results (e.g. conflicts or explosions, see Sect. 11.4.2) the logics that are going to be introduced in this Section block the application. Compared to the logic presented in Sect. 11.4.3 that makes use of a further restriction of (WRM), namely (WRM$_\star$), for the ALs that are presented in this section there is no need to add more auxiliary permission statements to the premise set. This accords with our goal to reduce the reasoning and interference of the user as much as possible.

We define the set of abnormalities

$$\Omega^r_d = \{\mathsf{O}(A \mid B) \wedge \mathsf{P}(A \wedge C \mid B) \wedge \neg\mathsf{O}(A \mid B \wedge C) \mid A, B, C \in \mathcal{W}_2\}$$

By propositional logic we have

$$\begin{aligned}\mathsf{O}(A \mid B), \mathsf{P}(A \wedge C \mid B) \vdash &\mathsf{O}(A \mid B \wedge C) \vee \\ &\big(\mathsf{O}(A \mid B) \wedge \mathsf{P}(A \wedge C \mid B) \wedge \neg\mathsf{O}(A \mid B \wedge C)\big),\end{aligned}$$

This enables us to derive $\mathsf{O}(A \mid B \wedge C)$ from $\mathsf{O}(A \mid B)$ and $\mathsf{P}(A \wedge C \mid B)$ on the condition $\{\mathsf{O}(A \mid B) \wedge \mathsf{P}(A \wedge C \mid B) \wedge \neg\mathsf{O}(A \mid B \wedge C)\}$.

Definition 11.5.2. Logic $\mathbf{A_r^{m/r}}$(**LLL**) is defined by the following triple:

$$\langle \mathbf{LLL}, \Omega^r_d, \mathbf{m}\text{inimal abnormality}/\mathbf{r}\text{eliability}\rangle$$

where **LLL** $\in \bigcup_{\alpha\in\{1,2'\}} \downarrow \ominus_{\mathrm{WRM}}$ **CDPM**.α**c** $\cup$ $\{\oplus_{\mathrm{CD}}\ominus_{\mathrm{WRM}}$**CDPM**.$\alpha$**c**$\}$.

Theorem 11.5.3. *Where* $\alpha \in \{1, 2'\}$ *and* $\mathbf{x} \in \{\mathbf{r}, \mathbf{m}\}$, *we have*

(i) $\mathbf{A_r^x}(\oplus_{\mathrm{CD}}\ominus_{\mathrm{WRM}}$**CDPM**.$\alpha$**c**) *satisfies (C‡).*
(ii) $\mathbf{A_r^x}(\oplus(\{\mathrm{CD}, \mathrm{PS}\}, \ominus(\{\mathrm{S}, \mathrm{WRM}\}$, **CDPM**.$\alpha$**c**))) *satisfies (C‡).*

Theorem 11.5.4. *Where* $\alpha \in \{1, 2'\}$, $\mathbf{x} \in \{\mathbf{r}, \mathbf{m}\}$, *and* **LLL** $\in \downarrow \ominus_{\mathrm{WRM}}$ **CDPM**.α**c**, $\mathbf{A_r^x}$(**LLL**) *does not validate* (CDEX-1), (CDEX-2), *and* (CDEX-3).

Let us take a look at a few examples.

Example 11.5.2. Let us return to the problematic asparagus example from Sect. 11.4.2 with lower limit logic $\ominus_{\mathrm{WRM}}$**CDPM.1c**. Let $!_r\mathsf{O}^B_A(C) =_{\mathrm{df}} \mathsf{O}(C \mid A) \wedge \mathsf{P}(C \wedge B \mid A) \wedge \neg\mathsf{O}(C \mid A \wedge B)$.

	1 $\mathsf{O}(\neg f \mid \top)$	PREM	$\emptyset$
	2 $\mathsf{P}(f \mid a)$	PREM	$\emptyset$
	3 $\mathsf{P}(\neg f \wedge a \mid \top)$	PREM	$\emptyset$
^{6}	4 $\mathsf{O}(\neg f \mid a)$	1,3; RC	$\{!_r\mathsf{O}^a_\top(\neg f)\}$
	5 $\neg\mathsf{O}(\neg f \mid a)$	2; RU	$\emptyset$
	6 $!_r\mathsf{O}^a_\top(\neg f)$	1,3,5; RU	$\emptyset$

This demonstrates that unwanted derivations are successfully blocked. In order to show that desired consequences are actually reached, the example needs to be extended. Let x stand for "being in a country C" and we know that eating with your fingers is strictly forbidden (no exceptions!) in C. Moreover, let n stand for "using a napkin".

7 $\mathsf{O}(\neg f \mid x)$	PREM	$\emptyset$
8 $\mathsf{O}(n \mid \top)$	PREM	$\emptyset$
9 $\mathsf{P}(\neg f \wedge a \mid x)$	PREM	$\emptyset$
10 $\mathsf{P}(n \wedge a \mid \top)$	PREM	$\emptyset$
11 $\mathsf{O}(\neg f \mid x \wedge a)$	7,9; RC	$\{!_r\mathsf{O}^a_x(\neg f)\}$
12 $\mathsf{O}(n \mid a)$	8,10; RC	$\{!_r\mathsf{O}^a_\top(n)\}$
13 $\mathsf{P}(n \wedge x \mid a)$	PREM	$\emptyset$
14 $\mathsf{O}(n \mid x \wedge a)$	12,13; RC	$\{!_r\mathsf{O}^a_\top(n), !_r\mathsf{O}^x_a(n)\}$

It can easily be shown that lines 11, 12 and 14 will not be marked in any extension of the proof.

Example 11.5.3. Also in the other problematic Example 11.4.1 the ALs block the undesired instances of (SA). Let the lower limit logic be $\ominus_{\mathrm{WRM}}$**CDPM**.α**c** where $\alpha \in \{1, 2'\}$.

	1 $\mathsf{O}(\neg s \mid h)$	PREM	$\emptyset$
	2 $\mathsf{P}(s \mid r)$	PREM	$\emptyset$
	3 $\mathsf{P}(s \mid h \wedge r)$	PREM	$\emptyset$
	4 $\mathsf{P}(\neg s \wedge r \mid h)$	PREM	$\emptyset$
^{8}	5 $\mathsf{O}(\neg s \mid h \wedge r)$	1,4; RC	$\{!_r\mathsf{O}^r_h(\neg s)\}$
^{8}	6 $\mathsf{O}(h \supset \neg s \mid r)$	5; S	$\{!_r\mathsf{O}^r_h(\neg s)\}$
	7 $\neg\mathsf{O}(\neg s \mid h \wedge r)$	3; Def	$\emptyset$
	8 $!_r\mathsf{O}^r_h(\neg s)$	1,4,7; Agg	$\emptyset$

Note that with **CDPM**.α**c** ($\alpha \in \{1, 2'\}$) the counter-intuitive $\mathsf{O}(\neg s \mid h \wedge r)$ is derivable from $\mathsf{O}(\neg s \mid h)$ and $\mathsf{P}(\neg s \wedge r \mid h)$ by (WRM) causing an explosion due to $\mathsf{P}(s \mid h \wedge r)$. Since line 5 gets marked in the adaptive proof this problem has evidently been overcome.

11.6 Combining the Adaptive Systems

This section offers insight into the way the various systems introduced above can be combined, allowing for different degrees of adaptiveness. This is desired since each of our logics so far only treated particular problems of the ones presented in Sect. 11.4. By combining them we are able to get the most powerful and intuitive systems. A natural first suggestion for an AL account of both inheritance and weak rational monotonicity is given by the following logic:

Definition 11.6.1. $\mathbf{A}^{\mathbf{m}}_{\mathbf{r,c}}(\mathbf{L})$ is defined by the following triple

$$\langle \mathbf{L}, \Omega^c_d \cup \Omega^r_d, \text{minimal abnormality} \rangle,$$

where $\mathbf{L} \in \bigcup_{\alpha \in \{1,2'\}} \downarrow \ominus_{\text{WRM}}\, \mathbf{CDPM}.\alpha\mathbf{c}$.

However, the following example shows that this logic is only suboptimal.

Example 11.6.1. Suppose the following obligations and permissions:

- If your friends Beth and Mike are around, you're supposed to serve coffee.—$\mathsf{O}(c \mid f)$
- If your friend Anna, who has a coffee allergy, is around, you're supposed not to serve coffee.—$\mathsf{O}(\neg c \mid a)$
- If your friends Beth and Mike are around, you're allowed to serve coffee and to have Anne around.—$\mathsf{P}(c \wedge a \mid f)$
- If your friend Anna is around, you're allowed to have Beth and Mike around and to not serve coffee.—$\mathsf{P}(\neg c \wedge f \mid a)$

The following proof is in $\mathbf{A}^{\mathbf{m}}_{\mathbf{r,c}}(\oplus_{\text{PS}}\ominus(\{\text{WRM}, \text{S}\}, \mathbf{CDPM.1c}))$.

1	$\mathsf{O}(c \mid f)$	PREM	$\emptyset$
2	$\mathsf{O}(\neg c \mid a)$	PREM	$\emptyset$
3	$\mathsf{P}(c \wedge a \mid f)$	PREM	$\emptyset$
4	$\mathsf{P}(\neg c \wedge f \mid a)$	PREM	$\emptyset$
^{9}5	$\mathsf{O}(c \mid f \wedge a)$	1, 3; RC	$\{!_r\mathsf{O}^a_f(c)\}$
^{9}6	$\mathsf{O}(\neg c \mid f \wedge a)$	2, 4; RC	$\{!_r\mathsf{O}^f_a(\neg c)\}$
7	$\mathsf{O}(c \mid f \wedge a) \vee \mathsf{O}(\neg c \mid f \wedge a)$	5; RU	$\{!_r\mathsf{O}^a_f(c)\}$
8	$\mathsf{O}(c \mid f \wedge a) \vee \mathsf{O}(\neg c \mid f \wedge a)$	6; RU	$\{!_r\mathsf{O}^f_a(\neg c)\}$
9	$!_c\mathsf{O}(c \mid f \wedge a) \vee !_r\mathsf{O}^a_f(c) \vee !_r\mathsf{O}^f_a(\neg c)$	5, 6; RU	$\emptyset$

The minimal choice sets are $\big\{\{!_c\mathsf{O}(c \mid f \wedge a)\}, \{!_r\mathsf{O}^a_f(c)\}, \{!_r\mathsf{O}^f_a(\neg c)\}\big\}$. Note that $\mathsf{O}(c \mid f \wedge a) \vee \mathsf{O}(\neg c \mid f \wedge a)$ is derivable. This is undesired. One possible solution is to use the reliability strategy instead of the minimal abnormality strategy. In this case lines 5–8 are marked at line 9. Note however that, as pointed out for the monadic case in Chap. 10, the reliability strategy is with respect to Ω^c_d suboptimal in the case of the life saver and similar examples since it does not allow to derive the obligations to bring about at least the disjunction of conflicting obligations.

A better solution to this problem is presented in the form of a sequential AL:

Definition 11.6.2. $\mathbf{A}^{\mathbf{x}}_{\mathbf{r}} \circ \mathbf{A_c}(\mathbf{LLL})$ (where $\mathbf{x} \in \{\mathbf{m}, \mathbf{r}\}$) is the combined AL defined by the consequence relation $Cn_{\mathbf{A}^{\mathbf{x}}_{\mathbf{r}}(\mathbf{LLL})}(Cn_{\mathbf{A_c}(\mathbf{LLL})}(\Gamma))$ where $\mathbf{LLL} \in \bigcup_{\alpha \in \{1,2'\}} \downarrow\ominus_{\text{WRM}}$ $\mathbf{CDPM}.\alpha\mathbf{c}$.

The intuition behind this system is to first interpret "locally" for each condition C the corresponding obligations $\mathsf{O}(A \mid C)$, $\mathsf{O}(B \mid C)$ etc. as non-conflicting as possible. Then, strengthening of the antecedent is applied adaptively.

We need to slightly adjust the marking conditions for abnormalities in Ω^r_d. We show this paradigmatically for the case $\mathbf{x} = \mathbf{r}$: for a more detailed explication of the proof theory for sequential combinations of ALs see Sect. 3.3.

Definition 11.6.3. We define $U^r_s(\Gamma) = \bigcup_I \Delta_i$ where $\{\mathsf{Dab}(\Delta_i) \mid i \in I, \Delta_i \subset \Omega^r_d\}$ is the set of all minimal Dab-formulas derived at stage s at unmarked lines on a condition Δ' such that $\Delta' \cap \Omega^r_d = \emptyset$.

Line i is marked at stage s iff, where Δ is its condition, $\Delta \cap U^r_s(\Gamma) \neq \emptyset$.

Now the following line can be added to the proof of Example (11.6.1.):

$$10 \; !_r\mathsf{O}^a_f(c) \vee !_r\mathsf{O}^f_a(\neg c) \qquad\qquad 9; \text{RC } \{!_c\mathsf{O}(c \mid f \wedge a)\}$$

Due to the reliability strategy we have to mark lines 5–8. Note that for the minimal abnormality strategy lines 7 and 8 would not be marked at line 10. Thus, reliability strategy is preferable. Note furthermore that, although the reliability strategy is employed for Ω^r_d and hence for the conditional applications of (WRM), we are free to use the minimal abnormality strategy for Ω^c_d and hence for the conditional applications of (RCPM).

Theorem 11.6.1. *Where* $\alpha \in \{1, 2'\}$ *and* $\mathbf{x} \in \{\mathbf{m}, \mathbf{r}\}$, *we have:*

(i) $\mathbf{A}^{\mathbf{x}}_{\mathbf{r}} \circ \mathbf{A_c}(\ominus_{\text{WRM}}\mathbf{CDPM}.\alpha\mathbf{c})$ *satisfies (C‡).*
(ii) $\mathbf{A}^{\mathbf{x}}_{\mathbf{r}} \circ \mathbf{A_c}(\oplus_{\text{PS}}\ominus(\{\text{WRM}, \text{S}\}, \mathbf{CDPM}.\alpha\mathbf{c}))$ *satisfies (C‡).*

Theorem 11.6.2. *Where* $\alpha \in \{1, 2'\}$, $\mathbf{x} \in \{\mathbf{m}, \mathbf{r}\}$, *and* $\mathbf{LLL} \in \downarrow \ominus_{\text{WRM}} \mathbf{CDPM}.\alpha\mathbf{c}$, $\mathbf{A}^{\mathbf{x}}_{\mathbf{r}} \circ \mathbf{A_c}(\mathbf{LLL})$ *does not validate* (CDEX-1), (CDEX-2), *and* (CDEX-3).

11.7 Conclusion

Lou Goble presented in [1, 2] dyadic deontic logics based on a restriction of the inheritance principle. These systems are able to deal with deontic conflicts in the sense that they are not explosive facing dilemmas and they block undesired derivations from these dilemmas. Furthermore, by including a restricted version of rational monotonicity Goble is able to give an intuitive account of a restricted strengthening the antecedent (SA) principle.

However, this chapter locates various shortcomings of Goble's logics. Most importantly the treatment of (SA) is counter-intuitive for some examples and can even lead to explosion. Furthermore, some of the principles employed by the **CDPM** systems do not behave well together.

An adaptive logic framework has been developed which enables a satisfactory treatment of these problems. The price to pay for going adaptive is the lack of monotonicity. However, although Goble's systems are monotonic logics, a defeasible approach is motivated by them due to the restrictiveness of some of their rules. Take for instance the restricted version of the inheritance principle. In order to apply the rule, in many cases the user needs to add permission statements to the premise sets. Especially for complicated premise sets this demands a great deal of reasoning by the user and can lead to explosive behavior in cases in which she is not able to foresee all consequences of the addition of certain premises. A defeasible approach is far more elegant. It not only avoids explosive behavior in such cases but also shifts the reasoning from the user to the logic. It is self-explanatory that one of the basic requirements for an adequate deontic logic is that it models as much reasoning with as less external interference as possible. The ALs are well-suited for this task. By interpreting a premise set as "normally as possible", the required additional permission statements are generated automatically as part of the proof dynamics. Furthermore these proof dynamics explicate our moral reasoning processes.

While this chapter demonstrated that going adaptive improves **CDPM** in terms of elegance, strength and intuitiveness, in Chap. 12 and in [4] I show that the adaptive logic approach on basis of **CDPM** is also able to give a solution to another deep problem of dyadic deontic logics, namely the lack of a proper treatment of (defeasible) detachment. The advantages of the logics presented in this chapter and in Chap. 12 resp. [4] can easily be assembled by forming combined systems in the same manner as it was done in Sect. 11.6.

Acknowledgments I thank Joke Meheus, Mathieu Beirlaen, Frederik Van De Putte, and the anonymous referees of the Journal of Logical Philosophy for valueable comments which helped to improve the paper.

References

1. Straßer, C.: An adaptive logic framework for conditional obligations and deontic dilemmas. Logic Log. Philos. **19**(1–2), 95–128 (2010)
2. Goble, L.: A proposal for dealing with deontic dilemmas. In: Lomuscio, A., Nute, D. (eds.) DEON. Lecture Notes in Computer Science, vol. 3065, pp. 74–113. Springer (2004)
3. Straßer, C., Meheus, J., Beirlaen, M.: Tolerating deontic conflicts by adaptively restricting inheritance. Logique et Analyse **219**, 477–506 (2012)
4. Straßer, C.: A deontic logic framework allowing for factual detachment. J. Appl. Logic **9**(1), 61–80 (2010)
5. Meheus, J., Beirlaen, M., Putte, F.V.D.: Avoiding deontic explosion by contextually restricting aggregation. In: Governatori, G., Sartor, G. (eds.) DEON (10th International Conference on Deontic Logic in Computer Science). Lecture Notes in Artificial Intelligence, vol. 6181, pp. 148–165. Springer (2010)
6. Meheus, J., Beirlaen, M., Putte, F.V.D., Straßer, C.: Non-adjunctive deontic logics that validate aggregation as much as Possible. (Forthcoming)

7. Goble, L.: A logic for deontic dilemmas. J. Appl. Logic **3**, 461–483 (2005)
8. Horty, J.F.: Nonmonotonic foundations for deontic logic. In: Nute, D. (ed.) Defeasible Deontic Logic, Studies in Epistemology, Logic, Methodology, and Philosophy of Science, vol 263, pp. 17–44. Kluwer (1997)
9. Chisholm, R.M.: Contrary-to-duty imperatives and deontic logic. Analysis **24**, 33–36 (1963)
10. Forrester, J.: Gentle murder, or the adverbial samaritan. J. Philos. **81**, 193–197 (1984)
11. Kraus, S., Lehmann, D.J., Magidor, M.: Nonmonotonic reasoning, preferential models and cumulative logics. Artif. Intell. **44**, 167–207 (1990)
12. Lehmann, D.J., Magidor, M.: What does a conditional knowledge base entail? Artif. Intell. **55**(1), 1–60 (1992)
13. van Fraassen, B.C.: The logic of conditional obligation. J. Philos. Logic **1**, 417–438 (1972)
14. Lewis, D.: Counterfactuals. Blackwell Publishers (2000)
15. Goble, L.: Preference semantics for deontic logic. Part I: Simple models. Logique et Analyse **183–184**, 383–418 (2003)

Chapter 12
A Deontic Logic Framework Allowing for Factual Detachment

Since our ethical and behavioral norms have a conditional form, it is of great importance that deontic logics give an account of deontic commitments such as "A commits you to do/bring about B". It is commonly agreed that monadic approaches are suboptimal for this task due to several shortcomings, for instance their falling short of giving a satisfactory account of "Strengthening the Antecedent" or their difficulties in dealing with contrary-to-duty paradoxes. While dyadic logics are more promising in these respects, they have been criticized for not being able to model "detachment": A and the commitment under A to do B implies the actual obligation to do B. Lennart Åqvist asks in his seminal entry on deontic logic in the Handbook of Philosophical Logic: "We seem to feel that detachment should be possible after all. But we cannot have things both ways, can we? This is the dilemma on commitment and detachment." (Lennart Åqvist in ([1], p. 199)).

In this chapter I answer Åqvist's question with "Yes, we can". I propose a general method to turn dyadic deontic logics in ALs allowing for a defeasible factual detachment while paying special attention to specificity and contrary-to-duty cases. I show that a lot of controversy about detachment can be resolved by analysing different notions of unconditional obligations. The logical modeling of detachment is paradigmatically realized on basis of one of Lou Goble's conflict tolerant **CDPM** logics.

12.1 Introduction

In this chapter I propose a way to adaptively enhance dyadic deontic logics in order to enable defeasible factual detachment. This is done by paying special attention to specificity and contrary-to-duty (CTD) cases. Paradigmatically the adaptive treatment is demonstrated by means of one of Lou Goble's powerful conflict-tolerant

---

A former version of the content of this chapter has been elaborated in [37].

C. Straßer, *Adaptive Logics for Defeasible Reasoning*, Trends in Logic 38,
DOI: 10.1007/978-3-319-00792-2_12,

CDPM logics.[1] But let me slowly situate this enterprise in the landscape of deontic logics and introduce the reader, by doing so, into the central notions and problems.

12.1.1 Monadic Approaches to Conditional Obligations

While monadic deontic logics, i.e., logics employing unary obligation and permission operators, are powerful tools to model normative notions such as obligations, duties, oughts, etc., there are several shortcomings concerning their abilities to model conditional obligations. Since most of our moral or behavioral norms are in a conditional form, it is important to develop deontic logics that adequately model them.

There are two canonical ways to represent of the conditional obligations of the type "Under condition A you're obliged/committed to bring about B" in unary logics: (i) by $A \supset \mathsf{O}B$ and (ii) by $\mathsf{O}(A \supset B)$, where $\mathsf{O}A$ is written for the obligation/the duty/etc. to bring about A (and analogously for permissions $\mathsf{P}A$). A lot of criticism of such ways of modeling commitment has been published (for a survey see [1]). Let me mention two of them.

(1) Some prominent arguments concern the appropriate representation of Strengthening the Antecedent (SA)—if A commits you to do B then A and C commit you to do B. In most situations the logic is obviously expected to validate SA. But consider the following case (A):

A1 In general we're supposed not to eat with our fingers.
A2 Being served asparagus, we are supposed to eat with our fingers.

If A1 is modeled by $\top \supset \mathsf{O}\neg f$, then $a \supset \mathsf{O}\neg f$ is derivable. This obviously is in conflict with A2, $a \supset \mathsf{O}f$. It would be counter-intuitive that, being served asparagus, we do have at the same time the obligation to eat and not to eat with our fingers. Similarly problematic is to use $\mathsf{O}(A \supset B)$ to represent the obligation B under condition A. In the example we can derive by the inheritance principle[2] $\mathsf{O}(a \supset \neg f)$ from $\mathsf{O}(\top \supset \neg f)$. However, this is counter-intuitive with respect to A2.

(2) Problems related to CTD obligations are commonly considered as the death-blow for any effort in modeling conditional obligations in a monadic way. A prominent example is given by the so-called Chisholm Paradox (C) (see [4]):

C1 John ought not to impregnate Diane.
C2 If John impregnates Diane, he ought to marry her.
C3 If John does not impregnate Diane, he ought not to marry her.
C4 John impregnates Diane.

---

[1] See [2, 3], where Goble presents a family of logics, **CDPM.xy** where $\mathbf{x} \in \{\mathbf{1}, \mathbf{2}\}$ and $\mathbf{y} \in \{\mathbf{a}, \mathbf{b}, \mathbf{c}\}$. I use in this chapter **CDPM** as a generic term for this family of logics.

[2] The inheritance principle, also often referred to as rule (K), is given by: If $\vdash A \supset B$ then $\vdash OA \supset OB$. It is validated by most deontic logics, above all by standard deontic logic.

As it has been shown for instance in [1], monadic approaches face severe difficulties in modeling such examples.[3]

It has been argued that deontic paradoxes are due to the lack of a modeling of the temporal aspect of obligations (see [11]). However, Prakken and Sergot have in [12] presented a version of Chisholm's paradox that is entirely independent of any temporal parameters (see Example 12.6.6, p. 322). This shows that any hopes to resolve all CTD puzzles by temporal deontic logics have to be abandoned. Furthermore, it motivates research in non-temporal deontic logics which are able to tackle CTD puzzles.

12.1.2 Dyadic Approaches and the Problem of Detachment

It is commonly accepted that dyadic approaches can lead to satisfying solutions with respect to the CTD problems and, as for instance in the case of Goble's **CDPM** [2, 3], there are also ways to tackle problems such as the ones related to (SA). In a dyadic deontic logic conditional obligations are modeled by a dyadic obligation operator. $\mathsf{O}(A \mid B)$ is written for the obligation to bring about A in case of B (and similarly for permissions $\mathsf{P}(A \mid B)$). I sometimes refer to A as being the conclusion and to B as being the antecedent or condition of the conditional obligation $\mathsf{O}(A \mid B)$.[4] Despite their merits, dyadic approaches are criticized for not giving a satisfactory account of detachment. Factual detachment (from now on shortly *detachment*) is, generally speaking, to derive the 'actual' obligation to bring about A from the commitment to bring about A under condition B and the fact B.[5] Formally:

$$\vdash \big(\mathsf{O}(A \mid B) \wedge B\big) \supset \mathsf{O}A \tag{FD}$$

The unconditional obligation $\mathsf{O}A$ should indicate that the obligation to bring about A is 'actual' for an agent in question and that it 'binds' her. This rough description is still somewhat ambiguous. It will be one of my tasks to analyze it further (see Sect. 12.2). That the modeling of detachment is not a trivial task can easily be demonstrated by the asparagus example (A). I write $\mathsf{O}(\neg f \top)$ for A1 and $\mathsf{O}(f \mid a)$ for $A2$. Now suppose

[3] It would go beyond the scope of this chapter to offer a detailed analysis of the example. However, this has been done before in the literature and I refer the interested reader to e.g. [1], Sect. 8. It should be mentioned that recent developments in reactive modal logics (see e.g., [5–10]) offer a way to overcome many of the usual shortcomings of monadic deontic logics in the context of Contrary-to-Duty examples such as the Chisholm Paradox.

[4] I restrict the discussion in this chapter to the case that the arguments of the obligation operator are propositional formulas, i.e., formulas without occurrences of obligation operators. The handling of detachment concerning obligations such as $\mathsf{O}(\mathsf{O}(A \mid B) \mid C)$ deserves a discussion in its own right. Such a discussion would have to answer questions concerning the proper way of dealing with nested obligations, e.g., should we infer $\mathsf{O}(A \mid B \wedge C)$ from $\mathsf{O}(\mathsf{O}(A \mid B) \mid C)$, or, to what extent, if any, can we make sense of detaching obligations from obligations.

[5] There is also *deontic detachment* (cp. [13]). I will discuss it in more detail in Sect. 12.6.1.

we are being served asparagus, a. However, we can apply (FD) to both premises A1 and A2 and thus get $\mathsf{O}\neg f$ and $\mathsf{O}f$, a deontic conflict. Confronted with such cases Åqvist pessimistically stated,

> We seem to feel that detachment should be possible after all. But we cannot have things both ways, can we? This is the dilemma on commitment and detachment. ([1], p. 199)

And indeed, most of the existing dyadic logics do not allow for detachment. Of course, then we have to ask with Van Eck rhetorically,

> How can we take seriously a conditional obligation if it cannot, by way of detachment, lead to an unconditional obligation? ([14], p. 263)

Indeed, normative reasoning is not just reasoning about conditional commitments, i.e., deriving other commitments from given ones, but to a high degree also applying a set of conditional norms to a factual situation in order to see what unconditional actual obligations bind us and guide our actions.

12.1.3 The Contribution and Structure of this Chapter

In this chapter I answer Åqvist's question whether it is possible to have both, commitment and detachment, with "Yes, we can". I propose a general method to turn dyadic deontic logics into ALs allowing for a defeasible factual detachment while paying special attention to specificity and CTD cases. This will equip the logics with the ability to model actual deontic reasoning/inferencing. That is to say, the logic is then able, given a set of factual information $\mathcal{F}$, to derive obligations that are actual and bind us (or some agents in question) in the very situation described by $\mathcal{F}$. I will proceed along the following steps:

Before giving a formal account of detachment, we have to ask, what kind of unconditional obligations do we want to detach. This is not a trivial question and there is a lot of controversy concerning CTD cases. In Sect. 12.2 I will point out that there is more than one sensible notion. For the formal modeling I paradigmatically settle for two notions that I consider to be very interesting for practical applications: instrumental and proper obligations.

Now, having a conceptual grip on the obligations that should be detached, I can proceed by giving a formal model. The central idea in this chapter is to apply detachment conditionally, namely on the condition that it can be considered safe to apply it. Therefore the language and axiomatization of a given dyadic deontic logic—let me call it the *base logic*—is enriched in such a way that it is expressible when it is unsafe to apply detachment (see Sect. 12.3). In such cases these obligations are labeled, $\bullet\mathsf{O}(A \mid B)$. Formally the conditional applications of detachment are realized by ALs. The idea of ALs is to interpret a premise set "as normally as possible" with respect to a certain standard of normality. They allow for some rules to be applied conditionally. In our case detachment is going to be applied "as much as possible".

The idea is to apply detachment to an obligation $\mathsf{O}(A \mid B)$ and B on the condition that $\bullet\mathsf{O}(A \mid B)$ can be considered to be false.

After having presented the generic framework I am going to apply the "adaptivization" to a concrete base logic. The choice is one of Lou Goble's **CDPM** systems (see [2, 3]) which I introduce in Sect. 12.5. This is a good choice in order to achieve an altogether very powerful logic. Let me highlight some of the strengths. Goble's logic has an intuitive handling of SA based on a weakening of the Rational Monotonicity principle (see e.g. [15]). Furthermore, by restricting the inheritance principle it is conflict-tolerant. That is, it is able to deal with deontic conflicts, i.e., situations in which we are committed to do A as well as to do not-A or some B which cannot be jointly realized with A. Standard deontic logic leads in such cases to deontic explosion, namely to the derivability of all obligations.

By going adaptive in Sect. 12.6 one more strength is added, namely an intuitive modeling of detachment. I will demonstrate the modus operandi and the strengths of the adaptive system by a number of standard benchmark examples.

Finally, in Sect. 12.8 I discuss some advantages of the adaptive handling of detachment and suggest some useful enhancements. In Appendix J I present the semantics, completeness and soundness proofs for the logics presented in this chapter.

12.2 What to Detach?

This section clarifies our intuitions about what kind of obligations to detach from conditional obligations. This is necessary since there is a variety of conceptions concerning the types of obligations involved in certain setups and concerning their relationships. Due to this, scholars disagree about how and when to apply detachment.

Preliminarily it can be said that what should be detached are obligations that are in some sense 'actual' and that in some sense 'bind us'. These are obligations whose condition is fulfilled, and that are not 'canceled' or 'destroyed' by other obligations. These points deserve a deeper analysis which I am going to provide in this section. I will emphasize two types of obligations that are especially interesting for detachment: *instrumental* and *proper* obligations.[6] In order to explicate these concepts I will first focus on two paradigmatical cases in which conditional obligations are in a sort of tension: specificity and CTD cases. This will provide a good basis to disentangle the discussion in the literature about when, or to what kind of obligations to apply detachment.

[6] This distinction is not exhaustive. There are other conceptions, such as Carmo and Jones' "ideal" obligation (see [16] and Sect. 12.2.5).

12.2.1 Exceptional Contexts and Specificity

Let us look at an example:

- When the plane takes off you ought to turn off electronic devices.—$\mathsf{O}(e\,|\,p)$
- The plane takes off.—p

It seems reasonable to apply detachment to the two premises in order to derive the 'actual' obligation to turn off electronic devices. However, we have to be careful. Norms and obligations often allow for exceptional contexts. Consider for instance our asparagus example (A1, A2), $\mathsf{O}(\neg f\,|\,\top)$ and $\mathsf{O}(f\,|\,a)$, or the weaker version, $\mathsf{O}(\neg f\,|\,\top)$ and $\mathsf{P}(f\,|\,a)$. Often scholars dub such settings *specificity*. The idea is that, where C is a more specific context than B, an obligation $\mathsf{O}(A\,|\,B)$ does not hold in the context C due to the presence of $\mathsf{P}(\neg A\,|\,C)$ or $\mathsf{O}(\neg A\,|\,C)$ if the context C can be considered as exceptional. In this case the obligation $\mathsf{O}(A\,|\,B)$ is excepted. For instance, the situation in which we eat asparagus is an exceptional context to the general obligation not to eat with our fingers and the latter is excepted in this context.

Let us get a more formal grip on this. $\langle C_1, \ldots, C_n\rangle$ is a *permissive sequence* from C_1 to C_n iff, for all $i < n$, (a) $\vdash C_{i+1} \supset C_i$ and (b) $\mathsf{P}(C_{i+1}\,|\,C_i)$. We say that C is a *permissible context* to B iff there is a permissive sequence from B to C.

Obviously a is a permissible context to $\top$ since $\mathsf{P}(a\,|\,\top)$. Moreover, also $f \wedge a$ is a permissible context to $\top$, although we do not have $\mathsf{P}(f \wedge a\,|\,\top)$.[7] However, we have $\mathsf{P}(a\,|\,\top)$ and $\mathsf{P}(f \wedge a\,|\,a)$ which constitute the permissive sequence $\langle\top, a, f \wedge a\rangle$.

C is an *exceptional context* to $\mathsf{O}(A\,|\,B)$ iff

EC1 C is a permissible context to B, and
EC2 there is an exceptional permission (resp. obligation), $\mathsf{P}(D\,|\,C)$ (resp. $\mathsf{O}(D\,|\,C)$), where $\vdash D \supset \neg A$.

The idea is that, if C is a permissible context to B and we have incompatible norms in the two contexts, e.g. $\mathsf{O}(A\,|\,B)$ and $\mathsf{O}(\neg A\,|\,C)$, then we have a case of specificity and $\mathsf{O}(A\,|\,B)$ is excepted. In Example (A) a situation in which we are being served asparagus, a, as well as the connected obligation $\mathsf{O}(f\,|\,a)$ are approved by $\mathsf{P}(a\,|\,\top)$. Thus, if a is the case, f should not be seen as a violation of $\mathsf{O}(\neg f\,|\,\top)$, but rather as an exception. Note that it would be suboptimal to use a more simplified definition of exceptional contexts in which EC1 is replaced by $\mathsf{P}(C\,|\,B)$. The reason is that according to the latter, simplified definition, $f \wedge a$ is not an exceptional context to $\mathsf{O}(\neg f\,|\,\top)$, which is counter-intuitive.

We say that $\mathsf{O}(A\,|\,B)$ is *excepted* in C iff C is an exceptional context to $\mathsf{O}(A\,|\,B)$. Hence, $\mathsf{O}(\neg f\,|\,\top)$ is excepted in the contexts a and $f \wedge a$. Cases such as the asparagus example are usually considered as obligations "overriding", "canceling", or "destroying" other obligations in certain contexts, i.e., $\mathsf{O}(f\,|\,a)$ overrides $\mathsf{O}(\neg f\,|\,\top)$.

[7] Note that it is rather problematic to add $\mathsf{P}(f \wedge a\,|\,\top)$ to the premises, since the latter is equivalent to $\neg\mathsf{O}(\neg f \vee \neg a\,|\,\top)$ and in most deontic logics $\mathsf{O}(\neg f \vee \neg a\,|\,\top)$ is entailed by $\mathsf{O}(\neg f\,|\,\top)$ (due to modal inheritance).

Let me contrast exceptional contexts and specificity with violation contexts and CTD obligations.

12.2.2 Violation Contexts and Contrary-to-Duty Obligations

C is a *violation context* to $\mathsf{O}(A|B)$ iff

VC1 $\vdash C \supset B$ (C is at least as specific as B),
VC2 C is not a permissible context to B, and
VC3 $\vdash A \supset \neg C$ (A and C are incompatible).

If C is the case then $\mathsf{O}(A|B)$ is (factually) violated by C. Note that $\mathsf{O}(A|B)$ is not excepted in the context C. In this sense violation contexts define sub-ideal situations, opposite to exceptional contexts, in which no obligation is violated. In order to give an example it is better to first introduce CTD obligations. A prominent case is given by the following "Gentle Murderer" example (G)[8]:

G1 Doe is in general obliged not to kill his mother.—$\mathsf{O}(\neg k|\top)$
G2 However, if Doe kills his mother, he ought to kill her gently.—$\mathsf{O}(g|k)$
G3 Doe kills his mother.—k

I presuppose $g \vdash k$. Obligations of this kind are usually dubbed *contrary-to-duty obligations*. Formally: $\mathsf{O}(D|C)$ is a *(strong) CTD obligation* to $\mathsf{O}(A|B)$ iff, (i) $\vdash A \supset \neg D$ and (ii) C is a violation context to $\mathsf{O}(A|B)$.[9] As is customary, I call $\mathsf{O}(A|B)$ the *primary* and $\mathsf{O}(D|C)$ the *secondary* obligation. Note that A is inconsistent, both with the antecedent and the conclusion of $\mathsf{O}(D|C)$. There is a certain sense in which the secondary obligation is dependent on the primary one. It specifies norms for the case that the primary obligation is violated. In the case of strong CTD obligations these norms are themselves incompatible with the primary obligation. In order to indicate this aspect, we say that the secondary obligation $\mathsf{O}(D|C)$ is *burdened* with the primary obligation $\mathsf{O}(A|B)$. The strong CTD obligation expresses that, if the primary obligation is being violated, then there are certain ways of violating it that are normatively preferable to others. If Doe already kills his mother, then he should at least do it gently. Evidently k is a violation context to $\mathsf{O}(\neg k|\top)$. Moreover, $\vdash \neg k \supset \neg g$. Hence, $\mathsf{O}(g|k)$ is a strong CTD obligation to $\mathsf{O}(\neg k|\top)$.

We say that $\mathsf{O}(D|C)$ is a *weak CTD obligation* to $\mathsf{O}(A|B)$ iff, C is a violation context to $\mathsf{O}(A|B)$ and $\mathsf{O}(D|C)$ is not a strong CTD obligation to $\mathsf{O}(A|B)$ (hence, $\nvdash A \supset \neg D$). Note that in this case $\mathsf{O}(D|C)$ is not burdened with $\mathsf{O}(A|B)$. Weak CTD obligations express that, if the primary obligation is being violated, then this violation introduces certain other consequential norms beside the strong CTD obligations that

[8] This version of the Forrester Paradox (cp. [17]) is taken from McNamara ([18], p. 243).

[9] It would be suboptimal to define (strong) CTD obligations by (i) together with $\vdash A \supset \neg C$, since then e.g. $\mathsf{O}(f|f \wedge a)$ would be a CTD obligation to $\mathsf{O}(\neg f|\top)$. However, this is counter-intuitive since $\mathsf{O}(\neg f|\top)$ is excepted in the context $f \wedge a$, as argued above.

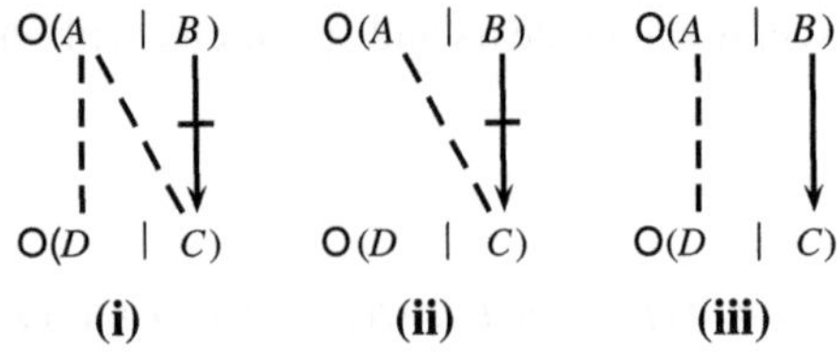

Fig. 12.1 (**i**) Strong CTD obligation, (**ii**) Weak CTD obligation, (**iii**) Specificity. The *dashed line* indicates an inconsistency between, for instance, A and D in (**i**). The (*striked-out*) *solid arrow* indicates that C is (not) a permissible context to B

specify the normatively optimal ways the violation should take place. An example (PA) is

PA1 In general you should not break your promise to attend the meeting.—$\mathsf{O}(\neg b|\top)$.

PA2 If you break your promise to attend the meeting, then you are supposed to apologize.—$\mathsf{O}(a|b)$.

Evidently b is a violation context to $\mathsf{O}(\neg b|\top)$. Note that $\mathsf{O}(a|b)$ is not a strong CTD obligation to $\mathsf{O}(\neg b|\top)$ and is hence not burdened by $\mathsf{O}(\neg b|\top)$, since its conclusion, i.e. to apologize, is not incompatible with the primary obligation. It does not inform us how to break the promise in the normatively preferable way but rather states a norm that is a consequence of the broken promise (and that is itself not incompatible with the primary obligation).

Figure 12.1 features a comparative overview of the introduced obligations types.

12.2.3 The Dissent in the Literature

There has been some discussion concerning detachment, especially with respect to (strong) CTD cases such as the Forrester Paradox (G). McNamara argues against detaching CTD obligations:

> So *carte blance* factual detachment seems to allow the mere fact that I will take an action in the future (killing my mother) that is horribly wrong and completely avoidable now to render obligatory another horrible (but slightly less horrible) action in the future (killing my mother gently). ([18], p. 268)

What McNamara has in mind in his criticism is the case that $\mathsf{O}(\neg k|\top)$ is violated (and not excepted): in the case that Doe cannot be excused for killing his mother since it was 'completely avoidable', but he undertakes the killing anyway, it seems absurd to detach an obligation to kill her gently, since, after all, the primary obligation, not to kill his mother, binds Doe despite him acting against it.

Wang in [19] disagrees. He points out the importance of "premise-dependency" (ibid., p. 13) when evaluating detachment inferences. He remarks:

> *the logic should construct the detachment inference in a way that makes the conclusion be evaluated in a premise-dependent way* (ibid., p. 12, Wang's emphasis)

In order to explicate this notion he reminds his readers of classical logic. There "B is a logical consequence of A, iff if A is true, then B is true, where the evaluation of B is given under the evaluation of A, i.e., B is evaluated in the models where A is true." (ibid., p. 12). In the Gentle Murderer case, the conclusion to kill the mother gently has the counter-intuitive appeal highlighted by McNamara mainly when interpreted in a premise-independent way, namely when we forget that the evaluation of the conclusion takes place in worlds/models in which Doe anyway kills her. Of course, isolated, i.e., independent from the evaluative constraint that he in fact kills her, the obligation to kill his mother gently is against our moral convictions. However, settling for a premise-dependent evaluation and accepting the premises G2 and G3, there is nothing which should cause any unease with our acceptance of the conclusion, that he should kill her gently. Similarly, accepting the conclusion "Doe will win the lottery tomorrow." from the premises (i) "If Doe knows the lottery numbers and he will play them in due time, then he will win." and (ii) "Doe knows the lottery numbers and he will play them in due time." is counter-intuitive only in the case that we forget about (ii). In general it is of course very unlikely for Doe to win the lottery. However, if we know that Doe knows the numbers already, it is perfectly correct to conclude that he will win.

Wang criticizes McNamara for giving a premise-independent interpretation of the conclusion for the Gentle Murderer: while

> *obligatory* in the premise [G2] is interpreted in the evaluative sense, i.e. given that Doe does kill his mother, it is better that Doe kills her gently, McNamara objects to the conclusion by a moral sense of the obligation based on what is morally right or wrong. (ibid., Footnote 12, p. 11).

He moreover points out that, opposite to the premise-dependent treatment of conclusions in the case of classical logic, "the evaluation should [in the case of detachment] be considered in a defeasible way." (ibid., p. 12). This is also clear looking at Example (G). Evidently, the antecedent of G1, $\top$, is fulfilled. However, the antecedent of the more specific G2 is also fulfilled, which leads in Wang's perspective to G1 being defeated.

A different view is offered by Van Der Torre and Tan in [20]. In cases such as the asparagus example they speak of an obligation, e.g. $\mathsf{O}(\neg f \mid \top)$, being *overridden* and *canceled* by a more specific obligation, e.g. $\mathsf{O}(f \mid a)$. However, in CTD situations they use the term *overshadowing*: for instance, $\mathsf{O}(\neg k \mid \top)$ is overshadowed by $\mathsf{O}(g \mid k)$ in a situation in which the killing takes place. The primary obligation, though being violated, is not in any way canceled or destroyed. Therefore, Van Der Torre and Tan argue that it would be intuitive to derive $O\neg k$, i.e., to apply detachment to $\mathsf{O}(\neg k \mid \top)$, while "[t]he consistency of $O\neg k \wedge Og$ is a solution that seems like overkill" ([21], p. 53). With respect to our terminology it seems to be the intuition of Van Der Torre and Tan that in most cases excepted obligations are being overridden while violated obligations are being overshadowed. It can be argued that Van Der Torre and

Tan circumvent the criticism by Wang: they evaluate the conclusion in a premise-dependent way and it is precisely the nature of CTD obligations that justifies and motivates that detachment is not applied to them, but rather to the more general primary obligation which is being violated. The intuition is that, whether Doe kills his mother or not, he has the obligation not to kill her and this obligation binds him. In the former case he violates this obligation.[10]

However, one might ask, what kind of conditional does G2 represent? Although it is of the form "if ..., then ...", a logic in McNamara's or Van Der Torre and Tan's sense would in no way allow to detach its conclusion. Nevertheless, in case of G3, G2 binds Doe and is actual in a certain way. In fact, both, G1 and G2 are of a binding nature in case of G3, despite the fact that both conclusions are incompatible. Thus, I suggest that a proper deontic logic allowing for detachment should be able to distinguish these cases and to offer an adequate detachment mechanism for both intuitions.

12.2.4 Instrumental and Proper Obligations

So, who is right, on the one hand Wang, or on the other hand McNamara, Van Der Torre and Tan? I suggest: all of them in their own way. This is due to the fact that different intuitions are in place which are all justified in their own respect once they are made more transparent.

Let us first look at Wang's intuition. The disagreement between him and the other scholars concerns mainly circumstances in which some (primary) obligations are factually violated. For instance, in the case in which Doe kills his mother, a fact that obviously violates his primary obligation not to kill her. However, in violation contexts as well as in any other context, there may still be certain obligations which guide our actions, despite the fact that some obligations are not realizable or excepted. This may for instance be obligations that tell us, if we already violate a primary obligation, how to do so in the morally most correct way. For Doe this means to at least kill his mother in a gentle way. There are obligations that tell us what we should and can bring about without considering already violated obligations. They tell us what *is* the right thing to do in a certain situation, but not what *would have been* the right thing to do in first place, for instance, to avoid a sub-ideal situation in which some obligations are violated.[11] I call these obligations from now on *instru-*

[10] Prakken and Sergot offer a similar view. A "key difference between contrary-to-duty and *prima facie* obligations" ([22], p. 224) is that, unlike prima facie obligations and opposite to Wang's view, the former do not satisfy any form of (defeasible) factual detachment. The only form of detachment they satisfy is the following strong detachment principle: $\models (\mathsf{O}(A \mid B) \wedge \Box B) \supset OA$, where $\Box$ is the necessity operator of an adequate modal logic. Detaching the obligation to kill gently would cause an inconsistency which is "counter to our intuitions". The obligation not to kill is not overridden, it is fully valid.

[11] Foot in [23] introduces a similar concept. She dubs 'obligations of type 2' obligations which "tell us *the right* thing to do" ([23], p. 385). They answer the question

mental obligations and write $\mathsf{O}^{\mathsf{i}}A$ for the instrumental obligation to bring about A. More formally the idea could be expressed as follows: given the factual information $\mathcal{F}$, the instrumental obligations amount to those obligations that an agent is expected to practically realize in the most normal and ideal worlds that validate $\mathcal{F}$. The latter are worlds in which as less violation and exceptional contexts are the case as possible, i.e., as is coherent with $\mathcal{F}$. For instance, in case of the asparagus example, if we do not have any specific factual information about the meal which is served, the most normal worlds are worlds in which asparagus is not served. Thus, we are expected not to eat with our fingers.

Instrumental obligations seem to cohere with Wang's intuition. Interpreting the factual premises as settled and immutable, all-things-considered it is the best thing to do for Doe to kill his mother gently, presupposing that he kills her. Thus, violated primary obligations are not candidates for detaching instrumental obligations. Rather are the secondary CTD obligations the ones which should be detached: in this case to kill the mother gently. Also, in case your promise is broken, the weak CTD obligation to apologize should be considered as an instrumental obligation to be detached, but not the primary obligation not to break the promise since the promise is already broken. Excepted obligations are not candidates for instrumental detachment: being served asparagus we are obviously exempted from the obligation not to eat with our fingers. The latter has to be considered as being overridden and the more specific obligation to eat with our fingers should be detached as instrumental obligation (see Table 12.1 for an overview).

Note that instrumental obligations are both, 'actual' (in our preliminary understanding of not being canceled or destroyed by another obligation) and binding in the sense of guiding our actions.

Instrumental obligations are not the only sensible notion for unconditional obligations which are interesting for detachment. The intuition behind Van Der Torre and Tan's approach may be linked to another concept which I call *proper obligations*. I write $\mathsf{O}^{\mathsf{p}}A$ for the proper obligation to bring about A. These obligations tell us, all-things-considered, which obligations are predominantly in force in a certain situation, where, (i) we also take into account obligations which have been violated but which still bind us, and, (ii) primary or stronger obligations are prioritized in

Table 12.1 What is detached? The columns are considered to be mutually exclusive. Ideal obligations are introduced in Sect. 12.2.5

	Violated obl.	Excepted obl.	Strong CTD obl.	Weak CTD obl.
Instr. obl.	×	×	✓	✓
Proper obl.	✓	×	×	✓
Ideal obl.	✓	×	×	×

(Footnote 11 continued)
"And what all things considered ought we to do?" (p. 386). Of course, it would have been the best thing to do for Doe not to kill his mother. However, interpreting the factual premises as unalterable facts, the best thing Doe may (still) do is to kill his mother gently (if he already kills her).

cases of conflicts. Hence, proper obligations are all obligations for a given context that are neither excepted nor burdened. For instance, all non-conflicted obligations conditioned for a given context are proper obligations. That proper obligations are 'actual' even in case they are violated is the reason why we judge a violator to be guilty, and it is witnessed by us being conscience-smitten after violating them.

In the case of example (G), the moral reasons for Doe not to kill his mother are stronger, or primary, compared to the reasons to kill her gently: after all, the obligation to kill her gently is only secondary for the case that he violates the primary obligation. It provides a practical guideline how to act in the situation in which the murder takes place. However, the obligation proper for Doe is not to kill. Thus, we are interested in detaching (not-excepted) violated primary obligations rather than their strong CTD counterparts. This is a key difference to the instrumental case. However, weak CTD obligations should be detached as being proper obligations since you are obliged to apologize in case you break your promise and this obligation is neither excepted nor burdened. Hence, it is not in conflict with a more general obligation that would bind you more. It is a proper obligation to apologize in the violation context, not "just" a norm informing you, if you already violate an obligation, how to do so in the normatively best way. Also the primary obligations in weak CTD cases should be detached as proper obligations, since neither are they excepted nor are they burdened. Analogous to the primary obligation in strong CTD cases they bind us as proper obligations despite the fact that they are violated.[12] Analogous to the instrumental case, excepted obligations should not be detached, but rather the more specific obligations which override them: it is our proper obligation to eat with our fingers in the case that asparagus is served (see Table 12.1 for an overview).

Concluding, it can be said that proper obligations bind us (in contrast to instrumental obligations) even in the case that they are violated by secondary obligations (if they are not excepted otherwise). They are 'actual' in our preliminary understanding since, also in the case of being violated they are not destroyed or canceled, or as Van Der Torre and Tan put it: they are merely overshadowed.

---

[12] Of course, in an 'ideal' world there is no need to apologize. Thus, Prakken and Sergot in [12] call it a 'pragmatic oddity' if, in the case of b, both obligations, to keep the promise and to apologize, are considered to be 'actual'. Van Der Torre and Tan ([21], p. 63) call it counter-intuitive. Let us see if our proposal causes an 'oddity'. An oddity could be given in two respects: (a) concerning ideality and (b) concerning a pragmatic aspect. Due to (a) there would be an oddity if we had $\mathsf{O}^{\mathsf{p}} A$ and $\mathsf{O}^{\mathsf{p}} B$ for incompatible A and B, and the intended reading of $\mathsf{O}^{\mathsf{p}} A$ would be that A is true in all "ideal worlds". This, however, is not the case with our analysis since for our proper obligation $\mathsf{O}^{\mathsf{p}} A$ the intended reading is not that this is an obligation ideal in the mentioned sense, but rather that it is an obligation actual in the given context that is neither excepted nor burdened. Furthermore, taking into account that, (i), some proper obligations may be violated while other proper obligations may still be realizable (the ones which are also instrumental obligations such as in our case the obligation to apologize) and, (ii), that the two obligations in question are consistent, I do not think that this is counter-intuitive. Due to (b) we would have a pragmatic oddity in case we would get two in some sense incompatible obligations which tell us what to do. This boils down to the concurrence of two incompatible instrumental obligations in our sense. However, again we do not have an oddity in the given example since we only have $\mathsf{O}^{\mathsf{i}} a$ and no other instrumental obligation with which it is incompatible.

12.2.5 Carmo and Jones' Proposal

Carmo and Jones propose in [16] a similar distinction of detachable obligations. They also reject the frequent reduction of an 'actual' unconditional obligation $\mathsf{O}A$ to the dyadic representation $\mathsf{O}(A|\top)$. Instead they define two notions of obligations, "actual" and "ideal" ones, each represented by a monadic operator, O^A and O^I. They realize detachment by introducing two modal operators, $\boxdot$ (and its dual $\Diamond\!\!\cdot$) for representing what "is *actually* fixed, or unalterable, given (among other factors) what the agents concerned have decided to do or not to do" (p. 286), and $\Box$ (and its dual $\Diamond$) for representing what is fixed and "could not have been avoided by the agents concerned, ... It is not even *potentially* possible for the agents to alter" (p. 287) these fixed facts. It is illuminating to take a look at their factual detachment principles (Table 12.2):

$$\big(\mathsf{O}(B|A) \wedge \boxdot A \wedge \Diamond\!\!\cdot B \wedge \Diamond\!\!\cdot \neg B\big) \supset \mathsf{O}^{\mathrm{A}} B \qquad (\mathsf{O}^{\mathrm{A}}\text{-FD})$$

$$\big(\mathsf{O}(B|A) \wedge \Box A \wedge \Diamond B \wedge \Diamond \neg B\big) \supset \mathsf{O}^{\mathrm{I}} B \qquad (\mathsf{O}^{\mathrm{I}}\text{-FD})$$

The intuition behind their "actual" obligations is very similar to our instrumental ones: in case an agent decides to act against a duty such as in the Gentle Murderer case, $\boxdot k$, and it is possible for him to either kill gently, $\Diamond\!\!\cdot g$, or to not kill gently $\Diamond\!\!\cdot \neg g$, he has, due to $\mathsf{O}(g|k)$, the "actual" obligation to kill gently, $\mathsf{O}^{\mathrm{A}} g$. There is a distinctive difference however concerning their "ideal" and our proper obligations. Take for instance example (PA): while in our analysis it is the proper obligation to apologize, it is not an "ideal" obligation in the sense of Carmo and Jones. As discussed above, our proper obligations are the non-excepted and non-burdened obligations that are valid for the given context. The only "ideal" obligation for (PA) is the obligation not to break the promise. This is also a proper obligation. Note however that $\mathsf{O}^{\mathrm{I}} a$ is not derivable if we consider the standard case for which $\Diamond b \wedge \Diamond \neg b$ is valid. Nevertheless, $\mathsf{O}^{\mathsf{p}} a$ is a proper obligation in example (PA) since it is not excepted and not burdened in the context b.

According to the distinction proposed by Carmo and Jones the treatment of strong and weak CTD cases is the same: in both cases the primary obligation is the "ideal"

Table 12.2 Overview of the Examples

Example	(A)	(G)	(PA)
Premises	$\mathsf{O}(\neg f|\top)$	$\mathsf{O}(\neg k|\top)$	$\mathsf{O}(\neg b|\top)$
	$\mathsf{O}(f|a)$	$\mathsf{O}(g|k)$	$\mathsf{O}(a|b)$
	$\mathsf{P}(a|\top)$		
	a	k	b
Proper	$\mathsf{O}^{\mathsf{p}} f$	$\mathsf{O}^{\mathsf{p}} \neg k$	$\mathsf{O}^{\mathsf{p}} \neg b$, $\mathsf{O}^{\mathsf{p}} a$
Instrumental	$\mathsf{O}^{\mathsf{i}} f$	$\mathsf{O}^{\mathsf{i}} g$	$\mathsf{O}^{\mathsf{i}} a$
Context	Exceptional	Violation	Violation
Over...	Overriding	Overshadowing	Overshadowing
Relationship	Specificity	CTD (strong)	CTD (weak)

one, while the secondary obligation is the "actual" one. Our distinction, however, mirrors the demarcation line between the two CTD cases: the difference is that in weak CTD cases the secondary obligation is treated as an instrumental and a proper obligation (since it is not excepted and it is not burdened) while in strong CTD cases it is only an instrumental but not a proper obligation (since it is burdened). Ideal obligations offer a third sensible notion of unconditional obligations which are interesting for detachment.

One of the interesting aspects of Carmo and Jones' approach is the idea to analyze the (actual and ideal) responsibilities of an agent by taking into account non-deontic modalities. If, for instance, $\Box b$ is the case, maybe due to the fact that our agent got forcefully hypnotized and thus cannot but break the promise, in Carmo and Jones' modeling the agent is not under the ideal obligation not to break the promise, since it is impossible to avoid breaking the promise, $\neg\Diamond\neg b$.

Due to the limited expressiveness of the standard language of dyadic deontic logics (such as used for this chapter), for these logics we need to presuppose that the deontic implications of the additional modal information are explicitly expressed in terms of the conditions of the obligations in question. In our case this means that, in order to express the example, we need to presuppose that our premises include $\mathsf{P}(b|h)$, i.e., that the agent is allowed to break a promise in case he is forcefully hypnotized, and that he is not obliged not to be forcefully hypnotized since it is beyond his power to avoid it, $\neg\mathsf{O}(\neg h|\top)$.[13] This remark hints at the main methodological difference between Carmo and Jones' and my approach: while their treatment of the detachment of "ideal" and "actual" obligations relies entirely on an analysis of certain types of possibilities concerning the conditions and conclusions of conditional obligations, my treatment of the detachment of proper and instrumental ("ideal") obligations relies entirely on an analysis of the relationships of the conditional obligations and the given facts, for instance, on the question if a certain obligation is overridden or overshadowed, resp. violated or excepted.[14]

There are certain shortcomings of the framework proposed by Carmo and Jones. Their paper is primarily concerned with CTD cases and thus they do not give a satisfying account of specificity. Note, that for the asparagus example we would

[13] Presupposing that the additional modal information modeled in Carmo and Jones' system is expressed in terms of conditional obligations, "ideal" obligations can be incorporated within the formal framework presented in this chapter as I will remark later (see Footnote 20). For the sake of conciseness I will though focus on instrumental and proper obligations.

[14] In this respect my approach is similar to the ones offered by defeasible deontic logics such as Rye's [24] or Horty's [25]. These accounts proceed in a similar way as default logic [26]: a given set of conditional obligations is enhanced to so-called extensions with respect to certain consistency criteria. One important feature is that by an analysis of the relationships between the conditional obligations the logics identify the "overridden" (in Horty's terminology) resp. "defeated" and "violated" (in Rye's terminology) obligations that are rejected as members of the extensions. In our terminology, excepted and violated obligations are sorted out. A consequence relation is then defined in terms of membership to these extensions.

get $\mathsf{O}^{I}\neg f$, by their handling, even if asparagus is served, since obviously we have $\Box\top \wedge \Diamond f \wedge \Diamond\neg f$. This is clearly counter-intuitive. In order to deal with specificity cases their framework needs to be complemented by the ability to analyze the relationships certain conditional obligations have in order to identify exceptional contexts. Furthermore, there seems to be no obvious way to express the detachment of our proper obligations in Carmo and Jones' way by modal operators $\Box$ and $\boxdot$.

Of course, given an enhanced modal framework such as proposed by Carmo and Jones nothing speaks against a hybrid framework that for instance inherits the defeasible character and the treatment of specificity from our framework and the treatment of "ideal" and "actual" obligations from Carmo and Jones.

12.3 Formally Realizing Detachment

In the remainder of the chapter I will present a logical framework that models detachment paradigmatically for instrumental and proper obligations. Following standard representations of dyadic deontic logics I presuppose a propositional calculus which is supplemented by a dyadic obligation operator O for conditional obligations $\mathsf{O}(A|B)$ where A and B are propositional formulas (without occurrences of O, see Footnote 4) and analogously a dyadic operator P for permissions.[15] Furthermore, two monadic obligation operators are used, O^{p} and O^{i}, for proper and instrumental obligations. Where $\wedge, \vee, \supset, \equiv$ and $\neg$ are the classical logical operators, two further symbols are added: $\bullet_{\mathsf{p}}$ and $\bullet_{\mathsf{i}}$. The language is restricted in such a way that $\bullet_{\mathsf{p}}$ and $\bullet_{\mathsf{i}}$ only precede conditional obligations $\mathsf{O}(A|B)$. The modal operators and the new symbols are going to be closed under substitution of equivalents. Where $\mathsf{x} \in \{\mathsf{i}, \mathsf{p}\}$, we have:

$$\text{If } \vdash A \equiv B, \text{ then } \vdash \mathsf{O}(C|A) \equiv \mathsf{O}(C|B) \qquad \text{(RCE)}$$

$$\text{If } \vdash A \equiv B, \text{ then } \vdash \mathsf{O}(A|C) \equiv \mathsf{O}(B|C) \qquad \text{(CRE)}$$

$$\text{If } \vdash A \equiv B, \text{ then } \vdash \mathsf{O}^{\mathsf{x}} A \equiv \mathsf{O}^{\mathsf{x}} B \qquad (\text{EO}^{\mathsf{x}})$$

$$\text{If } \vdash A \equiv B, \text{ then } \vdash \bullet_{\mathsf{x}}\mathsf{O}(A|C) \equiv \bullet_{\mathsf{x}}\mathsf{O}(B|C) \qquad \text{(CREx)}$$

$$\text{If } \vdash A \equiv B, \text{ then } \vdash \bullet_{\mathsf{x}}\mathsf{O}(C|A) \equiv \bullet_{\mathsf{x}}\mathsf{O}(C|B) \qquad \text{(RCEx)}$$

The intended meaning of $\bullet_{\mathsf{p}}\mathsf{O}(A|B)$ (resp. $\bullet_{\mathsf{i}}\mathsf{O}(A|B)$) is that the commitment to bring about A if B, is blocked from being detached as a proper (resp. instrumental) obligation.[16] Therefore the rule for factual detachment (FD) is altered as follows:

$$\vdash \big(\mathsf{O}(A|B) \wedge B \wedge \neg \bullet_{\mathsf{p}} \mathsf{O}(A|B)\big) \supset \mathsf{O}^{\mathsf{p}} A \qquad \text{(FDp)}$$

---

[15] The permission operator may also be defined by $\mathsf{P}(A|B) =_{\text{df}} \neg\mathsf{O}(\neg A|B)$.

[16] There is a similarity between the presented approach and defeasible deontic logics such as Rye's [24] or Horty's [25] (see Footnote 14) concerning the fact that excepted or violated obligations are in a sense incapacitated via an analysis of the relationships between the obligations. Of course,

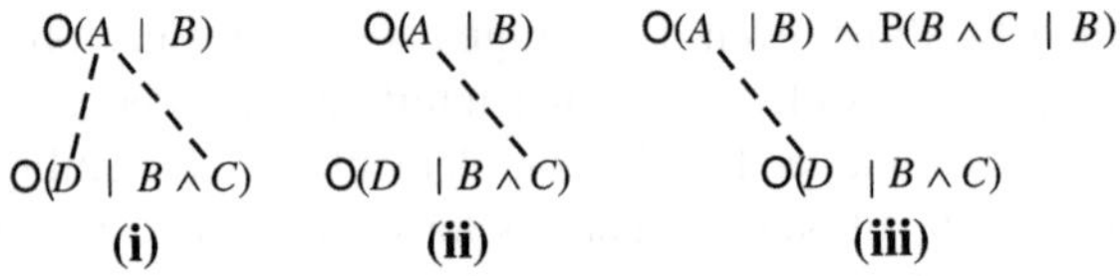

Fig. 12.2 The line indicates an inconsistency between, for instance, *A* and *D* in **(i)**. **(i)** Strong CTD obligation, **(ii)** weak CTD obligation, **(iii)** specificity

$$\vdash \big(\mathsf{O}(A|B) \wedge B \wedge \neg \bullet_{\mathsf{i}} \mathsf{O}(A|B)\big) \supset \mathsf{O}^{\mathsf{i}} A \tag{FDi}$$

By (FDp) and (FDi), detachment is applied to $\mathsf{O}(A|B)$ and B only in case also $\neg \bullet_{\mathsf{p}} \mathsf{O}(A|B)$ (resp. $\neg \bullet_{\mathsf{i}} \mathsf{O}(A|B)$) is derivable.

In order to provide a technically not too involving presentation I will introduce the adaptive handling of detachment in terms of a simplified notion of CTD and excepted obligations compared to the ones presented in Sect. 12.2 (see Fig. 12.2). This way I am able to focus more on the central issue of providing a formal modeling of defeasible factual detachment inferences. As will be demonstrated, for all the standard examples the simplified notions are sufficient in order to guarantee an apt modeling of detachment. However, for cases of nested permissible contexts the solution presented here is suboptimal. These are cases in which we have a permissible context C to B but not $\mathsf{P}(C|B)$. Hence, for the interested reader a generalized modeling that reflects faithfully the more realistic notions of CTD and excepted obligations from Sect. 12.2 is presented in Sect. 12.7.

In the following we treat an obligation $\mathsf{O}(A|B)$ to be *excepted* in the context $B \wedge C$ iff, (i) $\mathsf{P}(B \wedge C|B)$ and (ii) there is an exceptional permission (resp. obligation) $\mathsf{P}(D|B \wedge C)$ (resp. $\mathsf{O}(D|B \wedge C)$) where $\vdash D \supset \neg A$. Compared to the more generic definition of exceptional contexts in Sect. 12.2, (ii) is identical to EC2 and (i) is a simplification of EC1. Indeed, if $\mathsf{P}(B \wedge C|B)$ then $B \wedge C$ is a permissible context to B. However, what we do not take into account are nested permissible contexts. One example is in the asparagus case $f \wedge a$. Note that we do not have $\mathsf{P}(f \wedge a|\top)$ (see Footnote 7), but $\mathsf{P}(a|\top)$ and $\mathsf{P}(f \wedge a|a)$.[17]

We will use the following simplified notion of CTD obligations. $\mathsf{O}(D|B \wedge C)$ is a *(strong) CTD obligation* to $\mathsf{O}(A|B)$ iff $\vdash A \supset \neg D$ and $\vdash A \supset \neg C$. We define $\mathsf{O}(D|B \wedge C)$ to be a *weak CTD obligation* to $\mathsf{O}(A|B)$ iff only the latter is the case, i.e., $\vdash A \supset \neg C$. See Fig. 12.2 for an illustration.

(Footnote 16 continued)
the current approach does not "sort out" for instance overshadowed and overridden obligations by constructing extensions, but rather labels them by $\bullet_{\mathsf{i}}$ and $\bullet_{\mathsf{p}}$. Note that this allows us to stay within the (adaptively extended) standard proof theory of the given deontic base logic.

[17] Nevertheless, as will be shown later (see Example 12.6.2), the ALs which are going to be presented in the following model detachment as desired for this example.

Obviously, in case of Example (G), $\mathsf{O}(g|k)$ is a strong CTD obligation to $\mathsf{O}(\neg k|\top)$ according to this definition. Moreover, for Example (PA), $\mathsf{O}(a|b)$ is a weak CTD obligation to $\mathsf{O}(\neg b|\top)$.[18]

12.3.1 Blocking Proper Detachment

The following rule models specificity: excepted obligations are not to be detached as proper obligations.[19]

$$\text{If } \vdash D \supset \neg A, \text{ then } \big((\mathsf{P}(D|B \wedge C) \vee \mathsf{O}(D|B \wedge C)) \wedge B \wedge C \wedge \mathsf{P}(B \wedge C|B) \wedge \mathsf{O}(A|B)\big) \supset \bullet_{\mathsf{p}}\mathsf{O}(A|B) \qquad \text{(Ep)}$$

Note that in the asparagus example,—$\mathsf{O}(\neg f|\top), \mathsf{O}(f|a), \mathsf{P}(a|\top), a$—, $\bullet_{\mathsf{p}}\mathsf{O}(\neg f|\top)$ is derivable by (Ep).

Furthermore, proper detachment should be blocked in strong CTD cases. While in the case of specificity the more general primary obligation is blocked, the situation is now inverse:

$$\text{If } \vdash A \supset \neg C \text{ and } \vdash A \supset \neg D, \text{ then } \big(\mathsf{O}(D|B \wedge C) \wedge \mathsf{O}(A|B)\big) \supset \bullet_{\mathsf{p}}\mathsf{O}(D|B \wedge C) \qquad \text{(CTDR)}$$

Thus, in the case of the Gentle Murderer $\bullet_{\mathsf{p}}\mathsf{O}(g|k)$ is derivable from $\mathsf{O}(\neg k|\top)$ and $\mathsf{O}(g|k)$.[20]

---

[18] Again, the definition fails for cases that feature nested permissible contexts. For instance $\mathsf{O}(f \wedge a|a)$ is a strong CTD obligation to $\mathsf{O}(\neg f|\top)$ according to the simplified definition above, although, according to the refined notions in Sect. 12.2, $f \wedge a$ is an exceptional context to $\mathsf{O}(\neg f|\top)$.

[19] In logics verifying $\vdash \mathsf{O}(E|F) \supset \mathsf{P}(E|F)$ (e.g. standard deontic logic) the first condition of the antecedent, $\mathsf{P}(D|B \wedge C) \vee \mathsf{O}(D|B \wedge C)$, can be simplified to $\mathsf{P}(D|B \wedge C)$.

[20] It should be remarked at this place that our framework can easily be enhanced such as to model a notion similar to Carmo and Jones' "ideal" obligations. The language is enhanced similar as for instrumental and proper obligations by a unary "ideal" obligation operator O^{I} and by $\bullet_{\mathrm{I}}$. The factual detachment rule is defined analogous as for the instrumental and proper case, $\vdash \big(\mathsf{O}(A|B) \wedge B \wedge \neg \bullet_{\mathrm{I}} \mathsf{O}(A|B)\big) \supset \mathsf{O}^{\mathrm{I}}A$. An analogous rule to (Ep) is used where $\bullet_{\mathsf{p}}\mathsf{O}(A|B)$ is replaced by $\bullet_{\mathrm{I}}\mathsf{O}(A|B)$. The major change in comparison to the rules for the blocking of proper detachment is, as discussed in Sect. 12.2.4, with (CTDR) since in the "ideal" case we also block weak CTD obligations from detachment. Thus, the rule is altered to: If $\vdash D \supset \neg C$, then $\big(\mathsf{O}(A|B \wedge C) \wedge \mathsf{O}(D|B)\big) \supset \bullet_{\mathrm{I}}\mathsf{O}(A|B \wedge C)$. In order not to unnecessarily increase the complexity of the discussion, I will not follow this option further in this chapter. As remarked in Sect. 12.2.5, the language of standard deontic logic lacks the modal expressiveness of Carmo and Jones' proposal and the deontic implications of the additional modal information need to be explicitly expressed in terms of the conditions of obligations.

12.3.2 Blocking Instrumental Detachment

If $\neg A$ is the case we do not allow for the detachment of instrumental obligations such as $\mathsf{O}^{\mathsf{i}}A$, since conditional obligations with conclusion A are de facto violated:

$$\vdash \big(\mathsf{O}(A \mid B) \wedge \neg A \wedge B\big) \supset \bullet_{\mathsf{i}}\mathsf{O}(A \mid B) \tag{fV}$$

For instance, in the Gentle Murderer case (G) we have k and thus we get $\bullet_{\mathsf{i}}\mathsf{O}(\neg k \mid \top)$. Similarly, we get in Chisholm's example (C) $\bullet_{\mathsf{i}}\mathsf{O}(\neg i \mid \top)$ since we have i, where i stands for 'John impregnates Diane'.

Furthermore, we instrumentally prioritize more specific obligations over incompatible general norms[21]:

$$\text{If } \vdash D \supset \neg A, \text{ then } \big((\mathsf{P}(D \mid B \wedge C) \vee \mathsf{O}(D \mid B \wedge C)\big) \wedge B \wedge C \wedge \mathsf{O}(A \mid B)) \supset \bullet_{\mathsf{i}}\mathsf{O}(A \mid B) \tag{oV-Ei}$$

Note that if $\mathsf{O}(A \mid B)$ is excepted in the context $B \wedge C$, then by (oV-Ei), $\bullet_{\mathsf{i}}\mathsf{O}(A \mid B)$. Moreover, since the standard language of deontic logics lacks the means to express explicit preferences among incompatible obligations, I opt in this chapter heuristically for preferring instrumentally the more specific norm, $\mathsf{O}(D \mid B \wedge C)$ resp. $\mathsf{P}(D \mid B \wedge C)$, over the general obligation, $\mathsf{O}(A \mid B)$, even in cases in which $B \wedge C$ is not an exceptional context to $\mathsf{O}(A \mid B)$.

12.3.3 Realizing Detachment by an Adaptive Logic

We are now able to define the following enhancement of a given 'base logic' **L**:

Definition 12.3.1. We define $\mathbf{L}^+$ by adding the rules (EO^{i}), (EO^{p}), (CREi), (CREp), (RCEi), (RCEp), (FDi), (FDp), (Ep), (CTDR), (oV-Ei), and the axiom (fV) to the axiomatization of **L**.

However, $\mathbf{L}^+$ does not yet give a satisfactory account of detachment. In order to show this we take a look at two paradigmatic proofs: one for example (A) (to the left) and one for example (G) (to the right, we presuppose $g \vdash k$):

1	$\mathsf{O}(\neg f \mid \top)$	PREM	1	$\mathsf{O}(\neg k \mid \top)$	PREM
2	$\mathsf{O}(f \mid a)$	PREM	2	$\mathsf{O}(g \mid k)$	PREM
3	$\mathsf{P}(a \mid \top)$	PREM	3	k	PREM
4	a	PREM	4	$\bullet_{\mathsf{i}}\mathsf{O}(\neg k \mid \top)$	1,3; fV
5	$\bullet_{\mathsf{i}}\mathsf{O}(\neg f \mid \top)$	1,2,4; oV-Ei	5	$\bullet_{\mathsf{p}}\mathsf{O}(g \mid k)$	1,2; CTDR
6	$\bullet_{\mathsf{p}}\mathsf{O}(\neg f \mid \top)$	1,2,3,4; Ep			

[21] In logics verifying $\vdash \mathsf{O}(E \mid F) \supset \mathsf{P}(E \mid F)$ (e.g. standard deontic logic) the first condition of the antecedent, $\mathsf{P}(D \mid B \wedge C) \vee \mathsf{O}(D \mid B \wedge C)$, can be simplified to $\mathsf{P}(D \mid B \wedge C)$.

It is important to notice that, although the logic is able to block undesired applications of the detachment rules (FDi) (to $\mathsf{O}(\neg f|\top)$ resp. $\mathsf{O}(\neg k|\top)$) and (FDp) (to $\mathsf{O}(\neg f|\top)$ resp. $\mathsf{O}(g|k)$), detachment is not applied where desired. For instance in the case of the asparagus example (A) we are interested in applying (FDi) and (FDp) to $\mathsf{O}(f|a)$, and in the case of the Gentle Murderer (G) we are interested in applying (FDi) to $\mathsf{O}(g|k)$ and (FDp) to $\mathsf{O}(\neg k|\top)$. What is lacking are means to derive $\neg \bullet_{\mathsf{i}} \mathsf{O}(f|a)$ and $\neg \bullet_{\mathsf{p}} \mathsf{O}(f|a)$ in the former case and $\neg \bullet_{\mathsf{i}} \mathsf{O}(g|k)$ and $\neg \bullet_{\mathsf{p}} \mathsf{O}(\neg k|\top)$ in the latter case.

This is where ALs come in, since these logics allow for conditional applications of certain rules which enable them to interpret a premise set "as normally as possible" with respect to some given criterion for normality.

In our case we are interested in applying detachment to $\mathsf{O}(A|B)$ and B on the condition that $\bullet_{\mathsf{i}}\mathsf{O}(A|B)$ (resp. $\bullet_{\mathsf{p}}\mathsf{O}(A|B)$) can be assumed to be false.

Note that if we were able to derive $\neg \bullet_{\mathsf{i}} \mathsf{O}(A|B)$ (resp. $\neg \bullet_{\mathsf{p}} \mathsf{O}(A|B)$), then the restricted detachment rules (FDi) (resp. (FDp)) would be applicable to $\mathsf{O}(A|B)$ and B. For instance in the Gentle Murderer example it would be useful to extend the proof by the following lines:

6	$\mathsf{O}^{\mathsf{i}}g$	2, 3; cFDi	$\{\bullet_{\mathsf{i}}\mathsf{O}(g|k)\}$
7	$\mathsf{O}^{\mathsf{p}}\neg k$	1; cFDp	$\{\bullet_{\mathsf{p}}\mathsf{O}(\neg k|\top)\}$

At line 6 detachment is applied to $\mathsf{O}(g|k)$ and k on the condition $\{\bullet_{\mathsf{i}}\mathsf{O}(g|k)\}$ which is written in the last column. The idea is that, e.g. in the case that $\bullet_{\mathsf{i}}\mathsf{O}(g|k)$ is derived at a later stage of the proof, all lines that feature $\bullet_{\mathsf{i}}\mathsf{O}(g|k)$ as an element of the condition are marked. The second elements of such marked lines are not anymore considered to be derived. Similarly, $\mathsf{O}^{\mathsf{p}}\neg k$ is conditionally derived at line 7.

In order to technically realize the idea just described I introduce in Sect. 12.4 a generic format for an adaptive deontic logic that realizes factual detachment. Afterward, the adaptive approach is demonstrated by means of a concrete base logic.

12.4 A Generic Adaptive Logic for Detachment

Given that our enriched base logic $\mathbf{L}^+$ satisfies the criteria for lower limit logics as required in the standard format of ALs (see Sect. 2.2), we are able to define an AL $\mathbf{DL}^+$ which models detachment by the following triple:

1. the **LLL** is $\mathbf{L}^+$;
2. the abnormalities are $\Omega^d =_{\text{df}} \Omega^i \cup \Omega^p$ where $\Omega^x =_{\text{df}} \{\bullet_{\mathsf{x}}\mathsf{O}(A|B) : A, B \in \mathcal{P}\}$ for $\mathsf{x} \in \{\mathsf{i}, \mathsf{p}\}$, and $\mathcal{P}$ is the set of all propositional formulas;
3. the strategy is reliability.

Note that due to (FDp) and (FDi) we have, where $\mathsf{x} \in \{\mathsf{i}, \mathsf{p}\}$,

$$\mathsf{O}(A \mid B), B \vdash_{L^+} \mathsf{O}^{\mathsf{x}} A \vee \bullet_{\mathsf{x}} \mathsf{O}(A \mid B)$$

As argued above, this makes it possible to (conditionally) derive $\mathsf{O}^{\mathsf{x}} A$ in $\mathbf{DL}^+$ from $\mathsf{O}(A \mid B)$ and B on the condition $\{\bullet_{\mathsf{x}} \mathsf{O}(A \mid B)\}$. This is demonstrated by the following proof fragment:

1 $\mathsf{O}(A \mid B)$	PREM	$\emptyset$
2 B	PREM	$\emptyset$
3 $\mathsf{O}^{\mathsf{i}} A$	1,2; cFDi	$\{\bullet_{\mathsf{i}} \mathsf{O}(A \mid B)\}$
4 $\mathsf{O}^{\mathsf{p}} A$	1,2; cFDp	$\{\bullet_{\mathsf{p}} \mathsf{O}(A \mid B)\}$

We write cFDi and cFDp for these conditional derivations. If at a later stage of the proof e.g. $\bullet_{\mathsf{i}} \mathsf{O}(A \mid B)$ is derived as part of a minimal Dab-formula, say $\bullet_{\mathsf{i}} \mathsf{O}(A \mid B) \vee \bullet_{\mathsf{i}} \mathsf{O}(C \mid D)$, line 3 would be marked. If at a even later stage $\bullet_{\mathsf{i}} \mathsf{O}(C \mid D)$ is derived, then line 3 would be unmarked again since its condition is not anymore unreliable.

Note that the adaptive models of a given premise set do not validate any abnormalities that are not members of minimal Dab-consequences. As a look at rules (Ep), (CTDR), (fV), (oV-Ei) and at the discussion in Sect. 12.2 reveals, the abnormalities in Ω^d are caused by exceptional and violation contexts. Thus, the reliable models of Γ validate as less exceptional and violation contexts as possible.

In the remainder I will give concrete examples for such ALs by using variants of one of Lou Goble's **CDPM** logics (see [2, 3]) as lower limit logics.

12.5 A Lower Limit: Lou Goble's CDPM

In order to demonstrate the adaptive modeling of detachment by means of a concrete logic, I settle for this chapter on two variants of one of Lou Goble's **CDPM** systems (see [2, 3]) as base logics. This choice is not essential: it is possible to use other logics for this task. The choice however is not arbitrary either. **CDPM** was chosen due to its many nice properties. For instance it tolerates deontic conflicts while blocking counter-intuitive consequences from conflicting obligations. Furthermore, in the absence of deontic conflicts it offers a close approximation of (dyadic) standard deontic logic.[22] One of the main ideas behind these logics is to restrict the (conditional version of the) inheritance principle

$$\text{If } \vdash B \supset C, \text{ then } \vdash \mathsf{O}(B \mid A) \supset \mathsf{O}(C \mid A) \tag{RCM}$$

by adding a further permission statement:

$$\text{If } \vdash B \supset C, \text{ then } \vdash \mathsf{P}(B \mid A) \supset \big(\mathsf{O}(B \mid A) \supset \mathsf{O}(C \mid A)\big) \tag{RCPM}$$

[22] Goble defines a dyadic version **SDDL** of standard deontic logic in [27]. It is equivalent to van Fraassen's **CD** of [28] and David Lewis' **VN** of [29]. See Chap. 11, Sect. 11.2.

This principle ensures that the logic is non-explosive confronted with deontic conflicts.

Goble proposed two families of his **CDPM** systems: **CDPM.1**α and **CDPM.2**α where $\alpha \in \{\mathbf{a}, \mathbf{b}, \mathbf{c}\}$. Only the latter family verifies the rule (CP), $\vdash \neg\mathsf{O}(\bot \mid A)$. In order to ensure non-explosive behavior in **CDPM.2**α the aggregation principle has to be restricted in a similar way as inheritance.

$$\vdash \mathsf{P}(A \wedge B \mid C) \supset \big((\mathsf{O}(A \mid C) \wedge \mathsf{O}(B \mid C)\big) \supset \mathsf{O}(A \wedge B \mid C)\big) \qquad \text{(CPAND)}$$

For the conditional version Goble favors the second family, since in the first one a counter-intuitive kind of SA is valid.[23] I focus in this chapter on (a weakened version of) the system Goble dubs **CDPM.2c**. Goble gives an account of SA by restricting Rational Monotonicity[24] in the following way:

$$\vdash \big(\mathsf{O}(B \mid A) \wedge \mathsf{P}(B \wedge C \mid A)\big) \supset \mathsf{O}(B \mid A \wedge C) \qquad \text{(WRM)}$$

CDPM.2c is defined by adding to (RCE), (CRE), (CP), (RCPM), (CPAND), and (WRM) the axioms:

$$\vdash \mathsf{O}(\top \mid \top) \qquad \text{(CN)}$$

$$\vdash \mathsf{O}(B \mid A) \supset \mathsf{O}(A \mid A) \qquad \text{(QR)}$$

$$\vdash \mathsf{O}(A \mid B \wedge C) \supset \mathsf{O}(B \supset A \mid C) \qquad \text{(S)}$$

Axiom (QR) is in some cases severely counter-intuitive. Take for instance the Chisholm example (C) from p. 298. (QR) allows to derive from the commitment to marry Diane in case John impregnates her, $\mathsf{O}(m \mid i)$, the commitment to impregnate her in that very case, $\mathsf{O}(i \mid i)$. However, the fact that he impregnates her is a violation of the primary obligation not to impregnate her. I therefore abandon rule (QR) for our base logics.[25] Furthermore, the language is extended as discussed in Sect. 12.3. For reasons that are explicated later (see Sect. 12.6.1) I define two alternative lower limit logics for the adaptive systems which will be introduced in Sect. 12.6.

[23] It validates all instances of $\vdash \big(\mathsf{O}(B \mid A) \wedge \mathsf{P}(C \mid A) \wedge \mathsf{P}(B \mid A)\big) \supset \mathsf{O}(B \mid A \wedge C)$ (PRatMono). That (PRatMono) is counter-intuitive can be demonstrated by means of our asparagus example (namely $\mathsf{O}(\neg f \mid \top)$, $\mathsf{P}(a \mid \top)$ and $\mathsf{O}(f \mid a)$): if we add the intuitive premise $\mathsf{P}(\neg f \mid \top)$, then the counter-intuitive $\mathsf{O}(\neg f \mid a)$ is derivable. This defect is not fatal though: I proposed in [30] and in Sect. 11.4 a version of **CDPM.1** which overcomes this shortcoming.

[24] Rational Monotonicity (cp. [15]) can be stated in terms of the language of dyadic deontic logic used in this chapter as follows: $\vdash \big(\mathsf{O}(B \mid A) \wedge \mathsf{P}(C \mid A)\big) \supset \mathsf{O}(B \mid A \wedge C)$. It is verified in dyadic standard deontic logic. Goble is aware of the fact that his (WRM) leads to counter-intuitive, even explosive behavior in some cases. I offered an improvement based on the idea of conditionally applying SA within an AL which is able to avoid these problems (see [30] and Sect. 12.8).

[25] The reader might further object that (CN) (in a similar way as (QR) is not very intuitive. Goble's intention is to stay as close as possible to standard deontic logic. However, (CN) is neither an essential part of his logic nor in any way essential to the presented approach and may thus be disregarded as well.

Definition 12.5.1. **CDPM.2d$^+$** is defined by (CP), (RCPM), (CPAND), (WRM), (RCE), (CRE), (CN), (S), (FDp), (FDi), (Ep), (CTDR), (fV), (oV-Ei), (EO^i), (EO^p), (CREi), (CREp), (RCEi), and (RCEp). **CDPM.2e$^+$** is defined just as **CDPM.2d$^+$**, with the exception of (S) which is replaced by:

$$\vdash \big(\mathsf{O}(A|B \wedge C) \wedge \mathsf{P}(A|\neg B \wedge C)\big) \supset \mathsf{O}(B \supset A|C) \qquad \text{(PS')}$$

The semantics for these systems can be found in Appendix J. In Sect. 12.6.1 I will show that these systems handle the Chisholm example in different ways.

12.6 Handling Detachment Adaptively

As already explicated at the end of Sect. 12.3, the idea is to apply detachment to $\mathsf{O}(A|B)$ and B for proper (resp. instrumental) obligations on the condition that $\bullet_p\mathsf{O}(A|B)$ (resp. $\bullet_i\mathsf{O}(A|B)$) can be considered to be false. More precisely the condition is, with respect to the reliability strategy, that $\bullet_p\mathsf{O}(A|B)$ (resp. $\bullet_i\mathsf{O}(A|B)$) is not part of any minimal Dab-consequence. The following ALs are defined as suggested in Sect. 12.4, employing **CDPM.2d$^+$** (resp. **CDPM.2e$^+$**) as lower limit logic.

Definition 12.6.1. The AL in standard format **DCDPM.2**α^+, where $\alpha \in \{\mathbf{d}, \mathbf{e}\}$, is defined by the triple $\langle$**CDPM.2**α^+, Ω^d, reliability strategy$\rangle$.

Let me demonstrate the way the logics work by having a look at some examples. The proofs in this section are all for both logics, **DCDPM2.d$^+$** and **DCDPM2.e$^+$**.

Example 12.6.1 (G, see page 303). I have already stated a proof for the case of the Gentle Murderer on page 303.[26] Note that, for both strategies, there is no way to extend the proof in such a way that lines 6 and 7 are marked, i.e., there is no way to derive $\bullet_i\mathsf{O}(g|k)$ and $\bullet_p\mathsf{O}(\neg k|\top)$ as part of minimal Dab-consequences in our lower limits **CDPM.2**α^+ ($\alpha \in \{\mathbf{d}, \mathbf{e}\}$). Hence, $\bullet_i\mathsf{O}(g|k), \bullet_p\mathsf{O}(\neg k|\top) \notin U(\{\mathsf{O}(\neg k|\top), \mathsf{O}(g|k), k\})$. Thus, $\mathsf{O}^i g$ and $\mathsf{O}^p\neg k$ are finally derivable. As discussed, this is the desired outcome.

Example 12.6.2 (A, see page 298). For our asparagus example we have already derived $\bullet_i\mathsf{O}(\neg f|\top)$ and $\bullet_p\mathsf{O}(\neg f|\top)$ at lines 5 and 6. Now we apply detachment conditionally to $\mathsf{O}(f|a)$ and a in order to derive $\mathsf{O}^i f$ and $\mathsf{O}^p f$.

7 $\mathsf{O}^i f$	2,4; cFDi	$\{\bullet_i\mathsf{O}(f|a)\}$
8 $\mathsf{O}^p f$	2,4; cFDp	$\{\bullet_p\mathsf{O}(f|a)\}$

[26] Lines 1–5 are not stated in the form of an AL proof. It can be easily adjusted by adding the empty condition $\emptyset$ in a fourth column.

Note that the conditions $\bullet_{\mathsf{p}}\mathsf{O}(f|a)$ and $\bullet_{\mathsf{i}}\mathsf{O}(f|a)$ are not part of any minimal Dab-consequences and hence reliable. Thus, as desired, $\mathsf{O}^{\mathsf{i}}f$ and $\mathsf{O}^{\mathsf{p}}f$ are derivable.

Example 12.6.3 (PA, see page 304). Let us have a look at our weak CTD example.

1	$\mathsf{O}(\neg b|\top)$	PREM	$\emptyset$
2	$\mathsf{O}(a|b)$	PREM	$\emptyset$
3	b	PREM	$\emptyset$
^{5}4	$\mathsf{O}^{\mathsf{i}}\neg b$	1; cFDi	$\{\bullet_{\mathsf{i}}\mathsf{O}(\neg b|\top)\}$
5	$\bullet_{\mathsf{i}}\mathsf{O}(\neg b|\top)$	1,3; fV	$\emptyset$
6	$\mathsf{O}^{\mathsf{i}}a$	2, 3; cFDi	$\{\bullet_{\mathsf{i}}\mathsf{O}(a|b)\}$
7	$\mathsf{O}^{\mathsf{p}}a$	2, 3; cFDp	$\{\bullet_{\mathsf{p}}\mathsf{O}(a|b)\}$
8	$\mathsf{O}^{\mathsf{p}}\neg b$	1; cFDp	$\{\bullet_{\mathsf{p}}\mathsf{O}(\neg b|\top)\}$

Lines 4 and 6–8 feature conditional applications of detachment. Note that the first one gets marked, since its condition, $\bullet_{\mathsf{i}}\mathsf{O}(\neg b|\top)$, is derived on the empty condition at line 5. Hence, $\bullet_{\mathsf{i}}\mathsf{O}(\neg b|\top) \in U(\{\mathsf{O}(\neg b|\top), \mathsf{O}(a|b), b\})$ and there is no way to unmark line 4. The unconditional obligations derived at lines 6–8 are finally derived since neither of the resp. conditions is unreliable.

Example 12.6.4 (Reykjavik scenario (see [12]*)).* Let us look at another interesting example.

R1 The secret shall be told neither to Reagan nor to Gorbachev.—$\mathsf{O}(\neg r \wedge \neg g|\top)$
R2 If the secret is told to Reagan it shall also be told to Gorbachev.—$\mathsf{O}(g|r)$
R3 If the secret is told to Gorbachev it shall also be told to Reagan.—$\mathsf{O}(r|g)$

It is easy to see that if our factual knowledge is $\mathcal{F} = \emptyset$ then $\mathsf{O}^{\mathsf{i}}(\neg r \wedge \neg g)$ and $\mathsf{O}^{\mathsf{p}}(\neg r \wedge \neg g)$ are derivable. Let us see what happens if we add the premise g:

1	$\mathsf{O}(\neg r \wedge \neg g|\top)$	PREM	$\emptyset$
2	$\mathsf{O}(g|r)$	PREM	$\emptyset$
3	$\mathsf{O}(r|g)$	PREM	$\emptyset$
4	g	PREM	$\emptyset$
5	$\bullet_{\mathsf{p}}\mathsf{O}(g|r)$	1,2; CTDR	$\emptyset$
6	$\bullet_{\mathsf{p}}\mathsf{O}(r|g)$	1,3; CTDR	$\emptyset$
7	$\bullet_{\mathsf{i}}\mathsf{O}(\neg r \wedge \neg g|\top)$	1,4; fV	$\emptyset$
8	$\mathsf{O}^{\mathsf{p}}(\neg r \wedge \neg p)$	1; FDp	$\{\bullet_{\mathsf{p}}\mathsf{O}(\neg r \wedge \neg p|\top)\}$
9	$\mathsf{O}^{\mathsf{i}}r$	3,4; FDi	$\{\bullet_{\mathsf{i}}\mathsf{O}(r|g)\}$

Obviously we get the intuitive consequences: while $\mathsf{O}^{\mathsf{p}}(\neg r \wedge \neg g)$ informs us of our proper obligation not to tell the secret to either of the two politicians, we have in the violation context g the instrumental obligation to tell the secret to Reagan as well, $\mathsf{O}^{\mathsf{i}}r$.

12.6.1 Context Shifts

In this section it will be demonstrated how certain properties of deontic logics can shift the context of some conditional obligations to more general or entirely different contexts and in what way this effects detachment. While the variants of the Chisholm example in Examples 12.6.5 and 12.6.6 focus on deontic detachment, Example 12.6.7 focuses on the interplay of deontic detachment and SA.

I postponed a discussion as to why axiom (S) was restricted to (PS') for **CDPM**.**2e**. Taking a look at the Chisholm example will illuminate this point. In the following I will highlight the role of deontic detachment for factual detachment by showing that the choice between a stronger and a weaker version of deontic detachment manifests itself in two incompatible resolutions of the Chisholm example. It is easy to see that **DCDPM**.**2d**$^+$ allows for the following versions of deontic detachment (henceforth (DD))[27]:

$$\big(\mathsf{O}(A|C) \wedge \mathsf{P}(A \wedge B|C) \wedge \mathsf{O}(B|A \wedge C)\big) \supset \mathsf{O}(B|C) \qquad \text{(DDP1)}$$

$$\big(\mathsf{O}(A|\top) \wedge \mathsf{P}(A \wedge B|\top) \wedge \mathsf{O}(B|A)\big) \supset \mathsf{O}(B|\top) \qquad \text{(DDP}\top\text{1)}$$

Example 12.6.5 (C, *see p. 298*). Let i stand for John impregnating Diane, and m for him marrying her. We first take a look at a proof in **DCDPM2.d**$^+$:

1	$\mathsf{O}(\neg i|\top)$	PREM	$\emptyset$
2	$\mathsf{O}(\neg m|\neg i)$	PREM	$\emptyset$
3	$\mathsf{O}(m|i)$	PREM	$\emptyset$
4	i	PREM	$\emptyset$
5	$\mathsf{P}(\neg i \wedge \neg m|\top)$	PREM	$\emptyset$
6	$\mathsf{O}^{\mathsf{i}}m$	3, 4; cFDi	$\{\bullet_{\mathsf{i}}\mathsf{O}(m|i)\}$
7	$\mathsf{O}^{\mathsf{p}}\neg i$	1; cFDp	$\{\bullet_{\mathsf{p}}\mathsf{O}(\neg i|\top)\}$
8	$\bullet_{\mathsf{i}}\mathsf{O}(\neg i|\top)$	1,4; fV	$\emptyset$
9	$\mathsf{O}(\neg m|\top)$	1,2,5; DDP$\top$1	$\emptyset$
10	$\mathsf{O}(\neg i \wedge \neg m|\top)$	1,5,9; CPAND	$\emptyset$
11	$\bullet_{\mathsf{i}}\mathsf{O}(\neg i \wedge \neg m|\top)$	4,10; fV	$\emptyset$
12	$\bullet_{\mathsf{p}}\mathsf{O}(m|i)$	3,10; CTDR	$\emptyset$
[12]13	$\mathsf{O}^{\mathsf{p}}m$	3,4; cFDp	$\{\bullet_{\mathsf{p}}\mathsf{O}(m|i)\}$
14	$\bullet_{\mathsf{i}}\mathsf{O}(\neg m|\top)$	3,4,9; oV-Ei	$\emptyset$
15	$\mathsf{O}^{\mathsf{p}}\neg m$	9; cFDp	$\{\bullet_{\mathsf{p}}\mathsf{O}(\neg m|\top)\}$
16	$\mathsf{O}^{\mathsf{p}}(\neg i \wedge \neg m)$	10; cFDp	$\{\bullet_{\mathsf{p}}\mathsf{O}(\neg i \wedge \neg m|\top)\}$

What follows is a continuation of the proof from line 8 on in **DCDPM2.e**$^+$:

9′	$\mathsf{O}^{\mathsf{p}}m$	3,4; cFDp	$\{\bullet_{\mathsf{p}}\mathsf{O}(m|i)\}$

27 That (DDP$\top$1) is a consequence of (DDP1) can easily be shown. Proofs for the validity of (DDP1) and (DDP$\top$1) in **CDPM.2d** can be found in Appendix J.

In the case of the first proof we have $\bullet_p\mathsf{O}(m|i)$ at line 12, since (CTDR) is applied to $\mathsf{O}(\neg i \wedge \neg m|\top)$ and $\mathsf{O}(m|i)$. Thus, the conditional application of detachment at line 13 gets marked. In the second proof however we have no means to derive $\mathsf{O}(\neg i \wedge \neg m|\top)$. Therefore the conditional application of detachment to $\mathsf{O}(m|i)$ is not blocked and thus $\mathsf{O}^{\mathsf{p}}m$ is finally derivable at line 9′. In contrast, note that in the first proof $\mathsf{O}(\neg m|\top)$ is derivable and detachment is applied to this obligation: as a consequence we arrive at the proper obligation $\mathsf{O}^{\mathsf{p}}\neg m$ at line 15.

For an illustration of the two treatments of the Chisholm example by the two ALs see Fig. 12.3. Let me analyze it a bit more. What is the correct result? What about the solution offered by **DCDPM.2d**$^+$? (DDP⊤1) enables us to derive $\mathsf{O}(\neg m|\top)$ at line 9 from the given obligations $\mathsf{O}(\neg i|\top)$, $\mathsf{O}(\neg m|\neg i)$ and the harmless permission statement $\mathsf{P}(\neg i \wedge \neg m|\top)$ that is added at line 5. Note that the commitment $\mathsf{O}(\neg m|\neg i)$ not to marry Diane in the case that Doe doesn't impregnate her, is in this logic treated as the general obligation not to marry her, $\mathsf{O}(\neg m|\top)$. From this we immediately get $\mathsf{O}(\neg i \wedge \neg m|\top)$ at line 10 which makes $\mathsf{O}(m|i)$ a strong CTD obligation. Thus, in this logic the commitment not to marry Diane is a proper obligation which is in conflict with the instrumental CTD obligation to marry her. The latter is in force since its condition, him having impregnated her, is fulfilled (and it is not otherwise excepted). More generally speaking: the logic elevates, via deontic detachment, a commitment on a condition C stating that a proper obligation is fulfilled (such as $\mathsf{O}(\neg m|\neg i)$ where the condition $\neg i$ corresponds to the proper obligation $\mathsf{O}^{\mathsf{p}}\neg i$) to be a proper obligation itself (e.g. $\mathsf{O}^{\mathsf{p}}\neg m$), even in case the condition C is not fulfilled (e.g. i being the case). Deontic detachment has quite some intuitive appeal as well

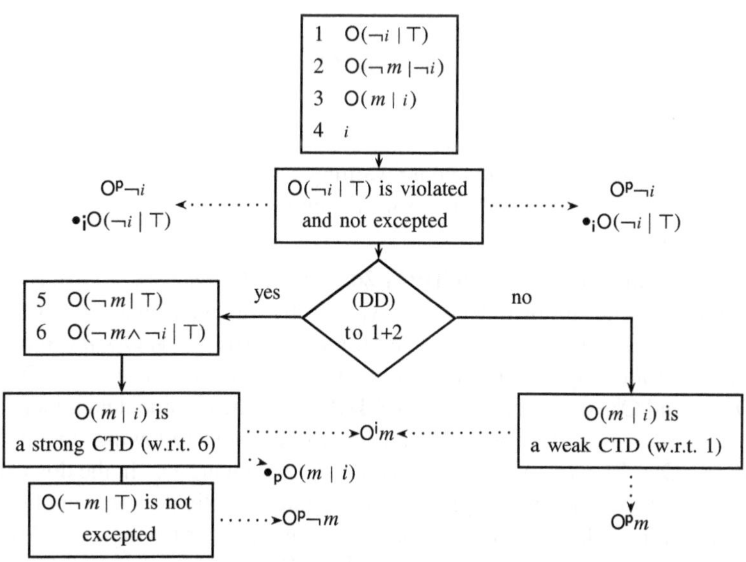

Fig. 12.3 Two ways of dealing with Chisholm's problem. *Left: the treatment in* **DCDPM.2d**$^+$, *Right: the treatment in* **DCDPM.2e**$^+$

as the thought that, if A is a proper obligation and you are committed to bring about B in case A, then also B should be a proper obligation.

We have a different result for the second logic. Indeed, there is an alternative intuition. In the case of **DCDPM.2e**$^+$ the deontic detachment that enabled us in **DCDPM.2d**$^+$ to derive $\mathsf{O}(\neg m|\top)$ is blocked, since this logic only supports a weaker form of (DD)[28]:

$$\big(\mathsf{O}(A|C) \wedge \mathsf{P}(A \wedge B|C) \wedge \mathsf{P}(B|\neg A \wedge C) \wedge \mathsf{O}(B|A \wedge C)\big) \supset \mathsf{O}(B|C) \qquad \text{(DDP2)}$$

$$\big(\mathsf{O}(A|\top) \wedge \mathsf{P}(A \wedge B|\top) \wedge \mathsf{P}(B|\neg A) \wedge \mathsf{O}(B|A)\big) \supset \mathsf{O}(B|\top) \qquad \text{(DDP}\top\text{2)}$$

In favor of this approach it may be argued that, after all, the obligation not to marry Diane was stated only on the condition that Doe doesn't impregnate her, $\mathsf{O}(\neg m|\neg i)$. This is clearly weaker than $\mathsf{O}(\neg m|\top)$, not to marry her in general. After all, $\mathsf{O}(m|i)$ is only a weak CTD obligation to $\mathsf{O}(\neg i|\top)$: thus to marry her in case of him impregnating her should not be considered as a violation of a primary obligation (such as the gentle killing in the Forrester paradox) but rather as a proper obligation in a violation context, similarly to our intuition which tells us in the (PA) example that it is not just an instrumental but rather a proper obligation to apologize in the sub-ideal situation of the broken promise. This logic reflects therefore the difference between stating $\mathsf{O}(\neg m|\top)$ and stating $\mathsf{O}(\neg m|\neg i)$ on the level of proper and instrumental obligations whereas in case of **DCDPM.2d**$^+$ this makes no difference since we get $\mathsf{O}^{\mathsf{p}}\neg m$ in both cases.

I leave it to the reader to settle for one of the two intuitions and to pick the corresponding logic.

Example 12.6.6. For another version of the Chisholm paradox (see [22]) the logic **DCDPM.2e**$^+$ seems clearly preferable:

W1 There must be no dog.—$\mathsf{O}(\neg d|\top)$
W2 If there is no dog, there must be no warning sign.—$\mathsf{O}(\neg s|\neg d)$
W3 If there is a dog, there must be a warning sign.—$\mathsf{O}(s|d)$
W4 There is a dog.—d

Analogously to the setup above **DCDPM.2d**$^+$ derives $\mathsf{O}(\neg d \wedge \neg s|\top)$ in case we add the in no way counter-intuitive $\mathsf{P}(\neg d \wedge \neg s|\top)$. This way we gain a strong CTD case between $\mathsf{O}(\neg d \wedge \neg s|\top)$ and $\mathsf{O}(s|d)$. With this we arrive at $\bullet_{\mathsf{p}}\mathsf{O}(s|d)$. Finally the logic has the following unconditional obligations as consequences: $\mathsf{O}^{\mathsf{i}}s$, $\mathsf{O}^{\mathsf{p}}(\neg d \wedge \neg s)$, $\mathsf{O}^{\mathsf{p}}\neg d$ and $\mathsf{O}^{\mathsf{p}}\neg s$. In contrast, for **DCDPM.2e**$^+$ the deontic detachment which leads to $\mathsf{O}(\neg d \wedge \neg s|\top)$ is blocked and we arrive at the following unconditional obligations: $\mathsf{O}^{\mathsf{i}}s$, $\mathsf{O}^{\mathsf{p}}\neg d$ and $\mathsf{O}^{\mathsf{p}}s$. As Prakken and Sergot argue: "In the dog example where there is a dog, having a sign does not violate an obligation that applies to the situation: no fine is due for having a warning sign, only for having a dog" ([22], p. 240). This clearly speaks against deriving $\mathsf{O}^{\mathsf{p}}\neg s$. But then it is clear that $\mathsf{O}^{\mathsf{p}}s$ should

[28] Proofs can be found in Appendix J.

be the case due to $\mathsf{O}(s|d)$, d and the fact that this obligation is neither excepted nor burdened. Thus, we treat $\mathsf{O}(s|d)$ as a weak CTD obligation.

Example 12.6.7. I add one more example (see [22]) to underline the role played by deontic detachment and also by (SA) in the derivation of proper and instrumental obligations.

F1 There must be no fence.—$\mathsf{O}(\neg f|\top)$
F2 There must be a white fence, if there is a fence.—$\mathsf{O}(w \wedge f|f)$
F3 There must be a fence, if there is a dog.—$\mathsf{O}(f|d)$
F4 It is allowed to have a dog.—$\mathsf{P}(d|\top)$
F5 There is a dog and no fence.—$d \wedge \neg f$

Let us take a look at a proof in **DCDPM.2d**$^+$ for this example:

1 $\mathsf{O}(\neg f|\top)$	PREM	$\emptyset$
2 $\mathsf{O}(w \wedge f|f)$	PREM	$\emptyset$
3 $\mathsf{O}(f|d)$	PREM	$\emptyset$
4 $\mathsf{P}(d|\top)$	PREM	$\emptyset$
5 $d \wedge \neg f$	PREM	$\emptyset$
6 $\bullet_{\mathsf{p}}\mathsf{O}(w \wedge f|f)$	1,2; CTDR	$\emptyset$
7 $\bullet_{\mathsf{p}}\mathsf{O}(\neg f|\top)$	1,3,4,5; Ep	$\emptyset$
8 $\bullet_{\mathsf{i}}\mathsf{O}(\neg f|\top)$	1,3,5; oV-Ei	$\emptyset$
9 $\mathsf{O}^{\mathsf{p}} f$	3,5; cFDp	$\{\bullet_{\mathsf{p}}\mathsf{O}(f|d)\}$

So far the logic derives the desired consequences. However, some readers might point out that also the obligation to bring about $w \wedge f$ should be detached as a proper obligation, since, after all, $\mathsf{O}(\neg f|\top)$ is excepted and $\mathsf{O}^{\mathsf{p}} f$ has been derived. However, we cannot apply detachment to $\mathsf{O}(w \wedge f|f)$ since its condition is not fulfilled and even if that were so, we derived $\bullet_{\mathsf{p}}\mathsf{O}(w \wedge f|f)$ since $\mathsf{O}(w \wedge f|f)$ is a (strong) CTD obligation with respect to $\mathsf{O}(\neg f|\top)$. Let us add the harmless premises $\mathsf{P}(w \wedge f|d)$ and $\mathsf{P}(w \wedge f \wedge d|f)$[29]:

10 $\mathsf{P}(w \wedge f|d)$	PREM	$\emptyset$
11 $\mathsf{P}(w \wedge f \wedge d|f)$	PREM	$\emptyset$
12 $\mathsf{O}(w \wedge f|f \wedge d)$	2,11; WRM	$\emptyset$
13 $\mathsf{O}(w \wedge f|d)$	3,10,12; DDP1	$\emptyset$
14 $\mathsf{O}^{\mathsf{p}}(w \wedge f)$	5,13; cFDp	$\{\bullet_{\mathsf{p}}\mathsf{O}(w \wedge f|d)\}$

The reader can see that by (WRM) at line 12 the obligation in question can be derived on the more restricted condition $f \wedge d$. This enables the application of deontic detachment in order to arrive at $\mathsf{O}(w \wedge f|d)$ at line 13. Now the detachment cannot be blocked anymore. Thus, despite the fact that $\mathsf{O}(w \wedge f|f)$ is a CTD obligation, this obligation can be detached since it can be derived also for the (exceptional) context

[29] In **DCDPM.2e**$^+$ we would have to add another additional premise, $\mathsf{P}(w \wedge f|d \wedge \neg f)$, in order to derive $\mathsf{O}(w \wedge f|d)$ analogously by (DDP2). Otherwise the proof is analogous.

d, which is the factual situation presented by the premises. This behavior of the logic is intuitive.

It is easy and left to the reader to prove that if F5 is replaced by $f \wedge \neg d$ we get $\mathsf{O}^{\mathsf{p}}\neg f$ and $\mathsf{O}^{\mathsf{i}}(w \wedge f)$ as consequences, and if F5 is replaced by $f \wedge d$ we get $\mathsf{O}^{\mathsf{p}} f$, $\mathsf{O}^{\mathsf{i}} f$, $\mathsf{O}^{\mathsf{i}}(f \wedge w)$. This is as expected, since "[w]hat is most striking about the fence example [...] is the observation that when the premise $\mathsf{O}(\neg f|\top)$ is violated by f, then the obligation for $\neg f$ should be derivable, but not when $\mathsf{O}(\neg f|\top)$ is overridden by the exception $f \wedge d$" ([21], p. 71).

Let me summarize this section. Given a logic **L** and a premise set Γ we may say that the *context of a conditional obligation* $\mathsf{O}(A|B) \in \Gamma$ *has shifted* if (i) $\Gamma \vdash_{\mathbf{L}} \mathsf{O}(A|D)$ and (ii) $\nvdash_{\mathbf{L}} B \equiv D$. Context shifts may cause the blocking of detachment for certain conditional obligations and may allow for additional detachments.

We have seen two examples for context shifts. First, with the Chisholm Example (C) it was demonstrated that the stronger form of deontic detachment validated by **DCDPM.2d**$^+$ causes that $\mathsf{O}(\neg m|\neg i)$ shifts to the more general context $\top$ since $\mathsf{O}(\neg m|\top)$ is derivable. This makes it on the one hand possible to detach $\mathsf{O}^{\mathsf{p}}\neg m$. On the other hand, aggregation with $\mathsf{O}(\neg i|\top)$ leads to the derivation of $\mathsf{O}(\neg m \wedge \neg i|\top)$ which makes $\mathsf{O}(m|i)$ a (strong) CTD obligation, and thus causes $\bullet_{\mathsf{p}}\mathsf{O}(m|i)$.

A second case of context shifting was demonstrated in example (F). There the interplay of deontic detachment and the weakened SA shifted the context of $\mathsf{O}(w \wedge f|f)$ to d. This made it possible to detach $\mathsf{O}^{\mathsf{p}}(w \wedge f)$ despite the fact that we had $\neg f$.

12.6.2 Deontic Conflicts

So far we did not take a look at the way the ALs handle detachment in face of deontic conflicts. Take for instance the situation that two persons are about to drown and, although it is possible to save one of them, it is not possible to save both. For each person individually we have in general the obligation to save his or her life, $\mathsf{O}(a|\top)$ and $\mathsf{O}(b|\top)$. However, it is intuitive that there is no obligation to save both of them, $\mathsf{O}(a \wedge b|\top)$, but rather to save at least one of them, $\mathsf{O}(a \vee b|\top)$. Following the intuition of the Kantian "ought implies can", it is also not desirable to derive the conjunction of the proper obligations, $\mathsf{O}^{\mathsf{p}}a \wedge \mathsf{O}^{\mathsf{p}}b$, since it is impossible to save the lives of both. Even more so is it counter-intuitive to have $\mathsf{O}^{\mathsf{i}}a \wedge \mathsf{O}^{\mathsf{i}}b$ since the instrumental obligation should tell us what to do. However, it would not be very useful to have two pragmatic instructions which are mutually exclusive. It is interesting that the following is true in our lower limit logics **CDPM2.d**$^+$ and **CDPM2.e**$^+$ (where $\mathsf{x} \in \{\mathsf{p}, \mathsf{i}\}$)[30]:

[30] In **CDPM.2d**$^+$ also the stronger "If $\vdash B \supset \neg A$ then $\big(\mathsf{O}(B|C) \wedge C \wedge \mathsf{O}(A|C)\big) \supset \bullet_{\mathsf{x}}\mathsf{O}(A|C)$" is valid. Note that in this version $\mathsf{P}(C|C)$ is not part of the antecedent. In view of this, in **CDPM2.e**$^+$ one may want to add the following rule in order to have a stricter handling of deontic conflicts: If $\vdash B \supset \neg A$ then $\big(\mathsf{O}(A|C) \wedge \mathsf{O}(B|C) \wedge C\big) \supset \bullet_{\mathsf{x}}\mathsf{O}(A|C)$.

$$\text{If } \vdash B \supset \neg A, \text{ then } \big(\mathsf{O}(B|\top) \wedge \mathsf{O}(A|\top)\big) \supset \bullet_{\mathsf{x}}\mathsf{O}(A|\top) \quad (\text{Cx}\top)$$

$$\text{If } \vdash B \supset \neg A, \text{ then } \big(\mathsf{O}(B|C) \wedge C \wedge \mathsf{P}(C|C) \wedge \mathsf{O}(A|C)\big) \supset \bullet_{\mathsf{x}}\mathsf{O}(A|C) \quad (\text{Cx})$$

In our example, presupposing $\vdash a \supset \neg b$, we get by (Cx⊤), $\bullet_{\mathsf{p}}\mathsf{O}(a|\top)$, $\bullet_{\mathsf{p}}\mathsf{O}(b|\top)$, $\bullet_{\mathsf{i}}\mathsf{O}(a|\top)$, $\bullet_{\mathsf{i}}\mathsf{O}(b|\top)$ and thus detachment is blocked for both obligations. This is the desired outcome as discussed above. Moreover, there is no way to block detachment from $\mathsf{O}(a \vee b|\top)$ and thus $\mathsf{O}^{\mathsf{p}}(a \vee b)$ and $\mathsf{O}^{\mathsf{i}}(a \vee b)$ are derivable (given $\mathsf{O}(a \vee b|\top)$). This is desired as well.[31]

12.7 Modeling Nested Permissible Contexts

As pointed out in Sect. 12.3, the generic enhancement $\mathbf{L}^+$ for deontic logics presented there is not able to model nested permissible contexts. These are cases in which we have a permissible context C to B but not $\mathsf{P}(C|B)$. The idea was there to focus on the explication of the adaptive handling of detachment and hence not to introduce additional complications. However, as will be demonstrated in this section, the logical framework can be enhanced with this ability by introducing some additional techniques.

Recall that $\langle C_1, \ldots, C_n \rangle$ is a *permissive sequence* from C_1 to C_n iff, for all $i < n$ (a) $\vdash C_{i+1} \supset C_i$ and (b) $\mathsf{P}(C_{i+1}|C_i)$. Moreover, C is a *permissible context* to B iff there is a permissive sequence from B to C.

We have already noticed that the permissive sequences characterizing permissible contexts have indeed sometimes a minimal length of more than 1. An instance was given by the asparagus example where we have $\mathsf{O}(\neg f|\top)$, $\mathsf{P}(a|\top)$ and $\mathsf{P}(f \wedge a|a)$, but not $\mathsf{P}(f \wedge a|\top)$ (see the discussion in Sect. 12.2.1). Evidently $f \wedge a$ describes a permissible context to $\top$.

12.7.1 Generalizing $\mathbf{L}^+$ *for Nested Permissible Contexts*

How can permissible sequences be formally modeled? The idea is to make use of an additional permission operator $\mathbb{P}(A \mid B)$ that expresses that A is a permissible context to B. It is axiomatized as follows:

$$\text{If } \vdash A \supset B, \text{ then } \mathsf{P}(A|B) \supset \mathbb{P}(A \mid B) \quad (\text{P-Ps})$$

[31] Furthermore, many of the adaptive strengthenings of **CDPM** defined in [30] are able to derive $\mathsf{O}(a \vee b|\top)$ from $\mathsf{O}(a|\top)$ and $\mathsf{O}(b|\top)$. Thus, forming a combined AL with one of these systems, analogous to the way it is sketched in Sect. 12.8, $\varphi = \mathsf{O}^{\mathsf{p}}(a \vee b) \wedge \mathsf{O}^{\mathsf{i}}(a \vee b)$ is derivable from $\mathsf{O}(a|\top)$ and $\mathsf{O}(b|\top)$, whereas for **DCDPM.2d**$^+$ and **DCDPM.2e**$^+$ we have to add the additional premise $\mathsf{O}(a \vee b|\top)$ in order to derive φ.

$$\vdash \big(\mathbb{P}(B \mid A) \wedge \mathbb{P}(C \mid B)\big) \supset \mathbb{P}(C \mid A) \qquad \text{(Ps-T)}$$

By these axioms we can derive $\mathbb{P}(f \wedge a \mid \top)$ from $\mathsf{P}(a|\top)$ and $\mathsf{P}(f \wedge a|a)$, as desired. More generally, we are able to derive $\mathbb{P}(C_n \mid C_1)$ from $\mathsf{P}(C_2|C_1)$, ..., $\mathsf{P}(C_n|C_{n-1})$ (where for all $i < n, \vdash C_{i+1} \supset C_i$) by multiple applications of (P-Ps) and (Ps-T).

Now we can adjust the axiomatization of our generic enhancement $\mathbf{L}^+$ of the base logic $\mathbf{L}$ from Sect. 12.3 so that it can model precisely the more general notions from Sect. 12.2.

$$\text{If } \vdash D \supset \neg A \text{ and } \vdash C \supset B, \text{ then } \big(\big(\mathsf{P}(D|C) \vee \mathsf{O}(D|C)\big) \wedge C \wedge \mathbb{P}(C \mid B) \wedge \mathsf{O}(A|B)\big) \supset \bullet_{\mathsf{p}}\mathsf{O}(A|B) \qquad \text{(Ep-g)}$$

$$\text{If } \vdash A \supset \neg D, \vdash A \supset \neg C, \text{ and } \vdash C \supset B, \text{ then } \big(\mathsf{O}(D|C) \wedge \mathsf{O}(A|B) \wedge \neg\mathbb{P}(C \mid B)\big) \supset \bullet_{\mathsf{p}}\mathsf{O}(D|C) \qquad \text{(CTDR-g)}$$

The idea behind (Ep-g) is that if $\mathsf{O}(A|B)$ is excepted in C, then the proper obligation to bring about A should not be detached from $\mathsf{O}(A|B)$. Hence, in this case $\bullet_{\mathsf{p}}\mathsf{O}(A|B)$ is derived. Rule (CTDR-g) concerns strong CTD obligations. Given that $\mathsf{O}(D|C)$ is a strong CTD obligation to $\mathsf{O}(A|B)$, the proper obligation to bring about D should not be detachable from $\mathsf{O}(D|C)$. Hence, $\bullet_{\mathsf{p}}\mathsf{O}(D|C)$ is derived.

The rules (fV) resp. (oV-ei) that manage the blocking of instrumental detachment in case an obligation is factually violated resp. in case there is a more specific obligation incompatible with it can remain as they were defined in Sect. 12.3, since permissible contexts do not play a role for them.

Definition 12.7.1. Given a base logic $\mathbf{L}$ we define $\mathbf{L}^+_{\mathbb{P}}$ to be $\mathbf{L}$ enriched by the axioms (P-Ps), (Ps-T), (Ep-g), (CTDR-g), (fV), (oV-ei), (CREi), (RCEi), (CREp), (RCEp), ($\mathrm{EO}^{\mathtt{i}}$), ($\mathrm{EO}^{\mathtt{p}}$), (FDp), and (FDi).

The underlying logic for the following examples is again an enriched **CDPM**.$\mathbf{2}\alpha$ where $\alpha \in\{\mathbf{d},\mathbf{e}\}$, i.e., **CDPM**.$\mathbf{2}\alpha^+_{\mathbb{P}}$.

Example 12.7.1. Let us again have a look at the asparagus example (PA). We enrich the premise set $\{\mathsf{O}(\neg f|\top), \mathsf{O}(f|a), \mathsf{P}(a|\top), a\}$ by the intuitive $\mathsf{O}(f|f \wedge a)$. One of the counter-intuitive consequences of **CDPM**.$\mathbf{2}\alpha^+$ is $\bullet_{\mathsf{p}}\mathsf{O}(f|f \wedge a)$ which is derivable by (CTDR) from $\mathsf{O}(\neg f|\top)$. Evidently, $\mathsf{O}(f|f \wedge a)$ is not a CTD obligation to $\mathsf{O}(\neg f|\top)$ since $f \wedge a$ is a permissible context to $\top$. It is easy to see that $\bullet_{\mathsf{p}}\mathsf{O}(f|f \wedge a)$ is not anymore derivable by **CDPM**.$\mathbf{2}\alpha^+_{\mathbb{P}}$. The reason is that $\mathbb{P}(f \wedge a \mid \top)$ is derivable (given $\mathsf{P}(f \wedge a|a)$) and hence (CTDR-g) is not applicable in such a way that $\bullet_{\mathsf{p}}\mathsf{O}(f|f \wedge a)$ is derivable.

There is still a drawback to the idea as it was presented so far. Take for instance the premises of the Forrester paradox: $\mathsf{O}(\neg k|\top)$ and $\mathsf{O}(g|k)$. Note that there are models[32] in which k is a permissible context to $\top$, that is to say, models in which $\mathbb{P}(k|\top)$ is verified. Take for instance the model that validates $\mathsf{P}(k \vee x|\top)$ and $\mathsf{P}(k|k \vee x)$.

[32] The semantics of **CDPM**.$\mathbf{2}\alpha^+_{\mathbb{P}}$ is defined by means of neighborhood frames similar as the semantics of **CDPM**.$\mathbf{2}\alpha^+$. This is spelled out in Appendix J.

Moreover, there is a model that validates $\mathbb{P}(k|\top)$ even if there is no permissive sequence from $\top$ to k. As a consequence, $\neg\mathbb{P}(k \mid \top)$ is not derivable and hence (CTDR-g) is not applicable in order to derive $\bullet_{\mathsf{p}}\mathsf{O}(g|k)$.

The reason for this is that all that is guaranteed by (P-Ps) and (Ps-T) is that if there is a permissive sequence from some A to some B then $\mathbb{P}(B \mid A)$. However, the other direction is not ensured. Moreover, there seem to be no simple axiomatic way of doing so. What would have to be expressed is that whenever we have $\mathbb{P}(B \mid A)$ then there is a natural number n such that there is a permissive sequence $\langle C_1, \dots, C_n\rangle$ where $A = C_1$ and $B = C_n$. However, without means to quantify over propositions and numbers this seems a hopeless enterprise.

Here is where ALs help us out another time. The idea is to interpret a premise set in such a way that B is a permissible context to A, i.e., $\mathbb{P}(B \mid A)$, iff there is an explicit permissive sequence from A to B. Our axioms (P-Ps) and (Ps-T) ensure the right-left direction. Hence, it is our task to ensure the left-right direction. In order to achieve this, we define the following AL:

Definition 12.7.2. Where the set of abnormalities $\Omega^{\mathbb{P}}$ is $\{\mathbb{P}(B \mid A) \mid A, B \in \mathcal{P}\}$, the adaptive logic $\mathbf{PL}^{+}_{\mathbb{P}}$ is defined by the triple $\langle \mathbf{L}^{+}_{\mathbb{P}}, \Omega_{\mathbb{P}}, \text{reliability strategy}\rangle$.

The reason why this realizes both directions is easy to see. If there is a permission sequence from A to B, then by (P-Ps) and (Ps-T), $\mathbb{P}(B \mid A)$. If there is no permissive sequence, then $\mathbb{P}(B \mid A)$ is not derivable by (P-Ps) and (Ps-T) and the AL $\mathbf{PL}^{+}_{\mathbb{P}}$ will take care of deriving $\neg\mathbb{P}(B \mid A)$, since $\mathbb{P}(B \mid A)$ is an abnormality. Obviously $\vdash_{\mathbf{L}^{+}_{\mathbb{P}}} \mathbb{P}(B \mid A) \vee \neg\mathbb{P}(B \mid A)$ and hence $\neg\mathbb{P}(B \mid A)$ is adaptively derivable on the condition $\{\mathbb{P}(B \mid A)\}$. In the remainder we indicate such conditional derivations by "RC$\mathbb{P}$" in the adaptive proofs. The following examples are formulated for $\mathbf{PCDPM.2}\alpha^{+}_{\mathbb{P}}$ where $\alpha \in \{\mathbf{d}, \mathbf{e}\}$.

Example 12.7.2. Let us take another look at the Gentle Murderer.

1	$\mathsf{O}(\neg k|\top)$	PREM	$\emptyset$
2	$\mathsf{O}(g|k)$	PREM	$\emptyset$
3	$\neg\mathbb{P}(k \mid \top)$	RC$\mathbb{P}$	$\{\mathbb{P}(k \mid \top)\}$
4	$\bullet_{\mathsf{p}}\mathsf{O}(g|k)$	1,2,3; CTDR-g	$\{\mathbb{P}(k \mid \top)\}$

It is easy to see that there is no way of extending the proof in such a way that lines 3 and 4 are marked. Hence, as desired, $\bullet_{\mathsf{p}}\mathsf{O}(g|k)$ is a finally derivable in $\mathbf{PCDPM.2}\alpha^{+}_{\mathbb{P}}$.

The following example features nested permissible contexts.

Example 12.7.3. Let $\vdash a_{i+1} \supset a_i$ where $1 \leq i < 3$.

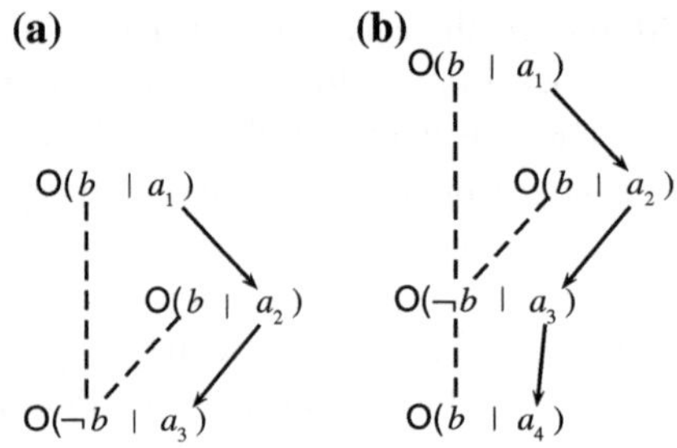

Fig. 12.4 The *dashed line* indicates an inconsistency between b and $\neg b$. The *solid arrow* indicates that e.g. a_2 is a permissible context to a_1

1 $\mathsf{O}(b\|a_1)$	PREM	$\emptyset$
2 $\mathsf{P}(a_2\|a_1)$	PREM	$\emptyset$
3 $\neg\mathsf{P}(a_3\|a_1)$	PREM	$\emptyset$
4 $\mathsf{O}(b\|a_2)$	PREM	$\emptyset$
5 $\mathsf{P}(a_3\|a_2)$	PREM	$\emptyset$
6 $\mathsf{O}(\neg b\|a_3)$	PREM	$\emptyset$
7 $\neg\mathsf{P}(a_3\|a_1)$	PREM	$\emptyset$
8 a_3	PREM	$\emptyset$
9 $\mathsf{O}(\neg a_3\|a_1)$	7; Def	$\emptyset$
10 $\mathbb{P}(a_2 \mid a_1)$	2; P-Ps	$\emptyset$
11 $\mathbb{P}(a_3 \mid a_2)$	5; P-Ps	$\emptyset$
12 $\mathbb{P}(a_3 \mid a_1)$	10,11; Ps-T	$\emptyset$
13 a_2	8; CL	$\emptyset$
14 $\bullet_{\mathsf{p}}\mathsf{O}(\neg a_3\|a_1)$	5,9,10,13; Ep-g	$\emptyset$
15 $\bullet_{\mathsf{i}}\mathsf{O}(\neg a_3\|a_1)$	8,9; fV	$\emptyset$
16 $\bullet_{\mathsf{p}}\mathsf{O}(b\|a_1)$	1,6,8,12; Ep-g	$\emptyset$
17 $\bullet_{\mathsf{p}}\mathsf{O}(b\|a_2)$	4,6,11,13; Ep-g	$\emptyset$
18 $\bullet_{\mathsf{i}}\mathsf{O}(b\|a_1)$	1,6,8; oV-Ei	$\emptyset$
19 $\bullet_{\mathsf{i}}\mathsf{O}(b\|a_2)$	4,6,8; oV-Ei	$\emptyset$

Note that $\mathbb{P}(a_3 \mid a_1)$ although $\neg\mathsf{P}(a_3|a_1)$. The two permissions $\mathsf{P}(a_2|a_1)$ and $\mathsf{P}(a_3|a_2)$ give rise to the nested permissible context a_3 to a_2 where a_2 is a permissible context to a_1. See for an illustration Fig. 12.4a. Note that $\bullet_{\mathsf{p}}\mathsf{O}(b|a_1)$ is not derivable by **CDPM.2**α^+. Evidently it is desired, since $\mathsf{O}(b|a_1)$ and $\mathsf{O}(b|a_2)$ are excepted in a_3 due to $\mathsf{O}(\neg b|a_3)$ and $\mathbb{P}(a_3 \mid a_1)$ (resp. $\mathbb{P}(a_3 \mid a_2)$).

Example 12.7.4. Let us extend the example from above. The reader may wonder what happens if the primary obligation to bring about b in the context a_1 gets reinstated at an even more specific level (see Fig. 12.4b). Suppose for the following that $\vdash a_4 \supset a_3$.

20 $\mathsf{O}(b\|a_4)$	PREM	$\emptyset$
21 a_4	PREM	$\emptyset$

22 $\mathsf{P}(a_4 \mid a_3)$	PREM	$\emptyset$
23 $\mathbb{P}(a_4 \mid a_3)$	22; P-Ps	$\emptyset$
24 $\bullet_{\mathsf{p}}\mathsf{O}(\neg b \mid a_3)$	6,20,21,23; Ep-g	$\emptyset$
25 $\bullet_{\mathsf{i}}\mathsf{O}(\neg b \mid a_3)$	6,20,21; oV-Ei	$\emptyset$

As desired, due to the second element of lines 24 and 25, proper and instrumental detachment is blocked from $\mathsf{O}(\neg b \mid a_3)$ since it is excepted in a_4 as we have $\mathsf{O}(b \mid a_4)$ and $\mathbb{P}(a_4 \mid a_3)$.

12.7.2 Adaptively Applying Detachment

In order to apply deontic detachment adaptively we can now proceed analogously to Sect. 12.4. Given a premise set Γ we first apply $\mathbf{PL}^+_{\mathbb{P}}$ and then $\mathbf{DL}^+_{\mathbb{P}}$. The latter logic is defined analogous to the definition of $\mathbf{DL}^+$ in Sect. 4.2 by the triple $\langle \mathbf{L}^+_{\mathbb{P}}, \Omega^d$ reliability strategy$\rangle$. The combination of the two logics is realized by a sequential AL (for a more general introduction and a detailed description of the proof theory see Chap. 3). **DPL** is characterized by the consequence relation

$$Cn_{\mathbf{DPL}}(\Gamma) = Cn_{\mathbf{DL}^+_{\mathbb{P}}}\left(Cn_{\mathbf{PL}^+_{\mathbb{P}}}(\Gamma)\right)$$

The marking for abnormalities in $\Omega_{\mathbb{P}}$ is analogous to Definition 2.4.2 (see Definition 12.7.3. below). We only need to slightly alter the marking for abnormalities in Ω^d. Since in the sequential case $\mathbf{DL}^+_{\mathbb{P}}$ operates on the consequence set of $\mathbf{PL}^+_{\mathbb{P}}$, Dab-formulas over abnormalities in Ω^d that are derived at unmarked lines on conditions that are subsets of $\Omega_{\mathbb{P}}$ have to be taken into account for the marking. Let me give an example.

Example (continues Example 12.7.2). We extend the proof above by the following lines:

5 k	PREM	$\emptyset$
$^{\ddagger(4)}$6 $\mathsf{O}^{\mathsf{p}} g$	2,5; cFDp	$\{\bullet_{\mathsf{p}}\mathsf{O}(g \mid k)\}$
7 $\mathsf{O}^{\mathsf{p}} \neg k$	1; cFDp	$\{\bullet_{\mathsf{p}}\mathsf{O}(\neg k \mid \top)\}$
8 $\bullet_{\mathsf{i}}\mathsf{O}(\neg k \mid \top)$	1,5; fV	$\emptyset$
9 $\mathsf{O}^{\mathsf{i}} g$	2,5; cFDi	$\{\bullet_{\mathsf{i}}\mathsf{O}(g \mid k)\}$

Note that the Dab-formula that is responsible for the marking of line 6 has been derived on the condition $\{\mathbb{P}(k \mid \top)\}$ at line 4. It is derivable in $\mathbf{PCDPM.2}\alpha^+_{\mathbb{P}}$ that k is not a permissible context to $\top$ (line 3). Given this, it follows further that $\mathsf{O}(g \mid k)$ is a strong CTD obligation to $\mathsf{O}(\neg k \mid \top)$ and hence $\bullet_{\mathsf{p}}\mathsf{O}(g \mid k)$ is derived at line 4. This, however, blocks the detachment at line 6.

Where $\mathsf{Dab}(\Delta_1), \ldots, \mathsf{Dab}(\Delta_m)$ are all minimal disjunctions of abnormalities in $\Omega^{\mathbb{P}}$ derived on the empty condition at stage s, we define $U^{\mathbb{P}}_s(\Gamma) = \Delta_1 \cup \ldots \cup \Delta_m$. The marking for $\mathbf{PL}^+_{\mathbb{P}}$ is defined as usual for the reliability strategy.

Definition 12.7.3. Line i is †-marked at stage s iff, where $\Delta \subseteq \Omega^{\mathbb{P}}$ is its condition, $\Delta \cap U_s^{\mathbb{P}}(\Gamma) \neq \emptyset$.

Where $\mathsf{Dab}(\Delta_1'), \ldots, \mathsf{Dab}(\Delta_n')$ are the minimal disjunctions of abnormalities in Ω^d derived at unmarked lines on conditions $\Theta \subseteq \Omega^{\mathbb{P}}$ at stage s, we define $U_s^d(\Gamma) = \Delta_1' \cup \ldots \cup \Delta_n'$.

Definition 12.7.4. Line i is ‡-marked at stage s iff, where Δ is its condition, $\Delta \cap U_s^d(\Gamma) \neq \emptyset$.

Example (continues Example 12.7.3). Prolonging the proof above nicely demonstrates the conditional applications of detachment for the case with nested exceptional contexts.

	20 a_1	8; CL	$\emptyset$
‡(16)	21 $\mathsf{O}^{\mathsf{p}} b$	1,20; cFDp	$\{\bullet_{\mathsf{p}}\mathsf{O}(b \mid a_1)\}$
‡(17)	22 $\mathsf{O}^{\mathsf{p}} b$	4,13; cFDp	$\{\bullet_{\mathsf{p}}\mathsf{O}(b \mid a_2)\}$
	23 $\mathsf{O}^{\mathsf{p}} \neg b$	6,8; cFDp	$\{\bullet_{\mathsf{p}}\mathsf{O}(\neg b \mid a_3)\}$
	24 $\mathsf{O}^{\mathsf{i}} \neg b$	6,8; cFDi	$\{\bullet_{\mathsf{i}}\mathsf{O}(\neg b \mid a_3)\}$

As expected, factual detachment is neither applicable to $\mathsf{O}(b \mid a_1)$ nor to $\mathsf{O}(b \mid a_2)$. Both are excepted in a_3. Hence, $\mathsf{O}^{\mathsf{p}} \neg b$ is derived at line 23 and $\mathsf{O}^{\mathsf{i}} \neg b$ at line 24.

Example (continues Example 12.7.4). The situation is different if we proceed with the enhanced premise set from Example 12.7.4.

‡(24)	26 $\mathsf{O}^{\mathsf{p}} \neg b$	6,8; cFDp	$\{\bullet_{\mathsf{p}}\mathsf{O}(\neg b \mid a_3)\}$
	27 $\mathsf{O}^{\mathsf{p}} b$	20,21; cFDp	$\{\bullet_{\mathsf{p}}\mathsf{O}(b \mid a_4)\}$
‡(25)	28 $\mathsf{O}^{\mathsf{i}} \neg b$	6,8; cFDi	$\{\bullet_{\mathsf{i}}\mathsf{O}(\neg b \mid a_3)\}$
	29 $\mathsf{O}^{\mathsf{i}} b$	20,21; cFDi	$\{\bullet_{\mathsf{i}}\mathsf{O}(b \mid a_4)\}$

In this case we are able to derive the proper and instrumental obligation to bring about b. This is intuitive since $\mathsf{O}(\neg b \mid a_3)$ is excepted in a_4.

12.8 Discussion

Let me begin this section by pointing out some advantages of the adaptive modeling of detachment.

First, it offers a very generic framework for detachment for dyadic deontic logics since ALs strengthen the enriched base logic, their lower limit logic, for which it is only required that it is a reflexive, transitive, monotonic and compact logic. Depending on the application the reader is free to use any base logic as long as it fulfills the mentioned requirements. Furthermore, since ALs have shown great unifying power representing nonmonotonic, defeasible logics, even deontic logics that do not fulfill the requirements may be represented by ALs.

By applying techniques of combining adaptive systems, the framework developed in this chapter may be applicable also in such cases.

Furthermore, the meta-theory of ALs in standard format is well-studied (see [31]). For instance, completeness and soundness are guaranteed by the completeness and soundness of the lower limit logic.

The internal dynamics modeled by the dynamic proof theory of ALs models the dynamic nature of human reasoning, more precisely, its diachronic defeasibility (cp. Pollock [32]). This type of defeasibility is based on the fact that we sometimes have to withdraw certain conclusion we drew before due to insights gained by analyzing the information at hand, even in cases in which we were not provided with any explicitly new information. Suppose we (conditionally) apply, for our asparagus example, detachment to $\mathsf{O}(\neg f \mid \top)$. However, analyzing our premise set further we come to the conclusion that we are in an exceptional context. In this case we revise the former derivation. In the adaptive proof the line at which detachment has been applied to $\mathsf{O}(\neg f \mid \top)$ is going to be marked and is hence considered not to be valid. This shows that the explication of the internal dynamics of the reasoning process is itself an integral part of the adaptive proof theory.

Yet another advantage of ALs is their ability to deal with external dynamics, i.e., with the synchronic defeasibility (see [32]) based on the introduction of new information. Again the markings of the dynamic proofs are able to model cases of specificity and CTD obligations which might be caused by new information. In the adaptive approach, despite the fact that new information might force us to withdraw certain conclusions, the proof dynamics via the marking procedure model in an accurate way the fact that we continue reasoning facing new information instead of beginning our reasoning process again from scratch.

Let me conclude this section by mentioning some possibilities to further strengthen the presented systems.

As pointed out in [30], one of the disadvantages of Goble's monadic logics **DPM** and of his dyadic generalizations **CDPM** is that, in order to make use of the restricted versions of inheritance, aggregation and SA, additional premises have to be added by the user of the logic. This is suboptimal in the sense that as much reasoning as possible should be performed by the logic and, inversely, as less as possible should be left to the user. Especially in complicated settings the adding of premises can be very bothersome for the user, since in such cases it is in no way a trivial question whether the addition of a certain permission statement leads to explosion or to other counter-intuitive consequences. In [30, 33] ALs on the basis of Goble's systems were developed which perform this task instead of the user: inheritance, aggregation and/or SA are applied conditionally (see Chap. 10 and Chap. 11). The way this is technically realized is very similar to the way detachment is realized in this chapter. By adjusting the marking conditions, it is technically straightforward to realize a sequential AL, with one of the logics in [30] resp. Chapter 11 being the first and one of our ALs for detachment being the second logic. This way the advantages of both systems can be combined in order to realize an altogether very powerful deontic logic.

Dyadic deontic logics are not the only field in which it is interesting to model detachment. For instance conditional logics of normality [34–36] offer promising ways to formalize default and common sense reasoning. Instead of our dyadic obligation operator there is a dyadic operator $\rightsquigarrow$, where $A \rightsquigarrow B$ is read as "A normally implies B". In order to realize actual default inferencing an adaptive approach to modus ponens may be developed. Techniques developed in this chapter for deontic logics may be transferred since also in default reasoning specificity occurs and limits the applicability of modus ponens (see Chap. 6).

12.9 Conclusion

In this chapter I proposed a generic way to turn dyadic deontic logics into ALs which model defeasible detachment. I proposed two intuitions concerning the question what obligations should be detached: proper and instrumental obligations. Obligations which are not excepted by more specific obligations bind us also in cases in which they are violated and are therefore considered to be proper obligations. Instrumental obligations are a more pragmatic concept: taking the factual premises as immutable, we can ask the question, what is the best thing to do in the given circumstances? For instance contrary-to-duty obligations indicate instrumental obligations, while the primary obligations which are violated do not cease to bind us as proper obligations. Paradigmatically I presented ALs based on one of Lou Goble's conflict-tolerant **CDPM** systems which implement these ideas. ALs are an excellent choice for defeasibly enabling detachment since they allow for certain rules to be applied "as much as possible". In this case, detachment with respect to proper and instrumental obligations is adaptively applied under special consideration of specificity and contrary-to-duty cases.

Acknowledgments I thank Joke Meheus, Mathieu Beirlaen, Frederik Van De Putte, Dunja Šešelja and the anonymous referees of the Journal for Applied Logic for valuable comments which helped to improve the paper.

References

1. Åqvist, L.: Deontic logic. In: Gabbay, D.M., Guenthner, F. (eds.) Handbook of Philosophical Logic, 2nd edn, vol. 8, pp. 147–264. Kluwer, Dordrecht (2002)
2. Goble, L.: Dilemmas in deontic logic (to appear)
3. Goble, L.: A proposal for dealing with deontic dilemmas. In: Lomuscio, A., Nute, D. (eds.) DEON. Lecture Notes in Computer Science, vol. 3065, pp. 74–113. Springer, Heidelberg (2004)
4. Chisholm, R.M.: Contrary-to-duty imperatives and deontic logic. Analysis **24**, 33–36 (1963)
5. Gabbay, D.M.: Completeness theorems for reactive modal logics (2010). Paper 392, July 2010. To appear in special issue of AMAI 2012

6. Gabbay, D.M.: Reactive Kripke Semantics and Arc Accessibility. Lecture Notes in Computer Science, pp. 292–341. Springer-Verlag, Berlin (2008)
7. Gabbay, D.M., Marcelino, S.: Modal logics of reactive frames. Studia Logica **93**, 403–444 (2009)
8. Gabbay, D.M.: Reactive Kripke models and contrary-to-duty obligations. In: van der Meyden, R., van der Torre, L. (eds.) DEON 2008, Deontic Logic in Computer Science, LNAI 5076, pp. 155–173. Springer, Berlin (2008)
9. Gabbay, D.M.: Temporal deontic logic for the generalised Chisholm set of contrary to duty obligations. In: DEON 2012, LNAI 7393, pp. 91–107. Springer, Heidelberg (2012)
10. Gabbay, D.M., Straßer, C.: Reactive standard deontic logic. J. Logic Comput. (2012)
11. Loewer, B., Belzer, M.: Dyadic deontic detachment. Synthese **54**, 295–318 (1983)
12. Prakken, H., Sergot, M.: Contrary-to-duty obligations. Studia Logica **57**(1), 91–115 (1996)
13. Greenspan, P.S.: Conditional ought and hypothetical imperatives. J. Philos. **72**, 259–276 (1975)
14. van Eck, J.: A system of temporally relative modal and deontic predicate logic and its philosophical applications. Logique at Analyse **99**, 249–290 (1982)
15. Lehmann, D.J., Magidor, M.: What does a conditional knowledge base entail? Artif. Intell. **55**(1), 1–60 (1992)
16. Carmo, J., Jones, A.J.I.: Deontic logic and contrary-to-duties. In: Gabbay, D.M., Guenthner, F. (eds.) Handbook of Philosophical Logic, 2nd edn., vol. 8, pp. 265–343. Kluwer, Dordrecht (2002)
17. Forrester, J.: Gentle murder, or the adverbial samaritan. J. Philos. **81**, 193–197 (1984)
18. McNamara, P.: Deontic logic. In: Gabbay, D., Woods, J. (eds.) Handbook of the History of Logic, pp. 197–288. Elsevier, Amsterdam (2007)
19. Wang, L.: The detachment problem of conditional obligations. http://goo.gl/EEiSmY (2008)
20. Torre, L.V.D., Tan, Y.H.: Cancelling and overshadowing: two types of defeasibility in defeasible deontic logic. In: Proceedings of the Fourteenth International Joint Conference on Artificial Intelligence (IJCAI'95), pp. 1525–1532 (1995)
21. van der Torre, L., Tan, Y.H.: Contrary-to-duty reasoning with preference-based dyadic obligations. Ann. Math. Artif. Intell. **27**(1–4), 49–78 (1999)
22. Prakken, H., Sergot, M.: Dyadic deontic logic and contrary-to-duty obligations. In: Nute, D. (ed.) Defeasible Deontic Logic, pp. 223–262. Kluwer, Dordrecht (1997)
23. Foot, P.: Moral realism and moral dilemma. J. Philos. **80**(7), 379–398 (1983)
24. Ryu, Y.U.: Conditional deontic logic augmented with defeasible reasoning. Data Knowl. Eng. **16**(1), 73–91 (1995)
25. Horty, J.F.: Deontic logic as founded on nonmonotonic logic. Ann. Math. Artif. Intell. **9**, 69–91 (1993)
26. Reiter, R.: A logic for default reasoning. Artif. Intell. **13**(1–2), 81–132 (1980)
27. Goble, L.: Preference semantics for deontic logic. Part I: simple models. Logique at Analyse **46**(183–184), 383–418 (2003)
28. van Fraassen, B.C.: The logic of conditional obligation. J. Philos. Logic **1**, 417–438 (1972)
29. Lewis, D.: Counterfactuals. Blackwell Publishers, Oxford (2000)
30. Straßer, C.: An adaptive logic framework for conditional obligations and deontic dilemmas. Logic Log. Philos. **19**(1–2), 95–128 (2010)
31. Batens, D.: A universal logic approach to adaptive logics. Logica Universalis **1**, 221–242 (2007)
32. Pollock, J.: Defeasible reasoning. Cogn. Sci. **11**(4), 481–518 (1987)
33. Straßer, C., Meheus, J., Beirlaen, M.: Tolerating deontic conflicts by adaptively restricting inheritance. Logique at Analyse **219**, 477–506 (2012)
34. Boutilier, C.: Conditional logics of normality: a modal approach. Artif. Intell. **68**(1), 87–154 (1994)
35. Kraus, S., Lehmann, D.J., Magidor, M.: Nonmonotonic reasoning, preferential models and cumulative logics. Artif. Intell. **44**, 167–207 (1990)
36. Lamarre, P.: S4 as the conditional logic of nonmonotonicity. In: Proceedings of the Second International Conference on the Principles of Knowledge Representation and Reasoning, pp. 357–367. Morgan Kaufmann (1991)
37. Straßer, C.: A deontic logic framework allowing for factual detachment. J. Appl. Logic 9(1), 61–80 (2010)

Appendix A
Some Generic Results on Choice Sets

In order to facilitate the reading we will shortcut $\min_{\subset}(X)$ by $X_{\min}$ in this section. Moreover, we will denote the set of choice sets of some set of sets Σ by $\mathsf{CS}(\Sigma)$. In the following results we will refer to X and Σ where X is an enumerable set while Σ is a set of finite subsets of X.

Lemma 5.5.1 (restated). *Where* $\varphi = \{A_i \mid i \in I\} \in \mathsf{CS}(\Sigma)$, *let* $\hat{\varphi} = \bigcap_{i \in I} \varphi_i$ *where* $\varphi_0 = \varphi$ *and (where* $i{+}1 \in I$*)*

$$\varphi_{i+1} = \begin{cases} \varphi_i & \textit{if} \text{ there is a } \Delta \in \Sigma \text{ such that } \varphi_i \cap \Delta = \{A_{i+1}\} \\ \varphi_i \setminus \{A_{i+1}\} & \textit{else} \end{cases}$$

we have: $\hat{\varphi} \in \mathsf{CS}_{\min}(\Sigma)$.

Proof. By the construction, (†) $\varphi_i \supseteq \varphi_{i+1}$ for all $i, i+1 \in I$. Note that, also by the construction, ($\star$) each φ_i is a choice set of Σ, i.e., $\varphi_i \in \mathsf{CS}(\Sigma)$ for each $i \in I$.

We now show that $\hat{\varphi} \in \mathsf{CS}(\Sigma)$. Assume otherwise and hence that there is a $\Delta \in \Sigma$ such that $\hat{\varphi} \cap \Delta = \emptyset$. Since Δ is finite and by ($\star$), $\Delta \cap \varphi_1 = \{B_1, \ldots, B_n\}$ for some $n \in \mathbb{N}$. By our assumption there is no $j \leq n$ such that $B_j \in \varphi_i \cap \Delta$ for all $i \in I$. Hence, for all $j \leq n$ there is a lowest $i_j \in I$ such that $B_j \notin \varphi_{i_j} \cap \Delta$. Take $k = \max(\{i_j \mid 1 \leq j \leq n\})$. Then, since due to (†) $\{B_1, \ldots, B_n\} \supseteq \varphi_i \cap \Delta \supseteq \varphi_{i+1} \cap \Delta$, also $B_j \notin \varphi_k \cap \Delta$ for all $j \leq n$ and thus $\varphi_k \cap \Delta = \emptyset$. However, this is a contradiction, since by ($\star$) $\varphi_k \in \mathsf{CS}(\Sigma)$. Thus, there is a $j \leq n$ such that $B_j \in \hat{\varphi}$. Hence, $\hat{\varphi} \in \mathsf{CS}(\Sigma)$.

Now assume $\hat{\varphi}$ is not minimal. Hence, there is a $A_i \in \hat{\varphi}$ (where $i \in I$) such that $\hat{\varphi} \setminus \{A_i\} \in \mathsf{CS}(\Sigma)$. By the construction, there is a $\Delta \in \Sigma$ such that $\varphi_{i-1} \cap \Delta = \{A_i\}$. By (†) and since $\hat{\varphi} \in \mathsf{CS}(\Sigma)$, $\hat{\varphi} \cap \Delta = \{A_i\}$,—a contradiction since then $\hat{\varphi} \setminus \{A_i\}$ cannot be a choice set of Σ. Hence, $\hat{\varphi} \in \mathsf{CS}_{\min}(\Sigma)$). □

Fact A.1. *Where* $\Delta \subseteq X$*: if there is no* $\Theta \in \Sigma$ *such that* $\Theta \subseteq \Delta$ *then* $\Delta \notin \mathsf{CS}(\mathsf{CS}_{\min}(\Sigma))$.

C. Straßer, *Adaptive Logics for Defeasible Reasoning*, Trends in Logic 38,
DOI: 10.1007/978-3-319-00792-2, © Springer International Publishing Switzerland 2014

Proof. Suppose the antecedent holds. Hence, for each $\Theta \in \Sigma$, $\Theta \setminus \Delta \neq \emptyset$. Let φ be any set in $\mathsf{CS}(\{\Theta \setminus \Delta \mid \Theta \in \Sigma\})$. Note that also $\varphi \setminus \Delta \in \mathsf{CS}(\{\Theta \setminus \Delta \mid \Theta \in \Sigma\})$ and moreover, $\varphi \setminus \Delta \in \mathsf{CS}(\Sigma)$. By Lemma 5.5.1, there is a $\psi \subseteq \varphi \setminus \Delta$ for which $\psi \in \mathsf{CS}_{\min}(\Sigma)$. Note that $\Delta \cap \psi = \emptyset$. Hence, $\Delta \notin \mathsf{CS}(\mathsf{CS}_{\min}(\Sigma))$. □

Fact A.2. *Where* $\Theta \in \Sigma$, $\Theta \in \mathsf{CS}(\mathsf{CS}_{\min}(\Sigma))$.

Proof. Let $\Theta \in \Sigma$ and $\varphi \in \mathsf{CS}_{\min}(\Sigma)$ arbitrary. Hence, $\varphi \in \mathsf{CS}(\Sigma)$ and thus, $\varphi \cap \Theta \neq \emptyset$. □

Lemma A.1. *Where* $\Delta \subseteq X$: $\Delta \in \Sigma_{\min}$ *iff* $\Delta \in \mathsf{CS}_{\min}(\mathsf{CS}_{\min}(\Sigma_{\min}))$.

Proof. Let $\Delta \in \Sigma_{\min}$. Obviously for each $\varphi \in \mathsf{CS}_{\min}(\Sigma_{\min})$, $\varphi \cap \Delta \neq \emptyset$. Hence, $\Delta \in \mathsf{CS}(\mathsf{CS}_{\min}(\Sigma_{\min}))$. Assume there is a $\Delta' \subset \Delta$ such that $\Delta' \in \mathsf{CS}(\mathsf{CS}_{\min}(\Sigma_{\min}))$. Hence, $\Delta' \notin \Sigma_{\min}$. By the contra-position of Fact A.1 there is a $\Theta \in \Sigma_{\min}$ such that $\Theta \subseteq \Delta'$. Thus, $\Theta \subset \Delta'$ since $\Delta' \notin \Sigma_{\min}$. But then $\Theta \subset \Delta$,—a contradiction to the fact that $\Delta \in \Sigma_{\min}$. Hence, our assumption is false. But that means that $\Delta \in \mathsf{CS}_{\min}(\mathsf{CS}_{\min}(\Sigma_{\min}))$.

Let now $\Delta \notin \Sigma_{\min}$. Assume $\Delta \in \mathsf{CS}_{\min}(\mathsf{CS}_{\min}(\Sigma_{\min}))$ and hence $\Delta \in \mathsf{CS}(\mathsf{CS}_{\min}(\Sigma_{\min}))$. By the contra-position of Fact A.1, there is a $\Theta \in \Sigma_{\min}$ such that $\Theta \subset \Delta$. By Fact A.2, $\Theta \in \mathsf{CS}(\mathsf{CS}_{\min}(\Sigma_{\min}))$,—a contradiction to our assumption. □

Lemma A.2. $\bigcup \Sigma_{\min} = \bigcup \mathsf{CS}_{\min}(\Sigma_{\min})$

Proof. Let $A \in \bigcup \Sigma_{\min}$. Hence, there is a $\Delta \in \Sigma_{\min}$ such that $A \in \Delta$. By Lemma A.1, $\Delta \in \mathsf{CS}_{\min}(\mathsf{CS}_{\min}(\Sigma_{\min}))$. Hence, for all $\varphi \in \mathsf{CS}_{\min}(\Sigma_{\min})$, $\Delta \cap \varphi \neq \emptyset$. Thus, $A \in \varphi$ for some $\varphi \in \mathsf{CS}_{\min}(\Sigma_{\min})$, else $\Delta \setminus \{A\} \in \mathsf{CS}(\mathsf{CS}_{\min}(\Sigma_{\min}))$ in contradiction to $\Delta \in \mathsf{CS}_{\min}(\mathsf{CS}_{\min}(\Sigma_{\min}))$.

Let $A \in \bigcup \mathsf{CS}_{\min}(\Sigma_{\min})$. Hence, there is a $\varphi \in \mathsf{CS}_{\min}(\Sigma_{\min})$ such that $A \in \varphi$. Suppose for all $\Delta \in \Sigma_{\min}$, $A \notin \Delta$. Then $\varphi \setminus \{A\} \in \mathsf{CS}(\Sigma_{\min})$ in contradiction to $\varphi \in \mathsf{CS}_{\min}(\Sigma_{\min})$. □

Applied to the adaptive logic context Lemma A.1 yields the following result which will be useful for proving results of Chap. 5: the set of minimal choice sets of the set of minimal choice sets of $\Sigma(\Gamma)$ is $\Sigma(\Gamma)$ itself. We use the notation of Chap. 5: $\Xi(\Gamma)$ $[\Xi_s(\Gamma)]$ denotes all $\varphi \subseteq \Omega$ such that φ is a choice set of $\Sigma(\Gamma)$ $[\Sigma_s(\Gamma)]$, and $\min^{\cup}_{\subset}(K) =_{\text{df}} \{\varphi \in K \mid \varphi \subseteq \bigcup \min_{\subset}(K)\}$ where K is a set of sets of abnormalities.

Corollary A.1. *Where* $\Gamma \subseteq \mathcal{W}^+$: $\Sigma(\Gamma) = \min_{\subset}(\mathsf{CS}(\min_{\subset}(\Xi(\Gamma)))) = \min_{\subset}(\mathsf{CS}(\Phi(\Gamma)))$.

Similarly, by Lemma A.2 we get:

Corollary A.2. *Where* $\Gamma \subseteq \mathcal{W}^+$:

(i) $\bigcup \Sigma(\Gamma) = \bigcup \min_{\subset}(\Xi(\Gamma)) = \bigcup \min^{\cup}_{\subset}(\Xi(\Gamma)) = \bigcup \Phi(\Gamma)$
(ii) $\bigcup \Sigma_s(\Gamma) = \bigcup \min_{\subset}(\Xi_s(\Gamma)) = \bigcup \min^{\cup}_{\subset}(\Xi_s(\Gamma)) = \bigcup \Phi_s(\Gamma)$

The following two lemmas follow by Lemma A.1 and are useful to prove the adequacy of the two marking definitions for normal selections in Sect. 2.8.

Lemma A.3. *Where $\Delta \subseteq X$: If there is a $\varphi \in \mathsf{CS}_{\min}(\Sigma_{\min})$ such that $\varphi \cap \Delta = \emptyset$ then there is no $\Theta \in \Sigma_{\min}$ such that $\Theta \subseteq \Delta$.*

Proof. Let $\varphi \in \mathsf{CS}_{\min}(\Sigma_{\min})$ such that $\varphi \cap \Delta = \emptyset$. Assume there is a $\Theta \subseteq \Delta$ such that $\Theta \in \Sigma_{\min}$. Thus, $\Theta \cap \varphi \neq \emptyset$. But then $\Delta \cap \varphi \neq \emptyset$,—a contradiction. □

Lemma A.4. *Where $\Delta \subseteq X$ is finite: If there is no $\Delta' \in \Sigma_{\min}$ such that $\Delta' \subseteq \Delta$ then there is a $\varphi \in \mathsf{CS}_{\min}(\Sigma_{\min})$ such that $\varphi \cap \Delta = \emptyset$.*

Proof. Suppose the antecedent is true. Hence, by Lemma A.1, there is no $\Delta' \in \mathsf{CS}_{\min}(\mathsf{CS}_{\min}(\Sigma_{\min}))$ such that $\Delta' \subseteq \Delta$. Since Δ is finite this means that $\Delta \notin \mathsf{CS}(\mathsf{CS}_{\min}(\Sigma_{\min}))$. Hence, there is a $\varphi \in \mathsf{CS}_{\min}(\Sigma_{\min})$ such that $\Delta \cap \varphi = \emptyset$. □

The following corollary is immediate in view of the two lemmas:

Corollary 2.8.1 (restated). *Where $\Delta \subseteq \Omega$ is finite and $\Gamma \subseteq \mathcal{W}^+$:*

(i) there is a $\varphi \in \Phi_s(\Gamma)$ such that $\Delta \cap \varphi = \emptyset$ iff there is no minimal Dab*-formula* $\mathsf{Dab}(\Theta)$ *at stage s such that $\Theta \subseteq \Delta$;*
(ii) there is a $\varphi \in \Phi(\Gamma)$ such that $\Delta \cap \varphi = \emptyset$ iff there is no minimal Dab*-consequence* $\mathsf{Dab}(\Theta)$ *such that $\Theta \subseteq \Delta$.*

The following insights are useful in Chap. 3 in order to prove the relationships between the criteria in Fig. 3.3.

Lemma A.5. *If every $\varphi \in \mathsf{CS}_{\min}(\Sigma_{\min})$ is finite, then $\Sigma_{\min}$ is finite.*

Proof. Suppose $\Sigma_{\min} = \{\Delta_i^0 \mid i \in \mathbb{N}\}$ is infinite. Obviously for each $i \in \mathbb{N}$ and each $j \in \mathbb{N}$ where $i \neq j$: $\Delta_i^0 \setminus \Delta_j^0 \supset \emptyset$. Note that for some $A \in \Delta_1^0$ the set

$$\Sigma_A^1 =_{\mathrm{df}} \{\Delta_i^0 \mid A \in \Delta_1^0 \setminus \Delta_i^0\}$$

is infinite. The reason is that (a) $\Sigma_{\min}$ is infinite and hence $\Sigma_{\min} \setminus \{\Delta_1^0\}$ is also infinite, (b) $\Sigma_{\min} \setminus \{\Delta_1^0\} = \bigcup_{A \in \Delta_1^0} \Sigma_A^1$, and (c) the latter is a finite union. Hence, some Σ_A^1 must be infinite. Let A_1 be such that the set $\Sigma_{A_1}^1$ is infinite. We rename the elements of $\Sigma_{A_1}^1$ in the following way:

$$\Sigma_{A_1}^1 = \{\Delta_i^1 \mid i \in \mathbb{N}\} = \{\Delta_i^0 \mid A_1 \in \Delta_1^0 \setminus \Delta_i^0\}$$

where, if $\Delta_k^0 = \Delta_l^1$ then $l \leq k$ (evidently this is possible).

By symmetric reasoning, there is a A_2 such that

$$\Sigma_{A_2}^2 =_{\mathrm{df}} \{\Delta_i^1 \mid A_2 \in \Delta_1^1 \setminus \Delta_i^1\}$$

is infinite. Again, we can rename the elements so that $\Sigma^2_{A_2} = \{\Delta^2_i \mid i \in \mathbb{N}\}$ and whenever $\Delta^0_k = \Delta^2_l$ then $l \leq k$.

Again, by an analogous argument there is a $A_3 \in \Delta^2_1$ such that $\Sigma^3_{A_3}$ is infinite. And so on. In this manner we can construct a set $\Theta = \{A_i \mid i \in \mathbb{N}\}$.

Note first that Θ is a choice set of $\Sigma_{\min}$ since by the construction, for each $i \in \mathbb{N}$ there is a $j \leq i+1$ such that either $\Delta^0_i = \Delta^{j-1}_1$ or j is minimal such that $\Delta^0_i \in \Sigma_{\min} \setminus \Sigma^j_{A_j}$. In each case $A_j \in \Delta^0_i$.

Note that for each A_i, for all $\Delta \in \Sigma^j_{A_j}$ and for all $j \geq i$, $A_i \notin \Delta$. Hence, no finite subset Θ' of Θ is a choice set of $\Sigma_{\min}$ since for all $A_i \in \Theta'$ and all $\Delta \in \Sigma^k_{A_k}$, $A_i \cap \Delta = \emptyset$ where $k = \max(\{l \in \mathbb{N} \mid A_l \in \Theta'\})$.

Hence, the minimal choice set $\hat{\Theta} \subseteq \Theta$ that is constructed as in Lemma 5.5.1 is infinite. □

Lemma A.6. *If $\Sigma_{\min}$ is finite, then $\mathsf{CS}_{\min}(\Sigma_{\min})$ is finite.*

Proof. Since $\Sigma_{\min}$ is finite, also $\bigcup \Sigma_{\min}$ is finite. By Lemma A.2, also $\bigcup \mathsf{CS}_{\min}(\Sigma_{\min})$ is finite. Evidently this means that every $\varphi \in \mathsf{CS}_{\min}(\Sigma_{\min})$ is finite. □

Corollary A.3. *Where $\Gamma \subseteq \mathcal{W}^+$: If each $\varphi \in \Phi(\Gamma)$ is finite then $\Phi(\Gamma)$ is finite.*

Corollary A.4. *Where $\Gamma \subseteq \mathcal{W}^+$: $U(\Gamma)$ is finite iff each $\varphi \in \Phi(\Gamma)$ is finite.*

Proof. Follows immediately by Lemma A.2. □

Appendix B
Appendix to Chapter 3

B.1 Proof of Lemma 3.2.7

The following fact states a useful insight. It follows immediately by way the semantic selections for reliability and minimal abnormality are defined.

Fact B.1.1. *Where* $\mathbf{AL} = \langle \mathbf{LLL}, \Omega, x\rangle$ *is an AL in standard format and* $M \in \mathcal{M}_{\mathbf{AL}}(\Gamma)$. *For all* $M' \in \mathcal{M}_{\mathbf{LLL}}(\Gamma)$ *for which* $Ab(M') \subseteq Ab(M)$, $M' \in \mathcal{M}_{\mathbf{AL}}(\Gamma)$.

Lemma 3.2.7 ***(restated).*** *Where* $i \in \{1, \ldots, n\}$, $\Phi^*(\Gamma)$ *is finite, and* $\Gamma \subseteq \mathcal{W}$:

(i) $\Phi^*\big(Cn^{\mathcal{L}^+}_{\mathbf{CAL_i}}(\Gamma)\big) \subseteq \Phi^*(\Gamma)$
(ii) *For all* $\varphi \in \boldsymbol{\Phi}^{\boldsymbol{i}}(\Gamma)$ *there is a* $\varphi' \in \Phi^*(\Gamma)$ *such that* $\varphi \subseteq \varphi'$.
(iii) $|\boldsymbol{\Phi}^{\boldsymbol{i}}(\Gamma)| \leq |\Phi^*(\Gamma)|$

Proof. Let $\{\varphi_1, \ldots, \varphi_m\} = \Phi^*(\Gamma)$ and **AL** be the AL in standard format characterized by the triple $\langle \mathbf{LLL}, \Omega^*, m\rangle$. I will show by an induction that (i)–(iii) hold for every $i \in \{1, \ldots, n\}$. It is enough to show that

$$\text{for each } M \in \mathcal{M}_{\mathbf{AL^{x_i}_i}}\left(Cn^{\mathcal{L}^+}_{\mathbf{CAL_{i-1}}}(\Gamma)\right) \text{ there is an } M' \in \mathcal{M}_{\mathbf{AL^m}}(\Gamma) \cap \mathcal{M}_{\mathbf{AL^{x_i}_i}}\left(Cn^{\mathcal{L}^+}_{\mathbf{CAL_{i-1}}}(\Gamma)\right) \text{ such that } \mathrm{Ab}(M') \subseteq \mathrm{Ab}(M) \qquad (\dagger)$$

Indeed, suppose that (†) holds, then

$$\left\{M \in \mathcal{M}_{\mathbf{AL^{x_i}_i}}\left(Cn^{\mathcal{L}^+}_{\mathbf{CAL_{i-1}}}(\Gamma)\right) \,\middle|\, M \text{ is minimally abnormal w.r.t. } \Omega^* \text{ in } \mathcal{M}_{\mathbf{AL^{x_i}_i}}\left(Cn^{\mathcal{L}^+}_{\mathbf{CAL_{i-1}}}(\Gamma)\right)\right\} \subseteq \mathcal{M}_{\mathbf{AL^m}}(\Gamma) \qquad (\ddagger)$$

C. Straßer, *Adaptive Logics for Defeasible Reasoning*, Trends in Logic 38,
DOI: 10.1007/978-3-319-00792-2, © Springer International Publishing Switzerland 2014

By Theorem 2.4.6, for every minimally abnormal (w.r.t. Ω_i) model $M \in \mathcal{M}_{\mathbf{AL}_i^{x_i}}\left(Cn^{\mathcal{L}^+}_{\mathbf{CAL_{i-1}}}(\Gamma)\right)$, $\text{Ab}(M) \cap \Omega_i \in \boldsymbol{\Phi}^i(\Gamma)$. By (†), there is an $M' \in \mathcal{M}_{\mathbf{AL^m}}(\Gamma) \cap \mathcal{M}_{\mathbf{AL}_i^{x_i}}\left(Cn^{\mathcal{L}^+}_{\mathbf{CAL_{i-1}}}(\Gamma)\right)$ such that $\text{Ab}(M') \subseteq \text{Ab}(M)$ and $\text{Ab}(M') \cap \Omega_i \subseteq \text{Ab}(M) \cap \Omega_i$. Due to the minimality of M, $\text{Ab}(M') \cap \Omega_i = \text{Ab}(M) \cap \Omega_i$. Since by Theorem 2.4.6, $\text{Ab}(M') \in \Phi^*(\Gamma)$, $\boldsymbol{\Phi}^i(\Gamma) \subseteq \{\varphi \cap \Omega_i \mid \varphi \in \Phi^*(\Gamma)\}$. Hence, for every $\varphi \in \boldsymbol{\Phi}^i(\Gamma)$ there is a $\varphi' \in \Phi^*(\Gamma)$ such that $\varphi \subseteq \varphi'$. Hence statement (ii) and (iii) hold.

Since by (iii), Lemma 3.2.6 and Lemma 3.1.3,

$$\left\{M \in \mathcal{M}_{\mathbf{AL}_i^{x_i}}\left(Cn^{\mathcal{L}^+}_{\mathbf{CAL_{i-1}}}(\Gamma)\right) \middle| M \text{ is minimally abnormal w.r.t. } \Omega^* \text{ in } \mathcal{M}_{\mathbf{AL}_i^{x_i}}\left(Cn^{\mathcal{L}^+}_{\mathbf{CAL_{i-1}}}(\Gamma)\right)\right\} = \left\{M \in \mathcal{M}_{\mathbf{LLL}}\left(Cn^{\mathcal{L}^+}_{\mathbf{CAL_i}}(\Gamma)\right) \middle| M \text{ is minimally abnormal w.r.t. } \Omega^* \text{ in } \mathcal{M}_{\mathbf{LLL}}\left(Cn^{\mathcal{L}^+}_{\mathbf{CAL_i}}(\Gamma)\right)\right\}$$

we get by (‡),

$$\left\{M \in \mathcal{M}_{\mathbf{LLL}}\left(Cn^{\mathcal{L}^+}_{\mathbf{CAL_i}}(\Gamma)\right) \middle| M \text{ is minimally abnormal w.r.t. } \Omega^* \text{ in } \mathcal{M}_{\mathbf{LLL}}\left(Cn^{\mathcal{L}^+}_{\mathbf{CAL_i}}(\Gamma)\right)\right\} \subseteq \mathcal{M}_{\mathbf{AL^m}}(\Gamma)$$

Hence, by Theorem 2.4.6, $\Phi^*\left(Cn^{\mathcal{L}^+}_{\mathbf{CAL_i}}(\Gamma)\right) \subseteq \Phi^*(\Gamma)$.

I now show by induction that (†) indeed holds.

"$i = 1$": Let $M \in \mathcal{M}_{\mathbf{AL}_1^{x_1}}(\Gamma)$. By the strong reassurance (Corollary 2.4.3ii) of **AL** there is an $M' \in \mathcal{M}_{\mathbf{AL^m}}(\Gamma)$ for which $\text{Ab}(M') \subseteq \text{Ab}(M)$. Hence, $\text{Ab}(M') \cap \Omega_1 \subseteq \text{Ab}(M) \cap \Omega_1$. Hence $M' \in \mathcal{M}_{\mathbf{AL}_1^{x_1}}(\Gamma)$ by Fact B.1.1.

"$i \Rightarrow i+1$": Let $M \in \mathcal{M}_{\mathbf{AL}_{i+1}^{x_{i+1}}}\left(Cn^{\mathcal{L}^+}_{\mathbf{CAL_i}}(\Gamma)\right)$. This is equivalent to

$$M \in \left\{M' \in \mathcal{M}_{\mathbf{LLL}}\left(Cn^{\mathcal{L}^+}_{\mathbf{CAL_i}}(\Gamma)\right) \middle| M' \text{ is a rel. } (x_{i+1} = r) \text{ resp. m.a. } (x_{i+1} = m) \text{ model in } \mathcal{M}_{\mathbf{LLL}}\left(Cn^{\mathcal{L}^+}_{\mathbf{CAL_i}}(\Gamma)\right)\right\} \tag{B.1}$$

By the induction hypothesis, $\boldsymbol{\Phi}^i(\Gamma)$ is finite and hence by Lemma 3.2.6,

$$\mathcal{M}_{\mathbf{LLL}}\left(Cn^{\mathcal{L}^+}_{\mathbf{CAL_i}}(\Gamma)\right) = \mathcal{M}_{\mathbf{AL}_i^{x_i}}\left(Cn^{\mathcal{L}^+}_{\mathbf{CAL_{i-1}}}(\Gamma)\right) \tag{$\star$}$$

Hence, (B.1) is equivalent to

$$M \in \Big\{ M' \in \mathcal{M}_{\mathbf{AL}_i^{x_i}} \left(Cn^{\mathcal{L}^+}_{\mathbf{CAL}_{i-1}}(\Gamma) \right) \Big| M' \text{ is a rel. } (x_{i+1} = r)$$

$$\text{resp. m.a. } (x_{i+1} = m) \text{ model in } \mathcal{M}_{\mathbf{AL}_i^{x_i}} \left(Cn^{\mathcal{L}^+}_{\mathbf{CAL}_{i-1}}(\Gamma) \right) \Big\} \qquad \text{(B.2)}$$

By the induction hypothesis there is an

$$M' \in \mathcal{M}_{\mathbf{AL^m}}(\Gamma) \cap \mathcal{M}_{\mathbf{AL}_i^{x_i}} \left(Cn^{\mathcal{L}^+}_{\mathbf{CAL}_{i-1}}(\Gamma) \right)$$

for which $\text{Ab}(M') \subseteq \text{Ab}(M)$ and hence $\text{Ab}(M') \cap \Omega_{i+1} \subseteq \text{Ab}(M) \cap \Omega_{i+1}$. Hence, by Fact B.1.1, $M' \in \mathcal{M}_{\mathbf{AL}_{i+1}^{x_{i+1}}} \left(Cn^{\mathcal{L}^+}_{\mathbf{CAL}_i}(\Gamma) \right)$. □

B.2 The Adequacy of the Proof Theory for CAL

In this Appendix we prove the adequacy of the proof theory for **CAL**. Let us first introduce some useful notions.

We say that $\mathsf{Dab}(\Delta)$ is a *minimal* Dab_i*-consequence* of Γ iff $\Delta \subseteq \Omega_i$, $\mathsf{Dab}(\Delta) \in Cn_{\mathbf{LLL}}(\Gamma)$, and for all $\Delta' \subseteq \Delta$: if $\mathsf{Dab}(\Delta') \in Cn_{\mathbf{LLL}}(\Gamma)$ then $\Delta' = \Delta$. Where $\mathsf{Dab}(\Delta_1), \mathsf{Dab}(\Delta_2), \ldots$ are the minimal Dab_i-consequences from Γ, let $\Sigma^i(\Gamma) =_{\text{df}} \{\Delta_1, \Delta_2, \ldots\}$. Let $\Phi^i(\Gamma)$ be the set of all minimal choice sets of $\Sigma^i(\Gamma)$ and $U^i(\Gamma) =_{\text{df}} \bigcup \Sigma^i(\Gamma)$.

In the remainder of this section, **AL** is a flat AL in standard format with lower limit logic **LLL** and the set of abnormalities $\Omega = \Omega_1 \cup \ldots \cup \Omega_n$. $\mathsf{Dab}(\Delta)$ is a *minimal* Dab*-consequence* of Γ iff $\Delta \subseteq \Omega$, $\mathsf{Dab}(\Delta) \in Cn_{\mathbf{LLL}}(\Gamma)$ and for all $\Delta' \subseteq \Delta$: if $\mathsf{Dab}(\Delta') \in Cn_{\mathbf{LLL}}(\Gamma)$ then $\Delta' = \Delta$. Where $\mathsf{Dab}(\Delta_1), \mathsf{Dab}(\Delta_2), \ldots$ are all the minimal Dab-consequences of Γ, $\Sigma(\Gamma) =_{\text{df}} \{\Delta_1, \Delta_2, \ldots\}$. $\Phi(\Gamma)$ is the set of minimal choice sets of $\Sigma(\Gamma)$ and $U(\Gamma) =_{\text{df}} \bigcup \Sigma(\Gamma)$.

In the following, $\check{\vee}\, \mathsf{Dab}(\Delta)$ denotes the empty string in case $\Delta = \emptyset$. $Cn_{\mathbf{CAL}_i}(\Gamma)$ denotes Γ if $i = 0$. For the sake of convenience we will sometimes speak about the empty proof, meaning the "proof" which consists of 0 lines. We denote this proof by $\mathcal{P}_\varepsilon$.

B.2.1 A Complete Proof Stage g

In the following it will be very useful to speak about the extension of a given (possibly empty) **AL**-, resp. **CAL**-proof $\mathcal{P}$ in which A is derived on the condition Δ whenever $\Gamma \vdash_{\mathbf{LLL}} A \check{\vee} \mathsf{Dab}(\Delta)$. We dub a corresponding stage $\mathbf{g}(\mathcal{P})$ a *complete stage*.

Note that each well-formed formula has a Gödel-number. From this it follows immediately that $Cn_{\mathbf{LLL}}(\Gamma)$ is enumerable, e.g. $Cn_{\mathbf{LLL}}(\Gamma) = \{B_1, B_2, \ldots\}$. Moreover, due to the compactness of **LLL**, for each $B_i \in Cn_{\mathbf{LLL}}(\Gamma)$ there are some $A_1, \ldots, A_m$ such that $A_1, \ldots, A_m \vdash_{\mathbf{LLL}} B_i$. Hence, for each $B_i \in Cn_{\mathbf{LLL}}(\Gamma)$ we have the following proof $\mathcal{P}_i$:

l_1^i	A_1	PREM	$\emptyset$
$\vdots$	$\vdots$	$\vdots$	$\vdots$
l_m^i	A_m	PREM	$\emptyset$
l_{m+1}^i	B_i	$l_1^i, \ldots, l_m^i$; RU	$\emptyset$

In case B_i is of the form $A \check{\vee} \mathsf{Dab}(\Delta)$ we add some further lines. Where $m' \leq m$ is the lowest natural number such that $\Delta \subseteq \Omega_1 \cup \ldots \cup \Omega_{m'}$ we add (where $\Delta_j = \Delta \cap \Omega_j$ for each $j \leq m'$):

l_{m+2}^i	$A \check{\vee} \mathsf{Dab}(\Delta_2 \cup \ldots \cup \Delta_{m'})$	l_{m+1}^i; RC	Δ_1
$\vdots$	$\vdots$	$\vdots$	$\vdots$
$l_{m+m'}^i$	$A \check{\vee} \mathsf{Dab}(\Delta_{m'})$	$l_{m+m'-1}^i$; RC	$\Delta_1 \cup \ldots \cup \Delta_{m'-1}$
$l_{m+m'+1}^i$	A	$l_{m+m'}^i$; RC	Δ

Where $\mathcal{P}$ consists of lines $l_1^0, l_2^0, \ldots$, we now combine the proofs $\mathcal{P}, \mathcal{P}_1, \mathcal{P}_2, \ldots$ to a proof $\mathcal{P}'$ that extends $\mathcal{P}$ to the stage $\mathbf{g}(\mathcal{P})$ by means of listing the respective lines as follows (and by renumbering the lines accordingly):

$$l_1^0, l_2^0, l_1^1, l_2^1, l_3^0, l_3^1, l_1^2, l_2^2, l_3^2, l_4^0, \ldots, l_4^2, l_1^3, \ldots, l_4^3, l_5^0, \ldots, l_5^3, l_1^4, \ldots, l_5^4, \ldots$$

Note that the marking at a stage is determined by the minimal Dab_i-formulas derived at this stage (where $i \leq n$). Since in $\mathbf{g}(\mathcal{P})$ every possible Dab_i-formula is derived on every possible condition, the marking remains stable from $\mathbf{g}(\mathcal{P})$ on. The following fact holds for the extension of an **AL** proof $\mathcal{P}$ to the stage $\mathbf{g}(\mathcal{P})$:

Fact B.2.1. *(i)* $\Sigma_{\mathbf{g}(\mathcal{P})}(\Gamma) = \Sigma(\Gamma)$ *and hence* $U_{\mathbf{g}(\mathcal{P})}(\Gamma) = U(\Gamma)$ *and* $\Phi_{\mathbf{g}(\mathcal{P})}(\Gamma) = \Phi(\Gamma)$. *(ii) Where* $i \leq n$: $\Sigma^i_{\mathbf{g}(\mathcal{P})}(\Gamma) = \Sigma^i(\Gamma)$ *and hence* $U^i_{\mathbf{g}(\mathcal{P})}(\Gamma) = U^i(\Gamma)$ *and* $\Phi^i_{\mathbf{g}(\mathcal{P})}(\Gamma) = \Phi^i(\Gamma)$.

Fact B.2.2. *If a line* l *is marked at stage* $\mathbf{g}(\mathcal{P})$*, then it is marked in every further extension. Hence, the markings remain stable from stage* $\mathbf{g}(\mathcal{P})$ *on.*

B.2.2 Some Results for Flat ALs

It is useful to first prove some lemmas about flat ALs. The following fact follows immediately by the reflexivity, the monotonicity, and the transitivity of **LLL**.

Fact B.2.3. **(Fixed point property for LLL).** $Cn_{\mathbf{LLL}}(Cn_{\mathbf{LLL}}(\Gamma)) = Cn_{\mathbf{LLL}}(\Gamma)$

The following two lemmas are known to hold where $\Gamma \subseteq \mathcal{W}$ (see Chap. 2). In what follows it is useful to show that they also hold where $\Gamma = Cn^{\mathcal{L}^+}_{\mathbf{LLL}}(\Gamma)$.[1]

Lemma B.2.1. *Where $\Gamma \subseteq \mathcal{W}$ or $\Gamma = Cn^{\mathcal{L}^+}_{\mathbf{LLL}}(\Gamma)$: $A \in Cn^{\mathcal{L}^+}_{\mathbf{AL^r}}(\Gamma)$ iff there is a $\Delta \subseteq \Omega \setminus U(\Gamma)$ for which $\Gamma \vdash_{\mathbf{LLL}} A \check{\vee} \mathsf{Dab}(\Delta)$.*

Proof. $A \in Cn^{\mathcal{L}^+}_{\mathbf{AL^r}}(\Gamma)$ iff [by Corollary 2.7.1] $\Gamma \Vdash_{\mathbf{AL^r}} A$ iff [by Corollary 2.4.1] there is a $\Delta \subseteq \Omega \setminus U(\Gamma)$ for which $\Gamma \Vdash_{\mathbf{LLL}} A \check{\vee} \mathsf{Dab}(\Delta)$ iff [by the soundness and completeness of **LLL**] there is a $\Delta \subseteq \Omega \setminus U(\Gamma)$ for which $\Gamma \vdash_{\mathbf{LLL}} A \check{\vee} \mathsf{Dab}(\Delta)$. □

Lemma B.2.2. *Where $\Gamma \subseteq \mathcal{W}$ or $\Gamma = Cn^{\mathcal{L}^+}_{\mathbf{LLL}}(\Gamma)$: $A \in Cn^{\mathcal{L}^+}_{\mathbf{AL^r}}(\Gamma)$ iff there is a $\Delta \subseteq \Omega \setminus U(\Gamma)$ for which $\Gamma \vdash_{\mathbf{AL^r}} A \check{\vee} \mathsf{Dab}(\Delta)$.*

Proof. "⇒": this follows by Lemma B.2.1 and since $Cn^{\mathcal{L}^+}_{\mathbf{LLL}}(\Gamma) \subseteq Cn^{\mathcal{L}^+}_{\mathbf{AL^r}}(\Gamma)$.

"⇐": Let $A \check{\vee} \mathsf{Dab}(\Delta) \in Cn^{\mathcal{L}^+}_{\mathbf{AL^r}}(\Gamma)$ for some $\Delta \subseteq \Omega \setminus U(\Gamma)$. Hence, by Lemma B.2.1, there is a $\Theta \subseteq \Omega \setminus U(\Gamma)$ such that $\Gamma \vdash_{\mathbf{LLL}} A \check{\vee} \mathsf{Dab}(\Delta \cup \Theta)$. Since $\Delta \cup \Theta \subseteq \Omega \setminus U(\Gamma)$, by Lemma B.2.1, $A \in Cn^{\mathcal{L}^+}_{\mathbf{AL^r}}(\Gamma)$. □

Lemma B.2.3. *Where $\Gamma \subseteq \mathcal{W}$ or $\Gamma = Cn^{\mathcal{L}^+}_{\mathbf{LLL}}(\Gamma)$: $A \in Cn^{\mathcal{L}^+}_{\mathbf{AL^m}}(\Gamma)$ iff for each $\varphi \in \Phi(\Gamma)$ there is a $\Delta \subseteq \Omega \setminus \varphi$ for which $\Gamma \vdash_{\mathbf{LLL}} A \check{\vee} \mathsf{Dab}(\Delta)$.*

Proof. $A \in Cn^{\mathcal{L}^+}_{\mathbf{AL^m}}(\Gamma)$ iff [by Corollary 2.7.1] $\Gamma \Vdash_{\mathbf{AL^m}} A$ iff [by Corollary 2.4.4] for each $\varphi \in \Phi(\Gamma)$ there is a $\Delta \subseteq \Omega \setminus \varphi$ such that $\Gamma \Vdash_{\mathbf{LLL}} A \check{\vee} \mathsf{Dab}(\Delta)$ iff [by the soundness and completeness of **LLL**] for each $\varphi \in \Phi(\Gamma)$ there is a $\Delta \subseteq \Omega \setminus \varphi$ such that $\Gamma \Vdash_{\mathbf{LLL}} A \check{\vee} \mathsf{Dab}(\Delta)$. □

Lemma B.2.4. *Where $\Gamma \subseteq \mathcal{W}$ or $\Gamma = Cn^{\mathcal{L}^+}_{\mathbf{LLL}}(\Gamma)$: $A \in Cn^{\mathcal{L}^+}_{\mathbf{AL^m}}(\Gamma)$ iff for each $\varphi \in \Phi(\Gamma)$ there is a $\Delta \subseteq \Omega \setminus \varphi$ for which $\Gamma \vdash_{\mathbf{AL^m}} A \check{\vee} \mathsf{Dab}(\Delta)$.*

Proof. "⇒: this follows by Lemma B.2.3 and since $Cn^{\mathcal{L}^+}_{\mathbf{LLL}}(\Gamma) \subseteq Cn^{\mathcal{L}^+}_{\mathbf{AL^m}}(\Gamma)$.

"⇐": Suppose for each $\varphi \in \Phi(\Gamma)$ there is a $\Delta_\varphi \subseteq \Omega \setminus \varphi$ such that $A \check{\vee} \mathsf{Dab}(\Delta_\varphi) \in Cn^{\mathcal{L}^+}_{\mathbf{AL^m}}(\Gamma)$. Hence, by Lemma B.2.3, for each $\varphi \in \Phi(\Gamma)$ and for each $\psi \in \Phi(\Gamma)$ there is a $\Delta^\psi_\varphi \subseteq \Omega \setminus \psi$ such that $\Gamma \vdash_{\mathbf{LLL}} A \check{\vee} \mathsf{Dab}(\Delta_\varphi \cup \Delta^\psi_\varphi)$. Since $\Delta_\varphi \cup \Delta^\varphi_\varphi \subseteq \Omega \setminus \varphi$ this implies by Lemma B.2.3 that $A \in Cn^{\mathcal{L}^+}_{\mathbf{AL^m}}(\Gamma)$. □

Lemma B.2.5. (**Dab-conservatism of AL**). *Where $\Gamma \subseteq \mathcal{W}$ or $\Gamma = Cn^{\mathcal{L}^+}_{\mathbf{LLL}}(\Gamma)$: If $\mathsf{Dab}(\Delta) \in Cn_{\mathbf{AL}}(\Gamma)$ then $\mathsf{Dab}(\Delta) \in Cn_{\mathbf{LLL}}(\Gamma)$.*

Proof. Let $\mathsf{Dab}(\Delta) \in Cn^{\mathcal{L}^+}_{\mathbf{AL}}(\Gamma)$. Case "minimal abnormality". By Lemma B.2.3, for each $\varphi \in \Phi(\Gamma)$ there is a $\Delta_\varphi \subseteq \Omega \setminus \varphi$ for which $\Gamma \vdash_{\mathbf{LLL}} \mathsf{Dab}(\Delta \cup \Delta_\varphi)$. Note that for each $\varphi \in \Phi(\Gamma)$ there is a $\Theta \in \Sigma(\Gamma)$ such that $\Theta \subseteq \Delta \cup \Delta_\varphi$. For each $\varphi \in \Phi(\Gamma)$, by Corollary 2.8.1, there is no $\psi \in \Phi(\Gamma)$ such that $\psi \cap (\Delta \cup \Delta_\varphi) = \emptyset$. This means that Δ is a choice set of $\Phi(\Gamma)$. Hence, by Corollary A.1, there is a $\Delta' \subseteq \Delta$ for which $\Delta' \in \Sigma(\Gamma)$. Hence, $\Gamma \vdash_{\mathbf{LLL}} \mathsf{Dab}(\Delta)$.

[1] Note that they do not hold for just any premise set that also contains formulas with "checked" symbols as the example in Sect. 2.7 shows.

Case "reliability". By Lemma B.2.1 there is a $\Theta \in \Omega \setminus U(\Gamma)$ such that $\Gamma \vdash_{\mathbf{LLL}}$ $\mathsf{Dab}(\Delta \cup \Theta)$. Hence, there is a $\Theta' \in \Sigma(\Gamma)$ such that $\Theta' \subseteq \Delta \cup \Theta$. By Corollary 2.8.1, for all $\psi \in \Phi(\Gamma)$, $\psi \cap (\Delta \cup \Theta) \neq \emptyset$. Since by Corollary A.2 $U(\Gamma) = \bigcup \Phi(\Gamma)$, $\psi \cap \Theta = \emptyset$ for all $\psi \in \Phi(\Gamma)$. Hence, Δ is a choice set of $\Phi(\Gamma)$. By Corollary A.1, there is a $\Delta' \subseteq \Delta$ for which $\Delta' \in \Sigma(\Gamma)$. Hence, $\Gamma \vdash_{\mathbf{LLL}} \mathsf{Dab}(\Delta)$. □

B.2.3 Proving the Adequacy of the CAL*-Proof Theory*

Corollary B.2.1. *Where $\Gamma \subseteq \mathcal{W}$: $A \in Cn^{\mathcal{L}^+}_{\mathbf{CAL_i}}(\Gamma)$ iff there is a $\Delta \subseteq \Omega_i$ such that $A \check{\vee} \mathsf{Dab}(\Delta) \in Cn^{\mathcal{L}^+}_{\mathbf{CAL_{i-1}}}(\Gamma)$ and*

(i) where $\mathbf{x_i} = \mathbf{r}$, $\Delta \cap U^i(Cn^{\mathcal{L}^+}_{\mathbf{CAL_{i-1}}}(\Gamma)) = \emptyset$, or
(ii) where $\mathbf{x_i} = \mathbf{m}$, there is a $\varphi \in \Phi^i(Cn^{\mathcal{L}^+}_{\mathbf{CAL_{i-1}}}(\Gamma))$ such that $\varphi \cap \Delta = \emptyset$ and for each $\varphi \in \Phi^i(Cn^{\mathcal{L}^+}_{\mathbf{CAL_{i-1}}}(\Gamma))$ there is a $\Theta \subseteq \Omega_i$ such that $A \check{\vee} \mathsf{Dab}(\Theta) \in Cn^{\mathcal{L}^+}_{\mathbf{CAL_{i-1}}}(\Gamma)$ and $\Theta \cap \varphi = \emptyset$.

Proof. We prove this by an induction for $j \leq n$:

"$j = 1$" follows directly Lemmas B.2.1 and B.2.3 and the fact that $Cn^{\mathcal{L}^+}_{\mathbf{CAL_1}}(\Gamma) = Cn^{\mathcal{L}^+}_{\mathbf{AL_1}}(\Gamma)$.

"$j \Rightarrow j+1$": By Lemma 3.1.3, $Cn^{\mathcal{L}^+}_{\mathbf{CAL_j}}(\Gamma) = Cn^{\mathcal{L}^+}_{\mathbf{LLL}}\left(Cn^{\mathcal{L}^+}_{\mathbf{CAL_j}}(\Gamma)\right)$. Thus, the corollary follows by Lemmas B.2.1 and B.2.3. □

The following corollary follows immediately by Lemma B.2.5 and Lemma 3.1.3.

Corollary B.2.2. (Dab-**conservatism of** $\mathbf{CAL_i}$.). *Where $\Delta \subseteq \Omega_i$: if $\mathsf{Dab}(\Delta) \in Cn^{\mathcal{L}^+}_{\mathbf{CAL_i}}(\Gamma)$ then $\mathsf{Dab}(\Delta) \in Cn^{\mathcal{L}^+}_{\mathbf{CAL_{i-1}}}(\Gamma)$.*

Lemma B.2.6. *Where $\Gamma \subseteq \mathcal{W}$ and $\mathcal{P}$ is a* **CAL**-*proof from Γ, we have for each $i \leq n$:*

(i) $\Sigma^i_{\mathbf{g}(\mathcal{P})}(\Gamma) = \Sigma^i(Cn^{\mathcal{L}^+}_{\mathbf{CAL_{i-1}}}(\Gamma))$ and hence $U^i_{\mathbf{g}(\mathcal{P})}(\Gamma) = U^i(Cn^{\mathcal{L}^+}_{\mathbf{CAL_{i-1}}}(\Gamma))$ and $\Phi^i_{\mathbf{g}(\mathcal{P})}(\Gamma) = \Phi^i(Cn^{\mathcal{L}^+}_{\mathbf{CAL_{i-1}}}(\Gamma))$;
(ii) there is a [$\leq i$]-line l with formula A and that is unmarked at stage $\mathbf{g}(\mathcal{P})$ iff $A \in Cn^{\mathcal{L}^+}_{\mathbf{CAL_i}}(\Gamma)$.

Proof. We show this by an induction on $i \leq n$.

"i=1": *Ad (i).* Trivial.

Ad (ii). Case $\mathbf{x_1} = \mathbf{r}$. There is a [$\leq$1]-line l with formula A and condition Δ that is unmarked iff [by the construction of stage $\mathbf{g}(\mathcal{P})$ and Definition 3.3.2] $\Gamma \vdash_{\mathbf{LLL}} A \check{\vee} \mathsf{Dab}(\Delta)$ and $\Delta \cap U^1_{\mathbf{g}(\mathcal{P})}(\Gamma) = \emptyset$, iff [by i.] $\Gamma \vdash_{\mathbf{LLL}} A \check{\vee} \mathsf{Dab}(\Delta)$ and $\Delta \cap U^1(\Gamma) = \emptyset$, iff [by Lemma B.2.1] $A \in Cn^{\mathcal{L}^+}_{\mathbf{AL_1}}(\Gamma)$, iff $A \in Cn^{\mathcal{L}^+}_{\mathbf{CAL_1}}(\Gamma)$.

Case $\mathbf{x_1} = \mathbf{m}$. The proof is similar and left to the reader.

"$i \Rightarrow i+1$": *Ad (i)*. Where $\Delta \subseteq \Omega_{i+1}$, we have: $\Delta \in \Sigma^{i+1}_{\mathbf{g}(\mathcal{P})}(\Gamma)$ iff $\mathsf{Dab}(\Delta)$ is derived at an unmarked [$\leq i$]-line and there is no $\Delta' \subset \Delta$ such that $\mathsf{Dab}(\Delta')$ is derived at an unmarked [$\leq i$]-line, iff [by (ii) and the induction hypothesis] $\mathsf{Dab}(\Delta) \in Cn^{\mathcal{L}^+}_{\mathbf{CAL_i}}(\Gamma)$ and for no $\Delta' \subset \Delta$, $\mathsf{Dab}(\Delta') \in Cn^{\mathcal{L}^+}_{\mathbf{CAL_i}}(\Gamma)$, iff [by Lemma 3.1.3] $\mathsf{Dab}(\Delta) \in Cn^{\mathcal{L}^+}_{\mathbf{LLL}}\left(Cn_{\mathbf{CAL_i}}(\Gamma)\right)$ and for no $\Delta' \subset \Delta$, $\mathsf{Dab}(\Delta') \in Cn^{\mathcal{L}^+}_{\mathbf{LLL}}\left(Cn_{\mathbf{CAL_i}}(\Gamma)\right)$, iff $\Delta \in \Sigma^{i+1}(Cn^{\mathcal{L}^+}_{\mathbf{CAL_i}}(\Gamma))$.

Ad (ii). *Case* $\mathbf{x_{i+1}} = \mathbf{r}$. Let l be some [$\leq i+1$]-line with formula A and condition Δ. Suppose line l is unmarked. If l is a j-line with $j \leq i$ we get $A \in Cn^{\mathcal{L}^+}_{\mathbf{CAL_{i+1}}}(\Gamma)$ due to the induction hypothesis and since $Cn^{\mathcal{L}^+}_{\mathbf{CAL_j}}(\Gamma) \subseteq Cn^{\mathcal{L}^+}_{\mathbf{CAL_{i+1}}}(\Gamma)$.

Thus, suppose l is an $i+1$-line with condition $\Delta_1 \cup \ldots \cup \Delta_{i+1}$ where $\Delta_j \subseteq \Omega_j$ for each $j \leq i+1$. We prove the statement by another induction on the number of steps j needed to derive A.

"$j = 1$": Only premises can be introduced in one inference step, but this does not lead to an $i+1$-line.

"$j = 2$": The proof looks as follows: A is derived by RC from some line l' at which some B is introduced as a premise and $B \vdash_{\mathbf{LLL}} A \check{\vee} \mathsf{Dab}(\Delta_{i+1})$ where $\Delta_{i+1} \subseteq \Omega_{i+1}$. Since l is unmarked at stage $\mathbf{g}(\mathcal{P})$, $\Delta_{i+1} \cap U^{i+1}_{\mathbf{g}(\mathcal{P})}(\Gamma) = \emptyset$ and hence by (i), $\Delta_{i+1} \cap U^{i+1}(Cn^{\mathcal{L}^+}_{\mathbf{CAL_i}}(\Gamma)) = \emptyset$. By Corollary B.2.1.i, $A \in Cn^{\mathcal{L}^+}_{\mathbf{CAL_{i+1}}}(\Gamma)$.

"$j \Rightarrow j+1$": Suppose A is derived with the justification $l_1, \ldots, l_m$; R where $R \in \{\text{RU}, \text{RC}\}$ and each line l_k (where $1 \leq k \leq m$) features a formula A_k and a condition $\Delta^k_1 \cup \ldots \cup \Delta^k_{i+1}$ where $\Delta^k_o \subseteq \Omega_o$ for each $o \leq i+1$. By the definition of RU and RC, (†) $A_1, \ldots, A_m \vdash_{\mathbf{LLL}} A \check{\vee} \mathsf{Dab}(\Delta'_{i+1})$ for some (possibly empty) $\Delta'_{i+1} \subseteq \Delta_{i+1} \subseteq \Omega_{i+1}$, and $\Delta'_{i+1} \cup \Delta^1_{i+1} \cup \ldots \cup \Delta^m_{i+1} = \Delta_{i+1}$. Since l is unmarked, (a) by Definition 3.3.2 and (i), $\Delta_{i+1} \cap U^{i+1}(Cn^{\mathcal{L}^+}_{\mathbf{CAL_i}}(\Gamma)) = \Delta_{i+1} \cap U^{i+1}_{\mathbf{g}(\mathcal{P})}(\Gamma) = \emptyset$, (b) by the definition of inh-marking each of the lines l_k is neither o-marked for any $o \leq i$ nor inh-marked, (c) neither line l_k is $i+1$-marked since $\Delta^k_{i+1} \subseteq \Delta_{i+1}$ and by (a). By our induction hypothesis, (b) and (c), $A_k \in Cn^{\mathcal{L}^+}_{\mathbf{CAL_{i+1}}}(\Gamma)$ and by (†) and Lemma 3.1.3 also $A \check{\vee} \mathsf{Dab}(\Delta'_{i+1}) \in Cn^{\mathcal{L}^+}_{\mathbf{CAL_{i+1}}}(\Gamma)$. By Lemma B.2.2, Lemma 3.1.3, and (a), $A \in Cn^{\mathcal{L}^+}_{\mathbf{CAL_{i+1}}}(\Gamma)$.

For the other direction suppose $A \in Cn^{\mathcal{L}^+}_{\mathbf{CAL_{i+1}}}(\Gamma)$. By Corollary B.2.1.i there is a $\Delta \subseteq \Omega_{i+1}$ for which $A \check{\vee} \mathsf{Dab}(\Delta) \in Cn^{\mathcal{L}^+}_{\mathbf{CAL_i}}(\Gamma)$ and $\Delta \cap U^{i+1}(Cn^{\mathcal{L}^+}_{\mathbf{CAL_i}}(\Gamma)) = \emptyset$. By the induction hypothesis there is an unmarked [$\leq i$]-line l at which $A \check{\vee} \mathsf{Dab}(\Delta)$ is derived on some condition $\Theta_1 \cup \ldots \cup \Theta_i$ where $\Theta_j \subseteq \Omega_j$ for each $j \leq i$. By the construction of stage $\mathbf{g}(\mathcal{P})$ there is a line l' with formula A, justification l; RC and condition $\Theta_1 \cup \ldots \cup \Theta_i \cup \Delta$. Since by i., $\Delta \cap U^{i+1}_{\mathbf{g}(\mathcal{P})}(\Gamma) = \emptyset$, line l' is not marked according to the $i+1$-marking with reliability. Moreover, since l is unmarked, l' is also not inh-marked.

Let $\mathbf{x_{i+1}} = \mathbf{m}$. The proof is similar and left to the reader. □

Since $A \in Cn^{\mathcal{L}^+}_{\mathbf{CAL}}(\Gamma)$ iff there is an $i \leq n$ such that $A \in Cn^{\mathcal{L}^+}_{\mathbf{CAL_i}}(\Gamma)$, we get by item 2 of the previous lemma:

Corollary B.2.3. *Where $\Gamma \subseteq \mathcal{W}$ and $\mathcal{P}$ is a* **CAL***-proof from Γ: $A \in Cn^{\mathcal{L}^+}_{\mathbf{CAL}}(\Gamma)$ iff A is derived at an unmarked line at stage $\mathbf{g}(\mathcal{P})$.*

Theorem B.2.1. *Where $\Gamma \subseteq \mathcal{W}$: if $\Gamma \vdash_{\mathbf{CAL}} A$ then $A \in Cn^{\mathcal{L}^+}_{\mathbf{CAL}}(\Gamma)$.*

Proof. Suppose $\Gamma \vdash_{\mathbf{CAL}} A$. Hence, there is a finite **CAL**-proof $\mathcal{P}$ in which A is finally derived at some line l. We extend $\mathcal{P}$ to stage $\mathbf{g}(\mathcal{P})$. By Definition 3.3.4 and Fact B.2.2, line l is unmarked and hence $A \in Cn^{\mathcal{L}^+}_{\mathbf{CAL}}(\Gamma)$ by Corollary B.2.3. □

Theorem B.2.2. *Where $\Gamma \subseteq \mathcal{W}$: if $A \in Cn^{\mathcal{L}^+}_{\mathbf{CAL}}(\Gamma)$ then $\Gamma \vdash_{\mathbf{CAL}} A$.*

Proof. Suppose $A \in Cn^{\mathcal{L}^+}_{\mathbf{CAL}}(\Gamma)$. Hence by Corollary B.2.3, (1) A is derived at an unmarked line at stage $\mathbf{g}(\mathcal{P}_\varepsilon)$ on a condition $\Theta_1 \cup \ldots \cup \Theta_m$ where $\Theta_j \subseteq \Omega_j$ for all $j \leq m$. The case where $\Theta_1 \cup \ldots \cup \Theta_m = \emptyset$ is trivial. Let's hence suppose $\Theta_1 \cup \ldots \cup \Theta_m \neq \emptyset$ and let m be minimal such that A is derived at an unmarked line at stage $\mathbf{g}(\mathcal{P}_\varepsilon)$ on a condition $\Theta_1 \cup \ldots \cup \Theta_m$. By Lemma B.2.6.ii this is to say, (2) m is minimal such that $A \in Cn_{\mathbf{CAL_m}}(\Gamma)$.

Since $\Gamma \vdash_{\mathbf{LLL}} A \check{\vee} \mathsf{Dab}(\Theta_1 \cup \ldots \cup \Theta_m)$ and by the compactness of **LLL**, there are $B_1, \ldots, B_o \in \Gamma$ such that $\{B_1, \ldots, B_o\} \vdash_{\mathbf{LLL}} A \check{\vee} \mathsf{Dab}(\Theta_1 \cup \ldots \cup \Theta_m)$. We now construct a **CAL**-proof $\mathcal{P}$ for A as follows:

1	B_1	PREM	$\emptyset$
$\vdots$	$\vdots$	$\vdots$	$\vdots$
o	B_o	PREM	$\emptyset$
$o{+}1$	$A \check{\vee} \mathsf{Dab}(\Theta_1 \cup \ldots \cup \Theta_m)$	$1, \ldots, o$; RU	$\emptyset$
$o{+}2$	$A \check{\vee} \mathsf{Dab}(\Theta_2 \cup \ldots \cup \Theta_m)$	$1, \ldots, o$; RU	Θ_1
$\vdots$	$\vdots$	$\vdots$	$\vdots$
$o{+}m$	$A \check{\vee} \mathsf{Dab}(\Theta_m)$	$o{+}m{-}1$; RC	$\Theta_1 \cup \ldots \cup \Theta_{m-1}$
$o{+}m{+}1$	A	$o{+}m$; RC	$\Theta_1 \cup \ldots \cup \Theta_m$

Let s be the stage of our proof. Since $\Gamma \subseteq \mathcal{W}$, the only Dab-formulas in $\{B_1, \ldots, B_o\}$ are abnormalities and hence (3) for every $j \leq n$, $\{B_1, \ldots, B_o\} \cap \Omega_j \subseteq U^j_{\mathbf{g}(\mathcal{P})}(\Gamma) = U^j_{\mathbf{g}(\mathcal{P}_\varepsilon)}(\Gamma)$; and for every $\varphi \in \Phi^j_{\mathbf{g}(\mathcal{P})}(\Gamma) = \Phi^j_{\mathbf{g}(\mathcal{P}_\varepsilon)}(\Gamma)$, $\{B_1, \ldots, B_o\} \cap \Omega_j \subseteq \varphi$.

Assume $A \in \Omega_m$. Then A is a Dab_m-formula. By Corollary B.2.2 and since $A \in Cn^{\mathcal{L}^+}_{\mathbf{CAL_m}}(\Gamma)$, $A \in Cn^{\mathcal{L}^+}_{\mathbf{CAL_{m-1}}}(\Gamma)$. This is a contradiction to (2). Hence, (4) $A \notin \Omega_m$.

By (3) and (4) we infer that line $o{+}m{+}1$ is unmarked.

Suppose line $o{+}m{+}1$ is marked in an extension of the proof resulting in the proof $\mathcal{P}'$. We can extend the proof further to stage $\mathbf{g}(\mathcal{P}')$. That line $o{+}m{+}1$ is unmarked is an immediate consequence of (1). □

B.2.4 Proving the Adequacy of the $\mathbf{CAL}^{\mathsf{ns}}$*-Proof Theory*

Lemma B.2.7. *$A \in Cn^{\mathcal{L}^+}_{\mathbf{CAL}^{\mathsf{ns}}}(\Gamma)$ iff there is a $\Delta \subseteq \Omega_n$ such that*

$$Cn^{\mathcal{L}^+}_{\mathbf{CAL_{n-1}}}(\Gamma) \vdash_{\mathbf{LLL}} A \check{\vee} \mathsf{Dab}(\Delta)$$

and for all $\Theta \in \Sigma^n\left(Cn^{\mathcal{L}^+}_{\mathbf{CAL_{n-1}}}(\Gamma)\right)$, $\Theta \not\subseteq \Delta$.

Proof. This follows by Corollary 3.1.2 and Theorem 2.8.2. □

Theorem B.2.3. *Where $\Gamma \subseteq \mathcal{W}$: $\Gamma \vdash_{\mathbf{CAL}^{\mathsf{ns}}} A$ implies $A \in Cn^{\mathcal{L}^+}_{\mathbf{CAL}^{\mathsf{ns}}}(\Gamma)$.*

Proof. Let $\Gamma \vdash_{\mathbf{CAL}^{\mathsf{ns}}} A$. Hence, there is a finite proof in which A is finally derived on a condition Δ on a line l. Suppose l is marked in an extension of the proof $\mathcal{P}$. We extend the proof further to the stage $\mathbf{g}(\mathcal{P})$. Since l is finally derived and since the markings remain stable from this point on, line l is unmarked at stage $\mathbf{g}(\mathcal{P})$. Hence, $\Delta \cap \Omega_n$ is such that for all $\Theta \in \Sigma^n_{\mathbf{g}(\mathcal{P})}(\Gamma)$, $\Theta \not\subseteq \Delta \cap \Omega_n$. Hence, by Lemma B.2.6, for all $\Theta \in \Sigma^n(Cn^{\mathcal{L}^+}_{\mathbf{CAL_{n-1}}}(\Gamma))$, $\Theta \not\subseteq \Delta \cap \Omega_n$.

In order to apply Lemma B.2.7 to conclude that $A \in Cn^{\mathcal{L}^+}_{\mathbf{CAL}}(\Gamma)$ we still have to show that $A \check{\vee} \mathsf{Dab}(\Delta \cap \Omega_n) \in Cn^{\mathcal{L}^+}_{\mathbf{CAL_{n-1}}}(\Gamma)$. Suppose that the justification of line l is $l_1, \ldots, l_m$; R where $R \in \{\text{RU, RC}\}$. Let A_k be the formula of line l_k and $\Delta^1_k \cup \ldots \cup \Delta^n_k$ its condition where $\Delta^o_k \subseteq \Omega_o$ for each $o \leq n$. By the definition of RU and RC, $\{A_1, \ldots, A_m\} \vdash_{\mathbf{LLL}} A \check{\vee} \mathsf{Dab}(\Delta'_n)$ for some (possibly empty) $\Delta'_n \subseteq \Delta \cap \Omega_n$ and $\Delta'_n \cup \Delta^n_1 \cup \ldots \cup \Delta^n_m = \Delta \cap \Omega_n$. For each line l_k there is an associated list of [$\leq n-1$]-lines $l^1_k, \ldots, l^{m_k}_k$ with formulas $B^1_k, \ldots, B^{m_k}_k$ such that A_k has been derived from these lines (possibly in more than one step: we can easily locate these lines by traversing backwards along the 'justification path' starting from the justification of line l_k). This holds due to the iterative character of the RC rule: see Remark 3.3.1. Note that $\{B^1_k, \ldots, B^{m_k}_k\} \vdash_{\mathbf{LLL}} A_k \check{\vee} \mathsf{Dab}(\Delta^n_k)$ and that the lines $l^1_k, \ldots, l^{m_k}_k$ are unmarked (otherwise l_k would be marked as well and as a consequence also l). By Lemma B.2.6 this shows that each $B^j_k \in Cn^{\mathcal{L}^+}_{\mathbf{CAL_{n-1}}}(\Gamma)$. Note that also $\{B^1_k, \ldots, B^{m_k}_k \mid k \leq m\} \vdash_{\mathbf{LLL}} A \check{\vee} \mathsf{Dab}(\Delta \cap \Omega_n)$. Since by Lemma 3.1.3 $Cn^{\mathcal{L}^+}_{\mathbf{LLL}}\left(Cn^{\mathcal{L}^+}_{\mathbf{CAL_{n-1}}}(\Gamma)\right) = Cn^{\mathcal{L}^+}_{\mathbf{CAL_{n-1}}}(\Gamma)$, this shows that $A \check{\vee} \mathsf{Dab}(\Delta \cap \Omega_n) \in Cn^{\mathcal{L}^+}_{\mathbf{CAL_{n-1}}}(\Gamma)$. Hence, our proof is finished. □

Theorem B.2.4. *Where $\Gamma \subseteq \mathcal{W}$: $A \in Cn^{\mathcal{L}^+}_{\mathbf{CAL}^{\mathsf{ns}}}(\Gamma)$ implies $\Gamma \vdash_{\mathbf{CAL}^{\mathsf{ns}}} A$.*

Proof. Let $A \in Cn^{\mathcal{L}^+}_{\mathbf{CAL}^{\mathsf{ns}}}(\Gamma)$. Hence, by Lemma B.2.7 there is a $\Delta \subseteq \Omega_n$ such that $Cn^{\mathcal{L}^+}_{\mathbf{CAL_{n-1}}}(\Gamma) \vdash_{\mathbf{LLL}} A \check{\vee} \mathsf{Dab}(\Delta)$ and for all $\Theta \in \Sigma^n(Cn^{\mathcal{L}^+}_{\mathbf{CAL_{n-1}}}(\Gamma))$, $\Theta \not\subseteq \Delta$. By Corollary 3.1.2 $A \check{\vee} \mathsf{Dab}(\Delta) \in Cn^{\mathcal{L}^+}_{\mathbf{CAL_{n-1}}}(\Gamma)$ and hence also $\check{\neg} A \check{\supset} \mathsf{Dab}(\Delta) \in Cn^{\mathcal{L}^+}_{\mathbf{CAL_{n-1}}}(\Gamma)$. By Corollary B.2.3, (†) it is derived at an unmarked line l in the complete extension of the empty proof on a condition $\Theta_1, \ldots, \Theta_{n-1}$

(where $\Theta_i \subseteq \Omega_i$). Thus, by the compactness of **LLL**, there are $B_1, \ldots, B_m \in \Gamma$ such that $\{B_1, \ldots, B_m\} \vdash_{\mathbf{LLL}} (\check{\neg}A \check{\supset} \textsf{Dab}(\Delta)) \check{\vee} \textsf{Dab}(\Theta_1 \cup \ldots \cup \Theta_{n-1})$. We construct now a proof of $\check{\neg}A \check{\supset} \textsf{Dab}(\Delta)$ as follows:

1	B_1	PREM	$\emptyset$
⋮	⋮	⋮	⋮
m	B_m	PREM	$\emptyset$
$m+1$	$(\check{\neg}A \check{\supset} \textsf{Dab}(\Delta)) \check{\vee} \textsf{Dab}(\Theta_2 \cup \ldots \cup \Theta_{n-1})$	1–m; RC	Θ_1
⋮	⋮	⋮	⋮
$m+n-2$	$(\check{\neg}A \check{\supset} \textsf{Dab}(\Delta)) \check{\vee} \textsf{Dab}(\Theta_{n-1})$	$m+n-3$; RC	$\Theta_1 \cup \ldots \cup \Theta_{n-2}$
$m+n-1$	$\check{\neg}A \check{\supset} \textsf{Dab}(\Delta)$	$m+n-3$; RC	$\Theta_1 \cup \ldots \cup \Theta_{n-1}$
$m+n$	A	$m+n-2$; RC	$\Theta_1 \cup \ldots \cup \Theta_{n-1} \cup \Delta$

Since $\Gamma \subseteq \mathcal{W}$ no Dab-formulas are derived at this stage of the proof. Thus, line $m+n$ is unmarked. Suppose the line is marked in an extension of the proof $\mathcal{P}$. We extend the proof further to $\mathbf{g}(\mathcal{P})$. That the line is unmarked is an immediate consequence of (†) and Fact B.2.1. □

Appendix C
Appendix to Chapter 5

C.1 $\Xi(\Gamma)$, $\Xi^{\text{sat}}(\Gamma)$, Density, and Smoothness

In this section we are going to prove that $\Xi^{\text{sat}}(\Gamma)$ is $\prec$-dense in $\Xi(\Gamma)$:

Theorem C.1.1. *Where $\Gamma \subseteq \mathcal{W}^+$ is* **LLL**-*non-trivial: $\Xi^{\text{sat}}(\Gamma)$ is $\prec$-dense in $\Xi(\Gamma)$.*

We will see that this implies the following corollaries: $\Xi^{\text{sat}}(\Gamma)$ and $\Xi(\Gamma)$ have the same $\prec$-minimal elements and $\Xi^{\text{sat}}(\Gamma)$ is smooth (relative to $\prec$) iff $\Xi(\Gamma)$ is smooth.

Corollary C.1.1. *Where $\Gamma \subseteq \mathcal{W}^+$ is* **LLL**-*non-trivial:* $\min_\prec(\Xi^{\text{sat}}(\Gamma)) = \min_\prec(\Xi(\Gamma))$.

Corollary C.1.2. *Where $\Gamma \subseteq \mathcal{W}^+$: If $\langle \Xi(\Gamma), \prec \rangle$ is smooth then also $\langle \Xi^{\text{sat}}(\Gamma), \prec \rangle$ is smooth.*

First it is useful to show that the following special case of Theorem C.1.1 holds:

Lemma 5.5.2 (restated). *Where Γ is* **LLL**-*non-trivial: $\Xi^{\text{sat}}(\Gamma)$ is $\subset$-dense in $\Xi(\Gamma)$.*

Before we prove this note that:

Lemma C.1.1. *Where $\Gamma \subseteq \mathcal{W}^+$ is* **LLL**-*non-trivial and $\varphi \in \Xi(\Gamma)$: $\Gamma \cup (\Omega \setminus \varphi)^{\check{\neg}}$ is* **LLL**-*non-trivial.*

Proof. Assume $\Gamma \cup (\Omega \setminus \varphi)^{\check{\neg}}$ is **LLL**-trivial. Hence, by the compactness of **LLL**, there is a finite and $\subset$-minimal $\Delta \subseteq \Omega \setminus \varphi$ such that $\Gamma \vdash_{\mathbf{LLL}} \mathsf{Dab}(\Delta)$. Hence, $\Delta \in \Sigma(\Gamma)$. Since φ is a choice set of $\Sigma(\Gamma)$, $\varphi \cap \Delta \neq \emptyset$,—a contradiction. □

Proof. (Proof of Lemma 5.5.2). Let $\varphi \in \Xi(\Gamma)$. Hence, by Lemma C.1.1, $\Gamma \cup (\Omega \setminus \varphi)^{\check{\neg}}$ is **LLL**-non-trivial. We construct $\psi = \bigcup_{i \in \mathbb{N}} \psi_i$ as follows: where $\psi_0 = \Gamma \cup (\Omega \setminus \varphi)^{\check{\neg}}$ and $\{A_0, A_1, \ldots\}$ is a list of the members of φ, let

DOI: 10.1007/978-3-319-00792-2, © Springer International Publishing Switzerland 2014

$$\psi_{i+1} = \begin{cases} \psi_i \cup \{A_i\} & \text{if } \psi_i \cup \{A_i\} \text{is } \mathbf{LLL}\text{-non-trivial} \\ \psi_i \cup \{\check{\neg} A_i\} & \text{else} \end{cases}$$

We can easily show by induction that each ψ_i is **LLL**-non-trivial. Since Γ is **LLL**-non-trivial, so is ψ_0. Suppose ψ_i is **LLL**-non-trivial and $\psi_i \cup \{A_i\}$ is **LLL**-trivial. Hence, by classical logic, $\psi_i \vdash_{\mathbf{LLL}} \check{\neg} A_i$. Since ψ_i is **LLL**-non-trivial, so is $\psi_i \cup \{\check{\neg} A_i\}$.

Assume ψ is **LLL**-trivial. By the compactness of **LLL**, there is a finite subset ψ' of ψ that is **LLL**-trivial. There is a $\psi_i \supseteq \psi'$. By the monotonicity of **LLL**, also ψ_i is **LLL**-trivial,—a contradiction. Hence, ψ is **LLL**-non-trivial.

Note that ψ can be written as $\Gamma \cup (\Omega \setminus \delta)^{\check{\neg}} \cup \delta$ for some $\delta \subseteq \varphi$.

Let now $\Delta \cup \Theta^{\check{\neg}} \in \Sigma^{\text{sat}}(\Gamma)$ and hence $\Gamma \vdash_{\mathbf{LLL}} \mathsf{Dabn}(\Delta, \Theta)$. Assume $\Delta \cap \delta = \emptyset = \Theta \cap (\Omega \setminus \delta)$. Hence, $\Delta \subseteq \Omega \setminus \delta$ and $\Theta \subseteq \delta$,—a contradiction to the **LLL**-non-triviality of ψ since by the monotonicity of **LLL** also $\psi \vdash_{\mathbf{LLL}} \mathsf{Dabn}(\Delta, \Theta)$. Hence, $\delta \in \Xi^{\text{sat}}(\Gamma)$. Since $\delta \subseteq \varphi$ this shows that $\Xi^{\text{sat}}(\Gamma)$ is $\sqsubset$-dense in $\Xi(\Gamma)$. □

Fact C.1.1. *Where X is $\prec$-dense in Y:* $\min_{\prec}(X) = \min_{\prec}(Y)$.

Proof. Assume $x \in \min_{\prec}(X) \setminus \min_{\prec}(Y)$. Since $X \subseteq Y$, $x \in Y$. Hence, there is a $y \in Y \setminus X$ for which $y \prec x$. Since X is $\prec$-dense in Y, there is a $z \in X$ for which $z \prec y$. By the transitivity of $\prec$, $z \prec x$,—a contradiction. Suppose $x \in \min_{\prec}(Y)$. Hence, there is no $y \in Y$ for which $y \prec x$. Since X is $\prec$-dense in Y, $x \in X$ and hence $x \in \min_{\prec}(X)$. □

Fact C.1.2. *Where $\prec_2 \subseteq \prec_1$: If X is $\prec_2$-dense in Y, then it is $\prec_1$-dense in Y.*

Proof. Let $x \in Y$. Hence, there is a $y \in X$ for which $y \preceq_2 x$. Since $\prec_2 \subseteq \prec_1$, also $y \preceq_1 x$. □

By Lemma 5.5.2, Fact C.1.2 and since $\sqsubset \subseteq \prec$, we immediately get Theorem C.1.1. Hence, by Fact C.1.1, also Corollary C.1.1 is immediate. Finally, also Corollary C.1.2 follows immediately by Theorem C.1.1.

C.2 Questions Concerning Membership in Υ

In this section we show that each of the following sets belongs to Υ: $\Xi_s^{\text{sat}}(\Gamma)$, $\Xi_s(\Gamma)$, and $\Xi(\Gamma)$. Furthermore we show that Υ is closed under $\cap$ and under $\min_{\sqsubset}^{\cup}$. The latter will help us to prove the adequacy of our characterization of the reliability strategy in terms of $\Lambda = \min_{\sqsubset}^{\cup}$.

Some notations for the next few results:

- $\Sigma^{\text{sat}}(\Gamma) =_{\text{df}} \{\mathsf{Dabn}(\Delta, \Theta) \mid \Delta \cup \Theta^{\check{\neg}} \in \Sigma^{\text{sat}}(\Gamma)\}$
- $\Sigma_s^{\text{sat}}(\Gamma) =_{\text{df}} \{\mathsf{Dabn}(\Delta, \Theta) \mid \Delta \cup \Theta^{\check{\neg}} \in \Sigma_s^{\text{sat}}(\Gamma)\}$
- $\Sigma(\Gamma) =_{\text{df}} \{\mathsf{Dab}(\Delta) \mid \Delta \in \Sigma(\Gamma)\}$
- $\Sigma_s(\Gamma) =_{\text{df}} \{\mathsf{Dab}(\Delta) \mid \Delta \in \Sigma_s(\Gamma)\}$

Lemma C.2.1. *Where $\Gamma \subseteq \mathcal{W}^+$: $\Xi_s^{\text{sat}}(\Gamma) = \mathsf{Ab}_{\mathbf{LLL}}^{\bullet\Sigma_s^{\text{sat}}(\Gamma)}$ and hence $\Xi_s^{\text{sat}}(\Gamma) \in \Upsilon$.*

Proof. "$\subseteq$": Assume $\varphi \in \Xi_s^{\text{sat}}(\Gamma) \setminus \mathsf{Ab}_{\mathbf{LLL}}^{\bullet\Sigma_s^{\text{sat}}(\Gamma)}$. Hence, $\bullet\Sigma_s^{\text{sat}}(\Gamma) \cup \varphi \cup (\Omega \setminus \varphi)^{\check{\neg}}$ is **LLL**-trivial. Since $\varphi \cup (\Omega \setminus \varphi)^{\check{\neg}}$ is a choice set of $\Sigma_s^{\text{sat}}(\Gamma)$, $\varphi \cup (\Omega \setminus \varphi)^{\check{\neg}} \vdash_{\mathbf{LLL}} \bullet\Sigma_s^{\text{sat}}(\Gamma)$. Hence, $\varphi \cup (\Omega \setminus \varphi)^{\check{\neg}}$ is **LLL**-trivial,—a contradiction to the **LLL**-contingency of Ω.

"$\supseteq$": Let $\varphi \in \mathsf{Ab}_{\mathbf{LLL}}^{\bullet\Sigma_s^{\text{sat}}(\Gamma)}$. Hence, obviously $\varphi \cup (\Omega \setminus \varphi)^{\check{\neg}}$ is a choice set of $\Sigma_s^{\text{sat}}(\Gamma)$. □

Lemma C.2.2. *Where $\Gamma \subseteq \mathcal{W}^+$: $\Xi_s(\Gamma) = \mathsf{Ab}_{\mathbf{LLL}}^{\bullet\Sigma_s(\Gamma)}$ and hence $\Xi_s(\Gamma) \in \Upsilon$.*

Proof. "$\subseteq$": Assume $\varphi \in \Xi_s(\Gamma) \setminus \mathsf{Ab}_{\mathbf{LLL}}^{\bullet\Sigma_s(\Gamma)}$. Hence, $\Sigma_s(\Gamma) \cup \varphi \cup (\Omega \setminus \varphi)^{\check{\neg}}$ is **LLL**-trivial. Since φ is a choice set of $\Sigma_s(\Gamma)$, $\varphi \cup (\Omega \setminus \varphi)^{\check{\neg}} \vdash_{\mathbf{LLL}} \bullet\Sigma_s(\Gamma)$. Hence, $\varphi \cup (\Omega \setminus \varphi)^{\check{\neg}}$ is **LLL**-trivial,—a contradiction to the **LLL**-contingency of Ω.

"$\supseteq$": Let $\varphi \in \mathsf{Ab}_{\mathbf{LLL}}^{\bullet\Sigma_s(\Gamma)}$. Hence, obviously φ is a choice set of $\Sigma_s(\Gamma)$. □

Lemma C.2.3. *Where $\Gamma \subseteq \mathcal{W}^+$: $\Xi(\Gamma) = \mathsf{Ab}_{\mathbf{LLL}}^{\bullet\Sigma(\Gamma)}$ and hence $\Xi(\Gamma) \in \Upsilon$.*

Proof. Analogous to the proof of Lemma C.2.2.

Lemma C.2.4. *$X, Y \in \Upsilon$ implies $X \cap Y \in \Upsilon$.*

Proof. Let $X, Y \in \Upsilon$. Hence, there are Γ_X and Γ_Y such that $X = \mathrm{Ab}_{\mathbf{LLL}}^{\Gamma_X}$ and $Y = \mathrm{Ab}_{\mathbf{LLL}}^{\Gamma_Y}$. By Lemma 5.3.2, $X = \Xi^{\text{sat}}(\Gamma_X)$ and $Y = \Xi^{\text{sat}}(\Gamma_Y)$. We will show that $\Xi^{\text{sat}}(\Gamma_X) \cap \Xi^{\text{sat}}(\Gamma_Y) = \Xi^{\text{sat}}(\bullet\Sigma^{\text{sat}}(\Gamma_X) \cup \bullet\Sigma^{\text{sat}}(\Gamma_Y))$ which is by Lemma 5.3.2 sufficient to prove the lemma.

Let $\varphi \in \Xi^{\text{sat}}(\Gamma_X) \cap \Xi^{\text{sat}}(\Gamma_Y)$. Assume $\varphi \notin \Xi^{\text{sat}}(\bullet\Sigma^{\text{sat}}(\Gamma_X) \cup \bullet\Sigma^{\text{sat}}(\Gamma_Y))$. Hence, there are Δ, Θ such that $\bullet\Sigma^{\text{sat}}(\Gamma_X) \cup \bullet\Sigma^{\text{sat}}(\Gamma_Y) \vdash_{\mathbf{LLL}} \mathsf{Dabn}(\Delta, \Theta)$ where $\varphi \cap \Delta = \emptyset = (\Omega \setminus \varphi) \cap \Theta$. However, by the supposition, φ is a choice set of $\Sigma^{\text{sat}}(\Gamma_X)$ and $\Sigma^{\text{sat}}(\Gamma_Y)$. Hence, $\varphi \cup (\Omega \setminus \varphi)^{\check{\neg}} \vdash_{\mathbf{LLL}} \bullet\Sigma^{\text{sat}}(\Gamma_x), \bullet\Sigma^{\text{sat}}(\Gamma_Y)$. By the transitivity of $\vdash_{\mathbf{LLL}}$ we have: $\varphi \cup (\Omega \setminus \varphi)^{\check{\neg}} \vdash_{\mathbf{LLL}} \mathsf{Dabn}(\Delta, \Theta)$. Hence, $\varphi \cup (\Omega \setminus \varphi)^{\check{\neg}}$ is **LLL**-trivial,—a contradiction to the **LLL**-contingency of Ω.

Let $\varphi \in \Xi^{\text{sat}}(\bullet\Sigma^{\text{sat}}(\Gamma_X) \cup \bullet\Sigma^{\text{sat}}(\Gamma_Y))$. Due to symmetry it is sufficient to show that $\varphi \in \Xi^{\text{sat}}(\Gamma_X)$. Assume thus that $\varphi \notin \Xi^{\text{sat}}(\Gamma_X)$. Hence, there are Δ, Θ such that $\mathsf{Dabn}(\Delta, \Theta) \in \bullet\Sigma^{\text{sat}}(\Gamma_X)$ for which $\Theta \subseteq \varphi$ and $\Delta \subseteq \Omega \setminus \varphi$. Note that $\bullet\Sigma^{\text{sat}}(\bullet\Sigma^{\text{sat}}(\Gamma_X) \cup \bullet\Sigma^{\text{sat}}(\Gamma_Y)) \vdash_{\mathbf{LLL}} \mathsf{Dabn}(\Delta, \Theta)$. But then $\varphi \notin \Xi^{\text{sat}}(\bullet\Sigma^{\text{sat}}(\Gamma_X) \cup \bullet\Sigma^{\text{sat}}(\Gamma_Y))$,—a contradiction. □

Fact C.2.1. *Where $\Gamma \subseteq \mathcal{W}^+$ is* **LLL***-trivial: (i) $\Xi(\Gamma) = \{\Omega\}$, (ii) $\Xi^{\perp}(\Gamma) = \{\Omega\}$, (iii) $\Xi^{\text{sat}}(\Gamma) = \emptyset$.*

Lemma C.2.5. *Where $\Gamma \subseteq \mathcal{W}^+$: Γ is* **LLL***-trivial iff $\Gamma \cup (\Omega \setminus \bigcup \Sigma(\Gamma))^{\check{\neg}}$ is* **LLL***-trivial.*

Proof. "⇒" follows by the monotonicity of **LLL**. "⇐": Suppose $\Gamma \cup (\Omega \setminus \bigcup \Sigma(\Gamma))^{\check{\neg}}$ is **LLL**-trivial. Assume Γ is **LLL**-non-trivial. Hence, there is a finite $\Delta \subseteq \Omega \setminus \bigcup \Sigma(\Gamma)$ such that $\Gamma \vdash_{\mathbf{LLL}} \mathsf{Dab}(\Delta)$. But then there is a $\Theta \subseteq \Delta$ such that $\Theta \in \Sigma(\Gamma)$,—a contradiction. □

Lemma C.2.6. *Where $\Gamma \subseteq \mathcal{W}^+$: $\varphi \in \Xi^{\mathrm{sat}}(\Gamma \cup (\Omega \setminus \bigcup \Sigma(\Gamma))^{\check{\neg}})$ iff (i) $\varphi \in \Xi^{\mathrm{sat}}(\Gamma)$ and (ii) $\varphi \subseteq \bigcup \Sigma(\Gamma)$.*

Proof. Note that by Lemma C.2.5 Γ is **LLL**-trivial iff $\Gamma \cup (\Omega \setminus \bigcup \Sigma(\Gamma))^{\check{\neg}}$ is **LLL**-trivial in which case (by Fact C.2.1) $\Xi^{\mathrm{sat}}(\Gamma) = \Xi^{\mathrm{sat}}(\Gamma \cup (\Omega \setminus \bigcup \Sigma(\Gamma))^{\check{\neg}}) = \emptyset$ and hence the Lemma holds trivially. Let thus, Γ be **LLL**-non-trivial.

Let $\varphi \in \Xi^{\mathrm{sat}}(\Gamma)$ and $\varphi \subseteq \bigcup \Sigma(\Gamma)$. Assume $\varphi \notin \Xi^{\mathrm{sat}}(\Gamma \cup (\Omega \setminus \bigcup \Sigma(\Gamma))^{\check{\neg}})$. Hence, there are $\Delta, \Theta \subseteq \Omega$ such that $\Delta \cup \Theta^{\check{\neg}} \in \Sigma^{\mathrm{sat}}(\Gamma \cup (\Omega \setminus \Sigma(\Gamma))^{\neg})$, $\Delta \cap \varphi = \emptyset = \Theta \cap (\Omega \setminus \varphi)$ and hence $\Delta \subseteq \Omega \setminus \varphi$ and $\Theta \subseteq \varphi \subseteq \bigcup \Sigma(\Gamma)$. Thus, $\Gamma \cup (\Omega \setminus \bigcup \Sigma(\Gamma))^{\check{\neg}} \vdash_{\mathbf{LLL}} \mathsf{Dabn}(\Delta, \Theta)$. By the compactness of **LLL**, there is a finite $\delta \subseteq \Omega \setminus \bigcup \Sigma(\Gamma)$ such that $\Gamma \vdash_{\mathbf{LLL}} \check{\neg}\check{\bigwedge}\delta^{\check{\neg}} \mathbin{\check{\vee}} \mathsf{Dabn}(\Delta, \Theta)$. Thus, $\Gamma \vdash_{\mathbf{LLL}} \mathsf{Dabn}(\Delta \cup \delta, \Theta)$. Hence, there are $\subset$-minimal $\Delta' \subseteq \Delta \cup \delta$ and $\Theta' \subseteq \Theta$ such that $\Gamma \vdash_{\mathbf{LLL}} \mathsf{Dabn}(\Delta', \Theta')$. Note though that $\varphi \cap \Delta' = \emptyset = \Theta' \cap (\Omega \setminus \varphi)$. This is a contradiction to $\varphi \in \Xi^{\mathrm{sat}}(\Gamma)$.

Let $\varphi \in \Xi^{\mathrm{sat}}(\Gamma \cup (\Omega \setminus \bigcup \Sigma(\Gamma))^{\check{\neg}})$. Evidently, $\varphi \subseteq \bigcup \Sigma(\Gamma)$. Assume $\varphi \notin \Xi^{\mathrm{sat}}(\Gamma)$. Hence, there are $\Delta, \Theta \subseteq \Omega$ for which $\varphi \cap \Delta = \emptyset = \Theta \cap (\Omega \setminus \varphi)$ and $\Delta \cup \Theta^{\check{\neg}} \in \Sigma^{\mathrm{sat}}(\Gamma)$. Hence, $\Gamma \vdash_{\mathbf{LLL}} \mathsf{Dabn}(\Delta, \Theta)$. By the monotonicity of **LLL**, also $\Gamma \cup (\Omega \setminus \bigcup \Sigma(\Gamma))^{\check{\neg}} \vdash_{\mathbf{LLL}} \mathsf{Dabn}(\Delta, \Theta)$. Hence, there are $\subset$-minimal $\Delta' \subseteq \Delta$ and $\Theta' \subseteq \Theta$ such that $\Delta' \cup \Theta'^{\check{\neg}} \in \Sigma^{\mathrm{sat}}(\Gamma \cup (\Omega \setminus \bigcup \Sigma(\Gamma))^{\check{\neg}})$. Note however that $\varphi \cap \Delta' = \emptyset = \Theta' \cap (\Omega \setminus \varphi)$. A contradiction to the supposition. □

Lemma C.2.7. *Where $\Gamma \subseteq \mathcal{W}^+$: $\Xi^{\mathrm{sat}}(\Gamma \cup (\Omega \setminus \bigcup \Sigma(\Gamma))^{\check{\neg}}) = \min_{\subset}^{\cup}(\Xi^{\mathrm{sat}}(\Gamma))$.*

Proof. Note that by Lemma C.2.5 Γ is **LLL**-trivial iff $\Gamma \cup (\Omega \setminus \bigcup \Sigma(\Gamma))^{\check{\neg}}$ is **LLL**-trivial in which case (by Fact C.2.1) $\Xi^{\mathrm{sat}}(\Gamma) = \Xi^{\mathrm{sat}}(\Gamma \cup (\Omega \setminus \bigcup \Sigma(\Gamma))^{\check{\neg}}) = \emptyset$ and hence the Lemma holds trivially. Let thus, Γ be **LLL**-non-trivial.

Let $\varphi \in \Xi^{\mathrm{sat}}(\Gamma \cup (\Omega \setminus \bigcup \Sigma(\Gamma))^{\check{\neg}})$. By Lemma C.2.6, $\varphi \subseteq \bigcup \Sigma(\Gamma)$. By Corollary A.2, $\bigcup \Sigma(\Gamma) = \bigcup \min_{\subset}(\Xi(\Gamma))$. By Corollary C.1.1, $\min_{\subset}(\Xi(\Gamma)) = \min_{\subset}(\Xi^{\mathrm{sat}}(\Gamma))$ and hence, $\bigcup \Sigma(\Gamma) = \bigcup \min_{\subset}(\Xi^{\mathrm{sat}}(\Gamma))$. Thus, (1) $\varphi \subseteq \bigcup \min_{\subset}(\Xi^{\mathrm{sat}}(\Gamma))$. By Lemma C.2.6, (2) $\varphi \in \Xi^{\mathrm{sat}}(\Gamma)$. By (1) and (2), $\varphi \in \min_{\subset}^{\cup}(\Xi^{\mathrm{sat}}(\Gamma))$.

Let $\varphi \in \min_{\subset}^{\cup}(\Xi^{\mathrm{sat}}(\Gamma))$. Hence, (1) $\varphi \in \Xi^{\mathrm{sat}}(\Gamma)$ and (2) $\varphi \subseteq \bigcup \min_{\subset}(\Xi^{\mathrm{sat}}(\Gamma))$. By Corollary C.1.1, (3) $\min_{\subset}(\Xi^{\mathrm{sat}}(\Gamma)) = \min_{\subset}(\Xi(\Gamma))$. By Corollary A.2, (2), (3), we have (4) $\varphi \subseteq \bigcup \Sigma(\Gamma)$. By (1), (4) and Lemma C.2.6, $\varphi \in \Xi^{\mathrm{sat}}(\Gamma \cup (\Omega \setminus \bigcup \Sigma(\Gamma))^{\check{\neg}})$. □

Corollary C.2.1. *Where $\Gamma \subseteq \mathcal{W}^+$: $\min_{\subset}^{\cup}(\mathsf{Ab}^{\Gamma}_{\mathbf{LLL}}) = \mathsf{Ab}^{\Gamma \cup (\Omega \setminus \bigcup \Sigma(\Gamma))^{\check{\neg}}}_{\mathbf{LLL}}$*

Corollary C.2.2. *Where $\Gamma \subseteq \mathcal{W}^+$: $\min_{\subset}^{\cup}(\Xi^{\mathrm{sat}}(\Gamma)) \in \Upsilon$ and $\min_{\subset}^{\cup}(\mathsf{Ab}^{\Gamma}_{\mathbf{LLL}}) \in \Upsilon$.*

Theorem C.2.1. *Where* $\mathbf{AL^r}$ *is the AL in standard format characterized by* $\langle \mathbf{LLL}, \Omega,$ *reliability*$\rangle$, $\mathbf{AL}_{\min^{\cup}_{\subset}}$ *is characterized* **LLL**, Ω *and the semantic selection* $\min^{\cup}_{\subset}$*:*

(i) $\Gamma \subseteq \mathcal{W}^+$*:* $\Gamma \Vdash_{\mathbf{AL^r}} A$ *iff* $\Gamma \Vdash_{\mathbf{AL}_{\min^{\cup}_{\subset}}} A$

(ii) $\Gamma \subseteq \mathcal{W}$*:* $\Gamma \vdash_{\mathbf{AL^r}} A$ *iff* $\Gamma \vdash_{\mathbf{AL}_{\min^{\cup}_{\subset}}} A$.

Proof. Ad (i): $\Gamma \Vdash_{\mathbf{AL^r}} A$, iff, for all reliable models M of Γ, $M \models A$, iff [note that $U(\Gamma) = \bigcup \Sigma(\Gamma)$], for all $M \in \mathcal{M}_{\mathbf{LLL}}(\Gamma)$ such that $\mathrm{Ab}(M) \subseteq \bigcup \Sigma(\Gamma)$, $M \models A$, iff, $\Gamma \cup (\Omega \setminus \bigcup \Sigma(\Gamma))^{\check{\neg}} \Vdash_{\mathbf{LLL}} A$, iff, for all $M \in \mathcal{M}_{\mathbf{LLL}}(\Gamma)$ for which $\mathrm{Ab}(M) \in \mathsf{Ab}_{\mathbf{LLL}}^{\Gamma \cup (\Omega \setminus \Sigma(\Gamma))^{\check{\neg}}}$, $M \models A$, iff [by Corollary C.2.1] for all $M \in \mathcal{M}_{\mathbf{LLL}}(\Gamma)$ for which $\mathrm{Ab}(M) \in \min^{\cup}_{\subset}(\mathsf{Ab}^{\Gamma}_{\mathbf{LLL}})$, $M \models A$, iff, $\Gamma \Vdash_{\mathbf{AL}_{\min^{\cup}_{\subset}}} A$.

Ad (ii): $\Gamma \vdash_{\mathbf{AL^r}} A$, iff, [by Corollary 2.4.2] $\Gamma \cup (\Omega \setminus \bigcup \Sigma(\Gamma))^{\check{\neg}} \vdash_{\mathbf{LLL}} A$, iff, [by the soundness and completeness of **LLL**] $\Gamma \cup (\Omega \setminus \bigcup \Sigma(\Gamma))^{\check{\neg}} \Vdash_{\mathbf{LLL}} A$, iff, for all $M \in \mathcal{M}_{\mathbf{LLL}}(\Gamma)$ for which $\mathrm{Ab}(M) \in \mathsf{Ab}_{\mathbf{LLL}}^{\Gamma \cup (\Omega \setminus \bigcup \Sigma(\Gamma))^{\check{\neg}}}$, $M \models A$, iff, [by Corollary C.2.1] for all $M \in \mathcal{M}_{\mathbf{LLL}}(\Gamma)$ for which $\mathrm{Ab}(M) \in \min^{\cup}_{\subset}(\mathsf{Ab}^{\Gamma}_{\mathbf{LLL}})$, $M \models A$, iff, $\Gamma \Vdash_{\mathbf{AL}_{\min^{\cup}_{\subset}}} A$, iff [by Corollary 5.4.2], $\Gamma \vdash_{\mathbf{AL}_{\min^{\cup}_{\subset}}} A$. □

C.3 Criteria for the Threshold Function Λ

In this section we investigate relationships among the various criteria for threshold functions introduced in the main text. Then we prove some properties of some specific threshold functions.

Fact C.3.1. *(i)* **C2** *implies* **C3***; (ii)* **C3** *implies* **C1***; (iii)* **C1** *implies* $\mathbf{DI}_{\subset}$*; (iv)* $\mathbf{DI}_{\subset}$ *implies* **C1***; (v)* **C4** *implies* $\mathbf{RA}_{\subset}$*; (vi)* $\mathbf{DI}_{\prec}$ *implies* $\mathbf{DI}_{\subset}$.

Proof. Ad (i): **C3** follows trivially. *Ad (ii):* **C1** follows immediately due to **T1**. *Ad (iii)*: Let X be $\subset$-dense in Y. By Fact C.1.1, $\min_{\subset}(X) = \min_{\subset}(Y)$. Hence, by **C1**, $\Lambda(X) \cap Y = \Lambda(Y) \cap X$. By **T1** and since $X \subseteq Y$, $\Lambda(X) = \Lambda(Y) \cap X$. Hence $\mathbf{DI}_{\subset}$ holds. *Ad (iv):* Suppose $\min_{\subset}(X) = \min_{\subset}(Y)$. We first show that $X \cap Y$ is $\subset$-dense in X and in Y. (Recall that $X \cap Y \in \Upsilon$ by Lemma C.2.4.) Without loss of generality we show the case for X. Let $x \in X$. By the smoothness of $\langle X, \subset\rangle$ (Theorem 5.5.3) there is a $y \in \min_{\subset}(X)$ such that $y \subseteq x$. Since $\min_{\subset}(X) = \min_{\subset}(Y)$, $y \in X \cap Y$. Hence, $X \cap Y$ is $\subset$-dense in X. Thus, by $\mathbf{DI}_{\subset}$, $\Lambda(X \cap Y) = \Lambda(X) \cap X \cap Y$ and by **T1**, $\Lambda(X \cap Y) = \Lambda(X) \cap Y$. By an analogous argument, $\Lambda(X \cap Y) = \Lambda(Y) \cap X$. Hence, $\Lambda(X) \cap Y = \Lambda(Y) \cap X$. *Ad (v):* This follows by Theorem 5.5.3. *Ad (vi):* Suppose X is $\subset$-dense in Y. Hence, by Fact C.1.2 and since $\subset \subseteq \prec$, X is $\prec$-dense in Y. Hence, due to $\mathbf{DI}_{\prec}$, $\Lambda(X) = \Lambda(Y) \cap X$. □

An overview of the relationship between the criteria is offered in Fig. 5.3.

Fact C.3.2. $\min_{\subset}(X) = \Psi_{\subset}(X)$

Proof. This follows immediately by Theorem 5.5.3 and Fact 5.2.1. □

Fact C.3.3. *Where X is $\prec$-dense in Y: $\Psi_{\prec}(X) = \Psi_{\prec}(Y) \cap X$. ($\Psi_{\prec}$ satisfies* **DI**$_{\prec}$*.)*

Proof. "$\subseteq$": Let $x \in \Psi_{\prec}(X)$. Assume $x \notin \Psi_{\prec}(Y) \cap X$. Since $\Psi_{\prec}$ is inclusive, $x \in X$ and thus $x \notin \Psi_{\prec}(Y)$. Hence, there is a $y \in \min_{\prec}(Y)$ for which $y \prec x$. However, by Fact C.1.1, $y \in \min_{\prec}(X)$ and hence $x \notin \Psi_{\prec}(X)$,—a contradiction. "$\supseteq$": Let $x \in \Psi_{\prec}(Y) \cap X$. Assume $x \notin \Psi_{\prec}(X)$. Since $x \in X$, there is a $y \in \min_{\prec}(X)$ for which $y \prec x$. However, by Fact C.1.1, $y \in \min_{\prec}(Y)$ and thus $x \notin \Psi_{\prec}(Y)$,—a contradiction. □

Fact C.3.4. *$\Psi_{\prec}(X)$ is $\prec$-dense in X. ($\Psi_{\prec}$ satisfies* **RA**$_{\prec}$*.)*

Proof. Let $x \in X$. Either there is a $y \in \min_{\prec}(X)$ such that $y \preceq x$ or not. In the second case $x \in \Psi_{\prec}(X)$. Since $\min_{\prec}(X) \subseteq \Psi_{\prec}(X)$, there is a $y \in \Psi_{\prec}(X)$ such that $y \preceq x$ also in the first case. □

Fact C.3.5. *$\Psi_{\prec}(X)$ is a $\prec$-lower set of X.*

Proof. Let $x \in \Psi_{\prec}(X)$ and $y \in X$ such that $y \prec x$. Hence, $x \notin \min_{\prec}(X)$. Hence, there is no $z \in \min_{\prec}(X)$ such that $z \prec x$. Hence, by the transitivity of $\prec$, there is no $z \in \min_{\prec}(X)$ such that $z \prec y$. Thus, $y \in \Psi_{\prec}(X)$. □

Fact C.3.6. *$\Psi_{[\prec_1,\ldots,\prec_n]}(X)$ is a $\prec_n$-lower set of X.*

Proof. We show this by induction on n. The case $n = 1$ is shown in Fact C.3.5. "$n-1 \Rightarrow n$": Let $x \in \Psi_{[\prec_1,\ldots,\prec_n]}(X)$ and $y \in X$ such that $y \prec_n x$. Hence, $x \in \Psi_{[\prec_1,\ldots,\prec_{n-1}]}(X)$ and, since $\prec_n \subseteq \prec_{n-1}$, $y \prec_{n-1} x$. Hence, $y \in \Psi_{[\prec_1,\ldots,\prec_{n-1}]}(X)$ by the induction hypothesis. The rest follows by Fact C.3.5. □

Fact C.3.7. *Where $\prec_2 \subseteq \prec_1$: If X is a $\prec_1$-lower set of Y, then X is a $\prec_2$-lower set of Y.*

Proof. Let $x \in X$ and $y \in Y$ such that $y \prec_2 x$. Hence, $y \prec_1 X$ and thus, $y \in X$ since X is a $\prec_1$-lower set of Y. □

Fact C.3.8. *$\Psi_{[\prec_1,\ldots,\prec_n]}$ (and hence also $\Psi_{\prec}$) is a threshold function.*

Proof. **T1** and **T3** are immediate. **T2** follows by Fact C.3.6, Fact C.3.7 and since $\subset \subseteq \prec_n$. □

The following related facts C.3.9–C.3.12 are especially useful in Sect. 5.5.1.

Fact C.3.9. *Where $\langle \prec_1, \ldots, \prec_n \rangle$ is an abstraction sequence:*

$$\min_{\prec_1}(X) \subseteq \min_{\prec_2}(\Psi_{\prec_1}(X)) \subseteq \ldots \subseteq \min_{\prec_n}(\Psi_{[\prec_1,\ldots,\prec_{n-1}]}(X))$$

Proof. We show this by induction. "$i = 1$": Let $x \in \min_{\prec_1}(X)$. Assume $x \notin \min_{\prec_2}(\Psi_{\prec_1}(X))$. Note that $\min_{\prec_1}(X) \subseteq \Psi_{\prec_1}(X)$ and hence $x \in \Psi_{\prec_1}(X)$. Thus, there is a $y \in \Psi_{\prec_1}(X)$ such that $y \prec_2 x$. But then $y \prec_1 x$,—a contradiction to the minimality of x.

"$i-1 \Rightarrow i$": Let $x \in \min_{\prec_i}(\Psi_{[\prec_1,\ldots,\prec_{i-1}]}(X))$. Hence, $x \in \Psi_{[\prec_1,\ldots,\prec_i]}(X)$. Assume there is a $y \in \Psi_{[\prec_1,\ldots,\prec_i]}(X)$ such that $y \prec_{i+1} x$. Then, $y \prec_i x$ and $y \in \Psi_{[\prec_1,\ldots,\prec_{i-1}]}(X)$,—a contradiction to the minimality of x. □

Fact C.3.10. *Where $\langle \prec_1, \ldots, \prec_n \rangle$ is an abstraction sequence, $x \in \Psi_{[\prec_1,\ldots,\prec_i]}(X)$ and $0 \leq i < j \leq n^2$: either $x \in \Psi_{[\prec_1,\ldots,\prec_j]}(X)$ or there is a $y \in \Psi_{[\prec_1,\ldots,\prec_j]}(X)$ for which $y \prec_{i+1} x$.*

Proof. Let $0 \leq i < n$. We show the fact by induction for all j such that $i < j \leq n$.

"$j = i+1$": Suppose $x \notin \Psi_{[\prec_1,\ldots,\prec_j]}(X)$. Since $x \in \Psi_{[\prec_1,\ldots,\prec_i]}(X)$ there is a $y \in \min_{\prec_j}(\Psi_{[\prec_1,\ldots,\prec_i]}(X)) \subseteq \Psi_{[\prec_1,\ldots,\prec_j]}(X)$ such that $y \prec_j x$.

"$j \Rightarrow j+1$": By the induction hypothesis there is a $x' \preceq_{i+1} x$ such that $x' \in \Psi_{[\prec_1,\ldots,\prec_j]}(X)$. Suppose $x' \notin \Psi_{[\prec_1,\ldots,\prec_{j+1}]}(X)$. Hence, there is a $y \in \min_{\prec_{j+1}}(\Psi_{[\prec_1,\ldots,\prec_j]}(X))$ such that $y \prec_{j+1} x'$. Hence, $y \prec_{i+1} x'$ and by the transitivity of $\prec_{i+1}$ also $y \prec_{i+1} x$. □

Fact C.3.11. *$\Psi_{[\prec_1,\ldots,\prec_n]}(X)$ is $\prec_1$-dense in X. ($\Psi_{[\prec_1,\ldots,\prec_n]}$ satisfies* **RA**$_{\prec_1}$.*)*

Proof. Follows by Fact C.3.10 (where $i = 0$ and $j = n$). □

Fact C.3.12. *Where $\Psi_{[\prec_1,\ldots,\prec_n]}(X) \subseteq Y \subseteq X$: $\Psi_{[\prec_1,\ldots,\prec_n]}(X) = \Psi_{[\prec_1,\ldots,\prec_n]}(Y)$. ($\Psi_{[\prec_1,\ldots,\prec_n]}$ satisfies* **CT** *and* **CM**.*)*

Proof. "$\supseteq$": Let $x \in \Psi_{[\prec_1,\ldots,\prec_n]}(Y)$. Note that (†) $x \in \Psi_{[\prec_1,\ldots,\prec_i]}(Y)$ for all $i \leq n$ and $x \in Y$. We show by induction that $x \in \Psi_{[\prec_1,\ldots,\prec_i]}(X)$ for every $i \leq n$.

"$i = 1$": Assume $x \notin \Psi_{\prec_1}(X)$. Hence, there is a $y \in \min_{\prec_1}(X)$ such that $y \prec_1 x$. By Fact C.3.9, $y \in \Psi_{[\prec_1,\ldots,\prec_n]}(X)$ and hence $y \in Y$. Since $x \in \Psi_{\prec_1}(Y)$, $y \notin \min_{\prec_1}(Y)$. Hence, there is a $z \in Y$ such that $z \prec_1 y$,—a contradiction to the minimality of y.

"$i \Rightarrow i+1$": By the induction hypothesis, $x \in \Psi_{[\prec_1,\ldots,\prec_i]}(X)$. Assume $x \notin \Psi_{[\prec_1,\ldots,\prec_{i+1}]}(X)$. Hence, there is a $y \in \min_{\prec_{i+1}}(\Psi_{[\prec_1,\ldots,\prec_i]}(X))$ such that $y \prec_{i+1} x$. Hence, by Fact C.3.9, $y \in \Psi_{[\prec_1,\ldots,\prec_n]}(X)$ and thus $y \in Y$. Since $\Psi_{[\prec_1,\ldots,\prec_{i+1}]}(Y)$ is a $\prec_{i+1}$-lower set of Y by Fact C.3.6 and by (†), $y \in \Psi_{[\prec_1,\ldots,\prec_{i+1}]}(Y)$. Hence, since $x, y \in \Psi_{[\prec_1,\ldots,\prec_{i+1}]}(Y)$ and $y \prec_{i+1} x$ there is a $z \in \Psi_{[\prec_1,\ldots,\prec_{i+1}]}(Y)$ such that $z \prec_{i+1} y$ (otherwise $y \in \min_{\prec_{i+1}}(\Psi_{[\prec_1,\ldots,\prec_i]}(Y))$ and hence $x \notin \Psi_{[\prec_1,\ldots,\prec_{i+1}]}(Y)$). Since $y \in \Psi_{[\prec_1,\ldots,\prec_{i+1}]}(X)$ and $\Psi_{[\prec_1,\ldots,\prec_{i+1}]}(X)$ is by Fact C.3.6 a $\prec_{i+1}$-lower set of X, also $z \in \Psi_{[\prec_1,\ldots,\prec_{i+1}]}(X)$,—a contradiction to the minimality of y.

"$\subseteq$": We show by induction that $x \in \Psi_{[\prec_1,\ldots,\prec_i]}(Y)$ for all $i \leq n$ and all $x \in \Psi_{[\prec_1,\ldots,\prec_n]}(X)$. Note that $x \in Y$ and $x \in \Psi_{[\prec_1,\ldots,\prec_i]}(X)$ for all $i \leq n$ and all $x \in \Psi_{[\prec_1,\ldots,\prec_n]}(X)$.

---

[2] For the special case $i = 0$ let $\Psi_{[\prec_1,\ldots,\prec_0]}(X)$ denote X.

"$i = 1$": Let $x \in \Psi_{[\prec_1,\ldots,\prec_n]}(X)$. Assume $x \notin \Psi_{\prec_1}(Y)$. Hence, there is a $y \in \min_{\prec_1}(Y)$ such that $y \prec_1 x$. Hence, there is a $z \in X$ such that $z \prec_1 y$ (otherwise $x \notin \Psi_{\prec_1}(X)$). By Fact C.3.11, there is a $z' \in \Psi_{[\prec_1,\ldots,\prec_n]}(X)$ such that $z' \preceq_1 z$. Since $z' \in Y$ and $z' \prec_1 y$ this is a contradiction to the minimality of y.

"$i \Rightarrow i+1$": By the induction hypothesis $x \in \Psi_{[\prec_1,\ldots,\prec_j]}(Y)$ for every $j \leq i$ and every $x \in \Psi_{[\prec_1,\ldots,\prec_n]}(X)$. Let $x \in \Psi_{[\prec_1,\ldots,\prec_n]}(X)$. Assume $x \notin \Psi_{[\prec_1,\ldots,\prec_{i+1}]}(Y)$. Since $x \in \Psi_{[\prec_1,\ldots,\prec_i]}(Y)$, there is a $y \in \min_{\prec_{i+1}}(\Psi_{[\prec_1,\ldots,\prec_i]}(Y))$ such that $y \prec_{i+1} x$. By Fact C.3.9, $y \in \Psi_{[\prec_1,\ldots,\prec_n]}(Y)$ and hence by "$\supseteq$", $y \in \Psi_{[\prec_1,\ldots,\prec_n]}(X)$. Hence, $y \in \Psi_{[\prec_1,\ldots,\prec_i]}(X)$. Since $y \prec_{i+1} x$ and $x \in \Psi_{[\prec_1,\ldots,\prec_i]}(X)$, there is a $z \in \Psi_{[\prec_1,\ldots,\prec_i]}(X)$ such that $z \prec_{i+1} y$ (otherwise $y \in \min_{\prec_{i+1}}(\Psi_{[\prec_1,\ldots,\prec_i]}(X))$ and hence $x \notin \Psi_{[\prec_1,\ldots,\prec_{i+1}]}(X)$). We know by Fact C.3.10 that there is a $z' \in \Psi_{[\prec_1,\ldots,\prec_n]}(X)$ for which $z' \preceq_{i+1} z$ and hence $z' \prec_{i+1} y$. By the induction hypothesis, $z' \in \Psi_{[\prec_1,\ldots,\prec_i]}(Y)$,—a contradiction to the minimality of y. □

Fact C.3.13. *Where X is $\prec_n$-dense in Y: $\Psi_{[\prec_1,\ldots,\prec_n]}(X) = \Psi_{[\prec_1,\ldots,\prec_n]}(Y) \cap X$. ($\Psi_{[\prec_1,\ldots,\prec_n]}$ satisfies $\mathbf{DI}_{\prec_n}$.)*

Proof. We show this by induction for each $n \in \mathbb{N}$. The case "$i = 1$" has been shown in Fact C.3.3. "$n \Rightarrow n+1$": By Fact C.1.2, X is $\prec_n$-dense in Y. Hence, by the induction hypothesis, $\Psi_{[\prec_1,\ldots,\prec_n]}(X) = \Psi_{[\prec_1,\ldots,\prec_n]}(Y) \cap X$. Let $x \in \Psi_{[\prec_1,\ldots,\prec_{n+1}]}(X)$. Thus, $x \in \Psi_{[\prec_1,\ldots,\prec_n]}(X) = \Psi_{[\prec_1,\ldots,\prec_n]}(Y) \cap X$.

Assume $x \notin \Psi_{[\prec_1,\ldots,\prec_{n+1}]}(Y)$. Hence, there is a $y \in \min_{\prec_{n+1}}(\Psi_{[\prec_1,\ldots,\prec_n]}(Y))$ such that $y \prec_{n+1} x$.

Assume now that $y \in X$. Then $y \in \Psi_{[\prec_1,\ldots,\prec_{n+1}]}(X)$ since by Fact C.3.6 $\Psi_{[\prec_1,\ldots,\prec_{n+1}]}(X)$ is a $\prec_{n+1}$-lower set. Hence, there is a $y' \in X$ such that $y' \prec_{i+1} y$ (since otherwise $y \in \min_{\prec_{n+1}}(\Psi_{[\prec_1,\ldots,\prec_n]}(X))$ and hence $x \notin \Psi_{[\prec_1,\ldots,\prec_{n+1}]}(X)$). But then, since by Fact C.3.6 $\Psi_{[\prec_1,\ldots,\prec_{n+1}]}(Y)$ is a $\prec_{n+1}$-lower set, also $y' \in \Psi_{[\prec_1,\ldots,\prec_{n+1}]}(Y)$ in contradiction to the minimality of y. Hence, our assumption is false and $y \notin X$.

Since X is $\prec_{n+1}$-dense in Y, there is a $x' \in X$ such that $x' \prec_{n+1} y$. Since by Fact C.3.6, $\Psi_{[\prec_1,\ldots,\prec_{n+1}]}(Y)$ is a $\prec_{n+1}$-lower set, also $x' \in \Psi_{[\prec_1,\ldots,\prec_{n+1}]}(Y)$,—a contradiction to the minimality of y. Hence, our first assumption is false and $x \in \Psi_{[\prec_1,\ldots,\prec_{n+1}]}(Y)$.

Let now $x \in \Psi_{[\prec_1,\ldots,\prec_{n+1}]}(Y) \cap X$. Hence, $x \in \Psi_{[\prec_1,\ldots,\prec_n]}(Y) \cap X = \Psi_{[\prec_1,\ldots,\prec_n]}(X)$. Assume that $x \notin \Psi_{[\prec_1,\ldots,\prec_{n+1}]}(X)$. Hence there is a $y \in \min_{\prec_{n+1}}(\Psi_{[\prec_1,\ldots,\prec_n]}(X))$ such that $y \prec_{n+1} x$. Hence, $y \prec_n x$. Since $y \in \Psi_{[\prec_1,\ldots,\prec_n]}(X)$, by the induction hypothesis also $y \in \Psi_{[\prec_1,\ldots,\prec_n]}(Y)$. Hence, there is a $y' \in \Psi_{[\prec_1,\ldots,\prec_{n+1}]}(Y)$ such that $y' \prec_{n+1} y$ (since otherwise $y \in \min_{\prec_{n+1}}(\Psi_{[\prec_1,\ldots,\prec_n]}(Y))$ in which case $x \notin \Psi_{[\prec_1,\ldots,\prec_{n+1}]}(Y)$). Since X is $\prec_{n+1}$-dense in Y, there is a $x' \in X$ such that $x' \preceq_{n+1} y'$. Since by Fact C.3.6, $\Psi_{[\prec_1,\ldots,\prec_{n+1}]}(X)$ is a $\prec_{i+1}$-lower set and $x' \prec_{i+1} x$, also $x' \in \Psi_{[\prec_1,\ldots,\prec_{n+1}]}(X)$,—a contradiction to the minimality of y. □

Remark C.3.1. $\mathbf{RA}_{\prec_n}$ does in general not hold for $\Psi_{[\prec_1,\ldots,\prec_n]}$. Take as an example $\Psi_{[\prec_c,\subset]}$ and the lower limit logic $\mathbf{L}_\circ$. Let $\Gamma = \{\circ p, \circ q, \circ r, \neg p \vee \neg q, \neg p \vee \neg r\}$. It is easy to see that $\Psi_{\prec_c}(\mathsf{Ab}^{\Gamma}_{\mathbf{L}_\circ})$ yields $\{\{\circ p \wedge \neg p\}\}$ and thus $\Psi_{[\prec_c,\subset]}(\mathsf{Ab}^{\Gamma}_{\mathbf{L}_\circ})$

$= \{\{\circ p \wedge \neg p\}\}$. The latter set is not $\sqsubset$-dense in $\mathsf{Ab}^{\Gamma}_{\mathbf{L}_\circ}$ since for instance there is no $\varphi \in \Psi_{[\prec_c, \sqsubset]}(\mathsf{Ab}^{\Gamma}_{\mathbf{L}_\circ})$ such that $\varphi \subseteq \{\circ q \wedge \neg q, \circ r \wedge \neg r\}$.

Remark C.3.2. Fact C.3.3 does not generalize to arbitrary abstraction sequences. Take as an example the set $\mathsf{Ab}^{\Gamma_a \cup \{\neg q, \neg r\}}_{\mathbf{L}^*_\circ}$ given our Example in Sect. 5.2.6. It is easy to check that $\mathsf{Ab}^{\Gamma_a \cup \{\neg q, \neg r\}}_{\mathbf{L}^*_\circ}$ is $\prec_{\mathsf{co}}$-dense in $\mathsf{Ab}^{\Gamma_a}_{\mathbf{L}^*_\circ}$. Note that $\Psi_{[\prec_{\mathsf{co}}, \sqsubset]}(\mathsf{Ab}^{\Gamma_a \cup \{\neg q, \neg r\}}_{\mathbf{L}^*_\circ}) = \{\mathsf{Ab}(M^{q,r}_i) \mid i \in \mathbb{N}\}$ while $\Psi_{[\prec_{\mathsf{co}}, \sqsubset]}(\mathsf{Ab}^{\Gamma_a}_{\mathbf{L}^*_\circ}) \cap \mathsf{Ab}^{\Gamma_a \cup \{\neg q, \neg r\}}_{\mathbf{L}^*_\circ} = \emptyset$. This shows that in general $\Psi_{[\prec_1, \ldots, \prec_n]}$ does not satisfy $\mathbf{DI}_{\prec_1}$.

However, for a more restricted class of abstraction orders we can guarantee $\mathbf{DI}_{\prec_1}$. This in shown in the remainder of this section.

Definition C.3.1. Given two partial orders $\prec$ and $\prec'$ we say that $\prec'$ is *$\prec$-transitivity preserving* iff (i) $x \prec y \prec' z$ implies $x \prec' z$, and (ii) $x \prec' y \prec z$ implies $x \prec' z$.

An abstraction sequence $\langle \prec_1, \ldots, \prec_n \rangle$ is *transitivity preserving* iff for each $i \leq n$, $\prec_i$ is $\prec_j$-transitivity preserving for all $j < i$.

Fact C.3.14. *Where X is $\prec_1$-dense in Y and the abstraction sequence $\langle \prec_1, \ldots, \prec_n \rangle$ is transitivity preserving: $\Psi_{[\prec_1, \ldots, \prec_n]}(X) = \Psi_{[\prec_1, \ldots, \prec_n]}(Y) \cap X$.*

Proof. We show this by induction for any $n \in \mathbb{N}$. By Fact C.3.3 this holds for $n = 1$. "$n \Rightarrow n + 1$":

"$\subseteq$": Let $x \in \Psi_{[\prec_1, \ldots, \prec_{n+1}]}(X)$. Hence, $x \in \Psi_{[\prec_1, \ldots, \prec_n]}(X)$, $x \in X$, and by the induction hypothesis, $x \in \Psi_{\prec_{n+1}}(\Psi_{[\prec_1, \ldots, \prec_n]}(Y) \cap X)$. Thus, $x \in \Psi_{[\prec_1, \ldots, \prec_n]}(Y)$. Assume $x \notin \Psi_{[\prec_1, \ldots, \prec_{n+1}]}(Y)$. Hence, there is a $y \in \min_{n+1}(\Psi_{[\prec_1, \ldots, \prec_n]}(Y))$ for which $y \prec_{n+1} x$. Since X is $\prec_1$-dense in Y, there is a $z \in X$ for which $z \preceq_1 y$. Hence, since $\prec_{n+1}$ is a transitivity preserving abstraction order of $\prec_1$ and $\prec_n$, $z \prec_{n+1} x$ and thus $z \prec_n x$. By Fact C.3.6 and since $x \in \Psi_{[\prec_1, \ldots, \prec_n]}(Y)$, $z \in \Psi_{[\prec_1, \ldots, \prec_n]}(Y) \cap X$. Now assume $z \notin \min_{\prec_{n+1}}(\Psi_{[\prec_1, \ldots, \prec_n]}(Y) \cap X)$. Hence, there is a $z' \in \Psi_{[\prec_1, \ldots, \prec_n]}(Y) \cap X$ for which $z' \prec_{n+1} z$. Since $\prec_{n+1}$ is a transitivity preserving abstraction order of $\prec_1$ and since $z \preceq_1 y$, $z' \prec_{n+1} y$,—a contradiction to the minimality of y. But then, $z \in \min_{\prec_{n+1}}(\Psi_{[\prec_1, \ldots, \prec_n]}(Y) \cap X)$. However, then $x \notin \Psi_{\prec_{n+1}}(\Psi_{[\prec_1, \ldots, \prec_n]}(Y) \cap X)$,—a contradiction. Hence, our first assumption is false and $x \in \Psi_{[\prec_1, \ldots, \prec_{n+1}]}(Y)$.

"$\supseteq$": Let $x \in \Psi_{[\prec_1, \ldots, \prec_{n+1}]}(Y) \cap X$. Hence, $x \in \Psi_{[\prec_1, \ldots, \prec_n]}(Y) \cap X$. Assume $x \notin \Psi_{[\prec_1, \ldots, \prec_{n+1}]}(X)$ and hence by the induction hypothesis, $x \notin \Psi_{\prec_{n+1}}(\Psi_{[\prec_1, \ldots, \prec_n]}(Y) \cap X)$. Hence, there is a $y \in \min_{\prec_{n+1}}(\Psi_{[\prec_1, \ldots, \prec_n]}(Y) \cap X)$ such that $y \prec_{n+1} x$. Since $x \in \Psi_{[\prec_1, \ldots, \prec_{n+1}]}(Y)$, $y \notin \min_{\prec_{n+1}}(\Psi_{[\prec_1, \ldots, \prec_n]}(Y))$. Thus, there is a $z \in \Psi_{[\prec_1, \ldots, \prec_n]}(Y) \setminus X$ such that $z \prec_{n+1} y$. Since X is $\prec_1$-dense in Y there is a $z' \in X$ such that $z' \prec_1 z$. Since $\prec_{n+1}$ is a transitivity preserving abstraction order of $\prec_1$, also $z' \prec_{n+1} y$ and thus $z' \prec_n y$. By Fact C.3.6 and since $y \in \Psi_{[\prec_1, \ldots, \prec_n]}(Y)$, $z' \in \Psi_{[\prec_1, \ldots, \prec_n]}(Y) \cap X$,—a contradiction to the minimality of y. □

C.4 Simplifying the Proof Theory: Some Meta-Proofs

Let in the following $Cn_{\mathbf{AL}^*_\Lambda}(\Gamma) =_{\text{df}} \{A \mid \Gamma \vdash_{\mathbf{AL}^*_\Lambda} A\}$. An $\mathbf{AL}^*_\Lambda$ proof is just like an $\mathbf{AL}_\Lambda$ proof except that in the marking definition we set $\Lambda^\Gamma_s = \Lambda(\Xi_s(\Gamma))$.

Fact C.4.1. $\Xi(\Gamma) \neq \emptyset \neq \Lambda(\Xi(\Gamma))$

Note that $\Xi(\Gamma) \neq \emptyset$ by definition. The rest follows by Fact 5.3.1.[3]

Lemma C.4.1. *Where $\Gamma \subseteq \mathcal{W}^+$ and $\Gamma \vdash_{\mathbf{AL}^*_\Lambda} A$:*

(i) A is derivable on a line l of a finite $\mathbf{AL}^*_\Lambda$ proof from Γ on a condition Δ such that $\Delta \cap \varphi = \emptyset$ for a $\varphi \in \Lambda(\Xi(\Gamma))$.
(ii) For every $\varphi \in \Lambda(\Xi(\Gamma))$ there is a finite $\Delta \subseteq \Omega \setminus \varphi$ for which $\Gamma \vdash_{\mathbf{LLL}} A \check{\vee} \mathsf{Dab}(\Delta)$.

Proof. Suppose $\Gamma \vdash_{\mathbf{AL}^*_\Lambda} A$. Hence, there is a finite $\mathbf{AL}^*_\Lambda$-proof $\mathcal{P}$ from Γ such that (1) A is derived in this proof on an unmarked line l with a condition Δ, and (2) every extension of the proof in which l is marked can be further extended such that l is unmarked again. We now extend our proof $\mathcal{P}$ to the complete stage $\mathbf{g}(\mathcal{P})$. Note that $\Lambda(\Xi(\Gamma)) = \Lambda\big(\Xi_{\mathbf{g}(\mathcal{P})}(\Gamma)\big) = \Lambda\big(\Xi_{s'}(\Gamma)\big)$ for every later stage s'.

Ad (i): By Fact C.4.1, $\Lambda(\Xi(\Gamma)) \neq \emptyset$. Assume there is no $\varphi \in \Lambda(\Xi(\Gamma))$ such that $\Delta \cap \varphi = \emptyset$. By Definition 5.3.4, line l is marked at stage $\mathbf{g}(\mathcal{P})$ and hence at every later stage s'—a contradiction to (2). *Ad (ii):* Assume there is a $\varphi \in \Lambda(\Xi(\Gamma))$ for which there is no $\Delta \subseteq \Omega$ such that $\Gamma \vdash_{\mathbf{LLL}} A \check{\vee} \mathsf{Dab}(\Delta)$ and $\Delta \cap \varphi = \emptyset$. By Definition 5.3.4.ii line l is marked at stage $\mathbf{g}(\mathcal{P})$ and hence at every later stage s'. This contradicts (2). □

Lemma C.4.2. *Where $\Gamma \subseteq \mathcal{W}$: If $\Gamma \vdash_{\mathbf{LLL}} A \check{\vee} \mathsf{Dab}(\Delta)$ and $\Delta \cap \varphi = \emptyset$ for a $\varphi \in \Lambda(\Xi(\Gamma))$, then there is a finite $\mathbf{AL}^*_\Lambda$-proof from Γ in which A is derived on the condition Δ at an unmarked line.*

Proof. Perfectly analogous to the proof of Lemma 5.4.3: just replace all occurrences of $\Xi^{\text{sat}}(\Gamma)$ by $\Xi(\Gamma)$ and all occurrences of $\Xi^{\text{sat}}_s(\Gamma)$ by $\Xi_s(\Gamma)$.

Lemma C.4.3. *Where $\Gamma \subseteq \mathcal{W}$: If for every $\varphi \in \Lambda(\Xi(\Gamma))$ there is a finite $\Delta_\varphi \subseteq \Omega \setminus \varphi$ such that $\Gamma \vdash_{\mathbf{LLL}} A \check{\vee} \mathsf{Dab}(\Delta_\varphi)$, then $\Gamma \vdash_{\mathbf{AL}^*_\Lambda} A$.*

Proof. Suppose the antecedent is true. By Lemma C.4.2, for every Δ_φ there is a finite $\mathbf{AL}^*_\Lambda$-proof from Γ in which A is derived on the condition Δ_φ at an unmarked line l. Given any such proof (by Fact C.4.1 $\Lambda(\Xi(\Gamma)) \neq \emptyset$ and hence there is such a Δ_φ), suppose the proof is extended to a stage s in which l is marked. Call this proof $\mathcal{P}$. We extend the proof further to the stage $\mathbf{g}(\mathcal{P})$. Note that for all $\varphi \in \Lambda(\Xi(\Gamma))$, A has been derived on the condition Δ_φ at this stage. By Definition 5.3.4, line l is unmarked at stage $\mathbf{g}(\mathcal{P})$. □

---

[3] Note that that Fact 5.3.1 is applicable since $\Xi(\Gamma) \in \Upsilon$ and $\Xi_s(\Gamma) \in \Upsilon$ by Lemma C.2.3 and Lemma C.2.2. We will in the following make use of these Lemmas without further notice.

By Lemma C.4.1 and Lemma C.4.3 we immediately get:

Theorem C.4.1. *Where* $\Gamma \subseteq \mathcal{W}$*:* $\Gamma \vdash_{\mathbf{AL}^{*}_{\Lambda}} A$ *iff for every* $\varphi \in \Lambda(\Xi(\Gamma))$*,* $\Gamma \cup (\Omega \setminus \varphi)^{\check{\neg}} \vdash_{\mathbf{LLL}} A$.

This representational result can be further strengthened:

Theorem C.4.2. *Where* $\Gamma \subseteq \mathcal{W}$*:* $\Gamma \vdash_{\mathbf{AL}^{*}_{\Lambda}} A$ *iff for every* $\varphi \in \Lambda(\Xi(\Gamma)) \setminus \Xi^{\perp}(\Gamma)$*,* $\Gamma \cup (\Omega \setminus \varphi)^{\check{\neg}} \vdash_{\mathbf{LLL}} A$.

Proof. Let $\varphi \in \Lambda(\Xi(\Gamma)) \cap \Xi^{\perp}(\Gamma)$. $\Gamma \cup (\Omega \setminus \varphi)^{\check{\neg}} \vdash_{\mathbf{LLL}} A$ iff [by the soundness and completeness of **LLL**] $\Gamma \cup (\Omega \setminus \varphi)^{\check{\neg}} \Vdash_{\mathbf{LLL}} A$ iff for all $M \in \mathcal{M}_{\mathbf{LLL}}\left(\Gamma \cup (\Omega \setminus \varphi)^{\check{\neg}}\right)$, $M \models A$ iff for all $M \in \mathcal{M}_{\mathbf{LLL}}(\Gamma)$ for which $\mathrm{Ab}(M) \subseteq \varphi$, $M \models A$ iff $\Gamma \cup (\Omega \setminus \psi)^{\check{\neg}} \Vdash_{\mathbf{LLL}} A$ for all $\psi \in \{\delta \in \mathsf{Ab}^{\Gamma}_{\mathbf{LLL}} \mid \delta \subseteq \varphi\}$ iff [by Lemma 5.3.2 and by the soundness and completeness of **LLL**] $\Gamma \cup (\Omega \setminus \psi)^{\check{\neg}} \vdash_{\mathbf{LLL}} A$ for all $\psi \in \{\delta \in \Xi(\Gamma) \setminus \Xi^{\perp}(\Gamma) \mid \delta \subset \varphi\}$ iff [by **T2** and since $\varphi \in \Lambda(\Xi(\Gamma))$] $\Gamma \cup (\Omega \setminus \psi)^{\check{\neg}} \vdash_{\mathbf{LLL}} A$ for all $\psi \in X_{\varphi}$ and $X_{\varphi} =_{\mathrm{df}} \{\psi \in \Lambda(\Xi(\Gamma)) \setminus \Xi^{\perp}(\Gamma) \mid \psi \subset \varphi\} \subseteq \Lambda(\Xi(\Gamma)) \setminus \Xi^{\perp}(\Gamma)$.

Hence, for each $\varphi \in \Lambda(\Xi(\Gamma)) \cap \Xi^{\perp}(\Gamma)$ there is a $X_{\varphi} \subseteq \Lambda(\Xi(\Gamma)) \setminus \Xi^{\perp}(\Gamma)$ such that (†) $Cn_{\mathbf{LLL}}\left(\Gamma \cup (\Omega \setminus \varphi)^{\check{\neg}}\right) = \bigcap_{\psi \in X_{\varphi}} Cn_{\mathbf{LLL}}\left(\Gamma \cup (\Omega \setminus \psi)^{\check{\neg}}\right)$.

By Theorem C.4.1,

$$Cn_{\mathbf{AL}^{*}_{\Lambda}}(\Gamma) = \bigcap_{\varphi \in \Lambda(\Xi(\Gamma))} Cn_{\mathbf{LLL}}\left(\Gamma \cup (\Omega \setminus \varphi)^{\check{\neg}}\right). \tag{1}$$

Obviously,

$$\bigcap_{\varphi \in \Lambda(\Xi(\Gamma))} Cn_{\mathbf{LLL}}\left(\Gamma \cup (\Omega \setminus \varphi)^{\check{\neg}}\right) = \bigcap_{\varphi \in \Lambda(\Xi(\Gamma)) \setminus \Xi^{\perp}(\Gamma)} Cn_{\mathbf{LLL}}\left(\Gamma \cup (\Omega \setminus \varphi)^{\check{\neg}}\right) \cap \bigcap_{\varphi \in \Lambda(\Xi(\Gamma)) \cap \Xi^{\perp}(\Gamma)} Cn_{\mathbf{LLL}}\left(\Gamma \cup (\Omega \setminus \varphi)^{\check{\neg}}\right) \tag{2}$$

By (†),

$$\bigcap_{\varphi \in \Lambda(\Xi(\Gamma)) \cap \Xi^{\perp}(\Gamma)} Cn_{\mathbf{LLL}}\left(\Gamma \cup (\Omega \setminus \varphi)^{\check{\neg}}\right) = \bigcap_{\varphi \in \Lambda(\Xi(\Gamma)) \cap \Xi^{\perp}(\Gamma)} \bigcap_{\psi \in X_{\varphi}} Cn_{\mathbf{LLL}}\left(\Gamma \cup (\Omega \setminus \psi)^{\check{\neg}}\right) \supseteq \bigcap_{\varphi \in \Lambda(\Xi(\Gamma)) \setminus \Xi^{\perp}(\Gamma)} Cn_{\mathbf{LLL}}\left(\Gamma \cup (\Omega \setminus \varphi)^{\check{\neg}}\right) \tag{3}$$

Hence, by (1), (2) and (3):

$$Cn_{\mathbf{AL}^{*}_{\Lambda}}(\Gamma) = \bigcap_{\varphi \in \Lambda(\Xi(\Gamma)) \setminus \Xi^{\perp}(\Gamma)} Cn_{\mathbf{LLL}}\left(\Gamma \cup (\Omega \setminus \varphi)^{\check{\neg}}\right)$$

□

Theorem 5.6.1 ***(restated).*** *Where $\Gamma \subseteq \mathcal{W}$: If Λ satisfies* **SIMP**, *then $\Gamma \vdash_{\mathbf{AL}_{\Lambda}} A$ iff $\Gamma \vdash_{\mathbf{AL}^{*}_{\Lambda}} A$ iff $\Gamma \Vdash_{\mathbf{AL}_{\Lambda}} A$.*

Proof. This follows immediately by Theorem C.4.2, Theorem 5.4.5 and Corollary 5.4.2. □

Lemma C.4.4. *Where $\Gamma \subseteq \mathcal{W}^{+}$ is* **LLL**-*trivial: $\Lambda(\Xi^{\text{sat}}(\Gamma)) = \Lambda(\Xi(\Gamma)) \setminus \Xi^{\perp}(\Gamma)$.*

Proof. By Fact C.2.1, $\Xi^{\text{sat}}(\Gamma) = \emptyset$ and hence by Fact 5.3.1 also (1) $\Lambda(\Xi^{\text{sat}}(\Gamma)) = \emptyset$. By Fact C.2.1, $\Xi(\Gamma) = \{\Omega\}$. Hence, by **T2** and **T3**, $\Lambda(\Xi(\Gamma)) = \{\Omega\}$. By Fact C.2.1, $\Xi^{\perp}(\Gamma) = \{\Omega\}$. Hence, (2) $\Lambda(\Xi(\Gamma)) \setminus \Xi^{\perp}(\Gamma) = \emptyset$. By (1) and (2), $\Lambda(\Xi^{\text{sat}}(\Gamma)) = \Lambda(\Xi(\Gamma)) \setminus \Xi^{\perp}(\Gamma)$. □

Theorem C.4.3. *Where $\Gamma \subseteq \mathcal{W}^{+}$: If Λ satisfies* **C1** *or* **DI**$_{\prec}$, *then $\Lambda(\Xi^{\text{sat}}(\Gamma)) = \Lambda(\Xi(\Gamma)) \setminus \Xi^{\perp}(\Gamma)$.*

Proof. Where Γ is **LLL**-trivial this has been shown in Lemma C.4.4. Let thus Γ be **LLL**-non-trivial.

Ad **C1**: this follows with the help of Corollary C.1.1 (where $\prec\ =\ \subset$): Since $\min_{\subset}(\Xi^{\text{sat}}(\Gamma)) = \min_{\subset}(\Xi(\Gamma))$, by **C1**, $\Lambda(\Xi^{\text{sat}}(\Gamma)) \cap \Xi(\Gamma) = \Lambda(\Xi(\Gamma)) \cap \Xi^{\text{sat}}(\Gamma)$. By **T1**, $\Lambda(\Xi^{\text{sat}}(\Gamma)) = \Lambda(\Xi(\Gamma)) \setminus \Xi^{\perp}(\Gamma)$. *Ad* **DI**$_{\prec}$: this follows by Fact C.3.1. □

Appendix D
Appendix to Chapter 6

D.1 Some proofs

Lemma D.1.1. *The core properties,* (rMP) *and* (Inh) *entail* $\big(A_1 \wedge (A_1 \leadsto \ldots \leadsto A_n \leadsto B)\big) \supset (B \vee \bullet A_n)$.

Proof. Due to A_1, $A_1 \leadsto A_2$ and (rMP) we have $A_2 \vee \bullet A_1$. Analogously we get $A_3 \vee \bullet A_2 \vee \bullet A_1$ and finally $B \vee \bullet A_n \vee \cdots \vee \bullet A_1$. By iterated applications of (Inh) we get $B \vee \bullet A_n$. □

Fact 6.3.1 (restated). (rMP), (Inh) *and the core properties entail*

$$\vdash \big(A \wedge (A \leadsto B \leadsto C) \wedge (A \leadsto \neg C)\big) \supset \bullet B \qquad \text{(Spe2)}$$

Proof. By means of A, $A \leadsto B \leadsto C$ and the lemma, (†) $C \vee \bullet B$. By means of A and $A \leadsto \neg C$, by (rMP) $\neg C \vee \bullet A$. Since $A \leadsto B$, by (Inh), $\bullet A \supset \bullet B$. Hence, (‡) $\neg C \vee \bullet B$. By (†) and (‡), $dotu B$. □

Fact 6.4.1 (restated). *The core properties, (*rMP*) and (*Inh*) imply (*Spe1*), (*Spe2*) and the following:*

$$\textit{If} \vdash A \supset B, \textit{ then} \vdash \big(A \wedge (B \leadsto C) \wedge (A \leadsto \neg C)\big) \supset \bullet B \qquad \text{(sSpe)}$$

$$\vdash \big(A \wedge (A \leadsto B_1 \leadsto \ldots \leadsto B_n \leadsto C) \wedge (A \leadsto \neg C)\big) \supset \bullet B_n \qquad \text{(SpeG)}$$

$$\begin{aligned}\vdash \big(&A \wedge (A \leadsto B_1 \leadsto \ldots \leadsto B_n \leadsto D) \wedge \\ &(A \leadsto C_1 \leadsto \ldots \leadsto C_m \leadsto \neg D) \wedge (B_n \leadsto \ldots \leadsto C_m)\big) \supset \bullet C_m\end{aligned} \qquad \text{(PreE)}$$

$$\textit{If} \vdash \neg \textstyle\bigwedge_I D_i, \textit{ then} \vdash \Big(A \wedge \textstyle\bigwedge_I (A \leadsto \ldots \leadsto B_i \leadsto D_i)\Big) \supset \textstyle\bigvee_I \bullet B_i \qquad \text{(Conf)}$$

Proof. "(Spe1)": this is trivial. "(SpeG)": By Lemma D.1.1, $C \vee \bullet B_n$ and $\neg C \vee \bullet A$. By multiple applications of (Inh), $\bullet A \supset \bullet B_n$. Hence $\neg C \vee \bullet B_n$. Thus, $\bullet B_n$. (Spe2)

C. Straßer, *Adaptive Logics for Defeasible Reasoning*, Trends in Logic 38,
DOI: 10.1007/978-3-319-00792-2, © Springer International Publishing Switzerland 2014

and (sSpe) follow immediately with (SpeG) and (CI). "(PreE)": By Lemma D.1.1, $D \vee \bullet B_n$ and $\neg D \vee \bullet C_m$. By multiple applications of (Inh), $\bullet B_n \supset \bullet C_m$. Hence, $D \vee \bullet C_m$. Thus, $\bullet C_m$. "(Conf)": By Lemma D.1.1, $D_i \vee \bullet B_i$. Due to $\neg \bigwedge_I D_i$ and by classical logic, $\bigvee_I \bullet B_i$. □

Fact 6.5.1 (restated). *The core properties,* (RM), (rMP) *and* (Inh) *imply*

$$\vdash \big(A \wedge (A \rightsquigarrow B \rightsquigarrow C) \wedge (A \not\rightsquigarrow C)\big) \supset \bullet B \qquad \text{(Spe')}$$

$$\vdash \big(A \wedge (A \rightsquigarrow B_1 \rightsquigarrow \ldots \rightsquigarrow B_n \rightsquigarrow C) \wedge (A \not\rightsquigarrow C)\big) \supset \bullet B_n \qquad \text{(SpeG')}$$

$$\vdash \big(A \wedge (A \rightsquigarrow B_1 \rightsquigarrow \ldots \rightsquigarrow B_n) \wedge (B_n \not\rightsquigarrow D) \wedge$$
$$(A \rightsquigarrow C_1 \rightsquigarrow \ldots \rightsquigarrow C_m \rightsquigarrow D) \wedge (B_n \rightsquigarrow \ldots \rightsquigarrow C_m)\big) \supset \bullet C_m \qquad \text{(PreE')}$$

Proof. "(Spe')": Suppose $\neg\bullet B$. Then, due to (Spe1), $B \not\rightsquigarrow \neg A$. By (RM), $(A \wedge B) \rightsquigarrow C$. But then by (RT), $A \rightsquigarrow C$,—a contradiction. "(SpeG')": Suppose $\neg\bullet B_n$. Hence, due to (Inh), $\neg\bullet B_i$ for all $i \rightsquigarrow n$ and $\neg\bullet A$. Hence, due to (Spe'), $A \rightsquigarrow B_i$ for all $i \leq n$ (otherwise, $\bullet B_i$). But then by (Spe'), $\bullet B_n$,—a contradiction. "(PreE')": similar and left to the reader. □

D.2 The Semantics

I focused in this chapter on the base logics $\mathbf{L} \in \{\mathbf{P}, \mathbf{R}\}$. There are many semantics around for the core properties (see Footnote 2). Paradigmatically I will extend the semantics based on preferential models (see [1]) for our lower limit logics **Lp**. Again there are various ways to enhance preferential models such as to serve as semantical representations of **Lp**. I am going to present versions which are technically straightforward. In this appendix I will cover the case for $\mathbf{L} = \mathbf{P}$ and hence the logic **Pp**. However, for **Rp** the semantics are defined analogously.

We call interpretations $\mathcal{W} \rightarrow \{0, 1\}$ which satisfy the classical truth conditions for $\wedge, \vee, \neg, \supset$ and $\equiv$ *classical propositional worlds* and write $\mathcal{U}$ for the set of all classical propositional worlds.

Definition D.2.1. Let $\prec$ be a partial order on a set U and $V \subseteq U$. We say that $x \in V$ is *minimal* in V iff there is no $y \in V$, such that $y \prec x$. We shall say that V is *smooth* iff for all $x \in V$, either there is a y minimal in V, such that $y \prec x$ or x is itself minimal.

Definition D.2.2. A *preferential model* M is a triple $\langle S, l, \prec\rangle$ where S is a set, the elements of which will be called states, $l : S \rightarrow \mathcal{U}$ assigns a classical propositional world to each state and $\prec$ is a strict partial order on S satisfying the following *smoothness condition*: forall $A \in \mathcal{W}$, the set of states $\hat{A} =_{\text{df}} \{s \mid s \in S, s \models\!\equiv A\}$ is smooth, where $\models$ is defined as $s \models A$ (read s satisfies A) iff $l(s)(A) = 1$. M validates $A \rightsquigarrow B$, in signs $M \models A \rightsquigarrow B$, iff, for any s minimal in $\hat{A}$, $s \models B$. For the classical connectives $\models$ is defined as usual:

$$M \models A \vee B \text{ iff } M \models A \text{ or } M \models B \quad \text{(S-}\vee\text{)}$$

$$M \models A \wedge B \text{ iff } M \models A \text{ and } M \models B \quad \text{(S-}\wedge\text{)}$$

$$M \models \neg A \text{ iff } M \not\models A \quad \text{(S-}\neg\text{)}$$

$$M \models A \supset B \text{ iff } M \models \neg A \vee B \quad \text{(S-}\supset\text{)}$$

$$M \models A \equiv B \text{ iff } M \models A \supset B \text{ and } M \models B \supset A \quad \text{(S-}\equiv\text{)}$$

where A and B are in the $(\wedge, \vee, \supset, \neg, \equiv)$-closure of $\mathcal{W}^{\rightsquigarrow}$ and $\mathcal{W}^{\rightsquigarrow}$ is the set of all conditionals.

Let $\mathcal{W}^{\bullet}$ be the set of all formulas of the form $\bullet A$. Let $\mathcal{P}$be the $(\wedge, \vee, \supset, \neg, \equiv)$-closure of $\mathcal{W} \cup \mathcal{W}^{\bullet} \cup \mathcal{W}^{\rightsquigarrow}$. We have two tasks in order to define the semantics for **Pp**. On the one hand, preferential models have to be generalized in order to allow for the modeling of factual premises. On the other hand, the new rules (rMP) and (Inh) have to be taken into account. We will realize both requirements by introducing an actual world to the preferential models defined above.

Definition D.2.3. *A preferential$_c$ model M with an actual world is defined by* $\langle S, l, \prec, @\rangle$ where $M' = \langle S, l, \prec\rangle$ is a preferential model and @ is an interpretation $\mathcal{P} \rightarrow \{0, 1\}$ such that the classical clauses (where now $A, B \in \mathcal{P}$) (M-$\vee$), (M-$\wedge$), (S-$\neg$), (M-$\supset$), and (S-$\equiv$) hold and the following requirements are satisfied:

$$M' \models A \rightsquigarrow B \text{ iff } @(A \rightsquigarrow B) = 1 \quad \text{(S-@)}$$

$$\text{If } @(A) = @(A \rightsquigarrow B) = 1 \text{ and } @(\bullet A) = 0, \text{ then } @(B) = 1 \quad \text{(S-rMP)}$$

$$\text{If } @(\bullet A) = @(A \rightsquigarrow B) = 1, \text{ then } @(\bullet B) = 1 \quad \text{(S-Inh)}$$

We define $M \models \varphi$ iff $@\ (\varphi) = 1$. We denote the corresponding semantic consequence relation by $\Vdash^{\mathbf{p}}_{\mathbf{P}}$ which is defined in the usual way: $\Gamma \Vdash^{\mathbf{p}}_{\mathbf{P}} \varphi$ iff all preferential$_c$ models M with an actual world that verify all members of Γ also verify φ.

Lemma D.2.1. *Let $\Gamma \subset \mathcal{P}$ be a* **Pp***-consistent premise set. There is a preferential$_c$ model M with an actual world for which $M \models \Gamma$.*

Proof. *(Sketch of the proof).* Let Γ' be a maximal consistent (w.r.t. **Pp**) extension of Γ. Take any preferential model M' of $\Gamma' \cap \mathcal{W}^{\rightsquigarrow}$. Obviously such a model exists since $\Gamma' \cap \mathcal{W}^{\rightsquigarrow}$ is **Pp**-consistent. Let @ be definedby $@\ (A) = 1$ iff $A \in \Gamma'$. Let

$M = \langle M', @\rangle$. Obviously @ fulfills the requirements (S-@), (S-rMP), (S-Inh), the classical properties and the core properties. □

Theorem D.2.1. *If* $\Gamma \vdash_{\mathbf{Pp}} \varphi$ *then* $\Gamma \Vdash^{\mathbf{p}}_{\mathbf{P}} \varphi$.

Proof. *(Sketch of the proof).* The proof proceeds via an induction over the derivative steps constituting a proof of φ.

"$n = 1$": If φ is derived by a core rule $\mathcal{R}$, then the antecedents of the rule are valid in all models $M = \langle M', @\rangle$ of Γ since they are in Γ and due to the fact that M' is a preferential model, φ is also valid in M'. By (S-@), φ is valid in M. If $\varphi = B$ has been derived by (rMP) from A, $A \rightsquigarrow B$, and $\neg\bullet A$, then $A, A \rightsquigarrow B, \neg\bullet A \in \Gamma$. By (S-rMP) and (S-@), B is valid in all models. For (S-Inh) and the classical rules the argument is similar.

"$n \rightarrow n + 1$": Let φ be derived by a core rule $\mathcal{R}$. All antecedents of the rule are valid in all models $M = \langle M', @\rangle$ of Γ and since M' is a preferential model, also the consequent of $\mathcal{R}$is valid in M'. By (S-@), φ is valid in M. If $\varphi = B$ has been derived by (rMP) from A, $A \rightsquigarrow B$, and $\neg\bullet A$, then $\Gamma \Vdash^{\mathbf{p}}_{\mathbf{P}} A, A \rightsquigarrow B, \neg\bullet A$. By (S-rMP), B is valid in all models. For (S-Inh) and the classical rules the argument is similar. □

Theorem D.2.2. *If* $\Gamma \Vdash^{\mathbf{p}}_{\mathbf{P}} \varphi$ *then* $\Gamma \vdash_{\mathbf{Pp}} \varphi$.

Proof. Suppose $\Gamma \nvdash_{\mathbf{Pp}} \varphi$, then $\Gamma \cup \{\neg\varphi\}$ is **Pp**-consistent. Thus, by Lemma D.2.1, there is a preferential$_c$ model with an actual world for $\Gamma \cup \{\neg\varphi\}$. □

So far I have presented the semantics for the ALs based on the core properties, i.e., based on **P**. For **Rp** the semantics are defined analogously. Instead of preferential models, ranked models are used. Ranked models are preferential models for which $\prec$ is modular (see [2] and Sect. 7.1.3 for details). The completeness and soundness results are shown analogously. The easy meta-proofs are left to the reader.

Appendix E
Appendix to Chapter 7

Fact 7.1.1 (restated). *Where* $\mid\!\sim$ *is a preferential consequence relation:*

(i) $\alpha < \beta$ *implies* $\alpha \mid\!\sim \alpha \wedge \neg\beta$, $\alpha \mid\!\sim \neg\beta$, *and* $\alpha \vee \beta \mid\!\sim \alpha$
(ii) $\alpha \mid\!\sim \beta$ *and* $\alpha < \gamma$ *imply* $\beta < \gamma$
(iii) $\alpha < \beta$ *and* $\beta < \gamma$ *imply* $\alpha < \gamma$

Proof. Ad (i): Suppose $\alpha < \beta$. Then, $\alpha \vee \beta \mid\!\sim \neg\beta$. Since also $\alpha \vee \beta \mid\!\sim \alpha \vee \beta$ and by (And), $\alpha \vee \beta \mid\!\sim \alpha \wedge \neg\beta$. By (Right Weakening), $\alpha \vee \beta \mid\!\sim \alpha$. Since by (Reflexivity) and (Right Weakening) also $\alpha \mid\!\sim \alpha \vee \beta$, we get by (Equivalence), $\alpha \mid\!\sim \alpha \wedge \neg\beta$ and $\alpha \mid\!\sim \neg\beta$ by (Right Weakening).

Ad (ii): Suppose $\alpha \mid\!\sim \beta$ and $\alpha < \gamma$. Hence, $\alpha \vee \gamma \mid\!\sim \neg\gamma$ and by (i), $\alpha \vee \gamma \mid\!\sim \alpha$. Evidently also $\alpha \mid\!\sim \alpha \vee \gamma$. Since $\alpha \mid\!\sim \beta$ by (EQ) also $\alpha \vee \gamma \mid\!\sim \beta$. Hence, by (Right Weakening), $\alpha \vee \gamma \mid\!\sim \beta \vee \gamma$. By (Cautious Monotonicity), $(\alpha \vee \gamma) \wedge (\gamma \vee \beta) \mid\!\sim \neg\gamma$. By (S), $\gamma \vee \beta \mid\!\sim (\alpha \vee \gamma) \supset \neg\gamma$ and hence $\gamma \vee \beta \mid\!\sim \neg\gamma$ which is the same as $\beta < \gamma$.

Ad (iii): Suppose (1) $\alpha < \beta$ and (2) $\beta < \gamma$. By (2), (i) and (Right Weakening), $\beta \vee \gamma \mid\!\sim \beta$ and $\beta \mid\!\sim \beta \vee \gamma$. Since $\beta \mid\!\sim \alpha \vee \beta$ and by (Equivalence), (3) $\beta \vee \gamma \mid\!\sim \alpha \vee \beta$. By (2) and (Cautious Monotonicity), $(\beta \vee \gamma) \wedge (\alpha \vee \beta) \mid\!\sim \neg\gamma$. By (S), $\alpha \vee \beta \mid\!\sim \neg(\beta \vee \gamma) \vee \neg\gamma$ and hence $\alpha \vee \beta \mid\!\sim \neg\gamma$. Since also $\beta \vee \gamma \mid\!\sim \neg\gamma$, by (OR), $\alpha \vee \beta \vee \gamma \mid\!\sim \neg\gamma$. By (Reflexivity) and (Right Weakening), $\alpha \vee \beta \mid\!\sim \alpha \vee \beta \vee \gamma$. Also, by (3) and since by (Reflexivity) $\alpha \vee \beta \mid\!\sim \alpha \vee \beta$, by (Or) $\alpha \vee \beta \vee \gamma \mid\!\sim \alpha \vee \beta$. By (1) and (i) $\alpha \vee \beta \mid\!\sim \alpha$. Thus, by (Equivalence), $\alpha \vee \beta \vee \gamma \mid\!\sim \alpha$. By (Reflexivity) and (Right Weakening), $\alpha \mid\!\sim \alpha \vee \beta \vee \gamma$. By (Reflexivity) and (Right Weakening), $\alpha \mid\!\sim \alpha \vee \gamma$. Hence, by (Equivalence), $\alpha \vee \beta \vee \gamma \mid\!\sim \alpha \vee \gamma$. By (Reflexivity) and (Right Weakening), $\alpha \vee \gamma \mid\!\sim \alpha \vee \beta \vee \gamma$. Since $\alpha \vee \beta \vee \gamma \mid\!\sim \neg\gamma$, by (Equivalence), $\alpha \vee \gamma \mid\!\sim \neg\gamma$. □

Fact 7.2.4 (restated). $\alpha \mid\!\sim_M l_0$

Proof. There are two possible cases: either $\sigma_M(\alpha) = \emptyset$ or $\sigma_M(\alpha) \neq \emptyset$. In the former case by definition $\alpha \mid\!\sim_M l_0$. In the latter case let $i = \mathsf{AbDeg}_{\prec}(\alpha)$. By (R) (i), $w \models l_0$ for all $w \in \min_{\prec}(\sigma_M(\alpha))$. Hence, $\alpha \mid\!\sim_M l_0$. □

Fact 7.2.5 (restated). $\mathsf{AbDeg}_{\prec}(l_i) = i$ *iff* $\sigma_M(l_i) \neq \emptyset$, *else* $\mathsf{AbDeg}_{\prec}(l_i) = \omega$.

C. Straßer, *Adaptive Logics for Defeasible Reasoning*, Trends in Logic 38,
DOI: 10.1007/978-3-319-00792-2, © Springer International Publishing Switzerland 2014

Proof. "⇒": By the definition of $\mathsf{AbDeg}_{\prec}$.

"⇐": Let $\sigma_M(l_i) \neq \emptyset$. Assume $\mathsf{AbDeg}_{\prec}(l_i) < i$. Then by (R) (ii), for all $w \in \min_{\prec}(\sigma_M(l_i))$, $w \models \neg l_i$,—a contradiction. By (R) (i), for all w for which $\mathsf{rank}_{\prec}(w) = i$, $w \models l_i$. Hence, $\mathsf{AbDeg}_{\prec}(l_i) = i$. The rest follows by the definition of $\mathsf{AbDeg}_{\prec}$. □

Fact 7.2.6 (restated). *Where $i, j \in \mathcal{N}$, if $i < j$ then $l_i \lessdot_M l_j$.*

Proof. Let $i, j \in \mathcal{N}$ such that $i < j$. There are two cases: (a) $\sigma_M(l_i \vee l_j) = \emptyset$ and (b) $\sigma_M(l_i \vee l_j) \neq \emptyset$. In the case (a) $l_i \vee l_j \mid\!\sim_M \neg l_j$ and hence $l_i \lessdot_M l_j$. Suppose now that (b) is the case. Let $k = \mathsf{AbDeg}_{\prec}(l_i \vee l_j)$. Suppose $k < i$. Then by (R) (ii), for all $w \in \min_{\prec}(\sigma_M(l_i \vee l_j))$, $w \models \neg l_i, \neg l_j$,—a contradiction. Hence $k \geq i$. Since by (R) (i) for all w for which $\mathsf{rank}_{\prec}(w) = i$, $w \models l_i$, $k = i$. Note that for all these w by (R) (ii), $w \models \neg l_j$. Hence $l_i \vee l_j \mid\!\sim_M \neg l_j$ and hence $l_i \lessdot_M l_j$. □

Fact 7.2.7 (restated). *For all $\alpha \in \mathcal{V}_p^+$ and all $i \in \mathcal{N}$,*

(i) $i < \mathsf{AbDeg}_{\prec}(\alpha)$ *iff* $l_i \lessdot_M \alpha$.
(ii) $i \leq \mathsf{AbDeg}_{\prec}(\alpha)$ *iff* $\alpha \mid\!\sim_M l_i$.
(ii) *where* $i < \mathsf{m}$, $\mathsf{AbDeg}_{\prec}(\alpha) = i$ *iff,* $\alpha \mid\!\sim_M l_i$ *and* $\alpha \not\mid\!\sim_M l_{i+1}$
(iv) $\mathsf{AbDeg}_{\prec}(\alpha) = i$ *iff, not* $l_i \lessdot_M \alpha$ *and* ($l_{i-1} \lessdot_M \alpha$ *or* $i = 0$).

Proof. Ad (ii): "⇒" Let $i \leq \mathsf{AbDeg}_{\prec}(\alpha)$. Let $\mathsf{AbDeg}_{\prec}(\alpha) = j$. If $j = \omega$ then $\sigma_M(\alpha) = \emptyset$ and hence $\alpha \mid\!\sim_M l_i$. If $j \neq \omega$ then by (R) (i) for all $w \in \min_{\prec}(\sigma_M(\alpha))$, $w \models l_i$. Hence, $\alpha \mid\!\sim_M l_i$. "⇐": Let $\alpha \mid\!\sim_M l_i$. Assume $\mathsf{AbDeg}_{\prec}(\alpha) < i$. Then by (R) (ii) for all $w \in \min_{\prec}(\sigma_M(\alpha))$, $w \models \neg l_i$,—a contradiction.

Ad (i): "⇐": Suppose $l_i \lessdot_M \alpha$. By Fact 7.2.5, $\mathsf{AbDeg}_{\prec}(l_i) \in \{i, \omega\}$. If $\mathsf{AbDeg}_{\prec}(l_i) = i$, then by Fact 7.1.2, $\mathsf{AbDeg}_{\prec}(\alpha) > i$. If $\mathsf{AbDeg}_{\prec}(l_i) = \omega$, then by Fact 7.1.2, $\mathsf{AbDeg}_{\prec}(\alpha) = \omega > i$. "⇒": Suppose now that $i < \mathsf{AbDeg}_{\prec}(\alpha)$. By (ii), $\alpha \mid\!\sim_M l_i$. By Fact 7.2.5, $\mathsf{AbDeg}_{\prec}(l_i) \in \{i, \omega\}$. If $\mathsf{AbDeg}_{\prec}(l_i) = i$ then by Fact 7.1.2, $l_i \lessdot_M \alpha$. If $\mathsf{AbDeg}_{\prec}(l_i) = \omega$ then $\sigma_M(l_i) = \emptyset$. Due to $\alpha \mid\!\sim_M l_i$, also $\sigma_M(\alpha) = \emptyset$. Hence $\sigma_M(\alpha \vee l_i) = \emptyset$. Hence $l \lessdot_M \alpha$.

Ad (iii): "⇒": Suppose $\mathsf{AbDeg}_{\prec}(\alpha) = i$. Then by (ii), $\alpha \mid\!\sim_M l_i$. Suppose that $i \neq \mathsf{m}$. By Fact 7.2.5, $\mathsf{AbDeg}_{\prec}(l_{i+1}) \in \{i + 1, \omega\}$. Hence, $\alpha \mid\!\sim_M \neg l_{i+1}$. Hence, $\alpha \not\mid\!\sim_M l_{i+1}$.

"⇐": Suppose $\alpha \not\mid\!\sim_M l_i$. By Fact 7.2.5, $\mathsf{AbDeg}_{\prec}(l_i) \in \{i, \omega\}$. Hence, $\mathsf{AbDeg}_{\prec}(\alpha) \geq i$. Suppose further $\alpha \not\mid\!\sim_M l_{i+1}$. By (ii), $\mathsf{AbDeg}_{\prec}(\alpha) < i + 1$. Hence, $\mathsf{AbDeg}_{\prec}(\alpha) = i$.

Ad (iv): "⇒": Suppose $\mathsf{AbDeg}_{\prec}(\alpha) = i$. By (i), not $l_i \lessdot_M \alpha$. Also by (i), if $i \neq 0$, $l_{i-1} \lessdot_M \alpha$. "⇐": Suppose now that not $l_i \lessdot_M \alpha$. Suppose further that $i > 0$ and $l_{i-1} \lessdot_M \alpha$. By (i), $i - 1 < \mathsf{AbDeg}_{\prec}(\alpha)$ and $i \not< \mathsf{AbDeg}_{\prec}(\alpha)$. Hence, $\mathsf{AbDeg}_{\prec}(\alpha) = i$. Suppose now $i = 0$. By (i), $0 \not< \mathsf{AbDeg}_{\prec}(\alpha)$. Hence, $\mathsf{AbDeg}_{\prec}(\alpha) = 0$. □

Fact 7.2.8 (restated). *If $l_{i-1} \lessdot_M \alpha$, then $\alpha \mid\!\sim_M l_j$ for all $j \leq i$ and $j \leq \mathsf{m}$.*

Proof. Let $l_{i-1} \lessdot_M \alpha$. By Fact 7.1.2, $\mathsf{AbDeg}_{\prec}(l_{i-1}) \leq \mathsf{AbDeg}_{\prec}(\alpha)$. If $\sigma_M(\alpha) = \emptyset$, $\alpha \mid\!\sim_M l_j$ for all $j \leq i$. Otherwise, by Fact 7.2.5, $\mathsf{AbDeg}_{\prec}(l_{i-1}) = i - 1$.

By Fact 7.1.2, $\mathsf{AbDeg}_{\prec}(l_{i-1}) < \mathsf{AbDeg}_{\prec}(\alpha)$. By (R) (i), for all $w \in \min_{\prec}(\sigma(\alpha))$, $w \models l_j$ for all $j \leq i$. Hence, $\alpha \mathrel{|\!\!\sim}_M l_j$ for all $j \leq i$. □

Fact 7.2.9 (restated). *If $l_{\mathsf{m}} \lessdot_M \alpha$, then $\alpha \mathrel{|\!\!\sim}_M \bot$.*

Proof. Let $l_{\mathsf{m}} \lessdot_M \alpha$. By Fact 7.2.5, $\mathsf{AbDeg}_{\prec}(l_{\mathsf{m}}) \in \{\mathsf{m}, \omega\}$. In the second case, by Fact 7.1.2, $\mathsf{AbDeg}_{\prec}(\alpha) = \omega$ and hence $\alpha \mathrel{|\!\!\sim}_M \bot$. In the first case, by Fact 7.1.2, $\mathsf{AbDeg}_{\prec}(\alpha) > \mathsf{m}$. By (R), for all $w \in W$, $\mathsf{rank}_{\prec}(w) \leq \mathsf{m}$. Hence, $\sigma_M(\alpha) = \emptyset$. Thus, $\alpha \mathrel{|\!\!\sim}_M \bot$. □

Lemma 7.2.1 (restated). *The following holds for the mapping λ:*

(i) $\lambda(M)$ satisfies requirement (R).
(ii) For all $\alpha, \beta \in \mathcal{V}_p$, $\alpha \mathrel{|\!\!\sim}_M \beta$ iff $\alpha \mathrel{|\!\!\sim}_{\lambda(M)} \beta$.
(iii) For all $w \in W$, $\mathsf{rank}_{\prec}(w) = \mathsf{rank}_{\prec'}(\pi(w))$.
(iv) $\lambda(M)$ is a rational$^+$ model.

Proof. Ad (iii): This can be shown by an induction on $\mathsf{rank}_{\prec}(w)$. Assume $\mathsf{rank}_{\prec}(w) = 0$. Suppose further that $\mathsf{rank}_{\prec'}(\pi(w)) \neq 0$. Hence, there is a w' such that $\mathsf{rank}_{\prec'}(\pi(w')) = 0$. However, then by the construction, $\mathsf{rank}_{\prec}(w') < \mathsf{rank}_{\prec}(w)$,—a contradiction. Let now $\mathsf{rank}_{\prec}(w) = n + 1$. By the definition of $\mathsf{rank}_{\prec}$ there is a sequence of worlds $w_0, \ldots, w_n$ such that $\mathsf{rank}_{\prec}(w_i) = i$ for all $i \leq n$. By the induction hypothesis, $\mathsf{rank}_{\prec'}(\pi(w_i)) = i$ for all $i \leq n$. Moreover, by the construction, $\mathsf{rank}_{\prec'}(\pi(w_n)) < \mathsf{rank}_{\prec'}(\pi(w))$ since $\mathsf{rank}_{\prec}(w_n) < \mathsf{rank}_{\prec}(w)$. But then, $\mathsf{rank}_{\prec'}(\pi(w)) > n$. Assume $\mathsf{rank}_{\prec'}(\pi(w)) > n + 1$. Hence, there is a w_{n+1} for which $\mathsf{rank}_{\prec'}(\pi(w_{n+1})) = n + 1$. Since $\mathsf{rank}_{\prec'}(\pi(w)) > \mathsf{rank}_{\prec'}(\pi(w_{n+1})) > \mathsf{rank}_{\prec'}(\pi(w_n)) = n$ also $\mathsf{rank}_{\prec}(w) > \mathsf{rank}_{\prec}(w_{n+1}) > \mathsf{rank}_{\prec}(w_n) = n$. A contradiction the fact that $\mathsf{rank}_{\prec}(w) = n+1$. Thus, our assumption was false and hence $\mathsf{rank}_{\prec'}(\pi(w)) = n + 1$.

Ad (i): By (iii) and Fact 7.1.6ii, for all $\pi(w) \in \pi(W)$, $\mathsf{rank}_{\prec'}(\pi(w)) \leq \mathsf{m}$. Let $\mathsf{rank}_{\prec'}(\pi(w)) = i$. By (iii), $\mathsf{rank}_{\prec}(w) = i$. By the definition of π, $\pi(w) \models l_j$ for all $j \leq i$ and $\pi(w) \models \neg l_j$ for all $j \in \mathcal{N}$ for which $j > i$. Hence, (R) is satisfied.

Ad (ii): By the definition of π and by (iii), $\min_{\prec}(\sigma_M(\alpha)) = \{w \mid \pi(w) \in \min_{\prec'}(\sigma_{\langle \pi(W), \prec' \rangle}(\alpha))\}$. Also due to the definition of π, $w \models \beta$ iff $\pi(w) \models \beta$.

Ad (iv): Follows by (i) and (iii). □

Lemma 7.2.2 (restated). *The following holds for the mapping μ:*

(i) $\mu(M)$ is a rational $\mathcal{L}$-model.
(ii) For all $\alpha, \beta \in \mathcal{V}_p$, $\alpha \mathrel{|\!\!\sim}_M \beta$ iff $\alpha \mathrel{|\!\!\sim}_{\mu(M)} \beta$.
(iii) For all $w \in W$, $\mathsf{rank}_{\prec}(w) \geq \mathsf{rank}_{\prec'}(\eta(w))$.

Proof. Ad (i): We have to show that $\prec'$ is modular. Define $\mathsf{D} =_{\mathrm{df}} \{\rho(w) \mid w \in W\}$ where $\rho(w)$ picks out an arbitrary member of $\min_{\prec}(\eta^{-1}(\{w\}))$. Define $\upsilon : \eta(W) \to \mathsf{D}$ by $\eta(w) \mapsto \rho(w)$. We will show that υ defines an isomorphism from $\mu(M) = \langle \eta(W), \prec' \rangle$ to $\langle \mathsf{D}, \prec_{\mathsf{D}} \rangle$ where $\prec_{\mathsf{D}} = \prec \cap (\mathsf{D} \times \mathsf{D})$. By definition υ is surjective. Let now $\eta(w) \neq \eta(w')$. Hence, there is an $\alpha \in \mathcal{P}$ such that $\eta(w) \models \alpha$ and $\eta(w') \not\models \alpha$. Hence, for all $w'' \in \eta^{-1}(\{w\})$, $w'' \models \alpha$ and for all $w''' \in \eta^{-1}(\{w'\})$, $w'' \not\models \alpha$. Hence,

$\upsilon(\eta(w)) \neq \upsilon(\eta(w'))$. This shows that υ is injective. We still need to show that υ is structure-preserving. Suppose $\eta(w) \prec' \eta(w')$. Then, by the definition of $\prec'$, for all $w'' \in \min_{\prec}(\eta^{-1}(\{w\}))$ and for all $w''' \in \min_{\prec}(\eta^{-1}(\{w'\}))$, $w'' \prec w'''$. Hence, $\upsilon(\eta(w)) \prec \upsilon(\eta(w'))$. Suppose now that $\upsilon(\eta(w)) \prec \upsilon(\eta(w'))$. Since $\upsilon(\eta(w)) \in \min_{\prec}(\eta^{-1}(\{w\}))$ and $\upsilon(\eta(w')) \in \min_{\prec}(\eta^{-1}(\{w'\}))$, $\eta(w) \prec' \eta(w')$.

Since $\prec$ is modular, also $\prec_{\mathsf{D}}$ is modular. Hence, due to the fact that υ is an isomorphism between $\mu(M)$ and $\langle \mathsf{D}, \prec_{\mathsf{D}} \rangle$, $\prec'$ is also modular.

Ad (iii): This can be shown by an induction on $\mathsf{rank}_{\prec'}(\eta(w))$. Suppose $\mathsf{rank}_{\prec'}(\eta(w)) = 0$. Evidently $\mathsf{rank}_{\prec}(w) \geq 0$. Suppose now that $\mathsf{rank}_{\prec'}(\eta(w)) = n+1$. By the definition of $\mathsf{rank}_{\prec'}$, there are $w_0, \ldots, w_n$ such that $\mathsf{rank}_{\prec'}(\eta(w_i)) = i$ for all $i \leq n$. Since $\eta(w_0) \prec' \cdots \prec' \eta(w_n) \prec' \eta(w)$, also $w'_0 \prec \cdots \prec w'_n \prec w'$ for all $w'_0 \in \min_{\prec}(\eta^{-1}(\{w_0\}))$, $\ldots$, $w'_n \in \min_{\prec}(\eta^{-1}(\{w_n\}))$ and all $w' \in \min_{\prec}(\eta^{-1}(\{w\}))$. Then $\mathsf{rank}_{\prec}(w') > n$ for all $w' \in \min_{\prec}(\eta^{-1}(\{w\}))$. Hence $\mathsf{rank}_{\prec}(w) > n$.

Ad (ii): Suppose first that $\alpha \mathrel{|\!\sim}_M \beta$. Assume $\alpha \mathrel{|\!\not\sim}_{\mu(M)} \beta$. Hence, there is a $\eta(w) \in \min_{\prec'}(\sigma_{\mu(M)}(\alpha))$ for which $\eta(w) \not\models \beta$. Let $w' \in \min_{\prec}(\sigma_M(\alpha))$. Since for all $w'' \in \eta^{-1}(\{\eta(w)\})$, (†) $w'' \models \alpha$, $\mathsf{rank}_{\prec}(w'') < \mathsf{rank}_{\prec}(w')$. Let $w'' \in \min_{\prec}(\eta^{-1}(\{\eta(w)\}))$. Note that $w' \in \min_{\prec}(\eta^{-1}(\{\eta(w')\}))$ since otherwise $w' \notin \min_{\prec}(\sigma_M(\alpha))$. By the definition of $\prec'$ and (†), $\mathsf{rank}_{\prec'}(\eta(w')) < \mathsf{rank}_{\prec'}(\eta(w'')) = \mathsf{rank}_{\prec'}(\eta(w))$. This is a contradiction to the minimality of $\eta(w)$ in $\min_{\prec'}(\sigma_{\mu(M)}(\alpha))$.

Suppose now that $\alpha \mathrel{|\!\sim}_{\mu(M)} \beta$. Assume $\alpha \mathrel{|\!\not\sim}_M \beta$. Hence, there is a $w \in \min_{\prec}(\sigma_M(\alpha))$ for which $\alpha \not\models \beta$. This also means that $\eta(w) \notin \min_{\prec'}(\sigma_{\mu(M)}(\alpha))$. Thus, there is a w' for which $\eta(w') \in \min_{\prec'}(\sigma_{\mu(M)}(\alpha))$ and $\eta(w') \prec' \eta(w)$. By the definition of $\prec'$, for all $w'' \in \min_{\prec}(\eta^{-1}(\{\eta(w')\}))$, $w'' \prec w$. Since $w'' \models \alpha$ this is a contradiction to $w \in \min_{\prec}(\sigma_M(\alpha))$. □

Appendix F
Appendix to Chapter 8

In Sect. F.1 we investigate the semantics of our core systems $\mathbf{L_A}$ and $\mathbf{L_C}$ and prove completeness and soundness. The representational results proven in Sect. F.2 for the logics for admissible and complete extensions, and in Sect. F.3 for the other extension types are given with respect to skeptical acceptance. We prove the representational results for credulous acceptance in Sect. F.4.

F.1 Semantics for the Core Systems $\mathbf{L_A}$ and $\mathbf{L_C}$

For the sake of clarity we recapitulate the definitions of the language and the axiomatization of logics $\mathbf{L_A}$ and $\mathbf{L_C}$ before we define the semantics.

F.1.1 The Language

We use a classical propositional language with an additional binary operator, $\twoheadrightarrow$, that represents the attack relation. Formally our language $\mathcal{W}_n$ (where n is a natural number) is defined in the following way:

$$\begin{aligned}
\mathcal{V}_n &:= p_1 \mid p_2 \mid p_3 \mid \cdots \mid p_n \\
\mathcal{W}_n^{\twoheadrightarrow} &:= \langle \mathcal{V}_n \rangle \twoheadrightarrow \langle \mathcal{V}_n \rangle \mid \bot \twoheadrightarrow \langle \mathcal{V}_n \rangle \\
\mathcal{W}_n &:= \bot \mid \langle \mathcal{V}_n \rangle \mid \langle \mathcal{W}_n^{\twoheadrightarrow} \rangle \mid \neg \langle \mathcal{W}_n \rangle \mid \langle \mathcal{W}_n \rangle \wedge \langle \mathcal{W}_n \rangle \mid \\
&\qquad \langle \mathcal{W}_n \rangle \vee \langle \mathcal{W}_n \rangle \mid \langle \mathcal{W}_n \rangle \supset \langle \mathcal{W}_n \rangle
\end{aligned}$$

$\mathcal{V}_n$ are the propositional letters of our language. We will in the remainder abbreviate $\neg(\alpha \twoheadrightarrow \beta)$ by $\alpha \not\twoheadrightarrow \beta$. Moreover, Greek letters α, β and γ are used as meta-variables for propositional letters, and φ and ψ are used for formulas in $\mathcal{W}_n$.

C. Straßer, *Adaptive Logics for Defeasible Reasoning*, Trends in Logic 38,
DOI: 10.1007/978-3-319-00792-2, © Springer International Publishing Switzerland 2014

F.1.2 The Syntactic Rules

Let us recapitulate the axiomatization of our core systems $\mathbf{L_A}$ and $\mathbf{L_C}$. Before we state the rules, we define:

$$\mathsf{def}\ \beta =_{\text{df}} \bigvee_{\alpha \in \mathcal{V}_n} (\alpha \wedge (\alpha \twoheadrightarrow \beta)) \qquad \text{(Def)}$$

The following rules are needed:

$$\frac{\alpha \quad \alpha \twoheadrightarrow \beta}{\neg\beta} \qquad (\text{R}{\twoheadrightarrow})$$

$$\frac{\alpha \quad \beta \twoheadrightarrow \alpha}{\mathsf{def}\ \beta} \qquad \text{(Rad)}$$

$$\frac{\bot \not\twoheadrightarrow \alpha}{\neg\alpha} \qquad (\text{R}\bot)$$

$$\frac{\bot \twoheadrightarrow \beta \bigwedge_{\alpha \in \mathcal{V}_n} ((\alpha \twoheadrightarrow \beta) \supset \mathsf{def}\ \alpha)}{\beta} \qquad \text{(RCo)}$$

Definition F.1.1. $\mathbf{L_A}$ is classical propositional logic enriched by rules (R↠), (Rad), and (R⊥). $\mathbf{L_C}$ is $\mathbf{L_A}$ enriched by (RCo).

F.1.3 The Semantics

We will define the semantics for logics $\mathbf{L}$ (where $\mathbf{L} \in \{\mathbf{L_A}, \mathbf{L_C}\}$) via an assignment function $v : \mathcal{V}_n \cup \mathcal{W}_n^{\twoheadrightarrow} \rightarrow \{0, 1\}$ and an $\mathbf{L}$-valuation $v_M^{\mathbf{L}} : \mathcal{W}_n \rightarrow \{0, 1\}$ determined by the assignment. A model M is defined by the assignment v. In Sect. F.1.3.1 we take a closer look at assignment functions, and in Sect. F.1.3.2 we define the $\mathbf{L}$-valuations.

Once the $\mathbf{L}$-valuations are defined, model validity and the semantic consequence relation can be defined in the usual way. We define $M \models_{\mathbf{L}} \varphi$ iff $v_M^{\mathbf{L}}(\varphi) = 1$. We say that a model M is an $\mathbf{L}$ model of $\Gamma \subset \mathcal{W}_n$ iff $M \models_{\mathbf{L}} \varphi$ for all $\varphi \in \Gamma$. We write $\mathcal{M}_{\mathbf{L}}(\Gamma)$ for the set of all $\mathbf{L}$-models of Γ. The semantic consequence relations $\Vdash_{\mathbf{L}}$ are defined in the usual way: $\Gamma \Vdash_{\mathbf{L}} \varphi$ iff for all $\mathbf{L}$-models M of Γ, $M \models_{\mathbf{L}} \varphi$.

F.1.3.1 The Assignment Function

We use an extended assignment function $v : \mathcal{V}_n \cup \mathcal{W}_n^{\twoheadrightarrow} \rightarrow \{0, 1\}$ that assigns truth values to both, propositional letters and 'attacks', i.e., formulas in $\mathcal{W}_n^{\twoheadrightarrow}$. A model M is defined by an assignment function v.

Some useful definitions for the remainder:

$$v(\mathsf{def}\,\alpha) =_{\mathrm{df}} \max_{\beta\in\mathcal{V}_n}\big(\min(v(\beta), v(\beta \twoheadrightarrow \alpha))\big)$$

$$v_{\mathrm{Rral}} =_{\mathrm{df}} 1 - \max_{\alpha,\beta\in\mathcal{V}_n}\big(\min(v(\alpha), v(\alpha \twoheadrightarrow \beta), v(\beta))\big)$$

$$v_{\mathrm{Rbot}} =_{\mathrm{df}} 1 - \max_{\alpha\in\mathcal{V}_n}\big(\min(v(\alpha), 1 - v(\bot \twoheadrightarrow \alpha))\big)$$

$$v_{\mathrm{Rad}} =_{\mathrm{df}} 1 - \max_{\alpha,\beta\in\mathcal{V}_n}\big(\min(v(\alpha), v(\beta \twoheadrightarrow \alpha), 1 - v(\mathsf{def}\,\beta))\big)$$

$$v_{\mathrm{RCo}} =_{\mathrm{df}} 1 - \max_{\beta\in\mathcal{V}_n}\Big(\min\big(v(\bot \twoheadrightarrow \beta), \min_{\alpha\in\mathcal{V}_n}(\max(1 - v(\alpha \twoheadrightarrow \beta), v(\mathsf{def}\,\alpha))), 1 - v(\beta)\big)\Big)$$

$$v_{\mathsf{i}}^{\mathbf{L_A}} =_{\mathrm{df}} \min(v_{\mathrm{Rral}}, v_{\mathrm{Rbot}}, v_{\mathrm{Rad}})$$

$$v_{\mathsf{i}}^{\mathbf{L_C}} =_{\mathrm{df}} \min(v_{\mathrm{Rral}}, v_{\mathrm{Rbot}}, v_{\mathrm{Rad}}, v_{\mathrm{RCo}})$$

Note that v_{Rral} corresponds to our syntactical rule (R$\twoheadrightarrow$) in the sense that $v_{\mathrm{Rral}} = 1$ if the assignment satisfies the semantic counterpart to (R$\twoheadrightarrow$). That is to say, $v_{\mathrm{Rral}} = 1$ iff v satisfies

$$\text{If } v(\alpha) = v(\alpha \twoheadrightarrow \beta) = 1, \text{ then } v(\beta) = 0. \qquad (\mathrm{S}\twoheadrightarrow)$$

The situation is analogous for v_{Rbot}, v_{Rad} and v_{RCo} with respect to the following properties:

$$\text{If } v(\alpha) = 1, \text{ then } v(\bot \twoheadrightarrow \alpha) = 1. \qquad (\mathrm{S}\bot)$$

$$\text{If } v(\alpha) = v(\beta \twoheadrightarrow \alpha) = 1, \text{ then there is a } \gamma \in \mathcal{V}_n \text{ for which } v(\gamma) = v(\gamma \twoheadrightarrow \beta) = 1. \qquad (\mathrm{Sad})$$

$$\text{If } v(\bot \twoheadrightarrow \beta) = 1 \text{ and for all } \alpha \in \mathcal{V}_n \text{ we have } \big[v(\alpha \twoheadrightarrow \beta) = 0 \text{ or (there is a } \gamma \in \mathcal{V}_n \text{ for which } v(\gamma) = v(\gamma \twoheadrightarrow \alpha) = 1)\big], \text{ then } v(\beta) = 1. \qquad (\mathrm{SCo})$$

We call an assignment v $\mathbf{L_A}$-*intended* iff $v_{\mathsf{i}}^{\mathbf{L_A}} = 1$, and we call it $\mathbf{L_C}$-*intended* iff $v_{\mathsf{i}}^{\mathbf{L_C}} = 1$. Obviously, every $\mathbf{L_C}$-intended assignment is $\mathbf{L_A}$-intended as well. The following lemma shows that intended assignments have the corresponding intuitive properties:

Lemma F.1.1. *Let v be an assignment function.*

(i) $v_{\mathsf{i}}^{\mathbf{L_A}} = 1$ *iff v satisfies* (S$\twoheadrightarrow$), (Sad) *and* (S$\bot$).
(ii) $v_{\mathsf{i}}^{\mathbf{L_C}} = 1$ *iff v satisfies* (S$\twoheadrightarrow$), (Sad), (S$\bot$), *and* (SCo).

Proof. Ad (i): "⇒": Let $v_{\mathsf{i}}^{\mathbf{L_A}} = 1$. Assume $v(\alpha) = 1$. Since $v_{\mathrm{Rbot}} = 1$, $\min(v(\alpha), 1 - v(\bot \twoheadrightarrow \alpha)) = 0$. Thus, $v(\bot \twoheadrightarrow \alpha) = 1$. Thus, (S⊥) is valid. Assume $v(\alpha) = v(\alpha \twoheadrightarrow \beta) = 1$. Since $v_{\mathrm{Rral}} = 1$, $\min(v(\alpha), v(\alpha \twoheadrightarrow \beta), v(\beta)) = 0$. Thus, $v(\beta) = 0$. Thus, (S↠) is valid. Assume $v(\alpha) = v(\beta \twoheadrightarrow \alpha) = 1$. Since $v_{\mathrm{Rad}} = 1$, $\min(v(\alpha), v(\beta \twoheadrightarrow \alpha), 1 - v(\mathsf{def}\ \beta)) = 0$. Hence, $v(\mathsf{def}\ \beta) = 1$. By definition, $\max_{\gamma \in \mathcal{V}_n}(\min(v(\gamma), v(\gamma \twoheadrightarrow \beta))) = 1$. Thus, there is a $\gamma \in \mathcal{V}_n$ for which $v(\gamma) = 1 = v(\gamma \twoheadrightarrow \beta)$. Thus, (Sad) is valid.

"⇐": Suppose $v_{\mathrm{Rbot}} = 0$. Then there is a $\alpha \in \mathcal{V}_n$ for which $\min(v(\alpha), 1 - v(\bot \twoheadrightarrow \alpha)) = 1$. Thus, $v(\alpha) = 1$ and $v(\bot \twoheadrightarrow \alpha) = 0$. Hence (S⊥) does not hold. The proof is similar for v_{Rral}, and v_{Rad}.

Ad (ii): "⇒": Let $v_{\mathsf{i}}^{\mathbf{L_C}} = 1$. In addition to what has been shown in (i), it has to be shown that (SCo) is valid. Assume $v(\bot \twoheadrightarrow \beta) = 1$ and for all $\alpha \in \mathcal{V}_n$, $v(\alpha \twoheadrightarrow \beta) = 0$ or there is a $\gamma \in \mathcal{V}_n$ for which $v(\gamma) = v(\gamma \twoheadrightarrow \alpha) = 1$. Thus, for all $\alpha \in \mathcal{V}_n$, $v(\alpha \twoheadrightarrow \beta) = 0$ or $v(\mathsf{def}\ \alpha) = 1$. Due to the fact that $v_{\mathrm{RCo}} = 1$, $\min\bigl(v(\bot \twoheadrightarrow \beta), \min_{\alpha \in \mathcal{V}_n}(\max(1 - v(\alpha \twoheadrightarrow \beta), v(\mathsf{def}\ \alpha))), 1 - v(\beta)\bigr) = 0$. Thus, $v(\beta) = 1$. Thus, (SCo) is valid.

"⇐": Suppose $v_{\mathrm{RCo}} = 0$. Then there is a $\beta \in \mathcal{V}_n$ for which $\min\bigl(v(\bot \twoheadrightarrow \beta), \min_{\alpha \in \mathcal{V}_n}(\max(1 - v(\alpha \twoheadrightarrow \beta), v(\mathsf{def}\ \alpha))), 1 - v(\beta)\bigr) = 1$ and thus $v(\beta) = 0$. Thus, for every $\alpha \in \mathcal{V}_n$ for which $v(\alpha \twoheadrightarrow \beta) = 1$, $v(\mathsf{def}\alpha) = 1$ and hence there is a $\gamma \in \mathcal{V}_n$ such that $v(\gamma) = v(\gamma \twoheadrightarrow \alpha) = 1$. Hence, (SCo) does not hold. □

F.1.3.2 The Valuation

Let us now take a look at valuation functions for our core logics $\mathbf{L_A}$ and $\mathbf{L_C}$. Let a model M be defined by an assignment v.

An $\mathbf{L_A}$-valuation $v_M^{\mathbf{L_A}} : \mathcal{W}_n \to \{0, 1\}$ determined by v is defined as follows (where $\alpha, \beta \in \mathcal{V}_n$; $\varphi, \varphi_1, \varphi_2 \in \mathcal{W}_n$; and $\mathbf{L} = \mathbf{L_A}$):

$$v_M^{\mathbf{L}}(\bot) = 0 \qquad \text{(s⊥)}$$

$$v_M^{\mathbf{L}}(\alpha \twoheadrightarrow \beta) = 1 \text{ iff } v(\alpha \twoheadrightarrow \beta) = 1 \qquad \text{(s↠)}$$

$$v_M^{\mathbf{L}}(\bot \twoheadrightarrow \alpha) = 1 \text{ iff } v(\bot \twoheadrightarrow \alpha) = 1 \qquad \text{(s⊥↠)}$$

$$v_M^{\mathbf{L}}(\alpha) = \min\bigl(v_{\mathsf{i}}^{\mathbf{L_A}}, v(\alpha)\bigr) \qquad \text{(sPA)}$$

$$v_M^{\mathbf{L}}(\varphi_1 \wedge \varphi_2) = \min\bigl(v_M^{\mathbf{L}}(\varphi_1), v_M^{\mathbf{L}}(\varphi_2)\bigr) \qquad \text{(s∧)}$$

$$v_M^{\mathbf{L}}(\varphi_1 \vee \varphi_2) = \max\bigl(v_M^{\mathbf{L}}(\varphi_1), v_M^{\mathbf{L}}(\varphi_2)\bigr) \qquad \text{(s∨)}$$

$$v_M^{\mathbf{L}}(\varphi_1 \supset \varphi_2) = \max\bigl(1 - v_M^{\mathbf{L}}(\varphi_1), v_M^{\mathbf{L}}(\varphi_2)\bigr) \qquad \text{(s⊃)}$$

$$v_M^{\mathbf{L}}(\neg\varphi) = 1 - v_M^{\mathbf{L}}(\varphi) \qquad \text{(s¬)}$$

In the case $v_{\mathsf{i}}^{\mathbf{L_A}} = 1$, i.e., in the case that the assignment is $\mathbf{L_A}$-intended, the valuation takes over the truth values from the assignment for all formulas in $\mathcal{V}_n$. However, if $v_{\mathsf{i}}^{\mathbf{L_A}} = 0$, the valuation assigns to all propositional letters the truth value 0. Note that for a given AF A the empty selection is always an admissible extension. Thus, the valuation on basis of a non-intended assignment corresponds to the empty extension.

$\mathbf{L_C}$-valuations are defined analogous to $\mathbf{L_A}$-valuations, with the exception of (sPA) which is replaced by:

$$v_M^{\mathbf{L_C}}(\alpha) = \max\big(\min\big(v_{\mathsf{i}}^{\mathbf{L_C}}, v(\alpha)\big), v_{\mathsf{g}}(\alpha)\big) \text{ where} \tag{sPC}$$

$$v_{\mathsf{g}}(\alpha) =_{\text{df}} \max_{i\geq 0}\big(v_{\mathsf{g}}^i(\alpha)\big), \text{ where}$$

$$v_{\mathsf{g}}^0(\alpha) =_{\text{df}} \min\big(v(\bot \twoheadrightarrow \alpha), 1 - \max_{\beta\in\mathcal{V}_n}(v(\beta \twoheadrightarrow \alpha))\big), \text{ and}$$

$$v_{\mathsf{g}}^i(\alpha) =_{\text{df}} \min\big(v(\bot \twoheadrightarrow \alpha), \mathsf{defended}^i(\alpha)\big), \text{ where } i > 0 \text{ and}$$

$$\mathsf{defended}^i(\alpha) =_{\text{df}} \min_{\beta\in\mathcal{V}_n}\big(\max(1 - v(\beta \twoheadrightarrow \alpha), \mathsf{defeated}^i(\beta))\big), \text{ where}$$

$$\mathsf{defeated}^i(\beta) =_{\text{df}} \max_{\gamma\in\mathcal{V}_n}\big(\min(v(\gamma \twoheadrightarrow \beta), v_{\mathsf{g}}^{<i}(\gamma))\big), \text{ where}$$

$$v_{\mathsf{g}}^{<i}(\gamma) =_{\text{df}} \max_{0\leq j<i}\big(v_{\mathsf{g}}^j(\gamma)\big)$$

Again, in the case that v is an $\mathbf{L_C}$-intended assignment, i.e., in the case $v_{\mathsf{i}}^{\mathbf{L_C}} = 1$, the valuation $v_M^{\mathbf{L_C}}$ takes over all truth values for all formulas in $\mathcal{V}_n$ from the assignment. However, in the case $v_{\mathsf{i}}^{\mathbf{L_C}} = 0$, the situation is more complicated than for the $\mathbf{L_A}$ case, since for a given AF A the empty selection may not correspond to a complete extension. In this case the valuation $v_M^{\mathbf{L_C}}$ verifies a propositional letter α iff $v_{\mathsf{g}}(\alpha) = 1$. As it will be shown, this way it is ensured that the models of the adaptive strengthening of $\mathbf{L_C}$ correspond to the complete extensions.

As the reader can see, everything except rules (s$\twoheadrightarrow$), (s$\bot\twoheadrightarrow$) and (sPA) (resp. (sPC)) is defined in the classical way. Note, that by (s$\twoheadrightarrow$) and (s$\bot\twoheadrightarrow$) the valuation takes over the assignment from v for formulas of the form $\alpha \twoheadrightarrow \beta$ and $\bot \twoheadrightarrow \alpha$. By (sPA) (resp. (sPC)) the valuation may have a different value for propositional letters than assigned by v. Note that although (sPA) and (sPC) are of a rather complex form, they are fully determined by the assignment v.

Our valuations satisfy the semantic properties corresponding to the rules (R$\twoheadrightarrow$), (Rad), (R$\bot$), (RCo) of logics $\mathbf{L_A}$ and $\mathbf{L_C}$.

Theorem F.1.1. *Let M be a model defined by the assignment v.*[4]

(i) $v_M^{\mathbf{L_A}}$ *satisfies* (S$\twoheadrightarrow$), (Sad) *and* (S$\bot$).
(ii) $v_M^{\mathbf{L_C}}$ *satisfies* (S$\twoheadrightarrow$), (Sad), (S$\bot$), *and* (S$\bot$).

[4] We postpone the proof of this theorem to page 375.

F.1.4 Soundness and Completeness

Lemma F.1.2. *Let v be an assignment. If $v_{\mathsf{i}}^{\mathbf{L_C}} = 1$, then $v_{\mathsf{g}}(\alpha) = 1$ implies $v(\alpha) = 1$.*

Proof. Suppose $v_{\mathsf{i}}^{\mathbf{L_C}} = v_{\mathsf{g}}(\alpha) = 1$. Then there is a $i \geq 0$ such that $v_{\mathsf{g}}^{i}(\alpha) = 1$. We prove the statement by an induction on i.

"$i = 0$": We have $v(\perp \twoheadrightarrow \alpha) = 1$ and for all $\beta \in \mathcal{V}_n$, $v(\beta \twoheadrightarrow \alpha) = 0$. Since $v_{\mathrm{RCo}} = 1, \min\big(v(\perp \twoheadrightarrow \alpha), \min_{\beta \in \mathcal{V}_n}\big(\max(1 - v(\beta \twoheadrightarrow \alpha), v(\mathsf{def}\,\beta))\big), 1 - v(\alpha)\big) = 0$. Thus, $1 - v(\alpha) = 0$ and hence, $v(\alpha) = 1$.

"$i \Rightarrow i+1$": Let $v_{\mathsf{g}}^{i+1}(\alpha) = 1$. Thus, $\min(v(\perp \twoheadrightarrow \alpha), \mathsf{defended}^{i+1}(\alpha)) = 1$ and hence $v(\perp \twoheadrightarrow \alpha) = \mathsf{defended}^{i+1}(\alpha) = 1$. Thus, $\min_{\beta \in \mathcal{V}_n}\big(\max(1 - v(\beta \twoheadrightarrow \alpha), \mathsf{defeated}^{i+1}(\beta))\big) = 1$. In the case that there is no $\beta \in \mathcal{V}_n$ for which $v(\beta \twoheadrightarrow \alpha) = 1$, $v_{\mathsf{g}}^{0}(\alpha) = 1$ and thus by induction hypothesis, $v(\alpha) = 1$. Suppose there is a β for which $v(\beta \twoheadrightarrow \alpha) = 1$. Then $\mathsf{defeated}^{i+1}(\beta) = 1$. Hence, $\max_{\gamma \in \mathcal{V}_n}\big(\min(v(\gamma \twoheadrightarrow \beta), v_{\mathsf{g}}^{<i+1}(\gamma))\big) = 1$. Thus, there is a γ_β for every such $\beta \in \mathcal{V}_n$ for which $v(\gamma_\beta \twoheadrightarrow \beta) = v_{\mathsf{g}}^{<i+1}(\gamma_\beta) = 1$. Hence there is a $j < i+1$ such that $v_{\mathsf{g}}^{j}(\gamma_\beta) = 1$. Thus, by induction hypothesis, $v(\gamma_\beta) = 1$. Thus, $v(\mathsf{def}\,\beta) = 1$ for all β for which $v(\beta \twoheadrightarrow \alpha) = 1$. Thus, due to the fact that $v_{\mathrm{RCo}} = 1$, $\min\big(v(\perp \twoheadrightarrow \alpha), \min_{\beta \in \mathcal{V}_n}(1 - v(\beta \twoheadrightarrow \alpha), v(\mathsf{def}\,\beta)), 1 - v(\alpha)\big) = 0$. Thus, we have $1 - v(\alpha) = 0$ and hence $v(\alpha) = 1$. □

Lemma F.1.3. *Let $v_M^{\mathbf{L}}$ be an $\mathbf{L}$-valuation (where $\mathbf{L} \in \{\mathbf{L_A}, \mathbf{L_C}\}$) with corresponding assignment v.*

(i) If $v_{\mathsf{i}}^{\mathbf{L}} = 1$ then $v_M^{\mathbf{L}}(\varphi) = v(\varphi)$ for all $\varphi \in \mathcal{V}_n \cup \mathcal{W}_n^{\twoheadrightarrow}$ and for all $\varphi = \mathsf{def}\,\alpha$ where $\alpha \in \mathcal{V}_n$.

(ii) Moreover, if $v_M^{\mathbf{L_A}}(\alpha) = 1$ for some $\alpha \in \mathcal{V}_n$ then $v_{\mathsf{i}}^{\mathbf{L_A}} = 1$.

Proof. Ad (i): For $\mathbf{L} = \mathbf{L_A}$: Since $v_M^{\mathbf{L}}(\alpha) = 1$ iff $v(\alpha) = v_{\mathsf{i}}^{\mathbf{L}} = 1$, $v_M^{\mathbf{L}}(\alpha) = 0$ iff $v_{\mathsf{i}}^{\mathbf{L}} = 0$ or $v(\alpha) = 0$. Since $v_{\mathsf{i}}^{\mathbf{L}} = 1$ the statement is true for all $\varphi \in \mathcal{V}_n$. For $\mathbf{L} = \mathbf{L_C}$: Since $v_M^{\mathbf{L}}(\alpha) = 1$ iff $v(\alpha) = v_{\mathsf{i}}^{\mathbf{L}} = 1$ or $v_{\mathsf{g}}(\alpha) = 1$, and $v_{\mathsf{g}}(\alpha) = 1$ implies $v(\alpha) = 1$ due to $v_{\mathsf{i}}^{\mathbf{L}} = 1$ by Lemma F.1.2, we have $v_M^{\mathbf{L}}(\alpha) = 1$ iff $v(\alpha) = 1$.

For $\varphi \in \mathcal{W}_n^{\twoheadrightarrow}$ the statement is true by (s$\twoheadrightarrow$) and (s$\perp\twoheadrightarrow$). Furthermore, $v_M^{\mathbf{L}}(\mathsf{def}\,\alpha) = v_M^{\mathbf{L}}(\bigvee_{\beta \in \mathcal{V}_n}((\beta \twoheadrightarrow \alpha) \wedge \beta)) = \max_{\beta \in \mathcal{V}_n}(\min(v_M^{\mathbf{L}}(\beta), v_M^{\mathbf{L}}(\beta \twoheadrightarrow \alpha))) = \max_{\beta \in \mathcal{V}_n}(\min(v(\beta), v(\beta \twoheadrightarrow \alpha))) = v(\mathsf{def}\,\alpha)$.

Ad (ii): this is the case due to (sPA). □

Lemma F.1.4. *Let v be an assignment function. If $v_{\mathsf{g}}(\beta) = 1$ then for all $\alpha \in \mathcal{V}_n$ for which $v(\alpha \twoheadrightarrow \beta) = 1$, $v_{\mathsf{g}}(\alpha) = 0$.*

Proof. Since $v_{\mathsf{g}}(\beta) = 1$ iff $\max_{i \geq 0}(v_{\mathsf{g}}^{i}(\beta)) = 1$, we have $v_{\mathsf{g}}^{i}(\beta) = 1$ for some $i \geq 0$. The proof is by induction on i.

"$i = 0$": In this case $\min(v(\perp \twoheadrightarrow \beta), 1 - \max_{\gamma \in \mathcal{V}_n}(v(\gamma \twoheadrightarrow \beta))) = 1$ and thus there is no $\gamma \in \mathcal{V}_n$ for which $v(\gamma \twoheadrightarrow \beta) = 1$.

"$i \Rightarrow i+1$": Let $v_{\mathsf{g}}^{i+1}(\beta) = 1$. Thus, $\min(v(\bot \twoheadrightarrow \beta), \mathsf{defended}^{i+1}(\beta)) = 1$. Thus, $\mathsf{defended}^{i+1}(\beta) = 1$. Thus, $\min_{\gamma \in \mathcal{V}_n}(\max(1 - v(\gamma \twoheadrightarrow \beta), \mathsf{defeated}^{i+1}(\gamma))) = 1$. Thus, for any γ for which $v(\gamma \twoheadrightarrow \beta) = 1$, $\mathsf{defeated}^{i+1}(\gamma) = 1$. In the case that there is such a γ, $\max_{\delta \in \mathcal{V}_n}\left(\min(v(\delta \twoheadrightarrow \gamma), v_{\mathsf{g}}^{<i+1}(\delta))\right) = 1$. Thus, there is a δ_γ for which $v(\delta_\gamma \twoheadrightarrow \gamma) = 1$ and $v_{\mathsf{g}}^{<i+1}(\delta_\gamma) = 1$. Thus, there is a $j < i+1$ such that $v_{\mathsf{g}}^{j}(\delta_\gamma) = 1$. Suppose $v_{\mathsf{g}}(\gamma) = 1$ then there is a k such that $v_{\mathsf{g}}^{k}(\gamma) = 1$. Note that $k > 0$ since $v(\delta_\gamma \twoheadrightarrow \gamma) = 1$. Thus $\mathsf{defended}^{k}(\gamma) = 1$. Hence, $\mathsf{defeated}^{k}(\delta_\gamma)$. Thus, there is an ϵ for which $v(\epsilon \twoheadrightarrow \delta_\gamma) = v_{\mathsf{g}}^{<k}(\epsilon) = 1$ and hence, $v_{\mathsf{g}}(\epsilon) = 1$. By induction hypothesis however, $v_{\mathsf{g}}(\epsilon) = 0$,—a contradiction. Thus, $v_{\mathsf{g}}(\gamma) = 0$. □

Proof (Proof of Theorem F.1.1). Ad (i): Let $\mathbf{L} = \mathbf{L_A}$. By Lemma F.1.3i , in the case that $v_{\mathsf{i}}^{\mathbf{L}} = 1$, $v_M^{\mathbf{L}}(\varphi) = v(\varphi)$ for all $\varphi \in \mathcal{V}_n \cup \mathcal{W}_n^{\twoheadrightarrow}$. By Lemma F.1.1i follows the rest. Assume thus that $v_{\mathsf{i}}^{\mathbf{L}} = 0$. Note that in this case by Lemma F.1.3ii $v_M^{\mathbf{L}}(\alpha) = 0$ for all $\alpha \in \mathcal{V}_n$. Hence, trivially (S$\twoheadrightarrow$), (Sad), and (S$\bot$) are valid.

Ad (ii): Let $\mathbf{L} = \mathbf{L_C}$. By Lemma F.1.3i, in the case $v_{\mathsf{i}}^{\mathbf{L}} = 1$, $v_M^{\mathbf{L}}(\varphi) = v(\varphi)$ for all $\varphi \in \mathcal{V}_n \cup \mathcal{W}_n^{\twoheadrightarrow}$. By Lemma F.1.1ii follows the rest. Assume thus that $v_{\mathsf{i}}^{\mathbf{L}} = 0$. Note that in this case $v_M^{\mathbf{L}}(\alpha) = 1$ iff $v_{\mathsf{g}}(\alpha) = 1$ for all $\alpha \in \mathcal{V}_n$. Assume $v_M^{\mathbf{L}}(\alpha) = 1$. Then $v_{\mathsf{g}}(\alpha) = 1$ and thus, by definition, $v(\bot \twoheadrightarrow \alpha) = v_M^{\mathbf{L}}(\bot \twoheadrightarrow \alpha) = 1$. Thus, (S$\bot$) is valid. Assume $v_M^{\mathbf{L}}(\alpha) = v_M^{\mathbf{L}}(\alpha \twoheadrightarrow \beta) = 1$. Hence, $v_{\mathsf{g}}(\alpha) = 1 = v(\alpha \twoheadrightarrow \beta)$. Thus, by Lemma F.1.4, $v_{\mathsf{g}}(\beta) = 0$ and hence, $v_M^{\mathbf{L}}(\beta) = 0$. Thus, (S$\twoheadrightarrow$) is valid. Assume $v_M^{\mathbf{L}}(\alpha) = v_M^{\mathbf{L}}(\beta \twoheadrightarrow \alpha) = 1$. Thus, $v_{\mathsf{g}}(\alpha) = 1 = v(\beta \twoheadrightarrow \alpha)$. Thus, there is an $i \geq 0$ such that $v_{\mathsf{g}}^{i}(\alpha) = 1$. Since $v(\beta \twoheadrightarrow \alpha) = 1$, $i > 0$. Since $\mathsf{defended}^{i}(\alpha) = 1$, $\mathsf{defeated}^{i}(\beta) = 1$. Thus, there is a $\gamma \in \mathcal{V}_n$ such that $v(\gamma \twoheadrightarrow \beta) = v_M^{\mathbf{L}}(\gamma \twoheadrightarrow \beta) = 1$ and $v_{\mathsf{g}}^{<i}(\gamma) = 1$. Thus, $v_{\mathsf{g}}(\gamma) = 1$ and thus $v_M^{\mathbf{L}}(\gamma) = 1$. Thus, (Sad) is valid. Assume $v_M^{\mathbf{L}}(\bot \twoheadrightarrow \beta) = 1$ and for all $\alpha \in \mathcal{V}_n$ either $v_M^{\mathbf{L}}(\alpha \twoheadrightarrow \beta) = 0$ or there is a $\gamma \in \mathcal{V}_n$ for which $v_M^{\mathbf{L}}(\gamma) = v_M^{\mathbf{L}}(\gamma \twoheadrightarrow \alpha) = 1$. Hence, $v(\bot \twoheadrightarrow \beta) = 1$ and for all $\alpha \in \mathcal{V}_n$ either $v(\alpha \twoheadrightarrow \beta) = 0$ or there is a $\gamma \in \mathcal{V}_n$ for which $v_{\mathsf{g}}(\gamma) = v(\gamma \twoheadrightarrow \alpha) = 1$. If for all $\alpha \in \mathcal{V}_n$, $v(\alpha \twoheadrightarrow \beta) = 0$ then by definition, $v_{\mathsf{g}}^{0}(\beta) = 1$ and thus $v_{\mathsf{g}}(\beta) = 1$. Hence, $v_M^{\mathbf{L}}(\beta) = 1$. Otherwise there is a $j > 0$ such that for all $\alpha \in \mathcal{V}_n$ for which $v(\alpha \twoheadrightarrow \beta) = 1$ there is a $\gamma_\alpha \in \mathcal{V}_n$ for which $v_{\mathsf{g}}(\gamma_\alpha) = v(\gamma_\alpha \twoheadrightarrow \alpha) = 1$, $v_{\mathsf{g}}^{<j}(\gamma_\alpha) = 1$. Then, $\mathsf{defended}^{j}(\beta) = 1$ and thus $v_{\mathsf{g}}^{j}(\beta) = 1$. Thus, $v_{\mathsf{g}}(\beta) = v_M^{\mathbf{L}}(\beta) = 1$. Thus, (SCo) is valid. □

Theorem F.1.2 ***(Soundness).*** *Let* $\mathbf{L} \in \{\mathbf{L_A}, \mathbf{L_C}\}$, $\Gamma \subseteq \mathcal{W}_n$ *and* $\varphi \in \mathcal{W}_n$. *If* $\Gamma \vdash_{\mathbf{L}} \varphi$, *then* $\Gamma \Vdash_{\mathbf{L}} \varphi$.

Proof. We use for the proof an induction on the derivation steps of a proof of φ analogous to the way soundness is usually proven for classical propositional logic (cp. e.g., [3] p. 40–43). If φ is introduced by premise introduction then trivially $\Gamma \Vdash_{\mathbf{L}} \varphi$, as $\varphi \in \Gamma$. If φ is obtained by aggregation $\frac{\varphi_1 \quad \varphi_2}{\varphi_1 \wedge \varphi_2}$ where $\varphi = \varphi_1 \wedge \varphi_2$, then by induction hypothesis, $\Gamma \Vdash_{\mathbf{L}} \varphi_1, \varphi_2$. Therefore we have for all $\mathbf{L}$-models of

$\Gamma, M \models_{\mathbf{L}} \varphi_1, \varphi_2$. But then by (s$\wedge$), $M \models_{\mathbf{L}} \varphi_1 \wedge \varphi_2$. The proof is analogous for the other classical rules.

We have to take a look at the non-classical rules: If $\varphi = \neg\beta$ and it has been derived by (R$\twoheadrightarrow$) from α and $\alpha \twoheadrightarrow \beta$, then by induction hypothesis we have $\Gamma \Vdash_{\mathbf{L}} \alpha, \alpha \twoheadrightarrow \beta$. Thus for all $\mathbf{L}$-models M of Γ, $M \models_{\mathbf{L}} \alpha, \alpha \twoheadrightarrow \beta$. Since by Theorem (F.1.1) $v_M^{\mathbf{L}}$ validates (S$\twoheadrightarrow$), $v_M^{\mathbf{L}}(\beta) = 0$. Thus, by (s$\neg$), $v_M^{\mathbf{L}}(\neg\beta) = 1$.

If $\varphi = \neg\alpha$ and it has been derived from $\bot \not\twoheadrightarrow \alpha$ by (R$\bot$), then by induction hypothesis for each $\mathbf{L}$-model M of Γ, $v_M^{\mathbf{L}}(\bot \not\twoheadrightarrow \alpha) = 1$ and hence by (s$\neg$), $v_M^{\mathbf{L}}(\bot \twoheadrightarrow \alpha) \neq 1$. By (S$\bot$), $v_M^{\mathbf{L}}(\alpha) \neq 1$ and hence $v_M^{\mathbf{L}}(\alpha) = 0$. Thus, by (s$\neg$), $v_M^{\mathbf{L}}(\neg\alpha) = 1$.

If $\mathsf{def}\,\beta$ has been derived from α and $\beta \twoheadrightarrow \alpha$ by (Rad), then by induction hypothesis for each $\mathbf{L}$-model M of Γ, $v_M^{\mathbf{L}}(\alpha) = v_M^{\mathbf{L}}(\beta \twoheadrightarrow \alpha) = 1$. By (Sad), there is a $\gamma \in \mathcal{V}_n$ for which $v_M^{\mathbf{L}}(\gamma) = v_M^{\mathbf{L}}(\gamma \twoheadrightarrow \beta) = 1$. By (s$\vee$), $v_M^{\mathbf{L}}(\mathsf{def}\,\beta) = 1$.

For $\mathbf{L} = \mathbf{L_C}$: If β has been derived from $\bot \twoheadrightarrow \beta$ and $\bigwedge_{\delta\in\mathcal{V}_n}\big((\delta \twoheadrightarrow \beta) \supset \mathsf{def}\,\delta\big)$ by (RCo), then by induction hypothesis, $v_M^{\mathbf{L}}(\bot \twoheadrightarrow \beta) = v_M^{\mathbf{L}}\big(\bigwedge_{\delta\in\mathcal{V}_n}((\delta \twoheadrightarrow \beta) \supset \mathsf{def}\,\delta)\big) = 1$ for each $\mathbf{L}$-model M of Γ. Thus, $\min_{\delta\in\mathcal{V}_n}(\max(1 - v_M^{\mathbf{L}}(\delta \twoheadrightarrow \beta), v_M^{\mathbf{L}}(\mathsf{def}\,\delta))) = 1$. Thus, for all $\delta \in \mathcal{V}_n$, $v_M^{\mathbf{L}}(\delta \twoheadrightarrow \beta) = 0$ or $v_M^{\mathbf{L}}(\mathsf{def}\,\delta) = 1$. In the latter case, by (s$\vee$), there is an $\epsilon_\delta \in \mathcal{V}_n$ for which $v_M^{\mathbf{L}}(\epsilon_\delta) = v_M^{\mathbf{L}}(\epsilon_\delta \twoheadrightarrow \delta) = 1$. By (SCo), $v_M^{\mathbf{L}}(\beta) = 1$. □

Let $\Gamma \subseteq \mathcal{W}_n$ and $\mathbf{L} \in \{\mathbf{L_A}, \mathbf{L_C}\}$. Γ is $\mathbf{L}$-*consistent* iff $\Gamma \nvdash_{\mathbf{L}} \bot$. Γ is *maximally* $\mathbf{L}$-*consistent* iff (a) Γ is $\mathbf{L}$-consistent and (b) if $\Gamma \subseteq \Gamma'$ and Γ' is $\mathbf{L}$-consistent, then $\Gamma = \Gamma'$. We say that Γ is $\mathbf{L}$-*inconsistent* in case it is not $\mathbf{L}$-consistent. The proofs of the following propositions are standard for classical propositional logic and can for instance be found in van Dalen [3] pp. 43–45. The proofs for our logics are analogous.

Lemma F.1.5. *Let $\Gamma \subseteq \mathcal{W}_n$, $\varphi \in \mathcal{W}_n$, and $\mathbf{L} \in \{\mathbf{L_A}, \mathbf{L_C}\}$. We have:*

(i) If $\Gamma \cup \{\neg\varphi\}$ is $\mathbf{L}$-inconsistent, then $\Gamma \vdash_{\mathbf{L}} \varphi$
(ii) If $\Gamma \cup \{\varphi\}$ is $\mathbf{L}$-inconsistent then $\Gamma \vdash_{\mathbf{L}} \neg\varphi$.
(iii) Each $\mathbf{L}$-consistent set Γ is contained in a maximally $\mathbf{L}$-consistent set Γ'.
(iv) If Γ is maximally $\mathbf{L}$-consistent, then $\varphi \in \Gamma$ iff $\neg\varphi \notin \Gamma$, and $\neg\varphi \in \Gamma$ iff $\varphi \notin \Gamma$.
(v) For an $\mathbf{L}$-consistent Γ and for all $\varphi \in \Gamma$, $\Gamma \nvdash_{\mathbf{L}} \neg\varphi$.

Lemma F.1.6. *Let $\mathbf{L} \in \{\mathbf{L_A},\mathbf{L_C}\}$. If $\Gamma \subseteq \mathcal{W}_n$ is $\mathbf{L}$-consistent, then there is an $\mathbf{L}$-model M of Γ.*

Proof. By Lemma F.1.5iii we know that Γ is contained in a maximally $\mathbf{L}$-consistent Γ'. We define an assignment $v : \mathcal{V}_n \cup \mathcal{W}_n^{\twoheadrightarrow} \to \{0, 1\}$ by $\varphi \mapsto 1$ iff $\varphi \in \Gamma'$. I will show that $v_M^{\mathbf{L}}(\varphi) = 1$ iff $\varphi \in \Gamma'$. In order to do so we first show that $v_{\mathsf{i}}^{\mathbf{L}} = 1$.

Suppose $v_{\text{Rbot}} = 0$ and thus $\max_{\delta\in\mathcal{V}_n}\big(\min(v(\delta), 1 - v(\bot \twoheadrightarrow \delta))\big) = 1$. Thus, there is an α for which $\min(v(\alpha), 1 - v(\bot \twoheadrightarrow \alpha)) = 1$. Thus, $v(\alpha) = 1$ and $v(\bot \twoheadrightarrow \alpha) = 0$. Hence, $\alpha \in \Gamma'$ and $\bot \twoheadrightarrow \alpha \notin \Gamma'$. Due to the maximal consistency of Γ', $\bot \not\twoheadrightarrow \alpha \in \Gamma'$. However, by (R$\bot$), $\neg\alpha \in \Gamma'$ and thus $\alpha \notin \Gamma'$,—a contradiction. Thus, $v_{\text{Rbot}} = 1$.

Suppose $v_{\text{Rral}} = 0$. Then $\max_{\gamma,\delta\in\mathcal{V}_n}(\min(v(\gamma), v(\gamma \twoheadrightarrow \delta), v(\delta))) = 1$. Thus, there are $\alpha, \beta \in \mathcal{V}_n$ such that $v(\alpha) = v(\alpha \twoheadrightarrow \beta) = v(\beta) = 1$. Hence, $\alpha, \beta, \alpha \twoheadrightarrow \beta \in \Gamma'$. But by (R$\twoheadrightarrow$) $\neg\beta \in \Gamma'$,—a contradiction. Thus, $v_{\text{Rral}} = 1$.

Suppose $v_{\text{Rad}} = 0$. Then $\max_{\gamma,\delta\in\mathcal{V}_n}(\min(v(\gamma), v(\delta \twoheadrightarrow \gamma), 1 - v(\mathsf{def}\,\delta)) = 1$. Thus, there are $\alpha, \beta \in \mathcal{V}_n$ such that $v(\alpha) = v(\beta \twoheadrightarrow \alpha) = 1$ and $v(\mathsf{def}\,\beta) = 0$. Thus, $\alpha, \beta \twoheadrightarrow \alpha \in \Gamma'$. By (Rad), $\mathsf{def}\,\beta \in \Gamma'$. Note that by definition $v(\mathsf{def}\,\beta) = \max_{\gamma\in\mathcal{V}_n}(\min(v(\gamma), v(\gamma \twoheadrightarrow \beta))) = 0$. Thus, for all $\gamma \in \mathcal{V}_n$, $v(\gamma) = 0$ or $v(\gamma \twoheadrightarrow \beta) = 0$. Thus due to the maximal consistency of Γ', for all $\gamma \in \mathcal{V}_n$, $\neg\gamma \in \Gamma'$, or $\gamma \not\twoheadrightarrow \beta \in \Gamma'$. Thus, due to the maximal consistency of Γ', $\neg\gamma \vee (\gamma \not\twoheadrightarrow \beta) \in \Gamma'$ for all $\gamma \in \mathcal{V}_n$. Thus, $\neg(\gamma \wedge (\gamma \twoheadrightarrow \beta)) \in \Gamma'$ for all $\gamma \in \mathcal{V}_n$. Thus, $\bigwedge_{\gamma\in\mathcal{V}_n} \neg(\gamma \wedge (\gamma \twoheadrightarrow \beta)) \in \Gamma'$ and hence, $\neg \bigvee_{\gamma\in\mathcal{V}_n}(\gamma \wedge (\gamma \twoheadrightarrow \beta)) \in \Gamma'$. Thus $\neg\mathsf{def}\,\beta \in \Gamma'$,—a contradiction. Thus, $v_{\text{Rad}} = 1$.

In the case of $\mathbf{L_C}$ also $v_{\text{RCo}} = 1$ has to be shown. Suppose $v_{\text{RCo}} = 0$. Then $\max_{\delta\in\mathcal{V}_n}\big(\min(v(\bot \twoheadrightarrow \delta), \min_{\alpha\in\mathcal{V}_n}(\max(1 - v(\alpha \twoheadrightarrow \delta), v(\mathsf{def}\,\alpha))), 1 - v(\delta))\big) = 1$. Thus, there is a $\beta \in \mathcal{V}_n$ for which $\min(v(\bot \twoheadrightarrow \beta), \min_{\alpha\in\mathcal{V}_n}(\max(1 - v(\alpha \twoheadrightarrow \beta), v(\mathsf{def}\,\alpha))), 1 - v(\beta)) = 1$. Thus, $v(\bot \twoheadrightarrow \beta) = 1$, $v(\beta) = 0$ and for all $\alpha \in \mathcal{V}_n$, $\max(1 - v(\alpha \twoheadrightarrow \beta), v(\mathsf{def}\,\alpha)) = 1$. Thus, $\bot \twoheadrightarrow \beta \in \Gamma'$ and $\beta \notin \Gamma'$ and hence, $\neg\beta \in \Gamma'$. Moreover, for all $\alpha \in \mathcal{V}_n$ either $v(\alpha \twoheadrightarrow \beta) = 0$ and hence $\alpha \not\twoheadrightarrow \beta \in \Gamma'$ or $v(\mathsf{def}\,\alpha) = 1$. In the latter case, $\max_{\gamma\in\mathcal{V}_n}(\min(v(\gamma), v(\gamma \twoheadrightarrow \alpha))) = 1$ and hence there is a $\gamma_\alpha \in \mathcal{V}_n$ for which $\min(v(\gamma_\alpha), v(\gamma_\alpha \twoheadrightarrow \alpha)) = 1$. Thus, $v(\gamma_\alpha) = v(\gamma_\alpha \twoheadrightarrow \alpha) = 1$. Thus, $\gamma_\alpha, \gamma_\alpha \twoheadrightarrow \alpha \in \Gamma'$. Hence, due to the maximal consistency of Γ', $\mathsf{def}\,\alpha \in \Gamma'$. Thus, for all $\alpha \in \mathcal{V}_n$, $\mathsf{def}\,\alpha \vee (\alpha \not\twoheadrightarrow \beta) \in \Gamma'$ and hence, $\bigwedge_{\alpha\in\mathcal{V}_n}\big((\alpha \twoheadrightarrow \beta) \supset \mathsf{def}\,\alpha\big) \in \Gamma'$. By (RCo), $\beta \in \Gamma'$,—a contradiction. Thus, $v_{\text{RCo}} = 1$.

We have shown that $v^{\mathbf{L}}_{\mathsf{i}} = 1$. We will show now that $v^{\mathbf{L}}_M(\varphi) = 1$ iff $\varphi \in \Gamma'$ by an induction on the length l of $\varphi \in \mathcal{W}_n$.

"$l = 0$": By Lemma F.1.3i, $v^{\mathbf{L}}_M(\varphi) = v(\varphi)$, for all $\varphi \in \mathcal{V}_n \cup \mathcal{W}^{\twoheadrightarrow}_n$. Thus, $v^{\mathbf{L}}_M(\varphi) = 1$ iff $\varphi \in \Gamma'$ for all $\varphi \in \mathcal{V}_n \cup \mathcal{W}^{\twoheadrightarrow}_n$.

"$l \Rightarrow l + 1$": Let $\varphi = \neg\varphi' \in \Gamma'$, then $\varphi' \notin \Gamma'$ and by the induction hypothesis, $v^{\mathbf{L}}_M(\varphi') = 0$. By (s$\neg$), $v^{\mathbf{L}}_M(\varphi) = 1$. Now let $v^{\mathbf{L}}_M(\varphi) = 1$, then by (s$\neg$), $v^{\mathbf{L}}_M(\varphi') = 0$ and by the induction hypothesis, $\varphi' \notin \Gamma'$. Due to the maximal consistency of Γ', $\varphi \in \Gamma'$.

For the other Boolean combinations the proof is analogous. Thus, $v^{\mathbf{L}}_M(\varphi) = 1$ iff $\varphi \in \Gamma'$. Since $\Gamma' \supseteq \Gamma$, we also have: if $\varphi \in \Gamma$ then $v^{\mathbf{L}}_M(\varphi) = 1$. □

Corollary F.1.1. *Let* $\mathbf{L} \in \{\mathbf{L_A},\mathbf{L_C}\}$ *and* $\Gamma \subseteq \mathcal{W}_n$. *If* $\Gamma \nvdash_{\mathbf{L}} \varphi$ *then there is an* $\mathbf{L}$*-model of* Γ *for which* $v^{\mathbf{L}}_M(\varphi) = 0$.

Proof. If $\Gamma \nvdash_{\mathbf{L}} \varphi$, then by Lemma F.1.5i, $\Gamma \cup \{\neg\varphi\}$ is $\mathbf{L}$-consistent. By Lemma F.1.6 there is an $\mathbf{L}$-model M of $\Gamma \cup \{\neg\varphi\}$. By (s$\neg$), $v^{\mathbf{L}}_M(\varphi) = 0$. □

Theorem F.1.3 (Completeness). *Let* $\mathbf{L} \in \{\mathbf{L_A},\mathbf{L_C}\}$ *and* $\Gamma \subseteq \mathcal{W}_n$. *If* $\Gamma \Vdash_{\mathbf{L}} \varphi$ *then* $\Gamma \vdash_{\mathbf{L}} \varphi$.

Proof. By Corollary F.1.1 we know that if $\Gamma \nvdash_{\mathbf{L}} \varphi$, then $\Gamma \nVdash_{\mathbf{L}} \varphi$. □

By Theorem F.1.2 and Theorem F.1.3, we get:

Corollary F.1.2. *Let* $\mathbf{L} \in \{\mathbf{L_A}, \mathbf{L_C}\}$ *and* $\Gamma \subseteq \mathcal{W}_n$. $\Gamma \Vdash_{\mathbf{L}} \varphi$ *iff* $\Gamma \vdash_{\mathbf{L}} \varphi$.

F.2 Semantics for Admissible and Complete Extensions

Let us recapitulate some definitions.

Definition F.2.1. Where $\Omega_{\twoheadrightarrow} =_{\text{df}} \{\alpha \twoheadrightarrow \beta \mid \alpha \in \mathcal{V}_n \cup \{\bot\}, \beta \in \mathcal{V}_n\}$, logic $\mathbf{AL_X}$ where $\mathbf{X} \in \{\mathbf{A}, \mathbf{C}\}$ is the AL in standard format defined by the triple $\langle \mathbf{L_X}, \Omega_{\twoheadrightarrow}, \text{simple strategy}\rangle$. For a model M, $\text{Ab}_{\twoheadrightarrow}^{\mathbf{L_X}}(M) =_{\text{df}} \{\varphi \in \Omega_{\twoheadrightarrow} \mid M \models_{\mathbf{L_X}} \varphi\}$. Furthermore, $\Gamma_{\mathsf{A}}^n =_{\text{df}} \{p_i \twoheadrightarrow p_j \mid (a_i, a_j) \in \rightarrow\} \cup \{\bot \twoheadrightarrow p_i \mid a_i \in \mathcal{A}\}$ where $\mathsf{A} = \langle \mathcal{A}, \rightarrow\rangle$ and $\mathcal{A} \subseteq \mathcal{A}_n$.

The following rather technical insights vastly simplify the proofs for the representational results for our systems (Theorem 8.4.1, Corollary 8.4.1). The proofs of the following results are very easy and are left to the reader:

Theorem F.2.1. *Let* $\mathbf{L} \in \{\mathbf{L_A}, \mathbf{L_C}\}$ *and* $\Gamma, \Gamma' \subseteq \mathcal{W}_n$.

(i) $\mathbf{L}$ *is reflexive, i.e.,* $\Gamma \subseteq Cn_{\mathbf{L}}(\Gamma)$.
(ii) $\mathbf{L}$ *is monotonic, i.e., if* $\Gamma \subseteq \Gamma'$ *then* $Cn_{\mathbf{L}}(\Gamma) \subseteq Cn_{\mathbf{L}}(\Gamma')$.
(iii) $\mathbf{L}$ *is transitive, i.e., if* $\Gamma' \subseteq Cn_{\mathbf{L}}(\Gamma)$, *then* $Cn_{\mathbf{L}}(\Gamma') \subseteq Cn_{\mathbf{L}}(\Gamma)$.
(iv) $\mathbf{L}$ *is compact, i.e., there is an* $\mathbf{L}$*-model of* Γ *iff for each finite* $\Gamma' \subseteq \Gamma$ *there is an* $\mathbf{L}$*-model of* Γ'.

Lemma F.2.1. *Let* $\mathsf{A} = \langle \mathcal{A}, \rightarrow\rangle$ *be an AF for which* $\mathcal{A} \subseteq \mathcal{A}_n$ *and* $\mathbf{X} \in \{\mathbf{A}, \mathbf{C}\}$.

(i) $\Gamma_{\mathsf{A}}^n \vdash_{\mathbf{AL_X}} \bot \not\twoheadrightarrow p_i$ *for all* $p_i \in \mathcal{V}_n \setminus \{p_l \mid a_l \in \mathcal{A}\}$, *otherwise* $\bot \twoheadrightarrow p_i \in \Gamma_{\mathsf{A}}^n$.
(ii) $\Gamma_{\mathsf{A}}^n \vdash_{\mathbf{AL_X}} p_i \twoheadrightarrow p_j$ *iff* $(a_i, a_j) \in \rightarrow$ *iff* $p_i \twoheadrightarrow p_j \in \Gamma_{\mathsf{A}}^n$, *otherwise* $\Gamma_{\mathsf{A}}^n \vdash_{\mathbf{AL_X}} p_i \not\twoheadrightarrow p_j$.
(iii) *For all* $M \in \mathcal{M}_{\mathbf{AL_X}}(\Gamma_{\mathsf{A}}^n)$, $Ab_{\twoheadrightarrow}^{\mathbf{L_X}}(M) = \Gamma_{\mathsf{A}}^n$.
(iv) *If* $p_i \in \mathcal{V}_n \setminus \{p_l \mid a_l \in \mathcal{A}\}$, *then* $\Gamma_{\mathsf{A}}^n \vdash_{\mathbf{AL_X}} \neg p_i$.

Proof. Let $S \subseteq \mathcal{A}$, in case $\mathbf{X} = \mathbf{A}$, be an admissible extension and, in case $\mathbf{X} = \mathbf{C}$, a complete extension. We construct an $\mathbf{L_X}$-model M of Γ_{A}^n on basis of the assignment $v : \mathcal{V}_n \cup \mathcal{W}_n^{\twoheadrightarrow} \rightarrow \{0, 1\}$, where

$$\varphi \mapsto \begin{cases} 1 \text{ if } \varphi = p_i \text{ where } a_i \in S & [1] \\ 1 \text{ if } \varphi = p_i \twoheadrightarrow p_j \text{ where } (a_i, a_j) \in \rightarrow & [2] \\ 1 \text{ if } \varphi = \bot \twoheadrightarrow p_i \text{ where } a_i \in \mathcal{A} & [3] \\ 0 \text{ else} & [4] \end{cases}$$

It is easy to check that the semantical properties (Sad), (S↠) and (S⊥) (resp. (Sad), (S↠), (S⊥) and (SCo) in the case $\mathbf{X} = \mathbf{C}$) are valid. Let for rule (S↠), $v(p_i) = v(p_i \twoheadrightarrow p_j) = 1$, then by definition of v, $a_i \in S$ and $(a_i, a_j) \in \rightarrow$. But then

$v(p_j) = 0$, as due the conflict-freeness of S, there is no $a_k \in S$ such that $(a_i, a_k) \in \rightarrow$. For rule (Sad) suppose $v(p_i) = v(p_j \twoheadrightarrow p_i) = 1$. By definition of v, $a_i \in S$ and $(a_j, a_i) \in \rightarrow$. Since S is admissible, there is an $a_k \in S$ such that $(a_k, a_j) \in \rightarrow$. But then, $v(p_k) = v(p_k \twoheadrightarrow p_j) = 1$. For (S$\bot$) let $v(p_i) = 1$. Then $a_i \in \mathcal{A}$ and thus $v(\bot \twoheadrightarrow p_i) = 1$. For (SCo) in the case $\mathbf{X} = \mathbf{C}$ let $v(\bot \twoheadrightarrow p_i) = 1$ and for all $p_k \in \mathcal{V}_n$ let $v(p_k \twoheadrightarrow p_i) = 0$ or there is a p_m such that $v(p_m) = v(p_m \twoheadrightarrow p_k) = 1$. Since $\{(a_j, a_l) \mid v(p_j \twoheadrightarrow p_l) = 1\} = \rightarrow$ and $\{a_l \mid v(p_l) = 1\} = S$, we know that a_i is defended by S. Thus, $a_i \in S$ and hence $v(p_i) = 1$.

Thus, since $v_{\mathsf{i}}^{\mathbf{Lx}} = 1$, $v_M^{\mathbf{Lx}}(\varphi) = v(\varphi)$ for all $\varphi \in \mathcal{V}_n \cup \mathcal{W}_n^{\twoheadrightarrow}$ by Lemma F.1.3i.

Note that by construction, $\mathrm{Ab}_{\twoheadrightarrow}^{\mathbf{Lx}}(M) = \Gamma_{\mathsf{A}}^n$. (i) – (iii) follow immediately. (iv) follows by (i) and (R$\bot$). □

Lemma F.2.1 shows that for our ALs all minimal disjunctions of abnormalities $\mathsf{Dab}(\Delta)$ (where $\Delta \subseteq \Omega_{\twoheadrightarrow}$) derivable by our lower limits $\mathbf{L_A}$ and $\mathbf{L_C}$ are such that Δ is a singleton. Therefore, the simple strategy defines in these cases the same consequence relation as the minimal abnormality strategy (or the reliability strategy, see Theorem 2.4.11).

Since all our ALs for skeptical acceptance are in the standard format and employ $\mathbf{L_A}$ resp. $\mathbf{L_C}$ as lower limit logics, the completeness and soundness of $\mathbf{L_A}$ and $\mathbf{L_C}$ provides us immediately with the completeness and soundness of the adaptive systems $\mathbf{AL_A}$ resp. $\mathbf{AL_C}$:

Theorem F.2.2. *We have $\Gamma_{\mathsf{A}}^n \Vdash_{\mathbf{AL_X}} \varphi$ iff $\Gamma_{\mathsf{A}}^n \vdash_{\mathbf{AL_X}} \varphi$ where $\mathbf{X} \in \{\mathbf{A}, \mathbf{C}\}$.*

Proof. This is due to Lemma F.2.1, Theorem F.2.1, Corollary F.1.2 and Theorem 2.4.12. □

Proof (Proof of Theorem 8.4.1i). Let $\mathsf{A} = \langle \mathcal{A}, \rightarrow \rangle$ be an AF for which $\mathcal{A} \subseteq \mathcal{A}_n$. "$\Leftarrow$": Let $M \in \mathcal{M}_{\mathbf{AL_A}}(\Gamma_{\mathsf{A}}^n)$. Note that $S = \{a_l \mid M \models_{\mathbf{L_A}} p_l\} \subseteq \mathcal{A}$, as for all $p_j \in \mathcal{V}_n \setminus \{p_l \mid a_l \in \mathcal{A}\}$ by Lemma F.2.1iv, $\Gamma_{\mathsf{A}}^n \vdash_{\mathbf{AL_A}} \neg p_j$ and thus $\Gamma_{\mathsf{A}}^n \Vdash_{\mathbf{AL_A}} \neg p_j$. S is conflict-free, as if $(a_i, a_j) \in \rightarrow$, then $p_i \twoheadrightarrow p_j \in \Gamma_{\mathsf{A}}^n$ and therefore $M \models_{\mathbf{L_A}} p_i \twoheadrightarrow p_j$. Hence, if $M \models_{\mathbf{L_A}} p_i$ then by (S$\twoheadrightarrow$) we have $M \models_{\mathbf{L_A}} \neg p_j$. Hence, either $a_i \notin S$ or $a_j \notin S$. For admissibility let $(a_j, a_i) \in \rightarrow$ and $M \models_{\mathbf{L_A}} p_i$. Hence, $p_j \twoheadrightarrow p_i \in \Gamma_{\mathsf{A}}^n$. Now by (Sad) there is a p_k such that $M \models_{\mathbf{L_A}} p_k, p_k \twoheadrightarrow p_j$. By Lemma F.2.1ii, $(a_k, a_j) \in \rightarrow$. Hence, S is admissible.

"$\Rightarrow$": Let $S \subseteq \mathcal{A}$ be an admissible set. Define an $\mathbf{AL_A}$-model M with respect to S as in Lemma F.2.1. Obviously $\{a_l \mid M \models_{\mathbf{L_A}} p_l\} = S$ and by construction $M \models_{\mathbf{L_A}} \Gamma_{\mathsf{A}}^n$. □

Proof (Proof of Theorem 8.4.1ii). Let $\mathsf{A} = \langle \mathcal{A}, \rightarrow \rangle$ be an AF for which $\mathcal{A} \subseteq \mathcal{A}_n$. "$\Leftarrow$": Let $M \in \mathcal{M}_{\mathbf{AL_C}}(\Gamma_{\mathsf{A}}^n)$. Note that $S = \{a_l \mid M \models_{\mathbf{L_C}} p_l\} \subseteq \mathcal{A}$, as for all $p_j \in \mathcal{V}_n \setminus \{p_l \mid a_l \in \mathcal{A}\}$ by Lemma F.2.1iv, $\Gamma_{\mathsf{A}}^n \vdash_{\mathbf{AL_C}} \neg p_j$ and thus $\Gamma_{\mathsf{A}}^n \Vdash_{\mathbf{AL_C}} \neg p_j$. S is conflict-free, as if $(a_i, a_j) \in \rightarrow$, then $p_i \twoheadrightarrow p_j \in \Gamma_{\mathsf{A}}^n$ and therefore $M \models_{\mathbf{L_C}} p_i \twoheadrightarrow p_j$. Hence, if $M \models_{\mathbf{L_C}} p_i$ then by (S$\twoheadrightarrow$) we have $M \models_{\mathbf{L_C}} \neg p_j$. Hence, either $a_i \notin S$ or $a_j \notin S$. For admissibility let $(a_j, a_i) \in \rightarrow$ and $M \models_{\mathbf{L_C}} p_i$. Hence, $p_j \twoheadrightarrow p_i \in \Gamma_{\mathsf{A}}^n$. Now by (Sad) there is a p_k such that $M \models_{\mathbf{L_C}} p_k, p_k \twoheadrightarrow p_j$. By

Lemma F.2.1ii, $(a_k, a_j) \in \rightarrow$. Hence, S is admissible. Let $a_i \in \mathcal{A}$ be such that S defends a_i. We have to show that $a_i \in S$. Note that, since $a_i \in \mathcal{A}$, $\bot \twoheadrightarrow p_i \in \Gamma^n_{\mathsf{A}}$. Let $\text{Att}_{a_i} =_{\text{df}} \{a_l \mid (a_l, a_i) \in \rightarrow\}$. If $\text{Att}_{a_i} = \emptyset$, then by Lemma F.2.1ii, for all p_j, $M \models_{\mathbf{L_C}} p_j \not\twoheadrightarrow p_i$ and thus by (SCo), $M \models_{\mathbf{L_C}} p_i$. Hence, $a_i \in S$. Let now $\text{Att}_{a_i} \neq \emptyset$. By Lemma F.2.1, $\{p_l \mid (a_l, a_i) \in \rightarrow\} = \{p_l \mid M \models_{\mathbf{L_C}} p_l \twoheadrightarrow p_i\}$. Since S defends a_i, there is an $a_k \in S$ for each $a_j \in \text{Att}_{a_i}$ for which $(a_k, a_j) \in \rightarrow$. Thus, there is a p_k such that $M \models_{\mathbf{L_C}} p_k, p_k \twoheadrightarrow p_j$ for each $p_j \in \{p_l \mid (a_l, a_i) \in \rightarrow\} = \{p_l \mid M \models_{\mathbf{L_C}} p_l \twoheadrightarrow p_i\}$. By (SCo), $M \models_{\mathbf{L_C}} p_i$ and thus, $a_i \in S$.

"$\Rightarrow$": Let $S \subseteq \mathcal{A}$ be a complete extension. We construct an $\mathbf{AL_C}$-model M of Γ^n_{A} such that $\{a_l \mid M \models_{\mathbf{L_C}} p_l\} = S$ as in Lemma F.2.1 (for the $\mathbf{L} = \mathbf{L_C}$ case). □

Proof (Proof of Corollary 8.4.1i,ii). Let $\mathbf{L} \in \{\mathbf{AL_A}, \mathbf{AL_C}\}$ and $\mathsf{A} = \langle \mathcal{A}, \rightarrow \rangle$ an AF for which $\mathcal{A} \subseteq \mathcal{A}_n$.

Let a_i be skeptically accepted w.r.t. admissible (in case $\mathbf{L} = \mathbf{AL_A}$) resp. complete (in case $\mathbf{L} = \mathbf{AL_C}$) extensions. Then $a_i \in S$ for all admissible (resp. complete) extensions S, $a_i \in \bigcap\{S \subseteq \mathcal{A} \mid S$ is an admissible (resp. complete) extension of $\mathsf{A}\}$. By Theorem 8.4.1i,ii we immediately get $\Gamma^n_{\mathsf{A}} \Vdash_{\mathbf{L}} p_i$ and due to Theorem F.2.2 we have $\Gamma^n_{\mathsf{A}} \vdash_{\mathbf{L}} p_i$.

Let $\Gamma^n_{\mathsf{A}} \vdash_{\mathbf{L}} p_i$. By Theorem F.2.2 we know that $\Gamma^n_{\mathsf{A}} \Vdash_{\mathbf{L}} p_i$. By Theorem 8.4.1i,ii we immediately get $a_i \in \bigcap\{S \subseteq \mathcal{A} \mid S$ is an admissible (resp. complete) extension by $\mathsf{A}\}$. □

F.3 Semantics for the Other Extension Types

F.3.1 Semantics for Preferred Extensions

Let us recapitulate some definitions.

Definition F.3.1. Where $\Omega_P =_{\text{df}} \{\neg\alpha \mid \alpha \in \mathcal{V}_n\}$ and $\mathbf{L} \in \{\mathbf{L_A}, \mathbf{L_C}\}$, $\text{Ab}^{\mathbf{L}}_P(M) =_{\text{df}} \{\varphi \in \Omega_P \mid M \models_{\mathbf{L}} \varphi\}$ for a model M. The sequential adaptive logic $\mathbf{AL_P}$ is defined by the following triple $\langle \mathbf{L_A}, [\Omega_{\twoheadrightarrow}, \Omega_P], [\text{simple strategy, minimal abnormality}]\rangle$.

Theorem F.3.1. *$\Gamma^n_{\mathsf{A}} \vdash_{\mathbf{AL_P}} \varphi$ iff $\Gamma^n_{\mathsf{A}} \Vdash_{\mathbf{AL_P}} \varphi$.*

Proof. This is due to Lemma F.2.1, Theorem F.2.1, Corollary F.1.2, Corollary 3.2.3, and Theorem 3.3.1. □

Lemma F.3.1. *Let $\mathsf{A} = \langle \mathcal{A}, \rightarrow \rangle$ be an AF for which $\mathcal{A} \subseteq \mathcal{A}_n$ and $\mathbf{L} \in \{\mathbf{L_A}, \mathbf{L_C}\}$. For all $M, N \in \mathcal{M}_{\mathbf{L}}\left(\Gamma^n_{\mathsf{A}}\right)$, if $Ab^{\mathbf{L}}_P(M) = Ab^{\mathbf{L}}_P(N)$ then $v^{\mathbf{L}}_M = v^{\mathbf{L}}_N$.*

Proof. Let $p_i \in \mathcal{V}_n$ and $M \models_{\mathbf{L}} p_i$. Then $N \models_{\mathbf{L}} p_i$ since otherwise $\neg p_i \in \text{Ab}^{\mathbf{L}}_P(N) \setminus \text{Ab}^{\mathbf{L}}_P(M)$. Let $\varphi \in \mathcal{W}^{\twoheadrightarrow}_n$. Then by Lemma F.2.1iii $M \models_{\mathbf{L}} \varphi$ iff $\varphi \in \Gamma^n_{\mathsf{A}}$ iff $N \models_{\mathbf{L}} \varphi$. It follows immediately by (s$\bot$), (s$\wedge$), (s$\vee$), (s$\supset$), and (s$\neg$) that if M and N valuate all formulas in $\mathcal{V}_n \cup \mathcal{W}^{\twoheadrightarrow}_n$ in the same way, then $v^{\mathbf{L}}_M = v^{\mathbf{L}}_N$. □

Proof (Proof of Theorem 8.4.1iii). Let $\mathsf{A} = \langle \mathcal{A}, \rightarrow \rangle$ be an AF for which $\mathcal{A} \subseteq \mathcal{A}_n$. "$\Leftarrow$": Let $M \in \mathcal{M}_{\mathbf{AL_P}}(\Gamma^n_{\mathsf{A}}) \subseteq \mathcal{M}_{\mathbf{AL_A}}(\Gamma^n_{\mathsf{A}})$ and $S = \{a_l \mid M \models_{\mathbf{L_A}} p_l\}$. By Theorem 8.4.1i we know that S is an admissible extension. Suppose there is a $S' \supset S$ such that S' is a preferred extension of A. Then S' is also admissible and therefore by Theorem 8.4.1i there is a $M' \in \mathcal{M}_{\mathbf{AL_A}}(\Gamma^n_{\mathsf{A}})$ such that $S' = \{a_l \mid M' \models_{\mathbf{L_A}} p_l\}$. But then $\{a_l \mid M' \models_{\mathbf{L_A}} p_l\} \supset \{a_l \mid M \models_{\mathbf{L_A}} p_l\}$ and hence $\mathrm{Ab}^{\mathbf{L_A}}_P(M') \subset \mathrm{Ab}^{\mathbf{L_A}}_P(M)$,—a contradiction. "$\Rightarrow$": Let $S \subseteq \mathcal{A}$ be a preferred extension. Therefore S is also an admissible extension and by Theorem 8.4.1i there is a $M \in \mathcal{M}_{\mathbf{AL_A}}(\Gamma^n_{\mathsf{A}})$ such that $S = \{a_l \mid M \models_{\mathbf{L_A}} a_l\}$. Suppose there is an $N \in \mathcal{M}_{\mathbf{AL_A}}(\Gamma^n_{\mathsf{A}})$ such that $\mathrm{Ab}^{\mathbf{L_A}}_P(N) \subset \mathrm{Ab}^{\mathbf{L_A}}_P(M)$. We know by Theorem 8.4.1i that there is an admissible extension T such that $T = \{a_l \mid N \models_{\mathbf{L_A}} p_l\}$. But then $T \supset S$,—a contradiction. □

Proof (Proof of Corollary 8.4.1iii). The proof is analogous to the proof of Corollary 8.4.1i. □

F.3.2 Semantics for Grounded Extensions

Let us recapitulate some definitions.

Definition F.3.2. Where $\Omega_G = \mathcal{V}_n$ and $\mathbf{L} \in \{\mathbf{L_A}, \mathbf{L_C}\}$, let for a model M, $\mathrm{Ab}^{\mathbf{L}}_G(M) =_{\mathrm{df}} \{\varphi \in \Omega_G \mid M \models_{\mathbf{L}} \varphi\}$. The prioritized adaptive logic $\mathbf{AL_G}$ is defined by the following triple: $\langle \mathbf{L_C}, [\Omega_{\twoheadrightarrow}, \Omega_G]$, [simple strategy, simple strategy]$\rangle$.

Let moreover, $\mathbf{AL^m_G}$ be the sequential AL defined by

$$\langle \mathbf{L_C}, [\Omega_{\twoheadrightarrow}, \Omega_G], [\text{simple strategy, minimal abnormality}]\rangle.$$

Lemma F.3.2. *$\Gamma^n_{\mathsf{A}} \vdash_{\mathbf{AL^m_G}} \varphi$ iff $\Gamma^n_{\mathsf{A}} \Vdash_{\mathbf{AL^m_G}} \varphi$.*

Proof. This is due to Lemma F.2.1, Theorem F.2.1, Corollary F.1.2, Corollary 3.2.3, and Theorem 3.3.1. □

Theorem F.3.2. *Let $\mathsf{A} = \langle \mathcal{A}, \rightarrow \rangle$ be an AF where $\mathcal{A} \subseteq \mathcal{A}_n$.*

(i) $\mathbf{AL^m_G}$ semantically represents grounded extensions for AFs with at most n arguments.
(ii) For all $M, N \in \mathcal{M}_{\mathbf{AL^m_G}}(\Gamma^n_{\mathsf{A}})$, $v^{\mathbf{L_C}}_M = v^{\mathbf{L_C}}_N$.
(iii) All minimal disjunctions $\mathsf{Dab}(\Delta)$ where $\Delta \subseteq \Omega_G$ in $Cn_{\mathbf{AL_C}}(\Gamma^n_{\mathsf{A}})$ are such that Δ is a singleton.
(iv) $\Gamma^n_{\mathsf{A}} \vdash_{\mathbf{AL^m_G}} \varphi$ iff $\Gamma^n_{\mathsf{A}} \vdash_{\mathbf{AL_G}} \varphi$.

Proof. Ad (i): "$\Rightarrow$": Let $M \in \mathcal{M}_{\mathbf{AL^m_G}}(\Gamma^n_{\mathsf{A}}) \subseteq \mathcal{M}_{\mathbf{AL_C}}(\Gamma^n_{\mathsf{A}})$. By Theorem 8.4.1ii we know that there is a complete extension S such that $S = \{a_l \mid M \models_{\mathbf{L_C}} p_l\}$. Suppose there is a $S' \subset S$ such that S' is a complete extension. By Theorem 8.4.1ii there is a $N \in \mathcal{M}_{\mathbf{AL_C}}(\Gamma^n_{\mathsf{A}})$ such that $\{a_l \mid N \models_{\mathbf{L_C}} p_l\} = S'$. But then $\{a_l \mid N \models_{\mathbf{L_C}} p_l\} \subset \{a_l \mid M \models_{\mathbf{L_C}} p_l\}$ and therefore $\mathrm{Ab}^{\mathbf{L_C}}_G(N) \subset \mathrm{Ab}^{\mathbf{L_C}}_G(M)$,—a contradiction.

"$\Leftarrow$": Let $S \subseteq \mathcal{A}$ be the grounded extension, then since S is complete there is by Theorem 8.4.1ii a $M \in \mathcal{M}_{\mathbf{AL_C}}(\Gamma^n_{\mathsf{A}})$ such that $\{a_l \mid M \models_{\mathbf{L_C}} p_l\} = S$. Now suppose there is a $N \in \mathcal{M}_{\mathbf{AL_C}}(\Gamma^n_{\mathsf{A}})$ such that $\mathrm{Ab}^{\mathbf{L_C}}_G(N) \subset \mathrm{Ab}^{\mathbf{L_C}}_G(M)$. But then $\{a_l \mid N \models_{\mathbf{L_C}} p_l\} \subset \{a_l \mid M \models_{\mathbf{L_C}} p_l\}$. Also by Theorem 8.4.1ii we know that there is a complete extension S' such that $S' = \{a_l \mid N \models_{\mathbf{L_C}} p_l\}$. But then $S' \subset S$,—a contradiction.

Ad (ii): This is due to the fact that there is only one grounded extension $S \subseteq \mathcal{A}$ and for all $M \in \mathcal{M}_{\mathbf{AL^m_G}}(\Gamma^n_{\mathsf{A}})$, $\{a_l \mid M \models_{\mathbf{L_C}} p_l\} = \{a_l \mid p_l \in \mathrm{Ab}^{\mathbf{L_C}}_G(M)\} = S$ by (i). Furthermore, by Lemma F.2.1ii, $\{\alpha \twoheadrightarrow p_l \mid M \models_{\mathbf{L_C}} \alpha \twoheadrightarrow p_l\} = \{\alpha \twoheadrightarrow p_l \mid N \models_{\mathbf{L_C}} \alpha \twoheadrightarrow p_l\} = \{\alpha \twoheadrightarrow p_l \mid \alpha \twoheadrightarrow p_l \in \Gamma^n_{\mathsf{A}}\}$ for all $M, N \in \mathcal{M}_{\mathbf{AL_C}}(\Gamma^n_{\mathsf{A}})$. By (s$\bot$), (s$\wedge$), (s$\vee$), (s$\supset$), and (s$\neg$), $v^{\mathbf{L_C}}_M = v^{\mathbf{L_C}}_N$.

Ad (iii): Follows immediately by (ii). Ad (iv): Follows immediately by (iii) and Theorem 2.4.11. □

Corollary F.3.1. $\Gamma^n_{\mathsf{A}} \vdash_{\mathbf{AL_G}} \varphi$ *iff* $\Gamma^n_{\mathsf{A}} \Vdash_{\mathbf{AL_G}} \varphi$.

Proof (Proof of Theorem 8.4.1iv). Follows immediately by Theorem F.3.2. □

Proof (Proof of Corollary 8.4.1iv). The proof is analogous to the proof of Corollary 8.4.1i. □

F.3.3 Semantics for (Semi)-Stable Extensions

Let us recapitulate some definitions.

Definition F.3.3. Where $\Omega_S =_{\mathrm{df}} \{\neg\alpha \wedge \neg\mathsf{def}\alpha \mid \alpha \in \mathcal{V}_n\}$ and $\mathbf{L} \in \{\mathbf{L_A}, \mathbf{L_C}\}$, let for a model M, $\mathrm{Ab}^{\mathbf{L}}_S(M) =_{\mathrm{df}} \{\varphi \in \Omega_S \mid M \models_{\mathbf{L}} \varphi\}$. The prioritized adaptive logic $\mathbf{AL_S}$ is defined by the following triple:

$$\langle \mathbf{L_C}, [\Omega_{\twoheadrightarrow}, \Omega_S], [\text{simple strategy, minimal abnormality strategy}]\rangle.$$

Theorem F.3.3. $\Gamma^n_{\mathsf{A}} \vdash_{\mathbf{AL_S}} \varphi$ *iff* $\Gamma^n_{\mathsf{A}} \Vdash_{\mathbf{AL_S}} \varphi$.

Proof. This is due to Lemma F.2.1, Theorem F.2.1, Corollary F.1.2, Corollary 3.2.3, and Theorem 3.3.1. □

In Footnote 5 in Chap. 8 we mentioned the following fact:

Fact F.3.1. *Let* $\mathsf{A} = \langle \mathcal{A}, \rightarrow\rangle$ *be an AF and* $S \subseteq \mathcal{A}$. *S is a semi-stable extension iff S is an admissible set of arguments for which there is no admissible set of arguments* $T \subseteq \mathcal{A}$ *such that* $T \cup T^+ \supset S \cup S^+$.

Proof. "$\Rightarrow$": S is admissible since it is complete. Suppose now that for an admissible T such that $T \cup T^+$ is maximal (w.r.t. $\subseteq$), $T \cup T^+ \supset S \cup S^+$. Thus T is not complete. Hence, $F(T) \neq T$. We have $T \subset F(T)$ since T is admissible. Therefore, there is an $\mathfrak{a} \in \mathcal{A} \setminus T$ such that T defends $\mathfrak{a}$. Let $T_{\mathfrak{a}} =_{\mathrm{df}} T \cup \{\mathfrak{a}\}$. Note that there is no argument in T attacking $\mathfrak{a}$ since T defends $\mathfrak{a}$ and is conflict-free. Furthermore, $\mathfrak{a}$ does not attack any argument in T, since, due to the fact that T is admissible, T defends itself against all attackers. Would $\mathfrak{a}$ attack T, then there would be an argument in T attacking $\mathfrak{a}$. But as we have just shown, this is not the case. Thus, $T_\mathfrak{a}$ is conflict-free. Suppose there is a $\mathfrak{b}$ attacking $T_\mathfrak{a}$. Then $\mathfrak{b}$ attacks T or it attacks $\mathfrak{a}$. Since T defends itself, in the first case there is an argument in T attacking $\mathfrak{b}$. In the second case there is also an argument in $T_\mathfrak{a}$ attacking $\mathfrak{b}$ since T defends $\mathfrak{a}$. Thus, $T_\mathfrak{a}$ is admissible.

Furthermore, $T_\mathfrak{a} \cup T_\mathfrak{a}{}^+ \supset T \cup T^+$ since obviously $T_\mathfrak{a}^+ \supseteq T^+$, $T_\mathfrak{a} \supset T$ and $\mathfrak{a} \notin (T \cup T^+)$,—a contradiction.

"$\Leftarrow$": Suppose there is an admissible extension $T \subseteq \mathcal{A}$ such that $S \subset T$. Then $T \cup T^+ \supset S \cup S^+$ since obviously $T^+ \supseteq S^+$, $T \supset S$ and for any $\mathfrak{a} \in T \setminus S$, $\mathfrak{a} \notin S \cup S^+$. Note that $\mathfrak{a} \notin S$, and $\mathfrak{a} \notin S^+$ since otherwise $\mathfrak{a} \in T^+$ which contradicts the conflictfreeness of T. However, that $T \cup T^+ \supset S \cup S^+$ is a contradiction since $S \cup S^+$ is maximal. Hence, S is a preferred extension. Since every preferred extension is complete, S is semi-stable. □

Lemma F.3.3. *Let* $\mathsf{A} = \langle \mathcal{A}, \rightarrow \rangle$ *be an AF for which* $\mathcal{A} \subseteq \mathcal{A}_n$ *and* $\mathbf{X} \in \{\mathbf{A}, \mathbf{C}\}$.

(i) Let $S \subseteq \mathcal{A}$ *be an admissible extension. For all* $M \in \mathcal{M}_{\mathbf{ALX}}(\Gamma^n_\mathsf{A})$ *such that* $\{a_l \mid M \models_{\mathbf{LX}} p_l\} = S$ *we have:* $a_i \in S \cup S^+$ *iff* $M \models_{\mathbf{LX}} p_i \vee \mathsf{def}\, p_i$.

(ii) Let $S, T \subseteq \mathcal{A}$ *be admissible extensions. For all* $M, N \in \mathcal{M}_{\mathbf{ALX}}(\Gamma^n_\mathsf{A})$ *where* $\{a_l \mid M \models_{\mathbf{LX}} p_l\} = S$, $\{a_l \mid N \models_{\mathbf{LX}} p_l\} = T$ *we have:* $Ab^{\mathbf{LX}}_S(M) \setminus Ab^{\mathbf{LX}}_S(N) = \{\neg p_l \wedge \neg \mathsf{def}\, p_l \mid a_l \in (T \cup T^+) \setminus (S \cup S^+)\}$.

Proof. Ad (i): Let $M \in \mathcal{M}_{\mathbf{ALX}}(\Gamma^n_\mathsf{A})$ such that $\{a_l \mid M \models_{\mathbf{LX}} p_l\} = S$. Let $M \models_{\mathbf{LX}} p_i \vee \mathsf{def}\, p_i$. If $M \models_{\mathbf{LX}} p_i$ then $a_i \in S$. If $M \models_{\mathbf{LX}} \mathsf{def}\, p_i$ then by rule (s$\vee$) and the definition of def, there is a p_j such that $M \models_{\mathbf{LX}} p_j, p_j \twoheadrightarrow p_i$. Therefore also $a_j \in S$. Then by Lemma F.2.1ii we have $(a_j, a_i) \in \rightarrow$ and therefore $a_i \in S^+$. Let now $a_i \in S \cup S^+$. If $a_i \in S$ then, by definition, $M \models_{\mathbf{LX}} p_i$. Let $a_i \in S^+$. Therefore there is a $a_j \in S$ such that $(a_j, a_i) \in \rightarrow$. Then $M \models_{\mathbf{LX}} p_j$ and $p_j \twoheadrightarrow p_i \in \Gamma^n_\mathsf{A}$. Thus, $M \models_{\mathbf{LX}} \mathsf{def}\, p_i$.

(ii) follows immediately by (i). □

Proof (Proof for Theorem 8.4.1v). Let $\mathsf{A} = \langle \mathcal{A}, \rightarrow \rangle$ be an AF for which $\mathcal{A} \subseteq \mathcal{A}_n$.

"$\Leftarrow$": Let $M \in \mathcal{M}_{\mathbf{AL_S}}(\Gamma^n_\mathsf{A}) \subseteq \mathcal{M}_{\mathbf{AL_C}}(\Gamma^n_\mathsf{A})$ and $S = \{a_l \mid M \models_{\mathbf{LC}} p_l\}$. Suppose there is a semi-stable extension $S' \subseteq \mathcal{A}$ such that $S' \cup S'^+ \supset S \cup S^+$. Then S' is also complete and therefore by Theorem 8.4.1ii there is a $N \in \mathcal{M}_{\mathbf{AL_C}}(\Gamma^n_\mathsf{A})$ such that $S' = \{a_l \mid N \models_{\mathbf{LC}} p_l\}$. Then by Lemma F.3.3ii, $\mathrm{Ab}^{\mathbf{LC}}_S(M) \supset \mathrm{Ab}^{\mathbf{LC}}_S(N)$,—a contradiction.

"$\Rightarrow$": Let $S \subseteq \mathcal{A}$ be a semi-stable extension. Due to the fact that S is complete there is by Theorem 8.4.1ii a $M \in \mathcal{M}_{\mathbf{AL_C}}(\Gamma^n_\mathsf{A})$ such that $\{a_l \mid M \models_{\mathbf{LC}} p_l\} = S$. Suppose there is a $N \in \mathcal{M}_{\mathbf{AL_C}}(\Gamma^n_\mathsf{A})$ such that $\mathrm{Ab}^{\mathbf{LC}}_S(N) \subset \mathrm{Ab}^{\mathbf{LC}}_S(M)$. By Theorem 8.4.1ii there is a complete extension $T \subseteq \mathcal{A}$ such that $\{a_l \mid N \models_{\mathbf{LC}} p_l\} = T$. By Lemma F.3.3ii we have $T \cup T^+ \supset S \cup S^+$,—a contradiction. □

Proof *(Proof of Corollary 8.4.1v).* The proof is analogous to the one for Corollary 8.4.1i. □

F.4 Semantics for Credulous Acceptance

F.4.1 Some Preliminary Results

We presented in Sect. 8.5 simplified definitions for our semantic consequence relations for the ALs for credulous acceptance (see Footnote 19). In this subsection we will show that we were justified in doing so. Therefore, we will first define the semantics in the way it is usually done for the normal selections strategy (see [4] and Sect. 2.8) which enables us to use the soundness and completeness results from Sect. 3.4. Then we show that the way we defined the semantics in Sect. 8.5 is equivalent.

Recall that the logics for credulous acceptance are defined as follows:

$$\mathbf{ALC_P} = \langle \mathbf{L_A}, [\Omega_{\twoheadrightarrow}, \Omega_P], \text{[simple strategy, normal selections]}\rangle$$
$$\mathbf{ALC_S} = \langle \mathbf{L_C}, [\Omega_{\twoheadrightarrow}, \Omega_S, \Omega_P], \text{[simple strategy, minimal abnormality strategy, normal selections]}\rangle$$

Where $\mathbf{X} \in \{\mathbf{P}, \mathbf{S}\}$, let $\mathbf{L[X]}$ be the lower limit logic of $\mathbf{ALC_X}$ and

$$\mathbf{K[X]} = \begin{cases} \mathbf{AL_A} & \text{if } \mathbf{X} = \mathbf{P} \\ \mathbf{AL_S} & \text{if } \mathbf{X} = \mathbf{S} \end{cases}$$

Let in the remainder $\mathbf{AP} = \langle \mathbf{L[X]}, \Omega_P, \text{normal selections}\rangle$.

The following Corollary follows with Theorem 3.4.2.

Corollary F.4.1. *$\varphi \in Cn_{\mathbf{AP}}\left(Cn_{\mathbf{K[X]}}\left(\Gamma^n_{\mathsf{A}}\right)\right)$ iff $\Gamma^n_{\mathsf{A}} \vdash_{\mathbf{ALC_X}} \varphi$.*

Let $\Vdash'_{\mathbf{ALC_X}}$ be defined as in Chap. 3 (Sect. 3.4) by: $\Gamma \vdash'_{\mathbf{ALC_X}} \varphi$ iff there is a $M \in \{M' \in \mathcal{M}_{\mathbf{K[X]}}(\Gamma) \mid \text{for all } M'' \in \mathcal{M}_{\mathbf{K[X]}}(\Gamma), Ab_P^{\mathbf{L[X]}}(M'') \not\subset Ab_P^{\mathbf{L[X]}}(M')\}$ such that for all $M' \in \mathcal{M}_{\mathbf{K[X]}}(\Gamma)$ for which $Ab_P^{\mathbf{L[X]}}(M'') = Ab_P^{\mathbf{L[X]}}(M')$, $M' \models_{\mathbf{L[X]}} \varphi$.

The following Corollary follows with Corollary 3.4.2 and Corollary F.4.1.

Corollary F.4.2. *$\Gamma^n_{\mathsf{A}} \vdash_{\mathbf{ALC_X}} \varphi$ iff $\Gamma^n_{\mathsf{A}} \Vdash'_{\mathbf{ALC_X}} \varphi$.*

Lemma F.4.1. *Where $\mathsf{A} = \langle \mathcal{A}, \rightarrow\rangle$ is an AF for which $\mathcal{A} \subseteq \mathcal{A}_n$: $\Gamma^n_{\mathsf{A}} \Vdash_{\mathbf{ALC_X}} \varphi$ iff $\Gamma^n_{\mathsf{A}} \Vdash'_{\mathbf{ALC_X}} \varphi$.*

Proof. $\Gamma^n_{\mathsf{A}} \Vdash_{\mathbf{ALC_X}} \varphi$, iff, there is a $M \in \{M' \in \mathcal{M}_{\mathbf{K[X]}}\left(\Gamma^n_{\mathsf{A}}\right) \mid$ there is no $M'' \in \mathcal{M}_{\mathbf{K[X]}}\left(\Gamma^n_{\mathsf{A}}\right)$ such that $Ab_P^{\mathbf{L[X]}}(M'') \subset Ab_P^{\mathbf{L[X]}}(M')\}$, $M \models \varphi$, iff [by Lemma F.3.1] there is a $M \in \{M' \in \mathcal{M}_{\mathbf{K[X]}}\left(\Gamma^n_{\mathsf{A}}\right) \mid$ there is no $M'' \in \mathcal{M}_{\mathbf{K[X]}}\left(\Gamma^n_{\mathsf{A}}\right)$ such that $\mathrm{Ab}_P^{\mathbf{L[X]}}(M'') \subset \mathrm{Ab}_P^{\mathbf{L[X]}}(M')\}$ such that for all $M' \in \mathcal{M}_{\mathbf{K[X]}}(\Gamma_{\mathsf{A}}^n)$ for which $\mathrm{Ab}_P^{\mathbf{L[X]}}(M) = \mathrm{Ab}_P^{\mathbf{L[X]}}(M')$, $M' \models \varphi$, iff, $\Gamma_{\mathsf{A}}^n \Vdash'_{\mathbf{ALC_X}} \varphi$. □

Corollary F.4.3. *Let* $\mathsf{A} = \langle \mathcal{A}, \rightarrow \rangle$ *be an AF for which* $\mathcal{A} \subseteq \mathcal{A}_n$*. The following statements are equivalent:*

(i) $\Gamma_{\mathsf{A}}^n \Vdash'_{\mathbf{ALC_X}} \varphi$
(ii) there is a $M \in \mathcal{M}_{\mathbf{ALC_X}}(\Gamma_{\mathsf{A}}^n)$ *for which* $M \models_{\mathbf{L[X]}} \varphi$
(iii) $\Gamma_{\mathsf{A}}^n \vdash_{\mathbf{ALC_X}} \varphi$
(iv) $\Gamma_{\mathsf{A}}^n \Vdash_{\mathbf{ALC_X}} \varphi$.

F.4.2 Semantics for Admissible, Complete, and Preferred Extensions

Corollary F.4.4. $\mathcal{M}_{\mathbf{ALC_P}}(\Gamma_{\mathsf{A}}^n) = \mathcal{M}_{\mathbf{AL_P}}(\Gamma_{\mathsf{A}}^n)$ *where* $\mathsf{A} = \langle \mathcal{A}, \rightarrow \rangle$ *is an AF for which* $\mathcal{A} \subseteq \mathcal{A}_n$.

Proof. Follows by the definitions. □

Fact F.4.1. *Where* A *is an AF, S is a preferred extension iff S is a maximal complete extension.*

Lemma F.4.2. *An argument* $\mathfrak{a}$ *is credulously accepted w.r.t. admissible (resp. complete) extensions iff* $\mathfrak{a}$ *is credulously accepted w.r.t. preferred extensions.*

Proof. If $\mathfrak{a}$ is credulously accepted w.r.t. the admissible (resp. complete) extension type, then it is an element of an admissible (resp. complete) extension S. Then there is a preferred extension $S' \supseteq S$. Therefore, $\mathfrak{a}$ is credulously accepted w.r.t. the preferred extension type. The other direction is clear, because every preferred extension is also an admissible (resp. complete) extension. □

Proof (Proof of Theorem 8.5.1i,ii). This follows immediately by Corollary F.4.4, Theorem 8.4.1iii and Fact F.4.1. □

Proof (Proof of Theorem 8.5.2i,ii,iii). Let $\mathsf{A} = \langle \mathcal{A}, \rightarrow \rangle$ be an AF for which $\mathcal{A} \subseteq \mathcal{A}_n$. Because of Lemma F.4.2 it is enough to show this for preferred extensions. If a_i is credulously accepted w.r.t. preferred extensions, then for a preferred extension S of A, $a_i \in S$. By Theorem 8.5.1ii there is a $M \in \mathcal{M}_{\mathbf{ALC_P}}(\Gamma_{\mathsf{A}}^n)$ such that $p_i \in \{p_l \mid M \models_{\mathbf{L_A}} p_l\}$. Hence, $\Gamma_{\mathsf{A}}^n \vdash_{\mathbf{ALC_P}} p_i$ by Corollary F.4.3.

If $\Gamma_{\mathsf{A}}^n \vdash_{\mathbf{ALC_P}} p_i$, then by Corollary F.4.3 there is a model $M \in \mathcal{M}_{\mathbf{ALC_P}}(\Gamma_{\mathsf{A}}^n)$ such that $M \models_{\mathbf{L_A}} p_i$. By Theorem 8.5.1ii there is a preferred extension S of Asuch that $S = \{a_l \mid M \models_{\mathbf{L_A}} p_l\}$. Therefore $a_i \in S$. □

F.4.3 Semantics for Grounded Extensions

Proof (Proof of Theorem 8.5.2iv). This is immediately clear in view of the fact that there is only one grounded extension and because of Corollary 8.4.1iv. □

F.4.4 Semantics for Stable and Semi-Stable Extensions

On first sight the AL defined by the triple

$$\mathbf{L} = \langle \mathbf{L_C}, [\Omega_{\twoheadrightarrow}, \Omega_S], [\text{simple strategy, normal selections}] \rangle$$

might to some readers seem to be a good candidate for a logical characterization of semi-stable extensions w.r.t. credulous acceptance. However, suppose we have the AF $\mathsf{A} = \langle \{a_1, a_2\}, \{(a_1, a_2), (a_2, a_1)\} \rangle$. In this case we have two stable extensions, namely $\{a_1\}$ and $\{a_2\}$. Our logic $\mathbf{AL_S}$ for language $\mathcal{W}_2$ has therefore two types of models, type (1) verifying p_1 and $\mathsf{def}\, p_2$ and type (2) verifying p_2 and $\mathsf{def}\, p_1$. Note that for models M of either type we have $\mathrm{Ab}_S^{\mathbf{Lc}}(M) = \emptyset$. Thus, it is easy to see that we have neither p_1 nor p_2 as a $\mathbf{L}$-consequence.

Note however, that we do have $\neg p_1 \vee \neg p_2$ as Ω_P-minimal Dab-consequence for $\mathbf{AL_S}$. If we apply normal selections w.r.t. Ω_P to the set of $\mathbf{AL_S}$-models we gain two selected sets of models: models of type (1) and models of type (2). Thus, this way we gain both, p_1 and p_2, as consequences.

Lemma F.4.3. *Where* $\mathsf{A} = \langle \mathcal{A}, \rightarrow \rangle$ *is an AF for which* $\mathcal{A} \subseteq \mathcal{A}_n$, $\mathcal{M}_{\mathbf{ALC_S}}(\Gamma^n_{\mathsf{A}}) = \mathcal{M}_{\mathbf{AL_S}}(\Gamma^n_{\mathsf{A}})$.

Proof. All $\mathbf{AL_S}$-models of Γ^n_{A} are Ω_P-minimally abnormal. To show this suppose there are $M, N \in \mathcal{M}_{\mathbf{AL_S}}(\Gamma^n_{\mathsf{A}})$ for which $\mathrm{Ab}_P^{\mathbf{Lc}}(M) \subset \mathrm{Ab}_P^{\mathbf{Lc}}(N)$. By Theorem 8.4.1v there are semi-stable extensions $S_M = \{a_l \mid M \models_{\mathbf{Lc}} p_l\}$ and $S_N = \{a_l \mid N \models_{\mathbf{Lc}} p_l\}$. Hence, $S_N \subset S_M$. But this is not possible, since both, S_M and S_N are also preferred extensions. □

Proof (Proof of Theorem 8.5.1iii). This is an immediate consequence of Lemma F.4.3 and Theorem 8.4.1v. □

Proof (Proof of Theorem 8.5.2v). Let $\mathsf{A} = \langle \mathcal{A}, \rightarrow \rangle$ be an AF for which $\mathcal{A} \subseteq \mathcal{A}_n$. If a_i is credulously accepted w.r.t. semi-stable extensions, then for a semi-stable extension S, $a_i \in S$. By Theorem 8.5.1iii there is an $M \in \mathcal{M}_{\mathbf{ALC_S}}(\Gamma^n_{\mathsf{A}})$, $M \models_{\mathbf{Lc}} p_i$. Hence, $\Gamma^n_{\mathsf{A}} \vdash_{\mathbf{ALC_S}} p_i$ by Corollary F.4.3.

If $\Gamma^n_{\mathsf{A}} \vdash_{\mathbf{ALC_S}} p_i$, then by Corollary F.4.3 there is a model $M \in \mathcal{M}_{\mathbf{ALC_S}}(\Gamma^n_{\mathsf{A}})$ such that $M \models_{\mathbf{Lc}} p_i$. By Theorem 8.5.1iii there is a semi-stable extension S such that $S = \{a_l \mid M \models_{\mathbf{Lc}} p_l\}$ and therefore $a_i \in S$. □

Appendix G
Appendix to Chapter 9

This Appendix provides the semantics to the logics defined in Chap. 9. Moreover, all representational results will be proven.

G.1 The Semantics

The semantics are defined analogous to the way they were defined for the logics for Dung's argumentation framework in Appendix F. Hence, the presentation in this section will be brief.

G.1.1 The Assignment

Let

$$\mathcal{W}_n^{\twoheadrightarrow,r} \quad := \quad \langle \mathcal{V}_n^{\wedge} \rangle \twoheadrightarrow \langle \mathcal{V}_n \rangle \mid \bot \twoheadrightarrow \langle \mathcal{V}_n \rangle$$

As in Appendix F I define the semantics for logics $\mathbf{L}$ (where $\mathbf{L} \in \{\mathbf{CL_A}, \mathbf{CL_C}\}$) via an assignment function $v : \mathcal{V}_n \cup \mathcal{W}_n^{\twoheadrightarrow,r} \to \{0, 1\}$ and an $\mathbf{L}$-valuation $v_M^{\mathbf{L}} : \mathcal{W}_n \to \{0, 1\}$ determined by the assignment. A model M is defined by the assignment v. We have to slightly alter our definitions in order to adjust them to the generalized framework. Let

C. Straßer, *Adaptive Logics for Defeasible Reasoning*, Trends in Logic 38, DOI: 10.1007/978-3-319-00792-2, © Springer International Publishing Switzerland 2014

$$v_\wedge\left(\bigwedge_I \alpha_i\right) =_{\text{df}} \min(\{v(\alpha_i) \mid i \in I\}) \text{ where } \alpha_i \in \mathcal{V}_n,$$

$$v_\vee\left(\bigvee_I \alpha_i\right) =_{\text{df}} \max(\{v(\alpha_i) \mid i \in I\}) \text{ where } \alpha_i \in \mathcal{V}_n,$$

$$v_{\mathsf{def}}\left(\mathsf{def} \bigwedge_I \alpha_i\right) =_{\text{df}} \max_{i \in I}\left(\max_{\beta \in \mathcal{V}_n^\wedge}\left(\min(v_\wedge(\beta), v(\beta \twoheadrightarrow \alpha_i))\right)\right) \text{ where } \bigwedge_I \alpha_i \in \mathcal{V}_n^\wedge,$$

$$v_\star(\alpha) =_{\text{df}} \begin{cases} v(\alpha) & \alpha \in \mathcal{V}_n \cup \mathcal{W}_n^{\twoheadrightarrow, r} \\ v_\wedge(\alpha) & \alpha \in \mathcal{V}_n^\wedge \setminus \mathcal{V}_n \\ v_\vee(\alpha) & \alpha \in \mathcal{V}_n^\vee \setminus \mathcal{V}_n \\ v_{\mathsf{def}}(\alpha) & \alpha = \mathsf{def}\,\beta, \beta \in \mathcal{V}_n^\wedge \end{cases}$$

We define

$$v_{\text{R}\twoheadrightarrow\text{C}} =_{\text{df}} 1 - \max_{\alpha \in \mathcal{V}_n^\wedge, \beta \in \mathcal{V}_n}\left(\min(v_\star(\alpha), v_\star(\alpha \twoheadrightarrow \beta), v_\star(\beta))\right)$$

$$v_{\text{R}\bot\text{C}} =_{\text{df}} 1 - \max_{\alpha \in \mathcal{V}_n}\left(\min(v_\star(\alpha), 1 - v_\star(\bot \twoheadrightarrow \alpha))\right)$$

$$v_{\text{RadC}} =_{\text{df}} 1 - \max_{\alpha \in \mathcal{V}_n, \beta \in \mathcal{V}_n^\wedge}\left(\min(v_\star(\alpha), v_\star(\beta \twoheadrightarrow \alpha), 1 - v_\star(\mathsf{def}\,\beta))\right)$$

$$v_{\text{RCoC}} =_{\text{df}} 1 - \max_{\beta \in \mathcal{V}_n}\Big(\min\big(v_\star(\bot \twoheadrightarrow \beta), \min_{\alpha \in \mathcal{V}_n^\wedge}(\max(1 - v_\star(\alpha \twoheadrightarrow \beta), v_\star(\mathsf{def}\,\alpha))), 1 - v_\star(\beta)\big)\Big)$$

$$v_{\mathsf{i}}^{\mathbf{CL_A}} =_{\text{df}} \min(v_{\text{R}\twoheadrightarrow\text{C}}, v_{\text{R}\bot\text{C}}, v_{\text{RadC}})$$

$$v_{\mathsf{i}}^{\mathbf{CL_C}} =_{\text{df}} \min(v_{\text{R}\twoheadrightarrow\text{C}}, v_{\text{R}\bot\text{C}}, v_{\text{RadC}}, v_{\text{RCoC}})$$

Note that $v_{\text{R}\twoheadrightarrow\text{C}}$ corresponds to our syntactical rule (R↠C) in the sense that $v_{\text{R}\twoheadrightarrow\text{C}} = 1$ if the assignment satisfies the semantic counterpart to (R↠C). That is to say, $v_{\text{R}\twoheadrightarrow\text{C}} = 1$ iff v has the following property (where $\mathsf{v} = v$):

$$\text{If } \mathsf{v}(\alpha) = \mathsf{v}(\alpha \twoheadrightarrow \beta) = 1, \text{ then } \mathsf{v}(\beta) = 0, \text{ where } \alpha \in \mathcal{V}_n^\wedge, \beta \in \mathcal{V}_n \tag{S↠C}$$

The situation is analogous for $v_{\text{R}\bot\text{C}}$, v_{RadC}, and v_{RCoC} with respect to the following properties:

$$\text{If } \mathsf{v}(\alpha) = 1, \text{ then } \mathsf{v}(\bot \twoheadrightarrow \alpha) = 1, \text{ where } \alpha \in \mathcal{V}_n. \tag{S⊥C}$$

$$\text{If } \mathsf{v}(\alpha) = \mathsf{v}\big((\bigwedge_I \beta_i) \twoheadrightarrow \alpha\big) = 1, \text{ then there is a } \gamma \in \mathcal{V}_n^\wedge \text{ for which } \mathsf{v}(\gamma) = \max_{i \in I}(\mathsf{v}(\gamma \twoheadrightarrow \beta_i)) = 1, \text{ where } \alpha, \beta_i \in \mathcal{V}_n. \tag{SadC}$$

$$\text{If } \mathsf{v}(\bot \twoheadrightarrow \beta) = 1 \text{ and for all } \bigwedge_I \alpha_i \in \mathcal{V}_n^\wedge \text{ we have } \big[\mathsf{v}\big((\bigwedge_I \alpha_i) \twoheadrightarrow \beta\big) = 0 \text{ or (there is a } \gamma \in \mathcal{V}_n^\wedge \text{ for which } \mathsf{v}(\gamma) = \max_{i \in I}(\mathsf{v}(\gamma \twoheadrightarrow \alpha_i)) = 1)\big], \text{ then } \mathsf{v}(\beta) = 1. \tag{SCoC}$$

We call an assignment v $\mathbf{CL_A}$-*intended* iff $v_{\mathsf{i}}^{\mathbf{CL_A}} = 1$, and we call it $\mathbf{CL_C}$-*intended* iff $v_{\mathsf{i}}^{\mathbf{CL_C}} = 1$. Obviously, every $\mathbf{CL_C}$-intended assignment is $\mathbf{CL_A}$-intended as well. The following Lemma shows that intended assignments have the corresponding intuitive properties:

Lemma G.1.1. *Let v be an assignment function.*

(i) $v_{\mathsf{i}}^{\mathbf{CL_A}} = 1$ *iff* $v_\star$ *satisfies* (S↠C), (SadC), (S⊥C).
(ii) $v_{\mathsf{i}}^{\mathbf{CL_C}} = 1$ *iff* $v_\star$ *satisfies* (S↠C), (SadC), (S⊥C), (SCoC).

Proof. Ad (i): "⇒": Let $v_{\mathsf{i}}^{\mathbf{CL_A}} = 1$. Assume $v_\star(\alpha) = 1$. Since $v_{\mathrm{R\bot C}} = 1$, $\min(v_\star(\alpha), 1 - v_\star(\bot \twoheadrightarrow \alpha)) = 0$. Thus, $v_\star(\bot \twoheadrightarrow \alpha) = 1$. Thus, (S⊥C) is valid.

Assume now $v_\star(\alpha) = v_\star(\alpha \twoheadrightarrow \beta) = 1$. Since $v_{\mathrm{R\twoheadrightarrow C}} = 1$, $\min(v_\star(\alpha), v_\star(\alpha \twoheadrightarrow \beta), v_\star(\beta)) = 0$. Thus, $v_\star(\beta) = 0$. Thus, (S↠C) is valid.

Assume now $v_\star(\alpha) = v_\star(\beta \twoheadrightarrow \alpha) = 1$ where $\beta = \bigwedge_I \beta_i$. Since $v_{\mathrm{RadC}} = 1$, $\min(v_\star(\alpha), v_\star(\beta \twoheadrightarrow \alpha), 1 - v_\star(\mathsf{def}\,\beta)) = 0$. Hence, $v_\star(\mathsf{def}\,\beta) = 1$. By definition, $\max_{i\in I} \max_{\gamma\in\mathcal{V}_n^\wedge}(\min(v_\star(\gamma), v_\star(\gamma \twoheadrightarrow \beta))) = 1$. Thus, there is a $\gamma \in \mathcal{V}_n^\wedge$ for which $v_\star(\gamma) = 1 = \max_{i\in I} v_\star(\gamma \twoheadrightarrow \beta_i)$. Thus, (SadC) is valid.

"⇐": Suppose that $v_{\mathrm{R\bot C}} = 0$ and $v_\star$ satisfies (S⊥C). Then there is a $\alpha \in \mathcal{V}_n^\vee$ for which $\min(v_\star(\alpha), 1 - v_\star(\bot \twoheadrightarrow \alpha)) = 1$. Thus, $v_\star(\alpha) = 1$ and $v_\star(\bot \twoheadrightarrow \alpha) = 0$. However, by (S⊥C), $v_\star(\bot \twoheadrightarrow \alpha) = 1$,—a contradiction. Thus, $v_{\mathrm{R\bot C}} = 1$. The proof is similar for $v_{\mathrm{R\twoheadrightarrow C}}$, and v_{RadC}.

Ad (ii): "⇒": Let $v_{\mathsf{i}}^{\mathbf{CL_C}} = 1$. In addition to what has been shown in (i), it has to be shown that (SCoC) is valid. Assume $v_\star(\bot \twoheadrightarrow \beta) = 1$ and for all $\alpha = \bigwedge_I \alpha_i \in \mathcal{V}_n^\wedge$, $v_\star(\alpha \twoheadrightarrow \beta) = 0$ or there is a $\gamma \in \mathcal{V}_n^\wedge$ for which $v_\star(\gamma) = \max_{i\in I}(v_\star(\gamma \twoheadrightarrow \alpha_i)) = 1$. Thus, for all $\alpha \in \mathcal{V}_n^\wedge$, $v_\star(\alpha \twoheadrightarrow \beta) = 0$ or $v_\star(\mathsf{def}\,\alpha) = 1$. Due to the fact that $v_{\mathrm{RCoC}} = 1$, $\min\big(v_\star(\bot \twoheadrightarrow \beta), \min_{\alpha\in\mathcal{V}_n^\wedge}(\max(1 - v_\star(\alpha \twoheadrightarrow \beta), v_\star(\mathsf{def}\,\alpha))), 1 - v_\star(\beta)\big) = 0$. Thus, $v_\star(\beta) = 1$. Thus, (SCoC) is valid.

"⇐": Suppose that $v_{\mathrm{RCoC}} = 0$ and $v_\star$ satisfies (SCoC). Then there is a $\beta \in \mathcal{V}_n$ for which $\min\big(v_\star(\bot \twoheadrightarrow \beta), \min_{\alpha\in\mathcal{V}_n^\wedge}(\max(1-v_\star(\alpha \twoheadrightarrow \beta), v_\star(\mathsf{def}\,\alpha))), 1-v_\star(\beta)\big) = 1$. Thus, $v_\star(\beta) = 0$ and $v_\star(\bot \twoheadrightarrow \beta) = 1$. Thus, for every $\alpha = \bigwedge_I \alpha_i \in \mathcal{V}_n^\wedge$ for which $v_\star(\alpha \twoheadrightarrow \beta) = 1$, $v_\star(\mathsf{def}\,\alpha) = 1$ and hence there is a $\gamma \in \mathcal{V}_n^\wedge$ such that $v_\star(\gamma) = \max_{i\in I}(v_\star(\gamma \twoheadrightarrow \alpha_i)) = 1$. But then, due to (SCoC), $v_\star(\beta) = 1$,—a contradiction. □

G.1.2 The Valuation

Let us now take a look at valuation functions for our core logics $\mathbf{CL_A}$ and $\mathbf{CL_C}$. Let a model M be defined by an assignment v.

An $\mathbf{CL_A}$-valuation $v_M^{\mathbf{CL_A}} : \mathcal{W}_n \to \{0, 1\}$ determined by v is defined as follows (where, if not indicated differently, $\alpha, \alpha_i, \beta \in \mathcal{V}_n$; $\varphi, \varphi_1, \varphi_2 \in \mathcal{W}_n$; and $\mathbf{L} = \mathbf{CL_A}$):

$$v_M^{\mathbf{L}}(\bot) = 0 \tag{s$\bot$}$$

$$v_M^{\mathbf{L}}(\alpha \twoheadrightarrow \beta) = v(\alpha \twoheadrightarrow \beta), \text{ where } \alpha \in \mathcal{V}_n^{\wedge}, \beta \in \mathcal{V}_n \tag{s$\twoheadrightarrow$}$$

$$v_M^{\mathbf{L}}(\bot \twoheadrightarrow \alpha) = v(\bot \twoheadrightarrow \alpha) \tag{s$\bot\twoheadrightarrow$}$$

$$v_M^{\mathbf{L}}\left(\left(\bigwedge_I \alpha_i\right) \twoheadrightarrow \left(\bigvee_J \alpha_j\right)\right) = v_M^{\mathbf{L}}\left(\bigwedge_J \left(\left(\bigwedge_I \alpha_i\right) \twoheadrightarrow \alpha_j\right)\right) \tag{s-CA}$$

$$v_M^{\mathbf{L}}\left(\left(\bigwedge_I \alpha_i\right) \twoheadrightarrow \left(\bigwedge_J \alpha_i\right)\right) = v_M^{\mathbf{L}}\left(\bigwedge_J \left(\left(\bigwedge_{I\cup J\setminus\{j\}} \alpha_i\right) \twoheadrightarrow \alpha_i\right)\right) \tag{s-IA}$$

$$v_M^{\mathbf{L}}(\alpha) = \min\left(v_{\mathrm{i}}^{\mathbf{L}}, v(\alpha)\right) \tag{sPA}$$

$$v_M^{\mathbf{L}}(\varphi_1 \wedge \varphi_2) = \min\left(v_M^{\mathbf{L}}(\varphi_1), v_M^{\mathbf{L}}(\varphi_2)\right) \tag{s$\wedge$}$$

$$v_M^{\mathbf{L}}(\varphi_1 \vee \varphi_2) = \max\left(v_M^{\mathbf{L}}(\varphi_1), v_M^{\mathbf{L}}(\varphi_2)\right) \tag{s$\vee$}$$

$$v_M^{\mathbf{L}}(\varphi_1 \supset \varphi_2) = \max\left(1 - v_M^{\mathbf{L}}(\varphi_1), v_M^{\mathbf{L}}(\varphi_2)\right) \tag{s$\supset$}$$

$$v_M^{\mathbf{L}}(\neg\varphi) = 1 - v_M^{\mathbf{L}}(\varphi) \tag{s$\neg$}$$

We define $M \models_{\mathbf{CL_A}} A$ iff $v_M^{\mathbf{L_A}}(A) = 1$.

In the case $v_{\mathrm{i}}^{\mathbf{CL_A}} = 1$, i.e., in the case that the assignment v is $\mathbf{CL_A}$-intended, the valuation takes over the truth values from the assignment for all formulas in $\mathcal{V}_n$. However, if $v_{\mathrm{i}}^{\mathbf{CL_A}} = 0$, the valuation assigns to all propositional letters the truth value 0. Note that for a given CAF A the empty selection is always an admissible extension. Thus, the valuation on basis of a non-intended assignment corresponds to the empty extension.

$\mathbf{CL_C}$-valuations are defined analogous to $\mathbf{CL_A}$-valuations, with the exception of (sPA) which is replaced by (where $\alpha \in \mathcal{V}_n$):

$$v_M^{\mathbf{L_C}}(\alpha) = \max\left(\min\left(v_{\mathrm{i}}^{\mathbf{L_C}}, v_\star(\alpha)\right), v_{\mathrm{g}}(\alpha)\right) \text{ where} \tag{sPC}$$

$$v_{\mathrm{g}}(\alpha) =_{\mathrm{df}} \max_{i\geq 0}\left(v_{\mathrm{g}}^i(\alpha)\right), \text{ where}$$

$$v_{\mathrm{g}}^0(\alpha) =_{\mathrm{df}} \min\left(v_\star(\bot \twoheadrightarrow \alpha), 1 - \max_{\beta\in\mathcal{V}_n^{\wedge}}\left(v_\star(\beta \twoheadrightarrow \alpha)\right)\right), \text{ and}$$

$$v_{\mathrm{g}}^i(\alpha) =_{\mathrm{df}} \min\left(v_\star(\bot \twoheadrightarrow \alpha), \mathsf{defended}^i(\alpha)\right), \text{ where } i > 0 \text{ and}$$

$$\mathsf{defended}^i(\alpha) =_{\mathrm{df}} \min_{\bigwedge_L \beta_l \in \mathcal{V}_n^{\wedge}}\left(\max\left(1 - v_\star\left(\left(\bigwedge_L \beta_l \twoheadrightarrow \alpha\right),\right.\right.\right.$$

$$\left.\left.\left.\mathsf{defeated}^i\left(\bigwedge_L \beta_l\right)\right)\right)\right), \text{ where}$$

$$\mathsf{defeated}^i\left(\bigwedge_L \beta_l\right) =_{\mathrm{df}} \max_{l\in L;\, \bigwedge_K \gamma_k \in \mathcal{V}_n^{\wedge}}\left(\min\left(v_\star\left(\left(\bigwedge_K \gamma_k\right) \twoheadrightarrow \beta_l\right),\right.\right.$$

$$\left.\left.v_{\mathrm{g}}^{<i}\left(\bigwedge_K \gamma_k\right)\right)\right), \text{ where}$$

$$v_{\mathrm{g}}^{<i}\left(\bigwedge_K \gamma_k\right) =_{\mathrm{df}} \min_{k\in K}\left(\max_{0\leq j<i}\left(v_{\mathrm{g}}^j(\gamma_k)\right)\right)$$

We define $M \models_{\mathbf{CL_C}} A$ iff $v_M^{\mathbf{L_C}}(A) = 1$. Again, in the case that v is an $\mathbf{CL_C}$-intended assignment, i.e., in the case $v_{\mathsf{i}}^{\mathbf{CL_C}} = 1$, the valuation $v_M^{\mathbf{L_C}}$ takes over all truth values for all formulas in $\mathcal{V}_n$ from the assignment (see Lemma G.2.1 below). However, in the case $v_{\mathsf{i}}^{\mathbf{CL_C}} = 0$, the situation is more complicated than for the $\mathbf{CL_A}$ case, since for a given AF A the empty selection may not correspond to a complete extension. In this case the valuation $v_M^{\mathbf{L_C}}$ verifies a propositional letter α iff $v_{\mathsf{g}}(\alpha) = 1$. As it will be shown, this way it is ensured that the models of the adaptive strengthening of $\mathbf{CL_C}$ correspond to the complete extensions.

As the reader can see, everything except rules (s-↠), (s⊥-↠), (s-CA), (s-IA), and (sPA) (resp. (sPC)) is defined in the classical way. Note, that by (s-↠) and (s⊥-↠) the valuation takes over the assignment from v for formulas of the form $\alpha \twoheadrightarrow \beta$ and $\bot \twoheadrightarrow \alpha$. By (sPA) (resp. (sPC)) the valuation may have a different value for propositional letters than assigned by v. Note that although (sPA) and (sPC) are of a rather complex form, they are fully determined by the assignment v.

Our valuations satisfy the semantic properties corresponding to the rules (R↠C), (RadC), (R⊥C), (RCoC) of logics $\mathbf{CL_A}$ and $\mathbf{CL_C}$.

Theorem G.1.1. *Let M be a model defined by the assignment v.*[5]

(i) $v_M^{\mathbf{CL_A}}$ *satisfies* (S↠C), (SadC), (S⊥C).
(ii) $v_M^{\mathbf{CL_C}}$ *satisfies* (S↠C), (SadC), (S⊥C), (SCoC).

G.2 Soundness and Completeness

Lemma G.2.1. *Let v be an assignment and $\alpha \in \mathcal{V}_n$. If $v_{\mathsf{i}}^{\mathbf{CL_C}} = 1$, then $v_{\mathsf{g}}(\alpha) = 1$ implies $v_\star(\alpha) = 1$.*

Proof. Suppose $v_{\mathsf{i}}^{\mathbf{CL_C}} = v_{\mathsf{g}}(\alpha) = 1$. Then there is a $i \geq 0$ such that $v_{\mathsf{g}}^{i}(\alpha) = 1$. We prove the statement by an induction on i.

"$i = 0$": We have $v_\star(\bot \twoheadrightarrow \alpha) = 1$ and for all $\beta \in \mathcal{V}_n^{\wedge}$, $v_\star(\beta \twoheadrightarrow \alpha) = 0$. Since $v_{\mathrm{RCoC}} = 1$, $\min\big(v_\star(\bot \twoheadrightarrow \alpha), \min_{\beta \in \mathcal{V}_n^{\wedge}}\big(\max(1 - v_\star(\beta \twoheadrightarrow \alpha), v_{\mathsf{def}}(\mathsf{def}\,\beta))\big), 1 - v_\star(\alpha)\big) = 0$. Thus, $1 - v_\star(\alpha) = 0$ and hence, $v_\star(\alpha) = 1$.

"$i \Rightarrow i+1$": Let $v_{\mathsf{g}}^{i+1}(\alpha) = 1$. Thus, $\min(v_\star(\bot \twoheadrightarrow \alpha), \mathsf{defended}^{i+1}(\alpha)) = 1$ and hence $v_\star(\bot \twoheadrightarrow \alpha) = \mathsf{defended}^{i+1}(\alpha) = 1$. Thus, $\min_{\beta \in \mathcal{V}_n^{\wedge}}\big(\max(1 - v_\star(\beta \twoheadrightarrow \alpha), \mathsf{defeated}^{i+1}(\beta))\big) = 1$. In the case that there is no $\beta \in \mathcal{V}_n^{\wedge}$ for which $v_\star(\beta \twoheadrightarrow \alpha) = 1$, $v_{\mathsf{g}}^{0}(\alpha) = 1$ and thus by induction hypothesis, $v_\star(\alpha) = 1$. Suppose there is a β for which $v_\star(\beta \twoheadrightarrow \alpha) = 1$. Then $\mathsf{defeated}^{i+1}(\beta) = 1$. Hence, where $\beta = \bigwedge_L \beta_l$, $\max_{l \in L; \gamma \in \mathcal{V}_n^{\wedge}}\big(\min(v_\star(\gamma \twoheadrightarrow \beta_l), v_{\mathsf{g}}^{<i+1}(\gamma))\big) = 1$. Thus, there is a γ_β for every such $\beta = \bigwedge_L \beta_l \in \mathcal{V}_n^{\wedge}$ for which $\max_L(v_\star(\gamma_\beta \twoheadrightarrow \beta_l)) = v_{\mathsf{g}}^{<i+1}(\gamma_\beta) = 1$. Hence, where $\gamma_\beta = \bigwedge_K \gamma_k$, for each $k \in K$ there is a $j_k < i+1$ such that

[5] We postpone the proof of this theorem to page 393.

$v_g^j(\gamma_k) = 1$. Thus, by induction hypothesis, $v_\star(\gamma_\beta) = 1$. Thus, $v_{\mathsf{def}}(\mathsf{def}\,\beta) = 1$ for all β for which $v_\star(\beta \twoheadrightarrow \alpha) = 1$. Thus, due to the fact that $v_{\mathrm{RCoC}} = 1$, $\min\big(v_\star(\bot \twoheadrightarrow \alpha), \min_{\beta \in \mathcal{V}_n^\wedge}(\max(1 - v_\star(\beta \twoheadrightarrow \alpha), v_{\mathsf{def}}(\mathsf{def}\beta))), 1 - v_\star(\alpha)\big) = 0$. Thus, we have $1 - v_\star(\alpha) = 0$ and hence $v_\star(\alpha) = 1$. □

Lemma G.2.2. *Let $v_M^{\mathbf{L}}$ be an* **L***-valuation (where $\mathbf{L} \in \{\mathbf{CL_A}, \mathbf{CL_C}\}$) with corresponding assignment v.*

(i) If $v_{\mathsf{i}}^{\mathbf{L}} = 1$ then $v_M^{\mathbf{L}}(\varphi) = v_\star(\varphi)$ for all $\varphi \in \mathcal{V}_n \cup \mathcal{V}_n^\vee \cup \mathcal{V}_n^\wedge \cup \mathcal{W}_n^{\twoheadrightarrow,r}$ and for all $\varphi = \mathsf{def}\,\alpha$ where $\alpha \in \mathcal{V}_n^\wedge$.
(ii) Moreover, if $v_M^{\mathbf{CL_A}}(\alpha) = 1$ for some $\alpha \in \mathcal{V}_n$ then $v_{\mathsf{i}}^{\mathbf{CL_A}} = 1$.

Proof. Ad (i): Let $\mathbf{L} = \mathbf{CL_A}$: Let $\alpha \in \mathcal{V}_n$. Since $v_M^{\mathbf{L}}(\alpha) = 1$ iff $v(\alpha) = v_{\mathsf{i}}^{\mathbf{L}} = 1$, $v_M^{\mathbf{L}}(\alpha) = 0$ iff $v_{\mathsf{i}}^{\mathbf{L}} = 0$ or $v(\alpha) = 0$. Since $v_{\mathsf{i}}^{\mathbf{L}} = 1$ the statement is true for all $\varphi \in \mathcal{V}_n$. Let $\mathbf{L} = \mathbf{CL_C}$ and $\alpha \in \mathcal{V}_n$: Since $v_M^{\mathbf{L}}(\alpha) = 1$ iff $v(\alpha) = v_{\mathsf{i}}^{\mathbf{L}} = 1$ or $v_g(\alpha) = 1$, and $v_g(\alpha) = 1$ implies $v(\alpha) = 1$ due to $v_{\mathsf{i}}^{\mathbf{L}} = 1$ by Lemma G.2.1, we have $v_M^{\mathbf{L}}(\alpha) = 1$ iff $v(\alpha) = 1$. For $\varphi \in \mathcal{V}_n^\vee \cup \mathcal{V}_n^\wedge$ the statement follows by the definition of $v_\star$.

For $\varphi \in \mathcal{W}_n^{\twoheadrightarrow,r}$ the statement is true by (s$\twoheadrightarrow$) and (s$\bot\twoheadrightarrow$) and the definition of $v_\star$. Furthermore, $v_M^{\mathbf{L}}(\mathsf{def}\bigwedge_I \alpha_i) = v_M^{\mathbf{L}}(\bigvee_{i \in I} \bigvee_{\beta \in \mathcal{V}_n^\wedge}((\beta \twoheadrightarrow \alpha_i) \wedge \beta)) = \max_{i \in I; \beta \in \mathcal{V}_n^\wedge}(\min(v_M^{\mathbf{L}}(\beta), v_M^{\mathbf{L}}(\beta \twoheadrightarrow \alpha_i))) = \max_{i \in I; \beta \in \mathcal{V}_n^\wedge}(\min(v_\star(\beta), v_\star(\beta \twoheadrightarrow \alpha_i))) = v_\star(\mathsf{def}\bigwedge_I \alpha_i)$.

Ad (ii): this is the case due to (sPA). □

Lemma G.2.3. *Le v be an assignment function and $\beta \in \mathcal{V}_n$. If $v_g(\beta) = 1$ then for all $\bigwedge_L \alpha_l \in \mathcal{V}_n^\wedge$ for which $v((\bigwedge_L \alpha_l) \twoheadrightarrow \beta) = 1$, $v_g(\alpha_l) = 0$ for some $l \in L$.*

Proof. Since $v_g(\beta) = 1$ iff $\max_{i \geq 0}(v_g^i(\beta)) = 1$, we have $v_g^i(\beta) = 1$ for some $i \geq 0$. The proof is by induction on i.

"$i = 0$": In this case $\min(v(\bot \twoheadrightarrow \beta), 1 - \max_{\gamma \in \mathcal{V}_n^\wedge}(v(\gamma \twoheadrightarrow \beta))) = 1$ and thus there is no $\gamma \in \mathcal{V}_n^\wedge$ for which $v(\gamma \twoheadrightarrow \beta) = 1$.

"$i \Rightarrow i + 1$": Let $v_g^{i+1}(\beta) = 1$. Thus, $\min(v(\bot \twoheadrightarrow \beta), \mathsf{defended}^{i+1}(\beta)) = 1$. Thus, $\mathsf{defended}^{i+1}(\beta) = 1$. Thus, $\min_{\gamma \in \mathcal{V}_n^\wedge}(\max(1 - v(\gamma \twoheadrightarrow \beta), \mathsf{defeated}^{i+1}(\gamma))) = 1$. Thus, for any $\gamma = \bigwedge_L \gamma_l$ for which $v(\gamma \twoheadrightarrow \beta) = 1$, $\mathsf{defeated}^{i+1}(\gamma) = 1$. In the case that there is such a γ, $\max_{l \in L; \delta \in \mathcal{V}_n^\wedge}\big(\min(v(\delta \twoheadrightarrow \gamma_l), v_g^{<i+1}(\delta))\big) = 1$. Thus, there are a δ_γ and a γ_l for which $v(\delta_\gamma \twoheadrightarrow \gamma_l) = 1$ and $v_g^{<i+1}(\delta_\gamma) = 1$. Thus, where $\delta_\gamma = \bigwedge_N \delta_n$, there is a $j_n < i + 1$ for every $n \in N$ such that $v_g^{j_n}(\delta_n) = 1$. Suppose $v_g(\gamma_l) = 1$. Then there is a k_l such that $v_g^{k_l}(\gamma_l) = 1$. Thus, $\mathsf{defended}^{k_l}(\gamma_l) = 1$. Hence, $\mathsf{defeated}^{k_l}(\delta_\gamma) = 1$. Thus there is an $\epsilon = \bigwedge_E \epsilon_e$ for which $\max_{n \in N}(v(\epsilon \twoheadrightarrow \delta_n)) = \min_E(v_g^{<k_l}(\epsilon_e)) = 1$. By the induction hypothesis however, $v_g(\epsilon_e) = 0$ for some $e \in E$,—a contradiction. Thus, $v_g(\gamma_l) = 0$. □

Proof (Proof of Theorem G.1.1). Ad (i): Let $\mathbf{L} = \mathbf{CL_A}$. By Lemma G.2.2i, in the case that $v_{\mathsf{i}}^{\mathbf{L}} = 1$, $v_M^{\mathbf{L}}(\varphi) = v(\varphi)$ for all $\varphi \in \mathcal{V}_n^\wedge \cup \mathcal{V}_n^\vee \cup \mathcal{W}_n^{\twoheadrightarrow}$. By Lemma G.1.1i follows the rest. Assume thus that $v_{\mathsf{i}}^{\mathbf{L}} = 0$. Note that in this case by Lemma G.2.2ii $v_M^{\mathbf{L}}(\alpha) = 0$ for all $\alpha \in \mathcal{V}_n$. Hence, trivially (S$\twoheadrightarrow$C), (SadC), and (S$\bot$C) are valid.

Ad (ii): Let $\mathbf{L} = \mathbf{CL_C}$. By Lemma G.2.2i, in the case $v_{\mathrm{i}}^{\mathbf{L}} = 1$, $v_M^{\mathbf{L}}(\varphi) = v(\varphi)$ for all $\varphi \in \mathcal{V}_n^{\wedge} \cup \mathcal{V}_n^{\vee} \cup \mathcal{W}_n^{\twoheadrightarrow}$. By Lemma G.1.1ii follows the rest.

Assume thus that $v_{\mathrm{i}}^{\mathbf{L}} = 0$. Note that for all $\alpha \in \mathcal{V}_n$ in this case $v_M^{\mathbf{L}}(\alpha) = 1$ iff $v_{\mathrm{g}}(\alpha) = 1$.

Assume $v_M^{\mathbf{L}}(\alpha) = 1$. Then $v_{\mathrm{g}}(\alpha) = 1$ and thus, by definition, $v(\bot \twoheadrightarrow \alpha) = v_M^{\mathbf{L}}(\bot \twoheadrightarrow \alpha) = 1$. Thus, (S⊥C) is valid.

Assume $v_M^{\mathbf{L}}(\alpha) = v_M^{\mathbf{L}}(\alpha \twoheadrightarrow \beta) = 1$. Hence, where $\alpha = \bigwedge_I \alpha_i$, for all $i \in I$, $v_M^{\mathbf{L}}(\alpha_i) = v_{\mathrm{g}}(\alpha_i) = 1 = v(\alpha \twoheadrightarrow \beta)$. Thus, by Lemma G.2.3, $v_{\mathrm{g}}(\beta) = 0$ and hence, $v_M^{\mathbf{L}}(\beta) = 0$. Thus, (S↠C) is valid.

Assume $v_M^{\mathbf{L}}(\alpha) = v_M^{\mathbf{L}}(\beta \twoheadrightarrow \alpha) = 1$. Thus, $v_{\mathrm{g}}(\alpha) = 1 = v(\beta \twoheadrightarrow \alpha)$. Thus, there is an $i \geq 0$ such that $v_{\mathrm{g}}^i(\alpha) = 1$. Note that this entails $\mathsf{defended}^i(\alpha)$. Since $v(\beta \twoheadrightarrow \alpha) = 1$, $i > 0$. Since $\mathsf{defended}^i(\alpha) = 1$, $\mathsf{defeated}^i(\beta) = 1$. Thus, where $\beta = \bigwedge_L \beta_l$, for some $l \in L$ and $\gamma = \bigwedge_G \gamma_g \in \mathcal{V}_n^{\wedge}$, $v(\gamma \twoheadrightarrow \beta_l) = v_M^{\mathbf{L}}(\gamma \twoheadrightarrow \beta_l) = 1$ and $\min_{g \in G}(v_{\mathrm{g}}^{<i}(\gamma_g)) = 1$. Thus, for all $g \in G$, $v_{\mathrm{g}}(\gamma_g) = 1$ and thus $v_M^{\mathbf{L}}(\gamma_g) = 1$. Hence $v_M^{\mathbf{L}}(\gamma) = 1$. Thus, (SadC) is valid.

Assume $v_M^{\mathbf{L}}(\bot \twoheadrightarrow \beta) = 1$ and for all $\alpha = \bigwedge_I \alpha_i \in \mathcal{V}_n^{\wedge}$ either $v_M^{\mathbf{L}}(\alpha \twoheadrightarrow \beta) = 0$ or there is a $\gamma = \bigwedge_G \gamma_g \in \mathcal{V}_n$ for which $\min_{g \in G}(v_M^{\mathbf{L}}(\gamma_g)) = \max_{i \in I}(v_M^{\mathbf{L}}(\gamma \twoheadrightarrow \alpha_i)) = 1$. Hence, $v(\bot \twoheadrightarrow \beta) = 1$ and for all $\alpha = \bigwedge_I \alpha_i \in \mathcal{V}_n$ either $v(\alpha \twoheadrightarrow \beta) = 0$ or there is a $\gamma = \bigwedge_G \gamma_g \in \mathcal{V}_n^{\wedge}$ for which $\min_{g \in G}(v_{\mathrm{g}}(\gamma_g)) = \max_{i \in I}(v(\gamma \twoheadrightarrow \alpha_i)) = 1$. If for all $\alpha \in \mathcal{V}_n^{\wedge}$, $v(\alpha \twoheadrightarrow \beta) = 0$ then by definition, $v_{\mathrm{g}}^0(\beta) = 1$ and thus $v_{\mathrm{g}}(\beta) = 1$. Hence, $v_M^{\mathbf{L}}(\beta) = 1$. Otherwise there is a $j > 0$ such that for all $\alpha = \bigwedge_I \alpha_i \in \mathcal{V}_n^{\wedge}$ for which $v(\alpha \twoheadrightarrow \beta) = 1$ there is a $\gamma_\alpha = \bigwedge_G \gamma_g \in \mathcal{V}_n^{\wedge}$ for which $\min_{g \in G}(v_{\mathrm{g}}^{<j}(\gamma_g)) = \max_{i \in I}(v(\gamma_\alpha \twoheadrightarrow \alpha_i)) = 1$. Then, $\mathsf{defended}^j(\beta) = 1$ and thus $v_{\mathrm{g}}^j(\beta) = 1$. Thus, $v_{\mathrm{g}}(\beta) = v_M^{\mathbf{L}}(\beta) = 1$. Thus, (SCoC) is valid. □

Theorem G.2.1 (Soundness). *Let* $\mathbf{L} \in \{\mathbf{CL_A}, \mathbf{CL_C}\}$, $\Gamma \subseteq \mathcal{W}_n$ *and* $\varphi \in \mathcal{W}_n$. *If* $\Gamma \vdash_{\mathbf{L}} \varphi$, *then* $\Gamma \Vdash_{\mathbf{L}} \varphi$.

Proof. We use for the proof an induction on the derivation steps of a proof of φ analogous to the way soundness is usually proven for classical propositional logic (cp. e.g., [3] p. 40–43). If φ is introduced by premise introduction then trivially $\Gamma \Vdash_{\mathbf{L}} \varphi$, as $\varphi \in \Gamma$. If φ is obtained by aggregation $\frac{\varphi_1 \quad \varphi_2}{\varphi_1 \wedge \varphi_2}$ where $\varphi = \varphi_1 \wedge \varphi_2$, then by induction hypothesis, $\Gamma \Vdash_{\mathbf{L}} \varphi_1, \varphi_2$. Therefore we have for all **L**-models of Γ, $M \models_{\mathbf{L}} \varphi_1, \varphi_2$. But then by (s∧), $M \models_{\mathbf{L}} \varphi_1 \wedge \varphi_2$. The proof is analogous for the other classical rules.

We have to take a look at the non-classical rules: If $\varphi = \neg\beta$ and it has been derived by (R↠C) from α and $\alpha \twoheadrightarrow \beta$, then by induction hypothesis we have $\Gamma \Vdash_{\mathbf{L}} \alpha, \alpha \twoheadrightarrow \beta$. Thus for all **L**-models M of Γ, $M \models_{\mathbf{L}} \alpha, \alpha \twoheadrightarrow \beta$. Since by Theorem G.1.1 $v_M^{\mathbf{L}}$ validates (S↠C), $v_M^{\mathbf{L}}(\beta) = 0$. Thus, by (s¬), $v_M^{\mathbf{L}}(\neg\beta) = 1$.

If $\varphi = \neg\alpha$ and it has been derived from $\bot \not\twoheadrightarrow \alpha$ by (R⊥C), then by induction hypothesis for each **L**-model M of Γ, $v_M^{\mathbf{L}}(\bot \not\twoheadrightarrow \alpha) = 1$ and hence by (s¬), $v_M^{\mathbf{L}}(\bot \twoheadrightarrow \alpha) \neq 1$. By (S⊥C), $v_M^{\mathbf{L}}(\alpha) \neq 1$ and hence $v_M^{\mathbf{L}}(\alpha) = 0$. Thus, by (s¬), $v_M^{\mathbf{L}}(\neg\alpha) = 1$.

If $\mathsf{def}\,\beta$ has been derived from α and $\beta \twoheadrightarrow \alpha$ by (RadC), then by induction hypothesis for each $\mathbf{L}$-model M of Γ, $v_M^{\mathbf{L}}(\alpha) = v_M^{\mathbf{L}}(\beta \twoheadrightarrow \alpha) = 1$. Where $\beta = \bigwedge_I \beta_i$, by (SadC), there is a $\gamma \in \mathcal{V}_n^{\wedge}$ and an $i \in I$ for which $v_M^{\mathbf{L}}(\gamma) = v_M^{\mathbf{L}}(\gamma \twoheadrightarrow \beta_i) = 1$. By (s$\vee$), $v_M^{\mathbf{L}}(\mathsf{def}\,\beta) = 1$.

Evidently, also the axioms (CA) and (IA) have the semantic counterparts (s-CA) and (s-IA) and hence the proof proceeds similarly.

For $\mathbf{L} = \mathbf{CL_C}$: If β has been derived from $\bot \twoheadrightarrow \beta$ and $\bigwedge_{\delta \in \mathcal{V}_n^{\wedge}} \big((\delta \twoheadrightarrow \beta) \supset \mathsf{def}\,\delta\big)$ by (RCoC), then by induction hypothesis, $v_M^{\mathbf{L}}(\bot \twoheadrightarrow \beta) = v_M^{\mathbf{L}}\big(\bigwedge_{\delta \in \mathcal{V}_n^{\wedge}}((\delta \twoheadrightarrow \beta) \supset \mathsf{def}\,\delta)\big) = 1$ for each $\mathbf{L}$-model M of Γ. Thus, $\min_{\delta \in \mathcal{V}_n^{\wedge}}(\max(1 - v_M^{\mathbf{L}}(\delta \twoheadrightarrow \beta), v_M^{\mathbf{L}}(\mathsf{def}\,\delta))) = 1$. Thus, for all $\delta = \bigwedge_I \delta_i \in \mathcal{V}_n$, $v_M^{\mathbf{L}}(\delta \twoheadrightarrow \beta) = 0$ or $v_M^{\mathbf{L}}(\mathsf{def}\,\delta) = 1$. In the latter case, by (s$\vee$), there is an $\epsilon_\delta \in \mathcal{V}_n^{\wedge}$ for which $v_M^{\mathbf{L}}(\epsilon_\delta) = \max_{i \in I}(v_M^{\mathbf{L}}(\epsilon_\delta \twoheadrightarrow \delta_i)) = 1$. Altogether, by (SCoC), $v_M^{\mathbf{L}}(\beta) = 1$. □

Let $\Gamma \subseteq \mathcal{W}_n$ and $\mathbf{L} \in \{\mathbf{CL_A}, \mathbf{CL_C}\}$. Γ is $\mathbf{L}$-*consistent* iff $\Gamma \nvdash_{\mathbf{L}} \bot$. Γ is *maximally* $\mathbf{L}$-*consistent* iff (a) Γ is $\mathbf{L}$-consistent and (b) if $\Gamma \subseteq \Gamma'$ and Γ' is $\mathbf{L}$-consistent, then $\Gamma = \Gamma'$. We say that Γ is $\mathbf{L}$-*inconsistent* in case it is not $\mathbf{L}$-consistent. The proofs of the following propositions are standard for classical propositional logic and can for instance be found in van Dalen [3] pp. 43–45. The proofs for our logics are analogous.

Lemma G.2.4. *Let $\Gamma \subseteq \mathcal{W}_n$, $\varphi \in \mathcal{W}_n$, and $\mathbf{L} \in \{\mathbf{CL_A}, \mathbf{CL_C}\}$. We have:*

(i) If $\Gamma \cup \{\neg\varphi\}$ is $\mathbf{L}$-inconsistent, then $\Gamma \vdash_{\mathbf{L}} \varphi$
(ii) If $\Gamma \cup \{\varphi\}$ is $\mathbf{L}$-inconsistent then $\Gamma \vdash_{\mathbf{L}} \neg\varphi$.
(iii) Each $\mathbf{L}$-consistent set Γ is contained in a maximally $\mathbf{L}$-consistent set Γ'.
(iv) If Γ is maximally $\mathbf{L}$-consistent, then $\varphi \in \Gamma$ iff $\neg\varphi \notin \Gamma$, and $\neg\varphi \in \Gamma$ iff $\varphi \notin \Gamma$.
(v) For an $\mathbf{L}$-consistent Γ and for all $\varphi \in \Gamma$, $\Gamma \nvdash_{\mathbf{L}} \neg\varphi$.

Lemma G.2.5. *Let $\mathbf{L} \in \{\mathbf{CL_A}, \mathbf{CL_C}\}$. If $\Gamma \subset \mathcal{W}_n$ is $\mathbf{L}$-consistent, then there is an $\mathbf{L}$-model M of Γ.*

Proof. By Lemma G.2.4v we know that Γ is contained in a maximally $\mathbf{L}$-consistent Γ'. We define an assignment $v : \mathcal{V}_n \cup \mathcal{W}_n^{\twoheadrightarrow, r} \to \{0, 1\}$ by $\varphi \mapsto 1$ iff $\varphi \in \Gamma' \cap (\mathcal{V}_n \cup \mathcal{W}_n^{\twoheadrightarrow, r})$. I will show that $v_M^{\mathbf{L}}(\varphi) = 1$ iff $\varphi \in \Gamma'$. In order to do so we first show that $v_{\mathsf{i}}^{\mathbf{L}} = 1$.

Suppose $v_{\mathrm{R\bot C}} = 0$ and thus $\max_{\delta \in \mathcal{V}_n}\big(\min(v(\delta), 1 - v(\bot \twoheadrightarrow \delta))\big) = 1$. Thus, there is an α for which $\min(v(\alpha), 1 - v(\bot \twoheadrightarrow \alpha)) = 1$. Thus, $v(\alpha) = 1$ and $v(\bot \twoheadrightarrow \alpha) = 0$. Hence, $\alpha \in \Gamma'$ and $\bot \twoheadrightarrow \alpha \notin \Gamma'$. Due to the maximal consistency of Γ', $\bot \not\twoheadrightarrow \alpha \in \Gamma'$. However, by (R$\bot$C), $\neg\alpha \in \Gamma'$ and thus $\alpha \notin \Gamma'$,—a contradiction. Thus, $v_{\mathrm{R\bot C}} = 1$.

Suppose $v_{\mathrm{R\twoheadrightarrow C}} = 0$. Then $\max_{\gamma \in \mathcal{V}_n^{\wedge}, \delta \in \mathcal{V}_n}(\min(v_\wedge(\gamma), v(\gamma \twoheadrightarrow \delta), v(\delta))) = 1$. Thus, there are $\alpha \in \mathcal{V}_n^{\wedge}$ and $\beta \in \mathcal{V}_n$ such that $v_\wedge(\alpha) = v(\alpha \twoheadrightarrow \beta) = v(\beta) = 1$. Hence, due to the maximal consistency of Γ', $\alpha, \beta, \alpha \twoheadrightarrow \beta \in \Gamma'$. But by (R$\twoheadrightarrow$C) $\neg\beta \in \Gamma'$,—a contradiction. Thus, $v_{\mathrm{R\twoheadrightarrow C}} = 1$.

Suppose $v_{\mathrm{RadC}} = 0$. Then $\max_{\gamma \in \mathcal{V}_n, \delta \in \mathcal{V}_n^{\wedge}}(\min(v(\gamma), v(\delta \twoheadrightarrow \gamma), 1 - v_{\mathsf{def}}(\mathsf{def}\,\delta)) = 1$. Thus, there are $\alpha \in \mathcal{V}_n$, $\beta \in \mathcal{V}_n^{\wedge}$ such that $v(\alpha) = v(\beta \twoheadrightarrow \alpha) = 1$ and

$v_{\mathsf{def}}(\mathsf{def}\,\beta) = 0$. Thus, $\alpha, \beta \twoheadrightarrow \alpha \in \Gamma'$. By (RadC), $\mathsf{def}\,\beta \in \Gamma'$. Note that, where $\beta = \bigwedge_I \beta_i$, by definition $v_{\mathsf{def}}(\mathsf{def}\,\beta) = \max_{i \in I; \gamma \in \mathcal{V}_n^{\wedge}}(\min(v_{\wedge}(\gamma), v(\gamma \twoheadrightarrow \beta_i))) = 0$. Thus, for all $\gamma \in \mathcal{V}_n^{\wedge}$, $v_{\wedge}(\gamma) = 0$ or $\max_{i \in I}(v(\gamma \twoheadrightarrow \beta_i)) = 0$. Thus due to the maximal consistency of Γ', for all $\gamma \in \mathcal{V}_n^{\wedge}$, $\neg\gamma \in \Gamma'$, or $\gamma \not\twoheadrightarrow \beta_i \in \Gamma'$ for all $i \in I$. Thus, due to the maximal consistency of Γ', $\neg\gamma \vee \bigwedge_I(\gamma \not\twoheadrightarrow \beta_i) \in \Gamma'$ for all $\gamma \in \mathcal{V}_n^{\wedge}$. Thus, $\bigwedge_I \neg(\gamma \wedge (\gamma \twoheadrightarrow \beta_i)) \in \Gamma'$ for all $\gamma \in \mathcal{V}_n^{\wedge}$. Thus, $\bigwedge_I \bigwedge_{\gamma \in \mathcal{V}_n^{\wedge}} \neg(\gamma \wedge (\gamma \twoheadrightarrow \beta_i)) \in \Gamma'$ and hence, $\neg \bigvee_I \bigvee_{\gamma \in \mathcal{V}_n^{\wedge}}(\gamma \wedge (\gamma \twoheadrightarrow \beta_i)) \in \Gamma'$. Thus $\neg\mathsf{def}\,\beta \in \Gamma'$,—a contradiction. Thus, $v_{\mathrm{RadC}} = 1$.

In the case of $\mathbf{CL_C}$ also $v_{\mathrm{RCoC}} = 1$ has to be shown. Suppose $v_{\mathrm{RCoC}} = 0$. Then $\max_{\delta \in \mathcal{V}_n}\big(\min(v(\bot \twoheadrightarrow \delta), \min_{\alpha \in \mathcal{V}_n^{\wedge}}(\max(1 - v(\alpha \twoheadrightarrow \delta), v_{\mathsf{def}}(\mathsf{def}\alpha))), 1 - v(\delta))\big) = 1$. Thus, there is a $\beta \in \mathcal{V}_n$ for which $\min(v(\bot \twoheadrightarrow \beta), \min_{\alpha \in \mathcal{V}_n^{\wedge}}(\max(1 - v(\alpha \twoheadrightarrow \beta), v_{\mathsf{def}}(\mathsf{def}\alpha))), 1 - v(\beta)) = 1$. Thus, $v(\bot \twoheadrightarrow \beta) = 1$, $v(\beta) = 0$ and for all $\alpha \in \mathcal{V}_n^{\wedge}$, $\max(1 - v(\alpha \twoheadrightarrow \beta), v_{\mathsf{def}}(\mathsf{def}\alpha)) = 1$. Thus, $\bot \twoheadrightarrow \beta \in \Gamma'$ and $\beta \notin \Gamma'$ and hence, $\neg\beta \in \Gamma'$. Moreover, for all $\alpha = \bigwedge_I \alpha_i \in \mathcal{V}_n^{\wedge}$ either $v(\alpha \twoheadrightarrow \beta) = 0$ and hence $\alpha \not\twoheadrightarrow \beta \in \Gamma'$ or $v_{\mathsf{def}}(\mathsf{def}\alpha) = 1$. In the latter case, $\max_I \max_{\gamma \in \mathcal{V}_n^{\wedge}}(\min(v_{\wedge}(\gamma), v(\gamma \twoheadrightarrow \alpha_i))) = 1$ and hence there is a $\gamma_\alpha \in \mathcal{V}_n^{\wedge}$ for which $\max_I \min(v_{\wedge}(\gamma_\alpha), v(\gamma_\alpha \twoheadrightarrow \alpha_i)) = 1$. Thus, $v_{\wedge}(\gamma_\alpha) = \max_I(v(\gamma_\alpha \twoheadrightarrow \alpha_i)) = 1$. Thus, $\gamma_\alpha, \gamma_\alpha \twoheadrightarrow \alpha_i \in \Gamma'$ for some $i \in I$. Hence, due to the maximal consistency of Γ', $\mathsf{def}\alpha \in \Gamma'$. Thus, for all $\alpha \in \mathcal{V}_n^{\wedge}$, $\mathsf{def}\alpha \vee (\alpha \not\twoheadrightarrow \beta) \in \Gamma'$ and hence, $\bigwedge_{\alpha \in \mathcal{V}_n^{\wedge}}\big((\alpha \twoheadrightarrow \beta) \supset \mathsf{def}\alpha\big) \in \Gamma'$. By (RCoC), $\beta \in \Gamma'$,—a contradiction. Thus, $v_{\mathrm{RCoC}} = 1$.

We have shown that $v_{\mathrm{i}}^{\mathbf{L}} = 1$. We will show now that $v_M^{\mathbf{L}}(\varphi) = 1$ iff $\varphi \in \Gamma'$ by an induction on the length l of $\varphi \in \mathcal{W}_n$.

"$l = 0$": By Lemma G.2.2i, $v_M^{\mathbf{L}}(\varphi) = v(\varphi)$, for all $\varphi \in \mathcal{V}_n \cup \mathcal{W}_n^{\twoheadrightarrow,r}$. Thus, $v_M^{\mathbf{L}}(\varphi) = 1$ iff $\varphi \in \Gamma'$ for all $\varphi \in \mathcal{V}_n \cup \mathcal{W}_n^{\twoheadrightarrow,r}$.

"$l \Rightarrow l + 1$": Let $\varphi = \neg\varphi' \in \Gamma'$, then $\varphi' \notin \Gamma'$ and by induction hypothesis, $v_M^{\mathbf{L}}(\varphi') = 0$. By (s$\neg$), $v_M^{\mathbf{L}}(\varphi) = 1$. Now let $v_M^{\mathbf{L}}(\varphi) = 1$, then by (s$\neg$), $v_M^{\mathbf{L}}(\varphi') = 0$ and by induction hypothesis, $\varphi' \notin \Gamma'$. Due to the maximal consistency of Γ', $\varphi \in \Gamma'$.

For the other Boolean combinations the proof is analogous.

Thus, $v_M^{\mathbf{L}}(\varphi) = 1$ iff $\varphi \in \Gamma'$. Since $\Gamma' \supseteq \Gamma$, we also have: if $\varphi \in \Gamma$ then $v_M^{\mathbf{L}}(\varphi) = 1$. □

Corollary G.2.1. *Let* $\mathbf{L} \in \{\mathbf{CL_A}, \mathbf{CL_C}\}$ *and* $\Gamma \subseteq \mathcal{W}_n$. *If* $\Gamma \nvdash_{\mathbf{L}} \varphi$ *then there is an* $\mathbf{L}$*-model of* Γ *for which* $v_M^{\mathbf{L}}(\varphi) = 0$.

Proof. If $\Gamma \nvdash_{\mathbf{L}} \varphi$, then by Lemma G.2.4i, $\Gamma \cup \{\neg\varphi\}$ is $\mathbf{L}$-consistent. By Lemma G.2.5 there is an $\mathbf{L}$-model M of $\Gamma \cup \{\neg\varphi\}$. By (s$\neg$), $v_M^{\mathbf{L}}(\varphi) = 0$. □

Theorem G.2.2 (Completeness). *Let* $\mathbf{L} \in \{\mathbf{CL_A}, \mathbf{CL_C}\}$. *If* $\Gamma \Vdash_{\mathbf{L}} \varphi$ *then* $\Gamma \vdash_{\mathbf{L}} \varphi$.

Proof. By Corollary G.2.1 we know that if $\Gamma \nvdash_{\mathbf{L}} \varphi$, then $\Gamma \nVdash_{\mathbf{L}} \varphi$. □

G.3 Representational Results for Admissible and Complete Extensions

Since all our ALs for skeptical acceptance are in the standard format and employ $\mathbf{CL_A}$ resp. $\mathbf{CL_C}$ as lower limit logics, the completeness and soundness of $\mathbf{CL_A}$ and $\mathbf{CL_C}$ provides us immediately with the completeness and soundness of the adaptive systems $\mathbf{ACL_A}$ resp. $\mathbf{ACL_C}$ (see Theorem 2.6.1).

Theorem G.3.1. *We have $\Gamma^n_{\mathsf{A}} \Vdash_{\mathbf{ACL_X}} \varphi$ iff $\Gamma^n_{\mathsf{A}} \vdash_{\mathbf{ACL_X}} \varphi$ where $\mathbf{X} \in \{\mathbf{A}, \mathbf{C}\}$.*

Proof. This follows by Theorem G.2.1, Theorem G.2.2 and Theorem 2.6.1. □

Lemma G.3.1. *Let $\mathsf{A} = \langle \mathcal{A}, \rightarrow \rangle$ be a sCAF for which $\mathcal{A} \subseteq \mathcal{A}_n$ and $\mathbf{X} \in \{\mathbf{A}, \mathbf{C}\}$.*

(i) $\Gamma^n_{\mathsf{A}} \vdash_{\mathbf{ACL_X}} \bot \not\twoheadrightarrow p_i$ *for all* $p_i \in \mathcal{V}_n \setminus \{p_l \mid a_l \in \mathcal{A}\}$, *otherwise* $\bot \twoheadrightarrow p_i \in \Gamma^n_{\mathsf{A}}$.
(ii) $\Gamma^n_{\mathsf{A}} \vdash_{\mathbf{ACL_X}} \left(\bigwedge_I p_i\right) \twoheadrightarrow p_j$ *iff* $(\{a_i \mid i \in I\}, a_j) \in \rightarrow$ *iff* $\left(\bigwedge_I p_i\right) \twoheadrightarrow p_j \in \Gamma^n_{\mathsf{A}}$, *otherwise* $\Gamma^n_{\mathsf{A}} \vdash_{\mathbf{ACL_X}} \left(\bigwedge_I p_i\right) \not\twoheadrightarrow p_j$.
(iii) *For all* $M \in \mathcal{M}_{\mathbf{ACL_X}}(\Gamma^n_{\mathsf{A}})$, $Ab^{\mathbf{CL_X}}_{\twoheadrightarrow}(M) = \Gamma^n_{\mathsf{A}}$.
(iv) *If* $p_i \in \mathcal{V}_n \setminus \{p_l \mid a_l \in \mathcal{A}\}$, *then* $\Gamma^n_{\mathsf{A}} \vdash_{\mathbf{ACL_X}} \neg p_i$.

Proof. Let $S \subseteq \mathcal{A}$, in case $\mathbf{X} = \mathbf{A}$, be an admissible extension and, in case $\mathbf{X} = \mathbf{C}$, a complete extension. We construct an $\mathbf{CL_X}$-model M of Γ^n_{A} on basis of the assignment $v : \mathcal{V}_n \cup \mathcal{W}_n^{\twoheadrightarrow, r} \to \{0, 1\}$, where

$$\varphi \mapsto \begin{cases} 1 \text{ if } \varphi = p_i \text{ where } a_i \in S & [1] \\ 1 \text{ if } \varphi = \left(\bigwedge_I p_i\right) \twoheadrightarrow p_j \text{ where } (\{a_i \mid i \in I\}, a_j) \in \rightarrow & [2] \\ 1 \text{ if } \varphi = \bot \twoheadrightarrow p_i \text{ where } a_i \in \mathcal{A} & [3] \\ 0 \text{ else} & [4] \end{cases}$$

It is easy to check that the semantical properties (SadC), (S↠C) and (S⊥C) (resp. (SadC), (S↠C), (S⊥C) and (SCoC) in the case $\mathbf{X} = \mathbf{C}$) are valid for $v_\star$. Let for property (S↠C), $v_\star(\bigwedge_I p_i) = v((\bigwedge_I p_i) \twoheadrightarrow p_j) = 1$, then by definition of v, $\{a_i \mid i \in I\} \subseteq S$ and $(\{a_i \mid i \in I\}, a_j) \in \rightarrow$. But then $v(p_j) = 0$, as due the conflict-freeness of S, there is no $S' \subseteq S$ such that $(S', a_k) \in \rightarrow$ for some $a_k \in S$. For property (SadC) suppose $v(p_i) = v((\bigwedge_J p_j) \twoheadrightarrow p_i) = 1$. By definition of v, $a_i \in S$ and $(\{a_j \mid j \in J\}, a_i) \in \rightarrow$. Since S is admissible, there is a $\{a_k \mid k \in K\} \subseteq S$ such that $(\{a_k \mid k \in K\}, a_j) \in \rightarrow$ for some $j \in J$. But then, $v_\wedge(\bigwedge_K p_k) = v((\bigwedge_K p_k) \twoheadrightarrow p_j) = 1$. For (S⊥C) let $v(p_i) = 1$. Then $a_i \in \mathcal{A}$ and thus $v(\bot \twoheadrightarrow p_i) = 1$. For (SCoC) in the case $\mathbf{X} = \mathbf{C}$ let $v(\bot \twoheadrightarrow p_i) = 1$ and for all $\bigwedge_K p_k \in \mathcal{V}_n^\wedge$ let $v((\bigwedge_K p_k) \twoheadrightarrow p_i) = 0$ or there is a $\bigwedge_M p_m$ such that $v_\wedge(\bigwedge_M p_m) = \max_K v((\bigwedge_M p_m) \twoheadrightarrow p_k) = 1$. Since $\left\{(\{a_j \mid j \in J\}, a_l) \mid v((\bigwedge_J p_j) \twoheadrightarrow p_l) = 1\right\} = \rightarrow$ and $\left\{a_l \mid v(p_l) = 1\right\} = S$, we know that a_i is defended by S. Thus, $a_i \in S$ and hence $v(p_i) = 1$.

Thus, since $v_{\mathsf{i}}^{\mathbf{CL_X}} = 1$, $v_M^{\mathbf{CL_X}}(\varphi) = v(\varphi)$ for all $\varphi \in \mathcal{V}_n \cup \mathcal{W}_n^{\twoheadrightarrow}$ by Lemma G.2.2i.

Note that by construction, $\mathrm{Ab}^{\mathbf{L_X}}_{\twoheadrightarrow}(M) = \Gamma^n_{\mathsf{A}}$. (i) – (iii) follow immediately. (iv) follows by (i) and (R⊥). □

Lemma G.3.1 shows that for our ALs all minimal disjunctions of abnormalities $\mathsf{Dab}(\Delta)$ (where $\Delta \subseteq \Omega_{\twoheadrightarrow}$) derivable by our lower limits $\mathbf{CL_A}$ and $\mathbf{CL_C}$ are such that Δ is a singleton. Therefore, the simple strategy defines in these cases the same consequence relation as the minimal abnormality strategy (or the reliability strategy, see Theorem 2.4.11).

Proof (Proof of Theorem 9.6.1i). Let $\mathsf{A} = \langle \mathcal{A}, \rightarrow \rangle$ be a sCAF for which $\mathcal{A} \subseteq \mathcal{A}_n$. "$\Leftarrow$": Let $M \in \mathcal{M}_{\mathbf{ACL_A}}(\Gamma^n_{\mathsf{A}})$. Note that $S = \{a_l \mid M \models_{\mathbf{CL_A}} p_l\} \subseteq \mathcal{A}$, as for all $p_j \in \mathcal{V}_n \setminus \{p_l \mid a_l \in \mathcal{A}\}$ by Lemma G.3.1iv, $\Gamma^n_{\mathsf{A}} \vdash_{\mathbf{ACL_A}} \neg p_j$ and thus $\Gamma^n_{\mathsf{A}} \Vdash_{\mathbf{ACL_A}} \neg p_j$. S is conflict-free, as if $(\{a_i \mid i \in I\}, a_j) \in \rightarrow$, then $(\bigwedge_I p_i) \twoheadrightarrow p_j \in \Gamma^n_{\mathsf{A}}$ and therefore $M \models_{\mathbf{CL_A}} \bigwedge_I p_i \twoheadrightarrow p_j$. Hence, if $M \models_{\mathbf{CL_A}} \bigwedge_I p_i$ and thus $M \models_{\mathbf{CL_A}} p_i$ for all $i \in I$, then by (S$\twoheadrightarrow$C) we have $M \models_{\mathbf{CL_A}} \neg p_j$. Hence, either $\{a_i \mid i \in I\} \nsubseteq S$ or $a_j \notin S$. For admissibility let $(\{a_j \mid j \in J\}, a_i) \in \rightarrow$ and $M \models_{\mathbf{CL_A}} p_i$. Hence, $(\bigwedge_J p_j) \twoheadrightarrow p_i \in \Gamma^n_{\mathsf{A}}$. Now by (SadC) there is a $\bigwedge_K p_k$ such that $M \models_{\mathbf{CL_A}} \bigwedge_K p_k, (\bigwedge_K p_k) \twoheadrightarrow p_j$ for some $j \in J$. By Lemma G.3.1ii, $(\{a_k \mid k \in K\}, a_j) \in \rightarrow$. Hence, S is admissible.

"$\Rightarrow$": Let $S \subseteq \mathcal{A}$ be an admissible set. Define an $\mathbf{ACL_A}$-model M with respect to S as in Lemma G.3.1. Obviously $\{a_l \mid M \models_{\mathbf{CL_A}} p_l\} = S$ and by construction $M \models_{\mathbf{CL_A}} \Gamma^n_{\mathsf{A}}$.

Due to Theorem 9.5.1 an analogous proof can be used to prove the property for CAFs. □

Proof (Proof of Theorem 9.6.1ii). Let $\mathsf{A} = \langle \mathcal{A}, \rightarrow \rangle$ be a sCAF for which $\mathcal{A} \subseteq \mathcal{A}_n$. "$\Leftarrow$": Let $M \in \mathcal{M}_{\mathbf{ACL_C}}(\Gamma^n_{\mathsf{A}})$. Note that $S = \{a_l \mid M \models_{\mathbf{CL_C}} p_l\} \subseteq \mathcal{A}$, as for all $p_j \in \mathcal{V}_n \setminus \{p_l \mid a_l \in \mathcal{A}\}$ by Lemma G.3.iv, $\Gamma^n_{\mathsf{A}} \vdash_{\mathbf{ACL_C}} \neg p_j$ and thus $\Gamma^n_{\mathsf{A}} \Vdash_{\mathbf{ACL_C}} \neg p_j$. S is conflict-free, as if $(\{a_i \mid i \in I\}, a_j) \in \rightarrow$, then $(\bigwedge_I p_i) \twoheadrightarrow p_j \in \Gamma^n_{\mathsf{A}}$ and therefore $M \models_{\mathbf{CL_C}} \bigwedge_I p_i \twoheadrightarrow p_j$. Hence, if $M \models_{\mathbf{CL_C}} \bigwedge_I p_i$ and thus $M \models_{\mathbf{CL_C}} p_i$ for all $i \in I$, then by (S$\twoheadrightarrow$C) we have $M \models_{\mathbf{CL_C}} \neg p_j$. Hence, either $\{a_i \mid i \in I\} \nsubseteq S$ or $a_j \notin S$. For admissibility let $(\{a_j \mid j \in J\}, a_i) \in \rightarrow$ and $M \models_{\mathbf{CL_C}} p_i$. Hence, $(\bigwedge_J p_j) \twoheadrightarrow p_i \in \Gamma^n_{\mathsf{A}}$. Now by (SadC) there is a $\bigwedge_K p_k$ such that $M \models_{\mathbf{CL_C}} \bigwedge_K p_k, (\bigwedge_K p_k) \twoheadrightarrow p_j$ for some $j \in J$. By Lemma G.3.1ii, $(\{a_k \mid k \in K\}, a_j) \in \rightarrow$. Hence, S is admissible.

Let $a_i \in \mathcal{A}$ be such that S defends a_i. We have to show that $a_i \in S$. Note that, since $a_i \in \mathcal{A}$, $\bot \twoheadrightarrow p_i \in \Gamma^n_{\mathsf{A}}$. If $\mathrm{Att}_{a_i} =_{\mathrm{df}} \{\{a_l \mid l \in L\} \mid (\{a_l \mid l \in L\}, a_i) \in \rightarrow\} = \emptyset$, then by Lemma G.3.1ii, for all $\bigwedge_J p_j$, $M \models_{\mathbf{CL_C}} (\bigwedge_J p_j) \not\twoheadrightarrow p_i$ and thus by (SCoC), $M \models_{\mathbf{CL_C}} p_i$. Hence, $a_i \in S$. Let now $\mathrm{Att}_{a_i} \neq \emptyset$. By Lemma G.3.1, $\{\bigwedge_L p_l \mid (\{a_l \mid l \in L\}, a_i) \in \rightarrow\} = \{\bigwedge_L p_l \mid M \models_{\mathbf{LC}} (\bigwedge_L p_l) \twoheadrightarrow p_i\}$. Since S defends a_i, there is an $\{a_k \mid k \in K\} \subseteq S$ for each $\{a_j \mid j \in J\} \in \mathrm{Att}_{a_i}$ for which $(\{a_k \mid k \in K\}, a_j) \in \rightarrow$ for some $j \in J$. Thus, there is a $\bigwedge_K p_k$ such that $M \models_{\mathbf{CL_C}} \bigwedge_K p_k, (\bigwedge_K p_k) \twoheadrightarrow p_j$ for some $j \in J$ for each $\bigwedge_J p_j \in \{\bigwedge_L p_l \mid (\{a_l \mid l \in L\}, a_i) \in \rightarrow\} = \{\bigwedge_L p_l \mid M \models_{\mathbf{CL_C}} (\bigwedge_L p_l) \twoheadrightarrow p_i\}$. By (SCoC), $M \models_{\mathbf{CL_C}} p_i$ and thus, $a_i \in S$.

"$\Rightarrow$": Let $S \subseteq \mathcal{A}$ be a complete extension. We construct an $\mathbf{ACL_C}$-model M of Γ^n_{A} such that $\{a_l \mid M \models_{\mathbf{CL_C}} p_l\} = S$ as in Lemma G.3.1 (for the $\mathbf{X} = \mathbf{C}$ case).

Due to Theorem 9.5.1 an analogous proof can be used to prove the property for CAFs. □

Proof (Proof of Corollary 9.6.1i,ii). Let $\mathbf{L} \in \{\mathbf{ACL_A}, \mathbf{ACL_C}\}$ and $\mathsf{A} = \langle \mathcal{A}, \rightarrow \rangle$ an sCAF (resp. $\mathsf{A} = \langle \mathcal{A}, \rightarrow_c, \rightarrow_i \rangle$ a CAF) for which $\mathcal{A} \subseteq \mathcal{A}_n$.

Let a_i be skeptically accepted with respect to admissible (in case $\mathbf{L} = \mathbf{ACL_A}$) resp. complete (in case $\mathbf{L} = \mathbf{ACL_C}$) extensions. Then $a_i \in S$ for all admissible (resp. complete) extensions S, $a_i \in \bigcap\{S \subseteq \mathcal{A} \mid S$ is an admissible (resp. complete) extension of $\mathsf{A}\}$. By Theorem 9.6.1i,ii we immediately get $\Gamma^n_{\mathsf{A}} \Vdash_{\mathbf{L}} p_i$ and due to Theorem 9.6.1 we have $\Gamma^n_{\mathsf{A}} \vdash_{\mathbf{L}} p_i$.

Let $\Gamma^n_{\mathsf{A}} \vdash_{\mathbf{L}} p_i$. By Theorem G.3.1 we know that $\Gamma^n_{\mathsf{A}} \Vdash_{\mathbf{L}} p_i$. By Theorem 9.6.1i,ii we immediately get $a_i \in \bigcap\{S \subseteq \mathcal{A} \mid S$ is an admissible (resp. complete) extension by $\mathsf{A}\}$. □

G.4 The Other Extension Types

The proofs are analogous to the proofs for standard AFs (see Appendix F.3).

Appendix H
Appendix to Chapter 10

In order to prove soundness and completeness with respect to our semantics for **DPM**, we will show that it is equivalent to Goble's original **DPM** semantics. Since Schröder and Pattinson have proven soundness and strong completeness for Goble's semantics in [5] this is sufficient.

Goble's original neighborhood semantics is very similar to the one presented here: the key difference is that we employ an actual world. Where frames are defined as before, an F-G-model M is a pair $\langle F, v \rangle$ where F is a frame and $v : \mathcal{S} \to \wp(W)$ as before. The essential difference concerns the definition of model-validity. While in our semantics it is defined in terms of validity with respect to the actual world, in Goble's semantics it is defined in terms of validity with respect to all given worlds: $M \models^G A$ iff $M, w \models A$ for all $w \in W$. All other definitions concerning validity are analogous. For a given frame $F = \langle W, \mathcal{O} \rangle$ and $\Gamma \subseteq \mathcal{W}$, $\Gamma \Vdash^G_F A$ iff for all F-G-models M and for all $w \in W$, if $M, w \models B$ for all $B \in \Gamma$, then $M, w \models A$. Moreover, where $\alpha \in \{1, 2'\}$, $\Gamma \Vdash^G_{\mathbf{DPM}.\alpha} A$ iff $\Gamma \Vdash^G_F A$ for all **DPM**.α-frames F. Schröder and Pattinson have shown the following strong completeness and soundness result in [5]:

Theorem H.1. *Where* $\alpha \in \{1, 2'\}$ *and* $\Gamma \subseteq \mathcal{W}$, $\Gamma \Vdash^G_{\mathbf{DPM}.\alpha} A$ *iff* $\Gamma \vdash_{\mathbf{DPM}.\alpha} A$.

Theorem H.2. *Where* $\alpha \in \{1, 2'\}$ *and* $\Gamma \subseteq \mathcal{W}$, $\Gamma \Vdash^G_{\mathbf{DPM}.\alpha} A$ *iff* $\Gamma \Vdash_{\mathbf{DPM}.\alpha} A$.

Proof. Let $\mathcal{F}$ be the class of **DPM**.α-frames. "$\Leftarrow$": Let $\Gamma \Vdash_{\mathbf{DPM}.\alpha} A$ and $F = \langle W, \mathcal{O} \rangle \in \mathcal{F}$. Suppose there is an F-G-model $M = \langle F, v \rangle$ and a world $w \in W$ for which $M, w \not\models A$ and $M, w \models B$ for all $B \in \Gamma$. Note that $M' = \langle F, v, w \rangle$ is a **DPM**.α-model of Γ for which $M' \not\models A$,—a contradiction.

"$\Rightarrow$": Let $\Gamma \Vdash^G_{\mathbf{DPM}.\alpha} A$. Suppose for some frame $F = \langle W, \mathcal{O} \rangle \in \mathcal{F}$ there is an F-model $M_w = \langle F, v, w \rangle$ of Γ for which $M_w \not\models A$. Let $M = \langle F, v \rangle$. Note that $M, w \models B$ for all $B \in \Gamma$ and $M, w \not\models A$,—a contradiction. □

Theorem 10.3.1 (restated). *Where* $\alpha \in \{1, 2, 2'\}$ *and* $\Gamma \subseteq \mathcal{W}$:

$$\Gamma \vdash_{\mathbf{DPM}.\alpha} A \textit{ iff } \Gamma \Vdash_{\mathbf{DPM}.\alpha} A$$

C. Straßer, *Adaptive Logics for Defeasible Reasoning*, Trends in Logic 38,
DOI: 10.1007/978-3-319-00792-2, © Springer International Publishing Switzerland 2014

Proof. Follows immediately by Theorem H.1 and Theorem H.2. □

Theorem 10.2.2 (restated). *Where* $\alpha \in \{1, 2'\}$, **DPM**.α *satisfies* $(\star)$.

Proof. For **DPM.1** this has already been shown by Goble in [6]. Note that (D) together with (PAND) results in (AND). Since **DPM.1** strengthened by (D) has the same corresponding consequence relation as **SDL**, it also validates (N). Thus, **DPM.1** strengthened by (D) and **DPM.2**′ strengthened by (D) have the same corresponding consequence relation. Thus, **DPM.2**′ strengthened by (D) has the same corresponding consequence relation as **SDL**. □

Theorem 10.8.1 (restated). *Where* $\alpha \in \{1, 2'\}$, *none of Goble's explosion principles (DEX), (DEX-1)–(DEX-3) is valid in* **ADPM**.α.

Proof. Let us first consider the case for **ADPM.2**′. Let $W = \wp(\mathcal{S})$ and p_1 and p_2 are sentential letters. We define

$$\begin{aligned} W_a &= \{w \in W \mid p_1 \notin w, p_2 \notin w\},\\ W_b &= \{w \in W \mid p_1 \notin w, p_2 \in w\},\\ W_c &= \{w \in W \mid p_1 \in w, p_2 \notin w\},\\ W_d &= \{w \in W \mid p_1 \in w, p_2 \in w\}. \end{aligned}$$

We define a frame $F = \langle W, \mathcal{O}\rangle$ where $\mathcal{O}_w = \{W_a \cup W_b, W_c \cup W_d, W\}$ for all $w \in W$. Note that F is a **DPM.2**′-frame. Let $M = \langle F, v, @\rangle$ where $v(p_i) = \{w \in W \mid p_i \in w\}$ and @ is any world in W. Note first that $M \models \mathsf{O}p_1, \mathsf{O}\neg p_1, \mathsf{O}\top, \mathsf{P}\top, \mathsf{P}(p_1 \wedge p_2)$ and $M \not\models \mathsf{O}(p_1 \wedge p_2)$. Thus, M models a counter-instance to (DEX), (DEX-1)–(DEX-3). Note furthermore that M is a minimally abnormal model of $\{\mathsf{O}p_1, \mathsf{O}\neg p_1\}$ and also of $\{\mathsf{O}p_1, \mathsf{O}\neg p_1, \mathsf{O}\top, \mathsf{P}\top\}$ since $\mathrm{Ab}(M) = \{\mathsf{O}A \wedge \mathsf{O}\neg A \mid\vdash A \equiv p_1\}$ and $\{\mathsf{O}p_1, \mathsf{O}\neg p_1\} \vdash_{\mathbf{DPM.2'}} A$ for all $A \in \mathrm{Ab}(M)$.

The proof for **ADPM.1** is similar. Where $W = \wp(\mathcal{S})$ and $v : \mathcal{S} \to \wp(W)$, $p_i \mapsto \{w \in W \mid p_i \in w\}$, we define a frame $F = \langle W, \mathcal{O}\rangle$ where for all $w \in W$,

$$\mathcal{O}_w = \{W' \mid W' \supseteq v(p_1)\} \cup \{\emptyset\}$$

Note that F is a **DPM.1**-frame. Let $M = \langle F, v, @\rangle$ where @ is any world in W. Evidently, $M \models \mathsf{O}p_1, \mathsf{O}\top, \mathsf{O}\bot, \mathsf{P}p_1, \mathsf{P}p_2$ and $M \not\models \mathsf{O}p_2$. Thus, M models a counter-instance to (DEX), (DEX-1)–(DEX-3). Also, M is a minimally abnormal model of $\{\mathsf{O}\top, \mathsf{O}\bot\}$ and of $\{\mathsf{O}\top, \mathsf{O}\bot, \mathsf{O}p_1, \mathsf{P}p_1\}$ since $\mathrm{Ab}(M) = \{!A \mid A$ has a subformula B for which $\vdash B \equiv \top\}$ and $\{\mathsf{O}\top, \mathsf{O}\bot\} \vdash_{\mathbf{DPM.1}} A$ for all $A \in \mathrm{Ab}(M)$. □

Theorem 10.2.1 (restated). *Where* $\mathbf{L} \in \{$**DPM.1**, **DPM.2**, **DPM.2**′$\}$, $\mathbf{L}$ *does not validate any of the explosion principles* (DEX), (DEX-1)–(DEX-3).

Proof. Due to the fact that all adaptively selected models are models of the lower limit logic, the counter-models to (DEX) and (DEX-1)—(DEX-3) constructed for

ADPM.1 and **ADPM.2**$'$ in the proof of Theorem 10.8.1 are also counter-models for **DPM.1** and **DPM.2**$'$. Theorem 10.2.1 was proven for **DPM.2** by Goble in [6]. □

Theorem 10.8.2 ***(restated)*** *Where* $\alpha \in \{1, 2'\}$, *the upper limit logic of* **ADPM.**α *is* **SDL.**

Proof. Given a $\Gamma \subseteq \mathcal{W}$ we have to show that $\Gamma \cup \{\neg(\mathsf{O}A \wedge \mathsf{O}\neg A) \mid A \in \mathcal{W}\} \vdash_{\mathbf{DPM}.\alpha} B$ iff $\Gamma \vdash_{\mathbf{SDL}} B$. Note that $\vdash_{\mathbf{DPM}.\alpha} \neg(\mathsf{O}A \wedge \mathsf{O}\neg A) \equiv (\mathsf{O}A \supset \mathsf{P}A)$. Thus, $\Gamma \cup \{\neg(\mathsf{O}A \wedge \mathsf{O}\neg A)\} \vdash_{\mathbf{DPM}.\alpha} B$ iff $\Gamma \vdash_{\mathbf{DDPM}.\alpha} B$ where **DDPM.**α is **DPM.**α enriched by (D). However, since by Theorem 10.2.2 **DDPM.**α has the same corresponding consequence relation as **SDL**, we are finished. □

Corollary 10.8.3 ***(restated).*** *Where* $\alpha \in \{1, 2'\}$, **ADPM.**α *satisfies* $(\star)$.

Proof. For **SDL**-consistent premise sets Γ this is an immediate consequence of Corollary 10.8.4. Let $\Gamma \subseteq \mathcal{W}$ be **SDL**-inconsistent. Where **DADPM.**α (resp. **DDPM.**α) is **ADPM.**α (resp. **DPM.**α) enriched by (D) and $\Gamma' = \Gamma \cup \{\mathsf{O}A \supset \mathsf{P}A \mid A \in \mathcal{W}\}$, note that $\mathcal{M}_{\mathbf{SDL}}(\Gamma) = \emptyset = \mathcal{M}_{\mathbf{DDPM}.\alpha}(\Gamma) = \mathcal{M}_{\mathbf{DPM}.\alpha}(\Gamma') \supseteq \mathcal{M}_{\mathbf{ADPM}.\alpha}(\Gamma') = \mathcal{M}_{\mathbf{DADPM}.\alpha}(\Gamma)$. □

Appendix I
Appendix to Chapter 11

I.1 Semantics

The interested reader can find semantics for the monadic systems in [7, 8] and Chap. 10. Semantics for some of the dyadic **CDPM** systems, such as for instance **CDPM.1** and **CDPM.1c**, are introduced by Goble in [9]. The semantics for the variations introduced in Chap. 11 can be defined along the same lines. The only difference to the original versions is that we employ an actual world. This makes the semantics philosophically more intuitive for our applications, since we are not only interested in modeling theoremhood but rather in defining a semantic consequence relation.

One of the basic ideas for the neighborhood semantics is that propositions are interpreted in terms of sets of worlds. Moreover, each world has associated with it pairs of propositions, i.e., pairs of sets of worlds. The idea is that an obligation $\mathsf{O}(A \mid B)$ is true at a world w, in case $\langle |B|, |A| \rangle$ is one of its associated pairs of propositions and where $|B|$ denotes the set of worlds representing the proposition B (analogous for A). Let us take a look at the formal details.

Let a dyadic neighborhood frame F be a pair $\langle W, \mathcal{O} \rangle$ where W is a set of worlds and $\mathcal{O}$ assigns each world $a \in W$ a set of ordered propositions, i.e. $\mathcal{O}_a \subseteq \wp(W) \times \wp(W)$. A model M on a frame F is a triple $\langle F, v, @ \rangle$ where $v(p) \subseteq W$ for each propositional letter p and $@ \in W$ is called the actual world. Where $\mathcal{S} = \{p_1, p_2, \dots\}$ is the set of sentential letters, we define:

$$\begin{aligned} M, a \models A \text{ iff } a \in v(A), \ \text{ where } A \in \mathcal{S} \\ M, a \models \neg A \text{ iff } M, a \not\models A \\ M, a \models A \wedge B \text{ iff } (M, a \models A \text{ and } M, a \models B) \\ M, a \models A \vee B \text{ iff } (M, a \models A \text{ or } M, a \models B) \\ M, a \models A \supset B \text{ iff } (M, a \models \neg A \text{ or } M, a \models B) \\ M, a \models \top \\ M, a \not\models \bot \end{aligned}$$

C. Straßer, *Adaptive Logics for Defeasible Reasoning*, Trends in Logic 38,
DOI: 10.1007/978-3-319-00792-2, © Springer International Publishing Switzerland 2014

Moreover, where $|A|_M =_{\mathrm{df}} \{a \in W \mid M, a \models A\}$, we define,

$$M, a \models \mathsf{O}(A|B) \text{ iff } \langle |B|_M, |A|_M \rangle \in \mathcal{O}_a$$

For a model $M = \langle W, \mathcal{O}, v, @\rangle$, $M \models A$ iff $M, @ \models A$. Where $\Gamma \subseteq \mathcal{W}_2$ and $M = \langle F, v, @\rangle$, we say that M is an F-model of Γ iff $M \models A$ for all $A \in \Gamma$. Moreover, for a frame F, $\Gamma \Vdash_F A$ iff for all F-models $M = \langle F, v, @\rangle$ of Γ, $M \models A$. For a class of frames $\mathcal{F}$, $\Gamma \Vdash_{\mathcal{F}} A$ iff $\Gamma \Vdash_F A$ for all $F \in \mathcal{F}$.

Semantics for **CDPM.1c** are defined by means of the following frame conditions. Where $F = \langle W, \mathcal{O}\rangle$ and $\overline{X} =_{\mathrm{df}} W \setminus X$, we require for all $a \in W$:

$$\langle W, W\rangle \in \mathcal{O}_a \tag{F-CN}$$

$$\text{If } \langle X, Y\rangle \in \mathcal{O}_a \text{ and } \langle X, Z\rangle \in \mathcal{O}_a, \text{ then } \langle X, Y \cap Z\rangle \in \mathcal{O}_a \tag{F-CAND}$$

$$\text{If } Y \subseteq Z \text{ and } \langle X, Y\rangle \in \mathcal{O}_a \text{ and } \langle X, \overline{Y}\rangle \notin \mathcal{O}_a \text{ then } \langle X, Z\rangle \in \mathcal{O}_a \tag{F-RCPM}$$

$$\text{If } \langle X, Y\rangle \in \mathcal{O}_a, \text{ for any } Y \subseteq W, \text{ then } \langle X, X\rangle \in \mathcal{O}_a \tag{F-QR}$$

$$\text{If } \langle X \cap Y, Z\rangle \in \mathcal{O}_a, \text{ then } \langle X, \overline{Y} \cup Z\rangle \in \mathcal{O}_a \tag{F-S}$$

$$\text{If } \langle X, Y\rangle \in \mathcal{O}_a \text{ and } \langle X, \overline{Y \cap Z}\rangle \notin \mathcal{O}_a, \text{ then } \langle X \cap Z, Y\rangle \in \mathcal{O}_a \tag{F-WRM}$$

Semantics for **CDPM.2′c** are defined by means of (F-CN), (F-RCPM), (F-QR), (F-S), (F-WRM) and the following frame conditions:

$$\text{If } \langle X, Y\rangle \in \mathcal{O}_a, \langle X, Z\rangle \in \mathcal{O}_a, \langle X, \overline{Y}\rangle \notin \mathcal{O}_a, \langle X, \overline{Z}\rangle \notin \mathcal{O}_a, \text{ then } \langle X, Y \cap Z\rangle \in \mathcal{O}_a \tag{F-CPAND′}$$

$$\langle X, \emptyset\rangle \notin \mathcal{O}_a \tag{F-CP}$$

Moreover, the following frame conditions are useful to define the semantics for some of our logical variants:

$$\text{If } \langle X, Y\rangle \in \mathcal{O}_a, \langle X, \overline{Y \cap Z}\rangle \notin \mathcal{O}_a \text{ and } \langle Z, Y \cup \overline{X}\rangle \in \mathcal{O}_a, \text{ then } \langle X \cap Z, Y\rangle \in \mathcal{O}_a \tag{F-WRM$_\star$}$$

$$\text{If } \langle X \cap Y, Z\rangle \in \mathcal{O}_a, \text{ and } \langle X, \overline{Z}\rangle \notin \mathcal{O}_a, \text{ then } \langle X, \overline{Y} \cup Z\rangle \in \mathcal{O}_a \tag{F-PS}$$

$$\text{If } \langle Y, Z\rangle \in \mathcal{O}_a, \text{ and } \langle Y, \overline{Z \cap X}\rangle \notin \mathcal{O}_a, \text{ then } \langle X, \overline{Y} \cup Z\rangle \in \mathcal{O}_a \tag{F-AWRM$_\star$}$$

$$\text{If } \langle X, Y\rangle \in \mathcal{O}_a, \text{ then } \langle X, \overline{Y}\rangle \notin \mathcal{O}_a \tag{F-CD}$$

Definition I.1.1. Let $\Psi \in \wp(\mathcal{P})$, $\mathbf{L} = \oplus(\Psi, \mathbf{CDPM}^-)$ and $\Gamma \subseteq \mathcal{W}_2$. Let $\mathcal{F}$ be the class of frames that meet the conditions in {F-P | P $\in \Psi$}, (F-CN) and (F-RCPM).

We define $\mathcal{M}_{\mathbf{L}}(\Gamma)$ to be the class of all F-models of Γ. Moreover $\Gamma \Vdash_{\mathbf{L}} A$ iff for all $M \in \mathcal{M}_{\mathbf{L}}(\Gamma)$, $M \models A$.

Fact I.1.1. *Let $\Psi \in \wp(\mathcal{P})$, $\mathbf{L} = \oplus(\Psi, \mathbf{CDPM}^-)$ and $\Gamma \subseteq \mathcal{W}_2$. Let $\mathcal{F}$ be the class of frames that meet the conditions in* {F-P | $P \in \Psi$}, *(F-CN) and (F-RCPM).*

$$\Gamma \Vdash_{\mathbf{L}} A \textit{ iff } \Gamma \Vdash_{\mathcal{F}} A.$$

Goble's original neighborhood semantics is very similar to the one presented here. As pointed out, the key difference is that we employ an actual world. Where frames are defined as before, an F-G-model M is a pair $\langle F, v\rangle$ where $F = \langle W, \mathcal{O}\rangle$ is a frame and $v(p) \subseteq W$ for each propositional letter p. The essential difference concerns the definition of model-validity. While in our semantics presented above it is defined in terms of validity with respect to the actual world, in Goble's semantics it is defined in terms of validity with respect to all given worlds: $M \models^G A$ iff $M, w \models A$ for all $w \in W$. All other definitions concerning validity are analogous. For a given frame $F = \langle W, \mathcal{O}\rangle$ and $\Gamma \subseteq \mathcal{W}_2$, $\Gamma \Vdash^G_F A$ iff for all F-G-models M and for all $w \in W$, if $M, w \models B$ for all $B \in \Gamma$, then $M, w \models A$. For a class of frames $\mathcal{F}$, $\Gamma \Vdash^G_{\mathcal{F}} A$ iff for all $F \in \mathcal{F}$, $\Gamma \Vdash^G_F A$.

Goble offered a rather involved proof of weak completeness and soundness for **CDPM.1c** in [9]. The authors in [5] have proven strong soundness and completeness for all rank-1 modal logics (i.e., logics which are axiomatized by formulas containing exactly one level of modal operators) with respect to their canonical neighborhood semantics, i.e. that is with respect to the way Goble defined his neighborhood semantics for them. Obviously all the logics defined in Chap. 11 are rank-1 modal logics. Hence, with the results of [5] we have:

Theorem I.1.1. *Where* $\Gamma \subseteq \mathcal{W}_2$, $\Psi \in \wp(\mathcal{P})$, $\mathbf{L} = \oplus(\Psi, \mathbf{CDPM}^-)$, *and* $\mathcal{F}$ *is the class of frames that meet the conditions in* {F-P | P $\in \Psi$}, (F-CN) *and* (F-RCPM), *we have:*

$$\Gamma \Vdash^G_{\mathcal{F}} A \textit{ iff } \Gamma \vdash_{\mathbf{L}} A$$

The following establishes the bridge between Theorem I.1.3 and Theorem I.1.1.

Theorem I.1.2. *Where* $\Gamma \subseteq \mathcal{W}_2$, $\Psi \in \wp(\mathcal{P})$, $\mathbf{L} = \oplus(\Psi, \mathbf{CDPM}^-)$, *and* $\mathcal{F}$ *is the class of frames that meet the conditions in* {F-P | P $\in \Psi$}, (F-CN) *and* (F-RCPM), *we have:*

$$\Gamma \Vdash_{\mathbf{L}} A \textit{ iff } \Gamma \Vdash_{\mathcal{F}} A \textit{ iff } \Gamma \Vdash^G_{\mathcal{F}} A$$

Proof. "$\Rightarrow$": Let $\Gamma \Vdash_{\mathcal{F}} A$ and $F = \langle W, \mathcal{O}\rangle \in \mathcal{F}$. Suppose there is an F-G-model $M = \langle F, v\rangle$ and a world $w \in W$ for which $M, w \not\models A$ and $M, w \models B$ for all $B \in \Gamma$. Note that $M' = \langle F, v, w\rangle$ is an F-model of Γ for which $M' \not\models A$,—a contradiction.

"$\Leftarrow$": Let $\Gamma \Vdash^G_{\mathcal{F}} A$. Suppose that for some frame $F = \langle W, \mathcal{O}\rangle \in \mathcal{F}$ there is an F-model $M_w = \langle F, v, w\rangle$ of Γ for which $M_w \not\models A$. Let $M = \langle F, v\rangle$. Note that $M, w \models B$ for all $B \in \Gamma$ and $M, w \not\models A$,—a contradiction. □

Theorem I.1.3. *Let* $\Psi \in \wp(\mathcal{P})$, $\mathbf{L} = \oplus(\Psi, \mathbf{CDPM}^-)$ *and* $\Gamma \subseteq \mathcal{W}_2$.

$$\Gamma \Vdash_{\mathbf{L}} A \textit{ iff } \Gamma \vdash_{\mathbf{L}} A$$

Proof. By Theorem I.1.2, $\Gamma \Vdash_{\mathcal{F}} A$ iff $\Gamma \Vdash^{G}_{\mathcal{F}} A$. By Theorem I.1.1, $\Gamma \Vdash^{G}_{\mathcal{F}} A$ iff $\Gamma \vdash_{\mathbf{L}} A$. Hence, $\Gamma \Vdash_{\mathcal{F}} A$ iff $\Gamma \vdash_{\mathbf{L}} A$. □

Corollary I.1.1.

(i) Where **LLL** $\in \downarrow$**CDPM.1c** $\cup \downarrow$**CDPM.2′c**, $\Gamma \vdash_{\mathbf{A_c}(\mathbf{LLL})} A$ *iff* $\Gamma \Vdash_{\mathbf{A_c}(\mathbf{LLL})} A$.
(ii) Where **LLL** $\in \bigcup_{\alpha \in \{1,2'\}} \downarrow$**CDPM.**$\alpha$**c** $\cup$ $\{\oplus_{\mathrm{CD}} \ominus_{\mathrm{WRM}}$ **CDPM.**α**c**$\}$ *and* $\mathbf{x} \in \{\mathbf{m}, \mathbf{r}\}$, $\Gamma \vdash_{\mathbf{A^x_r}(\mathbf{LLL})} A$ *iff* $\Gamma \Vdash_{\mathbf{A^x_r}(\mathbf{LLL})} A$.

Proof. Follows by Theorem 2.6.1 and Theorem I.1.3. □

The semantics of our sequential ALs from Sect. 11.6 are defined as follows.[6]

Definition I.1.2. Where **LLL** $\in \bigcup_{\alpha \in \{1,2'\}} \downarrow \ominus_{\mathrm{WRM}}$ **CDPM.**α**c** we define: $\Gamma \Vdash_{\mathbf{A^r_r} \circ \mathbf{A_c}(\mathbf{LLL})} A$ iff for all $M \in \mathcal{M}_{\mathbf{A^r_r}(\mathbf{LLL})}\left(Cn_{\mathbf{A_c}(\mathbf{LLL})}(\Gamma)\right)$, $M \models A$.

By Theorem 3.2.5 and Theorem 3.3.1 we have:

Corollary I.1.2. *Where* **LLL** $\in \bigcup_{\alpha \in \{1,2'\}} \downarrow \ominus_{\mathrm{WRM}}$ **CDPM.**α**c**, $\Gamma \Vdash_{\mathbf{A^r_r} \circ \mathbf{A_c}(\mathbf{LLL})} A$ *iff* $\Gamma \vdash_{\mathbf{A^r_r} \circ \mathbf{A_c}(\mathbf{LLL})} A$ *iff* $A \in Cn_{\mathbf{A^r_r} \circ \mathbf{A_c}(\mathbf{LLL})}(\Gamma)$.

I.2 Proofs

Theorem 11.3.1 (restated). *Where* $\alpha \in \{1, 2'\}$, **CDPM.**α**c** *satisfies (C⋆).*

Proof. Goble has proven the statement already for **CDPM.1c**. Note that from (CD) and (CPAND’), (CAND) is derivable. Since **CDPM.1c** together with (CD) validates (CN), **CDPM.2′c** together with (CD) results in the same consequence relation as the one of **CDPM.1c** together with (CD). Thus, **CDPM.2′c** together with (CD) results in the same consequence relation as the one of (**R**)**SDDL**. □

Lemma I.2.1. *Where* $\alpha \in \{1, 2'\}$ *and* $\Gamma \subseteq \mathcal{W}_2$*:*
M is a $\oplus(\{\mathrm{CD}, \mathrm{PS}\}, \ominus_{\mathrm{S}}$**CDPM.**$\alpha$**c**)*-model of* Γ *iff* M *is a* $\oplus_{\mathrm{CD}}$**CDPM.**α**c**)*-model of* Γ.

Proof. Let $M = \langle F, v, @\rangle$ be a $\oplus(\{\mathrm{CD}, \mathrm{PS}\}, \ominus_{\mathrm{S}}$**CDPM.**$\alpha$**c**)-model of Γ. It is enough to show that (F-S) is valid in F. Suppose there is a counter-instance to the frame condition (F-S) valid in F. Let thus

$$\langle X \cap Y, Z\rangle \in \mathcal{O}_a \tag{I.1}$$

and suppose

$$\langle X, \overline{Y} \cup Z\rangle \notin \mathcal{O}_a \tag{I.2}$$

[6] Since $\mathbf{A_c}$ uses minimal abnormality the semantics is not defined by the iterative procedure of Definition 3.2.1. Compare the discussion concerning Example 3.2.1 where the reasons are explained in detail.

By (F-CD) and (I.1), $\langle X \cap Y, \overline{Z}\rangle \notin \mathcal{O}_a$. By (F-PS), (I.1) and (I.2), $\langle X, \overline{Z}\rangle \in \mathcal{O}_a$. But then, by (F-WRM), $\langle X, \overline{Y} \cup Z\rangle \in \mathcal{O}_a$,—a contradiction. Hence F is a frame satisfying the frame conditions corresponding to $\oplus_{\mathrm{CD}}$**CDPM**.α**c**.

The other direction is trivial. □

Theorem 11.4.2 (restated). *Where* $\alpha \in \{1, 2'\}$, $\oplus_{\mathrm{PS}} \ominus_{\mathrm{S}}$ **CDPM**.α**c** *satisfies* ($C\star$).

Proof. This is an immediate consequence of Lemma I.2.1 and the fact that **CDPM**.α**c** satisfies (C$\star$). □

Lemma I.2.2. (**R**)**SDDL** *validates (WRM), (WRM$_\star$) and (AWRM$_\star$).*

Proof. Suppose $\mathsf{O}(B|A)$ and $\mathsf{P}(B \wedge C|A)$. Hence, $\neg\mathsf{O}(\neg B \vee \neg C|A)$. By (RCM), $\neg\mathsf{O}(\neg C|A)$ and hence $\mathsf{P}(C|A)$. By (RatMono), $\mathsf{O}(B|A \wedge C)$. Hence, $\vdash_{\mathbf{(R)SDDL}}$ $(\mathsf{O}(B|A \wedge \mathsf{P}B \wedge C|A)) \supset \mathsf{O}(B|A \wedge C)$. Hence (WRM) and (WRM$_\star$) are validated by (**R**)**SDDL**.

Suppose now $\mathsf{O}(B|A)$ and $\mathsf{P}(B \wedge C|A)$. Hence, $\neg\mathsf{O}(\neg B \vee \neg C|A)$. By (RCM), $\neg\mathsf{O}(\neg C|A)$ and hence $\mathsf{P}(C|A)$. Hence, by (RatMono), $\mathsf{O}(B|A \wedge C)$. By (S), $\mathsf{O}(A \supset B|C)$. Hence $\vdash_{\mathbf{(R)SDDL}}$ $\big(\mathsf{O}(B|A) \wedge \mathsf{P}(B \wedge C|A)\big) \supset \mathsf{O}(A \supset B|C)$. Hence, (**R**)**SDDL** validates (AWRM$_\star$). □

Theorem 11.4.1 (restated). *Where* $\alpha \in \{1, 2'\}$,
$\oplus$({WRM$_\star$, AWRM$_\star$} $\ominus_{\mathrm{WRM}}$ **CDPM**.α**c**) *satisfies* (C$\star$).

Proof. Let $\mathbf{L}' = \oplus$({WRM$\star$, AWRM$\star$}, $\ominus_{\mathrm{WRM}}$**CDPM**.α**c**). Since it validates all instances of (WRM$_\star$) and (AWRM$_\star$), it also validates all instances of (WRM). Hence, the consequence relation of $\mathbf{L}'$ is at least as strong as the consequence relation of **CDPM**.α**c**. Since $\oplus_{\mathrm{CD}}$**CDPM**.α**c** characterizes the same consequence relation as (**R**)**SDDL** and due to Lemma I.2.2, it also validates all instances of (WRM$_\star$) and (AWRM$_\star$). Hence, $\oplus_{\mathrm{CD}}\mathbf{L}'$ and $\oplus_{\mathrm{CD}}$**CDPM**.α**c** characterize the same consequence relation. □

Theorem 11.4.3 (restated). *Where* $\alpha \in \{1, 2'\}$,
$\oplus$({WRM$_\star$, AMRM$_\star$, PS}, $\ominus$({WRM, S}, **CDPM**.α**c**)) *satisfies (C$\star$).*

Proof. Let

$$\mathbf{L}' = \oplus(\{\text{WRM}_\star, \text{AMRM}_\star, \text{PS}\}, \ominus(\{\text{WRM}, \text{S}\}, \mathbf{CDPM}.\alpha\mathbf{c}))$$

Since $\mathbf{L}'$ validates all instances of (WRM$_\star$) and (AWRM$_\star$), it also validates all instances of (WRM). Hence, the consequence relation of $\mathbf{L}'$ is at least as strong as the consequence relation of $\oplus_{\mathrm{PS}} \ominus_S$ **CDPM**.α**c**. Due to Theorem 11.4.2, $\oplus$({CD, PS}, $\ominus_{\mathrm{S}}$**CDPM**.α**c**) satisfies (C$\star$). Hence, $\oplus_{\mathrm{CD}}\mathbf{L}'$ is at least as strong as (**R**)**SDDL**. Since (**R**)**SDDL** validates all instances of (WRM$_\star$) and (AWRM$_\star$), it characterizes the same consequence relation as $\oplus_{\mathrm{CD}}\mathbf{L}'$. □

Theorem 11.5.1 (restated). *Where* $\alpha \in \{1, 2'\}$, *we have:*

(i) $\mathbf{A_c}(\mathbf{CDPM}.\alpha\mathbf{c})$ *satisfies (C‡).*
(ii) $\mathbf{A_c}(\oplus_{\mathrm{PS}} \ominus_S \mathbf{CDPM}.\alpha\mathbf{c})$ *satisfies (C‡).*

Proof. Ad (i): Let Γ be a (**R**)**SDDL**-consistent premise set. As $\oplus_{\mathrm{CD}}\mathbf{CDPM}.\alpha\mathbf{c}$ is equivalent to (**R**)**SDDL**, $\mathcal{M}_{\oplus_{\mathrm{CD}}\ \mathbf{CDPM}.\alpha\mathbf{c}}(\Gamma) \neq \emptyset$. By the definition of Ω_d^c these are the minimally abnormal $\mathbf{CDPM}.\alpha\mathbf{c}$-models since for all these models M, $\mathrm{Ab}(M) = \emptyset$. Moreover, for all $M \in \mathcal{M}_{\mathbf{CDPM}.\alpha\mathbf{c}\Gamma} \setminus \mathcal{M}_{\oplus_{\mathrm{CD}}\ \mathbf{CDPM}.\alpha\mathbf{c}}(\Gamma)$, $\mathrm{Ab}(M) \neq \emptyset$ since M validates a counter-instance of (CD), $\mathsf{O}(A|B) \wedge \mathsf{O}(\neg A|B)$. Hence $\mathbf{A_c}(\mathbf{CDPM}.\alpha\mathbf{c})$ is equivalent to $\oplus_{\mathrm{CD}}\mathbf{CDPM}.\alpha\mathbf{c}$ for Γ. Therefore it is equivalent to (**R**)**SDDL** for all premise sets for which (**R**)**SDDL** is non-explosive.

Ad (ii): Due to Lemma I.2.1 this is proven analogously. □

Lemma I.2.3. *Where* $\alpha \in \{1, 2'\}$ *and* $\mathbf{x} \in \{\mathbf{m}, \mathbf{r}\}$, *for all premise sets for which* $\mathbf{CDPM}.\alpha\mathbf{c}$ *(resp.* $\oplus_{\mathrm{CD}}\mathbf{CDPM}.\alpha\mathbf{c}$ *resp.* $\oplus(\{\mathrm{CD}, \mathrm{PS}\}, \ominus_{\mathrm{S}}\mathbf{CDPM}.\alpha\mathbf{c})$*) is non-explosive,* $\mathbf{A_r^x}$ $(\ominus_{\mathrm{WRM}}\mathbf{CDPM}.\alpha\mathbf{c})$ *(resp.* $\mathbf{A_r^x}(\oplus_{\mathrm{CD}}\ominus_{\mathrm{WRM}}\mathbf{CDPM}.\alpha\mathbf{c})$ *resp.* $\mathbf{A_r^x}(\oplus(\{\mathrm{CD}, \mathrm{PS}\}, \ominus(\{\mathrm{WRM}, \mathrm{S}\}, \mathbf{CDPM}.\alpha\mathbf{c}))))$ *has the same consequence relation as* $\mathbf{CDPM}.\alpha\mathbf{c}$ *(resp.* $\oplus_{\mathrm{CD}}\mathbf{CDPM}.\alpha\mathbf{c}$ *resp.* $\oplus(\{\mathrm{CD}, \mathrm{PS}\}, \ominus_{\mathrm{S}}\mathbf{CDPM}.\alpha\mathbf{c})$*).*

Proof. It is immediately clear that all models in $\mathcal{M}_{\mathbf{CDPM}.\alpha\mathbf{c}}(\Gamma)$ are minimally abnormal $\ominus_{\mathrm{WRM}}\mathbf{CDPM}.\alpha\mathbf{c}$-models of Γ since they do not validate any abnormalities in Ω_d^r. For all $M \in \mathcal{M}_{\ominus_{\mathrm{WRM}}\mathbf{CDPM}.\alpha\mathbf{c}}(\Gamma) \setminus \mathcal{M}_{\mathbf{CDPM}.\alpha\mathbf{c}}(\Gamma)$ there is a counter-instance of (WRM). Therefore these M do validate abnormalities and are therefore not minimally abnormal. Hence $\mathcal{M}_{\mathbf{CDPM}.\alpha\mathbf{c}}(\Gamma)$ is the set of all minimally abnormal (and all reliable) models. The proof is similar for the other cases. □

Theorem 11.5.3 is an immediate consequence.

Lemma I.2.4. *Let* $\Gamma \subseteq \mathcal{W}_2$ *be a* (**R**)**SDDL**-*consistent premise set and* $\alpha \in \{1, 2'\}$.

(i) $M \in \mathcal{M}_{\mathbf{A_c}(\ominus_{\mathrm{WRM}}\mathbf{CDPM}.\alpha\mathbf{c})}(\Gamma)$ *iff* $M \in \mathcal{M}_{\oplus_{\mathrm{CD}}\ominus_{\mathrm{WRM}}\mathbf{CDPM}.\alpha\mathbf{c}}(\Gamma)$ *iff*
$M \in \mathcal{M}_{\ominus_{\mathrm{WRM}}\mathbf{CDPM}.\alpha\mathbf{c}}\left(Cn_{\mathbf{A_c}(\ominus_{\mathrm{WRM}}\mathbf{CDPM}.\alpha\mathbf{c})}(\Gamma)\right)$
(ii) $M \in \mathcal{M}_{\mathbf{A_c}(\oplus_{\mathrm{PS}}\ominus(\{\mathrm{WRM},\mathrm{S}\},\mathbf{CDPM}.\alpha\mathbf{c}))}(\Gamma)$ *iff*
$M \in \mathcal{M}_{\oplus(\{\mathrm{CD},\mathrm{PS}\},\ominus(\{\mathrm{WRM},\mathrm{S}\},\mathbf{CDPM}.\alpha\mathbf{c}))}(\Gamma)$ *iff*
$M \in \mathcal{M}_{\oplus_{\mathrm{PS}}\ominus(\{\mathrm{WRM},\mathrm{S}\},\mathbf{CDPM}.\alpha\mathbf{c})}\left(Cn_{\mathbf{A_c}(\oplus_{\mathrm{PS}}\ominus(\{\mathrm{WRM},\mathrm{S}\},\mathbf{CDPM}.\alpha\mathbf{c}))}(\Gamma)\right)$

Proof. Ad (i): Evidently, since Γ is (**R**)**SDDL**-consistent, there are $\oplus_{\mathrm{CD}}\ominus_{\mathrm{WRM}}$ $\mathbf{CDPM}.\alpha\mathbf{c}$-models of Γ, $\ominus_{\mathrm{WRM}}\mathbf{CDPM}.\alpha\mathbf{c}$-models of Γ and, due to the reassurance property (see Corollary 2.4.3i) also $\mathbf{A_c}(\ominus_{\mathrm{WRM}}\mathbf{CDPM}.\alpha\mathbf{c})$-models of Γ. Let M be an $\mathbf{A_c}(\ominus_{\mathrm{WRM}}\mathbf{CDPM}.\alpha\mathbf{c})$-model of Γ. Suppose $\mathrm{Ab}(M) \neq \emptyset$ (w.r.t. Ω_d^c). However, for every N that is a $\oplus_{\mathrm{CD}}\ominus_{\mathrm{WRM}}\mathbf{CDPM}.\alpha\mathbf{c}$-model of Γ, $\mathrm{Ab}(N) = \emptyset$ (w.r.t. Ω_d^c). Hence, M is not a minimally abnormal $\ominus_{\mathrm{WRM}}\mathbf{CDPM}.\alpha\mathbf{c}$-model of Γ,—a contradiction. Hence, $\mathrm{Ab}(M) = \emptyset$ (w.r.t. Ω_d^c). Hence, M validates all instances of (CD).

Thus, M is a $\oplus_{\mathrm{CD}}\ominus_{\mathrm{WRM}}\mathbf{CDPM}.\alpha\mathbf{c}$-model of Γ. Hence,

$$\mathcal{M}_{\mathbf{A_c}(\ominus_{\mathrm{WRM}}\mathbf{CDPM}.\alpha\mathbf{c})}(\Gamma) \subseteq \mathcal{M}_{\oplus_{\mathrm{CD}}\ominus_{\mathrm{WRM}}\mathbf{CDPM}.\alpha\mathbf{c}}(\Gamma).$$

By the completeness of $\mathbf{A_c}(\ominus_{\mathrm{WRM}}\mathbf{CDPM}.\alpha\mathbf{c})$, $\Gamma \vdash_{\mathbf{A_c}(\ominus_{\mathrm{WRM}}\mathbf{CDPM}.\alpha\mathbf{c})} \neg A$ for all $A \in \Omega_d^c$. Thus, $\Gamma \vdash_{\mathbf{A_c}(\ominus_{\mathrm{WRM}}\mathbf{CDPM}.\alpha\mathbf{c})} \mathsf{O}(B|C) \supset \mathsf{P}(B|C)$ for all $B, C \in \mathcal{W}_2$. Hence,

$$\mathcal{M}_{\ominus_{\text{WRM}}\mathbf{CDPM}.\alpha\mathbf{c}}\left(Cn_{\mathbf{ALc}(\ominus_{\text{WRM}}\mathbf{CDPM}.\alpha\mathbf{c})}(\Gamma)\right) \subseteq \mathcal{M}_{\oplus_{\text{CD}}\ominus_{\text{WRM}}\mathbf{CDPM}.\alpha\mathbf{c}}(\Gamma).$$

Let now M be a $\oplus_{\text{CD}}\ominus_{\text{WRM}}\mathbf{CDPM}.\alpha\mathbf{c}$-model of Γ. Hence, $\text{Ab}(M) = \emptyset$ (w.r.t. Ω_d^c). Hence, M is a minimally abnormal $\ominus_{\text{WRM}}\mathbf{CDPM}.\alpha\mathbf{c}$-model of Γ. Thus, $M \in \mathcal{M}_{\mathbf{A_c}(\ominus_{\text{WRM}}\mathbf{CDPM}.\alpha\mathbf{c})}(\Gamma)$. Moreover, trivially,

$$M \in \mathcal{M}_{\ominus_{\text{WRM}}\mathbf{CDPM}.\alpha\mathbf{c}}\left(Cn_{\mathbf{Ac}(\ominus_{\text{WRM}}\mathbf{CDPM}.\alpha\mathbf{c})}(\Gamma)\right).$$

Ad (ii): Analogous to part (i). □

Theorem 11.6.1 (restated). *Where* $\alpha \in \{1, 2'\}$ *and* $\mathbf{x} \in \{\mathbf{m}, \mathbf{r}\}$ *we have:*

(i) $\mathbf{A_r^x} \circ \mathbf{A_c}(\ominus_{\text{WRM}}\mathbf{CDPM}.\alpha\mathbf{c})$ *satisfies (C‡).*
(ii) $\mathbf{A_r^x} \circ \mathbf{A_c}(\oplus_{\text{PS}}\ominus(\{\text{WRM}, \text{S}\}, \mathbf{CDPM}.\alpha\mathbf{c}))$ *satisfies (C‡).*

Proof. Let Γ be (**R**)**SDDL**-consistent.
Ad (i): We show the case $\mathbf{x} = \mathbf{r}$. The case $\mathbf{x} = \mathbf{m}$ is analogous.

$$\mathcal{M}_{\mathbf{A_r^r}\circ\mathbf{A_c}(\ominus_{\text{WRM}}\mathbf{CDPM}.\alpha\mathbf{c})}(\Gamma) = \mathcal{M}_{\mathbf{A_r^r}(\ominus_{\text{WRM}}\mathbf{CDPM}.\alpha\mathbf{c})}\left(Cn_{\mathbf{A_c}(\ominus_{\text{WRM}}\mathbf{CDPM}.\alpha\mathbf{c})}(\Gamma)\right) = \left\{M \in \mathcal{M}_{\ominus_{\text{WRM}}\mathbf{CDPM}.\alpha\mathbf{c}}\left(Cn_{\mathbf{A_c}(\ominus_{\text{WRM}}\mathbf{CDPM}.\alpha\mathbf{c})}(\Gamma)\right) \mid M \text{ is reliable w.r.t. } \Omega_d^r\right\}$$

The latter is by Lemma I.2.4i identical to

$$\left\{M \in \mathcal{M}_{\oplus_{\text{CD}}\ominus_{\text{WRM}}\mathbf{CDPM}.\alpha\mathbf{c}}(\Gamma) \mid M \text{ is reliable w.r.t. } \Omega_d^r\right\} = \mathcal{M}_{\mathbf{A_r^r}(\oplus_{\text{CD}}\ominus_{\text{WRM}}\mathbf{CDPM}.\alpha\mathbf{c})}(\Gamma)$$

The rest follows with Theorem 11.5.3.
Ad (ii): Analogous to part (i). □

Theorem I.2.1. *All ALs defined in Sect. 11.5 and Sect. 11.6 with lower limit logics in* $\downarrow$**CDPM.1c** *and* $\downarrow$**CDPM.2′c** *falsify* (CDEX-1)–(CDEX-3).

Proof. Let $F = \langle W, \mathcal{O}\rangle$ where $W = \wp(\mathcal{S})$. We define $W' = \{w \in W \mid p_1 \in w\}$ and $\mathcal{O}_w$ is defined as follows for all $w \in W$:

$$\mathcal{O}_w = \bigcup_{X\in\Psi} \{\langle X, Y\rangle \mid Y \supseteq X \cap W'\} \cup \{\langle W, \emptyset\rangle\}$$

where $\Psi = \{W'' \subseteq W \mid W'' \cap W' \neq \emptyset\}$.

It is easy to see that F satisfies the frame conditions for **CDPM.1c** and hence also all frame conditions for logics in $\downarrow$**CDPM.1c**. Obviously $\langle W, W\rangle \in \mathcal{O}_w$ for every $w \in W$. Hence (F-CN) is valid.
Let now $\langle X, Y\rangle, \langle X, Z\rangle \in \mathcal{O}_w$. In the case that either $\langle X, Y\rangle$ or $\langle X, Z\rangle$ is $\langle W, \emptyset\rangle$ we have $Y \cap Z = \emptyset$ and hence $\langle X, Y \cap Z\rangle = \langle W, \emptyset\rangle \in \mathcal{O}_w$. Let now $Y, Z \supseteq X \cap W'$ and

$X \in \Psi$. Hence also $Y \cap Z \supseteq X \cap W'$. Hence, $\langle X, Y \cap Z\rangle \in \mathcal{O}_w$. Hence, (F-CAND) is valid.

Let $Y \subseteq Z$ and $\langle X, Y\rangle \in \mathcal{O}_w$ and $\langle X, \overline{Y}\rangle \notin \mathcal{O}_w$. Since $\langle W, W\rangle \in \mathcal{O}_w$, $\langle X, Y\rangle \neq \langle W, \emptyset\rangle$. Hence, $Y \supseteq X \cap W'$ and $X \in \Psi$. Hence $Z \supseteq X \cap W'$. Hence $\langle X, Z\rangle \in \mathcal{O}_w$. Hence, (F-RCPM) is valid.

Let us take a look at (F-QR). Let $\langle X, Y\rangle \in \mathcal{O}_w$. The case $\langle X, Y\rangle = \langle W, \emptyset\rangle$ is trivial, since $\langle W, W\rangle \in \mathcal{O}_w$. Let now $\langle X, Y\rangle \neq \langle W, \emptyset\rangle$. Then, since $X \supseteq X \cap W'$, also $\langle X, X\rangle \in \mathcal{O}_w$. Hence, (F-QR) is valid.

Let $\langle X \cap Y, Z\rangle \in \mathcal{O}_w$. In case $\langle X \cap Y, Z\rangle = \langle W, \emptyset\rangle$, we have $X = Y = W$. Then $\langle X, \overline{Y} \cup Z\rangle = \langle W, \emptyset\rangle \in \mathcal{O}_w$. Let now $\langle X \cap Y, Z\rangle \neq \langle W, \emptyset\rangle$. Hence, $X \cap Y \cap W' \neq \emptyset$. Hence, $X \cap W' \neq \emptyset$. Also, $Z \supseteq X \cap Y \cap W'$. Hence, $\overline{Y} \cup Z \supseteq X \cap W'$. Hence, $\langle X, \overline{Y} \cup Z\rangle \in \mathcal{O}_w$. Hence, (F-S) is valid.

Let $\langle X, Y\rangle \in \mathcal{O}_w$ and $\langle X, \overline{Y \cap Z}\rangle \notin \mathcal{O}_w$. Note that $\langle X, Y\rangle \neq \langle W, \emptyset\rangle$, since $\langle W, \overline{\emptyset \cap Z}\rangle = \langle W, W\rangle$ and $\langle W, W\rangle \in \mathcal{O}_w$. Hence, $X \cap W' \neq \emptyset$. Moreover, $\overline{Y \cap Z} \not\supseteq X \cap W'$. Hence there is a $x \in X \cap W'$ such that $x \in Y \cap Z$. Hence, $X \cap Z \cap W' \neq \emptyset$. Since $\langle X, Y\rangle \in \mathcal{O}_w$, $Y \supseteq X \cap W'$. Hence also $Y \supseteq X \cap Z \cap W'$. Altogether hence, $\langle X \cap Z, Y\rangle \in \mathcal{O}_w$. Thus, (F-WRM) is valid.

We define an F-model $M = \langle F, v, @\rangle$ as follows. The actual world @ is an arbitrary world in W. Moreover, $v : p_i \mapsto \{w \in W \mid p_i \in w\}$. The model validates all premise sets $\Gamma \in \big\{\{\mathsf{O}(\top|\top), \mathsf{O}(\bot|\top)\}, \{\mathsf{O}(\top|\top), \mathsf{O}(\bot|\top), \mathsf{P}(p_2|\top)\}, \{\mathsf{O}(\top|\top), \mathsf{O}(\bot|\top), \mathsf{O}(p_1|\top), \mathsf{P}(p_1|\top), \mathsf{P}(p_2|\top)\}\big\}$ and falsifies the following instance of (CDEX-3): $M \models \mathsf{O}(\top|\top), \mathsf{O}(\bot|\top), \mathsf{O}(p_1|\top), \mathsf{P}(p_1|\top), \mathsf{P}(p_2|\top)$ and $M \not\models \mathsf{O}(p_2|\top)$. Thus, it does also not validate (CDEX-1) and (CDEX-2). Furthermore the model is minimally abnormal for Γ with resepct to Ω_d^c and Ω_d^r and any lower limit $\mathbf{L} \in \downarrow$**CDPM.1c**. For Ω_d^c note that the only abnormalities are $\{!_c\mathsf{O}(A|\top)$: where $\vdash_{\mathbf{L}} \top \equiv A$ if $\mathbf{L}$ does not validate (CAND) resp. where A has a subformula B for which $\vdash_{\mathbf{L}} \top \equiv B$ if $\mathbf{L}$ validates (CAND)$\}$, but we also have $\Gamma \vdash_{\mathbf{L}} !_c\mathsf{O}(A|\top)$ for all these abnormalities. For Ω_d^r the set of abnormalities verified by M is empty. Therefore, given premises Γ, the model is selected in all ALs based on lower limits in $\downarrow$**CDPM.1c** that are defined in Sects. 11.5 and 11.6.

Let $F = \langle W, \mathcal{O}\rangle$ where $W = \wp(\mathcal{S})$. We define $W_a = \{w \in W \mid p_1 \notin w, p_2 \notin w\}$, $W_b = \{w \in W \mid p_1 \notin w, p_2 \in w\}$, $W_c = \{w \in W \mid p_1 \in w, p_2 \notin w\}$, $W_d = \{w \in W \mid p_1 \in w, p_2 \in w\}$, and $\mathcal{O}_w$ for all $w \in W$ as follows:

$$\mathcal{O}_w = \bigcup_{X \in \Psi} \{\langle X, Y\rangle \mid Y \supseteq X \cap (W_c \cup W_d)\} \cup \{\langle W, W_a \cup W_b\rangle\}$$

where $\Psi = \{W'' \subseteq W \mid W'' \cap (W_c \cup W_d) \neq \emptyset\}$.

It is easy to see that F is a **CDPM.2′c**-frame. Obviously $\langle W, W\rangle \in \mathcal{O}_w$ for every $w \in W$ and hence (F-CN) is valid.

Let $\langle X, Y\rangle, \langle X, Z\rangle \in \mathcal{O}_w$ and $\langle X, \overline{Y}\rangle, \langle X, \overline{Z}\rangle \notin \mathcal{O}_w$. Obviously neither $\langle X, Y\rangle$ nor $\langle X, Z\rangle$ is equal to $\langle W, W_a \cup W_b\rangle$, since $\langle W, \overline{W_a \cup W_b}\rangle = \langle W, W_c \cup W_d\rangle \in \mathcal{O}_w$. Hence, $Y, Z \supseteq X \cap (W_c \cup W_d)$ and $X \in \Psi$. Hence, $Y \cap Z \supseteq X \cap (W_c \cup W_d)$. Hence, $\langle X, Y \cap Z\rangle \in \mathcal{O}_w$. Hence, (F-CPAND′) is valid.

Let $\langle X, Y\rangle \in \mathcal{O}_w$ and $Y \subseteq Z$. Let moreover, $\langle X, \overline{Y}\rangle \notin \mathcal{O}_w$. Hence $\langle X, Y\rangle \neq \langle W, W_a \cup W_b\rangle$. Hence, $X \in \Psi$ and $Y \supseteq X \cap (W_c \cup W_d)$. Hence $Z \supseteq X \cap (W_c \cup W_d)$. Hence, $\langle X, Z\rangle \in \mathcal{O}_w$. Hence (F-RCPM) is valid.

Let $\langle X, Y\rangle \in \mathcal{O}_w$. Hence, $X \in \Psi$. Trivially $X \supseteq X \cap (W_c \cup W_d)$. Hence $\langle X, X\rangle \in \mathcal{O}_w$. Hence, (F-QR) is valid.

Let $\langle X \cap Y, Z\rangle \in \mathcal{O}_w$. In case $\langle X \cap Y, Z\rangle = \langle W, W_a \cup W_b\rangle$ we have $X = Y = W$. Then $\langle X, \overline{Y} \cup Z\rangle = \langle W, W_a \cup W_b\rangle \in \mathcal{O}_w$. Let now $\langle X \cap Y, Z\rangle \neq \langle W, W_a \cup W_b\rangle$. Hence, $X \cap Y \cap (W_c \cup W_d) \neq \emptyset$. Hence, $X \cap (W_c \cup W_d) \neq \emptyset$. Also, $Z \supseteq X \cap Y \cap (W_c \cup W_d)$. Hence, $\overline{Y} \cup Z \supseteq X \cap (W_c \cup W_d)$. Hence, $\langle X, \overline{Y} \cup Z\rangle \in \mathcal{O}_w$. Hence, (F-S) is valid.
Let $\langle X, Y\rangle \in \mathcal{O}_w$ and $\langle X, \overline{Y \cap Z}\rangle \notin \mathcal{O}_w$. Suppose that $\langle X, Y\rangle = \langle W, W_a \cup W_b\rangle$. Hence $\langle W, \overline{(W_a \cup W_b) \cap Z}\rangle \notin \mathcal{O}_w$. Hence, $\langle W, (W_c \cup W_d) \cup \overline{Z}\rangle \notin \mathcal{O}_w$,—a contradiction. Hence $\langle X, Y\rangle \neq \langle W, W_a \cup W_b\rangle$. Hence, $X \cap (W_c \cup W_d) \neq \emptyset$. Moreover, $\overline{Y \cap Z} \not\supseteq X \cap (W_c \cup W_d)$. Hence there is a $x \in X \cap (W_c \cup W_d)$ such that $x \in Y \cap Z$. Hence, $X \cap Z \cap (W_c \cup W_d) \neq \emptyset$. Since $\langle X, Y\rangle \in \mathcal{O}_w$, also $Y \supseteq X \cap (W_c \cup W_d)$. Hence also $Y \supseteq X \cap Z \cap (W_c \cup W_d)$. Altogether hence, $\langle X \cap Z, Y\rangle \in \mathcal{O}_w$. Thus, (F-WRM) is valid.

We define an F-model $M = \langle F, v, @\rangle$ as follows. The actual world @ is an arbitrary world in W. Moreover, $v : p_i \mapsto \{w \in W \mid p_i \in w\}$. The model validates all premise sets $\Gamma \in \big\{\{\mathsf{O}\top\top, \mathsf{O}\bot\top\}, \{\mathsf{O}\top\top, \mathsf{O}\bot\top, \mathsf{P}p_2\top\}, \{\mathsf{O}\top\top, \mathsf{O}\bot\top, \mathsf{O}p_1\top, \mathsf{P}p_1\top, \mathsf{P}p_2\top\}\big\}$ and falsifies the following instance of (CDEX-3): $M \models \mathsf{O}\top\top, \mathsf{O}\bot\top, \mathsf{O}p_1\top, \mathsf{P}p_1\top, \mathsf{P}p_2\top$ and $M \not\models \mathsf{O}p_2\top$. Thus, it does also not validate (CDEX-1) and (CDEX-2). Furthermore the model is minimally abnormal for Γ with resepct to Ω^c_d and Ω^r_d and any lower limit $\mathbf{L} \in \downarrow$**CDPM.1c**. For Ω^c_d note that the only abnormalities are $\{\mathsf{O}A\top \wedge \mathsf{O}\neg A\top \mid \vdash_{\mathbf{L}} \top \equiv A\}$, but we also have $\Gamma \vdash_{\mathbf{L}} \mathsf{O}A\top \wedge \mathsf{O}\neg A\top$ for all these abnormalities. For Ω^r_d the set of abnormalities verified by M is empty. Therefore, given premises Γ, the model is selected in all ALs based on lower limits in $\downarrow$**CDPM.1c** that are defined in Sects. 11.5 and 11.6. □

Theorem 11.5.2, Theorem 11.5.4 and Theorem 11.6.2 follow immediately by Theorem I.2.1.

Appendix J
Appendix to Chapter 12

Although the semantics that I introduce in this Appendix are very similar to Goble's semantics of **CDPM**.**2c** in [9, 6], they vary from the former in the following aspects:

- An actual world variant of the semantics is used here in order to model factual premises in an intuitive way.
- The semantics have to deal with a language enriched by modal operators O^{i} and O^{p}, symbols $\bullet_{\mathsf{i}}$ and $\bullet_{\mathsf{p}}$, and the additional principles characterizing them.
- The language is weaker than Goble's in the sense that nested occurrences of modal operators are not allowed.

J.1 Language

The language used for the logics defined in [10] is built up by propositional atoms, denoted by $\mathcal{A}$, the classical connectives, $\top$, $\bot$, a dyadic modal operator O, monadic modal operators $\mathsf{O}^{\mathsf{i}}, \mathsf{O}^{\mathsf{p}}$ and symbols $\bullet_{\mathsf{i}}, \bullet_{\mathsf{p}}$. We use for (classical) propositional formulas the letters A, B, C, D, E, F and denote by $\mathcal{P}$ the set of all propositional formulas. Let $\mathcal{L}'$ consist of all formulas of the form $\mathsf{O}(A \mid B)$, $\mathsf{O}^{\mathsf{i}} A$, $\mathsf{O}^{\mathsf{p}} A$, $\bullet_{\mathsf{i}}\mathsf{O}(A \mid B)$, $\bullet_{\mathsf{p}}\mathsf{O}(A \mid B)$ and the set of all propositional letters. Our set of wffs $\mathcal{L}$ is then defined by the $\langle \neg, \vee, \wedge, \supset \rangle$-closure of $\mathcal{L}'$. We use for formulas in $\mathcal{L}$ lower case greek letters. As usually done, we define $\mathsf{P}(A \mid B) =_{\mathrm{df}} \neg\mathsf{O}(\neg A \mid B)$.

J.2 Syntactic Characterization

First, in order to recapitulate the definitions from [10], we state again the syntactic rules used to define logics **CDPM**.**2d**$^{+}$ and **CDPM**.**2e**$^{+}$:

$$\text{If } \vdash A \equiv B, \text{ then } \vdash \mathsf{O}(C \mid A) \equiv \mathsf{O}(C \mid B) \qquad \text{(RCE)}$$

DOI: 10.1007/978-3-319-00792-2, © Springer International Publishing Switzerland 2014

$$\text{If } \vdash A \equiv B, \text{ then } \vdash \mathsf{O}(A|C) \equiv \mathsf{O}(B|C) \quad \text{(CRE)}$$

$$\text{If } \vdash A \equiv B, \text{ then } \vdash \mathsf{O}^{\mathsf{i}} A \equiv \mathsf{O}^{\mathsf{i}} B \quad (\text{EO}^{\mathsf{i}})$$

$$\text{If } \vdash A \equiv B, \text{ then } \vdash \mathsf{O}^{\mathsf{p}} A \equiv \mathsf{O}^{\mathsf{p}} B \quad (\text{EO}^{\mathsf{p}})$$

$$\text{If } \vdash A \equiv B, \text{ then } \vdash \bullet_{\mathsf{i}}\mathsf{O}(A|C) \equiv \bullet_{\mathsf{i}}\mathsf{O}(B|C) \quad \text{(CREi)}$$

$$\text{If } \vdash A \equiv B, \text{ then } \vdash \bullet_{\mathsf{i}}\mathsf{O}(C|A) \equiv \bullet_{\mathsf{i}}\mathsf{O}(C|B) \quad \text{(RCEi)}$$

$$\text{If } \vdash A \equiv B, \text{ then } \vdash \bullet_{\mathsf{p}}\mathsf{O}(A|C) \equiv \bullet_{\mathsf{p}}\mathsf{O}(B|C) \quad \text{(CREp)}$$

$$\text{If } \vdash A \equiv B, \text{ then } \vdash \bullet_{\mathsf{p}}\mathsf{O}(C|A) \equiv \bullet_{\mathsf{p}}\mathsf{O}(C|B) \quad \text{(RCEp)}$$

$$\text{If } \vdash B \supset C, \text{ then } \vdash \mathsf{P}(B|A) \supset \big(\mathsf{O}(B|A) \supset \mathsf{O}(C|A)\big) \quad \text{(RCPM)}$$

$$\text{If } \vdash D \supset \neg A, \text{ then } \vdash \big(\big(\mathsf{P}(D|B \wedge C) \vee \mathsf{O}(D|B \wedge C)\big) \wedge B \wedge C \wedge \mathsf{P}(B \wedge C|B) \wedge \mathsf{O}(A|B)\big) \supset \bullet_{\mathsf{p}}\mathsf{O}(A|B) \quad \text{(Ep)}$$

$$\text{If } \vdash A \supset \neg C \text{ and } \vdash A \supset \neg D, \text{ then } \vdash \big(\mathsf{O}(A|B \wedge C) \wedge \mathsf{O}(D|B)\big) \supset \bullet_{\mathsf{p}}\mathsf{O}(A|B \wedge C) \quad \text{(CTDR)}$$

$$\text{If } \vdash D \supset \neg A, \text{ then } \vdash \big(\big(\mathsf{P}(D|B \wedge C) \vee \mathsf{O}(D|B \wedge C)\big) \wedge B \wedge C \wedge \mathsf{O}(A|B)\big) \supset \bullet_{\mathsf{i}}\mathsf{O}(A|B) \quad \text{(oV-Ei)}$$

Furthermore, the following axioms are needed:

$$\vdash \mathsf{P}(\top|A) \quad \text{(CP)}$$

$$\vdash \mathsf{O}(C|A \wedge B) \supset \mathsf{O}(B \supset C|A) \quad \text{(S)}$$

$$\vdash \big(\mathsf{O}(A|C) \wedge \mathsf{O}(B|C) \wedge \mathsf{P}(A \wedge B|C)\big) \supset \mathsf{O}(A \wedge B|C) \quad \text{(CPAND)}$$

$$\vdash \big(\mathsf{O}(B|A) \wedge \mathsf{P}(B \wedge C|A)\big) \supset \mathsf{O}(B|A \wedge C) \quad \text{(WRM)}$$

$$\vdash \big(\mathsf{O}(A|B \wedge C) \wedge \mathsf{P}(A|\neg B \wedge C)\big) \supset \mathsf{O}(B \supset A|C) \quad \text{(PS')}$$

$$\vdash \big(\mathsf{O}(A|B) \wedge B \wedge \neg \bullet_{\mathsf{p}} \mathsf{O}(A|B)\big) \supset \mathsf{O}^{\mathsf{p}} A \quad \text{(FDp)}$$

$$\vdash \big(\mathsf{O}(A|B) \wedge B \wedge \neg \bullet_{\mathsf{i}} \mathsf{O}(A|B)\big) \supset \mathsf{O}^{\mathsf{i}} A \quad \text{(FDi)}$$

$$\vdash \big(\mathsf{O}(A|B) \wedge \neg A \wedge B\big) \supset \bullet_{\mathsf{i}}\mathsf{O}(A|B) \quad \text{(fV)}$$

Definition J.2.1. Logic **CDPM.2d**$^{+}$ is defined by all the rules and axioms stated above (with exception of (PS')),[7] **CDPM.2e**$^{+}$ is defined as **CDPM.2d**$^{+}$ with exception of (S), which is replaced by (PS'). Let $\mathbf{L}^{+}$ from now on be any of the two logics (if not specified beforehand).

J.3 The Neighborhood Semantics

One of the basic ideas for the neighborhood semantics is that propositions are interpreted in terms of sets of worlds. For each obligation type (such as O^{i}, O^{p}, ...) each world has associated with it propositions, i.e. sets of worlds. The idea is that an oblig-

[7] This rule follows directly from (S).

ation $\mathsf{O}^{\mathsf{i}}A$ is true at a world w, in case A is one of its associated propositions with respect to O^{i}. The generalization in terms of conditional obligations is canonical. In this case worlds are associated with ordered pairs of propositions representing conditional obligations. We are going to make use of an actual world. The reason is that we are going to work with premise sets containing propositional formulas representing given facts, which can be better modeled like that.

Let a dyadic neighborhood frame F be a tuple $\langle W, \mathcal{O}, \mathcal{N}^{\mathsf{i}}, \mathcal{N}^{\mathsf{p}}, \mathcal{O}^{\mathsf{i}}, \mathcal{O}^{\mathsf{p}} \rangle$ where W is a set of worlds and $\mathcal{O} : W \to \wp(\wp(W) \times \wp(W))$, $\mathcal{N}^{\mathsf{i}} : W \to \wp(\wp(W) \times \wp(W))$, $\mathcal{N}^{\mathsf{p}} : W \to \wp(\wp(W) \times \wp(W))$, $\mathcal{O}^{\mathsf{i}} : W \to \wp(\wp(W))$, $\mathcal{O}^{\mathsf{p}} : W \to \wp(\wp(W))$. Thus, $\mathcal{O}$, $\mathcal{N}^{\mathsf{i}}$ and $\mathcal{N}^{\mathsf{p}}$ assign to each world $w \in W$ a set of ordered propositions, i.e., $\mathcal{O}_w, \mathcal{N}^{\mathsf{i}}_w, \mathcal{N}^{\mathsf{p}}_w \subseteq \wp(W) \times \wp(W)$,[8] and $\mathcal{O}^{\mathsf{i}}$ and O^{p} assign to each world a proposition, i.e., $\mathcal{O}^{\mathsf{i}}_w, \mathcal{O}^{\mathsf{p}}_w \subseteq \wp(W)$. A model M on frame F is a triple $\langle F, @, v \rangle$ where $@ \in W$ is the actual world and $v : \mathcal{A} \to \wp(W)$. A propositional atom is mapped into the set of worlds in which it is supposed to hold. We define $M \models \varphi$ iff $M, @ \models \varphi$, $F \models \varphi$ iff for all models M defined on the basis of frame F, $M \models \varphi$, and $\mathcal{F} \models \varphi$ (where $\mathcal{F}$ is a set of frames) iff for all $F \in \mathcal{F}$, $F \models \varphi$. Furthermore, where $w \in W$, we have the following requirements for our models:

$$M, w \models p \text{ iff } w \in v(p), \text{ where } p \in \mathcal{A} \tag{M-p}$$

$$M, w \models \mathsf{O}(A|B) \text{ iff } \langle |B|_M, |A|_M \rangle \in \mathcal{O}_w \tag{M-$\mathcal{O}$}$$

$$M, w \models \bullet_{\mathsf{i}}\mathsf{O}(A|B) \text{ iff } \langle |B|_M, |A|_M \rangle \in \mathcal{N}^{\mathsf{i}}_w \tag{M-$\mathcal{N}^{\mathsf{i}}$}$$

$$M, w \models \bullet_{\mathsf{p}}\mathsf{O}(A|B) \text{ iff } \langle |B|_M, |A|_M \rangle \in \mathcal{N}^{\mathsf{p}}_w \tag{M-$\mathcal{N}^{\mathsf{p}}$}$$

$$M, w \models \mathcal{O}^{\mathsf{i}}A \text{ iff } |A|_M \in \mathcal{O}^{\mathsf{i}}_w \tag{M-$\mathcal{O}^{\mathsf{i}}$}$$

$$M, w \models \mathcal{O}^{\mathsf{p}}A \text{ iff } |A|_M \in \mathcal{O}^{\mathsf{p}}_w \tag{M-$\mathcal{O}^{\mathsf{p}}$}$$

where $|\varphi|_M =_{\text{df}} \{w \in W \mid M, w \models \varphi\}$. For the classical connectives the definitions are as usual:

$$M, w \models \neg\varphi \text{ iff } M, w \not\models \varphi \tag{M-$\neg$}$$

$$M, w \models \varphi \vee \psi \text{ iff } M, w \models \varphi \text{ or } M, w \models \psi \tag{M-$\vee$}$$

$$M, w \models \varphi \wedge \psi \text{ iff } M, w \models \varphi \text{ and } M, w \models \psi \tag{M-$\wedge$}$$

$$M, w \models \varphi \supset \psi \text{ iff } M, w \models \neg\varphi \vee \psi \tag{M-$\supset$}$$

$$M, w \models \top \tag{M-$\top$}$$

$$M, w \not\models \bot \tag{M-$\bot$}$$

We write $\overline{W'} =_{\text{df}} W \setminus W'$ where $W' \subseteq W$ for a given frame $F = \langle W, \mathcal{O}, \mathcal{N}^{\mathsf{i}}, \mathcal{N}^{\mathsf{p}}, \mathcal{O}^{\mathsf{i}}, \mathcal{O}^{\mathsf{p}} \rangle$. In order to define our **CDPM** systems we also need the following conditions on frames. For all $X, Y, Z \subseteq W$ and $w \in W$ we demand:

---

[8] We follow Goble's writing convention and write the argument of the mappings that constitute frames as subscripts.

$$\langle W, W\rangle \in \mathcal{O}_w \tag{F-CN}$$

$$\text{If } Y \subseteq Z;\ \langle X, Y\rangle \in \mathcal{O}_w \text{ and } \langle X, \overline{Y}\rangle \notin \mathcal{O}_w, \text{ then } \langle X, Z\rangle \in \mathcal{O}_w \tag{F-RCPM}$$

$$\text{If } \langle X \cap Y, Z\rangle \in \mathcal{O}_w, \text{ then } \langle X, \overline{Y} \cup Z\rangle \in \mathcal{O}_w \tag{F-S}$$

$$\text{If } \langle X, Y\rangle \in \mathcal{O}_w \text{ and } \langle X, \overline{Y \cap Z}\rangle \notin \mathcal{O}_w, \text{ then } \langle X \cap Z, Y\rangle \in \mathcal{O}_w \tag{F-WRM}$$

$$\begin{array}{r}\text{If } \langle X, Y\rangle \in \mathcal{O}_w;\ \langle X, Z\rangle \in \mathcal{O}_w \text{ and } \langle X, \overline{Y \cap Z}\rangle \notin \mathcal{O}_w,\\ \text{then } \langle X, Y \cap Z\rangle \in \mathcal{O}_w\end{array} \tag{F-CPAND}$$

$$\langle X, \emptyset\rangle \notin \mathcal{O}_w \tag{F-CP}$$

For the **e**-version of our system we need:

$$\text{If } \langle Y \cap Z, X\rangle \in \mathcal{O}_w \text{ and } \langle \overline{Y} \cap Z, \overline{X}\rangle \notin \mathcal{O}_w, \text{ then } \langle Z, \overline{Y} \cup X\rangle \in \mathcal{O}_w \tag{F-PS'}$$

In order to model detachment we are in need of the following conditions on frames:

$$\text{If } \langle Y, X\rangle \in \mathcal{O}_w;\ w \in Y;\ \text{and } \langle Y, X\rangle \notin \mathcal{N}^{\mathrm{i}}_w, \text{ then } X \in \mathcal{O}^{\mathrm{i}}_w \tag{F-FDi}$$

$$\text{If } \langle Y, X\rangle \in \mathcal{O}_w;\ w \in Y;\ \text{and } \langle Y, X\rangle \notin \mathcal{N}^{\mathrm{p}}_w, \text{ then } X \in \mathcal{O}^{\mathrm{p}}_w \tag{F-FDp}$$

$$\begin{array}{r}\text{If } w \in Y \cap Z;\ \langle Y, \overline{Y \cap Z}\rangle \notin \mathcal{O}_w;\ \langle Y, X\rangle \in \mathcal{O}_w;\ Z' \subseteq \overline{X};\ \text{and}\\ \big(\langle Y \cap Z, \overline{Z'}\rangle \notin \mathcal{O}_w \text{ or } \langle Y \cap Z, Z'\rangle \in \mathcal{O}_w\big), \text{ then } \langle Y, X\rangle \in \mathcal{N}^{\mathrm{p}}_w\end{array} \tag{F-Ep}$$

$$\begin{array}{r}\text{If } \langle Y \cap Z, X\rangle, \langle Y, Z'\rangle \in \mathcal{O}_w;\ Z' \subseteq \overline{Z};\\ \text{and } Z' \subseteq \overline{X}, \text{ then } \langle Y \cap Z, X\rangle \in \mathcal{N}^{\mathrm{p}}_w\end{array} \tag{F-CTDR}$$

$$\text{If } \langle Y, X\rangle \in \mathcal{O}_w;\ w \in Y;\ \text{and } w \notin X;\ \text{then } \langle Y, X\rangle \in \mathcal{N}^{\mathrm{i}}_w \tag{F-fV}$$

$$\begin{array}{r}\text{If } w \in Y \cap Z;\ \langle Y, X\rangle \in \mathcal{O}_w;\ \big(\langle Y \cap Z, \overline{Z'}\rangle \notin \mathcal{O}_w \text{ or}\\ \langle Y \cap Z, Z'\rangle \in \mathcal{O}_w\big);\ \text{and } Z' \subseteq \overline{X}, \text{ then } \langle Y, X\rangle \in \mathcal{N}^{\mathrm{i}}\end{array} \tag{F-oV-Ei}$$

J.4 Soundness

Note that for the proofs in the Appendix I sometimes write LHS $\stackrel{X}{=}$ RHS if the equation between LHS and RHS holds due to Lemma X. Obviously most of the following results and their proofs resemble results proven by Goble for his **CDPM** systems.

Lemma J.4.1. *For any model $M = \langle F, @, v\rangle$, where*

$$F = \langle W, \mathcal{O}, \mathcal{N}^{\mathrm{i}}, \mathcal{N}^{\mathrm{p}}, \mathcal{O}^{\mathrm{i}}, \mathcal{O}^{\mathrm{p}}\rangle,$$

(i) $|\varphi \wedge \psi|_M = |\varphi|_M \cap |\psi|_M$; (ii) $|\varphi \vee \psi|_M = |\varphi|_M \cup |\psi|_M$; (iii) $|\neg\varphi|_M = \overline{|\varphi|_M}$; (iv) $|\top|_M = W$; (v) $|\bot|_M = \emptyset$.

Proof. Ad (i): $|\varphi \wedge \psi|_M = \{w \in W \mid M, w \models \varphi \wedge \psi\} = \{w \in W \mid M, w \models \varphi, \psi\} = \{w \in W \mid M, w \models \varphi\} \cap \{w \in W \mid M, w \models \psi\} = |\varphi|_M \cap |\psi|_M$. Ad (ii):

analogous. Ad (iii): $|\neg\varphi|_M = \{w \in W \mid M, w \models \neg\varphi\} = \{w \in W \mid M, w \not\models \varphi\} = W \setminus \{w \in W \mid M, w \models \varphi\} = \overline{|\varphi|_M}$. Ad (iv): $|\top|_M = \{w \in W \mid M, w \models \top\} = W$. Ad (v): analogous. □

Lemma J.4.2. *For any model $M = \langle F, @, v\rangle$, where*

$$F = \langle W, \mathcal{O}, \mathcal{N}^{\mathrm{i}}, \mathcal{N}^{\mathrm{p}}, \mathcal{O}^{\mathrm{i}}, \mathcal{O}^{\mathrm{p}}\rangle,$$

(i), if $F \models \varphi \supset \psi$, then $|\varphi|_M \subseteq |\psi|_M$, and, (ii), if $F \models \varphi \equiv \psi$, then $|\varphi|_M = |\psi|_M$.

Proof. Ad (i): Suppose there is a $w \in W$ for which $M, w \models \varphi \wedge \neg\psi$. $M' = \langle F, w, v\rangle$ obviously satisfies the model conditions (M-p), (M-$\mathcal{O}$), (M-$\mathcal{N}^{\mathrm{i}}$), (M-$\mathcal{N}^{\mathrm{p}}$), (M-$\mathcal{O}^{\mathrm{i}}$), (M-$\mathcal{O}^{\mathrm{p}}$), (M-$\neg$), (M-$\vee$), (M-$\wedge$) and (M-$\supset$) since M satisfies them. But then $F \not\models \varphi \supset \psi$—a contradiction. Ad (ii): This is an immediate consequence of (i). □

Theorem J.4.1. $\mathbf{L}^+$ *is sound with respect to the class of frames $\mathcal{F}$ that meet the appropriate frame conditions. In case of* **CDPM.2d**$^+$ *the frame conditions are* (F-CN), (F-RCPM), (F-S), (F-WRM), (F-CPAND), (F-CP), (F-FDi), (F-FDp), (F-Ep), (F-CTDR), (F-fV) *and* (F-oV-Ei). *In case of* **CDPM.2e**$^+$ *we replace* (F-S) *by* (F-PS').

Proof. The proof is very simple: we thus show only for a few rules paradigmatically that they are valid in all models of the respective frames. Let $\mathcal{F}$ be our respective class of frames and let $M = \langle F, @, v\rangle$ be an arbitrary model on an arbitrary frame $F \in \mathcal{F}$.

We begin with (RCPM): Let $F \models B \supset C$. Assume that $M, @ \models \mathsf{P}(B|A) \wedge \mathsf{O}(B|A)$. Then $M, @ \models \mathsf{P}(B|A), \mathsf{O}(B|A)$ and thus, $M, @ \models \neg\mathsf{O}(\neg B|A), \mathsf{O}(B|A)$. Hence, $\langle |A|_M, |B|_M\rangle \in \mathcal{O}_@$ and $\langle |A|_M, |\neg B|_M\rangle \notin \mathcal{O}_@$. Thus by Lemma J.4.1 (iii), $\langle |A|_M, \overline{|B|_M}\rangle \notin \mathcal{O}_@$. Furthermore, by Lemma J.4.2 (i), $|B|_M \subseteq |C|_M$. Since $\mathcal{F}$ validates (F-RCPM), $\langle |A|_M, |C|_M\rangle \in \mathcal{O}_@$. Hence, $M, @ \models \mathsf{O}(C|A)$ and thus, $M, @ \models (\mathsf{P}(B|A) \wedge \mathsf{O}(B|A)) \supset \mathsf{O}(C|A)$. Hence, $M \models (\mathsf{P}(B|A) \wedge \mathsf{O}(B|A)) \supset \mathsf{O}(C|A)$. Since M and F were arbitrary, $\mathcal{F} \models \mathsf{P}(B|A) \supset \big(\mathsf{O}(B|A) \supset \mathsf{O}(C|A)\big)$.

For (WRM): Assume that $M, @ \models \mathsf{O}(B|A) \wedge \mathsf{P}(B \wedge C|A)$. Then $M, @ \models \mathsf{O}(B|A), \neg\mathsf{O}(\neg(B \wedge C)|A)$. Thus, $\langle |A|_M, |\neg(B \wedge C)|_M\rangle \overset{J.4.1iii}{=} \big\langle |A|_M, \overline{|B \wedge C|_M}\big\rangle \overset{J.4.1i}{=} \big\langle |A|_M, \overline{|B|_M \cap |C|_M}\big\rangle \notin \mathcal{O}_@$. Thus, since F validates (F-WRM), $\langle |A|_M \cap |C|_M, |B|_M\rangle \overset{J.4.1i}{=} \langle |A \wedge C|_M, |B|_M\rangle \in \mathcal{O}_@$. Hence, $M, @ \models \mathsf{O}(B|A \wedge C)$ and thus, $M, @ \models \big(\mathsf{O}(B|A) \wedge \mathsf{P}(B \wedge C|A)\big) \supset \mathsf{O}(B|A \wedge C)$. Hence, $M \models \big(\mathsf{O}(B|A) \wedge \mathsf{P}(B \wedge C|A)\big) \supset \mathsf{O}(B|A \wedge C)$. Since M and F were arbitrary, $\mathcal{F} \models \big(\mathsf{O}(B|A) \wedge \mathsf{P}(B \wedge C|A)\big) \supset \mathsf{O}(B|A \wedge C)$.

For (PS'): Assume that $M, @ \models \mathsf{O}(A|B \wedge C) \wedge \mathsf{P}(A|\neg B \wedge C)$. Thus, $M, @ \models \mathsf{O}(A|B \wedge C), \mathsf{P}(A|\neg B \wedge C)$. Thus, $M, @ \models \mathsf{O}(A|B \wedge C), \neg\mathsf{O}\neg A\neg B \wedge C$. Hence, $\langle |B \wedge C|_M, |A|_M\rangle \in \mathcal{O}_@$ and $\langle |\neg B \wedge C|_M, |\neg A|_M\rangle \notin \mathcal{O}_@$. Hence, by Lemma J.4.1, $\langle |B|_M \cap |C|_M, |A|_M\rangle \in \mathcal{O}_@$ and $\big\langle \overline{|B|_M} \cap |C|_M, \overline{|A|_M}\big\rangle \notin \mathcal{O}_@$. Since $\mathcal{F}$ satisfies (F-PS'), $\big\langle |C|_M, \overline{|B|_M} \cup |A|_M\big\rangle \in \mathcal{O}_@$. By Lemma J.4.1, $\langle |C|_M, |\neg B \vee A|_M\rangle \in \mathcal{O}_@$. By Lemma J.4.2 (ii), $\langle |C|_M, |B \supset A|_M\rangle \in \mathcal{O}_@$ and thus, $M, @ \models$

$\mathsf{O}(B \supset A|C)$. Hence, $M, @ \models (\mathsf{O}(A|B \wedge C) \wedge \mathsf{P}(A|\neg B \wedge C)) \supset \mathsf{O}(B \supset A|C)$. Thus, $M(\mathsf{O}(A|B \wedge C) \wedge \mathsf{P}(A|\neg B \wedge C)) \supset \mathsf{O}(B \supset A|C)$. Since F and M were arbitrary, $\mathcal{F} \models (\mathsf{O}(A|B \wedge C) \wedge \mathsf{P}(A|\neg B \wedge C)) \supset \mathsf{O}(B \supset A|C)$.

For (EO^{i}): Let $F \models A \equiv B$. Assume that $M, @ \models \mathsf{O}^{\mathsf{i}}A$. Thus, $|A|_M \in \mathcal{O}^{\mathsf{i}}_{@}$. By Lemma J.4.2 (ii), $|A|_M = |B|_M$. Thus, $|B|_M \in \mathcal{O}^{\mathsf{i}}_{@}$. Hence, $M, @ \models \mathsf{O}^{\mathsf{i}}B$. Thus, $M, @ \models \mathsf{O}^{\mathsf{i}}A \supset \mathsf{O}^{\mathsf{i}}B$. Hence, $M \models \mathsf{O}^{\mathsf{i}}A \supset \mathsf{O}^{\mathsf{i}}B$. Since M and F were arbitrary, $\mathcal{F} \models \mathsf{O}^{\mathsf{i}}A \supset \mathsf{O}^{\mathsf{i}}B$. The other direction is analogous.

For (FDi): Assume that $M, @ \models \mathsf{O}(A|B) \wedge B \wedge \neg \bullet_{\mathsf{i}} \mathsf{O}(A|B)$ and thus, $M, @ \models \mathsf{O}(A|B)$, B and $M, @ \not\models \bullet_{\mathsf{i}}\mathsf{O}(A|B)$. Hence, $\langle |B|_M, |A|_M \rangle \in \mathcal{O}_{@}$, $@ \in |B|_M$ and $\langle |B|_M, |A|_m \rangle \notin \mathcal{N}^{\mathsf{i}}_{@}$. Since $\mathcal{F}$ satisfies (F-FDi), $|A|_M \in \mathcal{O}^{\mathsf{i}}_{@}$ and thus, $M, @ \models \mathsf{O}^{\mathsf{i}}A$. Hence, $M, @ \models (\mathsf{O}(A|B) \wedge B \wedge \neg \bullet_{\mathsf{i}} \mathsf{O}(A|B)) \supset \mathsf{O}^{\mathsf{i}}A$. Hence, $M \models (\mathsf{O}(A|B) \wedge B \wedge \neg \bullet_{\mathsf{i}} \mathsf{O}(A|B)) \supset \mathsf{O}^{\mathsf{i}}A$. Since M and F were arbitrary, we have $\mathcal{F} \models (\mathsf{O}(A|B) \wedge B \wedge \neg \bullet_{\mathsf{i}} \mathsf{O}(A|B)) \supset \mathsf{O}^{\mathsf{i}}A$.

The other cases are shown in a similar way. □

J.5 Completeness

Completeness for our logics can be proven in a similar way as Goble proved completeness of his **(C)DPM** systems. We proceed in two steps:

1. We prove model-completeness by means of a canonical model $\dot{M}$ and adjusted model conditions. We show that for each non-theorem φ of $\mathbf{L}^+$ there is such a $\dot{M}$ falsifying φ.
2. Using filtration techniques on the canonical model $\dot{M}$ we arrive at an alternative model $\mathring{M}$ on a frame $\mathring{F}$ that satisfies the respective frame conditions. For each non-theorem φ of $\mathbf{L}^+$ we have an $\mathring{M}$ which falsifies φ. This suffices to prove completeness and decidability.

J.5.1 Model Completeness

First we define a frame for a canonical model for $\mathbf{L}^+$. Let $\dot{F} = \langle \dot{W}, \dot{\mathcal{O}}, \dot{\mathcal{N}^{\mathrm{i}}}, \dot{\mathcal{N}^{\mathrm{p}}}, \dot{\mathcal{O}^{\mathrm{i}}}, \dot{\mathcal{O}^{\mathrm{p}}} \rangle$. $\dot{W}$ contains all maximal $\mathbf{L}^+$-consistent sets of formulas in $\mathcal{L}$. We have the following assignments for all $w \in \dot{W}$:[9]

$$\dot{\mathcal{O}}_w = \{\langle X, Y \rangle \mid X \subseteq \dot{W}, Y \subseteq \dot{W},$$
$$\exists A \exists B(X = [A], Y = [B] \text{ and } \mathsf{O}(B|A) \in w\},$$
$$\dot{\mathcal{N}^{\mathrm{i}}}_w = \{\langle X, Y \rangle \mid X \subseteq \dot{W}, Y \subseteq \dot{W},$$

[9] For sake of readability we use from now on "∃" and "∀" in set descriptions in the canonical reading "there is" and "for all".

$$\exists A \exists B(X = [A], Y = [B] \text{ and } \bullet_{\mathsf{i}}\, \mathsf{O}(B|A) \in w\},$$
$$\dot{\mathcal{N}}^{\mathsf{p}}{}_w = \{\langle X, Y\rangle \mid X \subseteq \dot{W}, Y \subseteq \dot{W},$$
$$\exists A \exists B(X = [A], Y = [B] \text{ and } \bullet_{\mathsf{p}}\, \mathsf{O}(B|A) \in w\},$$
$$\dot{\mathcal{O}}^{\mathsf{i}}{}_w = \{X \subseteq \dot{W} \mid \exists A(X = [A] \text{ and } \mathsf{O}^{\mathsf{i}} A \in w\},$$
$$\dot{\mathcal{O}}^{\mathsf{p}}{}_w = \{X \subseteq \dot{W} \mid \exists A(X = [A] \text{ and } \mathsf{O}^{\mathsf{p}} A \in w\},$$

where $[\varphi] =_{\text{df}} \{w \in \dot{W} \mid \varphi \in w\}$. Now we can define a canonical model $\dot{M} = \langle \dot{F}, @, \dot{v}\rangle$. Let for every atomic formula p

$$\dot{v} : p \mapsto \{w \in \dot{W} \mid p \in w\}$$

Lemma J.5.1. *For any φ and ψ, (i) $[\varphi \wedge \psi] = [\varphi] \cap [\psi]$; (ii) $[\varphi \vee \psi] = [\varphi] \cup [\psi]$; (iii) $[\neg\varphi] = \overline{[\varphi]}$; (iv) $[\top] = \dot{W}$; (v) $[\bot] = \emptyset$.*

Proof. Ad (i): $[\varphi \wedge \psi] = \{w \in \dot{W} \mid \varphi \wedge \psi \in w\} \overset{(1)}{=} \{w \in \dot{W} \mid \varphi, \psi \in w\} = \{w \in \dot{W} \mid \varphi \in w\} \cap \{w \in \dot{W} \mid \psi \in w\} = [\varphi] \cap [\psi]$ where (1) is due to the fact that w is a maximal consistent extension. The other cases are shown in a similar way. □

Lemma J.5.2. *For any φ and ψ, (i) $[\varphi] \subseteq [\psi]$ iff $\vdash \varphi \supset \psi$, and (ii) $[\varphi] = [\psi]$ iff $\vdash \varphi \equiv \psi$.*

Proof. This was proven in an analogous way in Goble [6, 9]. For (i), suppose $[\varphi] \subseteq [\psi]$ but $\nvdash \varphi \supset \psi$. Then $\{\varphi, \neg\psi\}$ is consistent and so has a maximal consistent extension, w. $\varphi \in w$ so $w \in [\varphi]$. Hence $w \in [\psi]$, which is to say $\psi \in w$, contrary to the consistency of w since $\neg\psi \in w$. Therefore, $\vdash \varphi \supset \psi$. Further, if $\vdash \varphi \supset \psi$, then since maximal consistent extensions are closed under provable implications, it is automatic that for any $w' \in [\varphi]$, $w' \in [\psi]$, or $[\varphi] \subseteq [\psi]$. Part (ii) follows immediately from (i). □

Lemma J.5.3. *For all $\varphi \in \mathcal{L}$ and all $w \in \dot{W}$, $\dot{M}, w \models \varphi$ iff $\varphi \in w$ (or, $|\varphi|_{\dot{M}} = [\varphi]$).*

Proof. This is shown by induction over the length of φ. The case that φ is a propositional letter is trivial, since $\dot{M}, w \models \varphi$ iff $w \in \dot{v}(\varphi) = \{w' \in \dot{W} \mid \varphi \in w'\}$. Let φ now be a propositional formula. Suppose for the subformulas φ_1 and φ_2 of φ the equivalence holds. Now let $\varphi = \varphi_1 \wedge \varphi_2$. We have $\dot{M}, w \models \varphi$ iff $\dot{M}, w \models \varphi_1$ and $\dot{M}, w \models \varphi_2$ iff $\varphi_1, \varphi_2 \in w$ iff $\varphi_1 \wedge \varphi_2 \in w$ due to the fact that w is a maximal consistent extension. The argument is similar for $\varphi = \varphi_1 \vee \varphi_2$, $\varphi = \varphi_1 \supset \varphi_2$ and $\varphi = \neg\varphi_1$. Thus the equivalence holds for all propositional formulas φ ($\star$).

Now consider the other cases in $\mathcal{L}'$. Let $\varphi = \mathsf{O}(A|B)$. "$\Rightarrow$": In case $\dot{M}, w \models \mathsf{O}(A|B)$ we have $\langle |B|_{\dot{M}}, |A|_{\dot{M}}\rangle \in \dot{\mathcal{O}}_w$ and thus $\langle [B], [A]\rangle \in \dot{\mathcal{O}}_w$ by ($\star$). Hence there are A', B' such that $[A'] = [A]$ and $[B'] = [B]$ and $\mathsf{O}(A'|B') \in w$. By Lemma J.5.2 (ii), $\vdash A' \equiv A$ and $\vdash B' \equiv B$. Since w validates (RCE) and (CRE), $\mathsf{O}(A|B) \in w$. "$\Leftarrow$": Let $\mathsf{O}(A|B) \in w$, then $\langle [A], [B]\rangle \in \dot{\mathcal{O}}_w$. Thus, by ($\star$), $\langle |A|_{\dot{M}}, |B|_{\dot{M}}\rangle \in \dot{\mathcal{O}}_w$. Hence, $\dot{M}, w \models \mathsf{O}(B|A)$.

Let $\varphi = \bullet_{\mathsf{i}}\mathsf{O}(A|B)$. "$\Rightarrow$": In case $\dot{M}, w \models \bullet_{\mathsf{i}}\mathsf{O}(A|B)$ we have $\langle |B|_{\dot{M}}, |A|_{\dot{M}}\rangle \in \dot{\mathcal{N}}^{\mathrm{i}}{}_w$ and thus $\langle [B], [A]\rangle \in \dot{\mathcal{N}}^{\mathrm{i}}{}_w$ by $(\star)$. Hence there are A', B' such that $[A'] = [A]$ and $[B'] = [B]$ and $\bullet_{\mathsf{i}}\mathsf{O}(A'|B') \in w$. By Lemma J.5.2 (ii), $\vdash A' \equiv A$ and $\vdash B' \equiv B$. Since w validates (CREi) and (RCEi), $\bullet_{\mathsf{i}}\mathsf{O}(A|B) \in w$. "$\Leftarrow$": Let $\bullet_{\mathsf{i}}\mathsf{O}(A|B) \in w$, then $\langle [A], [B]\rangle \in \dot{\mathcal{N}}^{\mathrm{i}}{}_w$. Thus, by $(\star)$, $\langle |A|_{\dot{M}}, |B|_{\dot{M}}\rangle \in \dot{\mathcal{N}}^{\mathrm{i}}{}_w$. Hence, $\dot{M}, w \models \bullet_{\mathsf{i}}\mathsf{O}(B|A)$.

The case $\varphi = \bullet_{\mathsf{p}}\mathsf{O}(A|B)$ is analogous.

Let $\varphi = \mathsf{O}^{\mathsf{i}}A$. "$\Rightarrow$": In case $\dot{M}, w \models \mathsf{O}^{\mathsf{i}}A$ we have $|A|_{\dot{M}} \in \dot{\mathcal{O}}^{\mathrm{i}}{}_w$ and thus $[A] \in \dot{\mathcal{O}}^{\mathrm{i}}{}_w$. Hence there is a A' such that $[A'] = [A]$ and $\mathsf{O}^{\mathsf{i}}A' \in w$. By Lemma J.5.2, $\vdash A' \equiv A$. Since w validates (EO^{i}), $\mathsf{O}^{\mathsf{i}}A \in w$. "$\Leftarrow$": Let $\mathsf{O}^{\mathsf{i}}A \in w$, then $[A] \in \dot{\mathcal{O}}^{\mathrm{i}}{}_w$. Thus, by $(\star)$, $|A|_{\dot{M}} \in \dot{\mathcal{O}}^{\mathrm{i}}{}_w$. Hence, $\dot{M}, w \models \mathsf{O}^{\mathsf{i}}A$.

The case $\varphi = \mathsf{O}^{\mathsf{p}}A$ is analogous.

Now let $\varphi = \varphi_1 \wedge \varphi_2 \in \mathcal{L} \setminus (\mathcal{L}' \cup \mathcal{P})$. By induction hypothesis we suppose the equivalence to be valid for φ_1 and φ_2. We have $\dot{M}, w \models \varphi$ iff $\dot{M}, w \models \varphi_1$ and $\dot{M}, w \models \varphi_2$ iff $\varphi_1, \varphi_2 \in w$ iff $\varphi_1 \wedge \varphi_2 \in w$ due to the fact that w is a maximal consistent extension. The argument is similar for $\varphi = \varphi_1 \vee \varphi_2$, $\varphi = \varphi_1 \supset \varphi_2$ and $\varphi = \neg\varphi_1$. Thus the equivalence holds for all $\varphi \in \mathcal{L}$. □

In order to prove model-completeness we need to restrict our sets of worlds to sets corresponding to expressible propositions on $\dot{M}$. We define,[10] where $M = \langle F, @, v\rangle$ and $F = \langle W, \mathcal{O}, \mathcal{N}^{\mathrm{i}}, \mathcal{N}^{\mathrm{p}}, \mathcal{O}^{\mathrm{i}}, \mathcal{O}^{\mathrm{p}}\rangle$

$$\varepsilon_M =_{\mathrm{df}} \{X \subseteq W \mid \exists B(X = |B|_M)\}$$

Lemma J.5.4. *For all $w \in \dot{W}$ and $X, Y \in \varepsilon_{\dot{M}}$ there are A and B for which $[A] = X$ and $[B] = Y$ and we have for all such A and B:*

(i) $\langle X, Y\rangle \in \dot{\mathcal{O}}_w$ *iff* $\mathsf{O}(B|A) \in w$

(ii) $\langle X, Y\rangle \in \dot{\mathcal{N}}^{\mathrm{i}}{}_w$ *iff* $\bullet_{\mathsf{i}}\mathsf{O}(B|A) \in w$.

(iii) $\langle X, Y\rangle \in \dot{\mathcal{N}}^{\mathrm{p}}{}_w$ *iff* $\bullet_{\mathsf{p}}\mathsf{O}(B|A) \in w$.

(iv) $X \in \dot{\mathcal{O}}^{\mathrm{i}}{}_w$ *iff* $\mathsf{O}^{\mathsf{i}}A \in w$.

(v) $X \in \dot{\mathcal{O}}^{\mathrm{p}}{}_w$ *iff* $\mathsf{O}^{\mathsf{p}}A \in w$.

Proof. Let $w \in \dot{W}$ and $X, Y \in \varepsilon_{\dot{M}}$ By definition of $\varepsilon_{\dot{M}}$ there are A and B for which $X = |A|_{\dot{M}}$ and $Y = |B|_{\dot{M}}$. By Lemma J.5.3 we have $[A] = |A|_{\dot{M}} = X$ and $[B] = |B|_{\dot{M}} = Y$.

Ad (i) "$\Rightarrow$": Let $\langle X, Y\rangle \in \dot{\mathcal{O}}_w$. Thus, $\langle [A], [B]\rangle \in \dot{\mathcal{O}}_w$ and thus, $\langle |A|_{\dot{M}}, |B|_{\dot{M}}\rangle \in \dot{\mathcal{O}}_w$. Hence, by Lemma J.5.3, $\mathsf{O}(B|A) \in w$. "$\Leftarrow$": Let $\langle X, Y\rangle \notin \dot{\mathcal{O}}_w$. Suppose $\mathsf{O}(B|A) \in w$, then by Lemma J.5.3, $\dot{M}, w \models \mathsf{O}(B|A)$, then by (M-$\mathcal{O}$), $\langle |A|_{\dot{M}}, |B|_{\dot{M}}\rangle \in \dot{\mathcal{O}}_w$. Hence, $\langle [A], [B]\rangle \in \dot{\mathcal{O}}_w$ and thus, $\langle X, Y\rangle \in \dot{\mathcal{O}}_w$—a contradiction.

Ad (ii): "$\Rightarrow$": Let $\langle X, Y\rangle \in \dot{\mathcal{N}}^{\mathrm{i}}{}_w$. Then $\langle [A], [B]\rangle \in \dot{\mathcal{N}}^{\mathrm{i}}{}_w$ and thus, $\langle |A|_{\dot{M}}, |B|_{\dot{M}}\rangle \in \dot{\mathcal{N}}^{\mathrm{i}}{}_w$. Thus, by Lemma J.5.3, $\bullet_{\mathsf{i}}\mathsf{O}(B|A) \in w$. "$\Leftarrow$": Now let $\langle X, Y\rangle \notin \dot{\mathcal{N}}^{\mathrm{i}}{}_w$. Suppose

[10] Note that this definition differs from Goble's proposal to the extent that in our case B is a propositional formula, while in Goble's case it was any wff. The reason is, that we don't allow for nested modal operators in Chap. 12.

$\bullet_{\mathsf{i}}\mathsf{O}(B|A) \in w$, then by Lemma J.5.3, $\dot{M}, w \models \bullet_{\mathsf{i}}\mathsf{O}(B|A)$ and thus by (M-$\mathcal{N}^{\mathrm{i}}$), $\langle |A|_{\dot{M}}, |B|_{\dot{M}}\rangle \in \dot{\mathcal{N}}^{\mathrm{i}}{}_w$. Hence, $\langle [A], [B]\rangle \in \dot{\mathcal{N}}^{\mathrm{i}}{}_w$ and thus, $\langle X, Y\rangle \in \dot{\mathcal{N}}^{\mathrm{i}}{}_w$—a contradiction.

Ad (iii): analogous.

Ad (iv): "$\Rightarrow$": Let $X \in \dot{\mathcal{O}}^{\mathrm{i}}{}_w$ and thus $[A] \in \dot{\mathcal{O}}^{\mathrm{i}}{}_w$. Hence, $|A|_{\dot{M}} \in \dot{\mathcal{O}}^{\mathrm{i}}{}_w$. Thus, $\dot{M}, w \models \mathsf{O}^{\mathsf{i}}A$ and thus by Lemma J.5.3, $\mathsf{O}^{\mathsf{i}}A \in w$. "$\Leftarrow$": Now let $X \notin \dot{\mathcal{O}}^{\mathrm{i}}{}_w$. Suppose $\mathsf{O}^{\mathsf{i}}A \in w$. Then by Lemma J.5.3, $\dot{M}, w \models \mathsf{O}^{\mathsf{i}}A$ and thus by (M-$\mathcal{O}^{\mathrm{i}}$), $|A|_{\dot{M}} \in \dot{\mathcal{O}}^{\mathrm{i}}{}_w$. Hence, $[A] \in \dot{\mathcal{O}}^{\mathrm{i}}{}_w$ and thus, $X \in \dot{\mathcal{O}}^{\mathrm{i}}{}_w$—a contradiction.

Ad (v): analogous. □

We now modify the frame conditions to form conditions on models. Where $M = \langle F, @, v\rangle$ and $F = \langle W, \mathcal{O}, \mathcal{N}^{\mathrm{i}}, \mathcal{N}^{\mathrm{p}}, \mathcal{O}^{\mathrm{i}}, \mathcal{O}^{\mathrm{p}}\rangle$ we require:

$$\langle W, W\rangle \in \mathcal{O}_w \tag{M-CN}$$

$$\text{For all } X, Y, Z \in \varepsilon_M,\ \text{if } Y \subseteq Z \text{ and } \langle X, Y\rangle \in \mathcal{O}_w \text{ and } \langle X, \overline{Y}\rangle \notin \mathcal{O}_w \text{ then } \langle X, Z\rangle \in \mathcal{O}_w \tag{M-RCPM}$$

$$\text{For all } X, Y, Z \in \varepsilon_M,\ \text{if } \langle X \cap Y, Z\rangle \in \mathcal{O}_w,\ \text{then } \langle X, \overline{Y} \cup Z\rangle \in \mathcal{O}_w \tag{M-S}$$

$$\text{For all } X, Y, Z \in \varepsilon_M,\ \text{if } \langle X, Y\rangle \in \mathcal{O}_w \text{ and } \langle X, \overline{Y \cap Z}\rangle \notin \mathcal{O}_w,\ \text{then } \langle X \cap Z, Y\rangle \in \mathcal{O}_w \tag{M-WRM}$$

$$\text{For all } X, Y, Z \in \varepsilon_M,\ \text{if } \langle X, Y\rangle \in \mathcal{O}_w, \langle X, Z\rangle \in \mathcal{O}_w \text{ and } \langle X, \overline{Y \cap Z}\rangle \notin \mathcal{O}_w,\ \text{then } \langle X, Y \cap Z\rangle \in \mathcal{O}_w \tag{M-CPAND}$$

$$\text{For all } X \in \varepsilon_M, \langle X, \emptyset\rangle \notin \mathcal{O}_w \tag{M-CP}$$

$$\text{For all } X, Y, Z \in \varepsilon_M,\ \text{if } \langle Y \cap Z, X\rangle \in \mathcal{O}_w \text{ and } \langle \overline{Y} \cap Z, \overline{X}\rangle \notin \mathcal{O}_w,\ \text{then } \langle Z, \overline{Y} \cup X\rangle \in \mathcal{O}_w \tag{M-PS'}$$

$$\text{For all } X, Y \in \varepsilon_M,\ \text{if } \langle Y, X\rangle \in \mathcal{O}_w;\ w \in Y;\ \text{and } \langle Y, X\rangle \notin \mathcal{N}^{\mathrm{i}}_w,\ \text{then } X \in \mathcal{O}^{\mathrm{i}}_w \tag{M-FDi}$$

$$\text{For all } X, Y \in \varepsilon_M,\ \text{if } \langle Y, X\rangle \in \mathcal{O}_w;\ w \in Y;\ \text{and } \langle Y, X\rangle \notin \mathcal{N}^{\mathrm{p}}_w,\ \text{then } X \in \mathcal{O}^{\mathrm{p}}_w \tag{M-FDp}$$

$$\text{For all } X, Y, Z, Z' \in \varepsilon_M,\ \text{if } w \in Y \cap Z;\ \langle Y, \overline{Y \cap Z}\rangle \notin \mathcal{O}_w;\ \langle Y, X\rangle \in \mathcal{O}_w;\ Z' \subseteq \overline{X};\ \text{and } \big(\langle Y \cap Z, \overline{Z'}\rangle \notin \mathcal{O}_w \text{ or } \langle Y \cap Z, Z'\rangle \in \mathcal{O}_w\big),\ \text{then } \langle Y, X\rangle \in \mathcal{N}^{\mathrm{p}}_w \tag{M-Ep}$$

$$\text{For all } X, Y, Z, Z' \in \varepsilon_M,\ \text{if } \langle Y \cap Z, X\rangle, \langle Y, Z'\rangle \in \mathcal{O}_w;\ Z' \subseteq \overline{Z};\ \text{and } Z' \subseteq \overline{X},\ \text{then } \langle Y \cap Z, X\rangle \in \mathcal{N}^{\mathrm{p}}_w \tag{M-CTDR}$$

$$\text{For all } X, Y \in \varepsilon_M,\ \text{if } \langle Y, X\rangle \in \mathcal{O}_w;\ w \in Y;\ \text{and } w \notin X;\ \text{then } \langle Y, X\rangle \in \mathcal{N}^{\mathrm{i}}_w \tag{M-fV}$$

$$\text{For all } X, Y, Z, Z' \in \varepsilon_M,\ \text{if } w \in Y \cap Z;\ \langle Y, X\rangle \in \mathcal{O}_w;\ \big(\langle Y \cap Z, \overline{Z'}\rangle \notin \mathcal{O}_w \text{ or } \langle Y \cap Z, Z'\rangle \in \mathcal{O}_w\big);\ \text{and } Z' \subseteq \overline{X},\ \text{then } \langle Y, X\rangle \in \mathcal{N}^{\mathrm{i}} \tag{M-oV-Ei}$$

Theorem J.5.1. $\mathbf{L}^+$ *is sound and complete with respect to the class of models that meet conditions, as appropriate. In case of* **CDPM.2d**$^+$ *the appropriate conditions are* (M-CN), (M-RCPM), (M-S), (M-WRM), (M-CPAND), (M-CP), (M-FDi), (M-FDp), (M-Ep), (M-CTDR), (M-fV), *and* (M-oV-Ei). *In case of* **CDPM.2e**$^+$ (M-S) *is replaced by* (M-PS').

Proof. Soundness is trivial and is shown in a similar way as it was done in Theorem J.4.1. Some examples: Let $M = \langle F, @, v\rangle$ be a model that satisfies the required properties. For (WRM), Let $M \models \mathsf{O}(B|A) \wedge \mathsf{P}(B \wedge C|A)$, then $M, @ \models \mathsf{O}(B|A), \mathsf{P}(B \wedge C|A)$. Thus, $\langle |A|_M, |B|_M\rangle \in \mathcal{O}_@$ and $\langle |A|_M, \overline{|B \wedge C|_M}\rangle \overset{J.4.1i}{=} \langle |A|_M, \overline{|B|_M \cap |C|_M}\rangle \notin \mathcal{O}_@$. Since M fulfills (M-WRM), $\langle |B|_M \cap |C|_M, |A|_M\rangle \overset{J.4.1i}{=} \langle |B \wedge C|_M, |A|_M\rangle \in \mathcal{O}_@$ and hence $M, @ \models \mathsf{O}(A|B \wedge C)$. Hence $M \models \mathsf{O}(A|B \wedge C)$. For (RCPM) let $M \models \mathsf{P}(B|A), \mathsf{O}(B|A)$ and $\vDash B \supset C$. Then $M, @ \models \mathsf{P}(B|A), \mathsf{O}(B|A)$. Thus, $\langle |A|_M, |B|_M\rangle \in \mathcal{O}_@$ and $\langle |A|_M, \overline{|B|_M}\rangle \notin \mathcal{O}_@$. Hence, since M satisfies (M-RCPM) and since $|B|_M \subseteq |C|_M$, $\langle |A|_M, |C|_M\rangle \in \mathcal{O}_@$. Hence, $M, @ \models \mathsf{O}(A|C)$ and thus, $M \models \mathsf{O}(A|C)$. The other cases are shown analogously.

In order to show completeness let φ be a formula not provable in $\mathbf{L}^+$. Then $\{\neg\varphi\}$ is $\mathbf{L}^+$-consistent and there is, hence, a maximal consistent extension of all $\mathbf{L}^+$ theorems, $\dot{@} \in \dot{W}$, which verifies $\neg\varphi$. Let $\dot{M} = \langle \dot{W}, \dot{\mathcal{O}}, \dot{\mathcal{N}}^{\mathrm{i}}, \dot{\mathcal{N}}^{\mathrm{p}}, \dot{\mathcal{O}}^{\mathrm{i}}, \dot{\mathcal{O}}^{\mathrm{p}}, \dot{@}, \dot{v}\rangle$ be defined as above. We show now that $\dot{M}$ meets the respective model conditions via some paradigmatical examples.

For (M-WRM): Let $X, Y, Z \in \varepsilon_{\dot{M}}$, $\langle X, Y\rangle \in \dot{\mathcal{O}}_w$, and $\langle X, \overline{Y \cap Z}\rangle \notin \dot{\mathcal{O}}_w$. There are A, B such that $[A] = X$, $[B] = Y$ and $\mathsf{O}(B|A) \in w$. By Lemma J.5.3, $M, w \models \mathsf{O}(B|A)$ and thus, $\langle |A|_{\dot{M}}, |B|_{\dot{M}}\rangle \in \dot{\mathcal{O}}_w$. Furthermore there is a C such that $|C|_{\dot{M}} = Z$. Suppose, $\mathsf{O}(\neg(B \wedge C)|A) \in w$. Then, $\langle [A], [\neg(B \wedge C)]\rangle \in \dot{\mathcal{O}}_w$ and thus by Lemma J.5.3, $\langle |A|_{\dot{M}}, |\neg(B \wedge C)|_{\dot{M}}\rangle \in \dot{\mathcal{O}}_w$. But then, since $\langle |A|_{\dot{M}}, |\neg(B \wedge C)|_{\dot{M}}\rangle \overset{J.4.1iii}{=} \langle |A|_{\dot{M}}, \overline{|B \wedge C|_{\dot{M}}}\rangle \overset{J.4.1i}{=} \langle |A|_{\dot{M}}, \overline{|B|_{\dot{M}} \cap |C|_{\dot{M}}}\rangle = \langle X, \overline{Y \cap Z}\rangle$, $\langle X, \overline{Y \cap Z}\rangle \in \dot{\mathcal{O}}_w$— a contradiction. Thus, $\mathsf{O}(\neg(B \wedge C)|A) \notin w$ and thus $\mathsf{P}B \wedge CA \in w$. Since w validates (WRM), $\mathsf{O}(B|A \wedge C) \in w$. By Lemma J.5.3, $\dot{M}, w \models \mathsf{O}(B|A \wedge C)$. Thus, $\langle |A \wedge C|_{\dot{M}}, |B|_{\dot{M}}\rangle \overset{J.4.1i}{=} \langle |A|_{\dot{M}} \cap |C|_{\dot{M}}, |B|_{\dot{M}}\rangle = \langle X \cap Z, Y\rangle \in \dot{\mathcal{O}}_{\dot{M}}$.

For (M-RCPM) let $X, Y, Z \in \varepsilon_{\dot{M}}$, $Y \subseteq Z$, $\langle X, Y\rangle \in \dot{\mathcal{O}}_w$, and $\langle X, \overline{Y}\rangle \notin \dot{\mathcal{O}}_w$. There are A, B such that $[A] = X$, $[B] = Y$ and $\mathsf{O}(B|A) \in w$. Furthermore, there is a C such that $|C|_M = Z$. By Lemma J.5.3, $M, w \models \mathsf{O}(B|A)$ and hence $\langle |A|_{\dot{M}}, |B|_{\dot{M}}\rangle \in \dot{\mathcal{O}}_w$. Since, by Lemma J.5.3, $\overline{|B|_{\dot{M}}} = \overline{Y}$, by Lemma J.5.3 (iii), $|\neg B|_{\dot{M}} = \overline{Y}$. By Lemma J.5.4 (i), $\mathsf{O}(\neg B|A) \notin w$ since $\langle X, \overline{Y}\rangle \notin \dot{\mathcal{O}}_w$, and thus $\mathsf{P}BA \in w$. By Lemma J.5.2 (i), $\vdash B \supset C$ since $[B] \subseteq |C|_M \overset{J.5.3}{=} [C]$. Since w validates all $\mathbf{L}^+$-theorems and (RCPM), $\mathsf{O}(C|A) \in w$. Thus, by Lemma J.5.3, $\dot{M}, w \models \mathsf{O}(C|A)$ and thus, $\langle |A|_{\dot{M}}, |C|_{\dot{M}}\rangle \overset{J.5.3}{=} \langle X, Z\rangle \in \dot{\mathcal{O}}_w$.

For (M-FDi): Let $X, Y \in \varepsilon_{\dot{M}}$. By Lemma J.5.4 there are A and B such that $[A] = X$ and $[B] = Y$. Now let $\langle Y, X\rangle \in \dot{\mathcal{O}}_w$, $w \in Y$, and $\langle Y, X\rangle \notin \dot{\mathcal{N}}^{\mathrm{i}}{}_w$. Since $w \in Y$ we have $w \in [B]$ and thus $B \in w$. By Lemma J.5.4 (i), $\mathsf{O}(A|B) \in w$. By

Lemma J.5.4 (ii), $\bullet_{\mathsf{i}}\mathsf{O}(A|B) \notin w$ and thus $\neg \bullet_{\mathsf{i}} \mathsf{O}(A|B) \in w$ since w is maximal consistent. Thus, since w validates (FDi), $\mathsf{O}^{\mathsf{i}}A \in w$. By Lemma J.5.4 (iv), $X \in \dot{\mathcal{O}}^{\mathsf{i}}{}_{w}$.

For (M-FDp): the proof is analogous.

For (M-Ep): Let $X, Y, Z, Z' \in \varepsilon_{\dot{M}}$. By Lemma J.5.4 there are A, B, C and D for which $[A] = X, [B] = Y, [C] = Z$ and $[D] = Z'$. Suppose the antecedent of (M-Ep) is true. By Lemma J.5.1 (i) we have $[B] \cap [C] = [B \wedge C]$. Thus, since $w \in [B \wedge C]$, $B \wedge C \in w$. By Lemma J.5.1 (iii), $[\neg(B \wedge C)] = \overline{[B \wedge C]} = \overline{[B] \cap [C]} = \overline{Y \cap Z}$. Thus, by Lemma J.5.4 (i), $\mathsf{O}(\neg(B \wedge C)|B) \notin w$ since by hypothesis $\langle Y, \overline{Y \cap Z}\rangle \notin \dot{\mathcal{O}}_{w}$. By Lemma J.5.4 (i), $\mathsf{O}(A|B) \in w$ since $\langle Y, X\rangle \in \dot{\mathcal{O}}_{w}$. Since $Z' \subseteq \overline{X}$, $[D] \subseteq \overline{[A]}$ and thus by Lemma J.5.1 (iii), $[D] \subseteq [\neg A]$. Thus, by Lemma J.5.2 (i), $\vdash D \supset \neg A$. Now we have by hypotheses, (a) $\langle Y \cap Z, \overline{Z'}\rangle \notin \dot{\mathcal{O}}_{w}$, or, (b), $\langle Y \cap Z, Z'\rangle \in \dot{\mathcal{O}}_{w}$. Note that by Lemma J.5.1(iii), $\overline{[D]} = [\neg D]$. Thus in case (a) we have by Lemma J.5.4 (i), $\mathsf{O}(\neg D|B \wedge C) \notin w$, and thus $\neg\mathsf{O}(\neg D|B \wedge C) \in w$ which is equivalent to $\mathsf{P}(D|B \wedge C) \in w$. In case (b) we have by Lemma J.5.4 (i), $\mathsf{O}(D|B \wedge C) \in w$. Since w validates (Ep), $\bullet_{\mathsf{p}}\mathsf{O}(A|B) \in w$. By Lemma J.5.4 (iii), $\langle Y, X\rangle \in \dot{\mathcal{N}}^{\mathsf{p}}{}_{w}$.

For the remaining conditions the proofs are analogous.

Thus, our model $\dot{M}$ satisfies all the model-conditions. By Lemma J.5.3, $\dot{M}, \dot{@} \models \neg\varphi$ and thus, $\dot{M}, \dot{@} \not\models \varphi$. Hence, $\dot{M} \not\models \varphi$. By contraposition we have that if φ is valid in all models which meet the appropriate conditions, then φ is provable in $\mathbf{L}^{+}$. □

J.5.2 Frame Completeness and Decidability

As shown in [6, 9], the canonical models $\dot{M} = \langle \dot{F}, \dot{@}, \dot{v}\rangle$ do not suffice to prove frame completeness. The problem is that $\dot{F}$ does not in general satisfy the appropriate frame conditions (as demonstrated by Goble for the monadic case with the permitted inheritance principle). Let me demonstrate the problem by means of (F-RCPM): Let $X, Y, Z \subseteq \dot{W}$ such that $Y \subseteq Z$; $\langle X, Y\rangle \in \dot{\mathcal{O}}_{w}$ and $\langle X, \overline{Y}\rangle \notin \dot{\mathcal{O}}_{w}$. There are, by the definition of $\dot{\mathcal{O}}_{w}$, A and B for which $X = [A], Y = [B]$ and

$$\mathsf{O}(B|A) \in w \tag{J.1}$$

Now suppose $\mathsf{O}(\neg B|A) \in w$. Then $\langle [A], [\neg B]\rangle \in \dot{\mathcal{O}}_{w}$. However, by Lemma J.5.1 (iii), $\langle [A], [\neg B]\rangle = \langle [A], \overline{[B]}\rangle$. Thus, $\langle X, \overline{Y}\rangle \in \dot{\mathcal{O}}_{w}$—a contradiction. Thus,

$$\mathsf{O}(\neg B|A) \notin w \tag{J.2}$$

And hence due to the maximal consistency of w,

$$\mathsf{P}(B|A) \in w \tag{J.3}$$

Now, in case there would be a C such that $[C] = Z$ it would be easy to prove frame completeness. Since we have in that case $[B] \subseteq [C]$, we get, by Lemma J.5.2 (i),

$$\vdash B \supset C \tag{J.4}$$

Since w satisfies (RCPM) we have $\mathsf{P}(B|A) \supset (\mathsf{O}(B|A) \supset \mathsf{O}(C|A))$ due to (J.4). By (J.1) and (J.3) we get via modus ponens, $\mathsf{O}(C|A) \in w$. Hence, $\langle [A], [C] \rangle = \langle X, Z \rangle \in \dot{\mathcal{O}}_w$.

However, the problem is that we have no guarantee that there is such a C.

On the basis of a given model $\dot{M} = \langle \dot{W}, \dot{\mathcal{O}}, \dot{\mathcal{N}}^{\mathrm{i}}, \dot{\mathcal{N}}^{\mathrm{p}}, \dot{\mathcal{O}}^{\mathrm{i}}, \dot{\mathcal{O}}^{\mathrm{p}}, \dot{@}, \dot{v} \rangle$ we construct a model $\mathring{M} = \langle \mathring{F}, \mathring{@}, \mathring{v} \rangle$ on a frame $\mathring{F} = \langle \mathring{W}, \mathring{\mathcal{O}}, \mathring{\mathcal{N}}^{\mathrm{i}}, \mathring{\mathcal{N}}^{\mathrm{p}}, \mathring{\mathcal{O}}^{\mathrm{i}}, \mathring{\mathcal{O}}^{\mathrm{p}} \rangle$ by filtration in the following way.

Let Φ be a finite set of formulas closed under subformulas, i.e., if $\varphi \in \Phi$ and ψ is a subformula of φ, then $\psi \in \Phi$, and let $\top, \bot \in \Phi$. Furthermore, let $\hat{\Phi}$ be the closure of Φ under truth-functions, i.e., $\hat{\Phi}$ is the smallest set of formulas such that $\Phi \subseteq \hat{\Phi}$ and if $\varphi, \psi \in \hat{\Phi}$, then $\varphi \wedge \psi \in \hat{\Phi}$, $\varphi \vee \psi \in \hat{\Phi}$ and $\neg\varphi \in \hat{\Phi}$. Note that $\top, \bot \in \hat{\Phi}$, and that $\hat{\Phi}$ itself is closed under subformulas.

Now let $\Psi = \Phi \cap \mathcal{P}$ and $\hat{\Psi}$ be again the closure of Ψ under truth-functions.

We define an equivalence relation $\sim_{\Psi}^{\dot{M}}$ on $\dot{W}$ such that, for all $w, w' \in \dot{W}$:

$$w \sim_{\Psi}^{\dot{M}} w' \text{ iff } \forall\varphi\big(\text{if } \varphi \in \Psi \text{ then } (\dot{M}, w \models \varphi \text{ iff } \dot{M}, w' \models \varphi)\big).$$

Lemma J.5.5. *For all $w, w' \in \dot{W}$, if $w \sim_{\Psi}^{\dot{M}} w'$, then for all $A \in \hat{\Psi}$, ($\dot{M}, w \models A$ iff $\dot{M}, w' \models A$).*

Proof. Suppose $w \sim_{\Psi}^{\dot{M}} w'$. The proof is by induction on the length of A. For all $A \in \Psi$ the statement holds by definition. If $A = A_1 \wedge A_2$ or $A = A_1 \vee A_2$ or $A = \neg A_1$, for some A_1, A_2, then the result follows directly from the inductive hypothesis. □

It is important to note that $\sim_{\Psi}^{\dot{M}}$ partitions $\dot{W}$ into finitely many equivalence classes $[w]$ for $w \in \dot{W}$, where $[w] = \{w' \in \dot{W} \mid w' \sim_{\Psi}^{\dot{M}} w\}$.

We select now, for each equivalence class $[w]$, a member $\mathring{w} \in [w]$ in the following way: where $[w] = [\dot{@}]$ let $\mathring{w} = \dot{@}$ and for all $[w] \neq [\dot{@}]$ let $\mathring{w}$ be an arbitrary member of $[w]$ (not necessarily w itself). We define $\mathring{W}$ as the set of all these selected representants. The following fact follows directly from the definitions.

Lemma J.5.6. *(i) $\mathring{W} \subseteq \dot{W}$; (ii) $\mathring{W}$ is finite; (iii) for all $w' \in \dot{W}$ there is a $\mathring{w} \in \mathring{W}$ such that $w' \sim_{\Psi}^{\dot{M}} \mathring{w}$; (iv) for all $w, w' \in \mathring{W}$, if $w \neq w'$ then it is not the case that $w \sim_{\Psi}^{\dot{M}} w'$.*

Some more writing conventions: for $X \subseteq \dot{W}$, let $X{\downarrow} =_{\mathrm{df}} X \cap \mathring{W}$.

The assignments $\mathring{\mathcal{O}}, \mathring{\mathcal{N}}^{\mathrm{i}}, \mathring{\mathcal{N}}^{\mathrm{p}}, \mathring{\mathcal{O}}^{\mathrm{i}}, \mathring{\mathcal{O}}^{\mathrm{p}}$ fullfil the following conditions for each $w \in \mathring{W}$:

$$\begin{aligned}\langle X, Y \rangle \in \mathring{\mathcal{O}}_w \text{ iff } \exists A \exists B\big(&A, B \in \hat{\Psi} \text{ and } X = |A|_{\dot{M}}{\downarrow} \text{ and} \\ &Y = |B|_{\dot{M}}{\downarrow} \text{ and } \langle |A|_{\dot{M}}, |B|_{\dot{M}} \rangle \in \dot{\mathcal{O}}_w\big)\end{aligned} \qquad (\mathrm{D}\mathcal{O}\star)$$

$$\begin{aligned}\langle X, Y \rangle \in \mathring{\mathcal{N}}^{\mathrm{i}}{}_w \text{ iff } \exists A \exists B\big(&A, B \in \hat{\Psi} \text{ and } X = |A|_{\dot{M}}{\downarrow} \\ &\text{and } Y = |B|_{\dot{M}}{\downarrow} \text{ and } \langle |A|_{\dot{M}}, |B|_{\dot{M}} \rangle \in \dot{\mathcal{N}}^{\mathrm{i}}{}_w\big)\end{aligned} \qquad (\mathrm{D}\mathcal{N}^{\mathrm{i}}\star)$$

$$\langle X, Y\rangle \in \mathring{\mathcal{N}}^{\mathrm{p}}{}_w \text{ iff } \exists A \exists B\big(A, B \in \hat{\Psi} \text{ and } X = |A|_{\dot{M}}\downarrow \text{ and } Y = |B|_{\dot{M}}\downarrow \text{ and } \langle |A|_{\dot{M}}, |B|_{\dot{M}}\rangle \in \dot{\mathcal{N}}^{\mathrm{p}}{}_w\big) \tag{D$\mathcal{N}^{\mathrm{p}}\star$}$$

$$X \in \mathring{\mathcal{O}}^{\mathrm{i}}{}_w \text{ iff } \exists A\big(A \in \hat{\Psi} \text{ and } X = |A|_{\dot{M}}\downarrow \text{ and } |A|_{\dot{M}} \in \dot{\mathcal{O}}^{\mathrm{i}}{}_w\big) \tag{D$\mathcal{O}^{\mathrm{i}}\star$}$$

$$X \in \mathring{\mathcal{O}}^{\mathrm{p}}{}_w \text{ iff } \exists A\big(A \in \hat{\Psi} \text{ and } X = |A|_{\dot{M}}\downarrow \text{ and } |A|_{\dot{M}} \in \dot{\mathcal{O}}^{\mathrm{p}}{}_w\big) \tag{D$\mathcal{O}^{\mathrm{p}}\star$}$$

For all atomic formulas p we demand that $\mathring{v} : p \mapsto \dot{v}(p)\downarrow$.

Lemma J.5.7. *For all $\mathring{w} \in \mathring{W}$, there is a formula $B \in \hat{\Psi}$ such that $|B|_{\dot{M}}\downarrow = \{\mathring{w}\}$.*

Proof. The following proof only differs minimally from Goble's. First we have for all $\mathring{w}, \mathring{w}' \in \mathring{W}$, if $\mathring{w} \neq \mathring{w}'$ then there is a formula A such that $A \in \hat{\Psi}$ and $\mathring{w} \in |A|_{\dot{M}}$ and $\mathring{w}' \notin |A|_{\dot{M}}$. For suppose otherwise. Suppose $\mathring{w} \neq \mathring{w}'$ but for every $A \in \hat{\Psi}$ if $\mathring{w} \in |A|_{\dot{M}}$ then $\mathring{w}' \in |A|_{\dot{M}}$. Then $\mathring{w} \sim_{\Psi}^{\dot{M}} \mathring{w}'$, for consider any $B \in \Psi$, hence $B \in \hat{\Psi}$. If $\dot{M}, \mathring{w} \models B$, then $\mathring{w} \in |B|_{\dot{M}}$, so by supposition $\mathring{w}' \in |B|_{\dot{M}}$, and thus $\dot{M}, \mathring{w}' \models B$. Suppose then that $\dot{M}, \mathring{w}' \models B$, i.e., $\mathring{w}' \in |B|_{\dot{M}}$, but that it is not the case that $\dot{M}, \mathring{w} \models B$. Then $\dot{M}, \mathring{w} \models \neg B$ and $w \in |\neg B|_{\dot{M}}$. Since $\neg B \in \hat{\Psi}$, by supposition, $\mathring{w}' \in |\neg B|_{\dot{M}}$, or $\dot{M}, \mathring{w}' \models \neg B$. That means $\dot{M}, \mathring{w}' \not\models B$—a contradiction. Hence, if $\dot{M}, \mathring{w}' \models B$, then $\dot{M}, \mathring{w} \models B$, and so $\dot{M}, \mathring{w} \models B$ iff $\dot{M}, \mathring{w}' \models B$, which suffices for $\mathring{w} \sim_{\Psi}^{\dot{M}} \mathring{w}'$. But if $\mathring{w} \neq \mathring{w}'$ then it is not the case that $\mathring{w} \sim_{\Psi}^{\dot{M}} \mathring{w}'$, by Lemma J.5.6 (iv), a contradiction. Therefore, it must be the case that if $\mathring{w} \neq \mathring{w}'$, there is a $A \in \hat{\Psi}$ such that $\mathring{w} \in |A|_{\dot{M}}$ and $\mathring{w}' \notin |A|_{\dot{M}}$. For each $\mathring{w}'$ such that $\mathring{w}' \neq \mathring{w}$, select one such formula, and call it $A_{\mathring{w}'}$. Let $\Lambda = \{\gamma_i \mid i \in I\}$ be the set of all such formulas $A_{\mathring{w}'}$ for all $\mathring{w}' \neq \mathring{w}$. Λ is finite since $\mathring{W}$ is finite. Let $B_{\mathring{w}} = \bigwedge \Lambda = \bigwedge_I \gamma_i$ be the conjunction of all the members of Λ. $B_{\mathring{w}} \in \hat{\Psi}$ since each conjunct $\gamma_i \in \hat{\Psi}$ and $\hat{\Psi}$ is closed under truth-functions. We now show that $|B_{\mathring{w}}|_{\dot{M}}\downarrow = \{\mathring{w}\}$.

(i) Suppose $x \in |B_{\mathring{w}}|_{\dot{M}}\downarrow$. So $x \in |B_{\mathring{w}}|_{\dot{M}}$ and $x \in \mathring{W}$. Suppose $x \neq \mathring{w}$. Then there is a formula $A_x \in \Psi$ such that $\mathring{w} \in |A_x|_{\dot{M}}$ and $x \notin |A_x|_{\dot{M}}$. We have $|B_{\mathring{w}}|_{\dot{M}} = |\bigwedge_I \gamma_i|_{\dot{M}} \overset{J.4.1i}{=} \bigcap_I |\gamma_i|_{\dot{M}}$. Hence, $|B_{\mathring{w}}|_{\dot{M}} \subseteq |\gamma_i|_{\dot{M}}$ for all $i \in I$. Note that $A_x = \gamma_j$ for a $j \in I$. Since $x \in |B_{\mathring{w}}|_{\dot{M}}, x \in |A_x|_{\dot{M}}$—a contradiction. Therefore, if $x \in |B_{\mathring{w}}|_{\dot{M}}$, $x = \mathring{w}$ and so $x \in \{\mathring{w}\}$. Thus $|B_{\mathring{w}}|_{\dot{M}}\downarrow \subseteq \{\mathring{w}\}$.

(ii) Suppose $x \in \{\mathring{w}\}$, i.e., $x = \mathring{w}$. Thus $x \in \mathring{W}$. For all $\gamma_i \in \Lambda$, $x \in |\gamma_i|_{\dot{M}}$. Hence, $\dot{M}, x \models \gamma_i$ for all $i \in I$. Consequently, $\dot{M}, x \models \bigwedge_I \gamma_i$. But $\bigwedge_I \gamma_i = B_{\mathring{w}}$, hence $\dot{M}, x \models B_{\mathring{w}}$. That is to say, $x \in |B_{\mathring{w}}|_{\dot{M}}$, and therefore $x \in |B_{\mathring{w}}|_{\dot{M}}\downarrow$. Thus, $\{\mathring{w}\} \subseteq |B_{\mathring{w}}|\downarrow$. Therefore, by (i) and (ii) together, $|B_{\mathring{w}}|_{\dot{M}}\downarrow = \{\mathring{w}\}$, as required for the Lemma. □

Lemma J.5.8. *For all $w \in \mathring{W}$ we have: $\mathring{M}, w \models A$ iff $\dot{M}, w \models A$.*

Proof. This is shown by induction. Let $A \in \mathcal{A}$, then $\mathring{M}, w \models A$ iff $w \in \mathring{v}(A)$ iff $w \in \dot{v}(A)\downarrow$ iff $w \in \dot{v}(A) \cap \mathring{W}$ iff (since $w \in \mathring{W}$) $w \in \dot{v}(A)$ iff $\dot{M}, w \models A$. Now by induction hypothesis let the lemma hold for B and C. Let $A = B \wedge C$. Then $\mathring{M}, w \models A$ iff $\mathring{M}, w \models B, C$ iff $\dot{M}, w \models B, C$ iff $\dot{M}, w \models B \wedge C$. The argument is similar for $A = B \vee C$, $A = B \supset C$ and $A = \neg B$. □

Lemma J.5.9. $|A|_{\mathring{M}} = |A|_{\dot{M}}{\downarrow}$.

Proof. $|A|_{\mathring{M}} = \{w \in \mathring{W} \mid \mathring{M}, w \models A\} \overset{(1)}{=} \{w \in \mathring{W} \mid \dot{M}, w \models A\} = \{w \in \dot{W} \mid \dot{M}, w \models A\} \cap \mathring{W} = |A|_{\dot{M}}{\downarrow}$, where (1) is due to Lemma J.5.8. □

Lemma J.5.10. *For all $X \subseteq \mathring{W}$, there is a formula B such that $B \in \hat{\Psi}$ and $X = |B|_{\dot{M}}{\downarrow}$.*

Proof. Let $X \subseteq \mathring{W}$. Then $X = \{x_1, \ldots, x_n\}$ is finite, since $\mathring{W}$ is finite. By Lemma J.5.7, there is an $A_i \in \hat{\Psi}$ for each $x_i \in X$ such that $|A_i|_{\dot{M}}{\downarrow} = \{x_i\}$. Let $A_X = A_1 \vee \cdots \vee A_n$. Since $\hat{\Psi}$ is closed under classical connectives, $A_X \in \hat{\Psi}$. $|A_X|_{\dot{M}}{\downarrow} = |\bigvee_{i=1}^n A_i|_{\dot{M}}{\downarrow} \overset{J.5.9}{=} |\bigvee_{i=1}^n A_i|_{\mathring{M}} = \{w \in \mathring{W} \mid \mathring{M}, w \models \bigvee_{i=1}^n A_i\} = \{w \in \mathring{W} \mid \mathring{M}, w \models A_1 \text{ or } \ldots \text{ or } \mathring{M}, w \models A_n\} = \bigcup_{i=1}^n \{w \in \mathring{W} \mid \mathring{M}, w \models A_i\} = \bigcup_{i=1}^n |A_i|_{\mathring{M}} \overset{J.5.9}{=} \bigcup_{i=1}^n |A_i|_{\dot{M}}{\downarrow} = \bigcup_{i=1}^n \{x_i\} = X$. □

Lemma J.5.11. *For all $A, B \in \hat{\Psi}$, (i) if $|A|_{\dot{M}}{\downarrow} \subseteq |B|_{\dot{M}}{\downarrow}$, then $|A|_{\dot{M}} \subseteq |B|_{\dot{M}}$; (ii) if $|A|_{\dot{M}}{\downarrow} = |B|_{\dot{M}}{\downarrow}$, then $|A|_{\dot{M}} = |B|_{\dot{M}}$.*

Proof. Let $A, B \in \hat{\Psi}$ such that $|A|_{\dot{M}}{\downarrow} \subseteq |B|_{\dot{M}}{\downarrow}$. Take any $w \in |A|_{\dot{M}}$. By Lemma J.5.6 (iii), there is a $\mathring{w} \in \mathring{W}$ such that $w \sim_{\Psi}^{\dot{M}} \mathring{w}$. Since $\dot{M}, w \models A$, by Lemma J.5.5, $\dot{M}, \mathring{w} \models A$. Hence, $\mathring{w} \in |A|_{\dot{M}}$, and, since $\mathring{w} \in \mathring{W}$, $\mathring{w} \in |A|_{\dot{M}}{\downarrow}$. Thus, $\mathring{w} \in |B|_{\dot{M}}{\downarrow}$ and hence, $\mathring{w} \in |B|_{\dot{M}}$ or $\dot{M}, \mathring{w} \models B$. Thus, since $B \in \hat{\Psi}$ and $w \sim_{\Psi}^{\dot{M}} \mathring{w}$, $\dot{M}, w \models B$ by Lemma J.5.5. Thus, $w \in |B|_{\dot{M}}$. (ii) follows immediately. □

Lemma J.5.12. *(i) $|\varphi|_{\dot{M}}{\downarrow} \cap |\psi|_{\dot{M}}{\downarrow} = |\varphi \wedge \psi|_{\dot{M}}{\downarrow}$; (ii) $|\varphi|_{\dot{M}}{\downarrow} \cup |\psi|_{\dot{M}}{\downarrow} = |\varphi \vee \psi|_{\dot{M}}{\downarrow}$; (iii) $\overline{|\varphi|_{\dot{M}}{\downarrow}} = |\neg\varphi|_{\dot{M}}{\downarrow}$ (where the complement is interpreted w.r.t. frame $\mathring{F}$).*

Proof. Ad (i): $|\varphi \wedge \psi|_{\dot{M}}{\downarrow} = |\varphi \wedge \psi|_{\dot{M}} \cap \mathring{W} \overset{(1)}{=} (|\varphi|_{\dot{M}} \cap |\psi|_{\dot{M}}) \cap \mathring{W} = (|\varphi|_{\dot{M}} \cap \mathring{W}) \cap (|\psi|_{\dot{M}} \cap \mathring{W}) = |\varphi|_{\dot{M}}{\downarrow} \cap |\psi|_{\dot{M}}{\downarrow}$ where (1) is due to Lemma J.4.1 (i).

Ad (ii): $|\varphi \vee \psi|_{\dot{M}}{\downarrow} \overset{(2)}{=} (|\varphi|_{\dot{M}} \cup |\psi|_{\dot{M}}){\downarrow} = (|\varphi|_{\dot{M}} \cup |\psi|_{\dot{M}}) \cap \mathring{W} = (|\varphi|_{\dot{M}} \cap \mathring{W}) \cup (|\psi|_{\dot{M}} \cap \mathring{W}) = |\varphi|_{\dot{M}}{\downarrow} \cup |\psi|_{\dot{M}}{\downarrow}$ where (2) is due to Lemma J.4.1 (ii).

Ad (iii): $\overline{|\varphi|_{\dot{M}}{\downarrow}} = (\mathring{W} \setminus |\varphi|_{\dot{M}}{\downarrow}) \cap \mathring{W} = (\mathring{W} \setminus (|\varphi|_{\dot{M}} \cap \mathring{W})) \cap \mathring{W} = (\mathring{W} \setminus |\varphi|_{\dot{M}}) \cap \mathring{W} = (\dot{W} \setminus |\varphi|_{\dot{M}}) \cap \mathring{W} \overset{(3)}{=} |\neg\varphi|_{\dot{M}} \cap \mathring{W} = |\neg\varphi|_{\dot{M}}{\downarrow}$ where (3) is due to Lemma J.4.1 (iii). □

Lemma J.5.13. *If $\dot{M} = \langle \dot{W}, \dot{\mathcal{O}}, \dot{\mathcal{N}}^{\mathsf{i}}, \dot{\mathcal{N}}^{\mathsf{p}}, \dot{\mathcal{O}}^{\mathsf{i}}, \dot{\mathcal{O}}^{\mathsf{p}}, \dot{@}, \dot{v} \rangle$, defined as above, satisfies conditions* {M-X | X ∈ ***X***} *where* ***X*** ⊆ {CN, RCPM, S, WRM, CPAND, CP, PS', FD^{i}, FD^{p}, E^{p}, CTDR, fV, oV-E^{i}}, *then $\mathring{F}$ satisfies conditions* {F-X | X ∈ ***X***}.

Proof. We demonstrate the proof via some paradigmatical rules. For (F-PS'): Let $X, Y, Z \subseteq \mathring{W}$, $\langle Y \cap Z, X \rangle \in \mathring{\mathcal{O}}_w$ and $\langle \overline{Y} \cap Z, \overline{X} \rangle \notin \mathring{\mathcal{O}}_w$. To show: $\langle Z, \overline{Y} \cup X \rangle \in \mathring{\mathcal{O}}_w$. By (D$\mathcal{O}\star$), there are $E, F \in \hat{\Psi}$ for which $|E|_{\dot{M}}{\downarrow} = Y \cap Z$, $|F|_{\dot{M}}{\downarrow} = X$ and $\langle |E|_{\dot{M}}, |F|_{\dot{M}} \rangle \in \dot{\mathcal{O}}_w$. Furthermore, by Lemma J.5.10, there are $A, B, C \in \hat{\Psi}$ such that $|A|_{\dot{M}}{\downarrow} = X$, $|B|_{\dot{M}}{\downarrow} = Y$ and $|C|_{\dot{M}}{\downarrow} = Z$. By Lemma J.5.12 (i),

$|B|_{\dot{M}}{\downarrow} \cap |C|_{\dot{M}}{\downarrow} = |B \wedge C|_{\dot{M}}{\downarrow}$. Thus $|E|_{\dot{M}}{\downarrow} = |B \wedge C|_{\dot{M}}{\downarrow}$ and by Lemma J.5.11 and since $B \wedge C \in \hat{\Psi}$, $|E|_{\dot{M}} = |B \wedge C|_{\dot{M}}$. Also by Lemma J.5.11, $|F|_{\dot{M}} = |A|_{\dot{M}}$. Thus, $\langle |B \wedge C|_{\dot{M}}, |A|_{\dot{M}}\rangle \overset{J.4.1i}{=} \langle |B|_{\dot{M}} \cap |C|_{\dot{M}}, |A|_{\dot{M}}\rangle \in \dot{\mathcal{O}}_w$. Suppose $\langle \overline{|B|_{\dot{M}}} \cap |C|_{\dot{M}}, \overline{|A|_{\dot{M}}}\rangle \in \dot{\mathcal{O}}_w$. Note that $\langle \overline{|B|_{\dot{M}}} \cap |C|_{\dot{M}}, \overline{|A|_{\dot{M}}}\rangle \overset{J.4.1iii}{=}$ $\langle |\neg B|_{\dot{M}} \cap |C|_{\dot{M}}, |\neg A|_{\dot{M}}\rangle \overset{J.4.1i}{=} \langle |\neg B \wedge C|_{\dot{M}}, |\neg A|_{\dot{M}}\rangle \in \dot{\mathcal{O}}_w$. (iii) and (2) by Lemma J.4.1 (i). Now by (D$\mathcal{O}\star$), $\langle |\neg B \wedge C|_{\dot{M}}{\downarrow}, |\neg A|_{\dot{M}}{\downarrow}\rangle \in \mathring{\mathcal{O}}_w$. Note that $\langle |\neg B \wedge C|_{\dot{M}}{\downarrow}, |\neg A|_{\dot{M}}{\downarrow}\rangle \overset{J.5.12i}{=} \langle |\neg B|_{\dot{M}}{\downarrow} \cap |C|_{\dot{M}}{\downarrow}, |\neg A|_{\dot{M}}{\downarrow}\rangle \overset{J.5.12iii}{=} \langle \overline{|B|_{\dot{M}}{\downarrow}} \cap |C|_{\dot{M}}{\downarrow}, \overline{|A|_{\dot{M}}{\downarrow}}\rangle \in \mathring{\mathcal{O}}_w$. However, now we have $\langle \overline{Y} \cap Z, \overline{X}\rangle \in \mathring{\mathcal{O}}_w$—a contradiction. Hence, $\langle \overline{|B|_{\dot{M}}} \cap |C|_{\dot{M}}, \overline{|A|_{\dot{M}}}\rangle \notin \dot{\mathcal{O}}_w$. Since $\dot{M}$ satisfies (M-PS'), $\langle |C|_{\dot{M}}, \overline{|B|_{\dot{M}}} \cup |A|_{\dot{M}}\rangle \in \dot{\mathcal{O}}_w$. By Lemma J.4.1 (ii) and (iii), $\langle |C|_{\dot{M}}, |\neg B \vee A|_{\dot{M}}\rangle \in \dot{\mathcal{O}}_w$. Thus, by (D$\mathcal{O}\star$) and since $C, \neg B \vee A \in \hat{\Psi}$, $\langle |C|_{\dot{M}}{\downarrow}, |\neg B \vee A|_{\dot{M}}{\downarrow}\rangle \in \mathring{\mathcal{O}}_w$. By Lemma J.5.12 (ii) and (iii), $\langle |C|_{\dot{M}}{\downarrow}, \overline{|B|_{\dot{M}}{\downarrow}} \cup |A|_{\dot{M}}{\downarrow}\rangle \in \mathring{\mathcal{O}}_w$ and thus $\langle Z, \overline{Y} \cup X\rangle \in \mathring{\mathcal{O}}_w$.

For (F-RCPM): Let $X, Y, Z \subseteq \mathring{W}$, $Y \subseteq Z$, $\langle X, Y\rangle \in \mathring{\mathcal{O}}_w$, and $\langle X, \overline{Y}\rangle \notin \mathring{\mathcal{O}}_w$. By (D$\mathcal{O}\star$), there are $A, B \in \hat{\Psi}$ such that $|A|_{\dot{M}}{\downarrow} = X$, $|B|_{\dot{M}}{\downarrow} = Y$ and $\langle |A|_{\dot{M}}, |B|_{\dot{M}}\rangle \in \dot{\mathcal{O}}_w$. Furthermore, by Lemma J.5.10, there is a $C \in \hat{\Psi}$ for which $Z = |C|_{\dot{M}}{\downarrow}$. Suppose, $\langle |A|_{\dot{M}}, |\neg B|_{\dot{M}}\rangle \in \dot{\mathcal{O}}_w$. Then by (D$\mathcal{O}\star$), $\langle |A|_{\dot{M}}{\downarrow}, |\neg B|_{\dot{M}}{\downarrow}\rangle \overset{J.5.12iii}{=}$ $\langle |A|_{\dot{M}}{\downarrow}, \overline{|B|_{\dot{M}}{\downarrow}}\rangle = \langle X, \overline{Y}\rangle \in \mathring{\mathcal{O}}_w$—a contradiction. Thus, $\langle |A|_{\dot{M}}, |\neg B|_{\dot{M}}\rangle \overset{J.4.1iii}{=}$ $\langle |A|_{\dot{M}}, \overline{|B|_{\dot{M}}}\rangle \notin \dot{\mathcal{O}}_w$. By Lemma J.5.11, $|B|_{\dot{M}} \subseteq |C|_{\dot{M}}$, since $|B|_{\dot{M}}{\downarrow} \subseteq |C|_{\dot{M}}{\downarrow}$. Since $\dot{M}$ satisfies (M-RCPM), $\langle |A|_{\dot{M}}, |C|_{\dot{M}}\rangle \in \dot{\mathcal{O}}_w$. Thus, by (D$\mathcal{O}\star$), $\langle |A|_{\dot{M}}{\downarrow}, |C|_{\dot{M}}{\downarrow}\rangle = \langle X, Z\rangle \in \mathring{\mathcal{O}}_w$.

For (F-Ep): Let $X, Y, Z, Z' \in \mathring{W}$ such that, (a), $w \in Y \cap Z$, (b), $\langle Y, \overline{Y \cap Z}\rangle \notin \mathring{\mathcal{O}}_w$, (c), $\langle Y, X\rangle \in \mathring{\mathcal{O}}_w$, (d), $Z' \subseteq \overline{X}$, and either, (e), $\langle Y \cap Z, \overline{Z'}\rangle \notin \mathring{\mathcal{O}}_w$, or, (f), $\langle Y \cap Z, Z'\rangle \in \mathring{\mathcal{O}}_w$. To show: $\langle Y, X\rangle \in \mathcal{N}^{\mathsf{p}}{}_w$. By (D$\mathcal{O}\star$) and (c) there are $A, B \in \hat{\Psi}$ for which $|A|_{\dot{M}}{\downarrow} = X$, $|B|_{\dot{M}}{\downarrow} = Y$ and $\langle |B|_{\dot{M}}, |A|_{\dot{M}}\rangle \in \dot{\mathcal{O}}_w$. By Lemma J.5.10 there is a $C \in \hat{\Psi}$ such that $|C|_{\dot{M}}{\downarrow} = Z$. Suppose $\langle |B|_{\dot{M}}, \overline{|B \wedge C|_{\dot{M}}}\rangle \in \dot{\mathcal{O}}_w$, then by Lemma J.4.1 (iii), $\langle |B|_{\dot{M}}, |\neg(B \wedge C)|_{\dot{M}}\rangle \in \dot{\mathcal{O}}_w$. Now by (D$\mathcal{O}\star$) and since $B, \neg(B \wedge C) \in \hat{\Psi}$, $\langle |B|_{\dot{M}}{\downarrow}, |\neg(B \wedge C)|_{\dot{M}}{\downarrow}\rangle \in \mathring{\mathcal{O}}_w$. Then by Lemma J.5.12, $\langle |B|_{\dot{M}}{\downarrow}, |\neg(B \wedge C)|_{\dot{M}}{\downarrow}\rangle =$ $\langle |B|_{\dot{M}}{\downarrow}, \overline{|B \wedge C|_{\dot{M}}{\downarrow}}\rangle = \langle |B|_{\dot{M}}{\downarrow}, \overline{|B|_{\dot{M}}{\downarrow} \cap |C|_{\dot{M}}{\downarrow}}\rangle = \langle Y, \overline{Y \cap Z}\rangle \in \mathring{\mathcal{O}}_w$—a contradiction with (b). Thus, $\langle |B|_{\dot{M}}, \overline{|B \wedge C|_{\dot{M}}}\rangle \overset{J.4.1i}{=} \langle |B|_{\dot{M}}, \overline{|B|_{\dot{M}} \cap |C|_{\dot{M}}}\rangle \notin \dot{\mathcal{O}}_w$. By Lemma J.5.10 there is a $D \in \hat{\Psi}$ for which $|D|_{\dot{M}}{\downarrow} = Z'$. Thus $|D|_{\dot{M}}{\downarrow} \subseteq \overline{|A|_{\dot{M}}{\downarrow}}$ and thus by Lemma J.5.12 (iii), $|D|_{\dot{M}}{\downarrow} \subseteq |\neg A|_{\dot{M}}{\downarrow}$. By Lemma J.5.11 and since $D, \neg A \in \hat{\Psi}$, $|D|_{\dot{M}} \subseteq |\neg A|_{\dot{M}} \overset{J.4.1iii}{=} \overline{|A|_{\dot{M}}}$. Case (e): Suppose $\langle |B|_{\dot{M}} \cap |C|_{\dot{M}}, \overline{|D|_{\dot{M}}}\rangle \in \dot{\mathcal{O}}_w$, then by Lemma J.4.1 (i) and (iii), $\langle |B \wedge C|_{\dot{M}}, |\neg D|_{\dot{M}}\rangle \in \dot{\mathcal{O}}_w$. By (D$\mathcal{O}\star$) and since $B \wedge C, \neg D \in \hat{\Psi}$, $\langle |B \wedge C|_{\dot{M}}{\downarrow}, |\neg D|_{\dot{M}}{\downarrow}\rangle \in \mathring{\mathcal{O}}_w$. By Lemma J.5.12 (i) and (iii), $\langle |B|_{\dot{M}}{\downarrow} \cap |C|_{\dot{M}}{\downarrow}, \overline{|D|_{\dot{M}}{\downarrow}}\rangle = \langle Y \cap Z, \overline{Z'}\rangle \in \mathring{\mathcal{O}}_w$—a contradiction with (e). Thus, $\langle |B|_{\dot{M}} \cap |C|_{\dot{M}}, \overline{|D|_{\dot{M}}}\rangle \notin \dot{\mathcal{O}}_w$. Case (f): By (D$\mathcal{O}\star$) there are $E, F \in \hat{\Psi}$ such that $|E|_{\dot{M}}{\downarrow} = Y \cap Z$, $|F|_{\dot{M}}{\downarrow} = Z'$ and $\langle |E|_{\dot{M}}, |F|_{\dot{M}}\rangle \in \dot{\mathcal{O}}_w$. We have $|E|_{\dot{M}}{\downarrow} = |B|_{\dot{M}}{\downarrow} \cap |C|_{\dot{M}}{\downarrow} \overset{J.5.12i}{=} |B \wedge C|_{\dot{M}}{\downarrow}$. Thus, by Lemma J.5.11 and since

$E, B \wedge C \in \hat{\Psi}, |E|_{\dot{M}} = |B \wedge C|_{\dot{M}} \overset{J.4.1i}{=} |B|_{\dot{M}} \cap |C|_{\dot{M}}$. Thus, $\langle |B|_{\dot{M}} \cap |C|_{\dot{M}}, |F|_{\dot{M}} \rangle \in \dot{\mathcal{O}}_w$. Also by Lemma J.5.11, since $|F|_{\dot{M}}\downarrow = Z' = |D|_{\dot{M}}\downarrow$, $|F|_{\dot{M}} = |D|_{\dot{M}}$. Thus, $\langle |B \wedge C|_{\dot{M}}, |D|_{\dot{M}} \rangle \in \dot{\mathcal{O}}_w$.

Since $\dot{M}$ satisfies (M-Ep), $\langle |B|_{\dot{M}}, |A|_{\dot{M}} \rangle \in \dot{\mathcal{N}}^{\mathrm{p}}{}_w$. Hence, by (D$\mathcal{N}^{\mathrm{p}}\star$), $\langle Y, X \rangle \in \mathring{\mathcal{N}}^{\mathrm{p}}{}_w$.

For (F-WRM): Consider $X, Y, Z \in \mathring{W}$. Let $\langle X, Y \rangle \in \mathring{\mathcal{O}}_w$ and $\langle X, \overline{Y \cap Z} \rangle \notin \mathring{\mathcal{O}}_w$. By (D$\mathcal{O}\star$), there are $A, B \in \hat{\Psi}$ such that $|A|_{\dot{M}}\downarrow = X$, $|B|_{\dot{M}}\downarrow = Y$ and $\langle |A|_{\dot{M}}, |B|_{\dot{M}} \rangle \in \dot{\mathcal{O}}_w$. By Lemma J.5.10, there is a $C \in \hat{\Psi}$ such that $|C|_{\dot{M}}\downarrow = Z$. Suppose that $\langle |A|_{\dot{M}}, \overline{|B|_{\dot{M}} \cap |C|_{\dot{M}}} \rangle \in \dot{\mathcal{O}}_w$. Due to the fact that $\langle |A|_{\dot{M}}, \overline{|B|_{\dot{M}} \cap |C|_{\dot{M}}} \rangle \overset{J.4.1i}{=} \langle |A|_{\dot{M}}, \overline{|B \wedge C|_{\dot{M}}} \rangle \overset{J.4.1iii}{=} \langle |A|_{\dot{M}}, |\neg(B \wedge C)|_{\dot{M}} \rangle$, we have, $\langle |A|_{\dot{M}}, |\neg(B \wedge C)|_{\dot{M}} \rangle \in \dot{\mathcal{O}}_w$. By (D$\mathcal{O}\star$), $\langle |A|_{\dot{M}}\downarrow, |\neg(B \wedge C)|_{\dot{M}}\downarrow \rangle \in \mathring{\mathcal{O}}_w$. However, $\langle |A|_{\dot{M}}\downarrow, |\neg(B \wedge C)|_{\dot{M}}\downarrow \rangle \overset{J.5.12iii}{=} \langle |A|_{\dot{M}}\downarrow, \overline{|B \wedge C|_{\dot{M}}\downarrow} \rangle \overset{J.5.12i}{=} \langle |A|_{\dot{M}}\downarrow, \overline{|B|_{\dot{M}}\downarrow \cap |C|_{\dot{M}}\downarrow} \rangle = \langle X, \overline{Y \cap Z} \rangle$. Thus, $\langle X, \overline{Y \cap Z} \rangle \in \mathring{\mathcal{O}}_w$—a contradiction. Thus, $\langle |A|_{\dot{M}}, \overline{|B|_{\dot{M}} \cap |C|_{\dot{M}}} \rangle \notin \dot{\mathcal{O}}_w$. Since $\dot{M}$ satisfies (M-WRM), $\langle |A|_{\dot{M}} \cap |C|_{\dot{M}}, |B|_{\dot{M}} \rangle \overset{J.4.1i}{=} \langle |A \wedge C|_{\dot{M}}, |B|_{\dot{M}} \rangle \in \dot{\mathcal{O}}_w$. Since $A \wedge C \in \hat{\Psi}$ (since $\hat{\Psi}$ is closed under the classical connectives and $A, C \in \hat{\Psi}$) and $B \in \hat{\Psi}$, by (D$\mathcal{O}\star$), $\langle |A \wedge C|_{\dot{M}}\downarrow, |B|_{\dot{M}}\downarrow \rangle \overset{J.5.12i}{=} \langle |A|_{\dot{M}}\downarrow \cap |C|_{\dot{M}}\downarrow, |B|_{\dot{M}}\downarrow \rangle = \langle X \cap Z, Y \rangle \in \mathring{\mathcal{O}}_w$.

The other cases are shown in a similar way and are left to the reader. □

Now we show that $\dot{M}$ and $\mathring{M}$ are equivalent modulo $\hat{\Phi}$.

Lemma J.5.14. *For all $\psi \in \hat{\Phi}$ and all $w \in \mathring{W}$, $\dot{M}, w \models \psi$ iff $\mathring{M}, w \models \psi$.*

Proof. We show the equivalence by induction on the length of ψ. The equivalence holds for all propositional formulas ψ by Lemma J.5.8.

Let now $\psi = \mathsf{O}(A|B)$. Note that $A, B \in \hat{\Psi}$. $\mathring{M}, w \models \mathsf{O}(A|B)$ iff $\langle |B|_{\mathring{M}}, |A|_{\mathring{M}} \rangle \in \mathring{\mathcal{O}}_w$ iff (by Lemma J.5.9) $\langle |B|_{\dot{M}}\downarrow, |A|_{\dot{M}}\downarrow \rangle \in \mathring{\mathcal{O}}_w$. By (D$\mathcal{O}\star$), there are $A', B' \in \hat{\Psi}$ such that $|A'|_{\dot{M}}\downarrow = |A|_{\dot{M}}\downarrow$, $|B'|_{\dot{M}}\downarrow = |B|_{\dot{M}}\downarrow$ and $\langle |B'|_{\dot{M}}, |A'|_{\dot{M}} \rangle \in \dot{\mathcal{O}}_w$. Since $A, A', B, B' \in \hat{\Psi}$, by Lemma J.5.11, $|A|_{\dot{M}} = |A'|_{\dot{M}}$ and $|B|_{\dot{M}} = |B'|_{\dot{M}}$. Thus, $\langle |B|_{\dot{M}}, |A|_{\dot{M}} \rangle \in \dot{\mathcal{O}}_w$ and thus, $\dot{M}, w \models \mathsf{O}(A|B)$. Let now $\dot{M}, w \models \mathsf{O}(A|B)$. Then $\langle |B|_{\dot{M}}, |A|_{\dot{M}} \rangle \in \dot{\mathcal{O}}_w$ and thus by (D$\mathcal{O}\star$), $\langle |B|_{\dot{M}}\downarrow, |A|_{\dot{M}}\downarrow \rangle \in \mathring{\mathcal{O}}_w$. By Lemma J.5.9, $\langle |B|_{\mathring{M}}, |A|_{\mathring{M}} \rangle \in \mathring{\mathcal{O}}_w$ and thus, $\mathring{M}, w \models \mathsf{O}(A|B)$.

Let $\psi = \bullet_{\mathsf{i}}\mathsf{O}(A|B)$. Note that $A, B \in \hat{\Psi}$. $\mathring{M}, w \models \bullet_{\mathsf{i}}\mathsf{O}(A|B)$ iff $\langle |B|_{\mathring{M}}, |A|_{\mathring{M}} \rangle \in \mathring{\mathcal{N}}^{\mathrm{i}}{}_w$ iff (by Lemma J.5.9) $\langle |B|_{\dot{M}}\downarrow, |A|_{\dot{M}}\downarrow \rangle \in \mathring{\mathcal{N}}^{\mathrm{i}}{}_w$. By (D$\mathcal{N}^{\mathrm{i}}\star$), there are $A', B' \in \hat{\Psi}$ such that $|A'|_{\dot{M}}\downarrow = |A|_{\dot{M}}\downarrow$, $|B'|_{\dot{M}}\downarrow = |B|_{\dot{M}}\downarrow$ and $\langle |B'|_{\dot{M}}, |A'|_{\dot{M}} \rangle \in \dot{\mathcal{N}}^{\mathrm{i}}{}_w$. Since $A, A', B, B' \in \hat{\Psi}$, by Lemma J.5.11, $|A|_{\dot{M}} = |A'|_{\dot{M}}$ and $|B|_{\dot{M}} = |B'|_{\dot{M}}$. Thus, $\langle |B|_{\dot{M}}, |A|_{\dot{M}} \rangle \in \dot{\mathcal{N}}^{\mathrm{i}}{}_w$ and thus, $\dot{M}, w \models \bullet_{\mathsf{i}}\mathsf{O}(A|B)$. Let now $\dot{M}, w \models \bullet_{\mathsf{i}}\mathsf{O}(A|B)$. Then $\langle |B|_{\dot{M}}, |A|_{\dot{M}} \rangle \in \dot{\mathcal{N}}^{\mathrm{i}}{}_w$ and thus by (D$\mathcal{O}\star$), $\langle |B|_{\dot{M}}\downarrow, |A|_{\dot{M}}\downarrow \rangle \in \mathring{\mathcal{N}}^{\mathrm{i}}{}_w$. By Lemma J.5.9, $\langle |B|_{\mathring{M}}, |A|_{\mathring{M}} \rangle \in \mathring{\mathcal{N}}^{\mathrm{i}}{}_w$ and thus, $\mathring{M}, w \models \bullet_{\mathsf{i}}\mathsf{O}(A|B)$.

The case $\psi = \bullet_{\mathsf{p}}\mathsf{O}(A|B)$ is shown analogously.

Let $\psi = \mathsf{O}^{\mathsf{i}}A$. Note that $A \in \hat{\Psi}$. $\mathring{M}, w \models \mathsf{O}^{\mathsf{i}}A$ iff $|A|_{\mathring{M}} \in \mathring{\mathcal{O}}^{\mathrm{i}}{}_w$ iff (by Lemma J.5.9) $|A|_{\dot{M}}\downarrow \in \mathring{\mathcal{O}}^{\mathrm{i}}{}_w$. By (D$\mathcal{O}^{\mathrm{i}}\star$), there is a $A' \in \hat{\Psi}$ such that $|A'|_{\dot{M}}\downarrow = |A|_{\dot{M}}\downarrow$, and

$|A'|_{\dot{M}} \in \dot{\mathcal{O}}^{\mathsf{i}}{}_w$. Since $A, A' \in \hat{\Psi}$, by Lemma J.5.11, $|A|_{\dot{M}} = |A'|_{\dot{M}}$. Thus, $|A|_{\dot{M}} \in \dot{\mathcal{O}}^{\mathsf{i}}{}_w$ and thus, $\dot{M}, w \models \mathsf{O}^{\mathsf{i}}A$. Let now $\dot{M}, w \models \mathsf{O}^{\mathsf{i}}A$. Then $|A|_{\dot{M}} \in \dot{\mathcal{O}}^{\mathsf{i}}{}_w$ and thus by (D$\mathcal{O}^{\mathsf{i}}\star$), $|A|_{\dot{M}}\downarrow \in \mathring{\mathcal{O}}^{\mathsf{i}}{}_w$. By Lemma J.5.9, $|A|_{\mathring{M}} \in \mathring{\mathcal{O}}^{\mathsf{i}}{}_w$ and thus, $\mathring{M}, w \models \mathsf{O}^{\mathsf{i}}A$.

The case $\psi = \mathsf{O}^{\mathsf{p}}A$ is shown analogously.

We still have to show that our statement holds for $\varphi \in (\mathcal{L} \cap \hat{\Psi}) \setminus (\mathcal{P} \cup \mathcal{L}')$. As induction hypothesis, suppose that the equivalence holds for $\psi_1, \psi_2 \in \hat{\Phi}$. Let $\psi = \psi_1 \wedge \psi_2$. Then $\mathring{M}, w \models \psi_1 \wedge \psi_2$ iff $\mathring{M}, w \models \psi_1, \psi_2$ iff (by induction hypothesis) $\dot{M}, w \models \psi_1, \psi_2$ iff $\dot{M}, w \models \psi_1 \wedge \psi_2$. The cases $\psi = \psi_1 \vee \psi_2, \psi = \psi_1 \supset \psi_2$ and $\psi = \neg\psi_1$ are shown similarly. Thus, the equivalence holds for all formulas $\psi \in \hat{\Phi}$. □

Corollary J.5.1. *For all $\psi \in \hat{\Phi}$, $\dot{M} \models \psi$ iff $\mathring{M} \models \psi$.*

Proof. $\dot{M} \models \psi$ iff $\dot{M}, \dot{@} \models \psi$ iff (by Lemma J.5.14 and since $\mathring{@} = \dot{@}$) $\mathring{M}, \mathring{@} \models \psi$ iff $\mathring{M} \models \psi$. □

Theorem J.5.2. $\mathbf{L}^+$ *is complete with respect to the class of frames that satisfy the appropriate conditions.*

Proof. The proof is similar to the proof of Theorem J.5.1. Take again a formula ψ such that $\nvdash_{\mathbf{L}^+} \psi$. The model $\dot{M} = \langle \dot{F}, \dot{@}, \dot{v} \rangle$ constructed for Theorem J.5.1 meeting the respective model conditions was such that $\dot{M}, \dot{@} \not\models \psi$ and thus $\dot{M} \not\models \psi$. We choose now Φ to be the set of subformulas of ψ. We construct $\mathring{M}$ on basis of $\dot{M}$ as above. By Corollary J.5.1, $\mathring{M} \not\models \psi$. By Lemma J.5.13, $\mathring{F}$ satisfies the respective frame conditions. Therefore, there is a model in the respective class of frames that meets the respective frame conditions. By contraposition and generalization, if a formula ψ is valid with respect to that class, it must be provable in $\mathbf{L}^+$. □

The following two corollary follow immediately.

Corollary J.5.2. $\mathbf{L}^+$ *is sound and complete with respect to the class of all finite frames that meet the appropriate frame conditions.*

Corollary J.5.3. $\mathbf{L}^+$ *has the finite model property.*

Corollary J.5.4. $\mathbf{L}^+$ *is decidable.*

J.6 Dealing with (Finite) Premise Sets

In order to work with (finite) premise sets $\Gamma \subset \mathcal{L}$ we define:

$$M \models \Gamma \text{ iff for all } \varphi \in \Gamma, M \models \varphi$$
$$\Gamma \Vdash_F \varphi \text{ iff for all } M = \langle F, @, v \rangle : \text{if } M \models \Gamma, \text{ then } M \models \varphi$$
$$\Gamma \Vdash_{\mathcal{F}} \varphi \text{ iff for all } F \in \mathcal{F} : \Gamma \Vdash_F \varphi$$

$$\Gamma \vdash_{\mathbf{L}^+} \varphi \text{ iff } \vdash_{\mathbf{L}^+} \bigwedge \Gamma \supset \varphi.$$

Let $\Gamma \subset \mathcal{L}$ be finite and $\varphi \in \mathcal{L}$.

Lemma J.6.1. *$M \models \bigwedge \Gamma \supset \varphi$ iff (if $M \models \Gamma$, then $M \models \varphi$).*

Proof. "⇒": $M \models \bigwedge \Gamma \supset \varphi$ iff $M \models \neg(\bigwedge \Gamma) \vee \varphi$ iff ($M \models \neg(\bigwedge \Gamma)$ or $M \models \varphi$). Also, $M \models \Gamma$ iff $M \models \bigwedge \Gamma$ (due to (M-$\wedge$). Thus, if $M \models \Gamma$, then $M \models \varphi$.

"⇐": Suppose $M \not\models \bigwedge \Gamma \supset \varphi$, then $M \not\models \neg(\bigwedge \Gamma) \vee \varphi$. Then, by (M-$\vee$), it is not the case that ($M \models \neg(\bigwedge \Gamma)$ or $M \models \varphi$). Thus, $M \not\models \neg(\bigwedge \Gamma)$ and $M \not\models \varphi$, and hence, $M \models \bigwedge \Gamma$ and $M \not\models \varphi$—a contradiction. □

Theorem J.6.1. *Where $\mathbf{L}^+ \in \{\mathbf{CDPM.2d}^+, \mathbf{CDPM.2e}^+\}$ and $\mathcal{F}$ is the appropriate class of frames (with respect to the frame conditions that characterize $\mathbf{L}^+$, see Sect. J.3), $\Gamma \Vdash_{\mathcal{F}} \varphi$ iff $\Gamma \vdash_{\mathbf{L}^+} \varphi$.*

Proof. $\Gamma \Vdash_{\mathcal{F}} \varphi$ iff for all $F \in \mathcal{F} : \Gamma \Vdash_F \varphi$ iff for all $F \in \mathcal{F}$ and for all $M = \langle F, @, v\rangle$: if $M \models \Gamma$, then $M \models \varphi$ iff (Lemma J.6.1) for all $F \in \mathcal{F}$ and for all $M = \langle F, @, v\rangle : M \models \bigwedge \Gamma \supset \varphi$ iff (Theorem J.5.2) $\vdash_{\mathbf{L}^+} \bigwedge \Gamma \supset \varphi$ iff (by Definition) $\Gamma \vdash_{\mathbf{L}^+} \varphi$. □

J.7 Deontic Detachment

Recall the following deontic detachment principles:

$$\vdash \big(\mathsf{O}(A|C) \wedge \mathsf{P}(A \wedge B|C) \wedge \mathsf{O}(B|A \wedge C)\big) \supset \mathsf{O}(B|C) \tag{DDP1}$$

$$\vdash \big(\mathsf{O}(A|\top) \wedge \mathsf{P}(A \wedge B|\top) \wedge \mathsf{O}(B|A)\big) \supset \mathsf{O}(B|\top) \tag{DDPT1}$$

$$\vdash \big(\mathsf{O}(A|C) \wedge \mathsf{P}(A \wedge B|C) \wedge \mathsf{P}(B|\neg A \wedge C) \wedge \mathsf{O}(B|A \wedge C)\big) \supset \mathsf{O}(B|C) \tag{DDP2}$$

$$\vdash \big(\mathsf{O}(A|\top) \wedge \mathsf{P}(A \wedge B|\top) \wedge \mathsf{P}(B|\neg A) \wedge \mathsf{O}(B|A)\big) \supset \mathsf{O}(B|\top) \tag{DDPT2}$$

Theorem J.7.1. *In $\mathbf{CDPM.2d}^+$ (DDP1) and (DDP⊤1) are valid.*

Proof. By (S) and $\mathsf{O}(B|A \wedge C)$ we get $\mathsf{O}(A \supset B|C)$. $\mathsf{P}(A \wedge (A \supset B)|C)$ is a consequence of (CRE) and $\mathsf{P}(A \wedge B|C)$. By (CPAND), $\mathsf{O}(A|C)$, $\mathsf{P}(A \wedge (A \supset B)|C)$ and $\mathsf{O}(A \supset B|C)$ we have $\mathsf{O}(A \wedge (A \supset B)|C)$. Thus, by (CRE), $\mathsf{O}(A \wedge B|C)$. By this, (RCPM) and $\mathsf{P}(A \wedge B|C)$ we get $\mathsf{O}(B|C)$. (DDP⊤1) follows immediately. □

Theorem J.7.2. *In* **CDPM.2e**$^+$ *(DDP2) and (DDP$\top$2) are valid.*

Proof. The proof is similar to the one above. Since we don't have (S), but instead the weaker (PS'), we need the additional hypothesis $\mathsf{P}(B|\neg A \wedge C)$ in order to derive $\mathsf{O}(A \supset B|C)$ from $\mathsf{O}(B|A \wedge C)$. The rest of the proof is identical to the proof of Theorem J.7.1. □

J.8 The Semantics for $\mathbf{L}_{\mathbb{P}}^+$

The semantics for our enhanced new lower limit logic **CDPM.2**$\alpha_{\mathbb{P}}^+$ is defined in a similar way as the semantics of **CDPM.2**α^+. Neighborhood frames are now tuples $\langle W, \mathcal{O}, \mathcal{N}^{\mathrm{i}}, \mathcal{N}^{\mathrm{p}}, \mathcal{O}^{\mathrm{i}}, \mathcal{O}^{\mathrm{p}}, \mathcal{P}^{\star}\rangle$ where $W, \mathcal{O}, \mathcal{N}^{\mathrm{i}}, \mathcal{N}^{\mathrm{p}}, \mathcal{O}^{\mathrm{i}}$, and $\mathcal{O}^{\mathrm{p}}$ are defined as before and $\mathcal{P}^{\star} : W \to \wp(\wp(W) \times \wp(W))$ is used to characterize our new operator $\mathbb{P}$. We add the following requirement for all $w \in W$:

$$M, w \models \mathbb{P}(A \mid B) \text{ iff } \langle |B|_M, |A|_M\rangle \in \mathcal{P}^{\star}_w \qquad (\text{M-}\mathcal{P}^{\star})$$

We have to add two more frame conditions corresponding to the new rules (P-Ps) and (Ps-T), namely

$$\text{For all } X, Y \subseteq W, \text{ if } X \subseteq Y \text{ and } \langle Y, \overline{X}\rangle \notin \mathcal{O}_w, \text{ then } \langle Y, X\rangle \in \mathcal{P}^{\star}_w \qquad (\text{F-P-Ps})$$

$$\text{For all } X, Y, Z \subseteq W, \text{ if } \langle X, Y\rangle, \langle Y, Z\rangle \in \mathcal{P}^{\star}_w, \text{ then } \langle X, Z\rangle \in \mathcal{P}^{\star}_w \qquad (\text{F-Ps-T})$$

Moreover, the frame-conditions for the altered rules (Ep-g) and (CTDR-g) have to be adjusted.

$$\text{For all } X, Y, Z, Z' \subseteq W, \text{ if } X \subseteq \overline{Z'}, Z \subseteq Y, \left(\langle Z, \overline{Z'}\rangle \notin \mathcal{O}_w \text{ or } \langle Z, Z'\rangle \in \mathcal{O}_w\right), \\ w \in Z, \langle Y, Z\rangle \in \mathcal{P}^{\star}_w, \text{ and } \langle Y, X\rangle \in \mathcal{O}_w, \text{ then } \langle Y, X\rangle \in \mathcal{N}^{\mathrm{p}}_w \qquad (\text{F-Ep-g})$$

$$\text{For all } X, Y, Z, Z' \subseteq W, \text{ if } X \subseteq \overline{Z'}, X \subseteq \overline{Z}, Z \subseteq Y, \\ \langle Z, Z'\rangle, \langle Y, X\rangle \in \mathcal{O}_w \text{ and } \langle Y, Z\rangle \notin \mathcal{P}^{\star}_w, \text{ then } \langle Z, Z'\rangle \in \mathcal{N}^{\mathrm{p}}_w \qquad (\text{F-CTDR-g})$$

The soundness and completeness proofs offered for **CDPM.2**α^+ can be easily adjusted for the altered and additional frame conditions for **CDPM.2**$\alpha_{\mathbb{P}}^+$.

References

1. Kraus, S., Lehmann, D.J., Magidor, M.: Nonmonotonic reasoning, preferential models and cumulative logics. Artif. Intell. **44**, 167–207 (1990)
2. Lehmann, D.J., Magidor, M.: What does a conditional knowledge base entail? Artif. Intell. **55**(1), 1–60 (1992)
3. van Dalen, D.: Logic and Structure: Universitext. Springer (1980) (4th edn., 2004)
4. Batens, D., Meheus, J., Provijn, D.: An adaptive characterization of signed systems for paraconsistent reasoning (Unpublished)
5. Schröder, L., Pattinson, D.: Rank-1 modal logics are coalgebraic. J. Logic Comput. **5**(20), 1113–1147 (2010)
6. Goble, L.: A proposal for dealing with deontic dilemmas. In: Lomuscio, A., Nute, D. (eds.) DEON. Lecture Notes in Computer Science, vol. 3065, pp. 74–113. Springer, Heidelberg (2004)
7. Goble, L.: A logic for deontic dilemmas. J. Appl. Logic **3**, 461–483 (2005)
8. Straser, C., Meheus, J., Beirlaen, M.: Tolerating deontic conflicts by adaptively restricting inheritance. Logique at Analyse **219**, 477–506 (2012)
9. Straser, C.: A deontic logic framework allowing for factual detachment. J. Appl. Logic **9**(1), 61–80 (2010)
10. Goble, L.: Dilemmas in deontic logic (Unpublished)

Index

DOI: 10.1007/978-3-319-00792-2, © Springer International Publishing Switzerland 2014

MIX
Papier aus verantwortungsvollen Quellen
Paper from responsible sources
FSC® C105338
FSC
www.fsc.org

If you have any concerns about our products,
you can contact us on
ProductSafety@springernature.com

In case Publisher is established outside the EU,
the EU authorized representative is:
Springer Nature Customer Service Center GmbH
Europaplatz 3, 69115 Heidelberg, Germany

Printed by Libri Plureos GmbH
in Hamburg, Germany